W9-AXS-217

L. A. PATERNO

TRAFFIC ENGINEERING

TRAFFIC

LOUIS J. PIGNATARO

Head, Dept. of Transportation Planning and Engineering, Polytechnic Institute of Brooklyn Brooklyn, New York

ENGINEERING

theory and practice

with contributions by

Edmund J. Cantilli
John C. Falcocchio
Kenneth W. Crowley
William R. McShane
Roger P. Roess
Bumjung Lee

Prentice-Hall, Inc., Englewood Cliffs, New Jersey

Library of Congress Cataloging in Publication Data

PIGNATARO, LOUIS J
Traffic engineering: theory and practice.

Includes bibliographical references.
1. Traffic engineering. I. Cantilli, Edmund J.
II. Title.
HE333.P53 387.3'1 72-12683
ISBN 0-13-926220-2

10 9 8

Printed in the United States of America

Prentice-Hall International, Inc., *London*
Prentice-Hall of Australia, Pty. Ltd., *Sydney*
Prentice-Hall of Canada, Ltd., *Toronto*
Prentice-Hall of India Private Limited, *New Delhi*
Prentice-Hall of Japan, Inc., *Tokyo*

to **Edith** and **Thea**

Contents

Preface

This book is designed for use by students, practitioners, and researchers in traffic engineering. It is based on lecture notes which were primarily developed for courses in traffic engineering given at the Polytechnic Institute of Brooklyn. Although the format and style of the notes have been modified for the preparation of this book, extensive rewriting to provide a smoother exposition of subject material was not undertaken. It was felt, however, that this shortcoming would not detract from the book's overall value, and that it did not justify waiting one or two years before publishing the manuscript.

The book is divided into several parts which cover a traditional, current treatment of traffic engineering, as well as a series of special topics which the practicing traffic engineer encounters in the successful execution of his responsibilities. For each chapter there is an extensive list of references which will enable the reader to pursue any subject in as great a depth as he desires.

In the preparation of this book, the authors have drawn liberally from many papers and articles written by numerous individuals, and they are the first to be acknowledged. Without their contributions, this book would not have materialized. Special gratitude is expressed to the many organizations whose publications provided the primary sources of reference, particularly the Highway Research Board, the Institute of Traffic Engineers, the Eno Foundation for Transportation, the Federal Highway Administration, the American Association of State Highway Officials, the American Society of Civil Engineers, the American Road Builders Association, and the Automobile Manufacturers Association.

Support for making revisions, and for preparation of the manuscript, was partially provided by a grant from *Action for Transportation in New York State, Inc.*, administered by the New York State Science and Technology Foundation. Deep appreciation is expressed to both of these organizations, and particularly to Dr. Donald H. Davenport, the Executive Secretary of the Foundation.

LOUIS J. PIGNATARO

Introduction

The continuing expansion of the American city has made the daily movement of people and goods an ever-increasing complex problem. Urban growth is causing multiple transportation demands on inadequate facilities. Cities depend largely upon their street systems for transportation services. These systems are being overtaxed to meet increasing service demands for automobile traffic, commercial traffic, public transportation, access to abutting property, and parking.

Increasing urbanization gives rise to vexing problems of congestion. If metropolitan areas are to grow and prosper, it will be imperative to plan and build vast new facilities for public and private transport. These, as well as existing resources, must be operated so as to provide the largest possible free flow of traffic. But if a reasonable level of amenity is to be maintained, the added facilities must be planned to make a sparing, efficient use of land, to be convenient to use, and to make a positive esthetic contribution to the environment of both users and bystanders.

Society is ever more committed to these goals; it demands increasing care and professional competence in the planning and operation of highways, airports, public transit, and goods terminals. Federal, state, and local governments are responding by setting up and supporting appropriate authorities, planning groups, and research agencies devoted to transportation and land use planning.

The growth of large cities significantly affects social and economic activities in areas beyond the official borders of the city. Therefore, it is essential in planning and providing transportation facilities to consider not only the central city but also the surrounding areas that are directly affected by the city. This complete region is known as the metropolitan region or Standard Metropolitan Statistical Area (SMSA).

The U. S. Bureau of the Budget has established definite criteria to define an SMSA.[1] A metropolitan area is defined as a county or group of counties which contains at least one city of 50,000 population or more. Counties adjacent to the county in which the central city is located might be included within the metropolitan area depending upon specific requirements of population density and economic dependence on the central city. At the present time there are over 230 officially designated SMSA's.

Transportation facilities are a basic requirement for community growth and development. With the huge capital investment required to finance urban projects of all kinds, the consequences of not planning have become more acute than ever. Successful solutions to the complex mobility problems confronting urban areas throughout the nation require the full energies and imagination of many professionals, but particularly of the traffic engineer.

DEFINITION

Traffic engineering has been defined as "that phase of engineering which deals with the planning, geometric design and traffic operations of roads, streets and highways, their networks, terminals, abutting lands and relationships with other modes of transportation for the achievement of safe, efficient and convenient movement of persons and goods."[2] With the justified emphasis on comprehensive transportation planning, the definition might well be strengthened in its indication of the concern of traffic engineers with public transportation and the interface between different modes of transportation, including the pedestrian mode.

Traffic engineering, unlike most engineering disciplines, deals with problems which are not only dependent on physical factors but very often include the human behavior of the driver and the pedestrian and their interrelationships with the complexity of the environment.

Thus the traffic engineer is in a unique position among the family of professionals because it is necessary for him to broaden his background from purely functional considerations. The traffic engineer must instead be sensitive to a variety of disciplines, because what he does, and what he fails to do, affects very large numbers of people, whether they are road users or not.

HISTORICAL NOTES

Traffic engineering as it is known today has evolved with the advent of the motor vehicle. However, many of the profession's antecedents are rooted in ancient history. For example, one-way streets were known in ancient Rome, and special off-street parking facilities were provided to get chariots off the traveled way. Vehicles were prohibited from entering the business districts of large cities in the Roman Empire during certain hours of the day because of traffic congestion. It is most likely that similar traffic rules and regulations were necessary to control vehicular flow on the paved streets of Babylon in 2000 B.C. Modern traffic islands and rotaries have their origins in the monuments and public squares erected in roadways of centuries past. Pavement markings were used as early as 1600 A.D. on a road, leading from Mexico City, which had incorporated a built-in centerline of contrasting color.

According to the Institute of Traffic Engineers' Committee on Historic Development of Traffic Control Devices, the first centerline marking in modern times was applied in Wayne County, Michigan, in 1911.[3] The first traffic signal was installed in Houston, Texas, in 1921, and the first coordinated signal system was in operation in that same city in 1922.[4]

Years ago the engineer's work was finished with the completion of the roadway construction. But with the introduction of the automobile as a popular means of transportation, and the use of motor trucks for the transportation of goods, the new elements of high speeds and large volumes were injected into the picture. These developments created problems that were too complex for historic methods of police control and regulation. As a result, the engineer was called in to apply his science to the solution of the problems, and thus the profession of traffic engineering came into being.[5]

IMPORTANCE OF MOTOR VEHICLE TRANSPORTATION

The first gasoline driven motor vehicles available to the general public were perhaps those offered for sale in 1888 by the Connelly Motor Company of New York. The Daimler and Duryea were offered for sale in 1891 and 1892, respectively. Today the automobile has become a dominant factor in American life. The rapid growth in numbers of motor vehicles in the United States began to generate a tremendous demand for improved highways by the beginning of this century. The federal government, responding to this public concern, entered the field of highway construction through the establishment of policies that significantly affected national highway development.

Realization of the importance of motor vehicle transportation was reflected by the Federal-Aid Highway Act of 1934. Congress authorized the expenditure, by states, of up to $1\frac{1}{2}$ per cent of federal-aid funds for conducting highway planning studies and other investigations. This act motivated the first statewide highway planning studies, even though they were restricted to rural areas.

In 1956, a most significant governmental policy was adopted, marking the beginning of the largest peacetime public works program in history. This was the Federal-Aid Highway Act of 1956, a law which authorized construction of the 41,000-mile National System of Interstate and Defense Highways. Present estimates put the cost of this project at about $56.6 billion, 90 per cent of which is paid by the federal government and the remaining 10 per cent by the states. Because of the tremendous expenditures required for the program, the law also created a Highway Trust Fund in the Treasury Department. The Trust Fund was established to receive a substantial portion of federal automotive excise taxes and to disburse therefrom the funds authorized for all federal-aid highway systems.

Ever-growing motor vehicle travel has resulted in

increasing traffic congestion, with the greatest distress apparent in urban areas. The use of the motor vehicle—passenger car, truck, and bus—is such a significant part of the social, commercial, and industrial life of any community that the inability to move safely and efficiently results in anguish, inconvenience, and economic loss. Relief of traffic congestion in urban areas is most urgent and has been accorded high priority, as realized in the Federal-Aid Highway Act of 1962, according to which:

> "It is declared to be in the national interest to encourage and promote the development of transportation systems embracing various modes of transportation in a manner that will serve the states and local communities efficiently and effectively."

This section of the act applied to urbanized areas of over 50,000 population, and the states and these urban areas were given until July 1, 1965, to establish a comprehensive transportation plan.

The significance of the traffic safety problem was accentuated by the enactment of two related safety bills in September of 1966, the National Traffic and Motor Vehicle Safety Act of 1966, and the Highway Safety Act of 1966. The first of these laws deals with the establishment of minimum federal safety standards for new vehicles and tires. It directed the Secretary of Transportation to establish safety standards by January 31, 1967, with new or revised standards to be issued by January 31, 1968.

The other law, the Highway Safety Act, was intended to provide a coordinated, national highway safety program through federal grants to the states, and to include a greatly expanded program of research into the causes of accidents and the effectiveness of various corrective measures. This act requires the states to have uniform highway safety programs approved by the Secretary of Transportation. The uniform standards include provisions for accident record systems, thorough accident investigations after accidents occur to determine the causes of accidents, vehicle registration, operation, and inspection, highway design and maintenance, traffic control, vehicle codes and laws, traffic surveillance to detect and correct high accident locations, and emergency services.

The Federal-Aid Highway Act of 1968 had some very significant provisions.[6] Several important changes were contained in the law, which include authorizations for a new federal-aid program for fringe parking, a greatly liberalized program of relocation assistance for families and businesses displaced by highway construction, new legislative guidelines for equal employment opportunity in highway construction, and a 1500-mile extension of the Interstate System. The statutory length of the Interstate System was increased from 41,000 to 42,500 miles to provide for "missing links." The law provides for a demonstration program under which federal-aid funds could be used for the construction of fringe parking facilities in connection with mass transit facilities.

Another aspect of the law, which may have some far-reaching consequences, requires the states in requesting approval for highway locations to certify that public hearings will consider not only the economic impact of the proposed location (required under previous law) but also the social effects of such location, its impact on the environment, and its consistency with the goals and objectives of such urban planning as has been promulgated by the community.

The law also provides an authorization of $200 million for each of the two fiscal years 1970 and 1971, for the TOPICS programs (Traffic Operations Program to Increase Capacity and Safety). The program was initiated at the beginning of 1967 by the Bureau of Public Roads, and it provides for the use of federal-aid highway funds for the application of traffic engineering improvements to principal urban streets. TOPICS covers such improvements as the channelization of intersections, additional lanes at signalized intersections, pedestrian overpasses, traffic control systems, special lanes for buses, and elimination of spot safety hazards.

Public Law 91–605 was also of great importance with respect to highway legislation. It has three titles, including the "Federal-Aid Highway Act of 1970," the "Highway Safety Act of 1970," and the "Highway Trust Fund Extension Provision."[7] The highway act provides for the extension of the Interstate System completion date through fiscal year 1976 (formerly through fiscal year 1974); the creation of a new federal-aid urban highway system in urbanized areas of more than 50,000 population; the establishment of a 70-30 federal-state matching ratio (formerly 50-50) for financing the construction of all highway projects except the Interstate System, beginning in fiscal year 1974; the use of federal funds in urban areas to finance the construction of exclusive or preferential bus lanes, fringe parking facilities, and other bus transit ancillaries; the establishment of a National Highway Institute in the Federal Highway Administration to develop and administer training programs for governmental employees engaged in federal-aid highway work; and the authorization of expenditures from the general fund for the highway beautification program and creation of a commission on highway beautification to study the program.

The safety act created a National Highway Traffic Safety Administration (formerly National Highway Safety Bureau) in the Department of Transportation, and it required that each state have an agency to administer its highway safety program to the satisfaction of the Secretary of Transportation by the end of 1971. The act also provides that highway safety programs and research and development activities are to be financed two-thirds from the Highway Trust Fund and one-third from the general fund beginning in fiscal year 1972. Prior to this date the highway safety program was financed entirely from the general fund.

The trust fund title of the law provides for the extension of the termination date of the Highway Trust Fund and the postponement of the reduction date for certain federal excise taxes from October 1, 1972, to October 1, 1977.

PHASES OF TRAFFIC ENGINEERING

Studies of Traffic Characteristics

These include the methods of conducting traffic studies which are used to determine the character of traffic movement and an understanding of the basic characteristics of the driver, the vehicle, and the traffic stream, including studies of:

1. The road user
2. The vehicle
3. Speed, travel time, and delay
4. Traffic volume
5. Origin and destination
6. Capacity
7. Parking
8. Accidents
9. Public transit

Traffic Operations

Traffic operations include traffic regulatory measures and traffic control devices.

1. Regulatory measures
   a. Laws and ordinances for the purposes of driver, vehicle, and pedestrian controls.
   b. Regulations controlling operation of the vehicles in the traffic stream. The fundamental regulatory measures include intersection controls, speed controls, one-way streets, and parking controls.
2. Traffic control devices

   Fundamentals of the design, installation, operation, and maintenance of traffic signs, signals, pavement markings, and channelization devices to provide the basis for intelligent application of these devices to specific situations.

Prior to the adoption of a regulatory measure or the use of a traffic control device, it is necessary to examine the characteristics, advantages, disadvantages, warrants, and prerequisite studies to establish justification and legality.

Traffic Planning

This phase of traffic engineering encompasses the planning of traffic facilities. It covers the characteristics of urban travel, including public transportation, the conducting of major transportation studies, and the basic techniques in studying and evolving comprehensive transportation plans. Coordination between the traffic engineer and the urban planner is essential for the successful completion of objectives.

Geometric Design

This area includes street design, new arterial highway design and the improvement of existing ones, channelization and intersection design, and the design of off-street parking and terminal facilities.

Administration

An understanding of the essentials of the administrative and legal background of traffic control and regulation is an important adjunct to purely technical knowledge. The traffic engineer is concerned with the administrative framework and organization of the Traffic Engineering Department, programs of driver and public education and, to a lesser degree, the enforcement of traffic regulations.

TRAFFIC ENGINEERING ORGANIZATION

As traffic problems have grown since the advent of the automobile, so have traffic engineering functions, although not always under that title. 1924 marked the first time that a traffic engineering position was officially created.[70] With the automobile becoming ubiquitous in the United States, at least some form of traffic engineering has appeared at every level of government, in every part of the country, from the smallest towns, where the mayor or police chief handles the problems of traffic, to the federal government, where the U.S. Department of Transportation, with thousands of employees, is concerned with all phases of transportation.

Initially, traffic engineering dealt primarily with traffic control devices and traffic surveys. As the complexity of the field has increased, the functions of traffic engineering have expanded dramatically. This, of course, varies with the size of the area of jurisdiction (small town, city, county, state, federal) and the way in which these functions are discharged. The scope of traffic engineering functions also varies from one part of the nation to another.

Responsibilities of a Traffic Engineering Organization

A wide range of functions may be found within traffic engineering organizations which may have titles other than that of "traffic engineering," depending on the size or jurisdiction of the group. Those activities previously mentioned, plus those in the general category of "traffic safety," may be found.

Although *street lighting* is not usually under the jurisdiction of traffic engineering, proper placement of light standards, safety design (break-away supports, etc.), lighting of freeway entrances and exits, proper lighting for signs and intersections, are all important in traffic design, and the traffic engineering organization should at least be consulted in these areas.

Public Relations and Education

In addition to all of the technical activities of the traffic engineering body, there should be a direct link with the general public. It is desirable for the public to be aware of the major programs of the traffic department (or other jurisdictional body). The public should also be educated to help these programs work successfully. For example, if a street is to be closed to traffic during certain hours, or if a prohibition of on-street parking during certain hours is to be effective, the public should be informed and educated about the program, including the specific details and the reasons for, and goals of, the program.

Development of Laws, Ordinances, Codes, and Issuance of Permits

Traffic engineers are often called upon to help develop traffic laws and ordinances, and are also involved in the issuance of permits for construction on traffic facilities, movement of dangerous cargo or oversized vehicles, and other such considerations.

Types of Traffic Engineering Organizations

The functions of traffic engineering can be discharged in a variety of ways. One department may handle all functions, or these may be spread among several departments or divisions.

At the city level, traffic engineering functions may be contained under one separate Department of Traffic, parallel to other departments such as Sanitation, Police, Housing, and Fire. For smaller cities, the Traffic or Traffic Engineering Department usually handles motor vehicle traffic and related areas, often leaving transit to a separate department or bureau.

As the size of the city increases, the agency's structure increases in complexity. Los Angeles, one of the nation's most heavily auto-oriented cities, has a much more stratified organizational set-up compared with the smaller cities. In this case, there is a separation of operational and planning functions. Under *operations* are the placement and maintenance of signs, signals, markings, parking, etc. Under *planning* are the *design* of signs and signals, geometrics, traffic surveys, transportation planning, and research.

The ultimate in total transportation jurisdiction in cities occurs in New York City, where the Transportation Administration, one of the city's "super agencies," has supervision of the Department of Traffic, Department of Marine and Aviation, Department of Highways, and the Bureau of Parking Violations. Aside from functions of administration, legal affairs, and public relations, the Transportation Administration handles research and planning, rapid transit planning, coordination of mass transit, traffic, highways, ferries, bridges (excluding facilities run by the Port of New York Authority and the Triborough Bridge and Tunnel Authority), and land use coordination. This is one of the most comprehensive city transportation agencies in the country.

State governments usually are more involved with highways than are city traffic departments. States with small populations, such as North Dakota or Vermont, generally break responsibilities into planning-engineering, construction-operations, and management-administration areas. Vermont divides the department into engineering and administrative areas, with planning, design, bridges, and utilities forming one branch of *engineering*, and traffic, construction, materials, and district functions forming the other. Under *administration* are budget, accounting, personnel, etc., as one branch, and right-of-way, photography, safety, and highway garages as the other.

In Texas, fifteen major divisions of the State Highway Commission handle all aspects of highway planning, construction, operations, maintenance, and administration.

California has a Business and Transportation Agency, with a State Transportation Board which formulates policy, and has the California Highway Commission, Department of Public Works, and California Toll Bridge Authority under the Agency. The Highway Commission is concerned with the location of facilities and with allocation of funds for highway needs. Actual engineering is found in the Division of Highways of the Department of Public Works.

Many populous states are following the example of the federal government in placing *all* transportation modes under one roof. New York State was one of the first to do so, in 1967. In addition to the Highway, Traffic, and Waterway Divisions, the New York Department of Transportation also gives technical and financial assistance in aviation and mass transit, through demonstration projects, capital grants, and technical assistance programs.

Connecticut changed to the Department of Transportation arrangement, as did Pennsylvania, and Maryland is planning to do so. These are all important steps in recognizing and meeting the need for balanced transportation systems, with coordination, rather than competition, between modes. This coordination is greatly facilitated by placing administration of the various modes under one agency.

In many populous states, counties may have large populations, but not necessarily one large city. Nassau County, New York, has a population of approximately 1.5 million people, and an extensive highway and arterial network. The county has its own Department of Public Works, and the Traffic Engineering Division is within that department.

Dade County, Florida, has a population of over 1.2 million, and is also highly urbanized and suburbanized. It has one of the most comprehensive Traffic and Transportation Departments of any county in the country.

In the nation's capitol, the Federal Department of Transportation, in the President's Cabinet, has led the way in streamlining transportation organization, taking many diverse agencies and placing them under one well-coordinated agency. The U.S. Department of Transportation contains the National Transportation Safety Board, Federal Highway Administration (within which are the Bureau of Public Roads, Office of Highway Safety, and the Motor Carrier Safety Bureau), the Coast Guard, Federal Aviation Administration, Federal Railroad Administration, National Highway Traffic Safety Administration, Urban Mass Transportation Administration, and the Saint Lawrence Seaway Development Corporation. Under the National Highway Traffic Safety Administration are five major subdivisions, including planning and programming, motor vehicle programs, research and development, traffic safety programs, and administration.

The Appendix illustrates a number of the organization charts discussed, typical of various populations and jurisdictions.

THE TRAFFIC ENGINEER

The professional society for traffic engineers is the Institute of Traffic Engineers (ITE), which was founded in 1930, and its membership now exceeds 3600. A detailed survey of ITE membership was made in 1959,[8] and a preliminary study of 1964 membership was also published.[9] Based on the 1964 survey sample, 14 per cent of the membership had no engineering degree, 41 per cent had an engineering degree, and 45 per cent had one year or more of graduate study. Median annual salaries are commensurate with educational attainment and other engineering fields.

Traffic engineering staff, functions, and administration vary considerably with the size (population) of the area of jurisdiction.[10-25]

The traffic engineer's scope of responsibility has significantly increased in recent years with the advent of comprehensive transportation planning on a regional rather than on a local basis.[26-29]

Modern techniques for traffic control and regulation, the increasing emphasis and necessity for coordinated efforts between the traffic engineer, civil engineer, and urban planner, urban renewal, and the importance of research in traffic engineering have all contributed to the major role played by the traffic engineer in today's transportation problems.[30-58]

The traffic engineer is constantly being challenged with a dilemma in that congestion, particularly on city streets and urban freeways, will continue to become more acute, while it is increasingly more difficult to acquire the right-of-way needed to improve existing facilities or to build new facilities. However, the construction of new roads and streets is not always the solution for relieving congestion, and the traffic engineer has an obligation to get maximum utilization out of existing facilities by better control of traffic.

Multi-disciplinary interaction is essential to obtain balanced and mutually reinforcing solutions to transportation problems. Many major urban transportation projects are planned by using the design team concept. The team is usually composed of civil engineers, structural engineers, traffic engineers, architects, landscape architects, urban designers, city planners, sociologists, urban geographers, economists, applied mathematicians, lawyers, and market analysts.[59-64]

One of the ways in which the traffic engineer has become involved in the broader planning problem is through the multiple use of right-of-way. This concept, which has attracted much attention and research[65-67] in recent times, involves the use of air rights development over transport facilities, as well as the development of unused right-of-way lands. An example of such development is the multi-story apartment complex of four buildings, built over the approach highway to the George Washington Bridge in New York City. Application of this concept involves the traffic engineer in problems of land use planning and its sociological implications, and is a good indicator of his broadening scope in the planning field.

The traffic engineer is just one member of a team of professionals working to improve the human environment. His specialty is the optimization of traffic movement, but his work involves him in a complex set of interrelationships which go to make up human ecology and the urban environment.[61]

The dominant role played by the motor vehicle in our way of life will certainly not diminish within the relatively near future. Attitude surveys clearly indicate that the automobile is by far the most important mode of travel to the American household, and represents a significant factor in our way of life and general values, and it will become even more important in the immediate years ahead.[68-69]

However, greater attention must be paid to the effects of the automobile and its necessary roadway system, by traffic engineer and urban planner alike. Problems of congestion, air pollution, esthetic pollution, noise pollution, and the disposal of discarded vehicles, must be confronted by all those who have a hand in the management or planning of traffic. The future livability of city and suburb alike demands it.

I

Elements of Traffic Engineering

1

The Road User

Road users, including drivers and pedestrians, are one of the three main elements of automobile transportation. The vehicle and the road comprise the other two. Because the user is a major part of the system, human limitations and behavior must be understood and taken into account in all traffic engineering and design matters.

The success of traffic engineering measures depends heavily upon the user. An understanding of not only average physical and mental limitations, but of the range of user performance, is critical to proper exercise of traffic controls and operating measures.

In this chapter, several important user characteristics will be presented and discussed. However, consideration of the user is to be found throughout the subject material, since human factors affect every phase of traffic engineering. For example, traffic laws and ordinances must be made "reasonable," so as not to breed contempt for all laws among drivers. Control devices must be designed with consideration of their primary function as a medium of communication with drivers.

Because human performance varies from individual to individual, consideration must be given to atypical, or nonaverage, characteristics. For example, a traffic signal timed to permit an average pedestrian to cross the street safely may cause a severe hazard to the elderly and others whose capabilities fall below those of the "average" pedestrian. The sections following detail some primary human characteristics as they relate to road users, including the results of studies quantifying both the range and magnitude of these characteristics.

FUNDAMENTAL CHARACTERISTICS OF DRIVERS

Deficiencies in the human element contribute to almost every accident. Physical limitations, however, account for only a small proportion of accidents, since these limitations may be compensated for by slower or more careful driving.

Reaction to External Stimuli

Reaction involves a series of events which are closely related to human physical factors:

1. *Perception*, which involves seeing the stimuli along with the other perceived objects
2. *Identification or intellection*, which involves the identification and understanding of the stimuli
3. *Judgment or emotion*, which involves the decision-making process, in which a determination is made as to the proper course of action (to stop, swerve, pass, merge, cross, move laterally, or blow horn)

4. *Reaction or volition*, which involves the execution of the decision, which requires motor coordination

The total time required to perceive and complete a reaction to a stimulus is the sum of the times necessary for Perception, Identification, Emotion, and Volition. This total reaction time is often referred to as the PIEV time.

Visual Factors in Perception and Identification

Visual Acuity. This is a characteristic of great importance. Vision provides the driver with accurate information on the relationships among perceived objects and on traffic control messages. Good vision can permit earlier perception and identification, thus decreasing the total reaction (PIEV) time to any given stimulus. Visual acuity relates to the field of clearest vision. The most acute vision is within a narrow cone of 3 to 5 degrees; however, the limit of fairly clear sight is within a cone of 10 to 12 degrees. This must be kept in mind when placing traffic signs and symbols, because beyond the 0- to 12-degree cone vision becomes blurred. Visual acuity may refer to a variety of characteristics. Of interest to road user researchers are such factors as dynamic visual acuity (ability to see and perceive stimuli in a moving field), depth perception, "night vision," and glare recovery, in addition to the common standard of static visual acuity.

Peripheral Vision. This relates to the field of view within which an individual can see objects, but without clear detail or color. The angle of peripheral vision normally varies from 120 to 180 degrees. The peripheral area does not contribute to seeing clearly, but it does serve a warning function, and is important for road safety because it is sensitive to motion and brightness. "Tunnel vision" is the term often used to describe individuals who have an angle of peripheral vision as small as 40 degrees. There are few cases of "tunnel vision," and it can be compensated for by moving the head.

Based on studies of responses to flicker, it appears that a decisive change in peripheral vision occurs at an age of 60 years.[4,5] Peripheral visual stimulation results in more accurate estimates of the absolute speed of the vehicle in which one is traveling than stimulation of the frontal visual field.[6,62] The *frontal field* refers to the 25 degrees centered on the line of sight in the direction of motion; the *peripheral field* refers to the field subtended between 65 and 90 degrees.

There is a positive relationship between good dynamic visual acuity and a good driving record. There is not sufficient evidence of a similar relationship between static visual acuity (standard) and the driving record.[2] Poor vision is associated with a poor record. Vision tests include dynamic visual acuity, static visual acuity, field of vision, degree of cross-eyedness, low-illumination vision, and glare recovery. Of these, *dynamic visual acuity* is the most closely and consistently correlated with driving record, followed by static acuity, field of vision, and glare recovery.

Depth Perception. This visual factor is very important because of its relationship to the ability to estimate distances and speed. The speed-judgment factor is related to the increasing frequency of rear-end collisions on high-speed highways.[7]

Glare Vision and Recovery. These factors are very much affected by age. There is a marked change in sensitivity to glare over the age of 40.[8] In general, older people have much poorer night vision. Glare recovery time, that is, the time required to recover from the effects of glare after the light source is passed, is approximately six or more seconds when going from light to dark and about three seconds when going from dark to light.[9-13]

Color Vision. Generally, this is not of great importance in driving since color blindness can be compensated for by learning other means of recognizing signs and signals. In general, the eye is most sensitive to black and white and black and yellow combinations. Details of studies involving optimum color schemes for signs are included in Chapter 20.

Hearing in Perception

Hearing is needed to detect warning sounds, but lack of hearing acuity can be compensated for by the use of hearing aids. In matching samples of totally deaf drivers and nondeaf drivers, it was found that deaf females did not differ significantly from nondeaf females in traffic convictions or accidents. Deaf males, however, had a significantly greater number of accidents; the accident frequency was 1.8 times higher than for nondeaf males. With regard to traffic convictions, the males did not differ significantly from each other.[14]

Total Driver Response Time

Total driver response (PIEV) time increases with the number of choices and the complexity of the judgment required. PIEV times are involved in the determination of safe stopping distances, safe approach speeds at intersections, and the yellow clearance interval for traffic signals, as well as the necessity to react to any emergency situation.

Simple reaction time is the time required to respond to the simpler types of stimuli. Under laboratory conditions, a simple eye-to-finger reaction called for in depressing the horn usually takes about 0.15 to 0.25 second. The eye-to-foot reaction, called for in depressing the brakes, requires a longer period of about 0.5 second.

A limited study of reaction times initiated by light bulbs, under laboratory conditions, involved the time from the onset of the light until the horn or brake light received an electrical pulse:[15]

TASK	AVERAGE TIME (Sec.)
Honk horn, starting position of hand on horn ring.	0.38
Honk horn, starting position of hands on steering wheel.	0.56
Depress brake, starting position of left foot on brake.	0.39
Depress brake, starting position of right foot on depressed accelerator.	0.59

PIEV time, under laboratory conditions, varies from about 0.2 to 1.5 seconds, for different types of stimuli ranging from the simple to the complex. However, the element of indecision was removed from these tests because the subjects were expecting the signal to which they were to respond, whereas the average motor vehicle operator is not.

A laboratory study[56] was conducted to eliminate this factor, and measure brake reaction times under unexpected circumstances. A group of 320 drivers was studied under standard test conditions for an *expected* stimulus. A small control group was repeatedly studied for both expected and surprise stimuli. The control group results provided correction factors by which the major test group results could be adjusted. The mean brake reaction time was 0.9 second (adjusted), and 25 per cent of those tested had reaction times greater than 1.2 seconds.

Since the cone of clear vision has a maximum of about 10 degrees, we must note that, in addition to PIEV time, time is required to shift one's line of clear vision to perceive a situation. An average reaction time of about 0.2 second is needed for every movement of the eyes. Therefore, when a driver is faced with situations requiring simultaneous checking of hazards on two opposite sides (as at a four-way intersection), an average reaction time of 0.6 second is required (maximum reaction time is 1 second). This is made up of 0.2 second to shift the eyes from right to left, 0.2 second to fixate the gaze long enough to see at left, and 0.2 second to shift back to the right again.

The American Association of State Highway Officials (AASHO)[16] recommends that a PIEV time of 2.5 seconds be used in determining safe stopping distance for all ranges of speed, and a PIEV time of 2.0 seconds be used for intersection sight distances.

Under actual conditions, total PIEV time ranges from about 0.5 second to 4.0 seconds, depending on the complexity of the situation. PIEV time increases with age, fatigue, alcohol, and physical deficiencies. Design criteria are usually based upon the normal range of road users (85th percentile group) and not upon the abnormal.

Complexity of the Job of Driving

A driver is not only required to operate his vehicle, but also to watch other traffic and pedestrians, to see and comprehend traffic control devices, to keep on route, and to avoid distractions.[17]

It has been shown[18] that tension-causing events encountered in driving on two urban streets occurred, on the average, every 29.2 seconds and 41.4 seconds, respectively. The superiority of controlled-access design in reducing tensions has been established.[19] Total stress incurred in driving is a more important determinant of route choice between a toll expressway and a parallel primary rural highway than either operating costs or travel time costs.[20]

Physical, Mental, and Emotional Limitations

Such limitations account for many accidents.[21] Existing driver licensing requirements are not adequate to provide proper protection of road users from physically and mentally deficient drivers.[22] A review of the state-of-the-art of attitude analyses and testing as regards drivers and their attitude toward safety produced inconclusive results on the reliability of the accuracy of such methods.[56]

A sample of negligent drivers (legally categorized because of traffic law violations and/or accident involvement) in California was divided into two groups. One group (treatment) received an individual informal hearing, at which the individual's record was discussed and various suggestions for improvement were made. The other group (control) received no hearing. It was found that traffic law violations for the *treatment* group were significantly reduced for the first year following the hearing, but that no significant differences were found between the two groups with respect to accident involvement in either the first or second subsequent year.[23]

Accident and violation rates for a sample of drivers in California were compared with driving test scores obtained by these drivers when they were examined for licensing. The results indicated that although there was a tendency for drivers with low performance test scores to have higher accident rates, differences in violation rates were not significant.[24]

A televised driver reeducation program was tested for its effectiveness in Lexington-Fayette County, Kentucky. Three-minute segments were shown, illustrating driver errors, some leading to accidents, with the correct action also shown and explained. A before-and-after study showed a significant reduction in both accidents and violations.[52]

Comparative studies[25-28] between "good" and "bad" drivers, who were differentiated by their degree of involvement in traffic violations and accidents, reveal that psychological factors are more significant than psychophysical factors (reaction time, vision, strength, etc.). Convictions are a more stable measure of driving record than accident experience.

Accident and conviction frequencies increase with increasing mileage, are lower for females than for males, and are highest for the younger age groups.[3,57]

Variability of Drivers

The variability of drivers and their attitudes with respect to age, sex, attentiveness, knowledge and skill in driving, nervousness, and impatience are important factors, and must be taken into consideration in establishing design criteria. Design values are normally based upon satisfying the needs of the 85th percentile drivers, with consideration given to the variability of the 15 per cent of drivers who are poorer.[29-31]

Driver's Desires

A driver's desires govern most driver actions. Drivers and pedestrians often do not react to controls and regulations unless they appear reasonable. This must be considered when controls and regulations are established; however, some regulations that might appear to be unreasonable to the road user, but are necessary for safety, must be backed by proper enforcement.

Effects of Fatigue

Studies of the effects of fatigue on driver performance are not conclusive. Many studies[32,33] have shown that drivers who are fatigued by various conditions do not exhibit a decrease in driver performance.

The AASHO Road Test provided an opportunity to study driver vigilance under actual driving conditions.[34] *Vigilance* is the ability of a driver to discern environmental signals over a prolonged period. Drivers were required to respond to light signals after over seven hours of driving, and the results showed that there was no significant deterioration in performance with time.[35]

In a 1,200-mile round trip field test,[36] a two-man team drove a vehicle equipped with a drivometer (an instrument devised to measure driver performance.[37,38] For the outgoing trip, the team regularly changed positions from data recording to driving. For the return trip, both men were mentally tired at the start of the trip and changed positions only when they felt too tired to drive farther. The outgoing trip was normal with respect to driver performance, but the return trip showed the effects of fatigue on driver performance. The effects of fatigue were measured by analyzing steering wheel reversal rates, speed change rates, and average vehicle speed.[39]

A laboratory study[40] of driver vigilance and target detection revealed that control subjects, who did not undergo any type of fatiguing conditions, performed significantly better than those who were subjected to fatiguing conditions. The effects of mental fatigue were found to be more pronounced than skill fatigue. The vigilance task involved the detection of a flash of light, and the target detection task involved the detection of an odd or irregular letter from a background of similar letters.[41-43]

Effects of Illumination on Driver Performance

A series of tests was carried out with the cooperation of the Connecticut State Highway Department to determine the effects of nighttime illumination on driver performance. Although drivers could not visually perceive the difference between 0.22 and 0.62 footcandle of illumination, driver characteristics were affected. Speeds were slightly increased at the higher level of illumination, but the variability of speeds was markedly increased. While steering activity remained fairly constant, gas pedal activity increased at the higher level.[53]

It was found that while lighting conditions did affect driver performance, roadway geometry had a far greater effect. For this reason the study was unable to draw practical conclusions from the experiment.[46]

Dynamic Judgment

Several tests have been conducted which indicate that most drivers, even those with good vision, are unable to accurately judge the speed of oncoming vehicles, or the closing rate, when executing a passing maneuver. Other tests have shown the driver's inability to estimate passing and overtaking distance in similar situations.[47-51]

Experiments have also indicated that providing drivers with external knowledge of oncoming vehicle speed improves their passing performance.[47]

The time needed to detect speed changes of a lead vehicle by a following driver increases as the spacing between the vehicles increases, and decreases with the increasing lead vehicle acceleration.[54]

Distraction

In testing the distractive influence of music on driving performance,[61] the following background conditions were considered:

1. No music
2. Slow music
3. Lively music

Driving speed increased for both music types over the no-music condition, and "weaving" (changes in lateral placement) decreased for both music conditions. More control activity (steering reversals, pedal movements) occurred with lively music than with either slow or no music.

The Driver and the Vehicle

The physical placement of the accelerator and the brake pedals are apparently inefficient.[58] Time between release of the accelerator and the depression of the brake pedal was measured under various stimuli. The normal placement of a brake 6 inches higher than the accelerator is compared to pedals at the same level, for seat heights of 17 and 12 inches, with the results shown in Table 1-1.

TABLE 1-1. COMPARISON OF NORMAL BRAKE-ACCELERATOR LOCATION WITH LEVEL PLACEMENT[54]

	ACCELERATOR-TO-BRAKE TIMES (*Sec.*) *Brake height differential*	
SEAT HEIGHT (*in.*)	*6 in.*	*Level*
17	0.309	0.194
12	0.337	0.183

PEDESTRIAN FACTORS

Pedestrian control is greatly needed because of the high injury and fatality rates due to traffic accidents. Controls include sidewalks, crosswalks, special pedestrian signals, pedestrian barriers, safety zones and islands, pedestrian tunnels and overpasses, and highway lighting, as well as enforcement control.

A special traffic signal for blind pedestrians helped reduce accidents for those so afflicted.[44] The traffic signal is operated by turning a special key in a lock mounted on the signal post. This causes the signal to change to a red phase, 6 seconds after which a bell begins to ring, indicating that it is safe to cross. The bell continues to ring for 20 seconds. The special signal-lock key was distributed to all blind persons in the area.

Physical, mental, emotional, and variability factors of pedestrians must be compensated for by such controls. These factors are similar to those of drivers, but they are compounded by the fact that many pedestrians lack knowledge of the rules of the road, and some are illiterate or unable to read English.

Pedestrian walking speeds range from 3.0 to 4.5 feet per second, and their reaction times range from 4.0 to 5.0 seconds. Both of these physical factors must be taken into consideration in determining minimum phases for the timing of traffic signals.

Observations of pedestrian movements at midblock and intersection locations revealed that the mean travel rate was 4.80 feet per second and 4.72 feet per second, respectively. The differences between males and females was found to be statistically significant. At midblock locations, males averaged 4.93 and females 4.63 feet per second; at intersections, males averaged 4.93 and females 4.53 feet per second.[45]

2

The Vehicle

The number of motor vehicles registered throughout the world in 1969 was over 230 million. About 48 per cent of these vehicles were registered in the United States, although it contains only 5.8 per cent of the world's population.[1] In 1968, 79 per cent of the families in the United States owned at least one automobile, and the nationwide average was one vehicle per 2.1 persons. The world-wide average at the same time was one vehicle per 17 persons.

PURPOSE AND TYPE

Vehicles vary widely in form, characteristics, and the purposes for which they are used. Statistics on vehicle types, including both public and privately-owned vehicles, for two years are shown in Table 2-1.

Size and Weight

Design standards are in part controlled by the dimensions of vehicles using the roadway.[49,50] Maximum weights and dimensions of motor vehicles permitted on various types of highway in each state are established by the state legislature.[6]

A *design vehicle* is a selected motor vehicle of a designated type, whose weight, dimensions, and operating characteristics are used to establish highway design controls to accommodate vehicles of that type. AASHO has established four design vehicles to be used as controls in geometric design. (See Table 2-2.)

The dimensions and minimum turning path of a design vehicle are controls that strongly affect the radius and width of pavement in intersection areas. Since most urban arterial highways carry heavy truck and bus traffic and have frequent intersections, a decision must be made as to which design vehicle or vehicles will control the design. The choice of design vehicle depends on the expected amount of turning traffic, but generally it may be based upon the following:[7]

Freeways: Use the *WB*-50, *WB*-40, or *SU* design vehicle, whichever is of importance in the turning movements.

Expressways, at grade: Use the *SU*, *WB*-40, or *WB*-50 design vehicle, where these are of particular importance. In outlying areas, at locations where trucks turn only

TABLE 2-1. MOTOR VEHICLE REGISTRATIONS BY TYPE

TYPE	*1966*	*1968*
Passenger cars	78.1×10^6	82.8×10^6
Trucks	15.5×10^6	17.1×10^6
Buses	3.2×10^5	
Totals	94.2×10^6	99.9×10^6

TABLE 2-2. AASHO DESIGN VEHICLES[51]

DIMENSIONS	*PASSENGER VEHICLE P*	*SINGLE-UNIT TRUCK OR BUS SU*	*SEMITRAILER COMBINATION WB − 40*	*SEMITRAILER COMBINATION WB − 50*
Wheelbase	11 ft	20 ft	13 + 27 = 40 ft	20 + 30 = 50 ft
Front overhang	3	4	4	3
Rear overhang	5	6	6	2
Overall length	19	30	50	55
Overall width	7	8.5	8.5	8.5
Height		13.5	13.5	13.5

occasionally, the *P* design vehicle may be used in conjunction with speed change lanes.

Major Streets: Preferably use the *SU* design vehicle; the *P* design vehicle may be used at minor cross streets where turns are not in significant numbers, and at major cross streets where the inclusion of parking lanes and pedestrian crosswalks make small-radius layouts acceptable.

A design check should be made to insure that the largest vehicle expected at that location can negotiate the designated turns, particularly where pavements are curved.

It is vitally important that fire-fighting and other emergency equipment should be capable of maneuvering on all city streets.

OPERATING CHARACTERISTICS

The operating characteristics that influence design are turning radii, acceleration, and braking.

Turning Radii

There are two different situations:

1. *Low-speed turns* (at less than 10 miles per hour), at intersections where the radius is controlled by the minimum turning paths of the vehicles. For the sharpest turn, 90 degrees, the minimum turning radius and the minimum simple-curve, pavement-edge radius for the design vehicles are shown in Table 2-3.

TABLE 2-3. MINIMUM TURNING AND CURVE RADII

DESIGN VEHICLE	*P*	*SU*	*WB − 40*	*WB − 50*
Minimum turning radius	24 ft	42 ft	40 ft	45 ft
Minimum pavement-edge radius	30	50	*	*
Off-tracking	2.7	5.1		

* For semitrailer combinations it is not practicable to fit simple curves to their minimum design paths, and therefore compound curves are used.

When vehicles turn at low speeds, the rear wheels track the front wheels on a shorter radius, and the difference between the radii of the rear and front wheels is known as *off-tracking*. Off-tracking is dependent on the turning radius and the vehicle wheelbase. Combination vehicles involve two or more wheelbase lengths, and off-tracking data are normally obtained by using scale models.[8-12]

2. *High-speed turns* (at up to about 0.7 times design speed), at intersections where the radius is controlled by the amount of superelevation (to overcome centrifugal forces in turning) and by the side-friction factor between the tires and the pavement. The safe coefficient of side friction usually used in design varies from 0.32 at 15 miles per hour to 0.16 at 40 miles per hour.

The minimum turning radius may be calculated with the following formula:

$$R = \frac{V^2}{15(e + f)} \tag{2-1}$$

where R = radius of curve, in feet
V = turning speed, in miles per hour
e = rate of superelevation, in feet per foot
f = coefficient of side friction

When vehicles turn at high speed, there is a tendency for the rear wheels to slip toward the outside of the curve. The *slip angle* denotes the angle between the desired path of motion and the actual path. The slip angle normally used in design is 3 degrees. It is because of off-tracking and slip angle that curves must be widened.

Drivers accept higher side-friction factors in operating around curves at intersections than on open highway curves. This is reflected in the above values when compared with those recommended for open highway curves; these vary directly with the design speed, from 0.16 at 30 miles per hour to 0.11 at 80 miles per hour.

Acceleration

Acceleration data are used to determine:

1. The time to cross an intersection from a standing start

2. The distance required to pass another vehicle
3. The gap acceptance

The rate of acceleration of passenger cars is from 4 to 6 miles per hour per second, which is equal to 6 to 9 feet per second per second; for trucks it is from 1.5 to 2 miles per hour per second, or 2 to 3 feet per second per second. Engine horsepower and transmission operation control acceleration, as do speed and grade.

The acceleration rate of vehicles is a maximum at low speeds, and it changes little for speeds up to about 20 miles per hour. The values shown above represent rates for speeds less than 20 miles per hour.

Actual acceleration rates must be determined by tests. In a study of passenger cars,[13] the instantaneous acceleration rates, for the average vehicle, were found to be 2.5 miles per hour per second at speeds between 20 and 35 miles per hour, decreasing to 2.0 miles per hour per second at 65 miles per hour.[14-18]

Braking

The braking ability of the vehicle and the forward-friction factor between tires and pavement control the slowing and stopping abilities of motor vehicles.

1. The deceleration, or slowing rate, cannot exceed that which pavement friction will provide, since skidding occurs when frictional resistance is overcome.
2. A vehicle coasting to a stop with motor disengaged is being decelerated by tractive resistance alone. This is made up of rolling resistance, or drag factor (the friction between tires and pavement and the inherent friction of the vehicle), plus air resistance.
3. In addition to tractive resistance, engine resistance (when the motor is engaged) helps to decelerate a vehicle. Tests made to measure the deceleration due to tractive and engine resistance, without brakes being applied, indicate that the deceleration varies from 3.2 feet per second^2 at 70 miles per hour to 1.3 feet per second^2 at 20 miles per hour.[13,16,19-21]
4. When brakes are applied suddenly, the brakes tend to lock to the wheels, and the vehicle will skid. Skid marks will indicate the distance through which a vehicle decelerated by skidding. Vehicle braking distance or skidding distance, neglecting the tractive and engine resistances, is given by the following formulas:

   a. For level roadways:
   When the vehicle comes to a halt at the end of a skid:

$$S = \frac{V^2}{30F} \qquad (2\text{-}2)$$

where S = the braking distance or skidding distance, in feet
V = the initial speed (when skid starts), in miles per hour
F = the coefficient of forward friction when all tires skid, or the drag factor

   When the vehicle has a speed of U miles per hour at the end of a skid:

$$S = \frac{V^2 - U^2}{30F} \qquad (2\text{-}3)$$

   b. For roadways on an up- or downgrade:
   If the roadway on which skidding occurs is not level, the skid marks will be longer on downgrades and shorter on upgrades. Allowance for grade is made by decreasing (for downgrades), or increasing (for upgrades), the drag factor by the rate of grade expressed as a decimal.

$$S = \frac{V^2}{30(F \pm G)} \quad \text{or} \quad S = \frac{V^2 - U^2}{30(F \pm G)} \qquad (2\text{-}4)$$

   The coefficient of friction is dependent on the speed and type of vehicle, the type and condition of pavement, and the type and condition of tires. Values recommended by AASHO[51] are given in Table 2-4 for wet, level pavement surfaces for comfortable stops. The above formulas, (2-1) through (2-4), are used with the coefficients of friction in Table 2-4 to obtain braking distances without skidding, which are used in determining stopping sight distances.

Extensive research is being conducted to develop accurate, economical methods of improving pavement coefficients of friction. Attempts to improve such coefficients include tire design, the design of surface courses, and investigations into the effects of surface applications in overcoming inclement weather conditions.[16,22-38,50]

Example

A vehicle is known to have skidded on a level asphalt surface ($F = 0.50$), and then on the adjacent gravel shoulder ($F = 0.60$), where it finally came to a halt. The average

TABLE 2-4. SAFE COEFFICIENTS OF FRICTION[51]

DESIGN SPEED (mph)	*ASSUMED SPEED FOR CONDITION (mph)*	*SAFE COEFFICIENT OF FRICTION*
30	28	0.36
40	36	0.33
50	44	0.31
60	52	0.30
65	55	0.30
70	58	0.29
75	61	0.28
80	64	0.27

length of the skid marks on the asphalt surface was 120 feet, and on the gravel shoulder 40 feet. What was the speed of the vehicle at the beginning of the skid?

The speed at the beginning of the gravel portion of the skid was:

$$U^2 = 30F_GS_G = 30 \times 0.6 \times 40 = 720$$
$$U = 26.8 \text{ miles per hour}$$

Then, the speed at the beginning of the asphalt portion of the skid was:

$$V^2 = 30F_AS_A + U^2 = 30 \times 0.5 \times 120 + 720$$
$$V = 50.2 \text{ miles per hour}$$

VEHICLE DESIGN

Vehicle design, with regard to the driver's visibility and vehicle headlighting, must also be considered in determining the portion of the road open or lighted to the driver's view.

Fatalities, injuries, property damage, and economic loss due to motor vehicle accidents all emphasize the critical need to consider human engineering aspects in the design of vehicles. The dimensions of seats, the location and placement of controls, steering wheel dimensions and positioning, visibility, and other factors should be based on human sizing criteria. In the design of passenger cars, there is a wide variation in these features among manufacturers.[39]

Windshield visibility is an important factor to consider in vehicle design. Obstructions to vision contribute to motor vehicle accidents.[40] The effectiveness of tinted windshields on visual functions has been studied, and it was found that there is no advantage to their use.[41,42] It was also found that tinted contact lenses were a source of danger when driving at night.[43-49,52]

3

The Roadway and Geometric Design

The geometric design of highways includes the visible elements of the highway or street. It deals with the grade line or profile, horizontal alinement, the several components of the cross section, sight distances, and intersections.

Geometric design is intimately related to the capabilities and limitations of the roadway user and his vehicle. Who will use the road and how often is also most important. Toward that end, traffic volume, speed, and composition are three major items to be considered in striving to provide safe, efficient, and economic traffic operations.

In preparing the design of a new highway or the redesign of an old one, the highway engineer must give attention to the following basic considerations:

1. The design must be adequate for the estimated future traffic volume, both average daily traffic and design peak hour, for the character of vehicles, and for the design speed.
2. The design must be safe for driving and should instill confidence in the majority of drivers.
3. The design must be consistent, and must avoid surprise changes in alinement, grade, or sight distance.
4. The design must be complete. It must include the necessary roadside treatment and provide essential traffic control devices, such as markings and signs, and proper lighting.
5. The design must be as economical as possible relative to initial costs and maintenance costs.

There are a number of nonengineering considerations which are important components of the overall design process. In terms of these, the highway engineer should attempt to achieve:

1. A design that is esthetically pleasing to the user and to those who live along the highway.
2. A design that is beneficial to the social and community values of the adjacent area.
3. A design that is ecologically harmless.

HIGHWAY CLASSIFICATION[3]

Highway systems are grouped into a number of different classifications for administrative, planning, and design purposes. Examples of the various systems are: the federal-aid financing system; state, county, and city administrative systems; and commercial-industrial-residential systems.

In the most basic classification system for design work, highways and streets are grouped into *Interstate*, *Primary* (excluding *Interstate*), *Secondary*, and *Tertiary* road classifications in rural areas, and *Expressway*, *Arterial*, *Collector*, and *Local* road classifications in urban areas. These groupings usually carry with them a set of suggested minimum design standards which are in keeping with the importance of the system and are governed by the specific transportation services the system is expected to provide. The main considerations for classing roads into systems are the travel desires of the public, land service needs based on existing and expected future land use, and the overall continuity of the system.

Four basic purposes of urban street systems were proposed by the National Committee on Urban Transportation and have been generally accepted as a rational basis for grouping road use and service. They are:

1. *Freeway system* (including expressways and parkways): This provides for rapid and efficient movement of large volumes of through traffic between areas and across the urban area. It is not intended to provide land access service.
2. *Major arterial system:* This provides for through traffic movement between areas and across the city with direct access to abutting property. It is subject to required control of entrances, exits, and curb use.
3. *Collector street system:* This provides for traffic movement between major arterials and local streets, with direct access to abutting property.
4. *Local street system:* This provides for direct access to abutting land and for local traffic movement.

These basic purposes of city street systems are similar in function to those of rural Interstate, Primary, Secondary and Tertiary highways, respectively, insofar as the degrees of service to through traffic and land access are concerned. It is important to remember, however, that regional as well as national highway transportation requirements must also be met by rural highways. Table 3-1 compares the overall criteria of urban street and rural road systems.

DESIGN DESIGNATIONS[1]

Design designations is a term for the major controls or services for which a given highway is designed. It relates to highway classifications as they include certain elements of traffic requirements with which the design designation should not conflict. It should be concise enough to be readily understood, but sufficiently complete to show the major controls that characterize the whole design.

Such major controls are: traffic volume, character and composition of traffic, design speed, and control of access. These determine the principal geometric features of a highway. Other controls and criteria such as topography, physical features, capacity, safety, and economics are important, but are either reflected in the major controls previously mentioned, or have to do with more detailed design features which are not included in the general designation.

Traffic Volumes

This is the first major control and should include information relevant to both current and future traffic volumes. This is best expressed in terms of average daily traffic (ADT), with the current year and future (design) year noted. Most significant is the design hour volume

TABLE 3-1. CRITERIA FOR FACILITY CLASSIFICATION

CLASSIFICATION	*PRINCIPAL SERVICE FUNCTION*	*TYPICAL TRIP LENGTHS (mi)*	*ELEMENTS LINKED BY FACILITY*	*DESIRABLE SPACING (mi)*	*PER CENT OF SYSTEM*
URBAN:					
Expressway	Through movement exclusively	3	CBD, major generators	1–3	0–8
Arterial	Through movement, some land access	1	CBD, secondary generators	1	20–35
Collector	Through movement and land access	1	Local areas	½	
Local	Land access	½	Individual land sites	—	65–80
RURAL:					
Interstate	Through movement exclusively	—	Major cities	—	2
Primary	Through movement, some land access	—	Smaller cities	—	17
Secondary	Through movement and land access	—	Small cities and outlying regions	—	10
Tertiary	Land access, some through movement	—	Farm-to-market	—	71

(DHV), a two-way volume. Also of importance, particularly on multi-lane facilities, is the directional distribution of traffic during the design hour, expressed as the percentage (D) of DHV in the predominant direction of travel. The combination of these items designates the essentials of traffic volume data for a highway improvement. Both ADT and DHV should preferably be expressed in terms of mixed traffic, that is, passenger cars and trucks combined. In some instances it may be desirable to express these volumes in terms of equivalent passenger cars, since capacity determinations are sometimes calculated on this basis.

Character of Traffic

This factor should describe the proportion of trucks (excluding light delivery trucks) in the traffic stream. Since DHV is the controlling volume in geometric design, it follows that trucks should be expressed as a percentage of this volume.

Design Speed

Design speed is basic to the overall standards. Together with DHV and per cent of trucks, it is indicative of the speeds and type of operation to be expected.

The Degree of Access Control

Control of access is designated as *none*, *partial*, or *full*, and describes the extent of roadside interference, or restriction to freedom of movement, which affects so many elements of highway design.

A method suggested by AASHO for indicating the design designations for highway improvements is:

Control of access	(full, partial, none)
ADT (current)	(vehicles per day)
ADT (design year)	(vehicles per day)
DHV	(vehicles per hour, design hour volume)
D	(per cent of directional flow)
T	(per cent of trucks in DHV)
V	(design speed, in miles per hour)

DESIGN VEHICLES

Almost all highways carry both passenger cars and trucks, and design standards for different types of vehicles have been adopted by AASHO. The maximum dimensions, weights, and power of motor vehicles are the basic considerations involved in the design of highways. The specifications for the various AASHO design vehicles were detailed in Chapter 2.

HIGHWAY CROSS-SECTION ELEMENTS

The design of each of the elements of the highway cross section depends on the use to be made of the facility in question. Roads with higher design volumes and speeds require more lanes, flatter grades, and more gentle curves than those with lower volume, lower design speeds. Furthermore, for the former type of facility, consideration should be given to wider shoulders and medians, separate turning lanes, and control of access.

The highway cross section is made up of design elements which can be classified into three broad groups:

1. *The traveled way:* surface, width, cross slope, and number of traffic-bearing lanes.
2. *Road margins:* shoulders, sidewalks, curbs, guardrails, ditches, and the roadside.
3. *Traffic separation:* the median.

Together, these relevant elements determine the right-of-way required to provide a facility that will perform satisfactorily for its planned lifetime.

Pavement Surfaces

The surfaces of the traveled way are classified into three general types: *high*, *intermediate*, and *low*. The type of surfacing provided is determined by the volume and character of traffic, the availability of materials and of experienced contractors, the initial cost, and the extent and cost of maintenance.

Heavy traffic volumes call for a smooth riding surface with good all-weather anti-skid properties. The surface should be constructed to retain these qualities with a minimum of maintenance and interference to traffic operations for the life of the pavement. A smooth surface of this kind offers little frictional resistance to the flow of surface water, and may be designed with minimum cross slopes. Conversely, low-type rough surfaces must be crowned enough to drain well. Low-type surfaces tend to reduce operating speeds.

Pavement color has a direct bearing on traffic operations and safety. Light-colored pavements offer better visibility at night than do dark-colored pavements, with either headlight or streetlight illumination. This is because lighter pavements offer better reflective qualities and also provide a more contrasting background for dark-colored obstacles or objects upon the traveled way. Pavement color also may be used to contrast the shoulder with the driving lanes, to define the major roadway at a junction, or to indicate exit and entrance ramps at interchanges.

Lane Widths

These have a great influence on the safety and comfort of driving. Ten- to 12-foot lane widths are standard, and the tendency is to use the larger value with the continued upward trend in traffic volumes, vehicle speed, and widths of trucks. Lane widths narrower than 12 feet can adversely affect capacity and safety, so their use should be limited to other than high-speed, high-volume facilities.

Lane widths of 13 and 14 feet have been used on some high-speed, rural, two-lane roads. Attempts to use wider lanes lead to the practice of some drivers of using such roads as multi-lane facilities.

Normal Cross Slopes

These have a crown or high point in the middle and slope downward toward both edges to facilitate drainage. The downward cross slope may be a plane or curved section, or a combination of curve and plane. Curved cross slopes usually are parabolic. On divided highways each one-way pavement may be crowned separately, as on two-lane highways.

Since many highways are on a tangent or a flat-curve alinement, the rate of cross slope is an important element in cross-section design. Rates should be as low as practicable for vehicle operation, but they must be high enough for proper drainage. Where two or more lanes are inclined in the same direction on multi-lane pavements, each successive lane outward from the crown line should preferably have an increased slope. The lane adjacent to the crown line should be pitched at the normal minimum slope and on each successive lane outward the rate should be increased by $\frac{1}{16}$ inch per foot. Normal cross slopes vary from a minimum of $\frac{1}{8}$ inch per foot on high-type pavement surfaces to $\frac{1}{4}$ inch per foot.

Shoulders

Adjacent to the traveled way shoulders are provided for the accommodation of stopped vehicles, for emergency use, and for lateral support of the base and surface courses. The need for shoulders is determined by the type of highway, the traffic volume, the speed of traffic, traffic composition, and the type of terrain. Shoulders vary from a minimum width of 4 feet to a desirable width of 12 feet for heavily traveled and high-speed roads. Full-width usable shoulders (8 to 12 feet) should be provided on highways where the DHV exceeds 100 vehicles per hour.

An important element of shoulder design is its cross section. Normally shoulders are sloped to drain away from the traveled way, and are an important link in the lateral surface runoff system. Shoulders must be sloped sufficiently to remove surface water from the pavement areas, but not to the extent that vehicle use would be hazardous.

Sidewalks

In the city, sidewalks are accepted as integral parts of the streets. However, the need for sidewalks is great in many rural areas because high speeds and a general lack of adequate lighting make it hazardous to walk on the traveled way. The most pressing need for sidewalks in rural and suburban areas is at points of community development, such as schools, local businesses, industrial plants, etc. Justification for sidewalks in these areas is dependent upon the vehicle-pedestrian hazard, which is governed chiefly by the volume of pedestrian and vehicular traffic and the speed of vehicular traffic. In general, wherever the roadside and land development conditions are such that pedestrians regularly move along a main or high-speed highway, sidewalks should be provided.

Curbs

By their design and location curbs can appreciably affect drivers, and therefore the safety and utility of a highway. Curbs are used to control drainage, prevent vehicles from leaving the pavement at hazardous points, delineate the edge of the pavement, present a more finished appearance, and assist in the more orderly development of the roadside.

Curbs are needed most on highways in urban areas. In strictly rural areas there is extensive mileage without curbs, but curbs are frequently used in rural design for one or more of the purposes described.

There are two general classes of curbs: barrier and mountable. Barrier curbs are relatively high and steep-faced, designed to prevent vehicles from leaving the pavement. They range from 6 to 20 inches in height, and may have a two-step section. Barrier curbs are used on bridges, piers, sidewalks (maximum height of 8 inches to prevent scraping of bumpers), and in some instances on medians. Mountable curbs are low curbs or curbs with flat, sloping faces, or both, and are designed so that vehicles can cross them without a severe jolt. Mountable curbs are used primarily on medians, at the inside edge of shoulders, and to outline channelizing islands in intersection areas.

Guardrails and Guide Posts

These are used where vehicles accidentally leaving the highway would be subjected to considerable danger. Generally such points are fills on steep grades, long through fills, or fills on sharp curvatures. The need for guardrails or guide posts is related to the fill slope; they may be omitted where the fill slope is 1 on 4 or flatter, because a driver, forced onto such a slope, has a chance of regaining control of his vehicle.

While guardrails or guide posts are regularly used on fill sections, their need is also recognized at abrupt changes in shoulder width and at approaches to structures.

The choice of providing guardrails or guide posts is largely dependent on the hazard involved. Guardrails are designed to resist impact by deflecting the vehicle so that it continues to move at a reduced velocity along the guardrail. Any abrupt stop of a vehicle is dangerous, and a guide post, or a projection on guardrails, which might snag a moving vehicle, is not desirable. Guide posts are generally not intended to resist impact, and they are used primarily to delineate the direction of the road, particularly at night.

Side Slopes

On cut and fill sections side slopes should be flattened and liberally rounded where they intersect the ground surface. Effective erosion control, low-cost maintenance, stabilization of soil, and adequate drainage of the subgrade are dependent upon proper shaping of the side slopes.

Medians

Used to separate opposing lanes of travel, medians are an essential feature on a multi-lane highway. Almost all modern highways planned for four or more lanes in the final stage are being designed with a median.

Medians on rural highways should be of sufficient width to provide the desired freedom from interference of opposing traffic, to minimize headlight glare, to include space for safe operation of crossing and turning vehicles at intersections at grade, and to provide a haven in case of emergency.

Medians are provided primarily to separate opposing traffic streams, and should therefore be highly visible both day and night, contrasting definitely with the through-traffic lanes, and as wide as feasible. Medians range in width from a minimum of 4 feet, to 60 feet or more, and are grouped into several cross-sectional types depending upon width, treatment of the median area, and drainage arrangements. A median need not be of constant width, and the two pavements need not be at the same elevation.

Right-of-Way Widths

These should be sufficient to include all of the cross-section elements plus an adequate border. The border is the area between the roadway edge and the right-of-way (ROW) extremity line. These areas are used for the accommodation of pedestrians, the location of utilities, the placement of signs and other controls, and as a green-belt strip, enhancing the appearance of both the public highway and the property adjacent to it. It is advisable to purchase sufficient ROW to take care of the ultimate development of the highway, because as land development proceeds, additional ROW can only be purchased at high cost.

Minimum and desirable ROW widths vary considerably with the type of highway or street. For example, AASHO's recommendations for different types of rural highways are as follows:[1]

HIGHWAY TYPE	*RANGE OF ROW WIDTHS (feet)*
Two-lane	66–120 (minimum is for low-type highways)
Three-lane	100–140
Four-lane or more	90–310 (minimum is for construction in restricted areas)

HIGHWAY ALINEMENT

Highways may be considered as being made up of a number of "straight" and "curved" sections. The straight pieces are called *tangent sections* and the curved pieces are called *circular horizontal curves.* In profile there are grades (plus, minus), level sections, and parabolic vertical curves.

Horizontal Alinement

The horizontal alinement of a highway is shown on the plan view, and is a series of tangents connected by circular curves. It is good practice to interpose transitional or spiral curves between tangents and circular curves. Alinement must be consistent; sudden changes from flat to sharp curves, and long tangents followed by sharp curves, must be avoided to reduce accident hazards. The use of compound, broken-back curves and reverse curves is objectionable unless suitable transitions are provided between them. Long curves of very small degree are preferable at all times, as they are pleasing in appearance and riding qualities and decrease the possibility of future obsolescence.

For balance in highway design, all geometric elements

should, as far as economically practicable, be selected to provide safe, continuous operation at the design speed. In the design of highway curves, it is necessary to establish the proper relationship between design speed and curvature, and also their joint relationship with superelevation. While these may stem from the laws of mechanics, the actual values for use in design depend upon practical limits and factors determined more or less empirically.

Superelevation of Curves

This is necessary to overcome the tendency of vehicles to slide away from the center of the curve, or to overturn, due to centrifugal force. Centrifugal force acts above the roadway surface through the center of gravity of the vehicle, and creates an overturning moment about the points of contact between the outer wheels and the pavement. Opposing overturn is the stabilizing moment created by the weight acting downward through the center of gravity. Overturn can occur only when the overturning moment exceeds the stabilizing moment. Modern passenger cars have low centers of gravity, and consequently the overturning moment is relatively small. As a result, they will slide sidewise rather than overturn. Many trucks have high centers of gravity, so that relatively large overturning moments can be created; they may overturn before they slide.

On a flat circular curve, the only force resisting sliding is the frictional force between the pavement and the tires.

On a superelevated curve, both the tendency to slide and the tendency to overturn can be completely eliminated if friction is negated, by superelevating sufficiently so that the component of weight parallel to the roadway surface equals the component of centrifugal force parallel to the roadway surface.

Figure 3-1 shows the relationship of forces when a vehicle traverses a superelevated curve. To resist sliding and overturning (neglecting friction), W_h must equal C_h. Thus:

$$W \sin\theta = \frac{WV^2}{gR} \cos\theta \qquad (3\text{-}1)$$

Dividing both sides by $W \cos\theta$ gives

$$\tan\theta = \frac{V^2}{gR} \qquad (3\text{-}2)$$

But $\tan\theta = e$, so

$$e = \frac{V^2}{gR} \qquad (3\text{-}3)$$

Changing V feet per second to V miles per hour

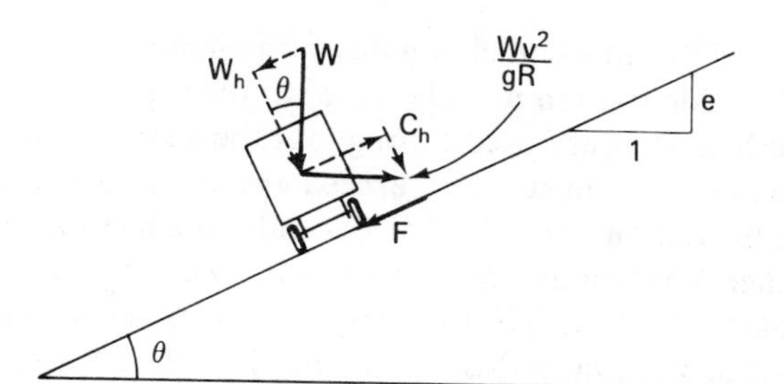

Legend
- W: Weight of vehicle, in pounds
- θ: Angle of pavement slope
- v: Velocity, in feet per second
- e: Superelevation = tan θ
- F: Side friction
  - F = fW cos θ where f is the coefficient of friction
- g: 32.2 feet per second²
- R: Radius of curve, in feet

Fig. 3-1. Conditions for vehicle traveling around curve

$$e = \frac{V^2}{15R} \qquad (3\text{-}4)$$

When a vehicle travels at a speed greater than that at which the superelevation balances all centrifugal force, the additional balancing force required comes from friction.

Considering both friction and superelevation, the resistance to sliding and overturning becomes

$$e + f = \frac{V^2}{15R} \qquad (3\text{-}5)$$

or

$$e = \frac{V^2}{15R} - f$$

Maximum coefficients of side friction on dry pavements range from 0.4 to 0.5. However, drivers become uncomfortable long before side friction approaches values at which slipping starts, because it is difficult to keep the vehicle in control. AASHO standards, based on studies of driver discomfort at different design speeds, recommend side-friction factors as shown in Table 3-2.

For practical reasons, the full theoretical rate of superelevation cannot be realized. When a vehicle is traveling at slower than the design speed, the centrifugal force is lessened, and there is a tendency to slide down the incline. For standing vehicles there is no centrifugal

TABLE 3-2. MAXIMUM SAFE SIDE-FRICTION FACTORS[1] (when e = .12)

Design speed (mph)	30	40	50	60	70	80
Maximum safe side-friction factors	0.16	0.15	0.14	0.13	0.12	0.11

force, and to prevent sliding down the incline the coefficient of side friction must be equal to the superelevation. Since highways are used throughout the year, maximum superelevation must never exceed the minimum coefficient of friction that will develop under the most adverse weather conditions. Maximum superelevation recommended by AASHO is 0.12 feet per foot. If snow and ice conditions prevail, this maximum is reduced to 0.08 feet per foot. On icy surfaces, the coefficient of side friction may be 0.05 or less; therefore skid-prevention measures must be taken by vehicle operators or maintenance personnel. In recent practice, superelevation up to 0.16 feet per foot has proved satisfactory on ramps at interchanges.

Circular Curves

These are described by giving either the radius or the degree of curve, which is the central angle subtended by a 100 foot length of curve. *Degree of curve* is inversely proportional to the radius, and the relationship between them is expressed as follows:

$$D = \frac{5729.58}{R} \qquad (3\text{-}6)$$

where D = degree of curve
R = radius of curve, in feet

The standards for curvature are intimately related to practical superelevation and side-friction factors. The relationship for superelevation can be expressed in terms of the radius or the degree of curve:

As previously noted, $$e + f = \frac{V^2}{15R} \qquad (3\text{-}5)$$

Thus, $$R = \frac{V^2}{15(e + f)} \qquad (3\text{-}7)$$

Note that the resulting equation (3-7) is the same equation as cited previously in Chapter 2.

Substituting the value for R given above,

$$D = \frac{85{,}950(e + f)}{V^2} \qquad (3\text{-}8)$$

For the maximum degree of curve allowed by AASHO standards, superelevation and coefficient of side friction are set at the maximum values.

For curves flatter than those requiring maximum permissible superelevation and side friction, either or both may be reduced. AASHO standards make no specific recommendations in this regard, and practice varies among agencies. Many agencies recommend that curves sharper than 1 degree shall be superelevated.

Spiral Easement Curves

These curves are provided to form a smooth transition between tangent and curved sections, since no superelevation is required on tangents, and full superelevation is required upon entering a circular curve. This insures the safety of the vehicle and the comfort of its occupants at high operating speeds. This transition is sometimes accomplished by means of a spiral curve. The effect of the spiral easement curve is to gradually change the radius from infinity on the tangent to that of the circular curve, so that centrifugal force also develops gradually. By careful application of superelevation along the spiral, a smooth and gradual application of centrifugal force can be effected.

The minimum length of easement curve recommended by AASHO for primary highways is given by the following expression:

$$L_s = 1.6\frac{V^3}{R} \qquad (3\text{-}9)$$

where L_s = length of easement curve, in feet
V = speed, in miles per hour
R = radius of circular curve, in feet

Easement curves are usually omitted from one-degree and flatter circular curves. AASHO standards for the Interstate system recommend that curves sharper than 2 degrees have spiral transitions.

Easement curves should also be used to separate reverse and compound curves (where degrees of curvature differ by more than 5 degrees).

Superelevation Runoff

The length required to effect the change in cross section from a normal crown section on tangents to the fully superelevated section required at the beginning of the horizontal curve, or vice versa, is called the superelevation runoff. To meet the requirements of comfort and safety, the superelevation runoff should be effected uniformly over a length adequate for the likely travel speeds.

The desired gradual change in cross section is provided by applying it over the entire length of the spiral, when it is used to accomplish the transition from tangent to circular section.

Pavement Widening on Curves

Pavement widening is used to make operating conditions on curves comparable to those on tangents. Pavement widening may be required on highway curves because of the manner in which the vehicle traverses the curve. As shown in Figure 3-2, the rear wheels of a vehicle will track around a curve on a shorter radius than do the

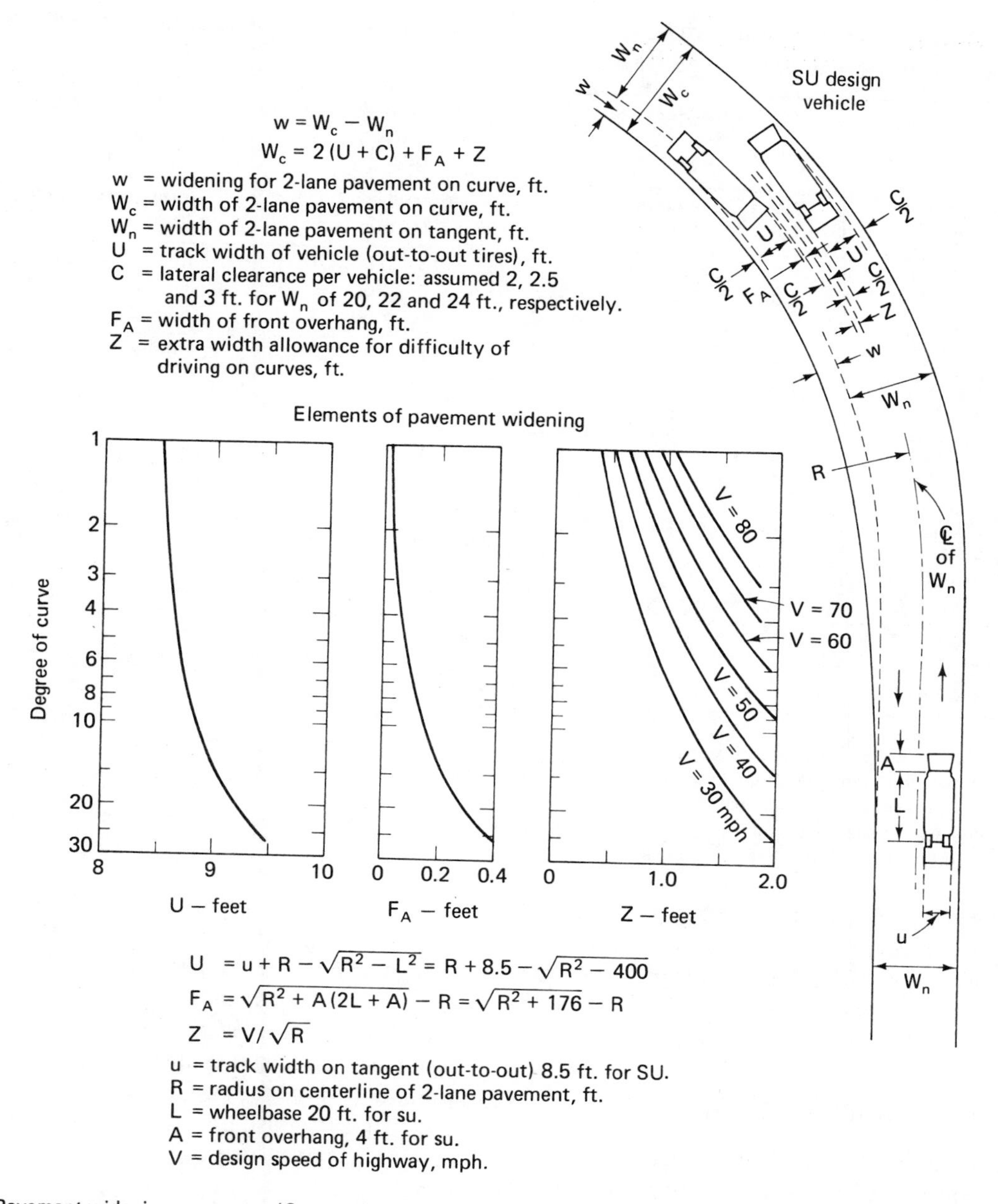

Fig. 3-2. Pavement widening on curves (*Source:* A Policy on Geometric Design of Rural Highways, *AASHO, 1965, Figure III-12*)

front wheels. Thus the vehicle occupies more pavement width than it does on a tangent section. When the curve is sharp and the vehicle relatively large, the extra width used may be appreciable.

There are numerous variations in applying widening to curves in current state highway department design practices. Several different formulas and other empirically derived methods are in use.

In general, two-lane pavement widening varies from 0 feet to 7 feet; maximum values in the range of 2 feet to 4 feet are the most widely used. For practical reasons (extra costs and details), and because of the rather empirical nature of the extra-width determination, design values for widening should be multiples of $\frac{1}{2}$ foot, and the minimum value should be 2 feet.

Values recommended by AASHO are calculated from a formula which takes into consideration the dimensions of the design vehicle, lateral clearance between vehicles, the radius of the curve, and design speed. This formula is given in Figure 3-2.

On four-lane undivided highways, the pavement widening should be double the values obtained from the formula. Generally, for two-lane and for four-lane divided highways, no widening is suggested for pave-

TABLE 3-3. CALCULATED AND DESIGN VALUES FOR PAVEMENT WIDENING ON OPEN HIGHWAY CURVES[1]
Two-Lane Pavements, One-Way or Two-Way

	WIDENING, IN FEET, FOR TWO-LANE PAVEMENTS, ON CURVES FOR WIDTH OF PAVEMENT ON TANGENT OF:														
	24 feet						*22 feet*					*20 feet*			
DEGREE OF CURVE	*Design Speed (mph)*						*Design Speed (mph)*					*Design Speed (mph)*			
	30	*40*	*50*	*60*	*70*	*80*	*30*	*40*	*50*	*60*	*70*	*30*	*40*	*50*	*60*
1	0.0	0.0	0.0	0.0	0.0	0.0	0.5	0.5	0.5	1.0	1.0	1.5	1.5	1.5	2.0
2	0.0	0.0	0.0	0.5	0.5	0.5	1.0	1.0	1.0	1.5	1.5	2.0	2.0	2.0	2.5
3	0.0	0.0	0.5	0.5	1.0	1.0	1.0	1.0	1.5	1.5	2.0	2.0	2.0	2.5	2.5
4	0.0	0.5	0.5	1.0	1.0		1.0	1.5	1.5	2.0	2.0	2.0	2.5	2.5	3.0
5	0.5	0.5	1.0	1.0			1.5	1.5	2.0	2.0		2.5	2.5	3.0	3.0
6	0.5	1.0	1.0	1.5			1.5	2.0	2.0	2.5		2.5	3.0	3.0	3.5
7	0.5	1.0	1.5				1.5	2.0	2.5			2.5	3.0	3.5	
8	1.0	1.0	1.5				2.0	2.0	2.5			3.0	3.0	3.5	
9	1.0	1.5	2.0				2.0	2.5	3.0			3.0	3.5	4.0	
10–11	1.0	1.5					2.0	2.5				3.0	3.5		
12–14.5	1.5	2.0					2.5	3.0				3.5	4.0		
15–18	2.0						3.0					4.0			
19–21	2.5						3.5					4.5			
22–25	3.0						4.0					5.0			
26–26.5	3.5						4.5					5.5			

NOTE: Values less than 2.0 may be disregarded.
Three-lane pavements: multiply above values by 1.5.
Four-lane pavements: multiply above values by 2.
Where semitrailers are significant, increase tabular values of widening by 0.5 for curves of 10 to 16 degrees, and by 1.0 for curves 17 degrees and sharper.

ments 24 feet wide; for 22-foot pavements, widening is not necessary where curves are 10 degrees or flatter.

Figure 3-2 gives the derivation of design values for pavement widening. Design values are given in Table 3-3.

Attainment of Widening on Curves

As with superelevation, the widening on curves is made gradually. Where spiral transition curves are used, widening may be placed on the inside, or divided equally between the inside and outside of the circular curve. On unspiraled curves, widening should be applied on the inside edge of pavement only. The maximum widening should be used for the entire length of the circular curve, and should vary from 0 at the point of spiral to the full amount at the point of circular curve. Regardless of how widening is applied, the final marked centerline should be placed midway between the edges of the widened pavement.

VERTICAL ALINEMENT

The grade line is shown on a profile taken along the centerline of the highway, and is a series of straight lines connected by parabolic vertical curves to which the straight lines are tangent. In establishing this grade line, the designer must insure economy by keeping earthwork quantities to the minimum consistent with meeting sight distance and other design requirements. In mountainous country, the grade may be set to balance excavation against fill, as an aim toward least overall cost. In flat or prairie country, the grade will be approximately parallel to the ground surface, but sufficiently above it to allow surface drainage and, where necessary, to permit the wind to clear drifting snow. Under all conditions, smooth, flowing grade lines should be the goal of the designer. When two grade lines having a difference in rate of more than $\frac{1}{2}$ per cent intersect, a vertical curve is used for a proper transition.

Maximum Grades

These are determined principally through consideration of vehicle operating characteristics on grades. Driving practices with respect to grades vary greatly, but there is general acceptance that nearly all passenger cars can readily negotiate grades as steep as 7 or 8 per cent, except for cars with high weight-to-horsepower ratios.

TABLE 3-4. RELATION OF MAXIMUM GRADES TO DESIGN SPEED FOR MAIN HIGHWAYS[1]

DESIGN SPEED (*mph*)	*MAXIMUM GRADE** (*per cent*)
30	6–9
40	5–8
50	4–7
60	3–6
70	3–5
80	3–4

* First value is for flat terrain, second value for mountainous terrain.

Studies show that operation on a 3 per cent upgrade, compared to that on a level road, has only a slight effect on passenger-car free speeds. On steeper grades, passenger-car speeds decrease progressively with an increase in grade. The effect of grades on truck speeds is much more pronounced than on passenger-car speeds. On upgrades, the maximum speed a commercial vehicle can maintain is dependent on the length and steepness of the grade and on the ratio of the gross vehicle weight to engine horsepower (weight-power ratio). It has been found that trucks with a weight-power ratio of about 400 have acceptable operating characteristics, in terms of being part of the traffic stream. Such a weight-power ratio will assure a minimum speed of about 15 mph on a 3 per cent upgrade.

Maximum grades are recommended on the basis of design speed control (Table 3-4).

Maximum grades for highways of a secondary nature are in the range of 1.2 to 1.5 times those shown in Table 3-4.

Critical Lengths of Grade for Design

In addition to the maximum allowable grade, critical lengths of grade for design must be considered. The term *critical length of grade* is used to describe the maximum length of a designated upgrade upon which a loaded truck can operate without an unreasonable reduction in speed. A reduction in average running speed of 15 mph is used by AASHO as a design control. AASHO has developed a curve relating critical lengths of grade for design to different per cents of upgrade for truck operation. This curve is shown in Figure 3-3.

Minimum grades are governed by drainage conditions. Flat or level grades on uncurbed pavements are virtually without objection when the pavement is adequately crowned to drain the surface water laterally. In cut sections, level pavement grades introduce the problem of obtaining sufficient slope in the side drainage channels. For this purpose grades should not be less than 0.5 per cent, though paved gutters will function with somewhat flatter grades. With curbed pavements longitudinal grades should be provided to facilitate surface drainage. A minimum grade for the usual case is 0.5 per cent, but a grade of 0.34 per cent may be used where there is a high-type pavement, accurately crowned.

Combined Curvature and Grade

These items are of particular concern in mountainous country, where the combination of sharp curves with steep grades occurs along the best located highway, causing a deleterious effect beyond that caused either by grade or curve alone. The practice of most agencies is

Fig. 3-3. Critical lengths of grade for design (*Source:* A Policy on Geometric Design of Rural Highways, *AASHO, 1965, Figure III-1*)

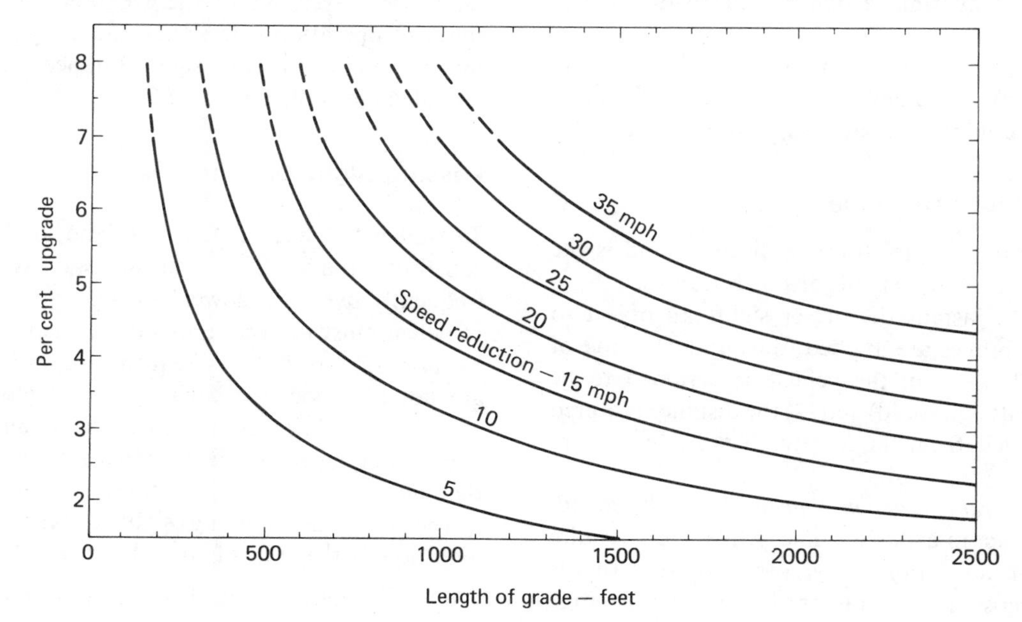

to reduce the grade for the length of curve, thus compensating for the increased resistance to vehicle operation caused by curvature. Many formulas have been used to determine the magnitude of this grade reduction. One such formula is

$$\text{Compensation in per cent of grade} = \frac{250}{R} \qquad (3\text{-}10)$$

where R = radius of curve, in feet

For example, on a curve of 500-foot radius, the grade would be reduced by $\frac{250}{500} = \frac{1}{2}$ per cent, say from 7 per cent to $6\frac{1}{2}$ per cent.

The use of grade compensation is commonly limited to curves sharper than 6 degrees and to grades of 5 per cent or greater. Another specification is that all grades over 5 per cent, where the radius of the horizontal curve is smaller than 1000 feet, should be compensated. These two specifications are essentially the same.

SIGHT DISTANCE

Sight distance is the length of highway ahead which is visible to the driver. The ability to see ahead is of the utmost importance in the safe and efficient operation of a highway. If safety is to be built into highways, the designer must provide a sight distance of sufficient length so that a driver can avoid striking an unexpected obstacle on the traveled way. This sight distance is called the *nonpassing sight distance* or the *minimum safe stopping distance*. In addition, the designer should provide, at frequent intervals along two- and three-lane highways, sufficient sight distance to enable drivers to pass vehicles without hazard. This sight distance is called the *passing sight distance*.

Sight distance at every point along every highway should be as long as possible, but at no point should it be less than the minimum safe stopping distance.

Nonpassing Sight Distance

This is the minimum safe stopping distance, and is the sum of two distances: (1) the distance traversed by a vehicle from the instant the driver sights an object to the instant the brakes are applied (during this period of perception and reaction, the vehicle is assumed to be traveling at the design speed) and (2) the distance required to stop the vehicle from the instant that the brakes are applied.

The average *brake reaction time* is about $\frac{1}{2}$ second. Some drivers react in a shorter time, while many require a full second or more. For safety, a reaction time that is sufficient for most drivers, rather than for the average driver, should be used in any determination of minimum sight distance. A brake reaction time of a full second is considered to satisfy this requirement. *Perception time* is the time required for a driver to come to the realization that the brakes must be applied. Research data on perception time are very limited. Most available information combines perception time with brake reaction time. From this it is evident that perception time is greater than brake reaction time. A perception time of 1.5 seconds is assumed to be large enough to include the time taken by nearly all drivers under most roadway conditions.

Therefore, according to AASHO, the total of brake reaction time and perception time is taken as 2.5 seconds and, based on available data, there is no marked difference in this time over the range of design speeds.

Nonpassing or minimum safe stopping distance (d, in feet), on level roadways or grades, may be calculated by use of the following formulas:

$$d = 1.47Vt + \frac{V^2}{30f} \qquad \text{on level roadway} \qquad (3\text{-}11)$$

$$d = 1.47Vt + \frac{V^2}{30(f \pm G)} \qquad \text{on grade} \qquad (3\text{-}12)$$

where

V = speed, in miles per hour
t = reaction time (2.5 seconds)
f = coefficient of skidding friction
G = grade, in per cent

For design calculations, a coefficient of friction equivalent to that on *wet pavements* is assumed for reasons of safety. However, under these conditions, the assumed speed of operation is less than the design speed. Values for minimum stopping sight distance which must be provided are shown in Table 3-5.

Passing Sight Distance

Two-Lane Highways. On two-lane highways, which constitute the bulk of our highway system, vehicles frequently overtake slower-moving vehicles, the passing of which must be accomplished on a lane regularly used by opposing traffic. If frequent passing opportunities are not provided, the capacity of a highway decreases and accident hazards increase. The minimum distance ahead that must be clear to permit passing is called the *minimum passing sight distance*.

The minimum passing sight distance for two-lane highways is the sum of four distances (AASHO):

d_1 = distance traversed during the preliminary delay

TABLE 3-5 MINIMUM STOPPING SIGHT DISTANCE[1]

Design speed (mph)	Assumed speed for condition (mph)	Perception and brake reaction: Time (sec)	Perception and brake reaction: Distance (ft)	Coefficient of friction, f	Braking distance on level (ft)	Stopping sight distance: Computed (ft)	Stopping sight distance: Rounded for design (ft)
			Design Criteria—WET PAVEMENTS				
30	28	2.5	103	.36	73	176	200
40	36	2.5	132	.33	131	263	275
50	44	2.5	161	.31	208	369	350
60	52	2.5	191	.30	300	491	475
65	55	2.5	202	.30	336	538	550
70	58	2.5	213	.29	387	600	600
75*	61	2.5	224	.28	443	667	675
80*	64	2.5	235	.27	506	741	750
			Comparative Values—DRY PAVEMENTS				
30	30	2.5	110	.62	48	158	
40	40	2.5	147	.60	89	236	
50	50	2.5	183	.58	144	327	
60	60	2.5	220	.56	214	434	
65	65	2.5	238	.56	251	489	
70	70	2.5	257	.55	297	554	
75	75	2.5	275	.54	347	622	
80	80	2.5	293	.53	403	696	

* Design speeds of 75 and 80 mph are applicable only to highways with full control of access or where such control is planned in the future.

period (the distance traveled during perception and reaction time and during the initial acceleration to the point of encroachment on the left lane)

d_2 = distance traveled while the passing vehicle occupies the left lane

d_3 = distance between the passing vehicle (at the end of its maneuver) and the opposing vehicle

d_4 = distance traversed by an opposing vehicle for $\frac{2}{3}$ of the time the passing vehicle occupies the left lane, or $\frac{2}{3} \times d_2$

Figure 3-4 illustrates these relationships and gives values for the necessary passing sight distance for various speeds of the *passing vehicle*. Design values are shown in Table 3-6.

TABLE 3-6. MINIMUM PASSING SIGHT DISTANCE FOR THE DESIGN OF TWO-LANE HIGHWAYS

Design Speed (mph)	30	40	50	60	70	80
Assumed Speeds						
Passed Vehicle (mph)	26	34	41	47	54	59
Passing Vehicle (mph)	36	44	51	57	64	69
Passing Sight Distance (ft)	1,100	1,500	1,800	2,100	2,500	2,740

Fig. 3-4. Elements of and total passing sight distance—2-lane highways (*Source:* A Policy on Geometric Design of Rural Highways, *AASHO, 1965, Figure III-2*)

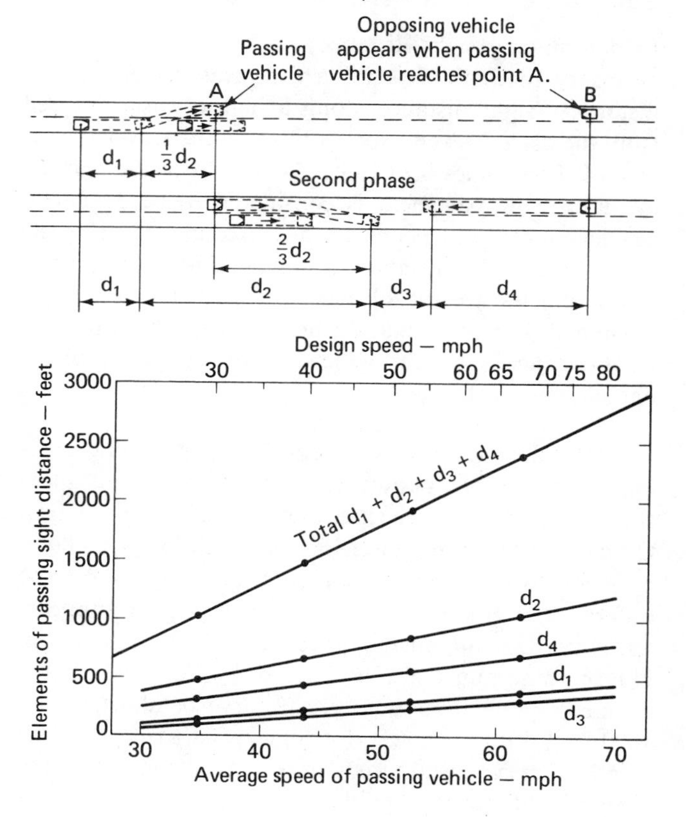

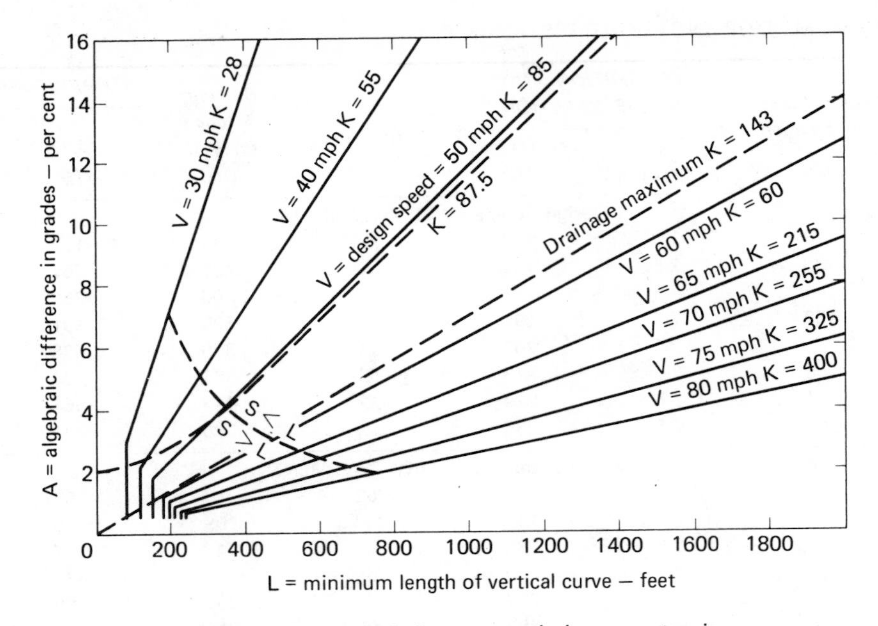

Fig. 3-5. Design controls for crest vertical curves, stopping sight distance (*Source:* A Policy on Geometric Design of Rural Highways, *AASHO, 1965, Figure III-19*)

Four-Lane Highways. On four-lane highways, only the minimum stopping sight distance must be provided because passing is accomplished in exclusive lanes.

Sight Distance on Vertical Curves

In determining sight distances on vertical curves, it is necessary to establish criteria for measurement of the distances. Sight distance along a highway is measured from the driver's eye to some object on the traveled way, when it first comes into view. The height of the driver's eye above the pavement is considered to be 3.75 feet. The height of the object or obstacle used in measuring the nonpassing sight distance is assumed to be 6 inches. Since vehicles are the objects that must be seen when passing, it is assumed that the height of object for passing sight distance is 4.5 feet, which is taken as the top of the oncoming vehicle.

Crest Vertical Curves. All vertical curves should be as long as conditions permit, and under no circumstances should they be shorter than the established minimums. Over crests these minimums are almost always dictated by sight distance requirements. On occasion, where the difference in grades is small, ease of riding and appearance may demand longer curves than does sight distance. No set minimums are prescribed by AASHO. Some engineers prefer that no vertical curve be shorter than 1000 feet. Others suggest, for high-speed roads, that the length of curve in hundreds of feet be not less than the algebraic difference in grades.

Nonpassing sight distance on crest vertical curves

Minimum lengths of crest vertical curves, as determined by sight distance requirements, are generally satisfactory from the standpoint of safety, comfort, and appearance. Basic formulas for the length of a parabolic vertical curve, in terms of the algebraic difference in grade and sight distance, are:

$$\text{When } S < L, \quad L = \frac{AS^2}{100(\sqrt{2h_1} + \sqrt{2h_2})^2} \tag{3-13}$$

$$\text{When } S > L, \quad L = 2S - \frac{200(\sqrt{h_1} + \sqrt{2h_2})^2}{A} \tag{3-14}$$

where
L = length of vertical curve, in feet
S = sight distance, in feet
A = absolute value of the algebraic difference in grades, in per cent
h_1 = height of eye above roadway surface, in feet
h_2 = height of object above roadway surface, in feet

When the height of the eye and the height of object are 3.75 feet and 6 inches, respectively, as used for nonpassing or stopping sight distance, the above formulas become:

$$\text{When } S_{np} < L, \qquad L = \frac{AS_{np}^2}{1398}$$

$$\text{When } S_{np} > L, \qquad L = 2S_{np} - \frac{1398}{A}$$

These equations may be simplified and represented in the form $L = KA$. These equations are plotted in Figure 3-5. Note that the equations are in the form to compute the minimum length of vertical curve necessary to provide safe stopping sight distance over the crest curve.

Passing sight distance on crest vertical curves

Design lengths of crest vertical curves for *passing* sight distance differ from those for *nonpassing* sight distance because of the different height of object criteria. The previous formulas apply. However, replacing the 6-inch value with the 4.5-foot height of object results in the following:

$$\text{When } S_p < L, \qquad L = \frac{AS_p^2}{3295} \tag{3-15}$$

$$\text{When } S_p > L, \qquad L = 2S_p - \frac{3295}{A} \tag{3-16}$$

Generally, it is impracticable to design crest vertical curves to provide for passing sight distance because of the high cost where crest cuts are involved, and because of the difficulty of fitting the required long vertical curves to the terrain, particularly for high-speed roads. Much longer vertical curves are required to provide for passing sight distance than for nonpassing sight distance. If a continuous passing opportunity is desired, a four-lane design may be cheaper than a two-lane one, with the saving in excavation more than offsetting the cost of extra lanes.

Sag Vertical Curves. Sight distance along sag vertical curves that do not pass through grade separation structures (underpasses) is usually determined by headlight reach distance, with the headlight beam directed into the roadway. The portion of highway lighted ahead is dependent upon the position of the headlight, and the direction of the light beam. Headlight height is generally taken to be 2.0 feet above the pavement, with a 1-degree upward divergence of the light beam from the longitudinal axis of the vehicle. The upward spread of the light beam provides some additional visible length, but it is generally ignored.

The following formulas show the S, L, and A relationships using S as the distance between the vehicle and the point where the 1-degree angle of light ray intersects the surface of the roadway:

$$\text{When } S < L, \qquad L = \frac{AS^2}{400 + 3.5S} \tag{3-17}$$

$$\text{When } S > L, \qquad L = 2S - \frac{400 + 3.5S}{A} \tag{3-18}$$

where L = length of sag vertical curve, in feet
S = light beam distance, in feet
A = absolute value of the algebraic difference in grades, in per cent

The length of vertical curve should be chosen such as to make S at least equal to the safe stopping distance (nonpassing sight distance). This relationship is plotted in Figure 3-6.

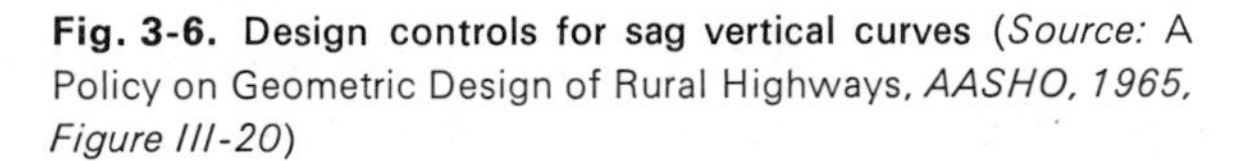

Fig. 3-6. Design controls for sag vertical curves (*Source:* A Policy on Geometric Design of Rural Highways, *AASHO, 1965, Figure III-20*)

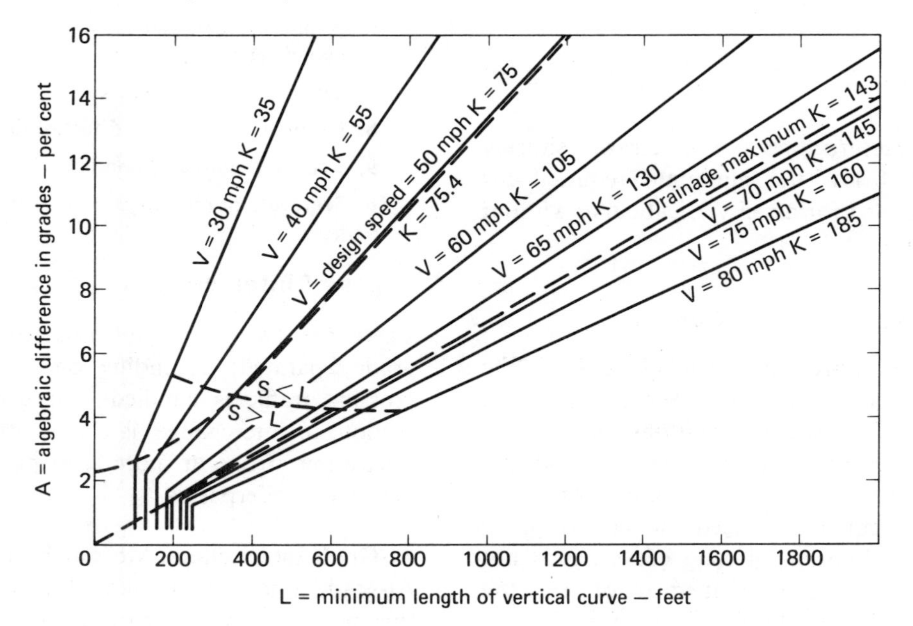

Sight Distances at Underpasses. The sight distances along sag vertical curves on the underpassing highway are influenced by the overpass structure. The following formulas show the S, L, and A relationships:

$$\text{When } S > L, \quad L = 2S - \frac{800}{A}\left(C - \frac{h + h_1}{2}\right) \qquad (3\text{-}19)$$

$$\text{When } S < L, \quad L = \frac{S^2 A}{800\left(C - \frac{h + h_1}{2}\right)} \qquad (3\text{-}20)$$

where
L = length of sag vertical curve, in feet
A = algebraic difference in grades (always positive), in per cent
S = sight distance, in feet
C = vertical clearance of underpass, in feet
h = vertical height of driver's eye above pavement, in feet
h_1 = vertical height of object, in feet

Generally, the desirable sight distance is chosen, and the minimum length of vertical curve that will provide this sight distance is then determined. The most critical conditions exist when C is taken as a minimum, and $h + h_1$ is taken as a maximum. The minimum vertical clearance is taken as 14.0 feet, the vertical height of the driver's eye above the pavement is taken as 6.0 feet (for truck drivers), and the object height is taken as 1.5 feet, assumed to be the taillight of the vehicle ahead, or the discernible portion of an oncoming vehicle.

Assuming the above values for C, h, and h_1, the above formulas become:

$$\text{When } S > L, \quad L = 2S - \frac{8200}{A} \qquad (3\text{-}21)$$

$$\text{When } S < L, \quad L = \frac{S^2 A}{8200} \qquad (3\text{-}22)$$

The overpass structure does not necessarily shorten the sight distance below the minimum required for stopping, as determined for a sag vertical curve without an overpass structure.

Sight Distance on Horizontal Curves

Another element of horizontal alinement design is the sight distance along the inside of curves. As a vehicle travels around a horizontal curve, any obstruction near the inside edge of the road blocks the driver's view ahead. If the design is to provide safe operation, the horizontal sight distance must equal or exceed the safe stopping distance. With the exception of long curves, it is not practical to provide passing sight distance, for the minimum passing sight distance for a two-lane highway is about four times greater than the stopping sight distance for the same design speed. Where possible, the obstruction (walls, cut slopes, wooded areas, buildings, hedgerows, high farm crops, etc.) should be removed to provide a better view ahead for the driver.

Horizontal sight distances may be scaled directly from the plans with a fair degree of accuracy, but to determine the radius of curve to provide a specified sight distance when the distance to the obstruction is known, the values derived from Figure 3-7 should be used.

INTERSECTIONS

An *intersection* is the area where two or more highways join or cross, and includes the roadway and roadside facilities for traffic movement in that area. The prime operational function of the intersection is to provide for changes in travel direction.

The intersection is an important part of the highway, since much of the efficiency, safety, speed, cost of operations, and capacity depend upon its design.

Principles of Intersection Design

The following ten principles[3] should serve as guides to intersection design:

1. Reduce the number of conflict points
2. Control the relative speed
3. Coordinate the design with the traffic controls
4. Consider alternate crossing geometries
5. Consider alternate turning geometries
6. Avoid multiple and compound merging and diverging maneuvers
7. Separate conflict points
8. Favor the heaviest and fastest flows
9. Reduce the area of conflict
10. Segregate nonhomogeneous flows

Types of Intersections

There are two classes of intersections: at-grade and grade-separated, depending on the manner in which crossing traffic is handled. The grade-separated intersection, or interchange, is characterized by the physical separation of one or more crossing maneuvers by overpasses or underpasses.

At-Grade Intersections. Most highways intersect at grade. At-grade intersections should provide for anticipated turning and crossing movements. Many factors enter

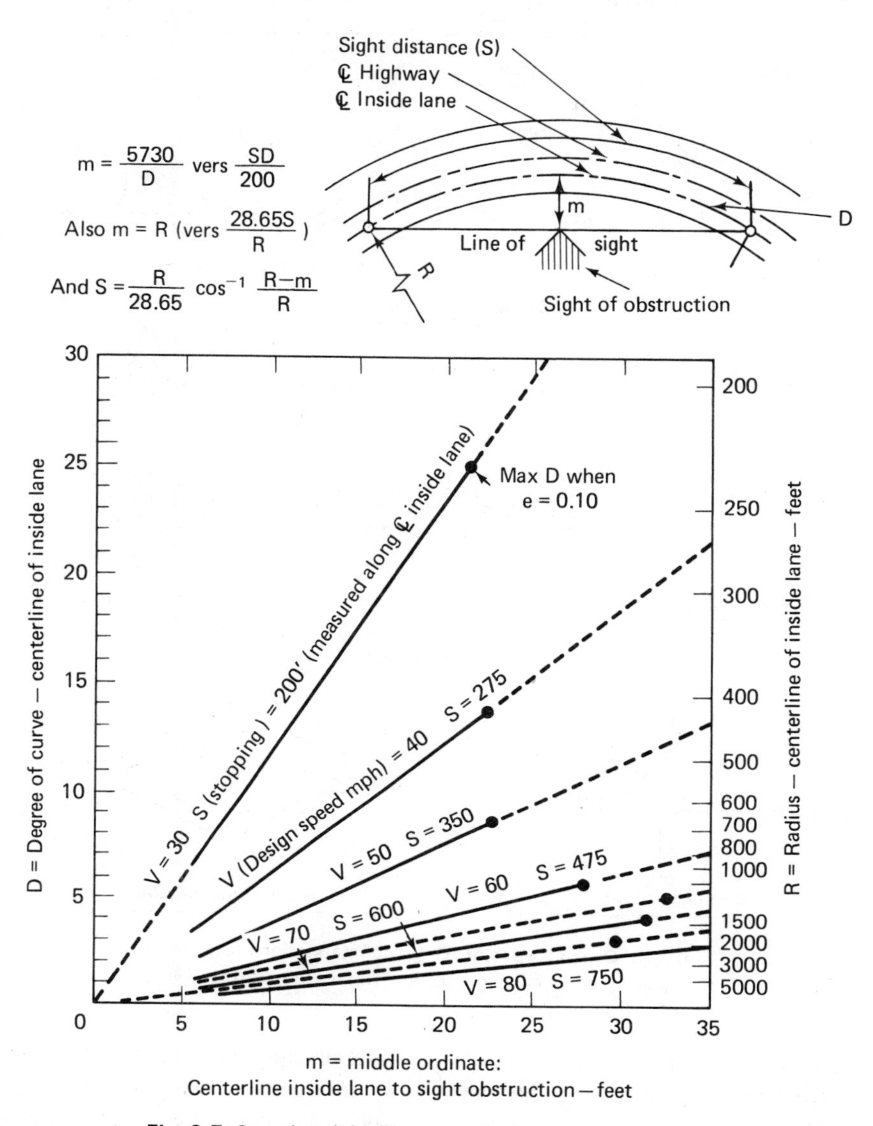

Fig. 3-7. Stopping sight distance on horizontal curves (*Source:* A Policy on Geometric Design of Rural Highways, *AASHO, 1965, Figure III-13*)

into the choice of type of intersection, but the principal controls are the design hour volume of traffic, the character of traffic (both through and turning), and the design speed. The character of traffic and the design speed affect many details of design, but in choosing the type of intersection they are not as significant as the traffic volumes involved in the various turning and through movements.

Basic types of at-grade intersections are T or Y (three-leg), four-leg, and rotary. In a particular case the type is determined primarily by the number of intersection legs, the physical controls of topography, the traffic pattern (traffic fluctuation), and the desired type of operation.

Any one basic intersection type can vary greatly in scope, shape, and degree of physical control or channelization.

Three-leg (T or Y) and four-leg intersections may be plain flared, or channelized. A *flared* intersection is one in which the number of traffic lanes or the pavement width exceeds the number of lanes or the width of the normal section of highway. A *rotary* intersection is one in which all traffic merges into, and emerges from, a one-way road around a central island. It is a form of channelized intersection.

Rotaries are not regularly included in newly designed roads, but they are still used where the amount of turning traffic approaches or exceeds the through volume, and where more than four intersection legs would otherwise produce complex arrangement or difficult operation. The use of rotaries should be restricted to design hour volumes (total entering the intersection from all legs) of 3000 vehicles or less. Rotaries are generally considered

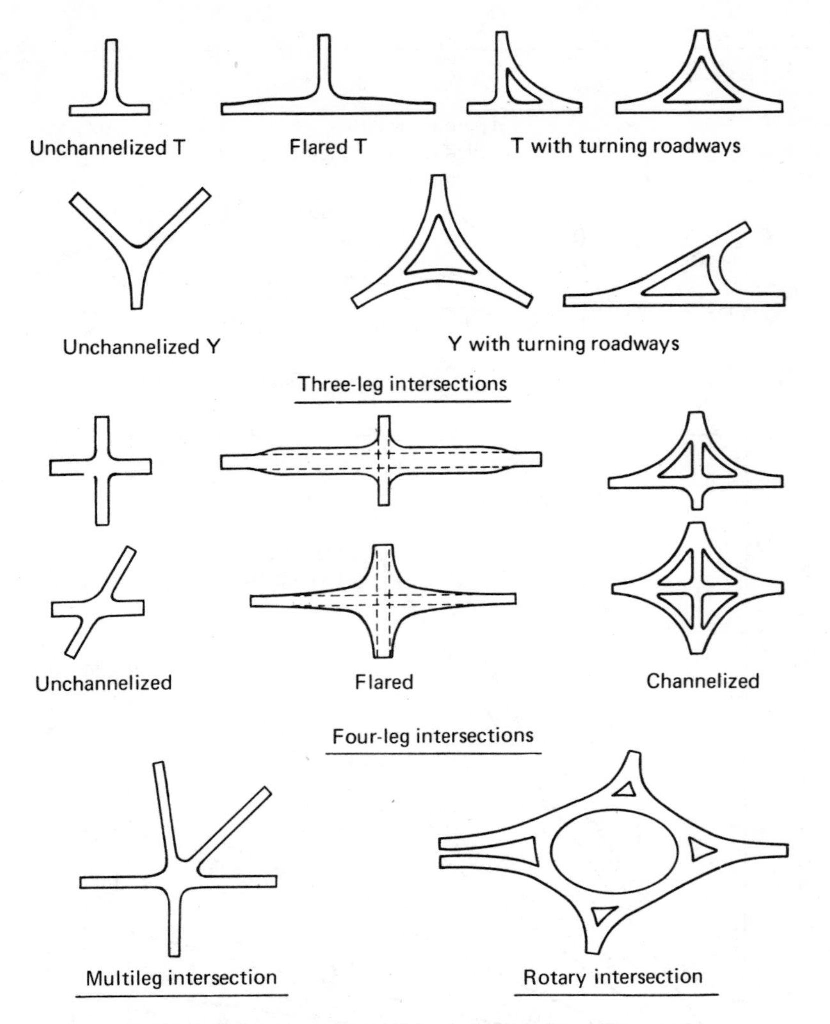

Fig. 3-8. General types of AT-grade intersections (*Source:* A Policy on Geometric Design of Rural Highways, *AASHO, 1965, Figure VIII-1*)

undesirable from an operational standpoint because of the high number of weaving movements which must be made in a relatively short distance.

Some at-grade intersections may have large paved areas (those with large corner radii and those with oblique angle crossings) which permit and encourage hazardous, uncontrolled vehicle movements, require long pedestrian crossings, and have unused pavement areas. Conflicts may be reduced in extent and intensity by layout design to include islands. An at-grade intersection in which traffic is directed into definite paths by islands is called a *channelized intersection.*

An *island* is a defined area between traffic lanes for the control of vehicle movements or for pedestrian refuge. Within an intersection a median or other separation is considered an island. An island has no single physical type; it may range from a raised area bordered by barrier curbs to a pavement area marked by paint.

Figure 3-8 shows a variety of general types of at-grade intersections; Figure 3-9 gives some examples of intersections where channelization is used.

Interchanges. The volume of traffic that can pass through an intersection can approach or equal the sum of the open-road capacities of two intersecting highways if the roadways are placed at different levels, thus enabling through traffic on both highways to flow without interruption. With connecting roadways for the turning movements, and adequate facilities for turning vehicles to slow down or speed up clear of through-traffic lanes, all traffic can proceed through the intersection with little or no interference.

The type of grade separation, or interchange type, and its design, is influenced by many factors, but those of greatest importance are: design hour volumes, character of traffic, design speed, topography, available right-of-way, and cost.

There are several basic interchange forms, or patterns of ramps, for turning movements at a grade separation. Their application to a particular site is determined by the number of intersection legs, the expected volumes of through and turning movements, and the physical controls of topography.

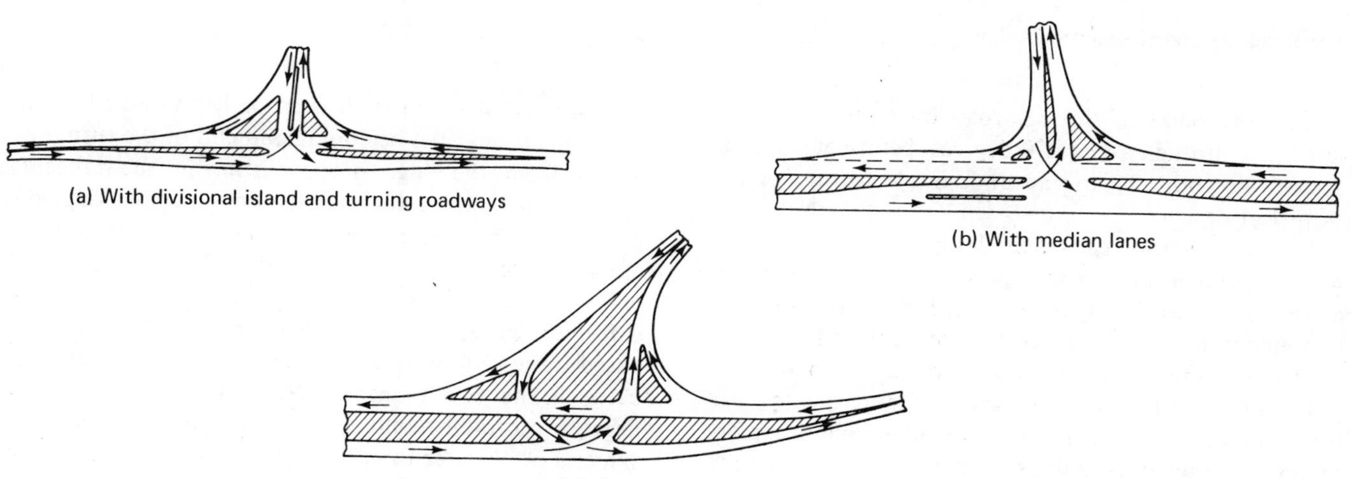

Fig. 3-9. T or Y intersections, channelized-high type (*Source:* A Policy on Geometric of Design Rural Highways, *AASHO, 1965, Figure VIII-28*)

Three-leg interchanges have the overall geometric form of a T (or trumpet). These are shown in general form as items (a) and (b) of Figure 3-10.

The most common four-leg interchange is the *cloverleaf.* The cloverleaf shown in Figure 3-10(d) has a full complement of ramps, with a separate one-way ramp for each turning movement. Direct left turns are not possible. Drivers desiring to turn left are required to travel

Fig. 3-10. General types of interchanges (*Source:* A Policy on Geometric Design of Rural Highways, *AASHO, 1965, Figure IX-17*)

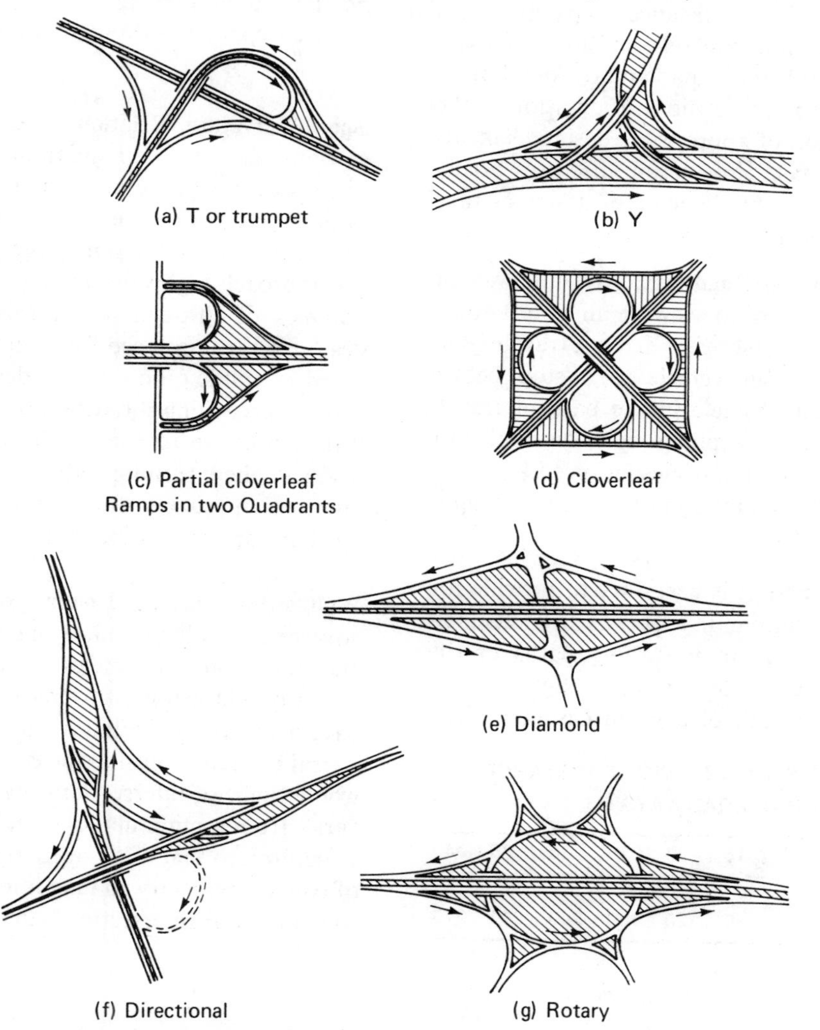

beyond the point of through-road intersection, and to turn right through about 270 degrees, before gaining the desired direction. This is an undesirable feature of the cloverleaf design.

The term *partial cloverleaf* designates those layouts with less than a full complement of ramps. In the partial cloverleaf shown in Figure 3-10(c), ramps are shown in two quadrants. All movements are provided for, but left turns at grade are required on the minor road.

The *diamond* type of interchange, Figure 3-10(e), has four one-way ramps. It is especially adaptable to major-minor highway intersections with limited right-of-way. However, left turns from and into the minor road must be executed at grade on minor roads.

Directional interchanges are those having ramps which tend to follow the natural direction of traffic movement. [See Figure 3-10(f) for one example.] One or more of the left turn movements is handled by a direct, or nearly direct, connection between major cross roadways.

Intersection Design Elements

Most of the roadway design elements: speed, horizontal and vertical alinement, sight distance, capacity, cross section, and superelevation, are relevant to the design of intersections. The problem is made more complicated because an intersection requires the consideration of the simultaneous application of appropriate design elements to two or more roadways. This section discusses briefly some of the more important elements of design as they relate to intersection development.

Sight Distance for Turning Roadways. Any highway or intersection should be open to view at all points for a sufficient distance to enable the driver, traveling at turning speed, to bring his vehicle to a stop before reaching an unexpected obstacle in the path of travel. Table 3-7 gives minimum stopping sight distances for turning roadways as a function of design speed.[1]

For a more detailed examination of intersection sight distances see Chapter 25.

Intersection Curves. Where it is necessary to provide for turning vehicles within minimum space, as at unchannelized intersections, the minimum turning paths of the design vehicles apply. In the design of the edge of pavement for the minimum path of a given design vehicle, it is assumed that the vehicle is properly positioned within the traffic lane at the beginning and end of the turn, i.e., 2 feet from the edge of pavement on the tangents approaching and leaving the intersection curve. Edge-of-pavement curve designs conforming to this assumption are shown in Table 3-8. They closely fit the inner wheel paths of the various design vehicles, clearing them by 2 feet or more, and at no point by less than 1 foot, based on a 12-foot lane. The angle of turn is the angle between the two tangents. With angles less than 90 degrees, the radii required to fit the minimum paths of vehicles are greater than those suggested for right-angle turns. With angles of turn greater than 90 degrees, the radii are decreased. Data are given for design vehicles operating at 10 miles per hour or less.

TABLE 3-7. MINIMUM STOPPING SIGHT DISTANCES FOR TURNING ROADWAYS

Design speed (mph)	15	20	25	30	35	40
Minimum stopping sight distance (ft)	80	120	160	200	240	275

Where the inner edges of pavement for right turns at intersections are designed to accommodate semitrailer combinations, or where the design permits passenger vehicles to turn at speeds of 15 miles per hour or more, the pavement area at the intersection may become excessively large for proper control of traffic. To avoid this, because it may lead to confusion of drivers and hazards to pedestrians, a corner island, placed within the unused portion of the turning roadway, should be provided to form a separate turning roadway, as shown in Figure 3-11.

Speed-Curvature Relations. Vehicles turning at intersections designed for minimum turns, as previously discussed, have to operate at low speeds (10 miles per hour or less). The speeds for which intersection curves should be designed depend largely upon vehicle speeds on the approach highways, the type of intersection, and the volumes of through and turning traffic. Generally, a desirable turning speed for design is the average running speed of traffic (about 0.7 of design speed) on the highways approaching the turn. Designs of such speeds offer little hindrance to smooth flow of traffic, and may be justified where turning traffic is high. Minimum radii for intersection curves for various speeds as given by AASHO are shown in Table 3-9.[1]

Widths for Turning Roadways. Pavement widths are governed by volumes of turning traffic, the turning speed, and the type of vehicles to be accommodated. The roadway width for a turning roadway, as distinct from pavement width, includes the shoulders or equivalent lateral clearance outside the edges of pavement. Over the whole range of intersections the required shoulder width varies from a minimum of 2 feet, as used on interchange structures, to that of an open highway cross section. It is, of course, necessary to take into consideration pavement widening on intersection curves. Required pavement

TABLE 3-8. MINIMUM EDGE OF PAVEMENT DESIGNS FOR TURNS AT INTERSECTIONS

DESIGN VEHICLE	ANGLE OF TURN (deg)	SIMPLE CURVE RADIUS (ft)	3-CENTERED COMPOUND CURVE, SYMMETRIC		3-CENTERED COMPOUND CURVE, ASYMMETRIC	
			Radii (ft)	Offset (ft)	Radii (ft)	Offset (ft)
P	30	60	—	—	—	—
SU		100	—	—	—	—
WB-40		150	—	—	—	—
WB-50		200	—	—	—	—
P	45	50	—	—	—	—
SU		75	—	—	—	—
WB-40		120	—	—	—	—
WB-50		170	200–100–200	3.0	—	—
P	60	40	—	—	—	—
SU		60	—	—	—	—
WB-40		90	—	—	—	—
WB-50		—	200–75–200	5.5	200–75–275	2.0–6.0
P	75	35	100–25–100	2.0	—	—
SU		55	120–45–120	2.0	—	—
WB-40		85	120–45–120	5.0	120–45–200	2.0–6.5
WB-50		—	150–50–150	6.0	150–50–225	2.0–10.0
P	90	30	100–20–100	2.5	—	—
SU		50	120–40–120	2.0	—	—
WB-40		—	120–40–120	5.0	120–40–200	2.0–6.0
WB-50		—	180–60–180	6.0	120–40–200	2.0–10.0
P	105	—	100–20–100	2.5	—	—
SU		—	100–35–100	3.0	—	—
WB-40		—	100–35–100	5.0	100–35–200	2.0–8.0
WB-50		—	180–45–180	8.0	150–40–210	2.0–10.0
P	120	—	100–20–100	2.0	—	—
SU		—	100–30–100	3.0	—	—
WB-40		—	120–30–120	6.0	100–30–180	2.0–9.0
WB-50		—	180–40–180	8.5	150–35–220	2.0–12.0
P	135	—	100–20–100	1.5	—	—
SU		—	100–30–100	4.0	—	—
WB-40		—	120–30–120	6.5	100–25–180	3.0–13.0
WB-50		—	160–35–160	9.0	130–30–185	3.0–14.0
P	150	—	75–18–75	2.0	—	—
SU		—	100–30–100	4.0	—	—
WB-40		—	100–30–100	6.0	90–25–160	3.0–11.0
WB-50		—	160–35–160	7.0	120–35–180	3.0–14.0
P	180	—	50–15–50	0.5	—	—
SU	U-TURN	—	100–30–100	1.5	—	—
WB-40		—	100–20–100	9.5	85–20–150	6.0–13.0
WB-50		—	130–25–130	9.5	100–25–180	6.0–13.0

(Source: *A Policy on Geometric Design of Rural Highways,* AASHO, 1965, Table VII-1)

widths as a function of radius of curve and type of roadway operation are tabulated in AASHO.[1]

Speed-Change Lanes. Speed-change lanes are warranted at important intersections on high-speed, high-volume highways. A speed-change lane is an auxiliary lane, including tapered areas, primarily for the acceleration or deceleration of vehicles entering or leaving through-traffic lanes. An *acceleration lane* is an added area 10 to 12 feet wide, of sufficient length to enable a turning vehicle to accelerate from the slower speed of a completed turn to a speed at which it is convenient and safe to merge with through traffic. A *deceleration lane* is an added area 10 to 12 feet wide, of sufficient length to enable a turning vehicle to slow down to the safe turning speed of the curve it approaches.

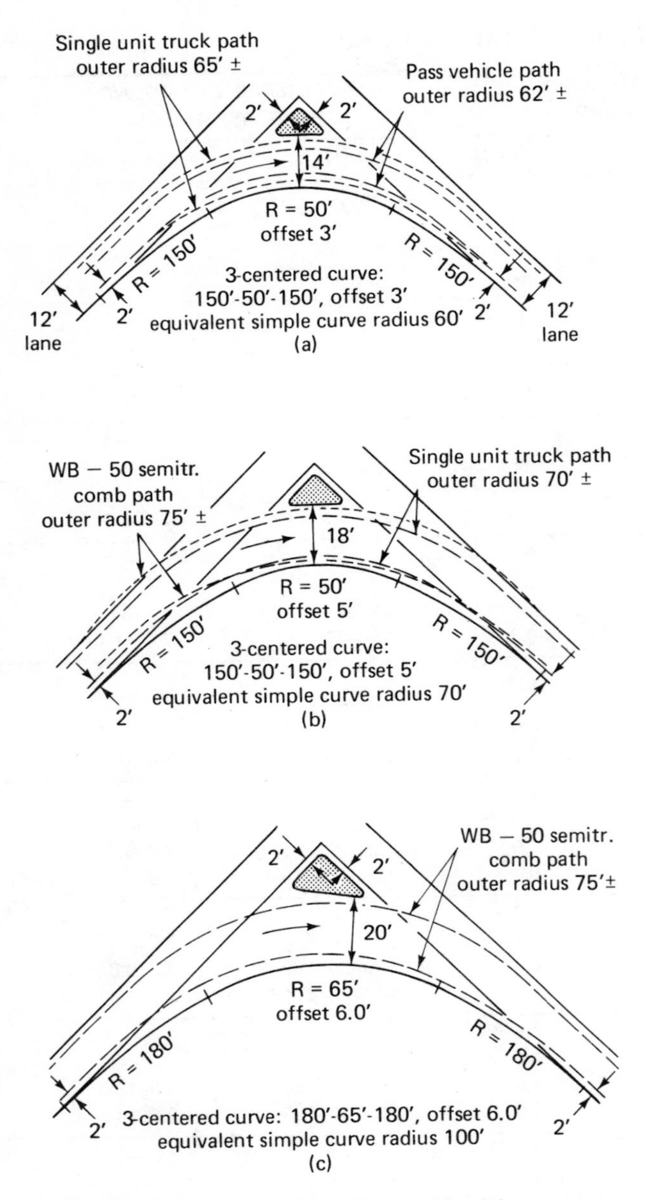

Fig. 3-11. Designs for turning roadways with minimum corner island, 90-degree right turn (*Source:* A Policy on Geometric Design of Rural Highways, *AASHO, 1965, Figure VII-6*)

TABLE 3-9. MINIMUM RADII FOR INTERSECTION CURVES

Design (turning) speed (V) (mph)	15	20	25	30	35	40
Side friction factor (f)	0.32	0.27	0.23	0.20	0.18	0.16
Assumed minimum superelevation (e)	0.00	0.02	0.04	0.06	0.08	0.09
Total $e + f$	0.32	0.29	0.27	0.26	0.26	0.25
Calculated minimum radius (R) (ft)	47	92	154	231	314	426
Suggested curvature for design:						
Radius: minimum (ft)	50	90	150	230	310	430
Degree of curve: maximum (deg)	—	64	38	25	18	13
Average running speed (mph)	14	18	22	26	30	34

Note: For design speeds of more than 40 mph, use values for open highway conditions.

Specific recommendations for the design of exit and entrance ramps are provided by AASHO[10,11]. They are summarized as follows:

Exit ramps

1. They should normally be one lane wide at the point of departure from the main roadway.
2. They should preferably have departure angles of 4–5 degrees.
3. They should leave the mainline on a tangent section of the mainline.
4. Right-hand exits are markedly superior to left-hand exits in traffic operations and safety.

Entrance ramps

1. They should, in general, have a single lane at the entrance nose.
2. They should take into account the effects of grades on trucks.
3. They should always enter from the right when serving local streets.

In addition, studies have shown[6] the value of designing entrance ramp profiles to maximize sight distance for vehicles on the ramp. This provides for a safer and more efficient use of the acceleration lane.[1]

Traffic Engineering Studies and Analyses

4

Introduction to Urban Transportation Planning

The purpose of this chapter is to describe the nature of the transportation planning mechanism, the relevant issues which guide its process, and the tremendous influence of this process on the pattern of urban land.

This complex objective will not be treated in a manner which will inform the reader on how to conduct a transportation analysis of an urban area, since this type of treatment would be too extensive. Rather, this chapter is aimed at informing the traffic engineer about some basic transportation principles and about the uses the traffic engineer can make of the information available from transportation planning studies.

DEFINITION

Transportation planning is a process whose objective, in a broad sense, is to develop a system of transport which will enable people and goods to travel safely and economically. In addition, the journeys should be comfortable and convenient. It is a dynamic process, in the sense that it must respond to time changes of land use, economic conditions, and travel patterns. Because of the vast amount of capital expenditures required in the implementation of transportation systems (either highway or rapid transit), these projects must inevitably radically influence the land use development in the region or area in which the transportation facilities are constructed. Because large-scale transportation projects have such long-lasting effects on the shape and character of land, it is important that they be developed with the objective of encouraging a desirable land use development in harmony with overall regional objectives.

THE EVOLUTION OF REGIONAL OR AREA-WIDE TRANSPORTATION PLANS

Prior to 1962, long-range transportation planning in most areas of the United States was accomplished primarily through independent community efforts. The sophistication of these efforts varied according to the professional expertise available in the communities, the quality of the consultants retained by the communities, or the funds available to do the job. The resulting plans were, in most part, neither comprehensive nor coordinated with the plans of adjacent communities, nor were the end results based on cooperative efforts between the various technical disciplines working toward the long-range plan.

In the United States, where highway construction was proceeding at such a fantastic pace (45,000 miles* per

* Excluding municipal mileage, including surfacing of existing nonsurfaced mileage. See Reference 8.

year over the period 1945–1960), the lack of a unified planning process cutting across county, city, or state lines was a serious one, indeed, for such lack of cooperation led to wasteful expenditures of resources. In view of these planning deficiencies, the federal government passed the Federal-Aid Highway Act of 1962, which stated: "It is declared to be in the national interest to encourage and promote the development of transportation systems embracing various modes of transportation in a manner that will serve the states and local communities efficiently and effectively." This section of the act applies to urban areas of over 50,000 population, and the states and urban areas were given until July 1, 1965, to establish a comprehensive transportation plan. Largely because of this persuasion of finance, the 1962 Highway Act resulted in the start of comprehensive, cooperative, and coordinated highway planning in the United States. Transportation studies were set up along the guidelines contained in the law, and for the first time rational approaches to long-range transportation planning took effect. Regional agencies which had the authority of acting as planning bodies for several political jurisdictions were set up. These include the Tri-State Transportation Commission, representing New York, New Jersey, and Connecticut, and the Delaware Valley Regional Planning Commission (originally set up as the Penn-Jersey Transportation Study), covering southern New Jersey and most of southeastern Pennsylvania.

THE NATURE OF TRAVEL

Travel can be defined, very broadly, as the motion of an object. The objects of interest to the transportation planner are people and goods. In order to plan transportation facilities it is extremely important to know *how* (by what mode) people or goods travel, and for what purposes.

People travel by the modes which are suitable to their needs and which are available to them. When short distances such as a few city blocks are involved, they may walk. As distances between their desired destinations and places of origin increase, they will require mechanical locomotion: the private auto, the mass-transit vehicle, the train, or the airplane. Eighty to 90 per cent of all trips made within an urban area are made by automobile.

People travel to satisfy their needs for various activities: work, shopping, recreation, etc. An example of how trip purposes are distributed in an urban area is given in Table 4-1 for the Philadelphia area.

Since most of these activities are located in buildings, it follows that travel may be thought of as the act of going from building to building. Buildings which house similar activities tend to be in the same neighborhood or locality, so it may be hypothesized that travel is a function of land use. The amount of travel destined to a particular parcel of land is not only dependent on the intensity of usage (represented in terms of building area to land area), but also on the location or accessibility of that land parcel to the rest of the region. Travel, therefore, is a function of many variables, some of which are rooted in the person making the trip, and some of which are characteristics of the physical and nonphysical environment.

TABLE 4-1. DESTINATION PURPOSE OF PERSONAL TRAVEL

To Home	40.0 per cent
To Work	19.6
To Shop	7.9
To School	3.9
Social Recreation	13.9
Personal Business	11.2
Other	3.5
Total	100.0 per cent

(Source: *Penn-Jersey Transportation Study*, Vol. 1, Philadelphia, Pa., 1964, p. 71)

To properly plan for the transportation needs of a region, it is important to understand and describe the type and amount of travel which takes place in that area, and also to clearly identify the goals and objectives which must be met by the transportation plan.

There are two major tasks which guide the development of transportation networks. One is *simulation*; the other is the *development of criteria* which are used in evaluating the effectiveness of a proposed transportation plan.

THE URBAN TRANSPORTATION PLANNING PROCESS

The transportation planning process as we know it today has been made possible only through widespread use of of large-storage electronic computers. The process consists of a series of interdependent steps which can be divided as shown in Figure 4-1. This flow chart illustrates the basic activities necessary to a comprehensive study. A brief description of each step follows.

Organization

The first task of the transportation planning process calls for the establishment of a framework within which policy decisions can be made, citizen participation can be assured, and a technical staff capable of undertaking the study can be assembled.

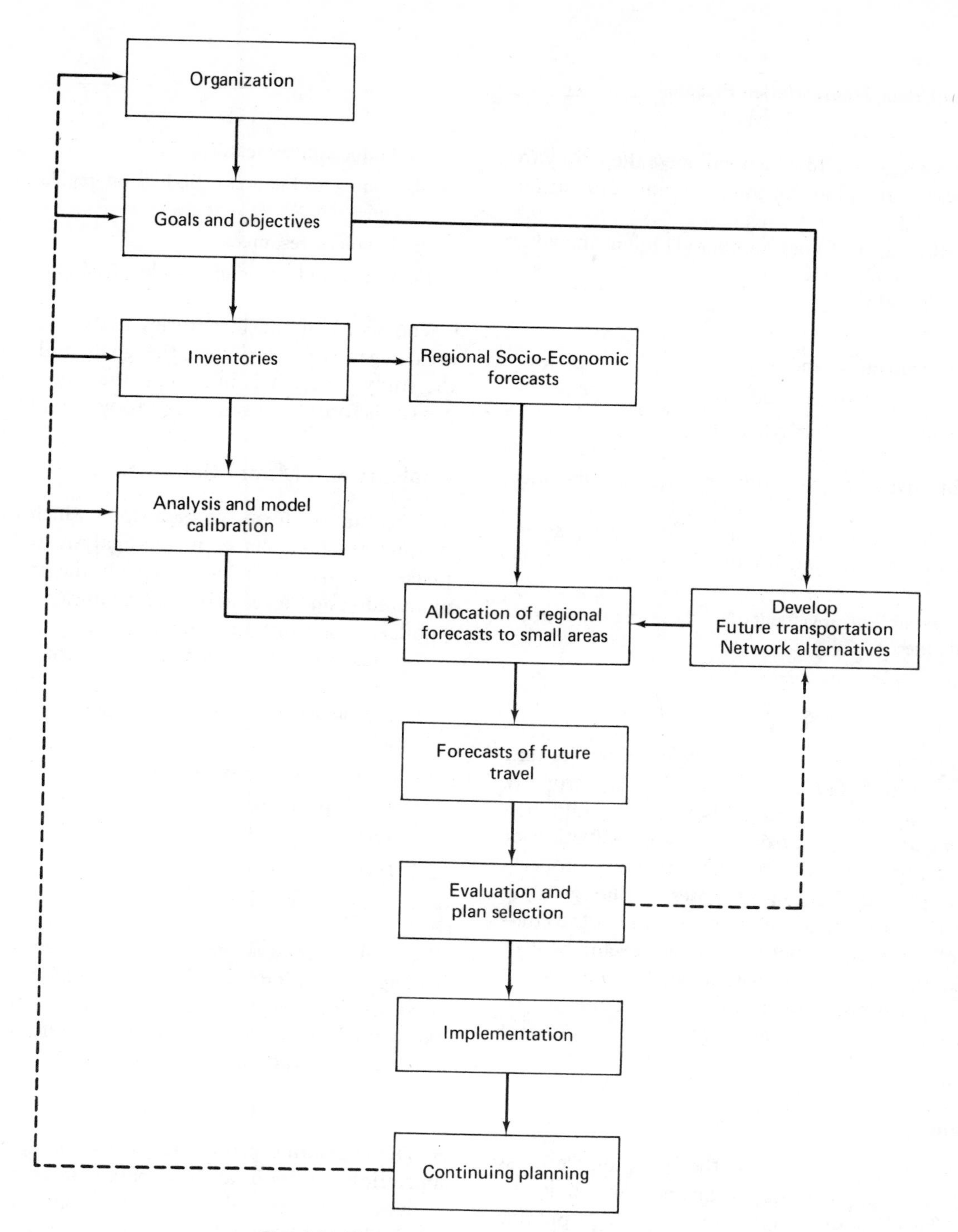

Fig. 4-1. The urban transportation planning process

The purpose of urban transportation planning is to develop, and continuously evaluate, transportation plans which meet the goals and objectives of the nation, the state, and the urban communities. Clear identification of the specific goals and objectives which the plans must satisfy is accomplished through the cooperative efforts of policy and citizens' advisory committees, with the assistance of the technical study staff.

Goals and Objectives

To guide the transportation planning process toward a successful outcome, it is customary to specify regional transportation goals and objectives which must be maximized by the "best" plan. For example, it might be stated that the transportation plan to be implemented should

1. Maximize mobility of people and goods
2. Enhance the urban environment.

These goals, as stated, are of little value in guiding the development of a plan that must be justified to others on the basis that it indeed maximizes these goals. Sets of more specific objectives are needed, which in turn can be related to each goal, and which will have the ability to be *quantified* in the plan formulation and evaluation phases.

Thus, the objectives of goal number (1) might include:

1. Minimize travel time
2. Minimize travel cost
3. Provide adequate system capacity
4. Provide adequate system safety
5. Provide adequate system reliability

And the objectives of goal number (2) might include:

1. Provide for equitable distribution of regional accessibility for employment, health, education, and shopping
2. Foster desirable arrangement of land use and transportation facilities
3. Minimize community disruption
4. Minimize air and noise pollution

The above objectives must be *measurable*, so that costable and quantifiable criteria may be properly expressed. Objectives containing this attribute could then be called *operational objectives*. A close examination of these sample objectives will show that they are properly suitable for plan evaluation purposes. Although it is not within the scope of this chapter to discuss the complex issue of what constitutes the proper development of criteria and standards of measurement, for the sake of completeness, it is useful to illustrate how some of the above objectives would be measured. This is done in Table 4-2.

Inventories

The comprehensive nature of the planning process requires that large amounts of diverse data must be gathered. The purpose of data collection is to provide the necessary elements for a basic understanding of the travel characteristics of the region. In addition, the type of data collected should be based on the requirements of the goals and objectives described above.

The typical elements for which inventories and analyses are required are:

1. Economic factors affecting development
2. Population
3. Land use
4. Transportation facilities
5. Travel patterns (discussed in Chapter 5)
6. Terminal and transfer facilities
7. Traffic control features
8. Zoning ordinances, subdivision regulations, building codes, etc.
9. Financial resources
10. Social and community-value factors

The scope of the inventories to be conducted in an area depends primarily on the goals and objectives of the study. Thus, it is important that these be identified clearly before the inventories are well under way.

Analysis and Model Calibration

The inventories form the base upon which the planning process can take shape. In the analysis and calibration phase, the data obtained through the inventories are analyzed and forecasting techniques are developed (model calibration) for estimating future travel demands.

The tasks usually included in this phase are:

1. Techniques for estimating population and economic activities
2. Land use forecasting techniques
3. Trip generation
4. Trip distribution
5. Modal split process
6. Traffic assignment

Regional Socioeconomic Forecasts: Techniques for Estimating Population and Economic Activities. This task involves the development of methods of forecasting population, employment, incomes, and car ownership for the total study area or region.

Population

The estimating procedures widely used in past transportation studies have been based on established techniques from the field of demography. The common methods used in population forecasting can be placed in three categories:

1. *Trend-based methods*, which employ graphical or mathematical projection of the curve of past population growth
2. *Ratio methods*, which are based on relationships of population growth in one area to that in other areas
3. *Component methods*, which are based on analyses of net migration and natural population increase

Each of these methods provides advantages and disadvantages to the user. A discussion of the most common techniques can be found in the literature.[11]

TABLE 4-2. OPERATIONAL OBJECTIVES

OBJECTIVES	*UNITS OF MEASUREMENT*
Minimize Travel Time	Hours
Minimize Travel Cost	Dollars
Provide System Capacity	Volume of Capacity (V/C) ratio
Provide Maximum Safety	Number of accidents
Provide Maximum Reliability	Variance of travel time during peak hour
Provide for Equitable Distribution of Accessibility to Regional Opportunities	Number of jobs, shopping areas (dollar sales), hospital beds, schools, etc., that can be reached from various socioeconomic areas at specified travel time intervals
Minimize Community Disruption	Number of housing units or persons, including dollar value per housing unit displaced
Minimize Business Disruption	Number of businesses displaced, including dollar value per business
Minimize Air Pollution Levels	Amounts of pollutants per mile of route as a function of traffic speed, volume, and composition, stratified by land use
Minimize Noise Levels	Decibels as a function of traffic volume stratified by land use
Minimize Construction Costs	Dollars per mile of route, stratified by area type and route type

Economic activities

The methods of forecasting economic activity for transportation study purposes may be categorized as follows:

1. *Trend line projection*, which employs time series data for the desired economic indicator which is to be forecast. The process involves fitting a trend line to historical data. The method of fit could be the "least squares" method, or some other technique.
2. *Step-down method*, which relates the economic activity in the study area to that of a larger economic region which includes the study area.
3. *Sector analysis*, which is similar in concept to the component method used in population forecasting. For example, instead of employment as a total unit, it stratifies this factor on a finer level of detail to represent subgroups which make up the total employment, that is, "heavy manufacturing," "light manufacturing," "wholesale and retail trade," etc.
4. *Economic base multiplier method*, which assumes that the local economy depends primarily on a group of "basic" sectors. These are the ones whose roles are primarily expanded to points outside the study area. It is assumed that changes in the basic sectors will generate changes in the "nonbasic" sectors of the local economy.

   Past data is analyzed to develop a "multiplier ratio" between total employment and basic employment. Thus, if future demands on the basic sectors are estimated, a forecast of the local economy can be made.
5. *The input-output model*, which is a method which explicitly relates the interactions of all major economic activities which affect one another. It consists of statements which equate the output (or product) of each subsystem to the inputs (or consumption) of the other subsystems.

   Although this model is very complex, it is nevertheless a gross simplification of reality.

Allocation of Regional Forecasts to Small Areas: Land Use Forecasting Techniques. The distribution of the end products of the regional population and economic forecasts to the small areas (analysis zones) located within the region is part of the land use forecast. Since travel is largely dependent on the spatial arrangement, intensity, and character of land use, this phase becomes an extremely important component of the transportation planning process. In the recent past much attention has been given to this subject by transportation and land use planners.

The development of sound land use models, however, is still in the evolutionary stage. This is due primarily to a lack of understanding of the complexity of the factors which influence the location decisions of people, industry, and other economic activities. Of the land use forecasting methods currently in use, residential land use models are by far the most advanced and reasonably acceptable. The overall pattern of residential development is much more regular than the seemingly erratic behavior of nonresidential activities. Retail land development tends to follow residential patterns, modified by zoning restrictions and the degree of accessibility of each of the alternative available sites. The location patterns of other commercial and manufacturing facilities are best predicted on an individual industry basis, with judgment supplemented by a forecasting model.

A number of techniques have been developed for forecasting land use. Some of these are:

Judgment or Trend Analysis. This involves simple visual extrapolation of data.

Density-Saturation Gradient Method. Used in the Chicago Area Transportation Study, this is a comparatively simple and intuitive approach. The area is divided into concentric ring zones about the high-value corner (HVC). Regularities in historical land development are observed, and curves which estimate these patterns are developed. One then makes intuitive judgments about the nature of further development, and extrapolates future land usage on the basis of historical patterns.

Accessibility Model. Unlike the above two methods, the accessibility model describes, in explicit mathematical form, some of the factors which influence land development. The general model is

$$G_i = G_t \frac{A_i^a V_i}{\sum_{i=1}^{n} A_i^a V_i} \tag{4-1}$$

where G_i = the forecast growth for zone i
G_t = total regional growth = $\sum G_i$
A_i = accessibility index for zone i
V_i = vacant land available for residences in zone i
a = empirically determined constant

The accessibility index (A_i) has the general form:

$$A_i = \sum_{j=1}^{n} E_j F_{ij} \tag{4-2}$$

where E_j = a measure of activity in zone j (for example, total employment)
F_{ij} = travel separation factor between zones i and j

The travel separation factor may be defined as

$$F_{ij} = \frac{1}{t_{ij}^b} \tag{4-3}$$

where t_{ij} = travel time from zone i to zone j
b = empirically determined constant

Intervening Opportunity Model. This is based on the hypothesis that the probability of a residential opportunity being chosen for development is a monotonically decreasing function of the number of intervening opportunities between a point within the center of the city and a location where a newcomer would find a site in which to build. This model is similar in concept to its namesake used in trip distribution, which is discussed in greater detail later.

The mathematical expression of the model is as follows:

$$A_j = A(e^{-l\theta} - e^{-l(\theta+\theta_j)}) \tag{4-4}$$

where A_j = number of households to be allocated to zone j
A = regional total number of households to be allocated
l = probability that a household will be located at a given opportunity
θ = the number of possible sites for locating a household, rank ordered up to, but excluding, zone j
θ_j = the number of possible sites in zone j

This formula produces results which are consistent with the objective of reducing travel friction, as shown by the fact that the probability of selecting a residential site per unit of available land is highest nearest the zone from which the outward search takes place.

Multiple Linear Regression Techniques. The idea is to develop an equation of the form

$$Y = a_0 + a_1x_1 + a_2x_2 + \cdots + a_nx_n \tag{4-5}$$

where Y = a measure of the change in the number of households occurring in a specified forecast time interval starting from the time for which the latest data are available
$x_1, x_2, \ldots, x_n$ = independent variables
$a_0, a_1, a_2, \ldots, a_n$ = regression coefficients

The independent variables may be:

1. Vacant land
2. Land value
3. Accessibility to employment
4. Residential zoning index (a number which is 1.0 if all of the land is zoned residentially, and which is 0.0 if no land is zoned residentially)
5. Measures of
   a. Zone size
   b. Zone population
   c. Amount of land devoted to varying uses
   d. Zone employment

Unlike the previous models, the regression techniques allow explicit inclusion of other important variables which enter into the decision process of activity allocation.

Trip Generation. Transportation studies have found that 80 to 90 per cent of all trips made by residents of an urban area originate or end at home. Residential land use, therefore, is a highly important trip generator. For this reason, and also because it exhibits a predictable pattern, much of the research on trip generation has been concentrated on residential land use. Nonresidential land use generation is usually estimated simply by measuring trip rates per unit activity during the survey year, intuitively modified to reflect changing conditions which might prevail in the forecast period.

To summarize:

1. Population and economic forecasts describe the *regional growth* in the area of study.
2. Land use forecasts describe *where* this growth will be allocated in the region.
3. Trip generation describes how many *trips* will be *generated* by these activities.

It has been found from empirical observation that people's travel (excluding walking trips) increases in proportion to their income and to their car ownership, and varies with location of residence. Other variables which may enter into the determination of people's travel characteristics are age of resident, race, and sex.

A typical model (equation) used in estimating trips emanating from a typical household in a given residential area is

$$Y = A + B_1X_1 + B_2X_2 + B_3X_3 + B_4X_4 \tag{4-6}$$

where Y = trips per household
X_1 = car ownership
X_2 = family income
X_3 = logarithm of net residential density
X_4 = family size

The above model is a multiple regression equation, and the parameters A, B_i must be determined through a calibration process which involves fitting the model to observed travel data. The model parameters are not likely, however, to be the same from area to area. This may be attributable not only to basic differences in life style from area to area, but also, in some degree, to the lack of a unified travel theory, which has thus far eluded transportation planners.

Once the structure of the trip generation model has been established, it is common practice to use it to obtain an estimate of future travel for some target periods. For this purpose one requires future estimates of each of the independent variables. However, while it may be impressive to build a multivariable model during the calibration phase, the increase in accuracy by the addition of an nth independent variable may dissipate at once when one considers the error introduced through the process of projecting this same variable into the target year.

The complexity of a transportation planning model, therefore, should be kept to a minimum when dealing with independent variables, which are subject to errors when they in turn must be forecast.

Trip Distribution

Just as it is necessary to determine how much travel will be generated in each land parcel of a region, it is equally important to obtain estimates of the direction of travel and the length of the trip originating from that land parcel. This information is obtained through the analysis of travel data collected by the origin-destination surveys, described in Chapter 5. For planning purposes it is necessary to reproduce these observed travel patterns by a mathematical model, so that the same model can be used to obtain estimates of travel distribution for a desired forecast year.

Two classes of models are currently available. One class contains the *growth factor models*, and the other contains *simulation models*. Of the growth factor models, the *Fratar model* is the most appealing because of its superior analytical strength. It is of the form:

$$T_{ij} = t_i F_i \left(\frac{t_{ij} F_j}{\sum_{j=1}^{n} t_{ij} F_j} \right) \tag{4-7}$$

where T_{ij} = future trips between i and j
t_{ij} = present trips between i and j
F_i = growth factor for zone i
F_j = growth factor for zone j
t_i = present trips to and from zone i

This method simply expands the observed travel patterns recorded in a survey year, in relation to the growth of each portion of the urban area under consideration. The method was developed in 1954, in connection with the Cleveland study.[12]

The process is repeated until the trips generated at any one zone equal the trips distributed to that zone from all other zones. Usually three to four iterations are required.

While appealing to the user because of its simplicity, this model has serious weaknesses, in that it is unable to forecast travel for those areas of the region which were predominantly undeveloped during the survey year, and it cannot differentiate the effects of changes in accessibility for various portions of the area.

Because of these serious limitations, this model is restricted to forecasts of trips which pass through the study area without stopping within it.

A more analytical approach to estimating the distribution of travel is the so-called *Gravity Model.* The general representation of this model is:

$$t_{ij} = P_i \frac{A_j/(D_{ij})^x}{\sum_{j=1}^{n} \frac{A_j}{(D_{ij})^x}} = P_i \frac{A_j/(D_{ij})^x}{\frac{A_1}{(D_{i_1})^x} + \frac{A_2}{(D_{i_2})^x} + \cdots + \frac{A_n}{(D_{i_n})^x}} \quad (4\text{-}8)$$

where: t_{ij} = Trips from zone i to zone j

P_i = Total trips originating in zone i

A_j = Attractive force in zone j (for example, if t_{ij} are work trips, A_j could be represented by the number of jobs located in zone j).

D_{ij} = Travel impedance between zones i and j (this could be *travel time* or travel *cost*).

x = The travel impedance exponent.

This model, related to Newton's law of gravity, assumes that trips emanating from a zone, i, are attracted to another zone, j, in proportion to the sizes of the two population groups and in inverse proportion to some power of the travel impedance between the zones. The analytical superiority of this model over the Fratar model lies in the fact that it enables forecasts of trips to areas which were only sparsely populated in the survey year, and it takes into account the changes in accessibility between zone pairs.

Another model with analytical strength is the *intervening opportunities model.* This is based on the principle that each zone in the study, containing a given activity of interest to the traveler, has an equal probability of being the traveler's destination. Because, however, the traveler is reluctant to travel any farther than he has to, the model assumes that zones closer to the traveler's origin have a greater likelihood of satisfying the traveler's desire than do zones located far away.

In its operational form the model is described as:

$$V_{ij} = V_i(e^{-lV_0} - e^{-l(V_0+V_j)}) \quad (4\text{-}9)$$

where V_{ij} = trips from zone i to zone j

V_i = total trips originating in zone i

l = probability of accepting a destination (this is considered constant for each zone, but varies according to trip purpose)

V_0 = number of destinations closer to zone i than zone j's destinations are to zone i

In both the *gravity* and *intervening opportunities* theories of trip distribution, it is necessary to find values for the exponent b and the probability of stopping l. These are determined during the calibration phase of the model with known data. They have been found to vary according to trip classification. Once calibrated, these models are used for forecasting purposes without changing the structure determined for the calibration period.

The Modal Split Process

The next step in the analysis of urban travel is to be able to reproduce traffic flows on the transportation networks. To accomplish this objective it is important to first estimate *how* people will travel. Most travel taking place in an urban area is made either by auto or mass transit. Apportioning person trips between mass transit and private transportation is called *modal split.*

The model split process is accomplished either before or after the trip distribution phase. The decision as to where to place it often depends upon the orientation of the transportation analyst or the region under study. If the level of mass transit service in a region is not expected to be significant to the extent that it will influence the growth structure of the area, then it may be well to employ the modal split *after* trip distribution. If, however, the study area is well served by mass transit, and the service is expected to continue in the future, it may be preferable to use the modal split model *before* trip distribution.

Some of the larger studies which have used either post- or pre-distribution modal split models are:

1. Post-Distribution
   a. Washington, D.C.
   b. Minneapolis-St. Paul, Minn.
   c. San Juan, P.R.
   d. Buffalo, N.Y.
2. Pre-Distribution
   a. Chicago, Ill.
   b. Pittsburgh, Pa.
   c. Erie, Pa.
   d. Puget Sound, R.I.
   e. Southeastern Wisconsin
   f. Philadelphia, Pa.

The methods used for determining the existing and future use of mass transportation in an urban area are based on developing quantitative relationships between the socio-economic characteristics of the traveler, the physical characteristics of the mass transit/highway system, and the characteristics of the land use arrangements in the area. These methods range from the simple development of travel-time ratios between the mass transit and the highway system, to more sophisticated and complex models such as the one developed by the Delaware Valley Regional Planning Commission.[3] A comprehensive summary of current techniques used in the modal split process is available.[7,3]

Like other models, the modal split model is based upon existing relationships between the variables used. For forecast purposes, these models assume that public policy toward mass transit will remain unchanged.

These assumptions may play a vital role in the future use of mass transit. Since the present techniques do not lend themselves to dealing with the effects of public policy, it is difficult to have the model predict the effect of these policies on mass transit riders. Nevertheless, mass transit planning must continue, and the results of more favorable fiscal public policies toward this mode may have to wait until they can be explicitly incorporated into a model's structure.

We have, however, historical evidence of the effects of a lack of a positive public policy toward mass transit. This is shown in the declining patronage of the mass transit system, attributable to the rapid increase in automobile ownership, accompanied by the slow but steady deterioration of mass transit service. If one measures present-day usage of mass transit, and projects this measured human behavior into the future, gross underestimates may result if the forecasts are left entirely to the model, with no allowance made for a more favorable climate for mass transit investment.

The modal split model is a very critical step in the transportation planning process because, on the results of the model, decisions are made which are vital to the region's transportation modal "balance."

Traffic Assignment

This is the "final" stage of the transportation planning process. Having divided total person travel between the private automobile and mass transit, one must allocate this travel to each major section (link) of the highway network or transit network. As a highway traffic engineer, one must be interested in highway traffic assignment. Required for the process are two basic data files: The *network file* and the *trip file*.

The network file contains information on the pertinent characteristics of each highway section, such as:

0. Type of route

1. Number of lanes

2. Capacity

3. Travel speed

4. Environmental area (central business district [CBD], urban, suburban, etc.)

5. Length

6. Others

The trip file contains:

1. Origin-destination of trips

2. Number of trips

3. Types of trips

4. Others

The objective of the traffic assignment process is to simulate the traffic flows on every highway section in the study area. The assignment process is first calibrated for existing conditions, then it is used for forecasting future traffic flows on future highway networks.

The accuracy of the traffic assignment is highest for high-volume (ADT of 30,000 vehicles or greater) highways, and lowest for ADT's of about 5,000–10,000 vehicles. The lowest standard error of estimate for high-volume roads is around 20 per cent of the average assigned volume, and for low-volume roads it can go higher than 100 per cent.[9] The highway or traffic engineer, therefore, should be aware of such variations when using traffic assignment outputs as his design volumes for future highway planning purposes.

Feedback is essential in evaluating the alternative transportation plans which have been forecast. The purpose of this feedback is to permit a plan evaluation which considers not only the level of service which the transportation network is expected to provide, but also whether the plan conforms with desirable community goals as specified by the initial set of policies which the elected community representatives formulated.

This is a painstaking process, subject to long, tedious work. It is through this process, however, that plans are realized and are ultimately implemented.

DEVELOPMENT OF ALTERNATIVE FUTURE TRANSPORTATION NETWORKS

Alternative future transportation networks are developed and tested through the process of system analysis. The purpose of developing various configurations of transportation networks is for (1) providing inputs in the travel forecasting processes, and (2) providing a range of highway networks to be tested for the purpose of recommending a plan which best meets the stated goals and objectives.

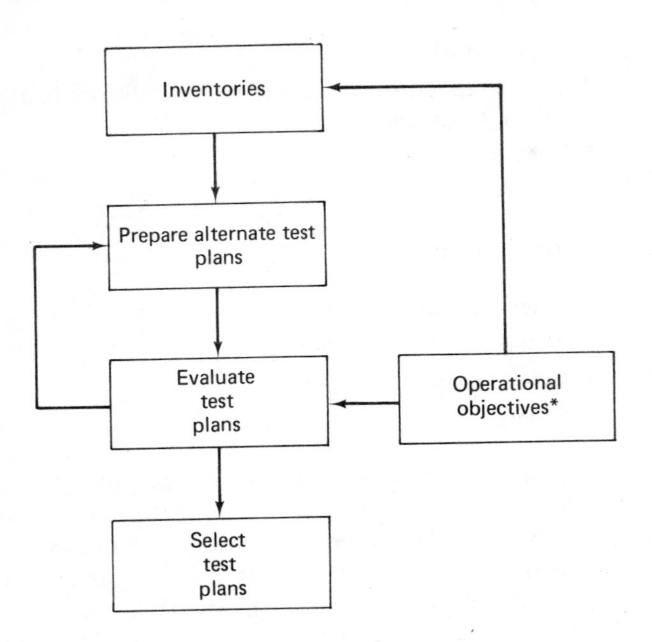

Fig. 4-2. Process for selecting alternative test plan

Current practice employs a great deal of subjective judgment, largely constrained by the (1) existing network, (2) proposed or committed network, and (3) political overtones. As a result, proposed alternative plans can be supported only through general beliefs or "theories." In addition, the existing linkage between the stated goals and objectives and a particular alternative plan is rather weak to be of use in relating plan design and its effectiveness or to satisfy the compatibility between the goals, the objectives, and the plan.

There is a strong need to strengthen the linkage between goals, objectives, and plan formulation. A suggested approach for accomplishing this is described in Figure 4-2.

It is not possible to include the complete set of goals and objectives at this level of analysis. Representative measures of travel time, construction costs, units of housing or businesses displaced, etc., could be effectively used as first-order measures in developing more rational highway test networks. These can then be subjected to more comprehensive and precise testing following the traffic assignment phase.

Before generating traffic demand volumes it may be feasible to measure the merits of a given alternate plan on the basis of the above measures, which can be reasonably estimated as a function of area type, route type, and other criteria (such as desirable levels of service) on the basis of existing known relationships.

This process will also aid in determining the amounts and type of data to be collected during the inventory phase for direct use in evaluating alternate test plans.

FORECASTS OF FUTURE TRAVEL

This task establishes the magnitude of travel demands which must be satisfied by the transportation networks. It is done by using the models calibrated on survey and historical data, as described.

EVALUATION AND PLAN SELECTION

Each alternative is subjected to a testing process in terms of the operational objectives. Each alternate transportation network is then evaluated on the basis of how well it meets the stated goals and objectives. Alternative plans are compared, and the plan which best meets the stated goals and objectives is recommended for implementation.

CONTINUING PLANNING

As already mentioned, transportation planning is a dynamic process. This implies that no future plan made today will be completely valid. Continuing monitoring of the region's pulse, therefore, is an important requirement. This involves updating the existing data files for purposes of detecting significant departures between forecast conditions and observed conditions. Whenever significant departures are identified, the plan should be modified to reflect the observed changes and reevaluated in terms of its original goals and possible changes in these goals. A sound transportation planning process must be cooperative, comprehensive, and continuing.

USES OF TRANSPORTATION PLANNING DATA BY THE TRAFFIC ENGINEER

The urban transportation planning process produces large amounts of data, much of which are of use to the traffic engineer. Thus, it is extremely desirable for him to keep in close contact with the local transportation study group, or state planning office, for access to these data.

The following is a sample list of information which could be of use:

1. Highway network
   a. Volume counts

b. Street widths
c. Travel times

2. Land use
   a. Location and size of major traffic generators
   b. Type of parking supply, usage, and demand
   c. Maps indicating job sites, business and combined areas, recreational land, etc.

3. Socioeconomic
   a. Auto ownership by traffic zones
   b. Household incomes by traffic zones

4. Travel characteristics
   a. Number and type of trips originating in, or destined to, an area
   b. Peak-hour factors, by route and area types
   c. Directional splits, by route and area types
   d. Commercial traffic, by route and area types

5

Origin and Destination Studies

The origin and destination (O & D) study establishes a measure of the patterns of movement of persons and goods within a particular area of interest. This kind of study estimates the travel characteristics observed for a typical day. The O & D study yields information regarding origins and destinations of trips, times of day in which trips are made, and mode of travel. In more comprehensive studies additional data is obtained. This includes trip purposes, land use at the beginning or end of the trip, and background social and economic data on the one making the trip.

The O & D study in which trip patterns for an entire metropolitan area are desired may be extensive in scope. For these studies, the area of interest must be divided into analysis zones, so that trips may be recorded as to zone of origin and zone of destination. A cordon line, representing the boundary of the study area, must be established. Comprehensive procedures have been developed for subdividing a study area into zones.[1,2]

O & D studies may also be smaller in scope. For example, the traffic engineer may want to study the origins and destinations of vehicles using an interchange, using a section of a roadway, or entering and leaving a CBD. These studies are less complex than the more sophisticated ones described above and, since they lend themselves to simpler study techniques, are conducted at considerably less cost.

USES OF ORIGIN AND DESTINATION DATA

The traffic engineer, concerned with providing safe and efficient movement of people in vehicles and on foot, finds it important to have knowledge of traffic patterns involving vehicles and pedestrians. O & D data enable the traffic engineer to determine:

1. Travel demand on existing or future transportation facilities
2. The adequacy of existing parking and other terminal facilities
3. The adequacy of existing mass transportation facilities
4. The most desirable location of new bridges and terminal facilities
5. The feasibility of bypass routes
6. The information needed for planning, locating, and designing new or improved street systems, expressway systems, and freeway systems
7. The information needed for planning, locating, and designing new or improved mass transportation systems
8. Through-traffic routes and truck routes

9. Estimates of the probable use of proposed new or improved routes, transit lines, and terminals
10. Travel characteristics from various types of land use
11. The means for estimating future travel patterns and requirements for transportation facilities
12. Construction priorities and economical solutions for improvement programs

DEFINITIONS OF SOME COMMONLY-USED TERMS IN TRANSPORTATION

Origin: the location of where a trip begins.

Destination: the location of where a trip ends.

Trip: a one-way movement between an origin and a destination, independent of length or distance.

Trip end: either a trip origin or a trip destination.

Internal or local trip: a trip having both origin and destination *within* the area under study.

Through-trip: a trip having both origin and destination *outside* the area under study.

Cordon: an imaginary line which defines the boundary of the study area.

Desire line: a straight line connecting zone centroids, representing travel taking place between zones. The width of the desire line is usually made proportional to the number of trips between the zones.

Screen line: a line established to divide the study area into parts for the purpose of checking the accuracy of survey data.

METHODS OF CONDUCTING O & D STUDIES

Roadside Interview[3]

Drivers of passenger cars and commercial vehicles are stopped at roadside locations and interviews are conducted. The data are recorded on prepared forms, and the information obtained may include part or all of the following:

1. Type of vehicle
2. Number of persons in vehicle
3. Origin and destination of trip
4. Purpose of trip
5. Parking location
6. Intermediate stops
7. Routes traveled

In addition to interviewing a sample of the drivers (in one or both directions of movement), other observers count and classify all traffic passing through the station. These volume counts are used to develop factors which permit expansion of the interview data to represent all vehicles. A trained interviewer can complete from 30 to 40 interviews per hour, provided there is sufficient traffic for continuous interviewing.

This method has an advantage in that direct and accurate information is obtained. It is well suited to conditions where personnel are limited because operations can be confined to one station per day, and the survey can be conducted over an extended period. However, the stopping of drivers can create congestion and antagonize the public. A significant disadvantage of using this method alone is that information about vehicular trips only is obtained.

Postcard[3,4]

There are two general methods of obtaining O & D data by the use of prepaid return postcards. The first involves the distribution of postcards to vehicle drivers at some location on their travel route, and the second involves Mailing the postcards to vehicle owners. The second method is generally referred to as a *controlled postcard survey*.

In the first method, postcards are distributed at roadside stations. If possible, these stations should be located at points where traffic flow is impeded for one reason or another, such as at toll booths, at locations controlled by a traffic signal, stop sign, or yield sign, and at spot locations which are congested. The information requested is similar to that obtained at the roadside interview stations. The postcards are precoded to indicate the station from which they were distributed. Vehicle classification may be obtained by a question on the postcard or by different colored cards. At each station, classified volume counts are made to permit expansion of the returns obtained to represent all vehicles.

In the controlled postcard procedure, a list is prepared of all owners of passenger cars, taxis, and trucks registered within the study area. Each person on the list is sent a postcard which provides space for recording the origin and destination of all vehicle trips made on a specified weekday. Prior to mailing, the cards are precoded for zone location and vehicle classification. The original listing contains the number of vehicles of each type, and permits expansion of the returns obtained to represent all vehicles.

Although a wider distribution of postcards is generally obtained by mailing postcards to vehicle owners as compared to handing out postcards at specific locations, more detailed trip information is normally obtained in the latter procedure. Regardless of which procedure is used, however, a return of at least 20 per cent is considered necessary for the data to be representative. Most of the postcards are usually returned within two weeks.

Use of postcards to obtain O & D information has the advantage of requiring less money, less time, and fewer trained personnel. However, information is only obtained about vehicular trips, and since all population classes are not equally cooperative, the results may be biased.

The controlled postcard procedure of obtaining O & D information within a study area was compared[6] with the home interview method. The results obtained by both methods were found to substantially agree.

License Plate[7]

Recording of license plate numbers as a method of obtaining O & D information has been applied to moving vehicles and to parked vehicles. As applied to moving vehicles, the last three or four digits of the license number of each passing vehicle are recorded by observers at various stations; the time of passage, type of vehicle, and direction of travel are noted. In analyzing the data, the zone of origin is assumed to be the station where the vehicle was first observed. The travel route is traced by the vehicle's successive appearances at a series of recording stations. The zone of destination is assumed to be the station where the vehicle was last observed.

The advantages of this method include:

1. Simplicity of field organization
2. Ease of tracing actual travel routes
3. No interference with moving traffic
4. Good likelihood of obtaining unbiased sample

The disadvantages include:

1. Extreme difficulty in analyzing data
2. Large numbers of observers required, since all stations must be operated simultaneously
3. large element of personal error in recording license numbers[8]

The method can be applied with some success to studies of single routes or facilities, but it produces no information relative to trip purpose, parking data, and trips by other modes of transportation. It is useful for limited area studies, such as at interchanges on freeways. Photographic techniques have been applied to this method of study, but data reduction remains a problem.

As applied to parked vehicles, the zone of destination is assumed to be the location of the parked vehicle, and the zone of origin is assumed to be the address of the vehicle owner as determined from the registration files of the motor vehicle department. The study is simple to make, requires a minimum of personnel, and permits complete coverage of the vehicles parked, but it does not provide any information relative to trip purpose, ultimate destination of driver, routes traveled, and trips by other modes of transportation.

"Lights On"

At a specified entrance location to the survey area, motorists are requested to turn on their headlights for a specific length of time or until they leave the study area. A volume count is made at this entrance location to determine the percentage of motorists complying with the lights-on request. This information will also aid in the expansion of the survey results. The duration of the volume count corresponds to the length of the study, which normally extends over a period of peak flow. Observers are stationed at all exit points of the survey area, and they record the number of vehicles passing which have headlights on. It is only possible to study one entrance location each day, and therefore the survey must be extended over a period of days until all entrances are studied under comparable flow conditions.

This method provides a very simple technique for tracing vehicular movements along a single route or through a complicated intersection. It does not interfere with traffic movement and can be applied during very heavy flow conditions. The same advantages and disadvantages associated with the moving-vehicle license plate method apply to the lights-on method, but the personal error involved in recording license numbers is eliminated.

The New York State Department of Public Works very successfully applied this method in the study of vehicular movements through a rotary intersection.[9] The method was also applied very satisfactorily to the study of traffic movements through complex interchange areas at three locations in the New Jersey–New York metropolitan region.[10]

Home Interview[1]

This method provides the most comprehensive procedure for obtaining travel characteristics within a study area. A representative sample of dwelling units is selected, and personal interviews are conducted to obtain travel characteristics for all members of the household by all modes of transportation for the previous day. In highly urbanized areas a competent interviewer can average about one interview per hour, including repeat calls and time used in going from one sample dwelling to the next.

The home interview method is superior to all previously discussed methods, since direct information is obtained about travel for all residents of the household. However, the method is costly, time-consuming, and requires a large work force.

License Renewal[11]

This method is predicated on the assumption that the license-renewal procedure for drivers requires them to appear in person on or about their birthday every so many years. At the time of license renewal, drivers are interviewed in the same manner as in the home interview method.

Although the license-renewal method has the advantage of lower cost, it is subject to several limitations. The procedure does not cover all travelers, because families without a licensed driver and family members who do not drive would be excluded. Duplication of travel characteristics may exist when the same household contains more than one licensed driver. The method can be applied only when it is a legal requirement to renew the license in person within the study area. However, the method can be used to keep O & D data up to date.

Telephone[12]

Interviews are conducted over the telephone in a manner similar to the home interview. A sample of households within the study area is randomly selected from the telephone book. These households are sent a pre-survey letter, describing the purpose of the study and the kinds of questions that will be asked. The letters are mailed two or three days before the telephone interviews. Calls are made during evening hours, and that day's travel behavior is requested.

This method has the advantage of obtaining a substantially higher sample of interviews at a considerably lower cost, but the results may be biased; households without a telephone could represent entirely different travel characteristics.

Photographic Techniques

The photographic techniques described for both volume and speed studies may also be used to obtain O & D data. Aerial photography can record vehicle movements, which can be traced through a given area and recorded for entry and exit points in the study section.

While this method has utility for limited area studies, as with a freeway interchange or intersection, data reduction is difficult, time-consuming, and costly. The method cannot be used for larger areas, because the height required to photograph the area makes accurate vehicle tracing difficult. However, sections as long as one mile have been photographed and analyzed using this technique.

TYPE OF STUDY VERSUS CITY SIZE

The National Committee on Urban Transportation has recommended different procedures for conducting O & D studies based upon the population of the urban area.[52] In smaller cities, where traffic patterns are less complex due to lower volumes, the smaller number of possible routes, and a single CBD concentration, the simpler, less costly techniques are adequate. These include postcard, lights-on, and license plate surveys. In larger cities, where volumes are higher, business activities may not be located in one centroid, and there are a multiplicity of routes, these methods are ineffective. For cities over 75,000 population, the National Committee on Urban Transportation recommends the *external cordon-home interview method.*[1] For conducting home interviews, the sample sizes given in Table 5-1 are recommended for proper accuracy.

The Bureau of Public Roads conducted a study of O & D results in 193 cities. The results are summarized in Table 5-2.

Table 5-2 shows that as cities increase in size, a greater proportion of approaching traffic has a destination within the city. These data apply only to vehicular trips approaching the city, and do not reflect the significant intracity traffic movements by all modes of transportation.

CHECKING THE ACCURACY OF O & D DATA

There are two methods available for checking the completeness with which internal vehicular trips are reported in internal and external surveys. (*Internal survey* refers to home interviews made within the survey

TABLE 5–1. SAMPLE SIZES FOR HOME INTERVIEW SURVEYS[1]

POPULATION	*SAMPLE SIZE*
under 50,000	1 in 5 dwelling units, or 20%
50,000–150,000	1 in 8 dwelling units, or 12.5%
150,000–300,000	1 in 10 dwelling units, or 10%
300,000–500,000	1 in 15 dwelling units, or 6.7%
500,000–1,000,000	1 in 20 dwelling units, or 5%
over 1,000,000	1 in 25 dwelling units, or 4%

TABLE 5–2. DESTINATIONS OF TRAFFIC APPROACHING VARIOUS SIZE CITIES[53]

POPULATION GROUP (THOUSANDS)	*NUMBER OF CITIES*	*% OF THROUGH (BY-PASSABLE) TRAFFIC*	*% OF TRAFFIC BOUND FOR CITY*	*% OF TRAFFIC BOUND FOR CBD*	*% BOUND FOR OTHER LOCATIONS*
less than 5	11	53.9	46.1	22.9	23.2
5–10	29	52.4	47.6	23.6	24.0
10–25	43	38.5	61.6	26.0	35.5
25–50	36	26.6	73.4	24.8	48.6
50–100	25	21.7	78.3	21.6	56.7
100–250	31	17.9	82.1	20.2	61.9
250–500	6	10.7	89.3	21.4	67.9
500–1000	9	8.6	91.4	12.5	78.9
over 1000	3	10.3	89.7	9.5	80.2

area; *external surveys* are studies of the traffic crossing the cordon line.)

Use of Screen Line Comparison

A screen line, intersecting the cordon line at two points, is established to divide the internal study area into two parts. Natural or man-made barriers such as rivers, railroads, or expressways provide excellent screen lines, since large volumes of traffic must cross them, and there are usually a limited number of such crossings. If two good screen lines are available, both should be adopted to check the adequacy of the internal trip data.

Classified, hourly volume counts are made at each crossing of the screen line. A comparison can then be made between the number of trips having their origin on one side of the screen line and their destination on the other, as actually counted, and that determined from an analysis of the expanded interview data from the internal and external surveys.

If the trip data and the actual ground counts show close agreement during morning and evening peak flow

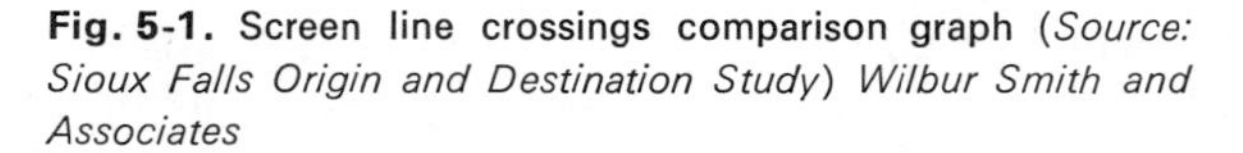

Fig. 5-1. Screen line crossings comparison graph (*Source: Sioux Falls Origin and Destination Study*) *Wilbur Smith and Associates*

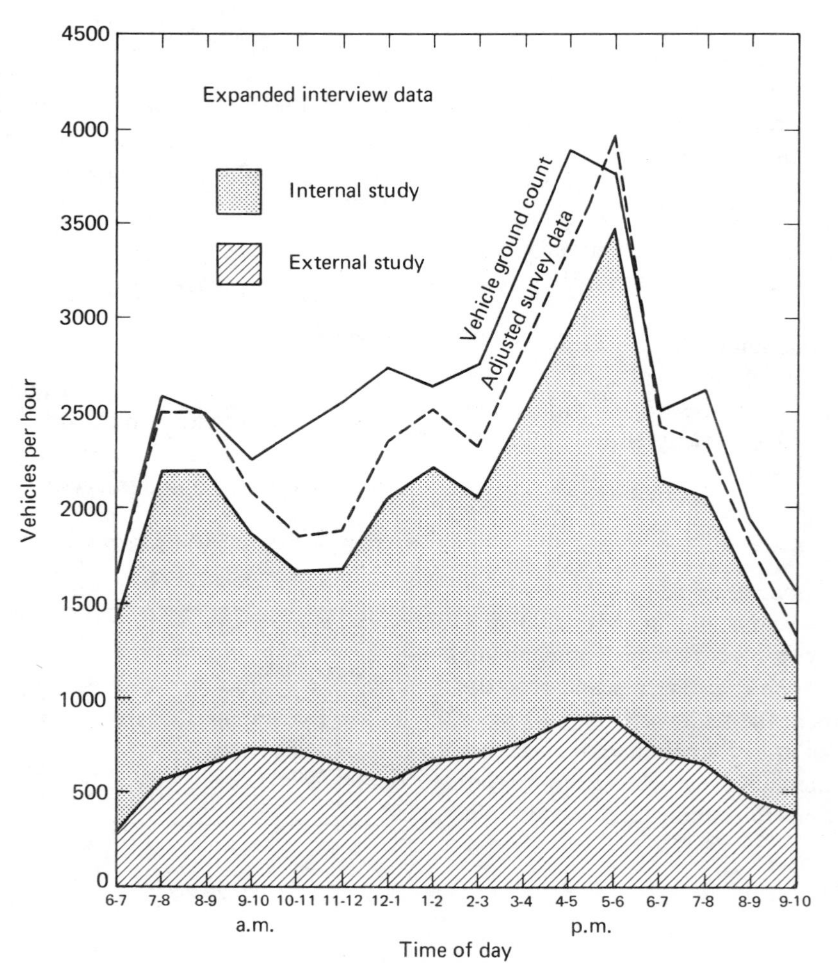

Comparison of passenger car trips at Screen Line					
Location________________					
Hour periods	Expanded trip data			Ground count	Per cent of total ground count
	Internal	External	Total		
A.M. 6:00-6:59 etc. P.M. 12:00-12:59 etc. 9:00-9:59					
Totals					

Fig. 5-2. Sample screen line data summary form

periods, agree reasonably well during the remaining periods of the day, and have a total number of trips for the 16-hour period that is at least 85 per cent of the total ground count for that period, then the trip data obtained in the survey may be considered adequate. However, if the trip data and ground counts fail to compare satisfactorily, some adjustment of the internal trip data is warranted.

Screen line comparisons are normally represented graphically, as shown in Figure 5-1. In this sample study, the comparison of the screen line ground counts with the vehicular trips between the two parts of the city as reported and expanded in the internal and external surveys for the 16-hour period (6:00 A.M. to 10:00 P.M.) gave a survey accuracy of 80 per cent. A correction factor of 1.2 was applied to the data obtained from the internal home interview and truck studies. The graph gives the hourly ground counts, the relative proportions of internal and external trips crossing the screen line, and the adjusted survey data. Caution, however, should be exercised in utilizing a uniform correction factor. Not all trip purposes may be equally under-reported and a thorough investigation of travel purposes should be made prior to reaching a decision of one or more correction factors.

Use of Cordon Line Comparison

A cordon line comparison concerns passenger-car trips by residents of the internal area and truck trips by trucks registered in the area which cross the external cordon line. The total number of such trips, as determined from the expanded *external* interview data, can be compared with the total number as determined from the expanded *internal* interview data. This check is considered to be only supplementary to the screen line.

Presentation and Summary of O & D Data

A vast amount of data are collected in O & D surveys, and a number of different methods are used to summarize information. These include various tabulations and graphic and pictorial representations.

Tabulations for Completeness of Data. The results of the methods discussed previously to evaluate survey accuracy (by comparison of trip data observed in the field with expanded internal and external interview trip data) must be analyzed first. This is necessary because any adjustments must be applied before any other tabulations are made.

Screen line comparisons are summarized by using separate tables for each vehicle classification (passenger cars, trucks, taxis, if sufficient in number) and for total vehicle trips. Using passenger cars as an example, the tables are arranged as shown in Figure 5-2. A table similar to the one in Figure 5-2 is prepared to summarize all vehicular trips. Such tables should be prepared for each screen line, grouping together the data for all crossings of each line.

Cordon line comparisons are summarized by using separate tables for passenger cars and trucks and a summary table of all vehicular trips. The data are grouped together for all external stations. Using passenger cars as an example, the tables are arranged as shown in Figure 5-3.

Comparison Graphs for Completeness of Data. After the above tables are compiled, the results are graphically presented, as previously discussed. If adjustment of trip data is warranted, the adjustment factor is applied according to trip purpose.

Exploratory and General Tabulations. These tabulations are general in nature and are used for two purposes: to form the basis for a study of travel habits, and to determine which items are of sufficient importance to be considered in subsequent analyses.

Fig. 5-3. Sample cordon-line summary form

Passenger car trips crossing Cordon Line			
Hour periods	Expanded internal	Expanded external	Per cent (internal/external)
6:00-6:59 A.M. etc. 9:00-9:59 P.M.			
Totals			

Vehicular trips

From (origin) zone	To (destination) zone			Σ of trips from zone
	1	2	3	
1	10	30	20	60
2	34	40	50	124
3	18	54	26	98
Σ of trips to zone	62	124	96	

Fig. 5-4. Rectangular table for internal origin and destination data

Dwelling unit summary
(only for home interview surveys)

A table is prepared which summarizes the expanded information recorded in the dwelling unit summary portion of the home interview form. The column headings for this table are as follows:

Tract or zone
Number of dwelling units
Number of passenger cars owned
Total number of persons
Number of persons 5 years of age or older
Number of persons 5 years of age or older making trips
Total number of trips

Mode and purpose of travel summary

1. Internal Survey

   Tables are prepared which summarize the expanded trip data recorded. A separate table for each mode of travel is compiled, showing the total number of trips by residents of the internal area and classified by trip purpose from origin to destination. A summary table is also prepared for all modes of travel. The column headings and stub headings are identical for these tables. The headings are:

   From or To
   Work
   Business
   Medical-Dental
   School
   Social-Recreation
   Change travel mode
   Eat meal
   Shopping
   Serve passenger
   Home
   Totals

   Another table is compiled from the expanded home interview data showing the average passenger-car occupancy by trip purpose. It is arranged as above, showing the average number of persons per vehicle for trips *from* each trip purpose *to* every other trip purpose.

2. External Survey

   Tables are prepared which summarize expanded passenger-car trip data as recorded in the roadside interviews at the external cordon stations. A separate table is compiled for local trips and through trips, showing the total number of passenger car trips passing each external station classified according to trip purpose. A summary table combining local and through trips is also prepared.

   Another table is compiled from the expanded external trip data which gives the average passenger-car occupancy by purpose of trip. The table is subdivided into local and through-trips.

Traffic Flow Tabulations. These tables form the basis for selecting the general route locations which will best serve traffic needs in the metropolitan area. The average daily vehicular or passenger traffic interzonal transfers are tabulated. Rectangular tables show one-way traffic movements from each zone to every other zone. A rectangular table with hypothetical interzone transfers, for a simple three-zone example, is shown in Figure 5-4. On an average day the trips into and out of a given zone must balance, otherwise vehicles would accumulate in the zone over a period of days.

The triangular table shown in Figure 5-5 is compiled by adding together the two directional movements between every pair of zones. The table does not show the direction of travel, and to obtain approximate one-way movements the number of trips between zones is halved.

Interzonal transfer tables are normally compiled separately to show the following movements:

1. Passenger cars
2. Trucks
3. Taxis
4. Summary of tables

Fig 5-5. Triangular table for internal origin and destination data

Vehicular trips

Between zones → ↓	1	2	3
1	20	64	38
2		80	104
3			52

Fig. 5-6. Desire lines of automobile drivers, traveling between the metropolitan traffic district and outside segments—1947 (*Source: B.A.M.T.S. 1946–47, California Division of Highways*)

5. Passengers carried by auto, truck and taxi
6. Passengers carried by bus

With the use of these tables, it is possible to estimate the volume of vehicles which would use any portion of a proposed route, as well as the direction and volume of interchange at any point on the route. These tables also provide data needed to synthesize traffic movements for developable land within the study area, as well as traffic movements for other cities.

It is not only possible to estimate the volume of vehicle movements into, through, and within the metropolitan area, but also the number of passenger trips from one zone to another made in auto, truck, taxi, or bus.

Pictorial Representation of Traffic Flow. The data in the interzonal transfer tables are used to prepare *desire line* charts. The two-directional movement between each pair of zones is represented by a straight line, extending between zone centroids, whose width is made proportional to the number of trips between zones. An example of a desire line chart is shown in Figure 5-6. *Desire lines* are the flows of traffic as they would proceed by the shortest distance between zones of origin and destination. An ideally located major street system, then, would closely correspond to the pattern of the desire lines.

A major desire line chart is prepared by summarizing individual desire lines, combining all those lines which follow approximately the same direction. These major directional patterns are very useful in determining the location of a proposed urban expressway system or to illustrate deficiencies in an existing system.

To represent desire lines of traffic flow for massive O & D data, such as in the Chicago Area Transportation Study, electronic data plotting equipment must be used.[54]

Parking Tabulations. These tables are used to estimate the demand for parking at different locations within the study area. Parking data are obtained at the internal study interviews, and this information can be used to prepare tables showing the kind of parking for each trip purpose in the metropolitan area as a whole, and for only those trips having destinations within the CBD. The formats of the tables for both summaries would be identical, with the *kind of parking* as stub headings and the *trip purpose* as column headings.

SUPPLEMENTARY O & D STUDIES

External-Internal Cordon Survey

For cities with populations between 5000 and 75,000, the National Committee on Urban Transportation recommends an external-internal cordon procedure for conducting O & D studies. This is on the condition that a large proportion of the traffic is headed for the CBD, or that the movement is predominately through traffic, and no major deficiencies exist outside the CBD.[55]

To obtain the required data, roadside interviews are conducted at two cordon lines, one near the edge of the urban developments and the other around the CBD.

External Cordon-Parking Survey

For cities with populations between 5000 and 75,000, the National Committee on Urban Transportation recommends the combination external cordon and parking procedure for conducting O & D studies. This is effective when the major problems are congestion and parking in the CBD.[55,56]

In addition to obtaining information from the external cordon survey, conducted as previously described, a comprehensive parking survey is made in the CBD. This study involves an inventory of existing parking facilities, interviews of parkers at parking stalls, and a cordon count of all traffic into and out of the CBD. Detailed procedures are available for conducting a comprehensive survey.[57]

External Cordon Survey

For cities with populations under 5000, the National Committee on Urban Transportation recommends the external cordon procedure for conducting O & D studies. This is adequate for cities of this size, because traffic approaching the city exerts the major influence on the traffic patterns within the study area.

6 Highway Economy Studies

Economy has been defined by Prof. Grant[1] as "getting the most for your money in the long run." He also gives a more elegant definition: "[Economy is] securing the highest possible ratio of utility to cost." The application of the principles of engineering economy to a proposed project will enable the engineer to select that type of construction or improvement which will serve the purpose at the lowest cost.

Highway economics, as distinguished from engineering economy, is concerned not only with the economic justification for certain improvements, but also with other factors that are of vital importance to highway development. For example, the use of the highways for commercial purposes has resulted in the adoption of certain methods of taxation, which may vary widely in their practical application. It is necessary, therefore, to examine the broad field of highway user taxation, to attempt to develop principles that should be followed in scheduling taxes for motor vehicles, and to consider the allocation of taxes among the various classes of users of the highways. The proper allocation of taxes among individual users is closely allied with the problem of costs. Before users can be taxed equitably for the costs of providing and maintaining highways, calculations must be made regarding these costs.

Modern methods of borrowing money to be used for highway construction embrace methods of paying the indebtedness. Amortization of bonded debt in an efficient manner must be considered in any discussion of highway economics.

The field of highway economics includes all elements and factors that affect the cost of transportation.

One of the objectives of traffic engineering is to provide safe, convenient, and economic transportation of persons and goods. Improvements in traffic facilities cost money, and the best way to evaluate and justify specific improvements is through an economic study. A factual approach to a problem can be defended more convincingly than mere opinion.

Highway planning, like private business planning, must be based on economic considerations and involves the establishment of long-range objectives for a period of years. These objectives must be logically programmed for execution over this period of time. A systematic scheduling of improvements is essential to assure the greatest benefits from the expenditure of funds.

PRINCIPLES OF ENGINEERING ECONOMY

A basic understanding of the principles of engineering economy is most desirable as a background for the study of highway economics.[1,2,3] A study of the methods for handling time differences is essential for the comparison of initial, ultimate, and recurring costs. Every engineer who is called upon to make engineering studies involving cost analyses should be familiar with the following derivations of mathematical formulas, which are useful in converting costs, dissimilar in time element, to comparable bases.

Note: In each derivation, n designates the number of compounding periods. For simplification, it is referred to throughout as the number of years, that is, a compounding period of one year is assumed. However, all equations hold regardless of whether n represents years, quarters, months, or other period. Of course, the correct interest rate i (annual, quarterly, monthly, etc.) must be chosen to correspond with the length of the compounding period.

Single Payment Formulas

Compound Amount. *Compound amount* is the amount S to which P dollars, deposited today at compound interest of rate i, will grow or accumulate in n years.

P = principal, or amount deposited at compound interest

$$P + Pi = P(1 + i) = \text{amount at end of first year}$$

$$P(1 + i) + P(1 + i)i = P(1 + i)(1 + i) = P(1 + i)^2 = \text{amount at end of second year}$$

$$P(1 + i)^2 + P(1 + i)^2 i = P(1 + i)^2(1 + i) = P(1 + i)^3 = \text{amount at end of third year}$$

Hence,

$$S = P(1 + i)^n = \textit{compound amount} \text{ at end of } n^{\text{th}} \text{ year} \qquad (6\text{-}1)$$

$$\text{Compound amount factor} = (1 + i)^n$$

Present Worth. The *present worth* of a future amount is the principal, or present deposit P, which, if placed now at compound interest i, will grow to an amount S in n years.

In the above formula, $S = P(1 + i)^n$, P is the principal or present deposit and S is the (future) compound amount. Therefore, solving for P,

$$P = \frac{S}{(1 + i)^n} = \text{present worth of the future amount of } S \qquad (6\text{-}2)$$

$$\text{Present worth factor} = \frac{1}{(1 + i)^n}$$

Uniform Annual Series Formulas

Sinking Fund. In order to accumulate a given sum S at the end of n years, a *sinking fund* is set up and uniform deposits R are paid into it periodically (say annually). The growing fund is kept at compound interest i.

S = the ultimate sum of the n compound amounts of the n annual deposits

Since

$R(1 + i)^{n-1}$ = compound amount at the end of n years, of the first deposit R made at the *end* of the first year

and

R = the deposit made at the end of the nth year, which draws no interest

then,

$$S = R(1 + i)^{n-1} + R(1 + i)^{n-2} + R(1 + i)^{n-3} + \cdots + R(1 + i)^2 + R(1 + i) + R \qquad (6\text{-}3)$$

To solve this series, a simple expedient that is very useful in handling definite series is the following:

First, multiply both sides of the equation by $(1 + i)$,

$$S(1 + i) = R(1 + i)^{n-1} + R(1 + i)^{n-2} + \cdots + R(1 + i)^3 + R(1 + i)^2 + R(1 + i)$$

Then subtract the former equation from the latter, thereby eliminating all but the end terms of the series.

$$S(1 + i - 1) - S = R(1 + i)^n - R$$

Simplifying,

$$S(1 + i - 1) = Si = R[(1 + i)^n - 1]$$

Solving for R,

$$R = \frac{Si}{(1 + i)^n - 1}$$

Sinking fund factor (of uniform annual series)

$$= \frac{i}{(1 + i)^n - 1}$$

Solving for S,

$$S = \frac{R[(1 + i)^n - 1]}{i} = \text{total amount in the sinking fund at the end of } n \text{ years}$$

Compound amount factor (of uniform annual series)

$$= \frac{(1+i)^n - 1}{i}$$

Solving for n,

$$\frac{Si}{R} + 1 = (1+i)^n$$

$$n \log (1+i) = \log \left(\frac{Si + R}{R} \right)$$

$$n = \frac{\log [(Si + R)/R]}{\log (1+i)}$$ = number of years (or periods) required to accumulate the sum S, if annual (or periodic) deposits of amount R are put into a growing fund, at compound interest i

Annual or Capital Recovery. To provide in advance for an annual withdrawal R at the end of each year for n years, i.e., for *recovery of capital* (including interest) at a uniform annual rate, an amount P is set aside at a compound interest rate i such that n annual withdrawals of R dollars each will exactly use up the fund.

$P(1+i) - R =$ amount in fund at end of first year.

$$[P(1+i) - R](1+i) - R$$
$$= P(1+i)^2 - R(1+i) - R$$
= amount in fund at end of second year

$$P(1+i)^3 - R(1+i)^2 - R(1+i) - R$$
= amount in fund at end of third year

$$P(1+i)^n - R(1+i)^{n-1} - R(1+i)^{n-2} - \cdots - R(1+i)^3 - R(1+i)^2 - R(1+i) - R$$
= amount in fund at end of nth year = 0 (by definition)

Multiplying by $(1+i)$,

$$P(1+i)^{n+1} - R(1+i)^n - R(1+i)^{n-1} - \cdots - R(1+i)^3 - R(1+i)^2 - R(1+i) = 0$$

Subtracting the former equation from the latter,

$$P(1+i)^{n+1} - P(1+i)^n - R(1+i)^n + R = 0$$

Simplifying,

$$P(1+i)^n(1+i-1) - R[(1+i)^n + 1] = 0$$

$$Pi(1+i)^n = R[(1+i)^n - 1]$$

Solving for R,

$$R = \frac{iP(1+i)^n}{(1+i)^n - 1}$$ = annual withdrawal, or annual amount of capital recovery

Capital recovery factor (of uniform annual series)

$$= \frac{i(1+i)^n}{(1+i)^{n-1}}$$

Solving for P,

$$P = \frac{R[(1+i)^n - 1]}{i(1+i)^n}$$ = amount to be set aside now, the purchase price or present worth of the annuity (6-4)

Present worth factor (of uniform annual series)

$$= \frac{(1+i)^n - 1}{i(1+i)^n}$$

Interest Tables

For convenience in the application of interest and annuity factors, values of these factors for various interest rates are tabulated. An abridgement of more complete tables (found in any book on engineering economy) is shown in Table 6-1.

COSTS OF HIGHWAY TRANSPORTATION

The costs of highway transportation may be divided into two broad categories:

1. Road User Costs: the costs incurred by the user in the operation of his vehicle
2. Highway Costs: the costs incurred in the construction and maintenance of the highways

Both are usually expressed on an annual basis. Therefore, highway costs may be generally defined as the annual costs accruing from the construction, maintenance and operation of an individual highway or highway system. Similarly, road user costs are those annual costs accruing from the ownership and operation of vehicles over the highway. Unit vehicle costs are generally expressed in terms of vehicle-miles, or ton-miles in the case of commercial vehicles. Generally speaking, since the same people pay both the highway costs and the road user costs, maximum economy is obtained only when the sum of these two costs is a minimum consistent with convenience and safety.

The costs of the highway are borne by the public as a whole through vehicle taxes, general taxes, and special assessments, whereas the costs of vehicle operation are borne by the individual owner of the vehicle.

TABLE 6-1. COMPOUND INTEREST AND ANNUITY FACTORS

		SINGLE PAYMENT		*UNIFORM ANNUAL SERIES*			
Rate	*Period*	*Compound Amount Factor*	*Present Worth Factor*	*Sinking Fund Factor*	*Capital Recovery Factor*	*Compound Amount Factor*	*Present Worth Factor*
i	n	$(1+i)^n$	$\frac{1}{(1+i)^n}$	$\frac{i}{(1+i)^n-1}$	$\frac{i(1+i)^n}{(1+i)^n-1}$	$\frac{(1+i)^n-1}{i}$	$\frac{(1+i)^n-1}{i(1+i)^n}$
3%	5	1.1593	0.8626	0.1884	0.2184	5.539	4.580
	10	1.3439	0.7441	0.0872	0.1172	11.464	8.530
	15	1.5580	0.6419	0.0538	0.0838	18.599	11.938
	20	1.8061	0.5537	0.0372	0.0672	26.870	14.877
	25	2.0938	0.4776	0.0274	0.0574	36.459	17.413
	30	2.4273	0.4120	0.0210	0.0510	47.575	19.600
4%	5	1.2167	0.8219	0.1846	0.2246	5.416	4.452
	10	1.4802	0.6756	0.0833	0.1233	12.006	8.111
	15	1.8009	0.5553	0.0499	0.0899	20.024	11.118
	20	2.1911	0.4564	0.0336	0.0736	29.778	13.590
	25	2.6658	0.3751	0.0240	0.0640	41.646	15.622
	30	3.2434	0.3083	0.0178	0.0578	56.085	17.292
5%	5	1.2763	0.7835	0.1810	0.2310	5.526	4.329
	10	1.6289	0.6139	0.0795	0.1295	12.578	7.722
	15	2.0789	0.4810	0.0463	0.0963	21.579	10.380
	20	2.6533	0.3769	0.0302	0.0802	33.066	12.462
	25	3.3864	0.2953	0.0210	0.0710	47.727	14.094
	30	4.3219	0.2314	0.0151	0.0651	66.439	15.372
6%	5	1.3382	0.7473	0.1774	0.2374	5.637	4.212
	10	1.7908	0.5584	0.0759	0.1359	13.181	7.360
	15	2.3966	0.4173	0.0430	0.1030	23.276	9.712
	20	3.2071	0.3118	0.0272	0.0872	36.786	11.470
	25	4.2919	0.2330	0.0182	0.0782	54.865	12 783
	30	5.7435	0.1741	0.0127	0.0727	79.058	13.765
7%	5	1.4026	0.7130	0.1739	0.2439	5.751	4.100
	10	1.9672	0.5083	0.0724	0.1424	13.816	7.024
	15	2.7590	0.3624	0.0398	0.1098	25.129	9.108
	20	3.8697	0.2584	0.0244	0.0944	40.995	10.594
	25	5.4274	0.1842	0.0158	0.0858	63.249	11.654
	30	7.6123	0.1314	0.0106	0.0806	94.461	12.409

Therefore, the selection of the program of highway improvement that insures the lowest overall cost requires consideration of both road costs and vehicle operating costs.

Road User Costs

The costs of motor vehicle operation are of extreme importance to the highway engineer, and more specifically to the highway economist, for several reasons. Most important are:

1. Comparing different forms of transportation
2. Determining road user benefits, which is an indication of the amount that should be spent on highway improvements
3. Aiding in deciding the priority of improvement
4. Helping to compare alternative highway locations

Cost Dependency. Certain items of cost in the operation

of a motor vehicle depend almost directly on the number of miles driven. Other items vary mainly with time, and still others are primarily dependent on speed.

The various items of cost might be grouped as:

Time-dependent

1. Interest on the original cost
2. Obsolescence, which is that portion of depreciation that results from inadequacy or from being out of date
3. Driver's license and registration fees
4. Garage rent
5. Insurance
6. Driver's wages, in the case of commercial vehicles
7. Taxes

Mileage-dependent

1. Fuel
2. Oil
3. Tires and tire maintenance
4. Maintenance and repairs
5. Depreciation, that portion attributable to wearing out

Speed-dependent

1. Value of the travel time of operator and rider (Charges on these vary inversely with speed.)
2. Fuel and oil consumption and tire wear (These increase as driving speed increases.)

Of the costs mentioned above, those that vary with mileage or speed are often affected by roadway improvement. These are of particular concern in highway economy studies, for justification of highway improvement depends largely on savings in operating costs to offset the expenditure required.

Actual costs of the above items for all types of vehicles are not available in the necessary detail. The significant study to date concerned with vehicle operating costs is the *Road User Benefit Analyses for Highway Improvements*, by AASHO.[4] The costs, even in this publication, are not complete. This study presents data in the necessary detail for operating costs for passenger cars in rural areas. However, similar information is required for trucks and buses in rural areas, and for passenger cars, trucks, and buses in urban areas. In addition, the unit cost values given in the 1960 edition of AASHO are based on vehicle performance data which were obtained from research conducted prior to 1950. The only significant difference between the 1960 and the 1952 editions of AASHO is that the former updated the costs of vehicle operation.

More current research data are available which can be used to supplement the information given in AASHO, and pertinent sources will be discussed at the end of the chapter. Also, the form of the unit costs developed by AASHO permits certain approximations that may be used to make relatively complete analyses, which take into consideration the effects of commercial vehicles and urban locations.

TABLE 6-2. RATIO OF TRUCK OPERATING COSTS TO THOSE OF PASSENGER CARS

TYPE OF TRUCK	*APPROXIMATE PERCENTAGE OF TOTAL ON MAIN HIGHWAYS*	*RATIO OF TRUCK COSTS TO THOSE OF CARS*
Single-unit trucks	70	2 to 4
Truck combinations	29	4 to 6
Buses	1	2 to 4
Composite truck and bus	100	2.5 to 4.5

(Source: *Road User Benefit Analyses for Highway Improvements*, AASHO, 1960)

Cost Criteria. The variable highway design features and conditions of vehicle operation considered by AASHO in developing unit road user costs are:

Type of vehicle

1. Passenger car
2. Truck or bus

To approximate the costs of operating commercial vehicles (trucks and buses) in rural areas, ratios ranging between 2 and 6 have been adopted. These ratios represent the relationship of commercial vehicle operating costs to those of passengers cars. Included in the passenger-car category are light delivery trucks, which are often classified as pickup or panel trucks. The commercial vehicles are divided into two categories: single-unit trucks, and combination trucks. Single-unit trucks are normally those which have dual tires on the rear wheels, and carry the engine and load on the same chassis. Combination trucks are vehicles consisting of a tractor unit and a semitrailer or full trailer.

Ratios for these two truck classifications and for buses are given in Table 6-2, which has been taken from AASHO. In practice, unit operating costs for passenger cars are applied to an equivalent number of passenger cars, which is a number obtained by multiplying the number of commercial vehicles by their appropriate ratios.

Type of area

1. Rural
2. Urban

AASHO recommends the following approximations for analyzing urban areas:

(a) On major streets and expressways where traffic flow is reasonably continuous, the same cost as for rural operation will apply directly.

(b) On major streets where traffic flow is not continuous but has only a moderate amount of interruption, the unit costs should be increased by 10 to 30 per cent.

(c) With stop-and-go operation, special field investigations are required to determine the number of stops and the average standing time. The analysis may then be carried out as in (a) above and should include the cost of stops.

(d) For conditions intermediate between (a) and (b) or between (b) and (c), values are determined by interpolation or measurement.

Type of highway

1. Two-lane
2. Divided

The type of highway is differentiated by AASHO primarily with respect to the number and arrangement of lanes. The main distinction is the one made between two-lane and divided highways. Unit costs for these two types of rural, tangent highways are given for various running speeds. Running speeds on three-lane and four-lane undivided highways are about equal to, or only slightly greater than those on two-lane highways for comparable conditions, and the effects of other variables are only slightly different. Therefore, unit costs on three-lane highways can be assumed to be nearly the same as costs on two-lane highways, and unit costs on undivided highways of four or more lanes can be approximated as values between those for two-lane highways and those for divided highways.

Type of operation

1. Free
2. Normal
3. Restricted

The traffic volume operating on a section of highway will decidedly affect the running speed and congestion, and therefore the unit cost of vehicle operation. Vehicle operation on rural highways is divided into three types of operation for computing unit costs: free, normal, and restricted. The type of operation is related to the 30th highest hourly volume, as follows:

TYPE OF OPERATION	RATIO OF 30th HIGHEST HOURLY TRAFFIC VOLUME TO PRACTICAL CAPACITY
R = Restricted	greater than 1.25
N = Normal	0.75 to 1.25
F = Free	less than 0.75

The concept of *level of service* has replaced *practical capacity* in the current edition of the *Highway Capacity Manual.* In the absence of more definitive directives, practical capacity may be thought to be equivalent to level of service B or C for rural highways, and to level of service C or D for urban highways.

Running speed

AASHO unit operating costs are expressed in terms of running speeds. Running speed is defined as the speed over a specified section of highway, being the total length of the section divided by the total running time (the time the vehicle is in motion). The running speed should be representative for the entire length of the analysis section. Traffic volume and geometric conditions can significantly affect running speed. When trucks and buses are included in the traffic stream in sufficient proportions, it is necessary to take into consideration the difference in running speed between such vehicles and passenger cars.

Gradient class

1. 0 to 3 percent
2. 3 to 5 percent
3. 5 to 7 percent
4. 7 to 9 percent

The above gradient classes are for an average grade along the analysis section, and the average grade can be determined by dividing the total rise and fall by the entire length of section under consideration.

Fig. 6-1. Profile A of highway to illustrate the calculation of average grade

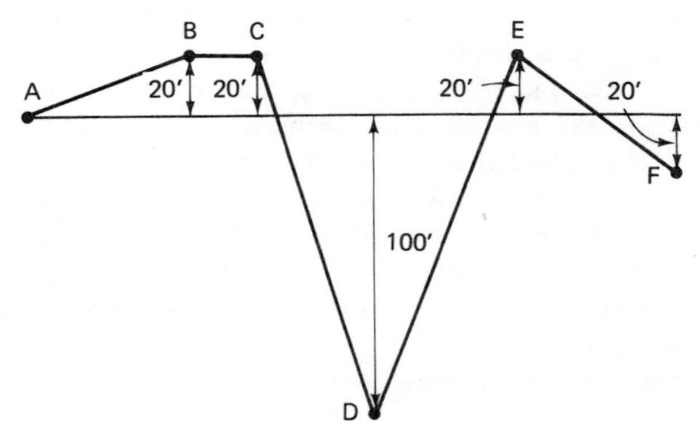

For example, assume the profile shown in Figure 6-1 is for a 10,000-foot length of highway.

Total rise and fall

$$\begin{array}{l} \quad\ AB \quad\ BC \quad\ CD \quad\ DE \quad\ EF \\ = 20 + 0 + 120 + 120 + 40 = 300 \text{ feet} \end{array}$$

$$\text{Average grade} = \frac{300}{10{,}000} \times 100 = 3 \text{ per cent}$$

Type of surface

1. Paved
2. Loose surface
3. Unsurfaced

Regardless of geometric design, the type of surface has a significant influence on unit costs. Unit road user costs are separated for types of surfaces: *paved surfaces*, either rigid or flexible; *loose surfaces*, primarily all-weather gravel; and *unsurfaced*. Unit costs are given for paved and loose surfaces in good condition. Costs for paved surfaces in fair to poor condition can be obtained by interpolation between the values for paved and loose conditions. Similarly, costs for gravel surfaces in poor condition can be estimated by interpolation between the values for loose and unsurfaced conditions.

Alinement

1. Tangent or flat curves
2. Curved

Unit costs are given for tangent alinement, and they must be adjusted to take into consideration the effects of curvature on vehicle operating costs. AASHO assumes that the effect of curvature is not pronounced providing the road is properly superelevated. Where proper superelevation is lacking, side friction is called upon to keep the vehicle in its curved path, and this results in an increase of unit costs for primarily fuel and tire costs. Normally the average degree of curvature and average rate of superelevation, weighted on the basis of length, are used to obtain the correction factor for curvature.

Other studies show that the effect of curvature with proper superelevation can increase operating costs, particularly with the combination of higher speeds and sharp curvature.

Components of Motor Vehicle Operating or Running Costs. These costs include the cost of fuel, oil, tires, maintenance, and repairs, and mileage depreciation.

Motor fuel costs

The cost of motor fuel is the most significant single item of vehicle operating cost. The amount of fuel consumed varies with each of the eight items discussed under cost criteria above.

Various studies have been conducted to determine the effect of grades on fuel consumption. All studies have shown that more fuel is used on upgrades, and less fuel is used on downgrades, than is used on level roadways. It is generally accepted that the overall effect of grades is to increase fuel consumption.

In considering the effect of curvature on fuel consumption, the rule that has been adopted by AASHO is that the excess of fuel required on curves is directly proportional, and numerically equal, to the coefficient of side friction developed. For example, if the coefficient of side friction developed is 0.10, the fuel used (as a percentage of that used on a tangent alinement) would be 110 per cent.

The type of surface has a significant effect on fuel consumption. Gasoline mileage on unsurfaced roads is about 3 miles per gallon less than on loose surface in good condition, and about 5 miles per gallon less than on paved surfaces in good condition.

Cost of oil

It is difficult to determine any definite relationship between oil consumption and the highway design features. Studies show that speed of travel and type of surface do have an appreciable effect on the amount of oil used. Oil consumption increases with speed, and progressively more oil is used as the type of surface changes from smooth and paved to unsurfaced.

Cost of tires

The rate of tire wear increases sharply with speed. Tire wear also increases with surface roughness, traffic congestion, steep grades, and improperly superelevated curves. Tire wear at a speed of 53 mph is about three times the wear at 33 mph.

Using normal operation as a base for comparison, it is assumed that for free operation tire wear will be about 90 per cent of that for normal, and for restricted operation it is about 125 per cent of that for normal. It has also been shown that continuous stop and go travel produces wear as much as seven times the wear produced at an equivalent uniform speed.

Little research has been conducted on the influence of grades on tire wear. However on the basis of the limited data available, AASHO assumed that, compared with operation on relatively level pavements, wear would be increased from 15 to 100 per cent as the grade increases from 3 to 9 per cent.

Driving around curves that are not properly superelevated increases tire wear. Wear is related to the coefficient of side friction that must be developed to keep the vehicle traveling in a circular path. Where the coefficient of friction that must be developed is about 0.15, the rate of tire wear is about two to three times that for operation at the same speed on tangent sections of highway.

Maintenance and repair costs

It is difficult to establish relationships between conditions of vehicle operation and expenditures for maintenance and repairs. Because of the lack of information, AASHO has related these costs only to the type of surface, and the cost for each type of surface is assumed constant for different operating conditions.

The cost of vehicle maintenance and repair for passenger cars on rural roads is assumed to be 1.20 cents per mile on paved surfaces, 1.80 cents on loose surfaces, and 2.40 cents on unsurfaced roads.

Depreciation

Passenger vehicles depreciate with the wear and tear of use, and they also depreciate with age. There is a considerable variety of opinion as to what portion of the depreciation should be assigned to wear and tear and what portion to age. AASHO arbitrarily assigns one-half of the depreciation to wear and tear and one-half to age. In making economy studies for road improvements, only the mileage depreciation (wear and tear) is considered, as the time portion is not influenced by the improvement.

To find the depreciated cost of passenger vehicles, the average undepreciated base cost of present-day vehicles is taken as $3,000. The average life of a vehicle is estimated to be about 10 years and 100,000 miles. For the 100,000 vehicle-miles of travel, the average charge for depreciation is, therefore,

$$\frac{\$1500 \times 100}{100{,}000} = 1.50 \text{ cents per mile}$$

Value of Time. A consequence of most roadway improvements is that the average speed of traffic is increased and the number of stops may be decreased, thereby resulting in a saving of travel time. There is general agreement that a saving in travel time for trucks, buses, and passenger vehicles used for business purposes has value that should be expressed in monetary units. However, it is difficult to assign an appropriate money value of time in many cases, as a savings of a few minutes in travel time may or may not result in a reduction in the costs associated with driver's wages, rental of equipment, and overhead. There will be a definite reduction in running costs (fuel and oil consumption, tire, vehicle repair, and maintenance).

There is a difference in opinion as to the method of including the value of time for passenger vehicles in economy studies. The questionable aspect is in not being concerned with time saving as a benefit to the road user, but with the monetary value assigned to it. There is little doubt that motor vehicle users attach importance to time saving, as evidenced by their willingness to pay for travel on facilities that result in a saving of time.[5,6]

AASHO assumes a value of time of $1.55 per hour as representative of current opinion for a logical and practical value for passenger cars. The typical passenger vehicle has 1.8 persons in it, and a time value of $0.86 per person per hour results in a vehicle total of $1.55 per hour. The conservative nature of this value reflects the controversial aspects of this assignment.

Different methods are advocated to evaluate the saving in time in highway economy studies.[7-17]

Comfort and Convenience. There is much more controversy over the value of comfort and convenience than over the value of time saving. As expressed by AASHO, an expressway which allows uninterrupted travel into the CBD of the city renders a service to the road user which has a value over and above the value of time and running cost savings. There is a value in the convenience of being able to go to one's destination without interference. There is a comfort value, over and above the saving in vehicle running cost, in being able to drive without frequent brake applications, stops and starts, or unexpected interference. There is a value in the conservation of health, through driving in a relaxed manner without the tension apparent when roadside interference is prevalent. The fact that some drivers place a value on comfort and convenience is demonstrated by their optional diversion to new and modern facilities, even though a greater mileage and little time difference may exist. After time saving, the most important reason stated by road users for using toll facilities over comparable free facilities was greater comfort and convenience.[5,18,19]

AASHO assumes that the value of comfort and convenience is related to the type of highway operation. On paved highways, the monetary values arbitrarily assigned for the cost of discomfort and inconvenience are:

TYPE OF OPERATION	*COST PER VEHICLE-MILE (cents)*
Free	0
Normal	0.5
Restricted	1.0

On unsurfaced roads, the conditions are considered to be comparable to a restricted type of operation, and a value of 1.0 cent per vehicle-mile is added. For loose surfaces, conditions are considered as intermediate to normal and restricted types of operation, and a value of 0.75 cent per vehicle-mile is added.

Since it is debatable as to whether comfort and convenience charges should be included in a listing of vehicle operating costs, AASHO has separated these charges, as well as those associated with time saving, from any composite figures of operating costs.

Tabulation of Unit Motor Vehicle Operating Costs. Prior to selecting the proper cost value for the various items included in road user costs, the following conditions must be evaluated for each analysis section.

1. Type of highway
2. Type of surface
3. Gradient class
4. Running speed
5. Type of operation

The tables in AASHO present unit operating costs, time cost, and value of comfort and convenience for passenger vehicles. Table 6-3 contains unit road user costs for tangent divided highways, pavement in good condition. Table 6-4 contains unit road user costs for tangent two-lane highways, pavement in good condition. Table 6-5 contains unit road user costs for tangent loose surface highways, in good condition. Table 6-6 contains unit road user costs for tangent unsurfaced roads. In each table the breakdown separates the factors of type of operation, running speed, and gradient class. Figure 6-2 presents the correction values for inclusion of curved alinement features.

Other studies have been conducted to obtain data on fuel consumption, travel time, operating characteristics for different classes of vehicles, and running costs of motor vehicles.[14,20-48]

A new series of tables has become available which includes running costs on level tangent, high-type pavements in good condition, plus and minus grades, horizontal curves, and for stops and 10 miles per hour speed changes. These costs are tabulated for the following vehicles: 4,000-pound passenger car, 5000-pound commercial delivery, 12,000-pound single-unit truck, 40,000-pound, gasoline powered combination vehicle, and 50,000-pound diesel powered combination vehicle. These tables are included in a book dealing with highway economics by Robley Winfrey.[183]

Cost of Stops. Traffic stops cost money. The costs of fuel, wear of tires and brakes, oil, and repairs are higher for stopping and starting than for continuous operation. It is also necessary to consider the amount of time that is consumed by stops. This can be divided into two parts, the standing delay and the delay due to deceleration and acceleration. The cost of delay due to standing, deceleration, and acceleration is based upon the assumed value of time.

AASHO has established additional costs per stop, taking into consideration the following items:

1. Additional operating costs, as compared to uninterrupted flow, due to stopping and starting. These costs are given in columns 2, 3, and 4 of Table 6-7 for various approach speeds.
2. Delay due to deceleration and acceleration. The cost of this portion of delay is computed on the basis of a time value of $1.55 per hour. This cost is given in column 5 of Table 6-7.
3. Standing delay as the vehicle stands waiting for a traffic light to change, or for some other cause of delay. This cost is also based upon a time value of $1.55 per hour. In addition to the value of time, it is necessary to consider the costs of fuel, oil, and repairs while the vehicle is idling.

The total additional cost per stop for various approach speeds, including different standing delay times, is given in Figure 6-3. The dashed curve represents the total cost per stop, considering just the additional costs due to operating costs and time for deceleration and acceleration. This curve depicts the data shown in column 6 of Table 6-7. The solid curves include the value of time for standing delay and the additional costs of fuel, oil, and repairs while the vehicle is idling.

Other studies have been conducted to determine the significance of additional fuel and time consumed due to stopping, slowdowns, speed changes, and congestion.[14,22-25,27,39-41,44,49-52]

Cost of Accidents. Any increase in safety is of benefit to the road user. However, it is difficult to measure increases in safety or to demonstrate actual reduction in accidents and to place a monetary value on them. Benefits resulting from accident reduction would be of extreme importance if it were possible to properly evaluate them. To do so, it would be necessary to have data available which would relate accident occurrence to highway operational and design features.

An appreciation of the total economic loss due to motor vehicle accidents can be obtained by examining statistics published by the National Safety Council

TABLE 6-3. ROAD USER COSTS FOR PASSENGER CARS IN RURAL AREAS TANGENT DIVIDED HIGHWAYS PAVEMENT IN GOOD CONDITION

Running Speed (mph)	Gradient Class (%)	Fuel	Tires	Oil	Maint. and Repairs	Depreciation	Subtot. Oper. Costs	Time	Comfort and Convenience	Total Costs
(1)	(2)	(3)	(4)	(5)	(6)	(7)	(8)	(9)	(10)	(11)
					USER COSTS, CENTS PER VEHICLE MILE FOR:					
					FREE OPERATION					
40	0–3	2.00	0.28	0.18	1.20	1.50	5.16	3.88	0	9.04
	3–5	2.10	0.33	0.18	1.20	1.50	5.31	3.88	0	9.19
	5–7	2.22	0.42	0.18	1.20	1.50	5.52	3.88	0	9.40
	7–9	2.53	0.57	0.18	1.20	1.50	5.98	3.88	0	9.86
44	0–3	2.09	0.34	0.21	1.20	1.50	5.34	3.52	0	8.86
	3–5	2.22	0.39	0.21	1.20	1.50	5.52	3.52	0	9.04
	5–7	2.35	0.51	0.21	1.20	1.50	5.77	3.52	0	9.29
	7–9	2.71	0.68	0.21	1.20	1.50	6.30	3.52	0	9.82
48	0–3	2.21	0.41	0.24	1.20	1.50	5.56	3.23	0	8.79
	3–5	2.35	0.47	0.24	1.20	1.50	5.76	3.23	0	8.99
	5–7	2.53	0.61	0.24	1.20	1.50	6.03	3.23	0	9.31
	7–9	2.95	0.81	0.24	1.20	1.50	6.70	3.23	0	9.93
52	0–3	2.34	0.47	0.29	1.20	1.50	5.80	2.98	0	8.78
	3–5	2.50	0.54	0.29	1.20	1.50	6.03	2.98	0	9.01
	5–7	2.72	0.71	0.29	1.20	1.50	6.42	2.98	0	9.40
	7–9	3.21	0.95	0.29	1.20	1.50	7.15	2.98	0	10.13
56	0–3	2.51	0.54	0.37	1.20	1.50	6.12	2.77	0	8.89
	3–5	2.71	0.62	0.37	1.20	1.50	6.40	2.77	0	9.17
	5–7	2.99	0.80	0.37	1.20	1.50	6.86	2.77	0	9.63
	7–9	3.58	1.07	0.37	1.20	1.50	7.72	2.77	0	10.49
60	0–3	2.73	0.56	0.52	1.20	1.50	6.51	2.58	0	9.09
	3–5	2.97	0.64	0.52	1.20	1.50	6.83	2.58	0	9.41
					NORMAL OPERATION					
32	0–3	1.85	0.23	0.15	1.20	1.50	4.93	4.84	0.50	10.27
	3–5	1.92	0.26	0.15	1.20	1.50	5.03	4.84	0.50	10.37
	5–7	2.01	0.34	0.15	1.20	1.50	5.20	4.84	0.50	10.54
	7–9	2.23	0.46	0.15	1.20	1.50	5.54	4.84	0.50	10.88
36	0–3	1.91	0.27	0.16	1.20	1.50	5.04	4.31	0.50	9.85
	3–5	2.00	0.31	0.16	1.20	1.50	5.17	4.31	0.50	9.98
	5–7	2.10	0.40	0.16	1.20	1.50	5.30	4.31	0.50	10.17
	7–9	2.34	0.53	0.16	1.20	1.50	5.73	4.31	0.50	10.54
40	0–3	2.00	0.32	0.18	1.20	1.50	5.20	3.88	0.50	9.58
	3–5	2.10	0.36	0.18	1.20	1.50	5.34	3.88	0.50	9.72
	5–7	2.22	0.47	0.18	1.20	1.50	5.57	3.88	0.50	9.95
	7–9	2.53	0.63	0.18	1.20	1.50	6.04	3.88	0.50	10.42
44	0–3	2.11	0.38	0.21	1.20	1.50	5.40	3.52	0.50	9.42
	3–5	2.23	0.44	0.21	1.20	1.50	5.58	3.52	0.50	9.60
	5–7	2.39	0.57	0.21	1.20	1.50	5.87	3.52	0.50	9.89
	7–9	2.75	0.76	0.21	1.20	1.50	6.42	3.52	0.50	10.44
48	0–3	2.27	0.45	0.24	1.20	1.50	5.66	3.23	0.50	9.39
	3–5	2.42	0.52	0.24	1.20	1.50	5.88	3.23	0.50	9.61
	5–7	2.62	0.67	0.24	1.20	1.50	6.23	3.23	0.50	9.96
	7–9	3.11	0.90	0.24	1.20	1.50	6.95	3.23	0.50	10.68
52	0–3	2.51	0.53	0.28	1.20	1.50	6.02	2.98	0.50	9.50
	3–5	2.71	0.60	0.28	1.20	1.50	6.29	2.98	0.50	9.77
	5–7	2.98	0.79	0.28	1.20	1.50	6.75	2.98	0.50	10.23
56	0–3	2.89	0.60	0.37	1.20	1.50	6.56	2.77	0.50	9.83
	3–5	3.18	0.68	0.37	1.20	1.50	6.93	2.77	0.50	10.20

TABLE 6-3. (continued)

Running Speed (mph)	Gradient Class (%)	Fuel	Tires	Oil	Maint. and Repairs	Depreciation	Subtot. Oper. Costs	Time	Comfort and Convenience	Total Costs
					USER COSTS, CENTS PER VEHICLE MILE FOR:					
					RESTRICTED OPERATION					
(1)	(2)	(3)	(4)	(5)	(6)	(7)	(8)	(9)	(10)	(11)
28	0–3	1.81	0.24	0.15	1.20	1.50	4.90	5.54	1.00	11.44
	3–5	1.89	0.27	0.15	1.20	1.50	5.01	5.54	1.00	11.55
	5–7	1.98	0.30	0.15	1.20	1.50	5.19	5.54	1.00	11.73
	7–9	2.22	0.48	0.15	1.20	1.50	5.55	5.54	1.00	12.09
32	0–3	1.85	0.28	0.15	1.20	1.50	4.98	4.84	1.00	10.82
	3–5	1.93	0.33	0.15	1.20	1.50	5.11	4.84	1.00	10.95
	5–7	2.02	0.43	0.15	1.20	1.50	5.30	4.84	1.00	11.14
	7–0	2.25	0.57	0.15	1.20	1.50	5.67	4.84	1.00	11.51
36	0–3	1.94	0.33	0.16	1.20	1.50	5.13	4.31	1.00	10.44
	3–5	2.03	0.38	0.16	1.20	1.50	5.27	4.31	1.00	10.58
	5–7	2.14	0.50	0.16	1.20	1.50	5.50	4.31	1.00	10.81
	7–9	2.38	0.67	0.16	1.20	1.50	5.91	4.31	1.00	11.22
40	0–3	2.08	0.39	0.18	1.20	1.50	5.35	3.88	1.00	10.23
	3–5	2.18	0.45	0.18	1.20	1.50	5.51	3.88	1.00	10.39
	5–7	2.32	0.59	0.18	1.20	1.50	5.79	3.88	1.00	10.67
	7–9	2.61	0.79	0.18	1.20	1.50	6.28	3.88	1.00	11.16
44	0–3	2.30	0.47	0.21	1.20	1.50	5.68	3.52	1.00	10.20
	3–5	2.41	0.55	0.21	1.20	1.50	5.87	3.52	1.00	10.39
	5–7	2.58	0.71	0.21	1.20	1.50	6.20	3.52	1.00	10.72
	7–9	2.96	0.95	0.21	1.20	1.50	6.82	3.52	1.00	11.34

(Source: *Road User Benefit Analyses for Highway Improvements,* AASHO, 1960)

TABLE 6-4. ROAD USER COSTS FOR PASSENGER CARS IN RURAL AREAS TANGENT 2-LANE HIGHWAYS PAVEMENT IN GOOD CONDITION

Running Speed (mph)	Gradient Class (%)	Fuel	Tires	Oil	Maint. and Repairs	Depreciation	Subtot. Oper. Costs	Time	Comfort and Convenience	Total Costs
					USER COSTS, CENTS PER VEHICLE MILE FOR:					
(1)	(2)	(3)	(4)	(5)	(6)	(7)	(8)	(9)	(10)	(11)
					FREE OPERATION					
32	0–3	1.85	0.21	0.15	1.20	1.50	4.91	4.84	0	9.75
	3–5	1.92	0.24	0.15	1.20	1.50	5.01	4.84	0	9.85
	5–7	2.01	0.31	0.15	1.20	1.50	5.17	4.84	0	10.01
	7–9	2.23	0.42	0.15	1.20	1.50	5.50	4.84	0	10.34
36	0–3	1.91	0.26	0.15	1.20	1.50	5.02	4.31	0	9.33
	3–5	2.00	0.30	0.15	1.20	1.50	5.15	4.31	0	9.46
	5–7	2.10	0.39	0.15	1.20	1.50	5.34	4.31	0	9.65
	7–9	2.34	0.52	0.15	1.20	1.50	5.71	4.31	0	10.02
40	0–3	2.00	0.32	0.18	1.20	1.50	5.20	3.88	0	9.02
	3–5	2.10	0.37	0.18	1.20	1.50	5.35	3.88	0	9.23
	5–7	2.22	0.48	0.18	1.20	1.50	5.58	3.88	0	9.46
	7–9	2.53	0.64	0.18	1.20	1.50	6.05	3.88	0	9.93

Table 6-4. (continued)

USER COSTS, CENTS PER VEHICLE MILE FOR:										
Running Speed (mph)	Gradient Class (%)	Fuel	Tires	Oil	Maint. and Repairs	Depreciation	Subtot. Oper. Costs	Time	Comfort and Convenience	Total Costs
(1)	(2)	(3)	(4)	(5)	(6)	(7)	(8)	(9)	(10)	(11)
FREE OPERATION (cont.)										
44	0–3	2.11	0.40	0.21	1.20	1.50	5.42	3.52	0	8.94
	3–5	2.23	0.46	0.21	1.20	1.50	5.60	3.52	0	9.12
	5–7	2.39	0.60	0.21	1.20	1.50	5.90	3.52	0	9.42
	7–9	2.75	0.80	0.21	1.20	1.50	6.46	3.52	0	9.98
48	0–3	2.27	0.50	0.24	1.20	1.50	5.71	3.23	0	8.94
	3–5	0.58	0.58	0.24	1.20	1.50	5.94	3.23	0	9.17
	5–7	2.62	0.75	0.24	1.20	1.50	6.31	3.23	0	9.54
	7–9	3.11	1.00	0.24	1.20	1.50	7.05	3.23	0	10.28
52	0–3	2.51	0.63	0.28	1.20	1.50	6.12	2.98	0	9.10
	3–5	2.71	0.72	0.28	1.20	1.50	6.41	2.98	0	9.39
	5–7	2.98	0.95	0.28	1.20	1.50	6.91	2.98	0	9.89
	7–9	3.65	1.26	0.28	1.20	1.50	7.89	2.98	0	10.87
56	0–3	2.91	0.75	0.37	1.20	1.50	6.73	2.77	0	9.50
	3–5	3.18	0.86	0.37	1.20	1.50	7.11	2.77	0	9.88
	5–7	3.60	1.12	0.37	1.20	1.50	7.79	2.77	0	10.56
60	0–3	3.64	0.84	0.52	1.20	1.50	7.71	2.58	0	10.29
	3–5	4.05	0.97	0.52	1.20	1.50	8.24	2.58	0	10.82
NORMAL OPERATION										
28	0–3	1.81	0.19	0.15	1.20	1.50	4.85	5.54	0.50	10.89
	3–5	1.89	0.22	0.15	1.20	1.50	4.96	5.54	0.50	11.00
	5–7	1.98	0.28	0.15	1.20	1.50	5.11	5.54	0.50	11.15
	7–9	2.22	0.38	0.15	1.20	1.50	5.45	5.54	0.50	11.49
32	0–3	1.85	0.23	0.15	1.20	1.50	4.93	4.84	0.50	10.27
	3–5	1.93	0.26	0.15	1.20	1.50	5.04	4.84	0.50	10.38
	5–7	2.02	0.34	0.15	1.20	1.50	5.21	4.84	0.50	10.55
	7–9	2.25	0.46	0.15	1.20	1.50	5.56	4.84	0.50	10.90
36	0–3	1.94	0.29	0.16	1.20	1.50	5.09	4.31	0.50	9.90
	3–5	2.03	0.33	0.16	1.20	1.50	5.22	4.31	0.50	10.03
	5–7	2.14	0.43	0.16	1.20	1.50	5.43	4.31	0.50	10.24
	7–9	2.38	0.58	0.16	1.20	1.50	5.82	4.31	0.50	10.63
40	0–3	2.08	0.36	0.18	1.20	1.50	5.32	3.88	0.50	9.70
	3–5	2.18	0.41	0.18	1.20	1.50	5.47	3.88	0.50	9.85
	5–7	2.32	0.54	0.18	1.20	1.50	5.74	3.88	0.50	10.12
	7–9	2.61	0.72	0.18	1.20	1.50	6.24	3.88	0.50	10.59
44	0–3	2.30	0.45	0.21	1.20	1.50	5.66	3.52	0.50	9.68
	3–5	2.41	0.52	0.21	1.20	1.50	5.84	3.52	0.50	9.86
	5–7	2.58	0.67	0.21	1.20	1.50	6.16	3.52	0.50	10.18
	7–9	2.96	0.90	0.21	1.20	1.50	6.77	3.52	0.50	10.79
48	0–3	2.71	0.56	0.24	1.20	1.50	6.21	3.23	0.50	9.94
	3–5	2.85	0.64	0.24	1.20	1.50	6.43	3.23	0.50	10.16
	5–7	3.13	0.84	0.24	1.20	1.50	6.91	3.23	0.50	10.64
	7–9	3.45	1.12	0.24	1.20	1.50	7.51	3.23	0.50	11.24
RESTRICTED OPERATION										
20	0–3	1.83	0.18	0.14	1.20	1.50	4.85	7.75	1.00	13.60
	3–5	1.91	0.21	0.14	1.20	1.50	4.96	7.75	1.00	13.71
	5–7	2.06	0.27	0.14	1.20	1.50	5.17	7.75	1.00	13.92
	7–9	2.34	0.36	0.14	1.20	1.50	5.54	7.75	1.00	14.29
24	0–3	1.81	0.21	0.14	1.20	1.50	4.86	6.46	1.00	12.32
	3–5	1.87	0.24	0.14	1.20	1.50	4.95	6.46	1.00	12.41
	5–7	1.99	0.31	0.14	1.20	1.50	5.14	6.46	1.00	12.60
	7–9	2.25	0.42	0.14	1.20	1.50	5.51	6.46	1.00	12.97

TABLE 6-4. (continued)

USER COSTS, CENTS PER VEHICLE MILE FOR:										
Running Speed (mph)	Gradient Class (%)	Fuel	Tires	Oil	Maint. and Repairs	Depreciation	Subtot. Oper. Costs	Time	Comfort and Convenience	Total Costs
(1)	(2)	(3)	(4)	(5)	(6)	(7)	(8)	(9)	(10)	(11)
RESTRICTED OPERATION (cont.)										
28	0–3	1.82	0.24	0.15	1.20	1.50	4.91	5.54	1.00	11.45
	3–5	1.89	0.28	0.15	1.20	1.50	5.02	5.54	1.00	11.56
	5–7	1.98	0.36	0.16	1.20	1.50	5.19	5.54	1.00	11.73
	7–9	2.23	0.48	0.15	1.20	1.50	5.56	5.54	1.00	12.10
32	0–3	1.87	0.29	0.15	1.20	1.50	5.01	4.84	1.00	10.85
	3–5	1.95	0.33	0.15	1.20	1.50	5.13	4.84	1.00	10.97
	5–7	2.03	0.43	0.15	1.20	1.50	5.31	4.84	1.00	11.15
	7–9	2.27	0.58	0.15	1.20	1.50	5.78	4.84	1.00	11.54
36	0–3	2.00	0.36	0.16	1.20	1.50	5.22	4.31	1.00	10.53
	3–5	2.09	0.41	0.16	1.20	1.50	5.36	4.31	1.00	10.67
	5–7	2.19	0.54	0.16	1.20	1.50	5.59	4.31	1.00	10.90
	7–9	2.45	0.72	0.16	1.20	1.50	6.03	4.31	1.00	11.34
40	0–3	2.27	0.45	0.18	1.20	1.50	5.60	3.88	1.00	10.48
	3–5	2.39	0.52	0.18	1.20	1.50	5.79	3.88	1.00	10.67
	5–7	2.53	0.67	0.18	1.20	1.50	6.08	3.88	1.00	10.96
	7–9	2.81	0.90	0.18	1.20	1.50	6.59	3.88	1.00	11.47

(Source: *Road User Benefit Analyses for Highway Improvements*, AASHO, 1960)

TABLE 6-5. ROAD USER COSTS FOR PASSENGER CARS IN RURAL AREAS TANGENT LOOSE SURFACE HIGHWAYS IN GOOD CONDITION

USER COSTS, CENTS PER VEHICLE MILE FOR:										
Running Speed (mph)	Gradient Class (%)	Fuel	Tires	Oil	Maint. and Repairs	Depreciation	Subtot. Oper. Costs	Time	Comfort and Convenience	Total Costs
(1)	(2)	(3)	(4)	(5)	(6)	(7)	(8)	(9)	(10)	(11)
20	0–3	2.07	0.44	0.18	1.80	1.50	5.99	7.75	0.75	14.49
	3–5	2.17	0.50	0.18	1.80	1.50	6.15	7.75	0.75	14.65
	5–7	2.37	0.63	0.18	1.80	1.50	6.48	7.75	0.75	14.98
	7–9	2.74	0.83	0.18	1.80	1.50	7.05	7.75	0.75	15.55
24	0–3	2.03	0.50	0.18	1.80	1.50	6.01	6.46	0.75	13.22
	3–5	2.12	0.57	0.18	1.80	1.50	6.17	6.46	0.75	13.38
	5–7	2.27	0.73	0.18	1.80	1.50	6.48	6.46	0.75	13.69
	7–9	2.61	0.95	0.18	1.80	1.50	7.04	6.46	0.75	14.25
28	0–3	2.03	0.57	0.19	1.80	1.50	6.09	5.54	0.75	12.38
	3–5	2.14	0.65	0.19	1.80	1.50	6.28	5.54	0.75	12.57
	5–7	2.25	0.83	0.19	1.20	1.50	6.57	5.54	0.75	12.86
	7–9	2.57	1.09	0.19	1.80	1.50	7.15	5.54	0.75	13.44
32	0–3	2.10	0.65	0.21	1.80	1.50	6.26	4.84	0.75	11.85
	3–5	2.19	0.75	0.21	1.80	1.50	6.45	4.84	0.75	12.04
	5–7	2.32	0.96	0.21	1.80	1.50	6.79	4.84	0.75	12.38
	7–9	2.63	1.26	0.21	1.80	1.50	7.40	4.84	0.75	12.99

TABLE 6-5. (continued)

		USER COSTS, CENTS PER VEHICLE MILE FOR:								
Running Speed (mph)	*Gradient Class (%)*	*Fuel*	*Tires*	*Oil*	*Maint. and Repairs*	*Depreciation*	*Subtot. Oper. Costs*	*Time*	*Comfort and Convenience*	*Total Costs*
(1)	(2)	(3)	(4)	(5)	(6)	(7)	(8)	(9)	(10)	(11)
36	0–3	2.21	0.75	0.22	1.80	1.50	6.48	4.31	0.75	11.54
	3–5	2.33	0.86	0.22	1.80	1.50	6.71	4.31	0.75	11.77
	5–7	2.47	1.11	0.22	1.80	1.50	7.10	4.31	0.75	12.16
	7–9	2.80	1.46	0.22	1.80	1.50	7.78	4.31	0.75	12.84
40	0–3	2.39	0.87	0.24	1.80	1.50	6.80	3.88	0.75	11.43
	3–5	2.53	0.99	0.24	1.80	1.50	7.06	3.88	0.75	11.69
	5–7	2.71	1.28	0.24	1.80	1.50	7.53	3.88	0.75	12.16
	7–9	3.12	1.69	0.24	1.80	1.50	8.35	3.88	0.75	12.98
44	0–3	2.69	1.01	0.29	1.80	1.50	7.29	3.52	0.75	11.56
	3–5	2.83	1.15	0.29	1.80	1.50	7.57	3.52	0.75	11.84
	5–7	3.09	1.49	0.29	1.80	1.50	8.17	3.52	0.75	12.44
	7–9	3.63	1.97	0.29	1.80	1.50	9.19	3.52	0.75	13.46

(Source: *Road User Benefit Analyses for Highway Improvements,* AASHO, 1960, Table 8)

TABLE 6-6. ROAD USER COSTS FOR PASSENGER CARS IN RURAL AREAS TANGENT UNSURFACED ROADS

		USER COSTS, CENTS PER VEHICLE MILE FOR:								
Running Speed (mph)	*Gradient Class (%)*	*Fuel*	*Tires*	*Oil*	*Maint. and Repairs*	*Depreciation*	*Subtot. Oper. Costs*	*Time*	*Comfort and Convenience*	*Total Costs*
(1)	(2)	(3)	(4)	(5)	(6)	(7)	(8)	(9)	(10)	(11)
16	0–3	2.70	0.53	0.25	2.40	1.50	7.38	9.69	1.00	18.07
	3–5	2.88	0.60	0.25	2.40	1.50	7.63	9.69	1.00	18.32
	5–7	3.31	0.77	0.25	2.40	1.50	8.23	9.69	1.00	18.92
	7–9	4.23	1.01	0.25	2.40	1.50	9.39	9.69	1.00	20.08
20	0–3	2.56	0.61	0.26	2.40	1.50	7.33	7.75	1.00	16.08
	3–5	2.72	0.69	0.26	2.40	1.50	7.57	7.75	1.00	16.32
	5–7	3.03	0.88	0.26	2.40	1.50	8.07	7.75	1.00	16.82
	7–9	3.69	1.16	0.26	2.40	1.50	9.01	7.75	1.00	17.76
24	0–3	2.51	0.71	0.27	2.40	1.50	7.39	6.46	1.00	14.85
	3–5	2.05	0.81	0.27	2.40	1.50	7.68	6.46	1.00	15.09
	5–7	2.88	1.04	0.27	2.40	1.50	8.09	6.46	1.00	15.55
	7–9	3.45	1.38	0.27	2.40	1.50	9.00	6.46	1.00	16.46
28	0–3	2.51	0.82	0.28	2.40	1.50	7.51	5.54	1.00	14.05
	3–5	2.66	0.94	0.28	2.40	1.50	7.78	5.54	1.00	14.32
	5–7	2.86	1.20	0.28	2.40	1.50	8.24	5.54	1.00	14.78
	7–9	3.39	1.59	0.28	2.40	1.50	9.16	5.54	1.00	15.70
32	0–3	2.62	0.96	0.30	2.40	1.50	7.78	4.84	1.00	13.62
	3–5	2.77	1.10	0.30	2.40	1.50	8.07	4.84	1.00	13.91
	5–7	2.96	1.41	0.30	2.40	1.50	8.57	4.84	1.00	14.41
	7–9	3.49	1.87	0.30	2.40	1.50	9.56	4.84	1.00	15.40
36	0–3	2.79	1.07	0.32	2.40	1.50	8.08	4.31	1.00	13.39
	3–5	2.98	1.22	0.32	2.40	1.50	8.42	4.31	1.00	13.73
	5–7	3.20	1.58	0.32	2.40	1.50	9.00	4.31	1.00	14.31
	7–9	3.79	2.09	0.32	2.40	1.50	10.10	4.31	1.00	15.41

(Source: *Road User Benefit Analyses for Highway Improvements,* AASHO, 1960, Table 9)

TABLE 6-7. EXTRA COST PER VEHICLE STOP NO STANDING DELAY

	ADDITIONAL COST PER STOP—CENTS				
Approach Speed (mph)	*Fuel*	*Tires and Brakes**	*Other Operating Costs*	*Time for Acceleration and Deceleration*	*Total*
(1)	(2)	(3)	(4)	(5)	(6)
10	0.06	0.03	0.05	0.02	0.16
20	0.12	0.07	0.10	0.09	0.38
30	0.19	0.16	0.18	0.21	0.74
40	0.28	0.33	0.31	0.40	1.32
50	0.41	0.56	0.46	0.64	2.07
60	0.55	0.80	0.68	0.91	2.94

(Source: *Road User Benefit Analyses for Highway Improvements,* AASHO, 1960, Table 1)

* Increase over 1952, approximately 25%

(NSC). For example, in 1966, the total economic loss due to motor vehicle accidents was subdivided as shown in Table 6-8.

Wage loss includes loss of wages due to temporary inability to work, lower wages when returned to work due to permanent partial disability, and the present value of anticipated future earnings for permanent total disability or death. *Medical expense* includes doctors' and hospitals' fees. *Overhead cost of insurance* includes all administrative, selling, and claim settlement expenses for insurance companies. It is essentially the difference between premiums paid to insurance companies and claims paid by them.

The NSC offers two methods for estimating the total economic loss due to traffic accidents: The first is used for large communities where the number of traffic deaths per year exceeds 10, and it involves the multiplication of the number of traffic deaths by a round sum which represents the loss due to 1 death, 36 personal injuries, and 235 property damage accidents. These figures represent the 1966 average ratios of the different types of accidents. The round sum based on 1966 figures is $190,000 which provides for an allowance of about $35,000 per death, $2,500 per injury, and $270 per property damage. The second method is to be used where the number of deaths in a community is less than 10, or the ratio of deaths to injuries varies from the above

TABLE 6-8. ECONOMIC LOSS DUE TO ACCIDENTS, 1966

TYPE	*AMOUNT*
Property damage	$ 4,300,000,000
Wage loss	6,700,000,000
Medical expense	2,700,000,000
Overhead cost of insurance	5,300,000,000
Total	19,000,000,000

(Source: *Accident Facts—1970 Edition,* NSC, 1970, p. 94)

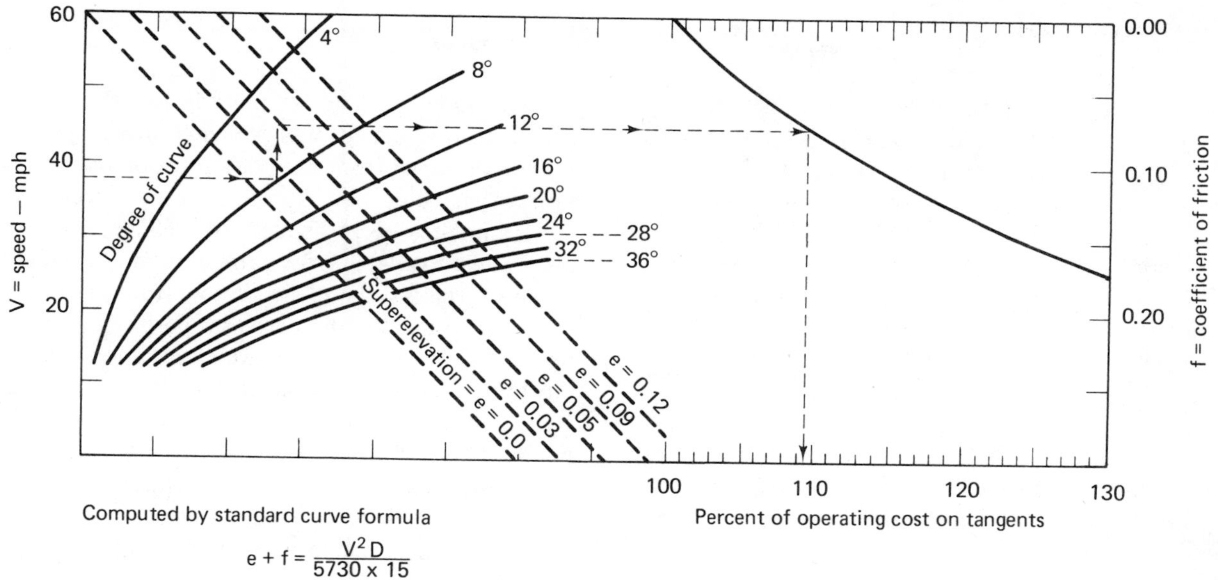

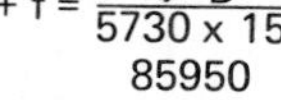

$$e + f = \frac{V^2 D}{\frac{5730 \times 15}{85950}}$$

Example: Assume 38 mph operation on 8° curve, 0.06 superelevation. Follow arrows and read 109.50% as the factor to apply to correct the tangent road user cost values.

Fig. 6-2. Relation between operating costs on curves and on tangents (*Source:* Road User Benefit Analysis for Highways, *AASHO, 1960, Figure 5*)

Fig. 6-3. Extra cost per vehicle-stop above that for constant speed operation (*Source:* Road User Benefit Analyses for Highway Improvement, *AASHO, 1960, Figure 21*)

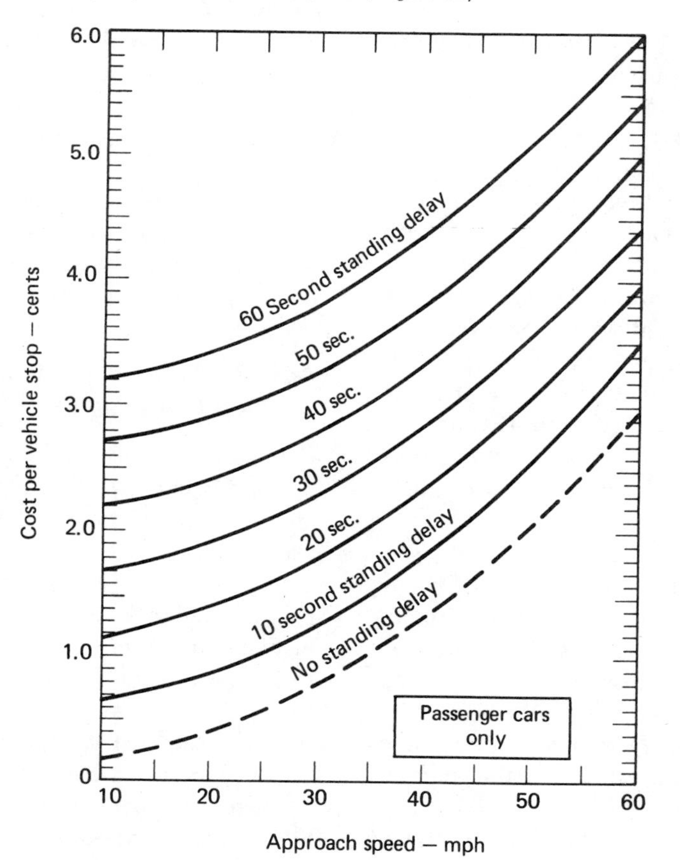

TABLE 6-9. ESTIMATING ECONOMIC LOSS DUE TO ACCIDENTS IN COMMUNITIES WITH LESS THAN 10 FATALITIES PER YEAR

AGE	COST PER DEATH	
	Male	*Female*
0–14	$24,000	$17,000
15–54	50,000	31,000
55 and older	10,000	8,000

(Source: *Accident Facts—1967 Edition,* NSC, 1967)

averages. This method is a little more precise in determining the economic loss due to deaths. The data given in Table 6-9 is currently used. Those figures for injury and property damage accidents are to be used as the unit costs for those types of accidents.

All of the expenses involved in motor vehicle accidents should not be considered as appropriate charges in determining the cost of accidents in economy studies. For example, the overhead cost of insurance is consistently about 35 per cent of the total economic loss due to traffic accidents, and the greatest portion of this overhead cost would continue regardless of accident reduction. Therefore, the cost of accidents could be more appropriately estimated by multiplying the above figures by 0.65. In addition, the proportion of wage loss which provides for the the present value of anticipated future earnings for permanent total disability or death is not considered to be a justified item to be included in the cost of accidents. Unfortunately, wage loss is not subdivided into its component parts.

It is normally agreed that the direct costs of accidents, for use in economy studies, should include the following items: property damage, medical expense, wage loss due to being temporarily out of work, damages awarded in excess of other direct costs, attorneys' services, and court fees. There is not a great deal of information available on direct costs of accidents, but several states such as Massachusetts,[54-56] Utah,[57] and Illinois,[60,61] and also Washington, D.C.,[58,59] are conducting comprehensive studies of the economic costs of motor vehicle accidents.

In the Illinois study, direct accident costs are divided for rural and urban areas, and the urban areas are further subdivided by population size. The data include direct costs for reported and unreported accident involvements. About 76 per cent of passenger car involvements and 80 per cent of truck involvements were not recorded in the official accident files of the state. Although most of these were minor accidents for which property damage costs were below the legal reporting minimum ($100), they accounted for 42 per cent of the total direct costs of passenger car accidents and 55 of the total direct costs of truck accidents. The magnitude of the unreported accident involvements was evaluated by expanding information obtained from a small sample of vehicle registrations that did not appear in the accident files. Questionnaires were mailed to vehicle owners. Procedures are available[44] for computation of accident costs based on the results of the Illinois study, but they were modified to include the overhead cost of insurance, and updated to June 1965 price levels. It is important to point out that the costs reported in the Illinois study are per vehicle involved, and not per accident, as was the case in the Massachusetts and Utah studies. The cost per involvement is applied to each vehicle involved in an accident. Therefore, a single vehicle colliding with a barrier constitutes one accident and one involvement. An accident in which three vehicles are involved constitutes one accident and three involvements. California has modified the Illinois accident study data and adopted it for computation of direct accident costs.[62] The modifications included an adjustment to 1964 prices and correlation with California accident characteristics. Illinois accident costs were increased 10 per cent to reflect 1964 prices, and the results in Table 6-10 were obtained.

Ideally, what is required is the effects on accident characteristics from individual changes in geometric design and operational conditions for various types of highway facilities. These correlations have not as yet been fully established; however, there is an excellent summary of the information available.[63] An attempt has been made to determine the relationship of motor vehicle accidents to rural highway types and highway design.[64]

Accident rates and characteristics related to different types of highway facilities provide valuable information in predicting the effects on accident experience by proposed changes in a highway facility.[53,65-73]

Additional User and Nonuser Benefits to be Derived from Highway Improvements

Practically all of the items discussed under costs of motor vehicle operation represent the direct benefits of highway improvements. These costs, which deal with the running costs of operating a vehicle, the cost of stops, and the cost of accidents, are all normally reduced as a result of highway improvements and they are the primary justification for improvements. The controversial items of the value of comfort and convenience and the value of time for noncommercial vehicle operation are purposely omitted from the category of direct monetary benefits but they are often considered in such a manner.

There are many significant benefits resulting from highway improvements that are not used in an economic analysis to justify the improvements. These normally

TABLE 6-10. CALIFORNIA ACCIDENT COSTS PER INVOLVEMENT

TYPE AND AREA	COST PER VEHICLE INVOLVED Fatal	Injury	PDO
Reported involvements			
Rural	$6,190	$1,563	$299
Urban	4,637	1,001	158
Unreported involvements			
Rural	*	585	143
Urban	*	508	86

(Source: Smith and Tamburri, "Direct Costs of California State Highway Accidents," in *HRR 255*, 1968, p. 13)
* All fatal accidents were reported
Note: To convert these costs to a per-accident basis, it is necessary to apply a conversion factor based on the number of vehicles per accident. In California, vehicle involvement rates for 1964 are given in Table 6-11.

TABLE 6-11. CALIFORNIA INVOLVEMENT RATES

CATEGORY	VEHICLE PER REPORTED ACCIDENT Fatal	Injury	PDO
Rural			
Nonfreeway	1.56	1.61	1.65
Freeway	1.56	1.59	1.50
Urban			
Nonfreeway	1.67	2.03	2.02
Freeway	1.66	2.02	1.75
Total, rural	1.56	1.60	1.64
Total, urban	1.67	2.03	1.93
Total, rural and urban	1.60	1.87	1.85

(Source: Smith and Tamburri, "Direct Costs of California State Highway Accidents," in *HRR 255*, 1968, p. 13)
Note: Multiplying the cost per vehicle involved in Table 6-10 by the appropriate involvement rate in Table 6-11 produces the cost per reported accident shown in Table 6-12 (the costs are rounded off for practical purposes).

TABLE 6-12. CALIFORNIA ACCIDENT COST PER ACCIDENT

AREA	COST PER REPORTED ACCIDENT Fatal	Injury	PDO	All
Rural	$9,700	$2,500	$500	$1,600
Urban	7,700	2,000	300	1,000
Total	9,000	2,200	400	1,200

(Source: Smith and Tamburri, "Direct Costs of California State Highway Accidents," in *HRR 255*, 1968, p. 18)
Note: The same factors were used to convert unreported involvements, which are shown in Table 6-13.

TABLE 6-13. CALIFORNIA ACCIDENT COST FOR UNREPORTED ACCIDENTS

AREA	COST PER UNREPORTED ACCIDENT Injury	PDO	All
Rural	$ 940	$230	$270
Urban	1,030	170	180
Total	980	180	200

(Source: Smith and Tamburri, "Direct Costs of California State Highway Accidents," in *HRR 255*, 1968, p. 16)
Note: Combined unit values, reflecting both reported and unreported accidents and expressed in terms of *reported accidents,* are shown in Table 6-14.

TABLE 6-14. CALIFORNIA ADJUSTED ACCIDENT COSTS

AREA	COST PER REPORTED ACCIDENT Fatal	Injury	PDO	All
Rural	$9,700	$2,600	$800	$1,800
Urban	7,700	2,100	600	1,200
Total	9,000	2,300	700	1,400

(Source: Smith and Tamburri, "Direct Costs of California State Highway Accidents," in *HRR 255*, 1968, p. 18)
Note: Direct unit costs for all accidents, in terms of cents per vehicle-mile, for California state highways for 1964, are shown in Table 6-15.

TABLE 6-15. CALIFORNIA TOTAL ACCIDENT COSTS, COSTS PER VEHICLE-MILE

TYPE AND CATEGORY	RURAL	URBAN	TOTAL
Reported			
Nonfreeway	0.41	0.55	0.47
Freeway	0.16	0.18	0.17
Total	0.35	0.30	0.32
Unreported			
Nonfreeway	0.06	0.14	0.09
Freeway	0.02	0.03	0.03
Total	0.05	0.07	0.06
Combined			
Nonfreeway	0.47	0.69	0.56
Freeway	0.18	0.21	0.20
Total	0.40	0.37	0.38

have to do with the social and economic development of the community, and as such they can have even greater significance than the direct benefits of highway improvement. However, it is difficult to establish generalized procedures to determine the monetary worth of these benefits.

A listing of claimed social and economic benefits of highway improvements is:[74]

1. Economic progress is aided through improved transportation, which permits regional specialization in large-scale production, thereby increasing efficiency by bringing together the materials needed and by distributing the products to expand markets.
2. Equalization and stabilization of prices is effected through improved transportation, which can effectively prevent the development of local monopolies.
3. There are increased land values and land use through improved transportation, which makes the remotest area accessible and which stimulates industrial growth in remote regions.
4. Suburbanization results from modern highway transportation, which permits filling of vacant areas near urban centers.
5. Population mobility is largely possible because of modern highway transportation. In any one year about 7 percent of the American people move their place of residence across county lines.
6. Blighted areas are reclaimed and blight is avoided by improved highway facilities.
7. There are increased employment opportunities through highway improvements, which have the effect of bringing additional employment opportunities within commuting distance.
8. There is improvement in public services through the availability of efficient highways and streets, which result in better police and fire protection, better postal and health services, and increased educational opportunities.
9. By making them more accessible through modern highway transportation there is greater utilization and development of recreational facilities.
10. The value and utilization of natural resources is increased by improved accessibility through modern highway transportation

There have been numerous studies and reports of the social and economic effects of highway improvements. There is also, of course, a growing awareness of and concern with some of the deleterious effects of highways on the specific environment they impinge upon and the general ecology. There is a need to translate these into numerical and financial terms for the purpose of these calculations.

Costs of Highways

The second item to be considered in arriving at the costs of highway transportation is the cost of highways. Road or highway costs include the costs incurred in the design, construction, maintenance, operation, and administration of a section of highway, of an individual highway, or of a highway system.

The construction of a new highway, or the improvement of an existing one, usually involves a direct expenditure of a considerable amount of money in a relatively short period of time. This initial expenditure constitutes what is commonly called the *first cost* or *capital investment*. This is generally regarded as consisting of:

1. Expenditures for engineering and design
2. Expenditures for acquiring rights-of-way
3. Expenditures for the physical structure itself

For the purposes of economy studies, these first costs are assigned as annual charges over their assumed useful life. If we add the annual expenditures for maintenance, operation, and administration to these costs, we find the total annual cost of the highway. To convert a first cost to an annual charge, we multiply it by the appropriate Capital Recovery Factor (CRF). In order to recover an investment in n years with interest rate i, the uniform annual amount necessary for capital recovery with interest is obtained by multiplying the investment by the CRF.

Capital Recovery Factor. The concept of capital recovery with interest is a controversial one, and different points of view exist:[1]

1. Interest rate should be zero for road improvements that are financed out of current taxation rather than by borrowing.
2. Interest rate should equal the rate paid on borrowings by the particular highway agency. If the proposed highway improvement is to be financed by borrowing. the interest rate should correspond to the cost of borrowing money. If the proposed highway improvement is to be financed from current income, the interest rate should correspond to that for longterm borrowing.
3. Interest rate should be a rate representing the minimum attractive return on invested money. This is the viewpoint normally taken by private enterprise. The cost of borrowed money is a primary factor in determining the minimum attractive return, but it is is not the sole factor to consider. A safety factor should also be included to reflect the risk involved, and therefore in most cases the appropriate minimum attractive return is somewhat higher than the cost of borrowed money.

In highway investments this risk is due to uncertainty as to the accuracy of assumed costs, service lives, and traffic forecasts over the analysis period.

The reasoning behind considering interest as a cost is the concept that if the funds absorbed in the capital investment for a highway had not been used for this purpose, they could have been invested to yield a safe and definite return. This return is thus lost to the public, and must be included as a part of the cost of the improvement. Essentially, this approach for highway investments is to use an interest rate that is approximately representative of the opportunity cost of capital, that is, the rate that could be earned by funds if privately invested rather than being collected by the governmental unit in the form of taxes. The opportunity cost of capital is sometimes referred to as the market rate of interest. There are three major premises involved in basing interest rates on the opportunity cost of capital:[14]

a. Nearly all resources are limited, including capital.
b. The future value to the investor of a limited resource for its most productive alternative use (its opportunity cost) is a proper estimate of its cost for the purposes under study.
c. Market prices are normally the best available measure of the value of capital.

In general, highway improvements in the United States are financed entirely by current highway user taxes and involve no public borrowing. This strengthens the arguments of the advocates of 0 per cent interest. However, the funds available in any year are limited by current tax collections, and the typical situation is that at any given time there are many economically justifiable projects that cannot be constructed because of the limitation on current funds.

On the policy level of determining the order of priority of projects competing for funds, it is essential that economy studies be made to compare the various alternate features of the design. If the investment of funds in a highway improvement resulted in benefits to highway users represented by a rate of return of 15 per cent on the highway investment, it is clearly not in the overall interest of highway users to have invested funds where the rate of return is only 3 per cent. In other words, where available funds are limited, the selection of an appropriate minimum attractive return calls for consideration of the prospective returns obtainable from alternate investments. This is as sound a principle in public works as it is in private enterprise.

If the time should ever be reached when economy studies show that all the highway funds available in any state cannot be used without undertaking a number of highway investments yielding very low returns, such as 2 per cent, a fair conclusion would be that highway user taxes should be lowered. Assuming the charge for interest is valid, which is the generally accepted viewpoint, the next point to consider is what rate should be used. AASHO currently has adopted a viewpoint somewhere between that specified in items 2 and 3 above. In their illustrative examples, interest rates range from $3\frac{1}{2}$ to 6 per cent and those who advocate a minimum attractive rate of return use rates ranging from 5 to 8 per cent.[20,75]

In response to a questionnaire which was sent to the highway departments of the fifty states and to the District of Columbia and Puerto Rico, the following average results were obtained relative to the choice of interest rate for economic analysis: about 45 per cent of the organizational units reported an interest rate of 0 per cent, about 22 per cent reported interest rates in the range of 2 to $3\frac{3}{4}$ per cent and about 33 per cent had interest rates in the range of 4 to 7 per cent.[76] The organizational units which reported interest rates in the range of 4 to 7 per cent are basing highway expenditures on a much sounder basis. The use of 0 per cent interest implies that all proposed increments of investment are equally attractive. However, this could hardly be the situation. Since there are many opportunities for productive investment of funds, each increment of investment in highway improvements that yields as little as 3 per cent has the effect of depriving some other investment of funds (in highways or elsewhere) that would be much more productive. The selection of a minimum attractive rate of return has a significant influence on the results of economy studies. This can be illustrated by the following example:[1] Assume that two alternate types of highway bridge are under consideration for the replacement of an existing timber structure. The first cost of a steel bridge is $500,000, and the first cost of a prestressed concrete bridge is $600,000. Maintenance costs for the steel bridge consist mainly of painting, at an average annual cost of $6000. Maintenance costs for the concrete bridge are assumed to be negligible. The estimated useful life of both structures is 40 years. A tabulation of annual costs with various rates of interest is given

TABLE 6-16. EXAMPLE OF THE EFFECT OF INTEREST RATE ON ECONOMY STUDIES

INTEREST RATE (%)	*ANNUAL COST*		*DIFFERENCE IN ANNUAL COST*	
	Steel	*Concrete*	*Favoring Steel*	*Favoring Concrete*
0	$18,500	$15,000		$3500
2	24,280	21,940		2340
4	31,260	30,310		950
5	35,140	34,970		170
6	39,230	39,880	$ 650	
8	47,930	50,320	2390	

in Table 6-16. If the minimum attractive rate of return is below 5.2 per cent, the concrete bridge is more economical. If it is above 5.2 per cent, the steel bridge is more economical.

Highway cost items to be considered in economy studies.[20,77-79] 1. The annual cost of a highway facility is made up of several items. Some of these cost items are not appropriate for inclusion in economy studies. A highway economy study is one of choosing between alternatives, and only the differences between alternatives are significant in their comparison. The alternatives must be clearly defined, so that the various highway and motor vehicle cost items associated with each alternative can be properly estimated.

Administrative expenses are usually affected to a very minor degree by the choice between alternatives. Similarly, expenditures for preliminary planning and operation are, for the most part, independent of alternatives. It is most likely that all of these expenditures would be incurred regardless of whether the proposed project is actually undertaken. Therefore, unless there is specific evidence that these costs are an important factor which could affect the economic comparison, they should be omitted from the calculations.

All relevant costs must be included in the analysis. These items generally include costs for the acquisition of rights-of-way, for construction, and for maintenance.

2. Some of the estimated combined disbursement items of all government agencies (including federal, state, country and township, and municipalities) for 1966 are given in Table 6-17.

Capital outlay includes costs for acquisition of rights-of-way (ROW), preliminary and construction engineering, construction of roads, construction of major structures, and installation of traffic service facilities. The expenditures for acquisition of ROW represented approximately 14 per cent of total capital outlay. *Maintenance* includes costs for physical maintenance of roads and structures, snow removal and sanding, traffic control and service facilities, and operation of roads and bridges. *Administration* includes costs for general administration and engineering, research investigation and planning, and land and buildings. *Highway police* includes costs for traffic law enforcement, safety education, and vehicle size and weight enforcement. The expenditures for administration constituted approximately 58 per cent of the combined costs for administration and highway police.

TABLE 6-17. HIGHWAY DISBURSEMENTS, 1966

DISBURSEMENTS	*MILLIONS OF DOLLARS*
Capital outlay	9,123
Maintenance	3,469
Administration and highway police	1,389
Total	13,981

3. Limited access highway costs have been developed that are representative of the Tri-State Region.[81] Cost data are expressed in terms of population density. Cost items are represented by regression equations separately for total cost per mile of road, total cost per lane-mile, total cost less interchange cost per lane-mile, right-of-way cost per mile of road, right-of-way cost per land unit (100 feet wide, 1 mile long), grading and surfacing cost per mile of road, and other costs per mile (traffic control devices, lighting, roadside elements, etc.). Additional cost items are tabulated, including cost of structures and interchange cost.

Cost of Right-of-Way. To build a highway, and in many cases to improve an existing highway, it is necessary to acquire the title to the land on which the highway is to be constructed (the right-of-way).

Contracts between a state highway agency and a property owner are often necessary in order to obtain the right-of-way for highway construction purposes. When land is vitally needed for a project, and no agreement can be reached for its purchase, the land may be taken under what is known as the right of *eminent domain.* This process of acquiring the property for public use is referred to as a condemnation proceeding, and follows a definite legal procedure. Resort to the courts, however, should be avoided if possible because of the extra cost and delay involved.

Rights-of-way represent one of the major costs of most highway improvements. In some cases it equals or exceeds the cost of construction. The cost of the right-of-way includes the cost of the land and buildings that lie within its limits. It may also include the damage to the remaining property caused by the taking of the right-of-way. For example, this damage may be caused by cutting the existing property into two pieces of such size that the saleable value of each is materially decreased. If the right-of-way taken destroys or interferes with the existing arrangement of a factory, plant, or other business, there may be further damages to pay, to compensate for the destruction or interference caused by the taking of the right-of-way.

A preliminary estimate of the cost of right-of-way may be made by applying the probable cost of the land per acre or square foot to the area of right-of-way to be acquired. This can usually be done where the right-of-way passes through unimproved land, and where the value per acre is fairly uniform and well established. Where the right-of-way covers improved property, a preliminary estimate can be made, basing it on the assessed values of the properties affected. However, the assessed values are generally much lower than the sales values, and it is necessary to apply a factor to the assessed values to obtain a proper estimate.

A closer estimate of cost may be made by engaging one or more appraisers with expert knowledge of real estate values in the vicinity of the right-of-way to make appraisals of the land to be taken.

The methods of acquiring right-of-way are often complex and varied, with different procedures in use throughout the country. In many cases complex legal problems are encountered.

Expenditures for the Physical Structure. These expenditures include all costs necessary to make the highway ready for use. They include the cost of excavation, fill, other grading operations, paving, drainage, culverts, bridges, and all other items entering into the construction of the highway.

The expenditures for engineering and design are generally calculated as a percentage of the direct costs of construction. In addition, engineering includes the costs of surveys, plans, tests, and inspection.

AASHO recommends the division of construction cost into the following categories:

1. Costs incurred for grading, drainage, minor structures, landscaping, and roadside development
2. Costs incurred for major structures
3. Costs incurred for pavement and associated items, such as lighting, guard rails, guide posts, traffic control devices, etc.

A preliminary estimate of the cost of construction may be made on the basis of the length of highway and the cost per mile of the construction of similar highways already built. To this estimate must be added the costs of culverts, bridges, and other special structures which also may be estimated by comparison with the costs of similar structures. Data from which construction costs may be estimated will, in large organizations like state highway departments, large county engineering departments, and city street departments, be available from the comprehensive records which are kept during the construction period and which show the cost of every item.

Economic Life of Highways. In order to assign the proper annual recovery charge for the various components making up the capital investment, it is necessary to estimate the average useful life of these components.

From the standpoint of economics, the pavement surface should be retired at the end of its economic life. Retirement may be accomplished by resurfacing, reconstruction, relocation, or abandonment. It is assumed that the surface is subjected to only routine maintainance during the period of its economic life. The average service life of a pavement surface is the average period of time after construction that the surface remains in service prior to being resurfaced, reconstructed, replaced, or otherwise taken out of service for any reason. Stated in another way, it is that period of time after construction during which the only operations performed on the pavement surface are those of maintenance as practiced by the various states.

For the purpose of expressing the average service life of different types of paving materials, surfaces are grouped as:

1. Low: includes soil and gravel or stone surfaces
2. Intermediate: includes bituminous-treated surfaces and mixed bituminous surfaces
3. High: includes bituminous concrete, bituminous penetration, portland cement concrete, brick, or block surfaces

The Bureau of Public Roads (BPR) has conducted studies of the average service lives of highway surfaces, and four reports have been issued. The first report was published in 1941[82] and covered 26 states, the second in 1949[83] and covered 16 states, the third in 1956[84] and covered 25 states and Puerto Rico, and the fourth in 1968[85] and covered 26 states and Puerto Rico. The 1956 study contains a comparison of the first three studies, and this comparison is shown in Table 6-18. Table 6-19 shows the distribution of how the various surfaces were retired at the ends of their average service lives.

Rights-of-way, grading, and major structures of modern highways are subject to less depreciation than are highway surfaces, and it is assumed that they have

TABLE 6-18. COMPARISON OF AVERAGE SERVICE LIVES PRESENTED IN THREE REPORTS[82,83,84]

SURFACE TYPE	*1941 STUDY AVERAGE LIFE (yrs)*	*1949 STUDY AVERAGE LIFE (yrs)*	*1956 STUDY AVERAGE LIFE (yrs)*
Low			
Soil surface	5.4	4.5	4.0
Gravel or stone	6.0	5.9	7.5
Intermediate			
Bituminous surface treated	11.4	10.2	11.7
Mixed bituminous	14.3	12.3	12.5
High			
Bituminous penetration	17.0	16.0	18.5
Bituminous concrete	17.9	18.1	20.3
Portland cement concrete	24.4	26.1	27.0
Brick or block	18.2	20.2	19.6

TABLE 6-19. PERCENTAGE DISTRIBUTION ACCORDING TO METHOD OF RETIREMENT

	METHOD OF RETIREMENT			
SURFACE TYPE	*Resurfaced* (%)	*Reconstructed* (%)	*Abandoned* (%)	*Transferred* (%)
Soil surface	37.5	58.1	1.2	3.2
Gravel or stone	58.0	30.4	2.5	9.1
Bituminous surface treated	53.3	36.7	2.6	7.4
Mixed bituminous	57.1	29.8	3.8	9.3
Bituminous penetration	51.8	31.4	3.6	13.2
Bituminous concrete	60.9	26.0	2.4	10.7
Portland cement concrete	66.4	21.3	2.1	10.2
Brick or block	45.3	44.1	1.3	9.3
Total	57.0	31.3	2.7	9.0

(Source: Gronberg and Blosser, "Lives of Highway Surfaces—Half Century Trends, in *DR*, June, 1956, Table 4)

TABLE 6-20. SERVICE LIFE RESULTS OF 1968 STUDY

SURFACE TYPE	*AVERAGE SERVICE LIFE* (*yrs*)
Bituminous surface treated	14.0
Mixed bituminous (pavement structure less than 7 in.)	12.0
Mixed bituminous (pavement structure greater than 7 in.)	17.5
Bituminous penetration (pavement structure less than 7 in.)	17.0
Bituminous penetration (pavement structure greater than 7 in.)	17.0
Bituminous concrete	17.0
Portland cement concrete	25.0

TABLE 6-21. METHODS OF RETIREMENT, 1968 STUDY

	METHOD OF RETIREMENT			
SURFACE TYPE	*Resurfaced* (%)	*Reconstructed* (%)	*Abandoned* (%)	*Transferred* (%)
Bituminous surface treated	58.5	32.6	2.5	6.4
Mixed bituminous (< 7 in.)	59.8	30.3	2.0	7.9
Mixed bituminous (> 7 in.)	56.6	30.3	3.9	9.2
Bituminous penetration (< 7 in.)	46.0	34.1	6.1	13.8
Bituminous penetration (> 7 in.)	45.4	39.9	3.1	11.6
Bituminous concrete	57.4	27.7	2.2	12.7
Portland cement concrete	66.0	22.8	1.8	9.4
Total	59.0	29.8	2.5	8.7

(Source: Winfrey and Howell, "Highway Pavements—Their Service Lives," in *HRR 252*, 1968, Table 9)

relatively long economic lives. The useful life for right-of-way is normally taken at 40 to 60 years, and for grading and structures at 30 to 40 years.

Costs of Maintenance and Operation. The expenditures for maintenance and operation include the sums spent to preserve the highway as nearly as possible in its original condition, as constructed or as subsequently improved, and for the operation of highway facilities and services to provide satisfactory and safe highway transportation.

1. The items normally included under *maintenance* are:
   a. Maintenance of wearing surface
   b. Maintenance of shoulders, ditches, cuts, and fills
   c. Maintenance of culverts and other drainage facilities
   d. Maintenance of bridges
   e. Maintenance of accessory structures
2. The items normally included under *operation* are:
   a. Roadside maintenance, including mowing, cutting and clearing vegetation, and planting
   b. Upkeep of guardrail, guide posts, roadway signs, and other traffic control devices
   c. Snow removal and ice control
   d. Clearing dirt and debris
   e. Highway lighting
   f. Traffic operations
   g. Portion of the cost of highway patrols and policing that is properly chargeable to traffic control.
3. Approximately $\frac{1}{4}$ or more of all highway funds go to maintenance and operation. The cost of maintenance and opeation for limited-access highways in rural areas ranges, on the average, from $6,000 to $7,500 per mile, and in urban areas from $10,000 to $20,000 per mile.
4. Maintenance and operation costs have increased from $1.42 billion in 1950 to $2.94 billion in 1963, and in 1973, the first full year after the Interstate System is expected to be completed, these costs have been estimated to be about $4.9 billion.[86]

   The main factors contributing to the increase in maintenance costs are:
   a. Expansion of maintenance activities and provision for more extensive and better services
   b. Overall increased use of highway facilities by all vehicle classifications
   c. Noneconomic or inefficient maintenance practices
   d. Inflation.

The 1968 BPR study was restricted to main rural highways, and the results are shown in Tables 6-20 and 6-21.

Some appreciation of the effect of inflation can be obtained by examination of cost trends over a period

TABLE 6-22. HIGHWAY MAINTENANCE AND OPERATION COST TRENDS
(1957–1959 base period = 100)

YEAR	*LABOR*	*MATERIAL*	*EQUIPMENT*	*OVERHEAD*	*TOTAL*
1950	66.44	81.15	72.77	70.95	70.49
1955	85.30	90.15	93.69	84.18	88.05
1960	108.28	103.09	109.77	104.66	107.65
1965	130.66	108.04	118.92	114.39	123.19
1966	149.46	108.53	122.03	119.64	134.32

(Source: *Highway Statistics—1966*, BPR, 1968)
Note: The largest increase in cost has been for labor, which, according to this index, in 1966 cost about 125 per cent more than it did in 1950. This is very significant since maintenance activities utilize labor to a large degree.[87-135]

of years. Highway maintenance and operation costs between 1950 and 1966 are shown in Table 6-22.

ECONOMY STUDIES

Economy studies in highway engineering are essential, since the highway engineers are charged with the responsibility for making the best possible use of a limited resource: the public funds available for highway purposes.

Even though enormous sums have been spent and will continue to be spent for highway improvements, they do not satisfy all needs and demands. Thus, highway engineers must continue to determine, by one method or another, the best uses and the priority among uses of the limited funds available. The expenditure of billions of dollars for the highway program is "big business" under any definition, and a sound economic approach must be used to a greater extent than it has been.

Textbooks in engineering economy develop four procedures for economy studies:

1. *Annual cost method.*
2. *Present worth method,* in which comparison is made between the present worth of all present and future expenditures
3. *Capitalized cost method,* in which comparison is made on a present worth basis, where expenditures are calculated as though they continue in perpetuity
4. *Rate of return method,* which involves the determination of the interest rate at which the alternates are equally attractive

To these four procedures, civil engineers involved with public works have added a fifth method.

5. *Benefit-cost ratio method,* which is often referred to as the benefit cost ratio

This is one of the most commonly used procedures for economy studies in highway engineering, but many engineers favor the rate of return method. However, the results obtained from any of the methods listed will lead to sound economic decisions.

A recent survey shows a sharp decrease in the use of the benefit-cost ratio, while the application of the total annual cost method shows a significant increase.[136] The results of the study, conducted by the South Dakota Department of Highways, were compared with an earlier study conducted by the BPR, and the results are shown in Table 6-23. The distribution of stated interest rates is given in Table 6-24.

The methods of *annual cost, present worth, rate of return,* and *benefit-cost ratio* will be discussed in detail, since they represent the procedures encountered in highway engineering economy studies.[137] The *present worth* method is receiving increased support by some engineers, but, not to the same degree as the other methods. The primary reason for preference of *annual cost, rate of return,* and *benefit-cost ratio* methods is that they are all based on annual cost comparisons. The average administrator is more accustomed to analyzing annual costs, and present worths are more difficult to interpret.

TABLE 6-23. USE OF ECONOMIC ANALYSIS METHODS

ITEM	*BPR ANALYSIS (1962)*	*SD ANALYSIS (1963–66)*
1. Per cent using road user analysis	73	89
2. Per cent of Item 1 using benefit-cost ratio method	72	49
3. Per cent of Item 1 using total annual cost method	28	51
4. Per cent of Item 2 using a second benefit ratio	11	20
5. Per cent of Item 1 using annual capital highway costs	63	92
6. Per cent of Item 1 including accident costs	5	10
7. Per cent of Item 1 including maintenance costs	55	79
8. Per cent of Item 1 including specific interest rate	67	77

TABLE 6-24. COMMONLY USED INTEREST RATES

INTEREST RATE %	*BPR ANALYSIS % REPORTS USING*	*SD ANALYSIS % REPORTS USING*
0	20	23
0.1–3.9	22	8
4.0–5.9	45	31
6.0–7.0	13	38
Over 7.0	0	0

PROCEDURES FOR ECONOMY STUDIES

Annual Cost Method

As already explained, the annual cost of a capital investment to be recovered in n years with interest is found by multiplying its first cost by the appropriate capital recovery factor. The uniform amount so determined, if charged at the end of each year for the assumed useful life, will exactly repay the initial investment with interest.

The total annual cost of a highway improvement is the sum of all annual costs of capital recovery, plus the annual costs of maintenance and operation, plus the annual costs of the road users, plus the annual costs of traffic accidents. Annual costs are computed for the existing facility and for each of the proposals for improvement. The alternative resulting in the lowest annual cost represents the best solution from an economic standpoint.

This method has a serious drawback, in that widely differing results are obtained for various assumed interest rates. Generally in the annual cost method, low interest rates favor those alternatives that combine large capital investments with low maintenance and operation or user costs, whereas high interest rates favor reverse combinations.

Present Worth Method

In making comparisons by the present worth method, all costs and benefits are reduced to present-day values. One difficulty encountered in the application of present worth comparisons is that equal periods of service must be analyzed. The following problem illustrates this point.

Two proposed structures are to be compared, using an interest rate of 6 per cent. Structure A has a first cost of $500 and an estimated life of 8 years. Structure B has a first cost of $1,000 and an estimated life of 20 years. Annual operating expenditures for both structures will be the same.

TABLE 6-25. PRESENT WORTH METHOD ILLUSTRATION

STRUCTURE A	
First cost	$ 500
Present worth of first renewal = 500 × 0.6274 =	314
Present worth of second renewal = 500 × 0.3936 =	197
Present worth of third renewal = 500 × 0.2470 =	124
Present worth of fourth renewal = 500 × 0.1550 =	78
Total present worth of cost of 40 years service =	$1,213
STRUCTURE B	
First cost	$1,000
Present worth of first renewal = 1,000 × 0.3118 =	312
Total present worth of cost of 40 years service =	$1,312

If a present worth comparison is to be made, the period of service under consideration must be the same for the two structures. Therefore, it is necessary to consider enough renewals of each structure to serve for the number of years which is the least common multiple of the estimated lives of the two structures, in this case it is 40 years. It is necessary to assume that renewal costs will be the same as the first cost. (See Table 6-25.) Therefore, according to Table 6-25, structure A would be selected as the best alternative.

In applying present worth to highway economy studies, comparison is made between the sum of the present worths of the benefits and costs for maintenance and operation during each year of the analysis period, and the total capital investment required to produce these benefits. This method tends to minimize the effects of large future benefits arising from projected increases in traffic volumes, and tends to place greater value upon benefits accruing in the near future.

Rate of Return Method

As explained above, the annual cost method of comparing alternatives involves converting each alternative into an equivalent uniform annual series of payments. This method is preceded by a decision on the appropriate interest rate of minimum attractive return to be used in the conversion.

Another method of comparing expenditure-time alternatives is to find the interest rate that makes them equivalent to one another. In this method the interest rate is the unknown in the problem. In the common situations of comparing two alternatives, one with a lower capital investment and higher maintenance and operation and road user costs, the other with a higher capital investment and lower maintenance and operation and road user costs, this unknown interest rate may be thought of as the prospective rate of return on the excess of investment in the alternative with the higher capital investment.

Therefore the rate of return method involves determining the rate of interest at which two alternatives have equal annual costs. This may be done by calculating annual costs for each alternate, for different rates of interest, and interpolating.

One alternative is taken as the base, each of the others is compared with it and the rate of return is computed. Those alternatives which do not show a rate of return sufficiently high or attractive are discarded. Economically speaking, the desirable alternatives are those that show a rate of return in excess of what is considered to be the minimum attractive return on investment. It is possible that some or all alternatives will be discarded as not yielding an attractive rate of return.

The rate of return method provides for selection of that project which will yield the most for the least expenditure of funds by giving priority to those projects yielding the highest rate. This is especially important in highway engineering where there are insufficient funds to finance all desirable or proposed projects.

Benefit-Cost Ratio Method

The benefit-cost method is extensively used for highway, waterway, and other public works projects. The method involves the determination of the ratio of estimated benefits to estimated costs, and the ratio is used as the major criterion in determining whether or not the proposed expenditures should be recommended.

For each alternative, the annual road user costs and the total annual cost for the highway are determined. The alternatives are then compared on the basis of benefit-cost ratios. The benefit-cost ratio is defined by:

$$\text{Benefit-cost ratio} = \frac{\text{benefits}}{\text{costs}}$$

$$= \frac{\text{difference in road user costs}}{\text{difference in highway costs}} = \frac{R - R_1}{H_1 - H}$$

where R = the total annual road user costs for the basic condition, usually the existing road

R_1 = the total annual road user costs for the proposed improvement

H = the total annual highway costs for the basic condition, usually the existing road

H_1 = the total annual highway costs for the proposed improvement

The same formula is used to compute the benefit ratio for all alternatives, each compared with the same basic condition. On this basis the relative values of the several ratios can be compared directly. A ratio greater than 1.0 indicates that the additional expenditure for the proposed improvement over the basic condition is justified. A ratio less than 1.0 indicates that the benefits are less than the costs, and in a road user benefit sense the base condition is to be preferred over the proposed improvement. The alternative with the highest ratio is usually the most acceptable. To conclusively establish the most attractive alternative, however, it is necessary to compare all alternatives with each other, and not just with the basic condition.

For comparable results it is essential that the same interest rates be used to evaluate the difference between annual costs of the various alternatives. The results obtained vary with the rate of interest assumed. In general, the interest rate only affects the denominator, and the higher the assumed interest rate the smaller the benefit-cost ratio.

DETERMINATION OF ADT FOR PERIOD OF ANALYSIS

In order to evaluate road user costs (RUC) it is necessary to determine the annual average daily traffic volume (ADT) for the period of analysis. AASHO recommends three steps to determine the appropriate ADT:

1. Estimate the ADT that will use the section upon its completion. This is the current traffic.
2. Determine the number of years (usually taken as 20) for which the analysis is to be made, and the expansion factor for traffic on the section during this period. The traffic forecast period is independent of the average lives assumed for the various components that make up the capital investment.
3. Calculate an expanded ADT that is a representative or average value for the period of analysis. This is the ADT to be used.

It is not entirely clear as to the proper meaning of the third step. If the expanded ADT is used for the analysis, it would result in an overstatement of road user costs. If the current ADT is used, it would result in an understatement of road user costs. It is proposed that a representative average value can be obtained by averaging the current ADT and the expanded ADT. But it is important to consider the appropriate ADT volume very carefully, as it has a significant influence on the analysis. This point is discussed further in the next section. It is necessary to separate traffic volume data by vehicle type, since different unit road user costs must be used for each.

In comparing alternatives, the question often arises as to how to treat induced traffic, which might be a part of the total forecasted traffic growth for a proposed improvement, but would not necessarily be part of the traffic growth for the existing facility. *Induced traffic* is used in the sense that it includes all traffic that would not have occurred without the new facility. The magnitude of induced traffic is very much dependent upon the type and location of the proposed improvement. For example, induced traffic will normally be greatest for a limited-access facility connecting the central city and the suburban area. On the other hand, induced traffic will be of little significance within urban or rural areas.

There are different approaches to the treatment of the problem which are best illustrated by the following simple, but extreme, example. Assume the conditions on an existing road and proposed new facility are:

Length of both alternatives = 10 miles

Unit road user cost on existing = 12.5 cents per vehicle-mile

Unit road user cost on proposed = 10.0 cents per vehicle-mile

Average ADT for analysis period for existing = 10,000 vehicles per day

Average ADT for analysis period for proposed = 15,000 vehicles per day

Annual cost of capital recovery for existing = 0

Annual cost of capital recovery for proposed (based on a capital investment of \$10 million ($n = 30$ years, $i = 6$ per cent) = \$726,000

Annual RUC for existing = $10{,}000 \times 10 \times 0.125 \times 365$ = \$4,562,500

Annual RUC for proposed = $15{,}000 \times 10 \times 0.10 \times 365 = \$5{,}475{,}000$

The annual RUC for the proposed improvement exceeds that for the existing road. When considering the different ADT volumes anticipated for each facility, a benefit-cost analysis would not justify the construction of the new facility.

Analysis based on the ADT given for the existing:

Annual RUC for proposed

$$= 10{,}000 \times 10 \times 0.10 \times 365 = \$3{,}650{,}000$$

$$\text{Benefit ratio} = \frac{912{,}500}{726{,}000} = 1.26$$

Analysis based on the ADT given for the proposed:

Annual RUC for existing

$$= 15{,}000 \times 10 \times 0.125 \times 365 = \$6{,}843{,}700$$

$$\text{Benefit ratio} = \frac{1{,}368{,}700}{726{,}000} = 1.88$$

Analysis based on total benefits to be derived from proposed:

Annual total benefits

$$= (0.125 - 0.10) \times 15{,}000 \times 10 \times 365 = \$1{,}368{,}700$$

This analysis leads to the same result as the analysis based on the ADT given for the proposed.

In the above example, when the same ADT is used in comparing alternatives, construction of the proposed facility would be justified.

Admittedly, there is no single answer to cover all situations, but it appears that the appropriate method of handling a problem like this is to base the comparison on the ADT given for the proposed improvement. The annual cost of capital recovery for the proposed improvement is significant in the comparison, and it is based on the capital investment required to provide a certain level of service for the traffic to be accommodated by the new facility. Therefore, the ADT on the proposed improvement is the more meaningful one to use in the analysis.

SENSITIVITY ASPECTS OF ECONOMY STUDIES[138,139]

Before an economy study can be undertaken, it is necessary in all methods of analysis, except the rate of return method, to adopt an interest rate. For all methods, *including* the rate of return, assumptions must be made for the useful life and the salvage value at the end of the useful life, for each component part of the highway improvement. In computing road user costs for all methods, it is first necessary to estimate an ADT volume. All of these factors are based on assumptions that can affect the final result of an economy study, to a greater or lesser degree. Where important decisions are involved, a series of parallel solutions encompassing a reasonable range of assumptions may be warranted.

Effect of Interest Rate

The selection of an interest rate, or minimum attractive rate of return, significantly affects the result of an economy study. A decided advantage of the rate of return method lies in the fact that no predetermined rate of interest need be selected. All of the precautions that might be taken in carrying out an economy study are meaningless if the interest rate is inappropriate for the conditions under which the decision is made.

Low interest rates favor alternatives with large capital investments, and high interest rates favor alternatives with small capital investments. Low interest rates give more significant weight to happenings in the more distant future, where uncertainties of prediction are greatest, while high interest rates tend to discount the effect of future happenings.

Effect of Assumed Useful Life

In an economy study, capital investments are uniformly spread over the assumed useful life of each element in the highway improvement. Where salvage value is not considered, this uniform annual charge for principal and interest is found by multiplying the capital investment of the element by the CRF.

As the assumed useful life increases, the CRF approaches the interest rate. At high interest rates this convergence occurs rapidly at low interest rates this convergence occurs slowly. Therefore, economy studies involving higher interest rates are relatively less sensitive to changes in assumed life, and at low or zero interest rates the sensitivity is correspondingly higher. For example, at 7 per cent the increase in the annual cost of capital recovery, when the assumed life is decreased from 30 to 20 years, is 17 per cent, while at 0 per cent the increase is 50 per cent.

Effect of Assumed Salvage Value

The salvage value of a component part of the highway improvement is its dollar worth at the end of its assumed useful life. The salvage value assumed is dependent upon the assumed useful life. At realistic interest rates and relatively long lives, the effect of salvage value is very small. Under these conditions the effect of including and excluding salvage value would result in an error of a magnitude that is considerably less than the expected error in other estimates. Therefore, for most highway facilities, or components of highway improvements, zero salvage value is appropriate in economy studies.

Effect of Assumed ADT Volume

Reductions in road user costs provide one of the major justifications for highway improvements, and therefore they have a significant influence on an economy study. The annual savings in road user costs is normally directly proportional to the estimated ADT for the analysis period. Projections of traffic on a new or improved facility may be considerably higher than traffic on the unimproved facility, and in many cases the projected ADT may be two or three times greater at the end of an analysis period of 20 to 30 years. Increased traffic volumes in the future may result in increased road user costs due to congestion. Obviously, if the facility is operating at near capacity at the beginning of the analysis period, road user costs can materially increase in the future if additional traffic causes greater congestion. The AASHO procedure of correlating vehicle running costs to type of highway operation is very desirable in this respect.

For important decisions two analyses may be warranted, one based on a pessimistic estimate of growth, and the other on an optimistic one. With this approach the range of variation in consequences of the improvement can be better evaluated.[14,44,140-144]

Effect of Inflation

Based on a comprehensive study[145] of different price changes and cost indices over a period of years, it was found that there was no single index completely satisfactory for measuring the general trend in prices, that is, inflation or deflation. The magnitude of future inflation is difficult to predict, but an approximate figure of 2 per cent per year, compounded, was thought to be reasonable for the long-term rate in the United States.

The report lists the following reasons why the majority of professional opinion is opposed to including an inflation rate in engineering economy studies:

1. Difficulties are inherent in forecasting.
2. The federal government is committed to price stabilization.
3. Federal programs, justified in part by inflating benefits, may contribute to inflation.
4. The gains received by debtors are offset by losses to creditors.
5. Future dollars to pay for future expenses will likewise be inflated, and there is no net change.
6. A bias toward capital-intensive and long-lived projects results, making adaptations to future changes more costly than otherwise.

EXAMPLES OF HIGHWAY ECONOMY STUDIES

Vehicle Costs, in Relation to Intersections

The significance of the cost of stops was previously discussed, and is now illustrated by an appropriate example.

Example

A grade separation is being considered at an intersection of a four-lane divided highway and a two-lane crossroad, where traffic signals now control vehicle movements. The ADT on the four-lane highway is 25,000 vehicles per day, of which 90 percent are passenger cars and light commerical vehicles, 5 per cent are single-unit trucks, and 5 per cent are combination vehciles. The ADT on the two-lane highway is 5,000 vehicles per day, and the classification of the traffic is similar to that on the four-lane highway. There is little interchange of traffic between the intersecting highways, and therefore no ramps need to be provided if a grade separation structure is considered economically feasible. The average intersection approach speeds for passenger cars and light commercial vehicles are 50 and 35 miles per hour on the four-lane and two-lane highways, respectively; the corresponding approach speeds for single-unit trucks and combination vehicles are 40 and 30 miles per hour on the two highways. A field study indicated that an average of 20 per cent of the vehicles on the four-lane highway are stopped for an average of 30 seconds per vehicle, and an average of 30 per cent of the vehicles on the two-lane highway are stopped for an average of 40 seconds per vehicle. The total costs incurred due to stopping are to be determined. That is done as follows:

1. Determination of equivalent passenger cars

From Table 6-2, passenger car equivalents of 2 and 4 are assumed, respectively, for single-unit trucks and combination vehicles. Therefore, the equivalent number of passenger cars becomes:

a. Four-lane highway

PC and LCV = 25,000 × 0.90
= 22,500 vehicles per day
SUT = 25,000 × 0.05 × 2
= 2,500
CV = 25,000 × 0.05 × 4
= 5,000
Total equivalent PC = 30,000 vehicles per day

b. Two-lane highway

PC and LCV = 5,000 × 0.90
= 4,500 vehicles per day
SUT = 5,000 × 0.05 × 2
= 500
CV = 5,000 × 0.05 × 4
= 1,000
Total equivalent PC = 6,000 vehicles per day

2. Cost per stop

The cost per stop per vehicle is determined for each of the intersecting highways, considering additional vehicle operating costs and time costs due to a vehicle coming to a stop, standing, and regaining initial speed. Figure 6-2 is used to obtain the total cost per stop for passenger cars.

a. Four-lane highway

With an approach speed of 50 miles per hour, and 30 seconds standing delay, the total cost per stop = 3.55 cents per stop.

With approach speed of 40 miles per hour, and 30 seconds standing delay, the total cost per stop = 2.85 cents per stop.

b. Two-lane highway

With approach speed of 35 miles per hour, and 40 seconds standing delay, the total cost per stop = 3.03 cost per stop.

With approach speed of 30 miles per hour, and 40 seconds standing delay, the total cost per stop = 2.79 cost per stop.

3. Number of vehicles stopped

The number of vehicles of each type that are stopped per day is next determined.

a. Four-lane highway

Number of PC and LCV stopped
= 22,500 × 0.20 = 4,500 vehicles per day
Number of equivalent SUT stopped
= 2,500 × 0.20 = 500 vehicles per day
Number of equivalent CV stopped
= 5,000 × 0.20 = 1,000 vehicles per day

b. Two-lane highway

Number of PC and LCV stopped
= 4,500 × 0.30 = 1,350
Number of equivalent SUT = 500 × 0.30 = 150
Number of equivalent CV = 1,000 × 0.30 = 300

4. Daily costs of stops

a. Four-lane highway

Cost of PC and LCV stops
= 4,500 × 0.0355 = $159.75 per day
Cost of SUT and CV stops
= 1,500 × 0.0285 = 42.75
Total = $202.50 per day

b. Two-lane highway

Cost of PC and LCV stops
= 1,350 × 0.0303 = $40.90 per day
Cost of SUT and CV stops
= 450 × 0.0279 = 12.55
Total = $53.45 per day

Therefore, the total cost of stops per day for both intersecting highways is 202.50 + 53.45 = $255.95, and the cost of stops per year is 255.95 × 365 = $93,420. The construction of the grade separation structure would be justified just on the basis of these costs, without considering the greater safety features of the grade separation, if the annual cost of building the grade separation was less than $93,420, which is almost a certainty.

Comparing Types of Pavements

Since the useful lives of various types of pavements differ, it is more appropriate to compare their annual costs over a period of time that exceeds their useful lives. The useful lives of major structures and ROW invariably exceed that of the surfacing, and almost 90 per cent of all pavements are either resurfaced or reconstructed at the end of their service lives, with only a very small percentage abandoned. Therefore, a more significant analysis period would be one that is closer, or equal to, the average service life of major structures. In this respect, an analysis period of 40 years is considered appropriate.

Example

To illustrate a procedure for comparing the relative economic value of two types of surfacing, assume the following conditions for a four-lane divided rural highway:

TYPE OF SURFACING	*INITIAL COST PER MILE*	*SERVICE LIFE (yrs)*	*RESURFACING COST PER MILE*	*ANNUAL MAINTENANCE COST PER MILE*
Asphaltic concrete	$100,000	18	$40,000	$600
Portland cement concrete	160,000	26	40,000	600

Resurfacing of both types of pavements will be done with asphaltic concrete, and the average service life of the resurfacings is assumed to be the same as that of the service life of the original asphaltic concrete, that is, 18 years. The annual cost of the pavement equals the appropriate capital recovery factor (CRF), times the summation of the initial cost, plus the present worth of the resurfacings required within the analysis period, minus the present worth of the residual value of the last resurfacing. To this is added the annual maintenance cost.

For the comparison, an interest rate of 6 per cent is used to determine the annual costs of the two pavements. Annual cost formula:

$$C = \text{CRF}_n\left[I + R_1 \times \text{PWF}_{n_1} + R_2 \times \text{PWF}_{n_1+n_2} - \frac{(n_1 + n_2 + n_3 - n)(R_1 \text{ or } R_2)\text{PWF}_n}{n_3}\right] + \text{M}$$

where
C = annual cost for pavement, per mile
n = analysis period in years
CRF = capital recovery factor
I = initial cost of pavement, per mile
R_1 = first resurfacing cost, per mile
n_1 = number of years between initial construction and first resurfacing
PWF = persent worth factor
R_2 = second resurfacing cost, per mile
n_2 = service life of first resurfacing
n_3 = service life of last resurfacing
$\frac{(n_1 + n_2 + n_3 - n)R_2 \times \text{PWF}_n}{n_3}$ = residual value of last resurfacing
M = annual maintenance cost

Annual cost of asphaltic concrete:

$n = 40$ years $I = \$100{,}000$ $R_1 = \$40{,}000$,
$n_1 = 18$ years $R_2 = \$40{,}000$ $n_2 = 18 = n_3$
$M = \$600$, $i = 6$ per cent

$$C = 0.0665\left[100{,}000 + 40{,}000 \times 0.3513 + 40{,}000 \times 0.1227 - \frac{(54 - 40)}{18} \times 40{,}000 \times 0.0972\right] + 600$$

$C = \$7{,}710$ per mile

Annual cost of Portland cement concrete:

$n = 40$ years $I = \$160{,}000$ $R_1 = \$40{,}000$
$n_1 = 26$ years $n_2 = 18$ years $M = \$600$
$i = 6$ per cent

$$C = 0.0665\left[160{,}000 + 40{,}000 \times 0.2198 - \frac{(44 - 40)}{18} \times 40{,}000 \times 0.0972\right] + 600$$

$C = \$11{,}170$ per mile

Based on the above analysis, the asphaltic concrete pavement would be more economical.

The above annual cost formula was used in an extensive study of annual costs to comparing different types of pavements.[146] A major purpose for making the study was to determine the effect and relative importance of each component in the formula. In addition, estimated costs of future resurfacing was increased by the application of a price trend factor, which was assumed as 2 per cent compound interest. However, the use of the price trend factor is not recommended if the factor represents an increase in future costs resulting from inflation. For example, if the cost per mile for resurfacing at the present time is \$20,000, and the estimated service life is 18 years, the resurfacing cost at the end of the 18th year would be $R_1 = 20{,}000 \times 1.428 = \$28{,}560$, where $1.428 = \text{CAF}$ for a single sum for $i = 2$ per cent and $n = 18$.

Some of the conclusions reached as a result of the investigation are:

1. The initial construction cost was found to be the most significant factor in a comparison of the annual costs of pavements. It overshadowed the variable effects of such factors as maintenance and resurfacing costs, rate of interest, pavement service life, and the analysis period.
2. For the majority of pavement sections analyzed, the initial construction costs constituted 85 to 95 per cent of the total annual costs, while maintenance costs amounted to less than 10 per cent of the total annual costs.
3. In general, the annual costs of rigid and flexible pavements designed for poor soil conditions were, for the most part, within the same price bracket, but the costs of the rigid pavements designed for excellent soil conditions were 25 to 40 percent higher than those of flexible pavements designed for the same conditions.[147, 148, 149]

Economics of Accident Reduction and Parking Prohibition

The cost of accident reduction and parking prohibition is illustrated by an example.

Example

The improvement of a 1.5-mile section of a two-way urban arterial is being considered. The existing conditions provide for two lanes for through traffic flow in each direction, parking on both sides, signalization of 10 of its 15 intersections (with the remaining intersections controlled by two-way stop signs), and grades that do not exceed 2 per cent. Accident experience at the intersections averages 1.5 injuries and 3.0 property-damage accidents per year; between intersections it averages 0.1 death, 3.0 injuries and 4.0 property-damage accidents per block per year.

The proposed improvements consist of left-turn channelization at eight intersections where heavy left-turn movements exist, and the prohibition of parking on both sides for the entire length of the section. With elimination of parking, no additional ROW is required, and the total cost of the improvements is estimated at \$500,000. The anticipated effect of the channelization is a 15 per cent reduction in all types of accidents at all intersections, and the expected effect of removing parking is a 30 per cent reduction in all types of accidents between intersections.

Traffic flow averages about 420 vehicles per hour per lane in each direction during peak periods, which constitute three hours of the day. The ratio of green time to cycle length on the arterial is 0.6, and it is assumed that speed characteristics will only differ significantly during the peak periods on weekdays for the parking and no parking conditions. An indication of the speed differentials for the two conditions can be obtained by using the following relations, which were developed from studies in Chicago:[150]

With parking $S = 22.78 - 0.0041q$

Without parking $S = 33.50 - 0.0066q$

where S = speed in miles per hour
q = rate of flow in vehicles per hour of green per lane

An analysis period of 20 years and 6 per cent interest will be used to determine if the proposed improvements are economically justified. This is done as follows.

1. Annual cost of accidents

Accident cost data in Table 6-14 will be used to estimate the cost of accidents. This table includes the cost of both reported and unreported accidents in terms of reported accidents.

a. At intersections

Existing conditions
$= (1.5 \times 2{,}100 + 3.0 \times 600)15 = \$74{,}200$

Proposed conditions
$= 0.85 \times 74{,}200 = \$63{,}100$

b. Between intersections

Existing conditions
$= (0.1 \times 7{,}700 + 3.0 \times 2{,}100 + 4.0 \times 600)14$
$= \$132{,}600$

Proposed conditions $= 0.70 \times 132{,}600 = \$92{,}800$.

c. Total annual cost of accidents

Existing conditions $= 74{,}200 + 132{,}600 = \$206{,}800$

Proposed conditions $= 63{,}100 + 92{,}800 = \$155{,}900$

2. Annual value of time

The AASHO-recommended value of time of \$1.55 per hour per passenger car will be used to estimate the cost of time. The number of vehicles which will be affected by any travel time savings in comparing the existing and proposed conditions are those during the three hours of peak period flows, on weekdays only. For two lanes in each direction, the total number of vehicles affected per weekday $= 4 \times 420 \times 3 = 5{,}040$.

a. Average overall speed

$$\text{Vehicles per hour of green per lane} = \frac{420}{0.6} = 700$$

Existing: $S = 22.78 - 0.0041 \times 700 = 19.9$ miles per hour

Proposed: $S = 33.50 - 0.0066 \times 700 = 28.9$ miles per hour

b. Annual value of time

$$\text{Existing} = \frac{1.55}{19.9} \times 5{,}040 \times 1.5(365 - 2 \times 52) = \$153{,}700$$

$$\text{Proposed} = \frac{1.55}{28.9} \times 5{,}040 \times 1.5(365 - 2 \times 52) = \$105{,}800$$

3. Annual running cost

Assuming that with parking the type of operation will be restricted, and without parking it will be normal, examination of Table 6-4 reveals little or no difference in operating costs for 0–4 per cent grade for restricted operation at 20 mph, and for normal operation at 28 mph.

4. Cost of stops

It is assumed that the number of stops will not be affected.

5. Annual road user costs

Existing $= 206{,}800 + 153{,}700 = \$360{,}500$.

Proposed $= 155{,}900 + 105{,}800 = \$261{,}700$

6. Annual cost of capital recovery

CRF ($n = 20$, $i = 6$ per cent) $= 0.0872$ (see Table 6-1)

Existing: capaital recovery $= \$0$

Proposed: capital recovery $= 0.0872 \times 500{,}000 = \$43{,}600$.

7. Benefit-cost ratio method

$$\text{Benefit-cost ratio} = \frac{360{,}500 - 261{,}700}{43{,}600 - 0} = 2.26$$

Therefore, the proposed improvement is justified since the benefit ratio is greater than 1.0.

8. Annual cost method

COST ITEM	EXISTING CONDITION	PROPOSED CONDITION
Road user costs	\$360,500	\$261,700
Capital recovery	0	43,600
Total annual cost	\$360,500	\$305,300

Therefore, the proposed improvement is desirable since it has the lowest annual cost.

9. Rate of return method

This method involves determining the interest

rate at which the annual costs are equal. This rate can be solved for directly, when the useful service life for each component of the capital investment is the same.

$$360{,}500 = 261{,}700 + \text{CRF} \times 500{,}000$$

$$\text{CRF} = 0.1976$$

The interest rate, expressed to the nearest tenth of a percent, corresponding to a CRF = 0.1976 for $n = 20$ years, is 19.1 per cent, which is obtained by interpolation in a table of capital recovery factors. Therefore the proposed improvement is better than the existing condition if the minimum attractive rate of return is equal to, or less than, 19.1 per cent.

10. Present worth method

COST ITEM	EXISTING CONDITION	PROPOSED CONDITION
Annual road user costs	$ 360,500	$ 261,700
PW of RUC at i = 6%, n = 20	4,134,900	3,001,700
Capital investment	0	500,000
Total Present Worth	$4,134,900	$3,501,700

Therefore, the proposed improvement is desirable, since it represents a propsective saving in present worth costs of $633,200.

An analysis of the economics of converting from two-way to one-way operation would be very similar to the above problem. The same procedure would be used, and travel time and accident reductions are of the same order of magnitude.

Example Number 1 from AASHO[4]

Assume a relocation project where, by heavy grading work on new alinement, it is possible to reduce the length between two points on an existing highway. The existing highway is to be abandoned. Present traffic on the existing route is 1,500 vehicle per day, and it is estimated that the average traffic for the next 20 years (the analysis period) will approximate 2,500 vehicle per day. This traffic is composed of passenger cars with a very small proportion of trucks. Due to the character of the trucks, a distinction between them and the passenger cars is not considered necessary. Assignment of type of operation is made on the basis of the indicated ratios of 30th highest hourly traffic volume to practical capacity. The proposed facility contemplates a pavement width of 20 feet, compared with 18 feet on the existing facility. This factor, together with improved alinement and grades, permits distinction as to type of operation, as follows:

ROUTE	30th HIGHEST HOUR	CAPACITY	RATIO	TYPE OF OPERATION
Existing	375	450	0.83	Normal
Proposed	375	600	0.63	Free

Other essential data follow:

ROUTE	DESIGN SPEED (mph)	RUNNING SPEED (mph)	No. LANES	LENGTH (miles)	GRADE CLASS (%)	TYPE OPERATION	CURVATURE
Existing	50	37	2	2.0	0–3	N	50% – 4°
Existing	50	34	2	0.4	3–5	N	50% – 4°
Proposed	60	42	2	1.5	0–3	F	Negligible

The curvature proposed on the new facility is generally flat, with a minor portion approaching the design speed limit. This is to be properly superelevated, so a correction for curvature can be ignored. However, the existing highway has a larger number of curves approaching the maximum for a 50 miles per hour design speed. It is estimated that 50 per cent of its length will have an average curvature of about 4 degrees, with superelevation negligible. From Figure 6-1 a correction of between 7 and 8 per cent is read for the speeds considered. Since only one-half of the roadway is curved, the correction is halved to 4 per cent, and applied for the whole of the route to the values from Table 6-4. The unit costs are as follows:

ROUTE	GRADE (%)	UNIT COSTS (cpvn)
Existing	0–3	9.85 × 1.04 = 10.24
Existing	3–5	10.20 × 1.04 = 10.61
Proposed	0–3	9.01 × 1.00 = 9.01

The average annual road user costs are calculated as follows for each section and added:

ROUTE (days)	×	A	×	L	×	U	=	$
Existing route								
365	×	2,500	×	2.0	×	0.1024	=	$186,900
365	×	2,500	×	0.4	×	0.1066	=	38,700
						Total		$225,600
Proposed route								
365	×	2,500	×	1.5	×	0.0901	=	$123,300

The average annual unit maintenance cost on the existing route is estimated to be $1,100 per mile, or a total of 2.4 × $1,100 = $2,640, and that of the proposed route is estimated to be $880 per mile, for a total of 1.5 × $880 = $1,320.

The total estimated cost of the proposed improvement is $550,000, and the prevailing local interest rate is 5 per cent. This interest rate is applied with reference to the cost and life expectancy of the individual items of improvement to compute the annual cost.

ITEM	ESTIMATED LIFE (yrs)	COST	CRF	ANNUAL COST
Pavement	20	$ 66,000	0.0802	$ 5,290
ROW	60	33,000	0.0528	1,740
Grading, drainage, and structures	40	451,000	0.0583	26,290
		Total Annual Cost =		$33,320

1. Benefit-cost ratio method

The benefit ratio for the proposed improvement is computed as follows:

$$\text{Benefit ratio} = \frac{\$225{,}600 - 123{,}000}{\$34{,}640 - 2{,}640} = 3.20$$

The analysis indicates that the annual benefits are over three times the annual project costs. From this it appears to be a worthy project, as far as road user benefits are concerned, and one that should be slated for early construction.

2. Annual cost method

SUMMARY OF ANNUAL COSTS

COST ITEM	*EXISTING ROUTE*	*PROPOSED ROUTE*
Road user costs	\$225,600	\$123,300
Maintenance	2,640	1,320
Capital recovery at i = 5%	0	33,320
Total annual cost	\$228,240	\$157,940

Based on the annual cost method, the proposed route is desirable since it has the lowest annual cost.

3. Rate of return method

The rate of return method involves determining the interest rate at which the annual costs are equal. This must be done by trial and error, as previously explained.

ANNUAL COST OF CAPITAL RECOVERY		*i = 15%*		*i = 20%*
Pavement (n = 20 yrs)				
	66,000 × 0.1598 =	\$10,550	× 0.2054 =	\$ 13,560
G and S (n = 40 yrs)				
	451,000 × 0.1506 =	67,920	× 0.2001 =	90,250
ROW (n = 60 yrs)				
	33,000 × 0.1501 =	4,950	× 0.2000 =	6,600
	Totals	\$83,420		\$110,410

At 15% interest

Annual cost of existing = \$228,240

Annual cost of proposed = 124,620 + 83,420 = \$208,040

At 20% interest

AC of existing = \$228,240

AC of proposed = 124,620 + 110,410 = \$235,030

The rate of return lies between 15 and 20 per cent. At 15 per cent interest, the existing route costs more, annually, than the proposed route by \$20,200, but at 20 per cent interest their relative positions are reversed, and the proposed route costs more by \$6,790. The rate of return is found by straight-line interpolation as follows:

$$i = 15 + 5 \times \frac{20{,}200}{20{,}200 + 6{,}790} = 18.7 \text{ per cent}$$

The proposed route is better than the existing route if the minimum attractive rate of return is equal to or less than 18.7 per cent, which obviously would be the case.

4. Present worth method

As previously discussed, in applying the present worth method, comparison is made between the sum of the present worths of the benefits and costs for maintenance and operation during each year of the analysis period, and the total capital investment required to produce these benefits.

SUMMARY OF PRESENT WORTHS (Analysis Period = 20 yrs.)

COST ITEM	*EXISTING ROUTE*	*PROPOSED ROUTE*
Annual RUC	\$ 225,600	\$ 123,300
Annual Maintenance	2,640	1,320
Total	228,240	124,620
PW at i = 5%, n = 20	\$2,844,330	\$1,553,010
Capital Investment	0	550,000
Total Present Worth	\$2,844,330	\$2,103,010

Based on the present worth method, the proposed route is desirable since as it represents a prospective saving in present worth costs of \$741,320.

Example Number 4 from AASHO[4]

Assume the case shown in Figure 6-4. The existing three-lane road A, which passes through a small town, is old and congested and is in need of improvement. A location and design study shows that the improvement of road A to modern standards is difficult and costly, due to both the terrain and the existing adjacent development. Two alternate locations, B and C, appear to be feasible for a Four-lane divided highway, which is considered necessary for the volumes involved.

For analysis purposes the following alternates are considered:

Alternate 1 (Basic condition): All traffic on the existing route, no new construction.

Alternate 2: Road B and two-lane connection D constructed; some traffic remaining on A.

Alternate 3: Road C constructed; some traffic remaining on A.

Fig. 6-4. (Illustration of differently located roads) (*Source: Road User Benefit Analyses for Highway Improvements, AASHO, 1960, Example 4*)

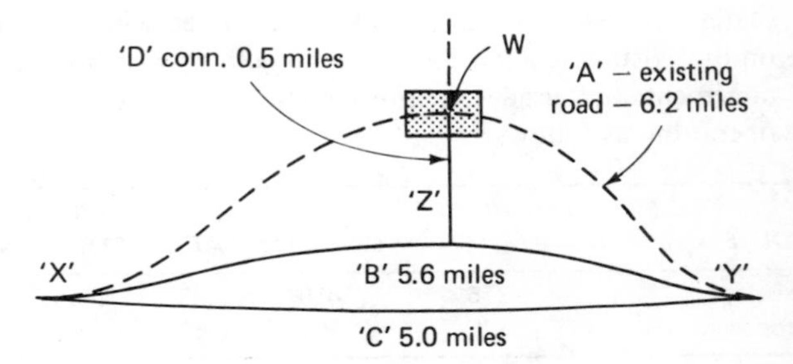

TABLE 6-26. ROAD USER UNIT COST DATA

		AVER. DAILY TRAFFIC VOL.							
Alt.	*Length (mi)*	A_p	*Equiv.* A_t	A_e	*Running Speed (mph)*	*Type of Oper.*	*Grade Class (%)*	*Pavmt. Condition*	*Unit Cost, (cents)*
1-A	6.2	12,000	2,400	19,200	28	R	0–3	Poor	11.92
2-A	6.2	4,000	400	5,200	32	N	0–3	Poor	11.06
2-B	5.6	8,000	2,000	14,000	44	F	0–3	Good	8.86
2-D	0.5	1,000	......	1,000	32	F	0–3	Good	9.75
3-A	6.2	4,500	400	5,700	32	N	0–3	Poor	11.06
3-C	5.0	7,500	2,000	13,500	44	F	0–3	Good	8.86

(Source: *Road User Benefit Analyses for Highway Improvements*, AASHO, 1960, Example 4)

Road user costs are:

Alternate 1: 365 × 19,200 × 6.2 × \$0.1192 = \$5,179,000

Alternate 2: 365 × 5,200 × 6.2 × \$0.1106 = \$1,301,000
365 × 14,000 × 5.6 × 0.0886 = 2,535,000
365 × 1,000 × 0.5 × 0.0975 = 18,000
Total \$3,854,000

Alternate 3: 365 × 5,700 × 6.2 × \$0.1106 = \$1,426,000
365 × 13,500 × 5.0 × 0.0886 = 2,183,000
Total \$3,609,000

TABLE 6-27. HIGHWAY COST ESTIMATES

ALTERNATE	*ESTIMATE LIFE*	*COST*	*K*	*INTEREST AND AMORTIZATION*
2-B ROW	60	\$ 572,000	0.0619	\$ 35,400
2-B Gr. and Drainage	40	1,474,000	0.0665	98,000
2-B Major Structures	40	726,000	0.0665	48,300
2-B Pavement	20	979,000	0.0872	85,400
2-D ROW	60	90,000	0.0619	5,600
2-D Gr. and Drainage	40	110,000	0.0665	7,300
2-D Major Structures	40	24,000	0.0665	1,600
2-D Pavement	20	88,000	0.0872	7,700
		Total Annual Cost =		\$289,300
3-C ROW	60	407,000	0.0619	25,200
3-C Gr. and Drainage	40	1,914,000	0.0665	127,300
3-C Major Structures	40	957,000	0.0665	63,600
3-C Pavement	20	946,000	0.0872	82,500
		Total Annual Cost =		\$298,600

(Source: *Road User Benefit Analyses for Highway Improvements*, AASHO, 1960, Example 4)

One truck is assumed to be equal to three passenger cars. Speeds are estimated from observations on similar facilities. Types of operation are computed from capacity data. Excessive curveture forms a very small portion of the total lengths, and is ignored. Because of the relatively narrow pavement and limited passing sight distance, unit costs on the existing three-lane highway are considered equivalent to that on two-lane highways. Unit road user costs are obtained from Tables 6-3 and 6-4 for the new facilities and by interpolation from Tables 6-4 and 6-5 for the old facilities. Complete road user unit cost data are shown in Table 6-26. The highway cost estimates shown is Table 6-27 were made from preliminary plans. An interest rate of 6 per cent was found applicable.

The annual maintenance cost for the existing road is estimated at \$3,850 per mile. This includes an amount for nominal widening to accommodate the estimated traffic movements. The maintenance on the new mileage of alternates 2 and 3 is estimated at \$1,320 per mile, and on the existing road at \$660 per mile, after B or C is built. The total annual highway cost becomes:

Existing road: (6.2 × \$3850) = \$23,900

Alternate 2: \$289,300 + (5.6 + 0.5) × 1320 + (6.2 × 660) = \$301,400

Alternate 3: \$298,600 + (5.0 × 1320) + (6.2 × 660) = \$309,300

1. Benefit-cost ratio method

The benefit ratios for alternates 2 and 3 are computed using the existing route (alternate 1) as the comparison base:

$$\text{Alternate 2:} \quad \frac{\$5{,}179{,}000 - 3{,}854{,}000}{\$\ 301{,}400 - \quad 23{,}900} = 4.77$$

$$\text{Alternate 3:} \quad \frac{\$5{,}179{,}000 - 3{,}609{,}000}{\$\ 309{,}300 - \quad 23{,}900} = 5.50$$

It was desired to know to what extent these expressions of desirability were due to the specific items making up the total operating costs. Unit costs were selected for the previously assigned speeds and type of operation from Tables 6-3 and 6-4 column *8* for subtotal operating cost; column *9* for time; and column *10* for comfort and convenience. Road user unit cost data are as given in Table 6-28.

The data in Table 6-28, extended by the traffic volume, road lengths, and days of the year give the annual road user cost data as shown in Table 6-29.

TABLE 6-28. ROAD USER COST DATA

ALTERNATE	SUBTOTAL OPERATING COST	TIME	COMFORT AND CONVENIENCE	TOTAL
1-A	5.50	5.54	0.88	11.92
2-A	5.60	4.84	0.62	11.06
2-B	5.34	3.52	0.00	8.86
2-D	4.91	4.84	0.00	9.75
3-A	5.60	4.84	0.62	11.6
3-C	5.34	3.52	0.00	8.86

(Source: *Road User Benefit Analyses for Highway Improvements*, AASHO, 1960, Example 4)

TABLE 6-29. ANNUAL ROAD USER COST DATA

ALTERNATE		SUBTOTAL OPERATING COST	TIME	COMFORT AND CONVENIENCE	TOTAL
1-A		$2,390,000	$2,407,000	$382,000	$5,179,000
2-A		$ 549,000	$ 569,000	73,000	1,301,000
2-B		1,528,000	1,007,000	0	2,535,000
2-D		9,000	9,000	0	18,000
	Total	$2,196,000	$1,585,000	$ 73,000	$3,854,000
3-A		$ 722,000	$ 624,000	$ 80,000	$1,426,000
3-C		1,316,000	867,000	0	2,183,000
	Total	$2,038,000	$1,491,000	$ 80,000	$3,609,000

(Source: *Road User Benefit Analyses for Highway Improvements*, AASHO, 1960, Example 4)

Use of the road user costs shown in Table 6-29 for each component of alternate 2 or 3 with the cost of the same component of alternate 1, determines the benefit attributable to that component, which is the numerator in the benefit-ratio. The highway costs (denominator) are the same as before, and the benefit ratios are as follows:

	BENEFIT RATIOS			
	Subtotal Operating Costs	Time	Comfort and Convenience	TOTAL
Alternate 2	0.70	2.96	1.11	4.77
Alternate 3	1.23	3.21	1.06	5.50

Based on the benefit-cost ratio method, either alternative represents an improvement over alternate 1, with alternate 3 apparently offering the best solution. However, to conclusively determine whether or not alternate 3 is the best choice, it is advisable to calculate a second benefit ratio. This is evaluated by eliminating the original basic condition, and using the alternate with the higher road user costs as the new basic condition. In this case, alternate 2 becomes the new basic condition.

$$\text{Second benefit ratio} = \frac{3{,}854{,}000 - 3{,}609{,}000}{309{,}300 - 301{,}400} = 31$$

Therefore, alternate 3 is definitely established as the best solution, since the benefits are greater than the costs.

2. Annual cost method

Cost Item	SUMMARY OF ANNUAL COSTS Alternate 1	Alternate 2	Alternate 3
Road user costs	$5,179,000	$3,854,000	$3,609,000
Maintenance	23,900	12,100	10,700
Capital recovery, i = 6%	0	289,300	298,600
Total Annual Costs	$5,202,900	$4,155,400	$3,918,300

Based on the annual cost method, alternate 3 is more desirable because it results in the lowest annual cost.

3. Rate of return method
   a. Comparing alternates 1 and 2.

Annual Cost of Capital Recovery	$i = 30\%$	$i = 40\%$
Pavement ($n = 20$ years)		
$1{,}067{,}000 \times 0.3016 =$	$ 321,800	
C & S ($n = 40$ years)		
$2{,}334{,}000 \times 0.3000 =$	700,200	
ROW ($n = 60$ years)		
$662{,}000 \times 0.300 =$	198,600	
Totals $4,063,000	$1,220,600 × 0.400 =	$1,625,200
Other annual costs	3,866,100	3,866,100
Total annual costs, alternate 2	$5,086,700	$5,491,300
Total annual costs, alternate 1	5,202,900	5,202,900

The rate of return lies between 30 and 40 per cent. At 30 per cent interest, alternate 1 costs more annually than alternate 2 by $116,200, but at 40 per cent interest alternate 2 costs more by $288,400. Therefore the rate of return is

$$i = 30 + 10 \times \frac{116{,}200}{116{,}200 + 288{,}400} = 32.9 \text{ per cent}$$

Alternate 2 is better than alternate 1 if the minimum attractive rate of return is 32.9 per cent or less.

   b. Comparing alternates 1 and 3

By following the identical procedure outlined above, the interest rate at which Alternates 1 and 3 have the same total annual costs is 37.5 per cent. Therefore, alternate 3 is better than alternate 1 if the minimum attractive rate of return is 37.5 per cent or less.

   c. Comparing alternates 2 and 3

Using the same procedure, the interest rate at which alternates 2 and 3 have the same total annual costs can be found. However, for this particular comparison, the solution need not be carried out, for the difference in annual costs (exclusive of capital recovery) between alternate 2 and alternate 3 is

$3,866,100 − $3,619,700 = $246,400. At 100 per cent interest (CRF = 1.0), the difference in capital recovery charges between alternates 2 and 3 is $4,224,000 − $4,063,000 = $161,000. Even at 100 per cent interest the annual cost of alternate 3 is less than that of alternate 2, and therefore it is obvious that alternate 3 represents the best solution.

4. Present worth method

SUMMARY OF PRESENT WORTHS (Analysis Period = 20 yrs.)

Cost Item	*Alternate 1*	*Alternate 2*	*Alternate 3*
Annual RUC	$ 5,179,000	$ 3,854,000	$ 3,609,000
Annual Maintenance	23,900	12,100	10,700
Total	$ 5,202,900	$ 3,866,100	$ 3,619,700
PW at i = 6%, n = 20	$59,677,260	$44,344,170	$41,517,960
Capital Investment	0	4,063,000	4,224,000
Total Present Worth	$59,677,260	$48,407,170	$45,741,960

Based on the present worth method, alternate 3 represents the best solution.

Comparison of Multiple Alternates[73,138]

A simplified hypothetical problem is employed to demonstrate the correct procedure for making an economic comparison among several alternates. The procedure involves an incremental analysis which was actually used in the previous example.

Example

Assume that an existing section of highway is inadequate. Several proposed designs at new locations, and proposed improvements of the existing facility, are all to be compared with a continuation of the present conditions on the existing section of highway. The existing facility if designated as alternate 1, and the several proposed improvements are designated as alternates 2 though 8.

To simplify the computations, it is assumed that the useful service lives for all components of the capital investment, required for each alternate, are the same. The useful life of the capital investment is assumed to be 30 years, and road user costs and benefits are assumed to be uniform through the 30-year period. The analysis is initially carried out by using a minimum attractive rate of return equal to 7 per cent.

1. Annual cost method

Table 6-30 contains the capital investment, estimated annual cost of maintenance and operation, and annual road user costs for the various alternates.

Table 6-31 contains a summary of annual costs for each alternate. The uniform cost of capital recovery is obtained by multiplying the capital investment by the CRF for $n = 30$ years and $i = 7$ per cent (CRF = 0.0806). All costs are shown in $1,000 units.

Based on a comparison of annual costs, alternate 7 is preferred over the other alternates, and all alternates for improvement result in a saving in annual costs over the existing facility.

2. Benefit-cost method

Table 6-32 contains the solutions for benefit-cost ratios for each of the proposed alternates, all compared with the existing facility as the basic condition. The values shown in column 2 of Table 6-32 are obtained by subtracting the road user costs for each alternate from the road user costs for alternate 1, which are shown in

TABLE 6-30. SUMMARY OF COSTS

ALTERNATE (1)	*CAPITAL INVESTMENT* ($1,000) (2)	*ANNUAL MAINTENANCE AND OPERATION* ($1,000) (3)	*ANNUAL ROAD USER COSTS* ($1,000) (4)
1	0	80	3,000
2	1,500	70	2,850
3	2,000	60	2,700
4	2,500	50	2,400
5	3,000	55	2,000
6	2,200	60	2,500
7	4,000	40	1,800
8	8,000	50	1,550

TABLE 6-31. SUMMARY OF ANNUAL COSTS

ALTERNATE (1)	*CAPITAL RECOVERY* (2)	*ANNUAL MAINTENANCE AND OPERATION* (3)	*HIGHWAY COSTS* (4)	*ROAD USER COSTS* (5)	*TOTAL ANNUAL COSTS* (6)
1	0	80	80	3,000	3,080.0
2	120.9	70	190.9	2,850	3,040.9
3	161.2	60	221.2	2,700	2,921.2
4	201.5	50	251.5	2,400	2,651.5
5	241.8	55	296.8	2,000	2,296.8
6	177.3	60	237.3	2,500	2,737.3
7	322.4	40	362.4	1,800	2,162.4*
8	644.8	50	694.8	1,550	2,244.8

* Alternate 7 is the lowest, and is therefore the preferred choice

TABLE 6-32. BENEFIT-COST RATIOS WITH ALTERNATE 1 AS THE BASIC CONDITION

ALTERNATE (1)	*ROAD USER BENEFITS OVER ALTERNATE 1* (2)	*EXTRA HIGHWAY COSTS OVER ALTERNATE 1* (3)	*BENEFIT-COST RATIO COLUMN 2/COLUMN 3* (4)
2	150	110.9	1.35
3	300	141.2	2.12
4	600	171.5	3.49
5	1,000	216.8	4.60
6	500	157.3	3.17
7	1,200	282.4	4.25
8	1,450	614.8	2.36

columns of Table 6-31. The values shown in column 3 of Table 6-32 are obtained by subtracting the highway costs for alternate 1 from those of each of the alternates, which are shown in column 4 of Table 6-31.

Based on a comparison of the benefit-cost ratios, it might appear that alternate 5 is to be preferred because it has the highest benefit-cost ratio. However, this is not necessarily true, because the benefit-cost ratios are obtained by comparing all alternates with alternate ones, and before a definite conclusion can be made it is necessary to compare all alternates with each other.

These comparisons are made by an incremental analysis, which involves the computation of ratios of increments of benefits to increments of costs. The computations are carried out in Table 6-33. The criterion used to carry out the analysis is that no increment of cost is justified unless it produces an increment of benefits at least equal to the increment of costs. All alternates have been compared with alternate 1 in Table 6-32, and alternate 5 resulted in the greatest increment of benefits to costs. Next, a comparison is made between alternates 2 and 5, using alternate 2 as the basic condition. The comparison favors alternate 5, because $850,000 of additional benefit is obtained through an additional cost of only $105,900; the former figure is obtained by subtracting the road user benefits for alternate 2 from those of alternate 5, which are shown in column 2 of Table 6-32, and the latter figure is obtained by subtracting the highway costs for alternate 2 from those of alternate 5 shown in column 3. Since the incremental benefit-cost ratio of 8.01 exceeds unity, Alternate 2 is eliminated from further comparisons.

Alternates 3, 4 and 6 are not as attractive when compared with alternate 5; therefore alternates 3, 4, and 6 need not be considered in further comparisons. A comparison of alternates 7 and 5 result in an incremental benefit-cost ratio of 3.05 in favor of alternate 7, thereby eliminating alternate 5 from further consideration.

Alternate 8 is not as attractive as alternate 7 because the incremental benefit-cost ratio as compared to alternate 7, is less than unity. Therefore, Table 6-33 leads to the selection of alternate 7 as the proposed improvement preferred over all others. This is true because, in comparing alternate 7 with any alternate having lower highway costs, the additional increment of benefits to be derived from alternate 7 is greater than the additional costs incurred. For all alternates having greater highway costs than alternate 7, the additional increments of benefits (as compared to alternate 7) are less than the additional costs incurred.

The solution based on a comparison of annual costs, in Table 6-31, also resulted in alternate 7 as the most desirable choice.

3. Rate of return method.

Table 6-34 contains the computations of prospective rates of return on investment for the various alternates, all of which are compared with the existing facility as the basic condition. A typical computation for the rate of return is shown below, when alternate 2 is compared with alternate 1:

AC of Alternate $1 = 3{,}000{,}000 + 80{,}000 =$ \$3,080,000; these values come from Table 6-30. Capital investment for alternate $1 = 0$.

AC of alternate 2, exclusive of capital recovery
$= 2{,}850{,}000 + 70{,}000$
$= 2{,}920{,}000$

Capital investment for alternate $2 = \$1{,}500{,}000$

The rate of return that makes both alternates equally attractive is obtained using the following relationship.

$$\text{CRF} \times 1{,}500 + 2{,}920 = 3{,}080$$

$$\text{CRF} \times 1{,}500 = 160 = \text{Reduction in AC, exclusive of capital recovery}$$

$$\text{CRF} = \frac{160}{1{,}500} = 0.10606$$

The interest rate, expressed to the nearest tenth of a per cent, corresponding to a CRF $= 0.10606$ for $n = 30$ years is 10.0 per cent, which is obtained by interpolation in a table of capital recovery factors. The rate of return will almost be as great as the CRF, because of the relatively long assumed service life.

The results shown in Table 6-34 can lead to the same erroneous conclusions as those associated with Table 6-32, which contains the benefit-cost ratios for all alternates compared with alternate 1. Alternate 5 again appears to be the most desirable solution, but before a definite choice can be made it is necessary to compare all alternates with each other. Actually, none of the alternates can be eliminated with the exception of alternate 1, because the rate of return on invested capital for each alternate compared with alternate 1 exceeds 7 per cent, which is assumed to be the minimum attractive rate of return.

Table 6-35 contains the computations for determining incremental rates of return, which are used to

TABLE 6-33. INCREMENTAL BENEFIT-COST RATIOS

ALTERNATES COMPARED	*INCREMENTS OF ROAD USER BENEFITS ($)*	*INCREMENTS OF HIGHWAY COSTS ($)*	*INCREMENTAL BENEFIT-COST RATIO*	*DECISION IN FAVOR OF*
Alternate 5 over alternate 1	1,000	216.8	4.60	Alternate 5
Alternate 5 over alternate 2	850	105.9	8.01	Alternate 5
Alternate 5 over alternate 3	700	75.6	9.26	Alternate 5
Alternate 5 over alternate 4	400	45.3	8.81	Alternate 5
Alternate 5 over alternate 6	500	59.5	8.40	Alternate 5
Alternate 7 over alternate 5	200	65.6	3.05	Alternate 7
Alternate 8 over alternate 7	250	332.4	0.75	Alternate 7

TABLE 6-34. RATES OF RETURN CONSIDERING ALTERNATE 1 AS THE BASIC CONDITION

ALTERNATE (1)	*REDUCTION IN AC EXCLUSIVE OF CR* (2)	*CAPITAL INVESTMENT* (3)	*CRF FOR n = 30 YEARS COLUMN 2/COLUMN 3* (4)	*RATE OF RETURN AS COMPARED TO ALTERNATE 1 (%)* (5)
2	160	1,500	0.10606	10.0
3	320	2,000	0.16000	15.8
4	630	2,500	0.25200	25.2
5	1,025	3,000	0.34167	34.2
6	520	2,200	0.23636	23.6
7	1,240	4,000	0.31000	31.0
8	1,480	8,000	0.18500	18.4

TABLE 6-35. INCREMENTAL ANALYSIS FOR RATES OF RETURN

ALTERNATES COMPARED (1)	*INCREMENT OF REDUCTION IN AC EXCLUSIVE OF CR* (2)	*INCREMENT OF CAPITAL INVESTMENT* (3)	*CRF FOR n = 30 YEARS COLUMN 2/COLUMN 3* (4)	*RATE OF RETURN ON INCREMENT OF INVESTMENT (%)* (5)	*DECISION IN FAVOR OF* (6)
Alternate 5 over alternate 1	1,025	3,000	0.34167	34.2	Alternate 5
Alternate 5 over alternate 2	865	1,500	0.57667	57.7	Alternate 5
Alternate 5 over alternate 3	705	1,000	0.70500	70.5	Alternate 5
Alternate 5 over alternate 4	395	500	0.79000	79.0	Alternate 5
Alternate 5 over alternate 6	505	800	0.63125	63.1	Alternate 5
Alternate 7 over alternate 5	215	1,000	0.21500	21.4	Alternate 7
Alternate 8 over alternate 7	240	4,000	0.06000	4.3	Alternate 7

TABLE 6-36. SUMMARY OF PRESENT WORTHS

ALTERNATE (1)	*CAPITAL INVESTMENT* (2)	*ANNUAL RUC* (3)	*ANNUAL MAINTENANCE AND OPERATION* (4)	*PRESENT WORTH RUC, MAINTENANCE AND OPERATION [(3) + (4)] × 12.409* (5)	*TOTAL PRESENT WORTH (2) + (5)* (6)
1	0	3,000	80	38,220	38,220
2	1,500	2,850	70	36,230	37,730
3	2,000	2,700	60	34,250	36,250
4	2,500	2,400	50	30,400	32,900
5	3,000	2,000	55	25,500	28,500
6	2,200	2,500	60	31,770	33,970
7	4,000	1,800	40	22,830	26,830*
8	8,000	1,550	50	19,850	27,850

* Alternate 7 is again the preferred alternate

compare alternates with one another. The procedure for carrying out the incremental analysis for rates of return is similar to that for the incremental analysis for benefit-cost ratios.

The incremental analysis leads to the selection of Alternate 7 as the most desirable solution.

4. Present worth method

Table 6-36 contains a summary of present worths for each alternate. An analysis period of 30 years is assumed, and the present worth of annual costs is obtained by multiplying the annual costs by the PWF for a uniform annual series for $n = 30$ years and $i = 7$ per cent (PWF = 12,409). All costs are shown in $1,000 units.

Based on a comparison of present worths, alternate 7 is preferred over the other alternates.

This conclusion coincides with the selection made by a comparison of annual costs (Table 6-31), by the incremental analysis of benefit-cost ratios (Table 6-33) and by the incremental analysis of rates of return (Table 6-35). This clearly demonstrates that economy studies conducted by different methods, properly applied, will lead to the same general conclusions.

5. Next best alternate

If it is desired to determine which alternate represents the next best solution, the same kinds of analyses are made, with alternate 7 eliminated, since it represents the best solution.

a. By Annual cost, Table 6-31, alternate 8 is preferred.

b. By incremental benefit-cost ratio, Table 6-33:

$$\text{Alternate 8 over alternate 5, BCR} = \frac{450}{398} = 1.13.$$

Therefore, alternate 8 is favored.

c. By incremental rate of return, Table 6-35:

$$\text{Alternate 8 over alternate 5, CRF} = \frac{455}{5000} = 0.09100.$$

Rate of return = 8.3 per cent, which is greater than 7 per cent.

Therefore, alternate 8 is favored.

d. By present worth, Table 6-36, alternate 8 is preferred.

There are a number of publications[45,44,143,151-162] which comment on methods of conducting economy studies and on the significance of such studies. There are also many applications of economy studies.[20,48,62,137,141,144,154,163-177] Publications containing computer-oriented techniques of economic analysis of highway system[51,160,178-182] also exist.

NEW SOURCES OF ROAD USER COST DATA

The deficiencies in AASHO's *Road User Benefit Analyses for Highway Improvements* relative to running costs of all types of vehicles in urban areas, and commercial vehicles and buses in rural areas, were previously noted.

TABLE 6-37. AUTOMOBILE FUEL CONSUMPTION AS AFFECTED BY SPEED AND GRADIENT—STRAIGHT HIGH-TYPE PAVEMENT AND FREE-FLOWING TRAFFIC*

UNIFORM SPEED (mph)	*GASOLINE CONSUMPTION (GPM) ON GRADES OF:*										
	LEVEL	*1%*	*2%*	*3%*	*4%*	*5%*	*6%*	*7%*	*8%*	*9%*	*10%*
	(*a*) Plus grades										
10	0.072	0.080	0.087	0.096	0.103	0.112	0.121	0.132	0.143	0.160	0.179
20	0.050	0.058	0.070	0.076	0.086	0.094	0.104	0.116	0.128	0.144	0.160
30	0.044	0.051	0.060	0.068	0.078	0.087	0.096	0.110	0.124	0.138	0.154
40	0.046	0.054	0.062	0.070	0.078	0.087	0.096	0.111	0.124	0.140	0.156
50	0.052	0.059	0.070	0.076	0.083	0.093	0.104	0.118	0.130	0.145	0.162
60	0.058	0.067	0.076	0.084	0.093	0.102	0.112	0.126	0.138	0.152	0.170
70	0.067	0.075	0.084	0.093	0.102	0.111	0.122	0.135	0.148	0.162	0.180
	(*b*) Minus grades										
10	0.072	0.060	0.045	0.040	0.040	0.040	0.040	0.040	0.040	0.040	0.040
20	0.050	0.040	0.027	0.022	0.021	0.021	0.021	0.021	0.021	0.021	0.021
30	0.044	0.033	0.022	0.016	0.014	0.013	0.013	0.013	0.013	0.013	0.013
40	0.046	0.035	0.025	0.018	0.014	0.012	0.012	0.012	0.012	0.012	0.012
50	0.052	0.041	0.030	0.025	0.021	0.018	0.014	0.013	0.010	0.010	0.008
60	0.058	0.048	0.036	0.037	0.030	0.027	0.022	0.018	0.014	0.011	0.008
70	0.067	0.058	0.048	0.043	0.039	0.036	0.031	0.027	0.022	0.016	0.013

(Source: Claffey, *Running Costs of Motor Vehicles as Affected by Road Design and Traffic,* NCHRP Report 111, 1971)

* The composite passenger car represented here reflects the following vehicle distribution: Large cars, 20 per cent; standard cars, 65 per cent; compact cars, 10 per cent; small cars, 5 per cent

The data presented in AASHO is in a form convenient for computation, analysis, and interpretation of road user costs, but it is in need of updating and supplementing. Two subsequent works have been published in the area of road user cost studies. While the form of the data presented and discussed in each does not lend itself to a straightforward updating of AASHO tables, each may be used to modify AASHO data in many cases. Two publications of note are:

1. *Economic Analysis for Highways*, by R. Winfrey[183]
2. *Running Costs of Motor Vehicles as Affected by Road Design and Traffic*, by Claffey, NCHRP Report 111[184]

NCHRP 111 is a result of a series of studies conducted during the years 1963 to 1969. The tables included in Winfrey's text represent a compilation of data from many sources, much of which is taken directly from the interim findings of the NCHRP study.

NCHRP Report Number 111

The tests conducted by Dr. Paul J. Claffey in this study include measurements of gasoline consumption, oil consumption, tire wear, maintenance costs, and depreciation. Studies of gasoline consumption are more detailed than those in other areas. Data is presented for a variety of vehicle types, including composite passenger car, pickup truck, single-unit truck (two-axle, six-tire), combination vehicles.

Gasoline Consumption. Table 6-37 shows the basic data for fuel consumption for passenger cars. Data is given for free-flowing traffic and is detailed for plus and minus grades separately. Gasoline consumption is given in gallons per mile. Table 6-38 gives correction factors to be applied to the data in Table 6-37 for curvature. The factors are valid only for grades of 3 per cent or less. Table 6-39 gives correction factors to be applied to the basic data for road surface type. These factors are also valid only for grades up to 3 per cent.

Correction factors were developed for various volume levels on a six-lane freeway and a six-lane arterial (base Table 6-37 is for free-flowing traffic). The correction factors are not as valuable as first appears for several major reasons. The combinations of speed and volume which are reported are not consistent with those expected for the various levels of service in the Highway Capacity Manual. On freeways, for example, test runs were made on sections free of side friction (ramps, weaving), a condition not consistent with normal situations. Unreasonable high speeds for high-volume circumstances may have occurred as a result. Another difficulty in the use of such corrections is the specification of "attempted speed" instead of a more definitive measure as a determinant. For these reasons, the volume-factor tables do not lend themselves to development of correction factors for various levels of service, as defined in the Highway Capacity Manual, or other volume-service measure, as would have been desirable.

TABLE 6-38. CORRECTION FACTORS TO ADJUST THE VALUES OF TABLE 6-37 FOR CURVATURE

DEGREE OF CURVE	*CORRECTION FACTORS BY UNIFORM SPEED OF AUTOMOBILES (MPH):* 10	20	30	40	50	60	70
0	1.000	1.000	1.000	1.000	1.000	1.000	1.000
1	1.000	1.001	1.005	1.025	1.025	1.040	1.060
2	1.001	1.002	1.010	1.031	1.054	1.080	1.120
3	1.002	1.003	1.016	1.048	1.090	1.132	1.182
4	1.002	1.004	1.022	1.065	1.120	1.200	1.300
5	1.003	1.005	1.028	1.082	1.180	1.300	—
6	1.004	1.006	1.034	1.120	1.250	1.500	—
7	1.005	1.007	1.040	1.170	1.430	1.900	—
8	1.005	1.008	1.080	1.230	1.610	—	—
9	1.006	1.010	1.140	1.340	1.820	—	—
10	1.008	1.030	1.200	1.480	2.070	—	—
11	1.010	1.070	1.280	1.620	2.200	—	—
12	1.020	1.110	1.360	1.800	2.500	—	—
90	1.130	2.000	—	—	—	—	—

(Source: Claffey, *Running Costs of Motor Vehicles as Affected by Road Design and Traffic,* NCHRP Report 111, 1971)

TABLE 6-39. CORRECTION FACTORS TO ADJUST THE VALUES OF TABLE 6-37 FOR ROUGH SURFACE

UNIFORM SPEED OF AUTOMOBILES (MPH)	*CORRECTION FACTORS BY ROAD SURFACE:* High-Type Concrete or Asphalt	Badly Broken and Patched Asphalt	Dry Well-Packed Gravel	Loose Sand
10	1.00	1.01	1.09	1.23
20	1.00	1.05	1.13	1.28
30	1.00	1.20	1.26	1.40
40	1.00	1.34	1.56	1.73
50	1.00	1.50	1.70	2.00

(Source: Claffey, *Running Costs of Motor Vehicles as Affected by Road Design and Traffic,* NCHRP Report 111, 1971)

Tables 6-40 and 6-41 give excess gallons of fuel consumed by passenger cars during a stop-go or speed change cycle. These tables do not include the cost of idling while stopped. Tables 6-42 through 6-44 give similar information for a single-unit truck. Tables are included in the study for other vehicle types, as well as those shown here. Fuel consumption while idling is given in Table 6-47 for all types of vehicles.

TABLE 6-40. EXCESS GALLONS OF GASOLINE CONSUMED PER STOP-GO SPEED CHANGE CYCLE—AUTOMOBILE

SPEED (MPH)	*EXCESS GASOLINE CONSUMED (GAL) BY DURATION OF STOPPED DELAY (SEC):* 0	30	60	90	120	150	180
10	0.0016	0.0021	0.0026	0.0031	0.0035	0.0040	0.0045
20	0.0066	0.0071	0.0076	0.0081	0.0085	0.0090	0.0095
30	0.0097	0.0102	0.0107	0.0112	0.0116	0.0121	0.0126
40	0.0128	0.0133	0.0138	0.0143	0.0147	0.0152	0.0157
50	0.0168	0.0173	0.0178	0.0183	0.0187	0.0192	0.0197
60	0.0208	0.0213	0.0218	0.0223	0.0228	0.0233	0.0238
70	0.0243	0.0248	0.0253	0.0258	0.0263	0.0268	0.0273

(Source: Claffey, *Running Costs of Motor Vehicles as Affected by Road Design and Traffic,* NCHRP Report 111, 1971)

TABLE 6-41. EXCESS GALLONS OF GASOLINE CONSUMED PER SLOWDOWN SPEED CHANGE CYCLE—AUTOMOBILE

SPEED (MPH)	*EXCESS GASOLINE CONSUMED (GAL) BY AMOUNT OF SPEED REDUCTION BEFORE ACCELERATING BACK TO SPEED (MPH):* 10	20	30	40	50	60
20	0.0032	—	—	—	—	—
30	0.0035	0.0062	—	—	—	—
40	0.0038	0.0068	0.0093	—	—	—
50	0.0042	0.0074	0.0106	0.0140	—	—
60	0.0046	0.0082	0.0120	0.0155	0.0190	—
70	0.0051	0.0090	0.0130	0.0167	0.0203	0.0243

(Source: Claffey, *Running Costs of Motor Vehicles as Affected by Road Design and Traffic,* NCHRP Report 111, 1971)

TABLE 6-43. CORRECTION FACTORS TO ADJUST THE VALUES OF TABLE 6-42 FOR CURVATURE*

DEGREE OF CURVE	*CORRECTION FACTORS BY SPEED OF TWO-AXLE SIX-TIRE TRUCKS (MPH):* 10	20	30	40	50	60	70
0	1.000	1.000	1.000	1.000	1.000	1.000	1.000
1	1.000	1.000	1.001	1.002	1.010	1.020	1.050
2	1.000	1.001	1.004	1.006	1.020	1.050	1.100
3	1.000	1.002	1.005	1.009	1.060	1.100	—
4	1.000	1.005	1.010	1.040	1.130	1.150	—
5	1.000	1.020	1.030	1.090	1.230	—	—
6	1.000	1.030	1.050	1.140	1.330	—	—
7	1.000	1.040	1.090	1.200	1.430	—	—
8	1.000	1.050	1.130	1.260	1.530	—	—
9	1.000	1.060	1.170	1.320	—	—	—
10	1.000	1.080	1.210	1.430	—	—	—
11	1.000	1.090	1.250	1.550	—	—	—
12	1.000	1.100	1.300	1.690	—	—	—
30	1.000	1.180	2.000	—	—	—	—

(Source: Claffey, *Running Costs of Motor Vehicles as Affected by Road Design and Traffic,* NCHRP Report 111, 1971)

* Correction factors determined for the vehicle distribution given in Footnote 1, Table 6-42. They apply for grades up to 1 per cent.

TABLE 6-42. TWO-AXLE SIX-TIRE TRUCK FUEL CONSUMPTION AS AFFECTED BY SPEED AND GRADIENT—STRAIGHT HIGH-TYPE PAVEMENT AND FREE-FLOWING TRAFFIC*

UNIFORM SPEED (mph)	*GASOLINE CONSUMPTION (GPM) ON GRADES OF:* LEVEL	1%	2%	3%	4%	5%	6%	7%	8%	9%	10%
					(a) Plus grades**						
10	0.074	0.094	0.120	0.143	0.175	0.195	0.225	0.255	0.289	0.324	0.357
20	0.059	0.080	0.112	0.140	0.167	0.190	0.214	0.254	0.295	0.344	0.394
30	0.067	0.094	0.121	0.150	0.181	0.206	0.232	0.268	0.305	—	—
40	0.082	0.112	0.141	0.173	0.210	0.228	—	—	—	—	—
50	0.101	0.130	0.159	0.194	—	—	—	—	—	—	—
60	0.122	0.150	—	—	—	—	—	—	—	—	—
					(b) Minus grades						
10	0.074	0.064	0.055	0.053	0.051	0.051	0.051	0.051	0.051	0.051	0.051
20	0.059	0.049	0.039	0.034	0.030	0.030	0.030	0.030	0.030	0.030	0.030
30	0.067	0.054	0.041	0.034	0.027	0.026	0.025	0.025	0.024	0.024	0.024
40	0.082	0.071	0.051	0.041	0.032	0.029	0.025	0.023	0.021	0.020	0.020
50	0.101	0.090	0.072	0.058	0.045	0.038	0.031	0.025	0.020	0.020	0.020
60	0.122	0.110	0.090	0.075	0.062	0.052	0.043	0.035	0.025	0.020	0.020

(Source: Claffey, *Running Costs of Motor Vehicles as Affected by Road Design and Traffic,* NCHRP Report 111, 1971)

* The composite two-axle six-tire truck represented here reflects the following vehicle distribution:

Two-axle trucks at 8,000 lb G.V.W. 50 per cent

Two-axle trucks at 16,000 lb G.V.W. 50 per cent

** Operation is in the highest gear possible for the grade and speed (No. 4, No. 3, or No. 2). When vehicle approach speed exceeds the maximum sustainable speed on plus grades, speed is reduced to this maximum as soon as the vehicle gets on the grade.

TABLE 6-45. EXCESS GALLONS OF GASOLINE CONSUMED PER STOP-GO SPEED CHANGE CYCLE—TWO-AXLE SIX-TIRE TRUCK

SPEED (MPH)	EXCESS GASOLINE CONSUMED (GAL) BY DURATION OF STOPPED DELAY (SEC): 0	30	60	90	120	150	180
10	0.0036	0.0090	0.0144	0.0198	0.0252	0.0306	0.0360
20	0.0097	0.0151	0.0205	0.0259	0.0313	0.0367	0.0421
30	0.0173	0.0227	0.0281	0.0335	0.0389	0.0443	0.0497
40	0.0242	0.0296	0.0350	0.0404	0.0458	0.0512	0.0566
50	0.0270	0.0326	0.0380	0.0434	0.0488	0.0542	0.0596

(Source: Claffey, *Running Costs of Motor Vehicles as Affected by Road Design and Traffic,* NCHRP Report 111, 1971)

TABLE 6-44. CORRECTION FACTORS TO ADJUST THE VALUES OF TABLE 6-42 FOR ROUGH SURFACE*

UNIFORM SPEED OF TWO-AXLE SIX-TIRE TRUCKS (MPH)	CORRECTION FACTORS BY SURFACE: High-Type Concrete or Asphalt	Badly Broken and Patched Asphalt	Well-Packed Gravel	Loose Sand
10	1.00	1 03	1.24	1.46
20	1.00	1.06	1.28	1.62
30	1.00	1.07	1.45	2.16
40	1.00	1.08	1.58	2.46
50	1.00	1.20	1.69	—

(Source: Claffey, *Running Costs of Motor Vehicles as Affected by Road Design and Traffic,* NCHRP Report 111, 1971)

* Correction factors determined for truck distribution of Footnote 1, Table 6-42. They apply on grades up to 1 per cent.

TABLE 6-46. EXCESS GALLONS OF GASOLINE CONSUMED PER SLOWDOWN SPEED CHANGE CYCLE—TWO-AXLE SIX-TIRE TRUCK

SPEED (MPH)	EXCESS GASOLINE CONSUMED (GAL) BY AMOUNT OF SPEED REDUCTION BEFORE ACCELERATING BACK TO SPEED (MPH): 10	20	30	40
20	0.0073	—	—	—
30	0.0080	0.0148	—	—
40	0.0096	0.0167	0.0226	—
50	0.0110	0.0168	0.0226	0.0266

(Source: Claffey, *Running Costs of Motor Vehicles as Affected by Road Design and Traffic,* NCHRP Report 111, 1971)

TABLE 6-47. IDLING FUEL CONSUMPTION RATES WITH VEHICLE STATIONARY

VEHICLE	CYLINDERS No.	Displ. (cu in.)	TRANSMISSION Type	Position	ENGINE SPEED (rpm)	FUEL CONSUMED (gph)
(a) Automobiles						
Composite	—	—	—	Neutral	—	0.63
				Drive	—	0.58
Chrysler	8	440	Auto.	Neutral	580	0.67
				Drive	530	0.61
Chevrolet	8	283	Auto.	Neutral	650	0.65
				Drive	500	0.61
Falcon	6	200	Auto.	Neutral	780	0.57
				Drive	630	0.52
Volkswagen	4	72	Man.	Engaged	550	0.34
(b) Trucks—gasoline						
Pickup	6	250	Man.	Engaged	500	0.45
Two-axle six-tire	6	351	Man.	Engaged	600	0.65
2-S2 combination	6	386	Man.	Enagged	600	0.80
2-S2 combination	6	503	Man.	Engaged	650	0.79
3-S2 combination	6	501	Man.	Engaged	650	0.89
(c) Trucks—diesel						
Two-axle six-tire	6	477	Man.	Engaged	—	0.38
2-S2 combination	6	672	Man.	Engaged	600	0.45
(d) Bus—gasoline						
Transit	6	260	Auto.	Drive	—	0.55
Transit	6	260	Auto.	Neutral	—	0.73

(Source: Claffey, *Running Costs Motor Vehicles as Affected by Road Design and Traffic,* NCHRP Report 111, 1971)

TABLE 6-48. AUTOMOBILE TIRE COST AS AFFECTED BY SPEED AND TYPE OF SURFACE—STRAIGHT ROAD AND FREE-FLOWING TRAFFIC*

UNIFORM SPEED (MPH)	*COST FOUR TIRES (CENTS/MI)***		
	High-Type Concrete	*High-Type Asphalt*	*Dry Well-Packed Gravel*
20	0.09	0.27	1.03
30	0.19	0.36	1.05
40	0.29	0.43	1.07
50	0.32	0.45	1.10
60	0.31	0.46	—
70	0.30	0.44	—
80	0.27	0.43	—

(Source: Claffey, *Running Costs of Motor Vehicles as Affected by Road Design and Traffic,* NCHRP Report 111, 1971)

* The composite passenger car represented here reflects the following vehicle distribution: large cars, 20 per cent; standard-size cars, 65 per cent; compact cars, 10 per cent; and small cars, 5 per cent.

** Tire costs were computed using a weighted average cost of $119 for a set of four new medium-quality tires based on the following unit tire costs by vehicle type (as noted in the northeastern states in 1969): Large cars, $35 per tire; standard-size cars, $30 per tire; compact cars, $25 per tire; and small cars, $15 per tire.

There are approximately 1,500 gm of usable tire tread in 80 per cent of passenger car tires. This weight of usable tire tread was also recorded for the tires used in the tire wear test.

TABLE 6-50. EXCESS TIRE COST PER SPEED CHANGE CYCLE—AUTOMOBILE

SPEED (MPH)	*COST OF FOUR TIRES (CENTS/CYCLE)*			
	Stop-Go Speed Change Cycles		*10-MPH Slowdown Cycles*	
	Concrete	*Asphalt*	*Concrete*	*Asphalt*
20	0.10	0.30	0.04	0.10
30	0.30	0.60	0.08	0.15
40	0.58	0.85	0.09	0.14
50	0.72	1.10	0.09	0.14
60	0.80	1.20	0.08	0.12
70	0.85	1.25	0.08	0.12

(Source: Claffey, *Running Costs of Motor Vehicles as Affected by Road Design and Traffic,* NCHRP Report 111, 1971)

TABLE 6-49. CORRECTION FACTORS TO ADJUST THE VALUES OF TABLE 6-48 FOR CURVATURE*,**

DEGREE OF CURVE	*CORRECTION FACTORS BY UNIFORM SPEED OF AUTOMOBILES (MPH):*					
	20	*30*	*40*	*50*	*60*	*70*
0	1.00	1.00	1.00	1.00	1.00	1.00
2	1.02	1.53	3.06	5.27	9.21	15.70
4	1.10	2.00	6.11	10.67	19.05	29.58
6	1.30	2.56	8.88	17.11	30.28	—
8	1.60	3.33	12.50	28.40	—	—
10	1.90	4.33	16.66	44.80	—	—
12	2.10	5.33	20.44	89.40	—	—
14	4.00	8.50	—	—	—	—
16	6.10	12.70	—	—	—	—
30	10.83	—	—	—	—	—

(Source: Claffey, *Running Costs of Motor Vehicles as Affected by Road Design and Traffic,* NCHRP Report 111, 1971)

* Correction factors apply on concrete, asphalt, and gravel surfaces

** Test operations were also carried out on a 90° curve at a speed of 20 mph with stops at 1-mi intervals. Tire wear was found to be approximately 1,000 times that on tangent.

Oil Consumption. Tabulations given for oil consumption are not in a form convenient for use. Consumption is given in terms of "oil consumed between oil changes," and does not include the cost or amounts of oil utilized in such periodic changes.

Tire Wear. Tire cost is shown in Table 6-48 for automobiles. Table 6-49 shows correction factors for curvature, and Table 6-50 shows excess tire cost per speed change cycle. Note that tire cost is not varied by grade. The report recognizes that grade is a significant factor, but adds that the measurement technique precluded accurate determination of the effect of grade. For these reasons the reported costs should be used with some degree of caution.

Depreciation. The depreciation data included in NCHRP 111 is somewhat sketchy. The report notes that its results are inconclusive and not sufficient for widespread use.

Maintenance. Maintenance costs were tabulated from secondary sources for various maintenance elements and total per mile cost. No attempt was made to vary the cost by traffic or roadway conditions. Table 6-51 shows the maintenance cost breakdown.

Other Elements. The NCHRP report does not concern itself with the elements of time cost and comfort and convenience costs, which make up a major portion of the AASHO tabulations. Should NCHRP cost tables be used, these elements must be added to determine total road user cost.

Winfrey Tables

The tables included in Robley Winfrey's text are the most finely stratified cost tables thus far in print. Unfortunately, the data shown are all from secondary sources (much of it from the above-mentioned NCHRP 111 interim findings). The precision breakdown of costs by $2\frac{1}{2}$ miles per hour speed ranges, grades, etc., was accomplished through much "judgment allocation," utilized to fill the many gaps in existing data. While these tables are useful because of their finely stratified format, they should be used with caution, since much of the data has been generated on the basis of the author's judgment. For example, Winfrey varies maintenance cost with speed, a variation entirely unsupported by data, and created entirely by "judgment allocation." These tables do not reflect the effect of volume on running costs.

TABLE 6-51. AVERAGE MAINTENANCE COST AS AFFECTED BY TRAVEL DISTANCE—STANDARD-SIZE PASSENGER CARS*

VEHICLE PART	*NO. OF VEHICLES FOR WHICH DATA WERE AVAILABLE*	*TRAVEL DISTANCE BEFORE REPAIRS ARE NECESSARY (1,000 MILES)*		*COST OF REPAIRS ($)*		*AVG. COST (CENTS/MILE)*
		RANGE	*AVG.*	*RANGE*	*AVG.*	
Automatic transmission	850	54–103	66	100–225	178	0.27
Engine block	395	50–90	70	65–130	93	0.13
Shock absorbers	1350	30–60	44	28–51	37	0.08
Brake system	1350	40–77	54	40–58	41	0.08
Distributor	1050	10–21	14	5–20	12	0.08
Exhaust	1350	30–56	39	18–33	26	0.07
Carburetor	1050	32–60	45	21–40	29	0.06
Universal	1050	30–54	44	20–31	28	0.06
Rear axle	50	100–113	106	54–75	66	0.06
Generator	1050	42–60	52	18–40	32	0.06
Water pump	1050	34–55	43	18–30	24	0.06
Springs	50	40–100	68	28–46	40	0.06
Fuel pump	50	44–64	52	12–20	15	0.03
Oil pump	50	92–138	109	16–28	21	0.02
Radiator	50	66–96	76	10–25	16	0.02
Fan belt	1050	40–68	51	3–6	4	0.01
Total						1.15

(Source: Claffey, *Running Costs of Motor Vehicles as Affected by Road Design and Traffic,* NCHRP Report 111, 1971)
* Based on responses to an inquiry submitted to the Chief Engineers of the highway departments of each of the 50 states and to each of the Division Engineers of the Bureau of Public Roads.

Comparisons Between AASHO, NCHRP, and Winfrey

Where data permits, it is valuable to compare these later cost studies to AASHO data, to generate a feeling for how running costs may have changed throughout the years.

NCHRP vs. AASHO. A comparison of these two data sources is difficult in many areas. Since the NCHRP report contains reliable data only for free-flow conditions, only those AASHO tables referring to the *free* condition may be used. Since AASHO contains detail only for passenger cars, only these may be used for comparisons. The computation of unit costs also precludes comparison in many areas.

Fuel consumption was compared for a passenger vehicle under free-flow conditions. To compare costs or consumption rates, the following steps are necessary:

1. Obtain average composite grade consumption rates for grade classifications (0–3, 3–5, 5–7, etc.) with NCHRP data in order to provide common format with AASHO.
2. Compute consumption rates from costs reported in AASHO by dividing the reported cost by the base cost of a gallon of gasoline used by AASHO (32 cents per gallon).

Comparisons are possible only for those speeds and grades which are common between both sources. Six such comparisons are possible: four at 40 miles per hour, and two at 60 miles per hour. The results are shown in Table 6-52.

Note that the later NCHRP results show consistently *lower* fuel consumption than that recorded by AASHO.

Comparison of oil consumption was not possible, since NCHRP tables do not include oil changes, and the AASHO tables do. No assumed average oil change frequency or amount is reported, which would have permitted conversion for comparison.

TABLE 6-52. COMPARISON OF COMPOSITE FUEL CONSUMPTION IN AASHO AND NCHRP FOR PASSENGER CARS

		FUEL CONSUMPTION (gal/mi)	
SPEED	GRADE CLASS	AASHO	NCHRP
40	0–3	0.062	0.045
	3–5	0.066	0.047
	5–7	0.069	0.055
	7–9	0.079	0.069
60	0–3	0.085	0.058
	3–5	0.093	0.062

Tire wear cannot be conveniently compared for several reasons. Of most difficulty is the fact that NCHRP reports considerable variation among costs for various surface types (all "good" surfaces), while AASHO reports only for "good" pavements. Also, NCHRP does not take grade into account, so that only level, or 0 per cent, grades could be used for comparison. Only two comparison points are possible, and considering the problem of surface type, such a comparison is not meaningful.

Maintenance costs agree quite well. AASHO uses a constant 1.20 cents per vehicle-mile, while NCHRP has found a value of 1.15 to be realistic. This difference is surprisingly small, considering the time span between the two studies.

NCHRP vs. Winfrey. Except for those values allocated by judgment, gasoline consumption for passenger cars, pickup trucks, and single-unit trucks in Winfrey agree quite closely to those shown in NCHRP. This is to be expected, because interim findings of the NCHRP study are cited by Winfrey as his major source of this data.

Both the NCHRP study and Winfrey cite a secondary source for fuel consumption of combination vehicles. Both used *Motor Transport Fuel Consumption Rates and Travel Time*, by Sawhill and Firey, but they do not agree with one another, because different vehicles were selected as the standard.

Tire and oil costs could not be compared for many of the same reasons as cited previously in the AASHO-NCHRP comparison. Winfrey includes the cost of oil changes where NCHRP does not. Base conditions for tire costs are not comparable.

As Winfrey allocated a variable maintenance cost (with speed), no meaningful comparison with the NCHRP constant cost is possible.

AASHO-Winfrey. Since NCHRP and Winfrey used the same gasoline consumption data, no further comparison is necessary. Of interest, however, is AASHO's inclusion of gasoline tax in the computation of cost, and Winfrey's exclusion of same, an issue of some significance.

An argument *against* the inclusion of at least the federal excise tax is that these taxes supply the Highway Trust Fund. Thus, in an economy study, these monies show up as *both* road user and highway costs, if they are included in the cost of gasoline. A strong argument *for* inclusion is the fact that a reduction in gasoline consumption represents a savings to the individual user related to the total gas price. Also, the vast majority of highway funds originate as user taxes in some form or

TABLE 6-53. A COMPARISON OF COST ELEMENTS AASHO VS. WINFREY
(cents per vehicle-mile)

	SPEED	TIRES	OIL	MAINTENANCE*	DEPRECIATION*
Winfrey	40	0.19	0.158	0.736	1.391
AASHO		0.28	0.18	1.20	1.50
Winfrey	60	0.353	0.143	0.913	1.208
AASHO		0.56	0.52	1.20	1.50

* Winfrey varies these with speed, on the basis of judgment allocation.

other. To deduct these from economy studies would result in virtual elimination of the consideration of highway cost, an unrealistic approach.

For level grades, passenger vehicles, and only those speeds common to both AASHO and Winfrey, cost elements other than fuel were compared (Table 6-53).

These comparisons are too sketchy to permit firm comparative statements. Tire costs are not directly comparable as a question of surface type arises, as previously discussed. Judgment allocations must be viewed cautiously, and unless a wide range of comparative values is available, no definitive statement should be made.

7

Travel Time and Delay Studies

Travel time varies inversely with travel speed. The travel time study provides data on the amount of time it takes to traverse a specified section of roadway. Such studies also provide speed and, usually, delay data.

Travel time and delay characteristics are good indicators of the level of service that is being provided, and can be used as a relative measure of efficiency of flow.

A delay study is made to determine the amount, cause, location, duration, and frequency of delays, as well as the overall travel and running speeds.

APPLICATIONS OF TRAVEL TIME AND DELAY DATA

1. *Congestion* can be properly evaluated when information is provided on the amount, location, and cause of delays. Such information is required for the selection of remedies to the congestion. These data also indicate the locations where other studies are needed to determine the proper remedy for a particular congestion problem. These other studies might include accidents, volumes, and driver and pedestrian observance of regulations and control devices.
2. *Sufficiency ratings*, *congestion indices*, or *quality indices* are all methods used to compare different roadways, and they are often based on travel time.[1-7,57]
3. *Before-and-after studies* utilize data on travel time and delay to determine the effectiveness of a change in parking restrictions or signal timing, new one-way streets, and similar changes.
4. *Assignment of traffic* to networks and to new or improved facilities is based upon relative travel time, in addition to other factors. This has a marked bearing on the physical plan and design of new facilities and on the nature of improvements to existing facilities.
5. *Economic studies*, such as *benefit-cost analyses*, use travel time data to evaluate the benefits of time saving.
6. *Trend studies* use travel time data to evaluate the level of service as it changes with the passage of time.

DEFINITIONS

Speed: See definitions 3 through 6 and 9 of Chapter 9.

Space mean speed: This is the speed corresponding to the average travel time over a given distance, or the average of speeds of vehicles within a given distance at a given instant.

$$\text{Space mean speed} = n \times \frac{\text{distance}}{\sum t} \qquad (7\text{-}1)$$

where n = number of test runs

TABLE 7-1. SAMPLE COMPUTATION FOR SMS AND TMS

RUN	*OVERALL TRAVEL TIME (min)*	*OVERALL TRAVEL SPEED (mph)*
1	2.0	60
2	2.4	50
3	3.0	40
Average	2.47	TMS = 50

$$\text{SMS} = \frac{2.0 \times 60}{2.47} = 48.6 \text{ mph}$$

The space mean speed can be converted directly to average travel time. However, the arithmetic mean speed (time mean speed) cannot be readily converted.

Time mean speed (TMS): This corresponds to the average of overall travel speeds, and it is higher than the space mean speed (SMS) for the same sample. SMS is affected more by slower-moving vehicles because they occupy the section of roadway for a longer period of time than faster-moving vehicles. To illustrate this difference, consider the simple example given in Table 7-1 for a roadway section 2.0 miles in length.

$$\text{Time mean speed} = \frac{\sum(\text{distance}/t)}{n} \qquad (7\text{-}2)$$

It has been shown that the relationship between TMS and SMS can be expressed as follows:[8]

$$\text{TMS} = \text{SMS} + \frac{\sigma^2}{\text{SMS}} \qquad (7\text{-}3)$$

σ = standard deviation of SMS

A regression analysis to establish a relationship between TMS and SMS resulted in the following equation:[9]

$$\text{SMS} = -1.88960 + 1.02619 \times \text{TMS}$$

The analysis revealed that, as speed increases, the difference between the two speed measurements becomes smaller.

Delay: This is the time lost while traffic is impeded by some element over which the driver has no control.

Operational delay: This is delay caused by interference between components of traffic, that is, the delay due to influences of other traffic. One type of operational delay is caused by other traffic movements that interfere with the stream flow (side frictions). This includes parking or unparking vehicles, pedestrians, stalled vehicles, double parking, and cross traffic. A second type of operational delay is caused by interferences within the traffic stream (internal frictions), this includes congestion due to high volumes, lack of roadway capacity, and merging or weaving maneuvers.

Fixed delay: This is delay caused by traffic control devices. It is the delay to which a vehicle is subjected regardless of the amount of traffic volume and interference present, and it occurs primarily at intersections. It may be caused by traffic signals, stop signs, yield signs, and railroad crossings.

Stopped time delay: This is the time period that a vehicle is actually standing still, due to any factor.

Travel time delay: This is delay caused by acceleration and deceleration, in addition to stopped time delay.

METHODS FOR CONDUCTING TRAVEL TIME OR DELAY STUDIES

Test-Car Techniques

These techniques utilize a test vehicle which is driven over the test section in a series of runs or trips. At least 12 trips should be made to adequately measure the average speed and delays for any one direction and set of conditions. The recommended number of test runs to obtain overall travel time (or speed), within certain limits of error, are given in Table 7-2 for different types of facilities.[10]

In the *floating car technique*, the observer's vehicle "floats" with traffic. The term "floating" refers to an attempt to pass as many vehicles as pass the test car. In another driving technique the driver selects a speed that, in his opinion, is the average rate of speed of the traffic stream, without trying to balance the passings.[10-13,58] Still another driving technique that has been advocated is to operate the test car at the posted speed limit, unless impeded by actual traffic conditions. A

TABLE 7-2. NUMBER OF TEST RUNS TO PREDICT OVERALL SPEED WITH 95% CONFIDENCE

TYPE OF FACILITY	*NUMBER OF RUNS REQUIRED TO PRODUCE AN ACCURACY OF:*	
	5%	*10%*
Signalized Urban Streets		
Two-lane, uncongested	30	8
Two-lane, congested	40	10
Multi-lane, uncongested	18	5
Multi-lane, congested	50	13
Rural Highways		
Two-lane, 1,130 vph	25	6
Two-lane, 1,440 vph	42	11

(Source: Berry, "Evaluation of Techniques for Determining Overall Travel Time," in *Proc. HRB*, 1949)

safe level of operation is maintained by observing minimum safe following distances, minimum passing distances, and reasonable acceleration and deceleration. The speed measured by this technique is the average maximum speed, and it is claimed that this speed provides a better base for measuring performance than does average speed. It is believed that the test car driven in this manner is subjected only to physical delays, while psychological delay due to the driver's mental attitude is minimized. The procedure is called the *maximum-car technique*.[14,15]

The most common method of recording data is with the use of two stop watches by the observer. The observer starts the first stop watch at the beginning of the test run, and he records the elapsed time at various control points along the route. The second watch is used to measure the length of individual stopped time delays. The time, location and cause of these delays are recorded.

Various instruments are available which graphically record a log of the relationship of vehicle speed and delay with regard to time. These instruments have the advantage of recording not only the duration of complete stops but also fluctuations in speed.[2,16-20]

In the case of delay studies, the cause of delay is usually a significant factor. Therefore it becomes necessary to supplement the instrument with an observer, to identify, tabulate, and connect the cause of delay to the graphical log being mechanically prepared.

License Plate Method

This method is useful only when travel time data are sufficient. Observers are posted at the entrance, at the exit, and, if necessary, at other strategic points of the test section where travel time is desired. Each observer records the last three or four numbers of each vehicle, along with the stop watch time at which the vehicle passes the observation post. A sample size of 50 license number matches usually provides good accuracy.[10,11,13,58] The recommended sample sizes to obtain overall travel time (or speed) with an error not exceeding 5 per cent are shown in Table 7-3 for different types of facilities.[10]

TABLE 7-3. SAMPLE SIZE TO PREDICT OVERALL SPEED WITH 95% CONFIDENCE

TYPE OF FACILITY	*NUMBER OF LICENSE MATCHINGS*
Signalized Urban Streets	
Two-lane, uncongested	32
Two-lane, congested	36
Multi-lane, uncongested	80
Multi-lane, congested	102
Rural Highways	
Two-lane, 1,130 vph	25
Two-lane, 1,440 vph	41
Four-lane, uncongested	30

(Source: Berry, "Evaluation Techniques for Measuring Overall Speeds in Urbran Areas" *Proc. HRB* 1949)

Fundamentally this method sacrifices accuracy in details of movement and delay to obtain a larger sample in travel time. It requires careful, long, and tedious office work to reduce the data manually but the use of a computer would greatly reduce this problem.

In a comparison of the license plate and test-car methods, the license plate method was found to be more accurate. However, it is also more expensive because of large manpower requirements in obtaining and analyzing data.[11]

Photographic Methods

These techniques are discussed under Spot Speed Studies in Chapter 8 and they are ideally suited for studies of the interrelationships of several factors such as speeds, time spacings, lane usage, acceleration rates, merging and crossing maneuvers, and delays. Photographic methods are suitable for short test sections such as intersections,[21,22] but they have also been applied for long test sections.[23]

Although these methods provide a means of obtaining a large sample of vehicles and a permanent record of the study, data analysis and equipment requirements raise costs, and such studies are limited to daylight hours and favorable atmospheric conditions.

Interview Method

This method involves interviewing selected individuals as to their travel time and delays experienced on trips. As an example, the employees of strategically located firms are asked to record their travel time to and from work on one particular day. With good cooperation the results obtained may be quite satisfactory. This method is useful where a large amount of data is required in a minimum of time, and at little expense for field observation.[13] A variation of this method was used in a transportation study[24] to obtain travel time data from zone to zone. The cooperation of major taxicab companies was enlisted. Taxi drivers were required to record origin and destination of trips as well as starting and completion times.

Elevated Observations

Observers are stationed at elevated vantage points. They select typical vehicles at random and record pertinent data regarding their progress through a section of roadway. This method is not practical for long-run observations, and it is further dependent on the availability of suitable observation posts.

Moving Vehicle Method of Estimating Volume and Travel Time

A test vehicle makes a series of test runs in each direction over the route under study. For reliable results, a minimum of six test runs should be made in each direction under comparable conditions. The method is applicable to two-way routes only. It has been found to be economical and to produce satisfactory, unbiased estimates of volume and travel time.[12,25-28] The test route is divided into sections which are as uniform as possible with respect to physical conditions (width, number of lanes, parking, etc.) and traffic conditions (volume, speed, type of traffic, etc.). The data required, which are recorded for each section along the route, include:

1. *Travel time*, obtained by a stop watch or other device
2. *Opposing traffic*, a manual count of the number of vehicles moving in the opposite direction that are met by the test car
3. *Overtaking traffic*, a count of vehicles, moving in the same direction, that overtake the test car
4. *Passed traffic*, a count of vehicles, moving in the same direction, that are passed by the test car

Computations for a typical set of data are shown below. The formulas for volume and average travel time each include the number of vehicles that overtake the test car and the number of vehicles passed by the test car. These values are necessary to compensate for irregular movement of the test car, for if the test car were traveling at the actual mean speed for the entire run, it would pass as many vehicles as pass it, and these values would cancel each other. In addition, the formula for volume contains the sum of the times to travel each direction. This is necessary because the volume for one time interval would be about one-half of that met by the test car, since the time required for the test car to travel to the midpoint of each direction is required for the vehicle met at that point to travel to the starting position of the test car.

Formulas. In the following computations, the test section is assumed to be a north-south street. The subscripts n and s refer to the direction the test car was traveling when the item was measured.

Hourly volume

Hourly volume for one directional flow on the section, under existing conditions, is determined by the following formula:

$$V_n = \frac{60(M_s + O_n - P_n)}{T_n + T_s} \tag{7-4}$$

where
V_n = volume per hour, northbound (for southbound volume all subscripts are reversed)
M_s = Opposing traffic count of vehicles met when the test car was traveling south
O_n = number of vehicles overtaking the test car while traveling north
P_n = number of vehicles passed by the test car while traveling north
T_n = travel time when traveling north, in minutes
T_s = travel time when traveling south, in minutes

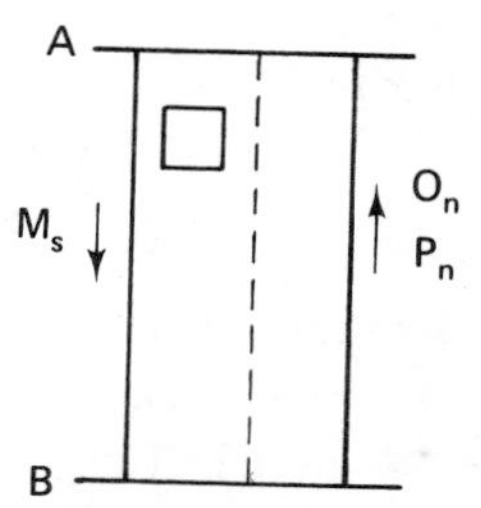

Fig. 7-1. Moving vehicle method

The test vehicle makes a round trip, essentially measuring the number of vehicles that will pass the starting point in the time it takes the vehicle to make a round trip from A to B and back to A again. Consider the diagram of Figure 7-1. The vehicle begins at A and proceeds in a southerly direction, counting all vehicles which pass it in the opposite direction (M_s). Certainly, all of these vehicles will pass point A in the time it takes the test vehicle to return to that spot. The test vehicle then turns around at point B. Any vehicle that passes the test car (O_n) will also arrive at A before the test car returns. Any vehicle overtaken by the test car has already been counted as part of M_s. However, these vehicles (P_n) will not arrive at A before the test vehicle. Therefore, the volume past point A, in a northerly direction, in the time it takes the test vehicle to make a round trip, is $M_s + O_n - P_n$, and the formula follows.

If the test car turned instantaneously at B, the count would be exact. However, there is a time loss while the vehicle turns, which may allow some error to occur. Also, one run may not be statistically representative of average conditions. For these reasons, the method is considered an estimate, and several runs are made and the results averaged.

Average travel time

The average travel time for one directional flow is determined by the following formula:

$$\bar{T}_n = T_n - \frac{60(O_n - P_n)}{V_n} \qquad (7\text{-}5)$$

where $\bar{T}_n$ = average travel time of all traffic northbound (for southbound travel time subscripts are reversed)

The value $(O_n - P_n)$ represents a correction accounting for the fact that the test car may not have been traveling at the average speed.

Space mean speed

The space mean speed for one directional flow is determined by the following formula:

$$S_n = \frac{60d}{\bar{T}_n} \qquad (7\text{-}6)$$

where S_n = space mean speed northbound, in miles per hour
d = length of test section, in miles

Illustrative Example

Data for test section length of 0.75 miles (major arterial) is given in Table 7-4.

$$V_n = \frac{60(M_s + O_n - P_n)}{T_n + T_s} = \frac{60(111.5 + 1.5 - 1.0)}{2.61 + 2.42}$$

= 1,336 vehicles per hour

$$V_s = \frac{60(M_n + O_s - P_s)}{T_s + T_n} = \frac{60(84.0 + 0.5 - 1.0)}{2.42 + 2.61}$$

= 996 vehicles per hour

$$\bar{T}_n = T_n - \frac{60(O_n - P_n)}{V_n} = 2.61 - \frac{60(1.5 - 1.0)}{1336}$$

= 2.59 minutes

$$\bar{T}_s = T_s - \frac{60(O_s - P_s)}{V_s} = 2.42 - \frac{60(0.5 - 1.0)}{996}$$

= 2.45 minutes

$$S_n = \frac{60d}{\bar{T}_n} = \frac{60 \times 0.75}{2.59} = 17.4 \text{ miles per hour}$$

$$S_s = \frac{60d}{\bar{T}_s} = \frac{60 \times 0.75}{2.45} = 18.4 \text{ miles per hour}$$

INTERSECTION DELAY STUDIES

Delay at intersections is a major factor in the analysis of congestion. There are several factors which should be considered in evaluating the efficiency or effectiveness of different types of intersection traffic control, but intersection delay is of primary importance. Some of the other factors include:[21,29]

1. The comparative accident experiences resulting from the use of different types of traffic control

TABLE 7-4. SAMPLE PROBLEM DATA FOR THE MOVING VEHICLES METHOD

NORTHBOUND TRIPS	T_n *(min)*	M_n	O_n	P_n
1N	2.65	85	1	0
2N	2.70	83	3	2
3N	2.35	77	0	2
4N	3.00	85	2	0
5N	2.42	90	1	1
6N	2.54	84	2	1
Total	15.66	504	9	6
Average	2.61	84.0	1.5	1.0
SOUTHBOUND TRIPS	T_s *(min)*	M_s	O_s	P_s
1S	2.33	112	2	0
2S	2.30	113	0	2
3S	2.71	119	0	0
4S	2.16	120	1	1
5S	2.54	105	0	2
6S	2.48	100	0	1
Total	14.52	669	3	6
Average	2.42	111.5	0.5	1.0

2. Motorists' desires: motorists may prefer one type of control over another, even though the average delay per vehicle may be greater. For example, the motorist may prefer the positive control of a traffic signal, even though the delay due to stop sign control may be less.
3. The comparative costs of installation and operation of different types of traffic control, and their effect on vehicle operating cost
4. The comparative effects on pedestrians
5. The physical conditions at the intersection area

Factors Affecting Intersection Delay[54]

1. *Physical factors*, such as the number of lanes, grades, width, access control, channelization and transit stops
2. *Traffic factors*, such as volume on each approach, turning movements, vehicle classification, driver characteristics, approach speeds, parking and pedestrians
3. *Traffic controls*, such as types and timing of signals, stop or yield signs, and turn and parking controls

Methods for Field Measurement of Intersection Delay

Travel time method. In intersection delay studies this method measures the travel time from a point in advance of the intersection to a point in or beyond the intersection. The methods used to obtain the travel time information include the following.

1. The test car is operated between the same two points.
2. License plate numbers and times are recorded at two points.

TABLE 7-5. SAMPLING PROCEDURE FOR INTERSECTION TRAVEL TIME

TIME (minute starting at)	*(5 MINUTES OF DATA FOR ONE APPROACH) INSTANTANEOUS DENSITY COUNTS (FOR TIME INTERVAL t) IN THE APPROACH AT TIME*			
	+0 sec.	+15 sec.	+30 sec.	+45 sec.
5:00 P.M.	0	4	5	7
5:01	3	8	4	2
5:02	5	0	6	1
5:03	5	3	6	6
5:04	6	7	4	7
Subtotal	19	22	25	23
Total Density = N		89		

3. Spaced aerial photographs are taken from a vantage point.[21]
4. The data is recorded by a 20-pen recorder, with road tubes at critical points and observers operating keys to aid in identifying vehicles passing through the test section.[30]
5. An observer, at a vantage point, traces vehicles through an intersection and records stop watch times between two points.[31]
6. A sampling procedure is used which involves the counting of vehicles occupying an intersection approach at successive time intervals (such as every 15 seconds), for a period of time (such as 10 minutes). Each successive count represents an instantaneous density (number of vehicles occupying the length of intersection approach per time interval). During the given time period, the number of vehicles leaving the intersection approach is also counted. This count represents the traffic volume. These counts permit estimating the average travel time for all vehicles traversing the intersection approach.[32] An example of 5 minutes of data is shown and analyzed in Table 7-5.

Total volume leaving intersection approach $= V = 70$ (during a time period of 5 minutes)

Average travel time

$$= T = \frac{Nt}{V} = \frac{89 \times 15}{70} = 19.1 \text{ seconds}$$

Stopped Time Delay Method. This is a second measure of intersection delay. This method involves the measurement of stopped time only, and does not include time losses due to deceleration and acceleration, which are included in travel time methods. The methods used to obtain stopped time delay include:

1. Spaced serial photographs are taken from a vantage point.
2. Delay meters are used which accumulate vehicle-seconds of stopped time, as an observer operates buttons or dials.[21]
3. A sampling procedure is used which involves the counting of the number of vehicles stopped in the intersection approach at successive intervals (such as every 15 seconds). This sampling, along with a volume count during the same time observations are made, permits estimating the vehicle-seconds of stopped time delay. When the study is conducted at an intersection controlled by fixed-time signals, the sampling interval should be selected so that repetitive sampling in the same parts of the signal cycle is avoided. Therefore, the sampling interval should not be an even subdivision of the cycle length.[21] An example of 5 minutes of data is shown and analyzed in Table 7-6.

Computations

Total delay

= total number observed × observation interval

$= 104 \times 15 = 1{,}560$ vehicle-seconds of delay

TABLE 7-6. SAMPLING PROCEDURE FOR INTERSECTION DELAY (5 minutes of data for one approach)

TIME (minute starting at)	*TOTAL NUMBER OF VEHICLES STOPPED IN THE APPROACH AT TIME*				*APPROACH VOLUME*	
	+0 sec.	+15 sec	+30 sec	+45 sec	*Number Stopping*	*Number Not Stopping*
5:00 P.M.	0	2	7	9	11	6
5:01	4	0	0	3	6	14
5:02	9	16	14	6	18	0
5:03	1	4	9	13	17	0
5:04	5	0	0	2	4	17
Subtotal	19	22	30	33	56	37
Total		104			93	

Note: A specific vehicle is counted more than once if it is stopped during more than one observation time. Because of this, it is necessary to count the number of vehicles stopping for volume determination.

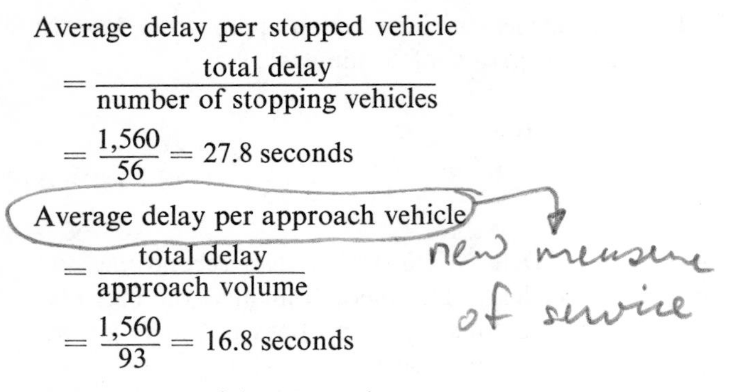

Average delay per stopped vehicle

$$= \frac{\text{total delay}}{\text{number of stopping vehicles}}$$

$$= \frac{1{,}560}{56} = 27.8 \text{ seconds}$$

Average delay per approach vehicle

$$= \frac{\text{total delay}}{\text{approach volume}}$$

$$= \frac{1{,}560}{93} = 16.8 \text{ seconds}$$

Per cent of vehicles stopped

$$= \frac{\text{number of stopping vehicles}}{\text{approach volume}} = \frac{56}{93} = 60.2 \text{ per cent}$$

In comparative studies of the above three methods for evaluating stopped time delay, it was found that both the delay meter and sampling procedures can give reliable results, but the spaced aerial photo method provides the most complete data.[21,22]

TRAVEL TIME (SPEED) AND DELAY AS MEASURES OF CONGESTION AND LEVEL OF SERVICE

Quantitative measurement of the economic loss to a community due to congestion is still problematical, but there is little doubt that congestion can adversely affect the economic life of a community. This is reflected by the detrimental effect on downtown business activities and property values, increased consumer prices, neighborhood disintegration, and increased road user costs.[33] Relief of traffic congestion results in decreased vehicle operating costs, reduced traffic accidents, reduced travel time, and substantial increase in driver comfort and convenience.

A suitable congestion index for urban facilities should embody a combination of three factors:[1]

1. *Operational characteristics*, which would entail measurements of speeds, delays, and overall travel times
2. *Volume-to-capacity characteristics*, which would entail determination of the relationship of actual traffic volume to the capacity of the facility
3. *Freedom-of-movement characteristics*, which would entail determination of the percentage of vehicles restricted from moving (free movement) and the durations of restrictions

There is no doubt that a congestion index embodying all of these factors would be most desirable in helping to program new construction or improvements, and in establishing levels of service. Recommended minimum desirable operating speeds (average overall travel speed) for different types of street systems, to provide a satisfactory level of service are shown in Table 7-7.[12,34]

TABLE 7-7. SUGGESTED STANDARDS FOR OVERALL SPEED AND RATE OF FLOW

TYPE OF STREET	OVERALL SPEED (mph)		RATE OF MOTION, (min/mi) PEAK HOUR
	Peak hour	Off peak	
Expressway:			1.71
1. Full Control	35	35–50	
2. Partial Control	35	35–50	
Major Arterial:			2.40
1. Divided	25	25–35	
2. Undivided	25	25–35	
Collector:			3.00
Local:			6.00
1. Business	10	10–20	
2. Industrial	10	10–20	
3. Residential	10	10–20	

Comparison of the results obtained in the illustrative example of the moving vehicle method of estimating volume and travel time and the above standards produce the following values:

Delay Rate

This is the difference between the *actual* rate of motion on the arterial and the *standard* rate of motion.

$$\textit{Actual} \text{ rate of motion} = \frac{60}{17.4} = 3.45 \text{ minutes per mile}$$

$$\textit{Standard} \text{ rate of motion} = 2.40 \text{ minutes per mile}$$

$$\text{Delay rate} = 3.45 - 2.40 = 1.05 \text{ minutes per mile}$$

Vehicle Delay Rate

This is the total time lost, in minutes per mile by vehicles in the traffic stream, because the arterial does not meet the standards. It is obtained by multiplying the peak-hour, one-direction volume by the delay rate.

Vehicle delay rate

$$= 1{,}336 \times 1.05 = 1{,}403 \text{ minutes per mile}$$

Primary factors used in the highway Capacity Manual to evaluate level of service are speed and travel time. Levels of service are established for each of the following types of facilities: freeways and other expressways, other multi-lane highways, two- and three-lane highways, urban arterial streets, and downtown streets.

Congestion Indices

Several congestion indices have been proposed for measuring the quality of traffic flow. The following brief summary represents the parameters upon which different methods are based:

1. The congestion index number for urban facilities is based on the ratio of the actual time a vehicle occupies a section of roadway to the optimum travel time (uncongested), which takes into consideration speed limits, safety, and prudent driving.[35]
2. Travel time is the basis of many methods of evaluating congestion and level of service.[5,6,36-38]
3. Speed inversion, applied to expressways, is the reduction in the speed of traffic which occurs at high volumes without a corresponding change in volume. Congestion is assumed to exist when a speed inversion of 10 miles per hour occurs.[23,39]
4. The quality index is the relationship of average overall speed to speed changes and frequency of speed changes per mile.[3,4,40,41,57]
5. The rating of congestion is the relationship of time loss to driver inconvenience and discomfort.[42]
6. The acceleration noise model is based on the disturbance of the vehicle's speed from a uniform speed.[43,44,45]

PRESENTATION OF TRAVEL TIME (SPEED) AND DELAY DATA

Time Contour Maps

Contour lines represent minutes of travel time from some central reference point. With the reference point as a center, a series of concentric circles are drawn, indicating distance from the central reference point. Time contour maps are particularly well suited for the comparison of the various routes leading from the central business district (CBD) and for estimating the time that may be saved by planning a new or improved facility.

Bar Charts

These charts indicate causes and distribution of delays. They depict at a glance all of the factors contributing to delays. Bar charts are also used to compare before-and-after results when changes have been made.

Speed Profiles

These may indicate the variation in average running speed and average overall speed (or travel times) of traffic along a given route, through different types of development, or on a block-by-block basis.

Frequency Distribution Curves

These curves depict the number or percentage of vehicles versus the travel time or delay.

Speed Zone Maps

These maps show the speed or travel time along different sections of various routes within an area.

Delay Zone Maps

These maps give the vehicle minutes of delay during different hours along different sections of various routes within an area.

SPEED-DELAY CHARACTERISTICS OF TRAFFIC

Various studies of speed-delay characteristics have been made. Several are summarized below:

1. Average overall speeds in urban areas range from 3 miles per hour in dense business areas to 35 to 40 miles per hour in limited lengths of routes in outlying residential areas. The overall speed on urban radial routes, from the center of the central business district (CBD) to the outer periphery of urban development, usually falls between 15 to 20 miles per hour.[46]

   The suggested minimum peak-hour standard rates of speed are as follows, for different types of facilities:

   Expressway or freeway: 35 miles per hour
   Major arterial: 25 miles per hour
   Collector street: 20 miles per hour
   Local streets: 10 miles per hour

2. Frequency of delays in urban areas for city-wide travel occur at an average of two to three times per mile, and as often as 10 to 12 times per mile in dense business districts.[46]
3. Under normal conditions, the duration of a single traffic delay is 15 to 24 seconds, ranging from 6 to 36 seconds. Shorter delays are usually caused by pedestrians and turning movements, and longer delays by traffic signals.[46]

4. Total time lost through delay ranges from 15 to 16 per cent for city-wide travel, and from 35 to 50 per cent for dense CBD's. Therefore, running time is about 85 per cent for city-wide travel on radial routes, and from 50 to 75 per cent in dense CBD's.[46]
5. A study of several arterial streets and some local streets in Pennsylvania cities revealed that for all locations combined the average speed was 16.3 miles per hour when traffic flow was *below* critical density, and 9.8 miles per hour when traffic flow was *above* critical density. Average speed for arterials within the CBD was 14.4 miles per hour; average speed for arterials within intermediate areas was 19.4 miles per hour; and average speed for local streets within intermediate areas was 14.2 miles per hour. A comparison of travel times along streets controlled by coordinated signal systems, and along streets controlled by signals that were *not* coordinated, revealed no significant differences in speeds.[36]
6. Based on a comparison of data obtained by the travel time and stopped time delay at two signalized intersections, the stopped time delay amounted to between 50 and 80 per cent of total delay. It was concluded that stopped time delay is a satisfactory method of comparison at signalized intersections. It was also concluded that when signalized intersections are spaced so that signals can be timed for progressive movement, the use of fixed-time coordinated timing can substantially reduce delay, as compared with fixed-time or traffic-actuated signals with no coordination in timing.[21,22]
7. A comparative study of different types of traffic control, including two-way stop, four-way stop, traffic-actuated signals, and fixed-time signals (not coordinated), was made at intersections located in different areas. In suburban areas, total delay experienced with various means of control was, in descending order, four-way stop, fixed-time, traffic-actuated, two-way stop. In urban areas, fixed-time signal control resulted in greater delay than two-way stop control. In rural areas, total delay experienced with various means of control was, in descending order, fixed-time, four-way stop, traffic actuated.[29,49-51]
8. A study of the effects of the installation of left-turn holding lanes and signals at intersections of high-volume left-turn traffic indicated that, while safety and left-turn capacity were increased, average vehicle delay to both turning and through vehicles also increased. Peak-hour delay increased over 100 per cent at most of the intersections studied.[52]
9. A study of the effects of parking turnover rate on delay, on an urban street, showed that this factor had no appreciable effect. The study showed that the level of service prevailing had a large effect on the percentage of vehicles affected by delay.[53]
10. Multi-variate analyses of travel speeds and delays on a two-lane highway resulted in the development of multiple linear regression equations to predict mean travel speeds and delays.[55] For uninterrupted flow sections (where traffic flow is not controlled at intersections by traffic lights or signs), the following equation was selected as the most valid functional relationship for estimating overall travel speed.

$$V_1 = 68.60 - 0.4541X_1 - 0.1775X_2 - 0.1007X_3 - 0.0150X_4 - 0.0301X_5$$

where
V_1 = mean overall travel speed, in miles per hour
X_1 = intersecting streets on both sides, in number per mile
X_2 = commercial establishments on both sides, in number per mile
X_3 = portion of section length where passing was not permitted, in per cent
X_4 = practical capacity, in vehicles per hour
X_5 = total traffic volume, in vehicles per 15 minutes

It should be noted that all five variables are negatively related to travel speed.

For interrupted flow sections (signalized intersections), the following equations were developed for overall travel speed and delay through the section which was considered to be a length of 1,000 feet (500 feet before and after the intersection).

$$V_2 = 28.595 - 0.4165X_6 - 0.2118X_7 - 0.0120X_8 - 0.0170X_9 + 29.4800X_{10}$$

where
V_2 = mean overall travel speed, in miles per hour
X_6 = average algebraic grade of intersection approach, per cent
X_7 = cycle length of traffic signal, in seconds
X_8 = traffic volume approaching the intersection in the direction of travel, in vehicles per 15 minutes
X_9 = total traffic volume entering the intersection on all four approaches, in vehicles per 15 minutes
X_{10} = green time-to-cycle length ratio

It should be noted that the test vehicle was not necessarily stopped at each signalized intersection, and that all variables except green time-to-cycle length ratio are negatively related to travel speed.

$$D = 11.951 + 0.0052X_{11} + 0.2299X_7 + 0.0135X_8 + 0.0168X_9 - 35.7935X_{10}$$

where
D = mean travel delay, in seconds
X_{11} = length of approach to special turning lane, in feet
X_7, X_8, X_9, X_{10} = defined as above

The effect of variables common to the above two equations is consistent in that whatever causes a decrease in travel speed would result in an increase in delay. The delay equation indicates that delay is directly related to length of approach to turning lane, and this is contrary to what might be expected. The length of approach, however, was associated with a high-volume intersection which had a relatively large number of turning movements. These conditions contributed to the increased delays.

See Spot Speed Characteristics, in Chapter 8 for further characteristics of travel time and delay.[29,47,56]

8

Spot Speed Studies

A spot speed study is a study of the speed of traffic at one point or spot on a trafficway. It consists of a series, or a sample, of observations of the individual speeds at which vehicles are approaching an intersection or passing a point at a nonintersection location. These observations are used to estimate the speed distribution of the entire traffic stream at that location, under the conditions prevailing at the time of the study.

APPLICATIONS OF SPOT SPEED DATA

Speed is a primary factor in all modes of transportation, and it is a basic measure of traffic performance. Spot speed data have many applications:

1. For trends in the operating speeds of different vehicle types, which are obtained from data collected through periodic sampling at selected locations
2. For speeds at problem locations, to determine whether speeds are too high and if complaints received are justified
3. For traffic operation (regulation and control)
   a. Establishing speed limits
   b. Determining safe speeds at curves and at approaches to intersections
   c. Establishing lengths of no-passing zones
   d. Locating traffic signs
   e. Locating and timing traffic signals
   f. School zone protection
   g. Establishing speed zones
4. For accident analysis, to determine the relationship of speed to accidents, which way help in developing corrective measures
5. For before-and-after studies, to evaluate the effect of some change in controls or conditions
6. For geometric design features
   a. Designs assume uniform speeds, and it is important to evaluate the effects of actual speed distributions on design features.
   b. Length of speed change lanes, curvature, superelevation, and sight distance are directly related to, and vary appreciably with, speed.
7. To evaluate capacity in relation to desired speeds. If all vehicles traveled at the same speed, capacity would be at a maximum and certain types of accidents, such as overtaking, passing, or rear-end collisions, would be eliminated.
8. As aids to enforcement
   a. To determine the effects of speed control measures
   b. To measure the effectiveness of changes in enforcement programs
9. Research studies

DEFINITIONS

Speed: The rate of movement of a vehicle, generally expressed in miles per hour.

Average spot speed: The arithmetic mean of the speeds of all traffic, or a component thereof, at a specified point.

Overall travel time: The time of travel, including stops and delays (except those off the traveled way).

Running time: The time the vehicle is in motion.

Overall travel speed: The speed over a specified section of highway that is the distance divided by the overall travel time. The average for all traffic, or a component thereof, is the summation of distances divided by the summation of overall travel times.

Running speed: The speed over a specified section of highway that is the distance divided by the running time. The average of all traffic, or a component thereof, is the summation of distances divided by the summation of running times.

Design speed: A speed determined for design as related to the physical features of a highway that might influence vehicle operation. It is the maximum safe speed that can be maintained over a specified section of highway when conditions are so favorable that the design features of the highway govern.

Eighty-five percentile speed: That speed below which 85 per cent of all traffic units travel, and above which 15 per cent travel.

Arithmetic mean speed or *time mean speed:* The speed obtained when the sum of all values is divided by the number of observations.

Median speed: The speed represented by a middle value when all speed values are arrayed in ascending order. Half the speed values will lie above the median, half below.

Modal speed: The speed value occurring most frequently. In a frequency distribution of speeds obtained in a speed study, the modal speed would be the value with the highest frequency of observations.

Pace: A given increment of speed that includes the greatest number of observations. It is usually taken in 10-mile increments.

Operating speed: The highest overall speed, exclusive of stops, at which a driver can travel on a given highway under prevailing conditions, without at any time exceeding the design speed.

WHERE TO MAKE SPOT SPEED STUDIES

1. Trend Locations
   (a) Straight, level, open sections of rural highways
   (b) Midblock locations on urban streets
2. Problem Locations for Specific Purposes
   (a) High accident frequency locations
   (b) At points where the installation of traffic signals and traffic signs is contemplated
   (c) Other locations that are important for traffic operations
3. Representative locations for basic data surveys
4. Locations where before-and-after studies are being conducted
5. The specific location for the speed study should be selected so as to reduce as much as possible the influence of the observer and of the measuring equipment.
   (a) Equipment should be concealed from, or made as inconspicuous as possible to, the approaching driver.
   (b) Recording of data should be as inconspicuous as possible.
   (c) Accumulation of onlookers should be avoided.

FACTORS WHICH AFFECT SPOT SPEEDS[44]

1. *The driver*, including the distance of his trip, the number of passengers he has, his sex, age, residence (rural or urban), in-state or out-of-state, and marital status,

2. *The vehicle*, including type, age, weight, manufacturer, and horsepower,

3. *The roadway*, including geographic location, type, curvature, grade, length of grade, sight distance, number of lanes, surface type, lane position, lateral clearance, frequency and spacing of intersections, and roadside development,

4. *The traffic*, including volume, density, passing maneuvers, opposing traffic, access control, traffic control devices, speed regulations, types of vehicles,

5. *The environment*, including time of day, month, season, and weather.

TIME AND LENGTH OF STUDY

It is recommended[62] that the study be divided into three parts during off-peak hours, and that observations be made for one hour, or not less than 50 motor vehicles for each period.

1. One hour between 9:00 and 12:00 A.M.
2. One hour between 3:00 and 6:00 P.M.
3. One hour between 8:00 and 10:00 P.M.

Statistically adequate samples must be obtained and the effect of volume on speed must be considered. Similar conditions must be present if results are to be comparable for before-and-after, trend, and characteristic studies.

TABLE 8-1. RECOMMENDED COURSE LENGTHS

AVERAGE SPEED OF TRAFFIC STREAM (mph)	*COURSE LENGTH (ft)*	*CHANGING SECONDS TO (mph)*
less than 25	88	60 ÷ seconds = mph
25 to 40	176	120 ÷ seconds = mph
greater than 40	264	180 ÷ seconds = mph

(Source: *Traffic Engineering Handbook* ITE, 1965, pg. 273)

METHODS FOR CONDUCTING SPOT SPEED STUDIES

There are two basic methods employed: one involves the measurement of time and distance, and the other takes advantage of the Doppler principle. The time versus measured distance method is the most commonly used; it involves measurement of the time required for a vehicle to traverse a measured course. There are a number of devices available to measure the time. The measured course is laid out at the location under study. When manual means (stop watch) are used to measure time, the length of the course is made some multiple of 1.467, so that speed may be conveniently expressed in miles per hour. For example, for a course length of 88 feet ($= 1.467 \times 60$) and a travel time of 1 second, the speed is $60/1.0 = 60$ miles per hour since 1.467 feet per second = 1 miles per hour. Therefore, to obtain the speed in miles per hour, it is only necessary to divide 60 by the number of seconds taken to travel the length of the course. The length of the course is dependent upon the average speed and the type of location under study. The shorter measured course (88 feet) is normally employed for intersection approaches, and the longer ones (176 or 264 feet) for other locations where speed is relatively more constant. (See Table 8-1.)

Methods for Measuring Time Over a Measured Distance

Stop Watch Methods

Pavement Markings. One method involves the use of transverse pavement markings which are placed at each end of the course, and the observer starts and stops the watch as the vehicle passes the markings.

Advantages: Minimum set-up time is needed initially, and none is needed for subsequent studies. The method is flexible and requires no maintenance costs; markings are easily renewed. Provided that the observer is able to station himself inconspicuously, driver behavior is not greatly influenced, since markings are in place over an extended period.

Disadvantages: It is possible to introduce a large error in the observations due to parallax. The extent of this error is largely dependent upon the observer's position. An inconspicuous vantage point is not always consistent with a position that will minimize the error because of parallax. Another source of error is the difference in reaction times of individual observers, and this factor becomes significant when subsequent studies are made by different observers. The method is not suitable for comprehensive, continuing studies because of human factors and the high personnel cost. The method is limited in accuracy and requires a systematic sampling procedure because the practical limit of observations is between 5 and 10 readings per minute.

Enoscope. Another method involves the use of an Enoscope (mirror box) at one end or both ends of the course to facilitate reading the vehicle entrance into, and exit from, the course. An Enoscope is an L-shaped box, open at both ends, with a mirror set inside at a 45-degree angle to the arms of the device. The device bends the line of sight of the observer so that it is perpendicular to the path of the vehicle at the limits of the measured course. If one Enoscope is used, the observer takes a position at one end of the course, with the Enoscope at the other end. It is convenient for the observer to line himself up opposite a tree or post on the other side of the roadway and to press the stop watch at the instant the vehicle breaks his line of sight to the tree or post. As a vehicle passes the Enoscope, a flash will be visible to the observer. The stop watch is started as the vehicle passes the Enoscope, and it is stopped when it passes the observer. If two Enoscopes are used, the observer stations himself between them. The Enoscope may also be employed at night by placing a light source on the opposite side of the roadway so that it shines into the mirror of the Enoscope.[62]

Advantages: The error, resulting from parallax, in judging the exact time that the vehicle crosses the limits of the course is eliminated. The method is flexible, equipment is of low cost, simple, easily set up, and fairly reliable.

Disadvantages: There are errors due to starting and stopping the stop watch manually. Considerable time is required to time one vehicle, which cuts down on the percentage of total traffic than can be timed. It is, therefore, not a satisfactory method for heavy, multi-lane traffic, but it is suitable for low-volume roads. The method is not well suited for continuing, long-term studies because of high personnel cost and human factors. It is difficult to conceal the Enoscope from passing traffic, which may influence the results obtained.

Methods Using Pneumatic Tubes

The limits of the measured course are defined by flexible rubber tubing fastened to the pavement and perpendicular to the path of vehicle travel. A vehicle crossing the tube compresses the air in it, actuating an air impulse switch. The switches operate the timing mechanism, which measures the time between the impulse from the first tube and the impulse from the second tube.

Pneumatic tubes are low in initial cost, simple to install, and easily maintained, but tubes are the most conspicuous detecting device and they may affect driver behavior and distort the speed distribution. In addition, they are subject to traffic hazards.

Speed Watch. One type of timing unit used in conjunction with pneumatic road tubes is the *speed watch.* The timing unit is a calibrated, spring-driven stop watch, which is operated by a solenoid actuated by the road tube impulses. The speed scale is calibrated against the time markings on the watch dial. The speed reading is held until the timing unit is manually reset.

Advantages: Human errors are reduced, and it is simple to operate.

Disadvantages: Road tubes are required and these are not suitable for continuous studies or for high-volume conditions.

20-Pen Graphic Recorder. This is another type of timing unit used in conjunction with pneumatic road tubes. The instrument records up to 20 items of information on a chart moving at constant speed. The speed of vehicles is determined by the distance between pips made by pens actuated by the road tubes. It is possible to separate data by lanes and directions. For example, a series of adjacent courses can be used near intersections to obtain the time spacing between vehicles (headways), acceleration or deceleration patterns, and similar kinds of data. The instrument is primarily used for research studies.

Advantages: A permanent record is obtained which includes every vehicle. Therefore, more comprehensive data are obtained.

Disadvantages: Road tubes are required, although pens may be actuated manually, which would then involve all of the errors due to human factors. The equipment is quite expensive. Data reduction is tedious and time consuming.

Electrically Operated Meter. This device is, in effect, an electric stop watch. The timing mechanism is a constant speed motor which runs continuously. Impulses from air switches engage and disengage an electric clutch, which connects the indicator needle to the motor shaft. Speeds are read from a calibrated dial and the device is equipped with a manual reset for indexing before checking the next vehicle.

Advantages: Human errors are reduced, and it is simple to operate under low volume conditions.

Disadvantages: Road tubes are required and these are not suitable for continuous studies or for high-volume conditions. Slight voltage changes will cause inaccuracies and the device is difficult to calibrate.

Electronic Meter. This is another timing unit used in conjunction with pneumatic road tubes. Timing is performed through a calibrated discharge circuit. A precision capacitor is charged to a set voltage by a battery. The impulse from the first road tube begins the discharge of the capacitor, and the impulse from the second tube stops the discharge. The remaining voltage left in the capacitor is proportional to the speed of the vehicle, and the speed is read directly in miles per hour. The meter must be read immediately and must be reset before timing another vehicle.

Advantages: Short courses can be used which permit the timing of individual vehicles during high-volume conditions. Course length is usually 11 feet. Initial cost is relatively low and the meter requires little maintenance.

Disadvantages: Road tubes are required. The meter is sensitive to temperature and humidity changes. The reading on the indicator will not hold for an infinite period.

Electronic Decade Meter. This timing unit is used together with electric contact switches or pneumatic tubes. Contacting detectors are placed at the limits of the measured course and they utilize vehicle axle weight to make an electrical contact between two conductors. A timer is started and stopped by impulses from the electric contact switches as a vehicle passes over the measured course. These contact detectors have advantages over the pneumatic detectors in that they do not require air impulse switches and are less conspicuous.

Timing is performed by the electronic decade through a continuously running oscillator which provides repetitive pulses of the time unit required ($\frac{1}{100}$ second, $\frac{1}{1000}$ second, etc.). As the vehicle passes the first contact detector, an electronic gate is opened and time pulses are accumulated by a series of decades until the vehicle passes the second contact detector, which shuts the gate. The vehicle's travel time over the course is obtained from the meter by recording the numerals on the decades, or the digits may be electronically transmitted to an automatic recorder. The decades must then be reset to zero in preparation for the next vehicle.

Advantages: The output of the meter is well suited to automatic digital recording. It requires little field maintenance and is reliable.

Disadvantages: The output indicates travel time and therefore requires a computation to obtain the speed. The initial cost is high.

The *traffic analyzer*[45,49] was developed by the BPR and is an elaborate assemblage of electronic and mechanical equipment which is capable of obtaining not only speed data in each lane, but also traffic volumes, headways, lane placement, and the effects of various highway design features on traffic operations. The traffic analyzer is mounted in a light truck. The speed detectors are spaced 12 to 36 feet apart and consist of pneumatic tubes or contact strips; the timing device is an electronic decade meter.

Multi-Parameter Detectors. Magnetic loop detectors have been successfully used in conjunction with computer processing of data to obtain a variety of traffic parameters such as speed, volume, space headway, time headway, vehicle length, and lane changing.

A crystal-controlled oscillator is connected to turns of wire installed below the wearing surface of the pavement (or taped to the surface) to form an inductance loop. When the front of a vehicle passes over the loop, it produces a phase shift in the voltage across the loop and a relay is activated. When the back of the vehicle leaves the loop, the relay is opened. The duration of relay closure varies directly with the length of the vehicle and inversely with its velocity. Vehicle velocity and length cannot be determined independently using a single-loop detector. By assuming an average car length, the speed can be determined by dividing the length by the duration of relay closure. A pair of loop detectors can be used to obtain speed and vehicle length data.[50] Assume a vehicle (length L) traverses a pair of loops (loop length d) separated by a distance (D) with a constant velocity (V). The magnetic field extends slightly beyond the loop boundaries, and the effective field length (d') depends on the sensitivity of the detector. Each of the two detectors produces a pulse, and the time delay between pulses is designated as Δt. Then,

$$V = \frac{D}{\Delta t} \tag{8-1}$$

The pulse width (T) is the time required for the vehicle to traverse the field (d') along its entire length (L). Then,

$$T = \frac{d' + L}{V} \tag{8-2}$$

$$L = \frac{D}{\Delta t} T - d'$$

Pulse lengths recorded on magnetic tape, and a computer program, can provide for the direct processing of data to obtain various traffic parameters.[51]

Loop detectors must be calibrated, and a malfunction may result if the loop is installed near a large mass of buried metal. Passage of multi-section vehicles will yield more than one pulse, as though these sections were separate vehicles. The same error results when vehicles follow too closely (tailgating).

Photographic Techniques

The photographic technique involves distance and time relations, as in all the above methods for conducting spot speed studies, but results are obtained from camera film rather than from various types of meters or stop watches. These methods are normally limited to research studies.

Time-Lapse Photography.[46] This method requires a motion-picture type camera, equipped with a time-lapse mechanism (which consists of a device to move the film a single frame at a time at a given time interval) and a frame-numbering device. When the developed film is projected onto a screen, a vehicle's movement along the roadway is readily determined. The spot speed equals the distance traversed by the vehicle from frame to frame, divided by the time between the same frames.

Advantages: A permanent record is obtained for all traffic flow characteristics, providing information about speeds, volumes, vehicle classification, spacing between vehicles, and lateral vehicle maneuvers.

Disadvantages: This method is not suitable for continuing studies. In most instances a high vantage point is necessary. There is a great possibility of malfunctioning of camera and auxiliary appurtenances. Analysis is time-consuming and must be done at a considerable time after the data are taken because the film must be developed.

Continuous-Strip Photography.[47,48,53,54] This method involves the taking of continuous-strip photographs from an airplane over the area to be surveyed. Based on the principles of photogrammetry, mathematical relations have been developed to obtain speed data and volume data, and other traffic flow characteristics can be obtained from the photographs.

The advantages and disadvantages outlined under time-lapse photography apply in general to continuous-strip photography, but it is claimed that automatic processing of the data could result in very early analysis of all kinds of traffic flow characteristics. This would eliminate the need to make separate studies to evaluate different characteristics.

Meters Based on the Doppler Principle

These meters are *radar* and *ultrasonic*, and they direct a radar or audio beam of a certain frequency at the moving vehicle. The reflected signal is shifted in frequency, and the difference in frequency is proportional to the speed of the vehicle.

Radar Meter. Radar meters operate on the Doppler principle that the speed of a moving target is proportional to the change in frequency between the radio beam transmitted to the target and the reflected radio beam. The equipment measures this difference and converts it to a direct reading in miles per hour. Most radar meters are designed so that graphic recorders may be attached to them, to provide a permanent record. The accuracy of speed determinations is ± 2 miles per hour up to 100 miles per hour.

The meter should be set up to keep the angle between the direction of vehicle travel, and a straight line from the transmitter-receiver to the vehicle, at a minimum. Radar units should be located so that they are aimed as near as possible to the line of vehicle travel. When the angle differs from zero, an error is introduced into the speed reading, but this error is always such as to cause the meter reading to be *lower* than the true speed of the vehicle. The magnitude of the error is directly related to the cosine of the angle, and it is not very large, the cosine function does not deviate much from 1.0 except for large angles. For example, an angle of 15 degrees produces an error of approximately 3.5 per cent. The meter should be positioned at the edge of the roadway, at an angle of about 15 degrees with the centerline. In this position, the meter will show speeds for vehicles traveling in either direction in the adjacent two or three lanes.

Advantages: In this method, no road tubes are required. The equipment is easily set up and operated and can be inconspicuous, reducing any influence on driver behavior.

Disadvantages: Since the equipment involves the transmission of radio waves, a Federal Communications Commission (FCC) license is required. Accuracy is not sufficiently great for all types of speed studies, but this difficulty can be overcome by proper calibration. The inherent accuracy of radar produces better results than many of the meters previously discussed. It is, however, difficult to distinguish a single vehicle being observed in heavy and/or multi-lane traffic.

Ultrasonic Meter. This device operates on the Doppler principle. A high-energy audio tone is directed at on coming vehicles, and the return signal is changed in frequency by the movement of the vehicle. This change is proportional to the speed of the vehicle. The transmitter-receiver is mounted over the centerline of a lane, and it is directed downward, at a 45-degree angle toward approaching traffic. The detection zone is very sharp, and all vehicles are detected at approximately the same point on the roadway. Data are usually transmitted from the field equipment to a central point by standard telephone circuits. Accuracy is equivalent to that obtained with radar meters.

Advantages: The advantages are similar to those for radar meters.

Disadvantages: The equipment is permanent. It requires telephone lines or other data transmission devices.

Remote Sensing Techniques.[55] Research is under way to determine the feasibility of using remote sensors (infrared, radar, and multi-channel and multi-band instruments) for conducting area-wide traffic surveys. Such surveys would provide traffic volumes, vehicle classification, speed, spacing, origin and destination data, and regional land use.

Weather and lighting conditions limit the use of aerial photography for collecting data, and the area of coverage by this method is also restricted. A potential solution for obtaining coverage of large regional areas is in the use of satellites. Where weather or light restricts photographic coverage, radar sensors might be used. Remote sensors also offer advantages for obtaining detailed traffic information over limited areas. Both infrared and radar sensors offer the possibility of obtaining data at any time of day, and in the case of radar, in almost any kind of weather in which an airplane can fly. The use of lasers and television cameras also offer potentially fruitful areas for research.[1]

ANALYSIS AND PRESENTATION OF SPOT SPEED DATA

Characteristically, the engineer collects data on a sampling basis. This is most certainly the case where the collection of spot speed data is concerned, where only a percentage of the vehicles at any given location may be measured. From a measured sample, the engineer attempts to characterize the entire population, which, in this case, consists of the entire traffic stream.

Because of the uncertainty involved in the characterization of an entire population by variables based upon a sample, and because vehicles in a traffic stream are not traveling at uniform speeds but follow a distribution of speeds within a comparatively wide range, the mathematics of statistics must be intoduced into the analysis of spot speed data.

TABLE 8-2. SAMPLE ANALYSIS OF SPOT SPEED DATA FREQUENCY DISTRIBUTION TABLE

1	2	3	4	5
SPEED GROUP (mph)	*MEAN SPEED OF GROUP, V (mph)*	*NUMBER OF VEHICLES IN GROUP, f*	*PER CENT OF TOTAL OBSERVATIONS IN GROUP*	*CUMULATIVE PER CENT OF TOTAL OBSERVATIONS*
10–14.9	12.5	0	0	0
15–19.9	17.5	6	2.0	2.0
20–24.9	22.5	8	2.7	4.7
25–29.9	27.5	29	9.7	14.4
30–34.9	32.5	60 $\sum = 103$	20.0	34.4
35–39.9	37.5	63	21.0	55.4
40–44.9	42.5	74	24.7	80.1
45–49.9	47.5	29	9.7	89.8
50–54.9	52.5	19	6.3	96.1
55–59.9	57.5	10	3.3	99.4
60–64.9	62.5	2	0.7	100.1
65–69.9	67.5	0	0	
Totals		$300 = n$	100.1	

Statistics addresses, mathematically, problems in two general areas:

1. Given a display of data, statistics may be utilized to determine a mathematical model or equation which best describes the variation pattern displayed.
2. Given that the display of data is generated by the measurement of a small fraction (sample) of a population, statistics may be employed to evaluate the mathematical uncertainty in characterizing the entire population by a display of data representing only a sample.[2-7]

In the analysis of spot speed data, it is convenient to make use of statistics in addressing both of the above problem areas.

Tabular Arrangement and Collection of Spot Speed Data

It is not convenient to record the speed of each vehicle separately, particularly when a large number of measurements is required in a limited amount of time. For this reason, speed groupings, or ranges (usually of 1 or 2

Fig. 8-1. Frequency histogram

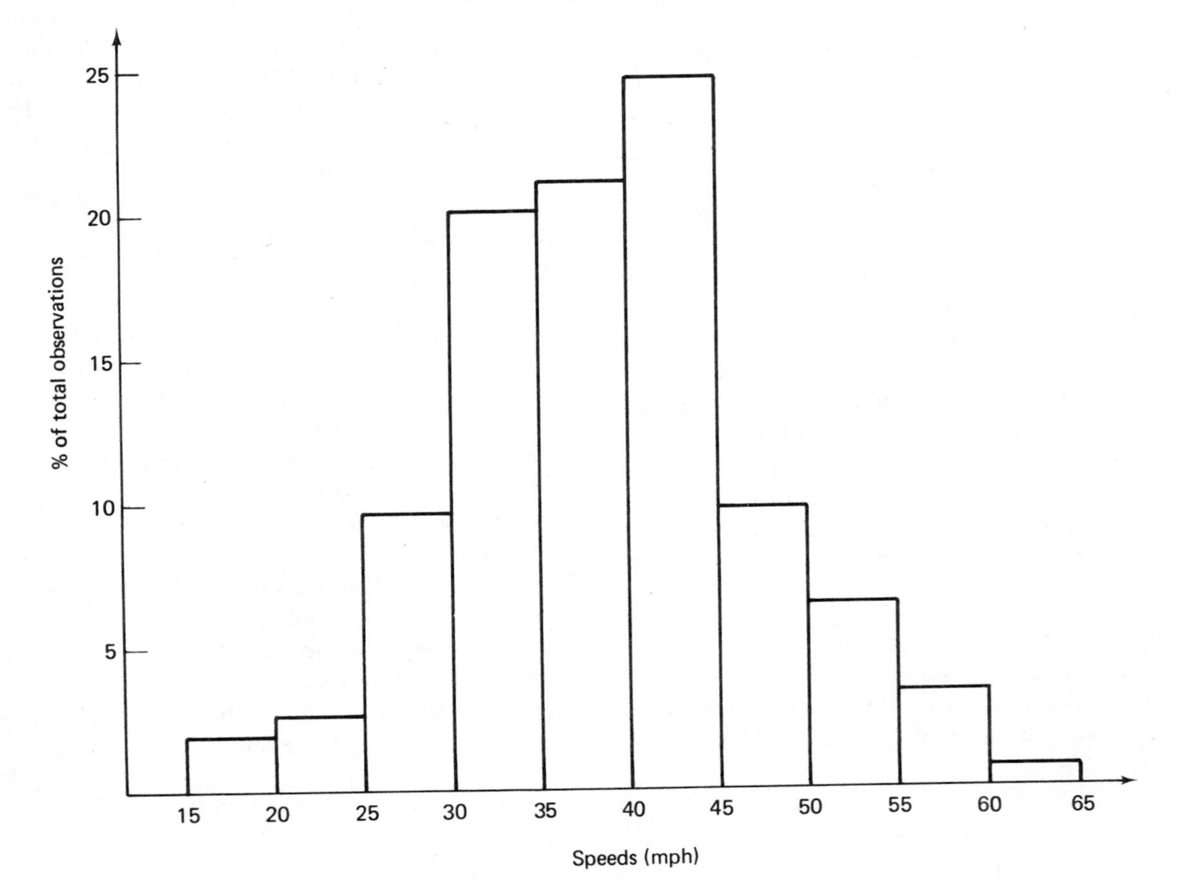

miles per hour) are established, and data is recorded by indicating the numbers of vehicles which are observed to be in each speed category.

This data is displayed in a *frequency distribution spot speed table*, an example of which is shown in Table 8-2. In the example of Table 8-2, the speed groupings are 5 miles per hour, and the total number of observations (or total frequency) was 300.

Graphical Presentation of Spot Speed Data Representative Values

Frequency Histogram. The data from columns 1 and 4 of the frequency distribution table are used to construct the *frequency histogram* (Figure 8-1). The speed groups form the bases of the frequency rectangles, and the respective per cents of total observations define the altitudes.

Frequency Distribution Curve. The data from columns 2 and 4 of the frequency distribution table are used to form the *frequency distribution curve* (Figure 8-2). Points are obtained by plotting the percentage of total observations versus the mean speed for each group. The points are then joined by a smooth curve.

A *unit area* is defined by a rectangle whose base is one speed grouping and whose altitude is 100 per cent on the frequency histogram. Therefore, the total area under the frequency histogram is unity, or 100 per cent. Similarly, the total area under the frequency distribution curve is

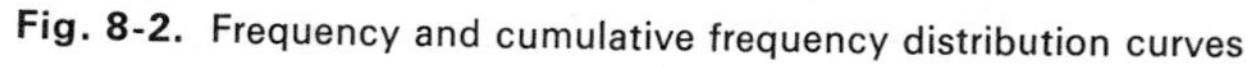

Fig. 8-2. Frequency and cumulative frequency distribution curves

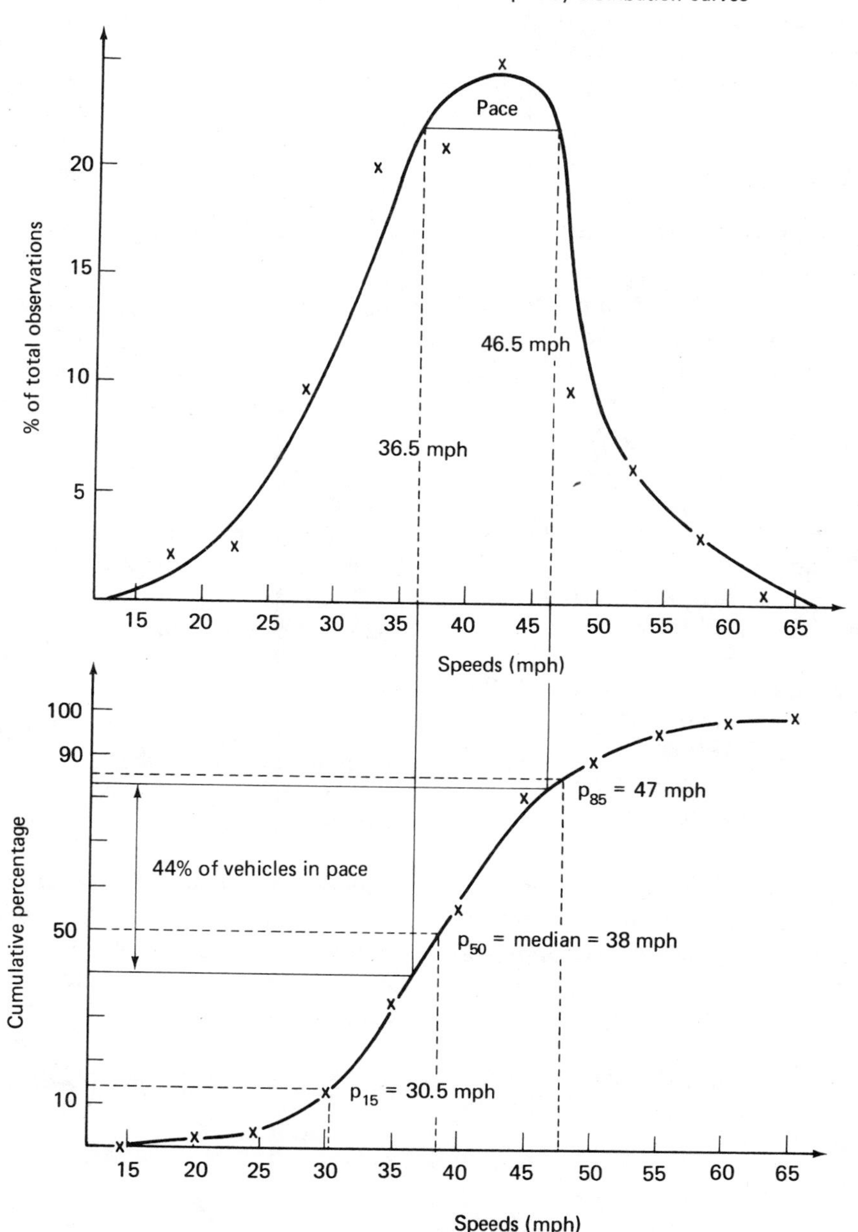

also unity, since this is merely the joining of the mid-points of the several frequency rectangles by a smooth curve.

Cumulative Frequency Distribution Curve. The data from columns 1 and 5 are used to plot the *cumulative frequency distribution curve* (Figure 8-2). Because the cumulative percentage refers to the percentage of vehicles traveling at or below a given speed, the cumulative percentages are plotted versus the *upper limit* of each speed group. The points are then joined by a smooth curve.

Representative Values

The engineer is interested in mathematically describing the speed characteristics of a particular flow in as clear and succinct a fashion as possible. It is cumbersome and inconvenient to refer to the entire distribution of speeds each time the speed characteristics of a traffic stream are to be described. For this reason, the engineer chooses several significant values which will adequately describe the distribution in question.

Measures of Central Tendency.

Arithmetic mean speed (time mean speed)

The *arithmetic mean* is simply the average speed of all observed vehicles. It is computed by multiplying the mean speed of each group by the number of observations for that group, summing the products, and dividing by the total number of observations, or

$$\text{Arithmetic mean speed } (\bar{x}) = \frac{\sum fV}{n} \qquad (8\text{-}3)$$

where f = frequency of observations in a particular speed group
V = mean speed for each speed group
n = number of observed speeds

The arithmetic mean speed is an estimate of the expected speed of any random vehicle at the point where the study was conducted. The arithmetic mean of the speed sample is a statistically unbiased estimator of the true (but unknown) mean of the population.

Median speed

The *median* is the middle value in a distribution whose values have been arranged in ascending or descending order. One-half the observed values are higher than the median and one-half are lower.

The median may be identified as the 50th percentile speed on the cumulative frequency distribution curve, the abscissa of the vertical line which divides the frequency distribution curve into two equal areas, or may be found by interpolation in the frequency distribution table according to the formula:

$$\text{Median} = L + \left(\frac{(n/2) - f_L}{f_m}\right)C \qquad (8\text{-}4)$$

where L = lower bound of the group in which the median lies
n = total number of observations
f_L = cumulative number of observations up to the lower bound of the group in which median lies
f_m = number of observations in the group in which the median lies
C = speed interval of group in which the median lies

Of these methods for determination, the use of the 50th percentile on the cumulative distribution curve is the most straightforward.

As the median is a positional value, it is less affected by extreme values than the arithmetic mean.

Modal speed

The *mode* or *modal value* is the most frequently occurring value in a distribution. It is that speed at which more vehicles pass the point of observation than any other. The mode is difficult to identify precisely when data is collected in groups. The *interval* in which the mode occurs can be identified as the group with the highest number of observations. The mode is sometimes estimated as the speed corresponding to the peak point on the frequency distribution curve.

If the frequency distribution curve is symmetrical around its vertical axis, the mean, the mode, and the median will all be the same.

The mode is less affected by extreme values than either the mean or the median.

Pace

The *pace* is defined as that 10-mile range in speeds in which the highest number of observations were recorded. It is identified by a 10-mile speed increment which "cuts off" the peak of the frequency distribution curve.

Standard Deviation—Measure of Dispersion. The description of central tendency is not sufficient to define a distribution. A measure of the dispersion, or spread, of values is needed. The standard deviation (STD, s) is computed from the following formula:

$$\text{STD}(s) = \sqrt{\frac{\sum f(V - \bar{x})^2}{n - 1}} \qquad (8\text{-}5)$$

observing that $\bar{x} = \sum fV/n$,

$$\begin{aligned}\sum f(V-\bar{x})^2 &= \sum fV^2 - \sum 2fV\left(\frac{\sum fV}{n}\right) + \sum f\frac{(\sum fV)^2}{n^2}\\ &= \sum fV^2 - \frac{2}{n}(\sum fV)^2 + \frac{1}{n}(\sum fV)^2\\ &= \sum fV^2 - \frac{1}{n}(\sum fV)^2\end{aligned}$$

and substituting,

$$\text{STD}(s) = \sqrt{\frac{\sum fV^2}{(n-1)} - \frac{(\sum fV)^2}{n(n-1)}}$$

or:

$$\text{STD}(s) = \sqrt{\frac{\sum fV^2}{(n-1)} - \left(\frac{n}{n-1}\right)(\bar{x})^2}$$

The sample standard deviation (s) is an unbiased estimator of the true (but unknown) standard deviation of the population (σ).

It is sometimes convenient to express the standard deviation as a per cent or fraction of the mean, to obtain a relative measure of dispersion. The *coefficient of variation* is defined as follows:

$$\text{CV} = \frac{s}{\bar{x}} \tag{8-6}$$

A rough estimate of the standard deviation may be obtained using values from the cumulative frequency distribution curve and applying the equation

$$\text{STD}\ (s_{\text{est}}) = \frac{P_{85} - P_{15}}{2.0} \tag{8-7}$$

where P_{85} = the 85th percentile speed on the cumulative distribution curve (that speed at which 85 per cent of the vehicles travel below)

P_{15} = the 15th percentile speed on the cumulative distribution curve

The values for the sample problem of Table 8-2 are summarized below. All values taken from the frequency or cumulative frequency distribution curves are illustrated in Figure 8-2.

$$\text{Arithmetic mean} = \bar{x} = \frac{\sum fV}{n} = \frac{11{,}615}{300} = 38.72 \text{ miles per hour}$$

Median = P_{50} = 38 miles per hour (cumulative frequency distribution curve)

Mode = in 40 to 44.9 miles per hour speed group (frequency distribution table)

Pace = 36.5 – 46.5 miles per hour (frequency distribution curve)

Per cent of vehicles in pace = 44 (cumulative frequency distribution curve)

$$s = \sqrt{\frac{472.250}{300-1} - \frac{300}{299}(38.72)^2} = 8.6 \text{ miles per hour}$$

$$s_{\text{est}} = \frac{P_{85} - P_{50}}{2.0} = \frac{47 - 30.5}{2} = \frac{16.5}{2} = 8.25 \text{ miles per hour}$$

The Normal Distribution

This is a mathematical expression which may be used to describe the probable outcomes of certain processes. If the assumption is made that spot speed data may be represented by an equation of the normal form, the mathematics of statistics enables certain statements and analyses to be made which would be otherwise impossible. For the remainder of this section, spot speed data will be assumed normally distributed. A statistical test for the accuracy of this assumption is treated later.

The *normal distribution* is described by an equation of the form:

$$f(x) = \frac{1}{\sigma\sqrt{2\pi}}e^{-(x-\mu)^2/2\sigma^2} \tag{8-8}$$

where μ = true mean of the population ($\bar{x}$ is a *sample* mean, which is an unbiased estimate of μ)

σ = true standard deviation (s is an unbiased estimate of σ)

σ^2 = variance (the square of the standard deviation)

Since the equation involves only two variable parameters (μ, σ), the specification of a mean and variance completely defines the distribution. For this reason, normal distributions are often referred to by a standard notation of the form $N[\mu, \sigma^2]$, that is, normally distributed with a mean of μ and a variance of σ^2. $N[25, 100]$ therefore represents a normal distribution with a mean of 25 and a variance of 100. The standard deviation is σ, or $\sqrt{100} = 10$. The form of the normal distribution is illustrated in Figure 8-3. Significant properties of the normal distribution include:

1. As with the frequency distribution curve, the total area under the normal distribution is unity, or 100 per cent.
2. The probability of an occurrence between the values x_1 and x_2 is given by the area under the normal distribution curve between the values x_1 and x_2.
3. The normal distribution is symmetrical around the mean and is asymptotic to the x-axis.

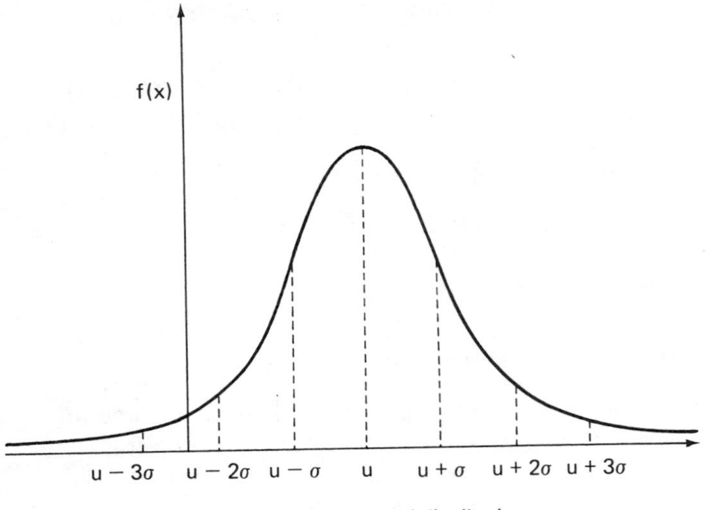

Fig. 8-3. A normal distribution

4. The area between $\mu + \sigma$ and $\mu - \sigma$ is 0.6827.
   The area between $\mu + 1.96\sigma$ and $\mu - 1.96\sigma$ is 0.9500.
   The area between $\mu + 2\sigma$ and $\mu - 2\sigma$ is 0.9545.
   The area between $\mu + 3\sigma$ and $\mu - 3\sigma$ is 0.9971.
   The area between $\mu + \infty$ and $\mu - \infty$ is 1.0000.

The last property enables the engineer to make certain valuable quantitative statements. Consider a distribution with a known true mean (μ) of 40 miles per hour and a true standard deviation (σ) of 5 miles per hour. It is now possible to say that 68.3 per cent of all vehicles can be expected to have a speed between $40 + 5 = 45$ miles per hour and $40 - 5 = 35$ miles per hour. If the speed of a random vehicle were estimated to be in this range, the estimate would be correct 68.3 per cent of the time and incorrect 31.7 per cent of the time. If the speed of a random vehicle were estimated to be the mean value, 40 miles per hour, the estimate would be wrong by no more than 5 miles per hour (or $\frac{5}{40} \times 100 = 12.5$ per cent) 68.3 per cent of the time. The error would be more than 5 miles per hour, 12.5 per cent, in 31.7 per cent of the estimates so made. Similarly, the error in estimating the speed of a random vehicle as the mean (40 miles per hours), would be in error at most $2 \times 5 = 10$ miles per hour (or 25 per cent), 95.5 per cent of the time, and at most $3 \times 5 = 15$ miles per hour (or 37.5 per cent), 99.7 per cent of the time.

These statements may be made under the assumption that the spot speed data represented a normal distribution with a *true mean* of 40 miles per hour and a *true standard deviation* of 5 miles per hour. Additional errors are introduced if sample means ($\bar{x}$) and standard deviations (s) are used. The error limits must be multiplied by $\sqrt{1 + (1/n)}$, where n is the sample size, and where the *sample* rather than *true* characteristics are employed.

Accuracy of Sampling

When evaluating the results of a speed sample, a single value, the mean speed, is generally taken as representative of the true mean of all vehicles in the traffic stream. The question is validly posed: How good is this estimate, and what are the probable limits of error?

The accuracy of any given sample depends upon two factors. The first is *pure chance*. That is, the sample observed may not be representative of the entire traffic stream due to the chance observation of an inordinate number of faster-than-average or slower-than-average drivers. This kind of error can be accounted for statistically, and the methodology for doing so is discussed in the following paragraphs. The larger the sample size, the smaller the chance that an estimated mean ($\bar{x}$) is significantly different than the true mean (μ).

The second kind of error occurs when vehicles are selected for measurement in a non-random way, thereby biasing the sample in one direction. The most common types of such sampling errors are:[8]

1. Selection of the first vehicle in a platoon. This is generally the easiest vehicle to observe, but it is most often the slowest vehicle in the platoon. Vehicles of differing platoon position must be selected to avoid biasing the sample towards lower speeds.
2. Selection of too large a percentage of trucks. Trucks are generally slower than other vehicle types and should be measured in the same proprotion as their percentage of the traffic stream.
3. Tendency to try and "catch" extremely high- or slow-speed vehicles. This must be avoided. It is not necessary to determine the "fastest" or "slowest" vehicle in a spot speed study.

Assuming that the sample is randomly selected, it is possible to examine the probable limits of chance error.

The spot speed study has resulted in a mean value ($\bar{x}$) which was computed from a sample of n vehicles. It is possible to examine the limits of error in estimating the speed of a single random vehicle to be the mean, by referring to the distribution of values x, that is, the normal distribution with an estimated mean of $\bar{x}$ and an estimated standard deviation s. To examine the limits of error in estimating the average speed of all vehicles to be $\bar{x}$, the mathematical distribution of values ($\bar{x}_n$) must be utilized, where n is the sample size used to compute $\bar{x}$.

It can be shown that if a variable x is normally distributed with a mean μ and variance σ^2(STD $= \sigma$), that is,

$$x = N[\mu, \sigma^2]$$

then a variable $\bar{x}$, which is the average value of x for samples of size n, is also normally distributed with a mean μ and a variance σ^2/n (STD $= \sigma/\sqrt{n}$), that is,

$$\bar{x} = N\left[\mu, \frac{\sigma^2}{n}\right]$$

The standard deviation of this distribution is defined as the standard error of the mean:

$$E = \frac{\sigma}{\sqrt{n}} \tag{8-9}$$

At this point, probable limits of error may be defined in terms of the applicable normal distribution. Consider the illustrative problem discussed earlier in which $\bar{x} = 38.7$ miles per hour and $\sigma = 8.6$ miles per hour E may be computed as $8.6/\sqrt{300} = 0.5$ miles per hour. It may now be stated that the true mean of the population lies between $\bar{x} \pm E$ or 38.7 ± 0.5 miles per hour with 68.3 per cent certainty. Further, $\mu = 38.7 \pm (2 \times 0.5)$ with 95.5 per cent certainty, and $\mu = 38.7 \pm (3 \times 0.5)$ with 99.7 per cent certainty. It is 31.7 per cent probable that the error in the estimate of the true mean as 38.7 miles per hour ($\bar{x}$) is at least 0.5 miles per hour, 4.5 per cent probable that the error is at least 1.0 miles per hour, and only 0.3 per cent probable that the error is as large as 1.5 miles per hour.

It is possible to use these statistical statements to determine the necessary size of a sample to yield any given accuracy with any given level of confidence using the equation:

TABLE 8-3. DETERMINATION OF SAMPLE SIZE FOR THE ESTIMATION OF THE MEAN

DESIRED LEVEL OF CONFIDENCE (per cent)	*EQUATION*
68.3	$n = \frac{\sigma^2}{e^2}$
95.0	$n = \frac{3.84\sigma^2}{e^2}$
95.5	$n = \frac{4\sigma^2}{e^2}$
99.7	$n = \frac{9\sigma^2}{e^2}$

Note: A nomograph for solution of these equations is presented in Figure 8-4.

$$e = K\frac{\sigma}{\sqrt{n}} \tag{8-10}$$

where
n = sample size
e = limits of tolerable error
$\frac{\sigma}{\sqrt{n}}$ = standard deviation of values $\bar{x}$ computed from samples of size n (the standard error)
K = the number of standard deviations referring to the desired confidence level

Fig. 8-4. Nomograph for confidence bounds

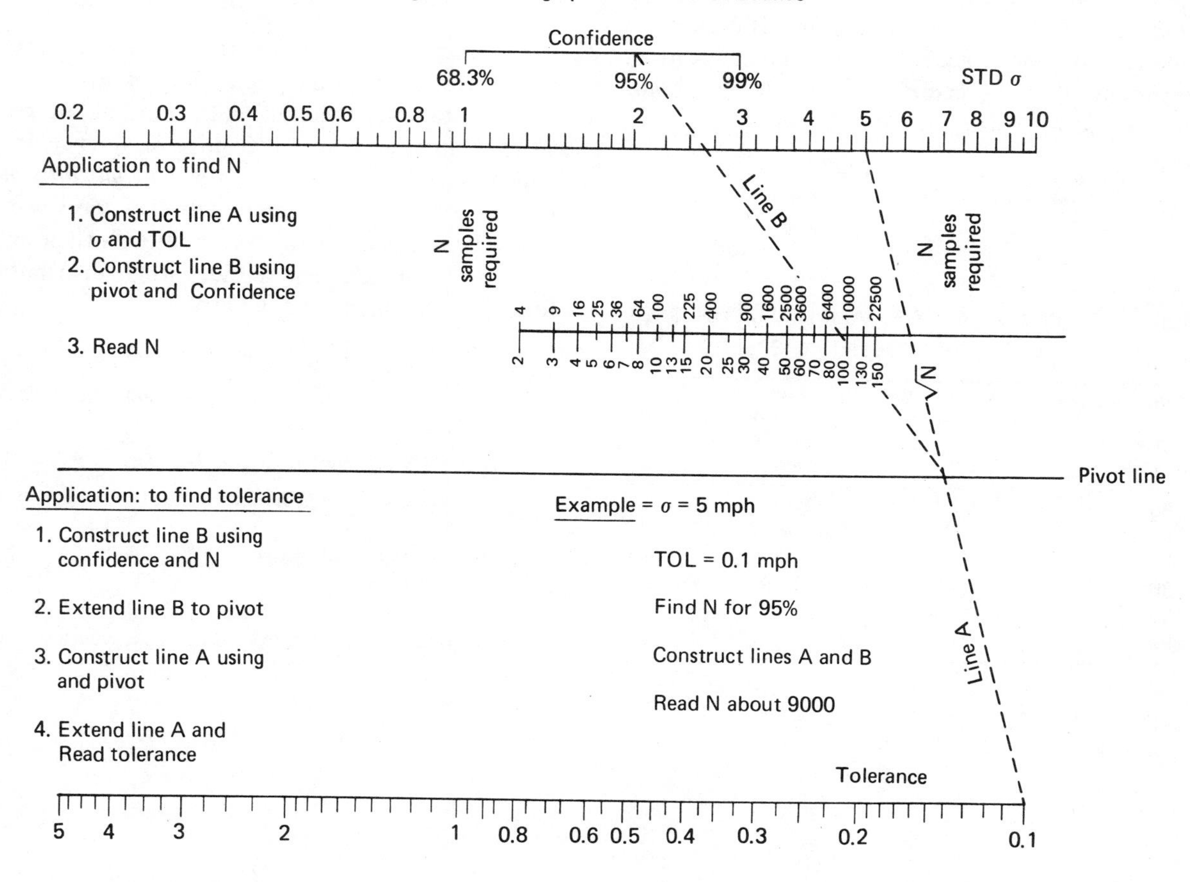

Therefore, if 95 per cent confidence were desired, the equation becomes:

$$e = 1.96\frac{\sigma}{\sqrt{n}}$$

Solving for n,

$$n = \frac{3.84\sigma^2}{e^2} \tag{8-11}$$

Similar equations may be derived for any given level of confidence. These are shown in Table 8-3.

Another similar set of equations has been developed to compute the necessary sample size to estimate any desired percentile speed (not just the mean).[9] The equation is

$$n = \frac{K^2\sigma^2(2 + U^2)}{2e^2} \tag{8-12}$$

where n = sample size
K = normal deviate corresponding to the desired confidence level
U = normal deviate corresponding to the desired percentile speed

Values of K and U are given in Table 8-4. The value σ may be known from a previous speed study at a particular location, or may be estimated as 5 miles per hour. This value was adopted after an extensive investigation of speeds on various facilities and under various conditions.[10] It was found that 5 miles per hour is a good estimate for σ in most cases.

For example, a 95.5 per cent confidence is desired on an estimate of the mean speed at a particular location. The maximum allowable error is ± 1.0 miles per hour. The necessary sample size is computed as:

$$n = \frac{4\sigma^2}{e^2}(\text{Table 8-3}) = \frac{4(5)^2}{(1.0)^2} = 100$$

TABLE 8-4. DETERMINATION OF SAMPLE SIZE IN THE ESTIMATION OF PERCENTILE SPEEDS

DETERMINATION OF K		*DETERMINATION OF U*	
Confidence Level (per cent)	*K*	*Percentile Speed*	*U*
68.3	1.000	50th	0.00
90.0	1.645	15th or 85th	1.04
95.0	1.960	7th or 93th	1.48
95.5	2.000	5th or 95th	1.67
99.0	2.575		
99.7	3.000		

Before-and-After Studies

It is often necessary to examine the significance of a difference in mean speeds observed at a particular location at which some control measure has been instituted. For example, assume that speeds had been recorded at a high accident-frequency location. A reduction had been observed when speeds were again surveyed after the placement of a new speed limit sign. The engineer must evaluate whether the observed difference in mean speeds was purely due to chance, or whether it reflected a significant change in driving habits due to the erection of the speed limit.

Assuming that all conditions, except for the placement of the speed sign, are the same in both the before and the after speed study, there are a variety of statistical tests which aid in making such evaluations.

Normal Approximation. It can be shown that if the sample size is sufficiently large ($n \geq 30$) and both $\bar{x}_1$ and $\bar{x}_2$ (the sample means from the *before* and *after* studies, respectively) are mean values of samples from populations having the *same distribution* (i.e., no significant difference in before or after population), the value $(x_1 - x_2)$ is asymptotically normal with a mean of 0 and a standard deviation of σ_D. That is to say that

$$(\bar{x}_1 - \bar{x}_2) = N[0, \sigma_D^2] \text{ for } n \geq 30$$

where

$$\sigma_D = \sqrt{\frac{s_1^2}{n_1} + \frac{s_2^2}{n_2}} \tag{8-13}$$

The implicit assumption that $\bar{x}_1$ and $\bar{x}_2$ are sample means from the same population or distribution is a vital one to note. When using the normal approximation to evaluate the significance in a difference in means, it is assumed that there is *no significant difference*, and that they are in fact *representative of the same population*. The accuracy of this assumption is then examined.

Using the properties of the normal distribution, the following statements can be made:

1. It is 68.3 per cent probable that $(\bar{x}_1 - \bar{x}_2) = 0 \pm \sigma_D$.
2. It is 95.0 per cent probable that $(\bar{x}_1 - \bar{x}_2) = 0 \pm 1.96\sigma_D$.
3. It is 95.5 per cent probable that $(\bar{x}_1 - \bar{x}_2) = 0 \pm 2\sigma_D$.
4. It is 99.7 per cent probable that $(\bar{x}_1 - \bar{x}_2) = 0 \pm 3\sigma_D$

If the observed value of $(\bar{x}_1 - \bar{x}_2)$ is $1.96\sigma_D$ or greater, it is possible to say:

1. If the before and after populations are actually the same,

a difference in means as high as $1.96\sigma_D$ would occur only 5 per cent of the time.

2. Given that $(\bar{x}_1 - \bar{x}_2)$ is greater than $1.96\sigma_D$, it is at most 5 per cent probable that this difference occurred by chance, if the two samples are representative of the same distribution.
3. The assumption that $\bar{x}_1$ and $\bar{x}_2$ are from the same distribution is, therefore, rejected, and the difference observed is taken as significant.

The general specification for certification of the significance of an observed difference (that is, rejection of the assumption that they are the same) is that it must be no more than 5 per cent probable that the difference in means was due to pure chance. This is sometimes referred to as the 95 per cent point, since 95 per cent of all chance occurrences would show mean differences less than $1.96\sigma_D$ if they were from the same distribution. The specification of the 95 per cent point as the decision point is somewhat arbitrary, and may be adjusted to suit engineering judgment. For example, if it is 10 per cent probable that the difference in means occurred by chance (the 90 per cent point), the engineer may wish to conduct further studies rather than reject the significance of the observed difference.

The 95 per cent specification may be restated in the following way: If $(\bar{x}_1 - \bar{x}_2) \geq 1.96\sigma_D$, the difference in means is significant.

Consider the following example:

$\bar{x}_1 = 38.72$ miles per hour $\quad \bar{x}_2 = 41.22$ miles per hour

$\sigma_1 = 8.6$ miles per hour $\quad \sigma_2 = 7.9$ miles per hour

$n_1 = 300 \quad n_2 = 330$

$$\sigma_D = \sqrt{\frac{(8.6)^2}{300} + \frac{(7.9)^2}{330}} = 0.66$$

The actual difference in means is

$$41.22 - 38.72 = 2.50 \text{ miles per hour}$$
$$\geq 1.96(0.66) = 1.30 \text{ miles per hour}$$

The difference in means is therefore significant.

The example could have been solved in another manner. The standard normal distribution, where the standard normal is defined as a normal distribution with mean 0 and standard deviation 1 ($N[0, 1]$), is tabulated for use (Table 8-5). Any normal distribution may be expressed in terms of the standard normal by shifting the axis to zero and dividing by the standard deviation. For example, consider a normally distributed variable:

$$x = N[\mu, \sigma^2]$$

It is required to evaluate the probability of the variable x falling between two values a and b, that is,

$$\text{Prob } a \leq x \leq b$$

Since tables are not available for $N[\mu, \sigma^2]$, it is desirable to utilize the standard normal. The equivalent probability on the standard normal is given by:

$$\text{Prob } \frac{a - \mu}{\sigma} \leq x' \leq \frac{b - \mu}{\sigma}$$

where $\quad x' = N[0, 1]$

In the sample problem, the actual difference in means was 2.50 miles per hour. The standard normal may be used to evaluate the probability of such a difference occurring by chance. The probability of $-2.50 \leq \bar{x}_1 - \bar{x}_2 \leq +2.50$ is desired, where $(\bar{x}_1 - \bar{x}_2) = N[0, \sigma_D^2]$, $\sigma_D = 0.66$. Shifting to the standard normal,

$$\text{Prob } \left(\frac{-2.50 - 0}{0.66}\right) \leq (\bar{x}_1 - \bar{x}_2)' \leq \left(\frac{+2.50 - 0}{0.66}\right)$$

where $\quad (x_1 - x_2)' = N[0, 1]$

$$\frac{2.50}{0.66} = 3.79$$

Referring to Table 8-5 (standard normal), the area to the left of $z = 3.79$ (probability that $z \leq +3.79$) is greater than 0.9998 (the table does not have values of $z > 3.49$). For the sake of continuing the illustrative analysis, assume the value is exactly 0.9998. This value represents the probability of a difference $(\bar{x}_1 - \bar{x}_2)$ being less than +2.50 miles per hour (or a difference $(\bar{x}_1 - \bar{x}_2)'$ being less than +3.79). The desired probability is:

$$\text{Prob } -3.79 \leq (\bar{x}_1 - \bar{x}_2)' \leq +3.79$$

The conversion is easily made. As the standard normal is symmetric about its mean (0), the probability that $z \leq -3.79$ is the same as the probability that $z \geq +3.79$, or $1.0000 - 0.9998 = 0.0002$. Therefore,

$$\text{Prob } -3.79 \leq (\bar{x}_1 - \bar{x}_2)' \leq +3.79$$
$$= 0.9998 - 0.0002 = 0.9996$$

or

$$\text{Prob } -2.50 \leq (\bar{x}_1 - \bar{x}_2) \leq +2.50 = 0.9996$$

This is interpreted to mean that the probability of a difference less than ± 2.50 miles per hour occurring by chance is 99.96 per cent. The probability of a difference as large as or larger than ± 2.50 miles per hour arising by chance is only 00.04 per cent. It is 99.96 per cent certain that this difference arose due to a significant difference in the characteristics of the measured driving population. It is only 00.04 per cent probable that this difference arose due to pure chance. Therefore, the assumption of equal means is rejected and *the significance of the difference is established.*

The test outlined above computes the probability of a difference in means being between $\pm v$ miles per hour. Often, this is impractical. If a speed law had been instituted as an attempt to reduce speeds at that location, a value of $(\bar{x}_2 - \bar{x}_1)$ which was negative would be entirely contradictory. That would tell the engineer that his attempt to reduce speeds resulted in increased speeds. No statistical analysis is necessary to establish the failure of such an effort. In most practical cases, only one side of the equation is of interest. A slight modification in the analysis is needed.

In the example discussed previously, the probability of $(\bar{x}_1 - \bar{x}_2)$ being less than $+2.50$ miles per hour is the only value with any meaning. This was read directly from Table 8-5 as greater than 0.9998. No further adjustment is needed. It is more than 99.98 per cent probable that a difference in means of $+2.50$ miles per hour (3.79 on the standard normal) arose due to a significant change in driving behavior, and less than 00.02 per cent probable that the difference arose by chance. *The significance of the difference is again established.*

In the first case, the significance of a difference in means is evaluated without regard for whether speed has increased or decreased. In the latter case only the difference of interest is considered (in this case, a decrease).

The *t*-Test. Where sample sizes are small ($n < 30$), the normal approximation is not valid. However, it can be shown that the variable

$$t = \frac{\bar{x}_1 - \bar{x}_2}{s_p\sqrt{(1/n_1) + (1/n_2)}} \tag{8-14}$$

is distributed according to the *t*-distribution, assuming that $\bar{x}_1$ and $\bar{x}_2$ are sample means from the same population. Like the standard normal, the *t*-distribution is tabulated (Table 8-6). As in the use of the normal approximation, the *t*-test is conducted by assuming that both $\bar{x}_1$ and $\bar{x}_2$ are equal (sample means from the same population), and then evaluating the accuracy of that assumption.

In the value t:

$\bar{x}_1$ = sample mean, *before* population

$\bar{x}_2$ = sample mean, *after* population

n_1 = *before* sample size

n_2 = *after* sample size

s_p = pooled standard deviation

where
$$s_p = \sqrt{\frac{(n_1 - 1)s_1^2 + (n_2 - 1)s_2^2}{n_1 + n_2 - 2}}$$

The use of the *t*-test requires that both σ_1 and σ_2 are, in actuality, equivalent. This assumption should be checked statistically. A test for doing so, the *F*-test, is discussed below.

The use of tabulated values of the *t*-distribution (Table 8-6) requires the determination of a value f, *degrees of freedom.* The number of degrees of freedom corresponds to the number of independent pieces of data. In the case of before-and-after studies, this is equivalent to $N_1 + N_2$. However, two degrees of freedom are lost. Note that $\bar{x}_1$ and $\bar{x}_2$ are inputs into the calculation of t. If the average $\bar{x}_1$ is taken as known, then only the first $n - 1$ observations on x_1 are truly independent. Consider the numbers 4, 6, 4. If it were known that the average of four numbers was 5, the fourth number must be 6. Therefore, one degree of freedom is lost for the assumption of each $\bar{x}_1$ and $\bar{x}_2$ in the computation of t. The number of degrees of freedom utilized in the *t*-test is, therefore,

$$f = n_1 + n_2 - 2$$

For small samples (that is, $n_1 < 30$ and/or $n_2 < 30$), the *t*-test rather than the normal approximation must be used to evaluate the significance of an observed difference in means. For these small samples, the use of the improper test may lead to an erroneous conclusion. Consider the following example:

Example

Given:

$x_1 = 50$ $\quad x_2 = 42$

$\sigma_1 = 6.48$ $\quad \sigma_2 = 7.48$

$n_1 = 4$ $\quad n_2 = 8$

By the normal approximation

$$\sigma_D = \sqrt{\frac{(6.48)^2}{4} + \frac{(7.48)^2}{8}} = 4.15$$

$$\frac{50 - 42}{4.15} = 1.92 \text{ standard deviations}$$

Referring to Table 8-5 (normal curve), the probability of a difference as large as +1.92 standard deviations occurring by chance is 1.00 − 0.9725 = 0.274 (or the 97.26 per cent point). The significance is, therefore, statistically established by Normal Approximation. Using the t-Test:

$$s_p = \sqrt{\frac{(4-1)(6.48)^2 + (8-1)(1.48)^2}{4+8-2}} = 7.20$$

$$t = \frac{50-42}{7.20\sqrt{\frac{1}{4}+\frac{1}{8}}} = 1.810$$

$$f = 4 + 8 - 2 = 10$$

Referring to Table 8-6, the probability of a value t as large as 1.810 occurring by chance is slightly more than 5 per cent (1.812 corresponds directly to 5 per cent). Thus, value of t is slightly below the 95 per cent point. As 95 per is the acceptance criteria, *the significance of the observed difference is not established.*

The above exercise illustrates the necessity of using the t-test for small samples (n_1 and/or $n_2 < 30$). At values of >30, the t-test differs from the normal curve only slightly (in the third decimal), and the normal approximation may be used. However, as n gets larger, the normal approximation becomes more accurate.

The *F*-Test. As the use of the t-test involves the assumption of equivalent variances (σ^2), the F-test should be used to investigate the accuracy of that assumption. It can be shown that the variable

$$F = \frac{\sigma_1^2}{\sigma_2^2} \tag{8-15}$$

is distributed according to the F-distribution if σ_1^2 and σ_2^2 are sample variances from the same population. The F-distribution is tabulated in Table 8-7 and involves the use of the value f_1 degrees of freedom for the before study and the value f_2 degrees of freedom for the after population, where

$$f_1 = n_1 - 1$$
$$f_2 = n_2 - 1$$

In the use of the F-test, it is customary to utilize the larger value of σ^2 in the numerator, regardless of whether it is the before or after value.

Applying the F-test to the example of the previous section, we have the following.

Example

$$\sigma_1^2 = (6.48)^2 = 41.8$$
$$\sigma_2^2 = (7.48)^2 = 55.3$$
$$f_2 = 8 - 1 = 7 \text{ (numerator)}$$
$$f_1 = 4 - 1 = 3 \text{ (denominator)}$$
$$F = \frac{\sigma_2^2}{\sigma_1^2} = \frac{55.3}{41.8} = 1.32$$

Note: σ_2^2, The larger value, is used in the numerator. Referring to Table 8-7, for $f_{\text{numerator}} = 7$, $f_{\text{denominator}} = 3$, $F = 5.27$ for $P(F) = 0.10$, $F = 8.89$ for $P(F) = 0.05$, and $F = 14.62$ for $P(F) = 0.025$. As the computed value of F is less than 5.27, suggesting that the probability of occurrence by chance is greater than 10 per cent, *the difference between the variances is not significant.*
Note that in Table 8-7, the values f_1 and f_2 refer to *numerator* and *denominator*, not *before* and *after*.

Goodness of Fit

Much of the previous discussion was predicated on the assumption that spot speed survey results could indeed be represented by a normal distribution. This assumption should be checked using the *Chi-Square* (χ^2) *Goodness of Fit Test*. This test may be used to check the characterization of a given data set by any mathematical distribution. The value of chi-square may be computed as

$$\chi^2 = \sum_{n=1}^{K} \frac{(f_o - f_t)^2}{f_t} \tag{8-16}$$

where
f_o = observed frequencies of the various speed groups
f_t = theoretical frequencies of the various speed groups if the data were perfectly normal
K = number of speed groups

The number of degrees of freedom f for a χ^2 distribution is given by $K - 3$ when a fit to a normal curve is being tested. Three degrees of freedom are lost, since the mean $\bar{x}$, the standard deviation σ, and the total frequency f_t are utilized in the calculation of theoretical frequencies, reducing the number of independent speed groups by 3.

The χ^2 distribution is not accurate when the theoretical frequencies are small. The general rule is not to have theoretical frequencies fewer than 5. This is accomplished by keeping the number dealt with as large as possible. However, should low theoretical frequencies occur in any event, it is desirable to rearrange the table by combining groups, or, in certain cases, by eliminating the low-frequency groups entirely.

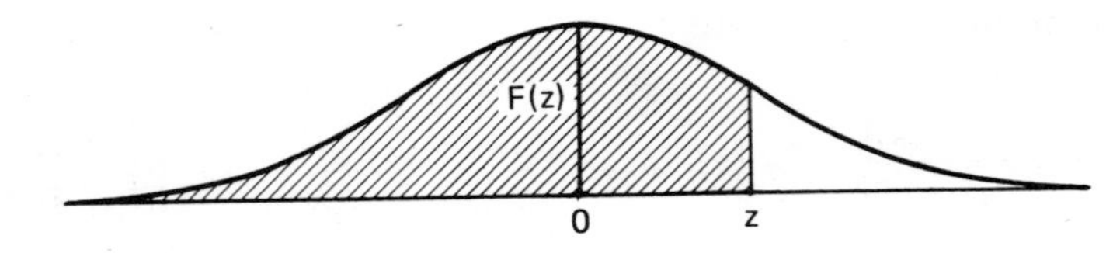

TABLE 8-5. CUMULATIVE NORMAL DISTRIBUTION

$$F(z) = \int_{-\infty}^{z} \frac{1}{\sqrt{2\pi}} e^{-z^2/2}\, dz$$

z	.00	.01	.02	.03	.04	.05	.06	.07	.08	.09
.0	.5000	.5040	.5080	.5120	.5160	.5199	.5239	.5279	.5319	.5359
.1	.5398	.5438	.5478	.5517	.5557	.5596	.5636	.5675	.5714	.5753
.2	.5793	.5832	.5871	.5910	.5948	.5987	.6026	.6064	.6103	.6141
.3	.6179	.6217	.6255	.6293	.6331	.6368	.6406	.6443	.6480	.6517
.4	.6554	.6591	.6628	.6661	.6700	.6736	.6772	.6808	.6844	.6879
.5	.6913	.6950	.6985	.7019	.7054	.7083	.7123	.7157	.7190	.7224
.6	.7257	.7291	.7324	.7357	.7389	.7422	.7454	.7486	.7517	.7549
.7	.7580	.7611	.7642	.7673	.7704	.7734	.7764	.7794	.7823	.7852
.8	.7881	.7910	.7939	.7967	.7995	.8023	.8051	.8078	.8106	.8133
.9	.8159	.8186	.8212	.8238	.8264	.8289	.8315	.8340	.8365	.8389
1.0	.8413	.8438	.8461	.8485	.8508	.8531	.8554	.8577	.8599	.8621
1.1	.8643	.8665	.8686	.8708	.8729	.8749	.8770	.8790	.8810	.8830
1.2	.8849	.8869	.8888	.8907	.8925	.8944	.8962	.8980	.8997	.9015
1.3	.9032	.9049	.9066	.9082	.9099	.9115	.9131	.9147	.9162	.9177
1.4	.9192	.9207	.9222	.9236	.9251	.9265	.9279	.9292	.9306	.9319
1.5	.9332	.9345	.9357	.9370	.9382	.9394	.9406	.9418	.9429	.9441
1.6	.9432	.9463	.9474	.9484	.9495	.9505	.9515	.9525	.9535	.9545
1.7	.9554	.9564	.9573	.9582	.9591	.9599	.9608	.9616	.9625	.9633
1.8	.9641	.9649	.9656	.9664	.9671	.9678	.9686	.9693	.9699	.9706
1.9	.9713	.9719	.9726	.9732	.9738	.9744	.9750	.9756	.9716	.9767
2.0	.9772	.9778	.9783	.9788	.9793	.9798	.9803	.9808	.9812	.9817
2.1	.9812	.9826	.9830	.9834	.9838	.9842	.9846	.9854	.9854	.9857
2.2	.9861	.9864	.9868	.9871	.9875	.9878	.9881	.9884	.9887	.9890
2.3	.9893	.9896	.9898	.9901	.9904	.9906	.9909	.9911	.9913	.9916
2.4	.9918	.9920	.9922	.9925	.9927	.9929	.9931	.9932	.9934	.9936
2.5	.9938	.9940	.9941	.9943	.9945	.9946	.9948	.9949	.9951	.9952
2.6	.9953	.9955	.9956	.9937	.9959	.9960	.9961	.9962	.9963	.9964
2.7	.9965	.9966	.9967	.9968	.9969	.9970	.9971	.9972	.9973	.9974
2.8	.9974	.9975	.9976	.9977	.9977	.9978	.9979	.9979	.9980	.9981
2.9	.9981	.9982	.9982	.9983	.9984	.9984	.9985	.9985	.9986	.9986
3.0	.9987	.9987	.9987	.9988	.9988	.9989	.9989	.9989	.9990	.9990
3.1	.9990	.9991	.9991	.9991	.9992	.9992	.9992	.9992	.9993	.9993
3.2	.9993	.9993	.9994	.9994	.9994	.9994	.9994	.9995	.9995	.9995
3.3	.9995	.9995	.9995	.9996	.9996	.9996	.9996	.9996	.9996	.9997
3.4	.9997	.9997	.9997	.9997	.9997	.9997	.9997	.9997	.9997	.9998

(Source: Crow, Davis, and Maxfield, *Statistics Manual*, U. S. Naval Ordinance Test Station, Dover, N. J., 1960, Table 1)

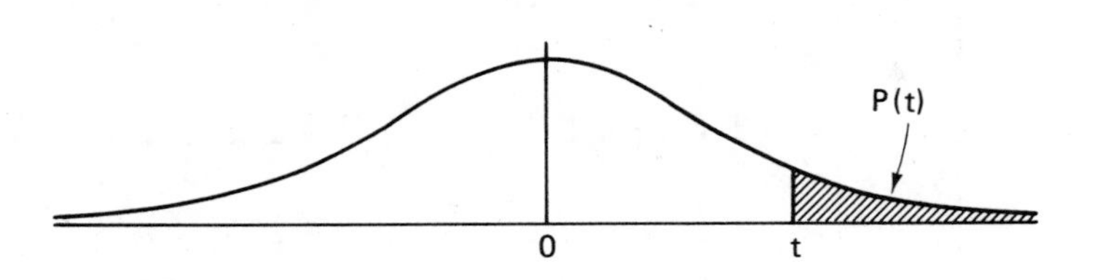

TABLE 8-6. UPPER PERCENTAGE POINTS OF THE *t*-DISTRIBUTION

$$P(t) = \int_t^\infty \frac{(f-1/2)!}{(f-2/2)!\sqrt{\pi f}} (1+t^2/f)^{+(f+1)/2}\, dt$$

$P(t)$ / f	.40	.30	.25	.20	.15	.10	.05	.025	.01	.005	.0003
1	.325	.727	1.000	1.376	1.963	3.078	6.314	12.706	31.821	63.657	636.619
2	.289	.617	.816	1.061	1.386	1.886	2.920	4.303	6.965	9.925	31.598
3	.277	.584	.765	.978	1.250	1.638	2.353	3.182	4.541	5.841	12.924
4	.271	.569	.741	.941	1.190	1.533	2.132	2.776	3.747	4.604	8.610
5	.287	.559	.727	.920	1.156	1.476	2.015	2.571	3.365	4.032	6.869
6	.265	.553	.718	.906	1.134	1.440	1.943	2.447	3.143	3.707	5.959
7	.263	.549	.711	.896	1.119	1.415	1.895	2.365	2.998	3.499	5.408
8	.262	.546	.706	.889	1.108	1.397	1.860	2.306	2.896	3.355	5.401
9	.261	.546	.703	.883	1.100	1.383	1.833	2.262	2.821	3.250	4.781
10	.260	.542	.700	.879	1.093	1.372	1.812	2.228	2.764	3.169	4.587
11	.260	.540	.697	.876	1.088	1.363	1.796	2.201	2.718	3.106	4.437
12	.259	.539	.695	.873	1.083	1.356	1.782	2.179	2.681	3.055	4.318
13	.259	.538	.694	.870	1.079	1.350	1.771	2.160	2.650	3.012	4.221
14	.258	.537	.692	.866	1.076	1.345	1.761	2.143	2.624	2.977	4.140
15	.258	.536	.691	.866	1.074	1.341	1.753	2.131	2.602	2.947	4.073
16	.258	.535	.690	.865	1.071	1.337	1.746	2.120	2.583	2.921	4.015
17	.257	.534	.689	.865	1.069	1.333	1.740	2.110	2.567	2.898	3.965
18	.257	.534	.688	.862	1.067	1.330	1.734	2.101	2.552	2.878	3.922
19	.257	.533	.688	.861	1.066	1.328	1.729	2.093	2.539	2.861	3.883
20	.257	.533	.687	.860	1.064	1.325	1.725	2.086	2.528	2.845	3.850
21	.257	.532	.686	.859	1.063	1.323	1.721	2.080	2.518	2.831	3.819
22	.256	.532	.686	.858	1.061	1.321	1.717	2.074	2.508	2.819	3.792
23	.256	.532	.685	.858	1.060	1.319	1.714	2.069	2.500	2.807	3.767
24	.256	.531	.685	.857	1.059	1.318	1.711	2.064	2.492	2.797	3.745
25	.256	.531	.684	.856	1.058	1.316	1.708	2.060	2.485	2.787	3.725
26	.256	.531	.684	.856	1.058	1.315	1.706	2.056	2.479	2.779	3.707
27	.256	.531	.684	.855	1.057	1.314	1.703	2.052	2.473	2.771	3.690
28	.256	.530	.683	.855	1.056	1.313	1.701	2.048	2.467	2.763	3.674
29	.256	.530	.683	.854	1.055	1.311	1.699	2.045	2.462	2.756	3.659
30	.256	.530	.683	.854	1.055	1.310	1.697	2.042	2.457	2.750	3.646
40	.255	.529	.681	.851	1.050	1.303	1.684	2.021	2.423	2.704	3.551
60	.254	.527	.679	.848	1.046	1.296	1.671	2.000	2.390	2.660	3.460
120	.254	.526	.677	.845	1.041	1.289	1.658	1.980	2.358	2.617	3.373
∞	.253	.524	.674	.842	1.036	1.282	1.645	1.960	2.326	2.576	3.291

(Source: Crow, Davis, and Maxfield, *Statistics Manual,* U. S. Naval Ordinance Test Station, Dover, N. J., 1960, Table 3)

TABLE 8-7. CRITICAL POINTS ON THE *F*-DISTRIBUTION

$$P(F) = \int_F^{\infty} \frac{(f_1+f_2-2)/2!}{(f_1-2)/2!\,(f_2-2)/2!} f_1^{f_1/2} f_2^{f_2/2} F^{(f_1-2)/2}(f_2+f_1F)^{-(f_1+f_2)/2}\,dF$$

NOTE: The number of degrees of freedom for the numerator is f_1*, for the denominator,* f_2.

$P(F) = 0.10$

f_2 \ f_1	1	2	3	4	5	6	7	8	9	10	12	15	20	24	30	40	60	120	∞
1	39.86	49.50	53.59	55.83	57.24	58.20	58.91	59.44	59.86	60.20	60.70	61.22	61.74	62.00	62.26	62.53	62.79	63.06	63.33
2	8.53	9.00	9.16	9.24	9.29	9.33	9.35	9.37	9.38	9.39	9.41	9.42	9.44	9.45	9.46	9.47	9.47	9.48	9.49
3	5.54	5.46	5.39	5.34	5.31	5.28	5.27	5.25	5.24	5.23	5.22	5.20	5.18	5.18	5.17	5.16	5.15	5.14	5.13
4	4.54	4.32	4.19	4.11	4.05	4.01	3.98	3.95	3.94	3.92	3.90	3.87	3.84	3.83	3.82	3.80	3.79	3.78	3.76
5	4.06	3.78	3.62	3.52	3.45	3.40	3.37	3.34	3.32	3.30	3.27	3.24	3.21	3.19	3.17	3.16	3.14	3.12	3.10
6	3.78	3.46	3.29	3.18	3.11	3.04	3.01	2.98	2.96	2.94	2.90	2.87	2.84	2.82	2.80	2.78	2.76	2.74	2.72
7	3.59	3.26	3.07	2.96	2.88	2.83	2.78	2.75	2.72	2.70	2.67	2.63	2.59	2.58	2.56	2.54	2.51	2.40	2.47
8	3.46	3.11	2.92	2.81	2.73	2.67	2.62	2.59	2.56	2.54	2.50	2.46	2.42	2.40	2.38	2.30	2.34	2.32	2.29
9	3.36	3.01	2.81	2.69	2.61	2.55	2.51	2.47	2.44	2.42	2.38	2.34	2.30	2.28	2.25	2.23	2.21	2.18	2.16
10	3.28	3.92	2.73	2.61	2.52	2.46	2.41	2.38	2.35	2.32	2.28	2.24	2.20	2.18	2.16	2.13	2.11	2.08	2.06
11	3.23	2.86	2.68	2.54	2.45	2.39	2.34	2.30	2.27	2.25	2.21	2.17	2.12	2.10	2.08	2.05	2.03	2.00	1.97
12	3.18	2.81	2.61	2.48	2.39	2.33	2.28	2.24	2.21	2.19	2.15	2.10	2.06	2.04	2.01	1.99	1.96	1.93	1.90
13	3.14	2.76	2.56	2.43	2.35	2.28	2.23	2.20	2.16	2.14	2.10	2.05	2.01	1.98	1.96	1.93	1.90	1.88	1.85
14	3.10	2.73	2.52	2.39	2.31	2.24	2.19	2.15	2.12	2.10	2.05	2.01	1.96	1.94	1.91	1.89	1.86	1.83	1.80
15	3.07	2.70	2.49	2.36	2.27	2.21	2.16	2.12	2.09	2.06	2.02	1.97	1.92	1.90	1.87	1.85	1.82	1.79	1.76
16	3.05	2.67	2.46	2.33	2.24	2.18	2.13	2.09	2.06	2.03	1.99	1.94	1.89	1.87	1.84	1.81	1.78	1.75	1.72
17	3.03	2.64	2.44	2.31	2.22	2.15	2.10	2.06	2.03	2.00	1.96	1.91	1.86	1.84	1.81	1.78	1.75	1.72	1.60
18	3.01	2.62	2.42	2.29	2.20	2.13	2.08	2.04	2.00	1.98	1.93	1.89	1.84	1.81	1.78	1.75	1.72	1.69	1.66
19	2.99	2.61	2.40	2.27	2.18	2.11	2.06	2.02	1.98	1.96	1.91	1.86	1.81	1.79	1.76	1.73	1.70	1.67	1.63
20	2.97	2.59	2.38	2.25	2.16	2.09	2.04	2.00	1.96	1.94	1.89	1.84	1.79	1.77	1.74	1.71	1.68	1.64	1.61
21	2.96	2.57	2.38	2.23	2.14	2.08	2.02	1.98	1.95	1.92	1.88	1.83	1.78	1.75	1.72	1.69	1.66	1.62	1.59
22	2.95	2.56	2.35	2.22	2.13	2.06	2.01	1.97	1.93	1.90	1.86	1.81	1.76	1.73	1.70	1.67	1.64	1.60	1.57
23	2.94	2.55	2.34	2.21	2.11	2.05	1.99	1.95	1.92	1.89	1.84	1.80	1.74	1.72	1.69	1.66	1.62	1.59	1.55
24	2.93	2.54	2.33	2.19	2.10	2.04	1.98	1.94	1.91	1.88	1.83	1.78	1.73	1.70	1.67	1.64	1.61	1.57	1.53
25	2.92	2.53	2.32	2.18	2.09	2.02	1.97	1.93	1.89	1.87	1.82	1.77	1.72	1.69	1.66	1.63	1.59	1.56	1.52
26	2.91	2.52	2.31	2.17	2.08	2.01	1.96	1.92	1.88	1.86	1.81	1.76	1.71	1.68	1.65	1.61	1.58	1.54	1.50
27	2.90	2.51	2.30	2.17	2.07	2.00	1.95	1.91	1.87	1.85	1.80	1.75	1.70	1.67	1.64	1.60	1.57	1.53	1.49
28	2.89	2.50	2.29	2.16	2.06	2.00	1.94	1.90	1.87	1.84	1.79	1.74	1.69	1.66	1.63	1.59	1.56	1.52	1.48
29	2.89	2.50	2.28	2.15	2.06	1.99	1.93	1.89	1.86	1.83	1.78	1.73	1.68	1.65	1.62	1.58	1.55	1.51	1.47
30	2.89	2.49	2.28	2.14	2.05	1.98	1.93	1.88	1.85	1.82	1.77	1.72	1.64	1.64	1.61	1.57	1.54	1.50	1.46

TABLE 8-7. (contin'd)

$P(F) = 0.10$

f_2 \ f_1	1	2	3	4	5	6	7	8	9	10	12	15	20	24	30	40	60	120	∞
40	2.84	2.44	2.23	2.09	2.00	1.93	1.87	1.83	1.79	1.76	1.71	1.66	1.61	1.57	1.54	1.51	1.47	1.42	1.38
60	2.79	2.39	2.18	2.04	1.95	1.87	1.82	1.77	1.74	1.71	1.66	1.60	1.54	1.51	1.48	1.44	1.40	1.35	1.29
120	2.75	2.35	2.13	1.99	1.90	1.82	1.77	1.72	1.68	1.65	1.60	1.54	1.48	1.45	1.41	1.37	1.32	1.26	1.19
∞	2.71	2.30	2.08	1.94	1.85	1.77	1.72	1.67	1.63	1.00	1.55	1.49	1.42	1.39	1.34	1.30	1.24	1.17	1.00

$P(F) = 0.05$

f_2 \ f_1	1	2	3	4	5	6	7	8	9	10	12	15	20	24	30	40	60	120	∞
1	161,45	199.50	215.71	224.56	230.16	233.99	236.77	238.88	240.54	241.88	243.91	245.95	248.01	249.05	250.09	251.14	252.20	253.25	254.32
2	18.51	19.00	19.16	19.25	19.30	19.33	19.35	19.37	19.38	19.40	19.41	19.43	19.45	19.45	19.46	19.47	19.48	19.49	19.50
3	10.13	9.55	9.28	9.12	9.01	8.94	8.89	8.85	8.81	8.79	8.74	8.70	8.66	8.64	8.62	8.59	8.57	8.55	8.53
4	7.71	6.94	6.59	6.39	6.26	6.16	6.00	6.04	6.00	5.96	5.91	5.86	5.80	5.77	5.75	5.72	5.69	5.66	5.63
5	6.61	5.79	5.41	5.19	5.05	4.95	4.88	4.82	4.77	4.74	4.68	4.62	4.56	4.53	4.50	4.46	4.43	4.40	4.36
6	5.99	5.14	4.76	4.53	4.39	4.28	4.21	4.15	4.10	4.06	4.00	3.94	3.87	3.84	3.81	3.77	3.74	3.70	3.67
7	5.59	4.74	4.35	4.12	3.97	3.87	3.79	3.73	3.68	3.64	3.57	3.51	3.44	3.41	3.38	3.34	3.30	3.27	3.23
8	5.32	4.46	4.07	3.84	3.69	3.58	3.50	3.44	3.39	3.35	3.28	3.22	3.15	3.12	3.08	3.04	3.01	2.97	2.93
9	5.12	4.26	3.86	3.63	3.48	3.37	3.29	3.23	3.18	3.14	3.07	3.01	2.94	2.90	2.86	2.83	2.79	2.75	2.71
10	4.96	4.10	3.71	3.48	3.33	3.22	3.14	3.07	3.02	2.98	2.91	2.84	2.77	2.74	2.70	2.66	2.62	2.58	2.54
11	4.84	3.98	3.59	3.36	3.20	3.09	3.01	2.95	2.90	2.85	2.79	2.72	2.65	2.61	2.57	2.53	2.49	2.45	2.40
12	4.75	3.89	3.49	3.26	3.11	3.00	2.91	2.85	2.80	2.75	2.69	2.62	2.54	2.51	2.47	2.43	2.38	2.34	2.30
13	4.67	3.81	3.41	3.18	3.03	2.92	2.83	2.77	2.71	2.67	2.60	2.53	2.46	2.42	2.38	2.34	2.30	2.25	2.21
14	4.60	3.74	3.34	3.11	2.96	2.85	2.76	2.70	2.65	2.60	2.53	2.46	2.39	2.35	2.31	2.27	2.22	2.18	2.13
15	4.54	3.68	3.29	3.06	2.90	2.79	2.71	2.64	2.59	2.54	2.48	2.40	2.33	2.29	2.25	2.20	2.16	2.11	2.07
16	4.49	3.63	3.24	3.01	2.85	2.74	2.66	2.59	2.54	2.49	2.42	2.35	2.28	2.24	2.19	2.15	2.11	2.06	2.01
17	4.45	3.59	3.20	2.96	2.81	2.70	2.61	2.55	2.49	2.45	2.38	2.31	2.23	2.19	2.15	2.10	2.06	2.01	1.96
18	4.41	3.55	3.16	2.93	2.77	2.66	2.58	2.51	2.46	2.41	2.34	2.27	2.19	2.15	2.11	2.06	2.02	1.97	1.92
19	4.38	3.52	3.13	2.90	2.74	2.63	2.54	2.48	2.42	2.38	2.31	2.23	2.16	2.11	2.07	2.03	1.98	1.93	1.88
20	4.35	3.49	3.10	2.87	2.71	2.60	2.51	2.45	2.39	2.35	2.28	2.20	2.12	2.08	2.04	1.99	1.95	1.90	1.84
21	4.32	3.47	3.07	2.84	2.68	2.57	2.49	2.42	2.37	2.32	2.25	2.18	2.10	2.05	2.01	1.96	1.92	1.87	1.81
22	4.30	3.44	3.05	2.82	2.66	2.55	2.46	2.40	2.34	2.30	2.23	2.15	2.07	2.03	1.98	1.94	1.89	1.84	1.78
23	4.28	3.42	3.03	2.80	2.64	2.53	2.44	2.37	2.32	2.27	2.20	2.13	2.05	2.00	1.96	1.91	1.86	1.81	1.76
24	4.26	3.40	3.01	2.78	2.62	2.51	2.42	2.36	2.30	2.25	2.18	2.11	2.03	1.98	1.94	1.89	1.84	1.79	1.73
25	4.24	3.39	2.99	2.76	2.60	2.49	2.40	2.34	2.28	2.24	2.16	2.09	2.01	1.96	1.92	1.87	1.82	1.77	1.71
26	4.23	3.37	2.98	2.74	2.59	2.47	2.39	2.32	2.27	2.22	2.15	2.07	1.99	1.95	1.90	1.85	1.80	1.75	1.69
27	4.21	3.35	2.96	2.73	2.57	2.46	2.37	2.31	2.25	2.20	2.13	2.06	1.97	1.93	1.88	1.84	1.79	1.73	1.67
28	4.20	3.34	2.95	2.71	2.56	2.45	2.36	2.29	2.24	2.19	2.12	2.04	1.96	1.91	1.87	1.82	1.77	1.71	1.65
29	4.18	3.33	2.93	2.70	2.55	2.43	2.35	2.28	2.22	2.18	2.10	2.03	1.94	1.90	1.85	1.81	1.75	1.70	1.64
30	4.17	3.32	2.92	2.69	2.53	2.42	2.33	2.27	2.21	2.16	2.09	2.01	1.93	1.89	1.84	1.79	1.74	1.68	1.62
40	4.08	3.23	2.84	2.61	2.45	2.34	2.25	2.18	2.12	2.08	2.00	1.92	1.84	1.74	1.79	1.69	1.64	1.58	1.51
60	4.00	3.15	2.76	2.53	2.37	2.25	2.17	2.10	2.04	1.99	1.92	1.84	1.75	1.70	1.65	1.59	1.53	1.47	1.39
120	3.92	3.07	2.68	2.48	2.29	2.18	2.09	2.02	1.96	1.91	1.83	1.75	1.66	1.61	1.55	1.50	1.43	1.35	1.23
∞	3.84	3.00	2.60	2.37	2.21	2.10	2.01	1.94	1.88	1.83	1.75	1.67	1.57	1.52	1.46	1.39	1.32	1.22	1.00

TABLE 8-7. (contin'd)

$P(F) = 0.025$

f_2 \ f_1	1	2	3	4	5	6	7	8	9	10	12	15	20	24	30	40	60	120	∞
1	647.79	799.50	864.16	899.58	921.85	937.11	948.22	956.66	963.28	968.63	976.71	984.10	993.10	997.25	1001.4	1005.6	1009.8	1014.0	1018.3
2	38.51	39.00	39.16	39.25	39.30	39.33	39.36	39.37	39.37	39.40	39.42	39.43	39.45	39.46	39.46	39.47	39.48	39.49	39.50
3	17.44	16.04	15.44	15.10	14.88	14.74	14.62	14.54	14.47	14.42	14.34	14.25	14.17	14.12	14.08	14.04	13.99	13.95	13.90
4	12.22	10.65	9.98	9.60	9.36	9.20	9.07	8.98	8.90	8.84	8.75	8.66	8.56	8.51	8.46	8.41	8.34	8.31	8.26
5	10.01	8.43	7.76	7.39	7.15	6.98	6.85	6.76	6.68	6.62	6.52	6.43	6.33	6.28	6.23	6.18	6.12	6.07	6.02
6	8.81	7.26	6.60	6.23	5.99	5.82	5.70	5.60	5.52	5.46	5.37	5.27	5.17	5.12	5.07	5.01	4.06	4.90	4.85
7	8.07	6.54	5.89	5.52	5.29	5.12	4.99	4.90	4.82	4.76	4.67	4.57	4.47	4.42	4.36	4.31	4.25	4.20	4.14
8	7.57	6.06	5.42	5.05	4.82	4.65	4.53	4.43	4.36	4.30	4.20	4.10	4.00	3.95	3.89	3.84	3.78	3.73	3.67
9	7.21	5.71	5.08	4.72	4.48	4.32	4.20	4.10	4.03	3.96	3.87	3.77	3.67	3.61	3.50	3.51	3.45	3.39	3.33
10	6.94	5.46	4.83	4.47	4.24	4.07	3.95	3.85	3.78	3.72	3.62	3.52	3.42	3.37	3.31	3.26	3.20	3.14	3.08
11	6.72	5.26	4.63	4.28	4.04	3.88	3.76	3.66	3.59	3.53	3.43	3.33	3.23	3.17	3.12	3.06	3.00	2.94	2.88
12	6.55	5.10	4.47	4.12	3.89	3.73	3.61	3.51	3.44	3.37	3.28	3.18	3.07	3.02	2.96	2.91	2.85	2.79	2.72
13	6.41	4.97	4.35	4.00	3.77	3.60	3.48	3.39	3.31	3.25	3.15	3.05	2.95	2.89	2.84	2.78	2.72	2.66	2.60
14	6.36	4.86	4.24	3.89	3.66	3.50	3.38	3.29	3.21	3.15	3.05	2.95	2.84	2.79	2.73	2.67	2.61	2.55	2.49
15	6.20	4.76	4.15	3.80	3.58	3.41	3.29	3.20	3.12	2.96	2.96	2.86	2.76	2.70	2.58	2.58	2.52	2.46	2.40
16	6.12	4.69	4.08	3.73	3.50	3.34	3.22	3.12	3.05	2.99	2.89	2.79	2.68	2.63	2.57	2.51	2.45	2.38	2.32
17	6.04	4.62	4.01	3.66	3.44	3.28	3.16	3.06	2.98	2.92	2.82	2.72	2.62	2.56	2.50	2.44	2.38	2.32	2.25
18	5.98	4.56	3.95	3.61	3.38	3.22	3.10	3.01	2.93	2.87	2.77	2.67	2.56	2.50	2.44	2.38	2.32	2.26	2.19
19	5.92	4.51	3.90	3.56	3.33	3.17	3.05	2.96	2.88	2.82	2.72	2.62	2.51	2.45	2.39	2.33	2.27	2.20	2.13
20	5.87	4.46	3.86	3.51	3.29	3.13	3.01	2.91	2.84	2.77	2.68	2.57	2.46	2.41	2.35	2.29	2.22	2.16	2.09
21	5.83	4.42	3.82	2.48	3.25	3.09	2.97	2.87	2.80	2.73	2.64	2.53	2.42	2.37	2.31	2.25	2.18	2.11	2.04
22	5.79	4.38	3.78	3.44	3.22	3.05	2.93	2.84	2.76	2.70	2.60	2.50	2.39	2.33	2.27	2.21	2.14	2.08	2.00
23	5.75	4.35	3.75	3.41	3.18	3.02	2.90	2.81	2.73	2.67	2.57	2.47	2.36	2.30	2.24	2.18	2.11	2.04	1.97
24	5.72	4.32	3.72	3.38	3.15	2.99	2.87	2.78	2.70	2.64	2.54	2.44	2.33	2.27	2.21	2.15	2.08	2.01	1.94
25	5.69	4.29	3.69	3.35	3.13	2.97	2.85	2.75	2.68	2.61	2.51	2.41	2.30	2.24	2.18	2.12	2.05	1.98	1.91
26	5.66	4.27	3.67	3.33	3.10	2.94	2.82	2.73	2.65	2.59	2.49	2.39	2.28	2.22	2.16	2.09	2.03	1.95	1.88
27	5.63	4.24	3.65	3.31	3.08	2.92	2.80	2.71	2.63	2.57	2.47	2.36	2.25	2.19	2.13	2.07	2.00	1.93	1.85
28	5.61	4.22	3.63	3.29	3.06	2.90	2.78	2.69	2.61	2.55	2.45	2.34	2.23	2.17	2.11	2.05	1.98	1.91	1.83
29	5.59	4.20	3.61	3.27	3.04	2.88	2.76	2.67	2.59	2.53	2.43	2.32	2.21	2.15	2.09	2.03	1.96	1.89	1.81
30	5.57	4.18	3.59	3.25	3.03	2.87	2.75	2.65	2.57	2.51	2.41	2.31	2.20	2.14	2.07	2.01	1.94	1.87	1.79
40	5.42	4.05	3.46	3.13	2.90	2.74	2.62	2.53	2.45	2.39	2.29	2.18	2.07	2.01	1.94	1.88	1.80	1.72	1.64
60	5.29	3.93	3.34	3.01	2.79	2.63	2.51	2.41	2.33	2.27	2.17	2.06	1.94	1.88	1.82	1.74	1.67	1.58	1.48
120	5.15	3.80	3.23	2.89	2.67	2.52	2.39	2.30	2.22	2.16	2.05	1.94	1.82	1.76	1.69	1.61	1.53	1.43	1.31
∞	5.02	3.69	3.12	2.79	2.57	2.41	2.29	2.19	2.11	2.05	1.94	1.83	1.71	1.64	1.57	1.48	1.39	1.27	1.00

The test for goodness of fit to a normal distribution curve, using the chi-square test, is carried out for the illustrative spot speed study in Table 8-8. The *area method* of fitting a normal curve to the observed distribution is used. The area included within any number of standard deviations may be obtained from prepared tables of area of the normal curve.

Columns of Table 8-8 are described below. The χ^2 distribution is listed in Table 8-9. As in previous tests, a 95 per cent assurance that the data is significantly different from a normal distribution is required to establish that difference.

Column 1: 5-miles per hour speed groups
2: the number of speeds recorded in each speed group, i.e., observed frequencies f_o
3: deviations of the furthest interval limit from the mean value
4: deviations of interval limits from the mean, expressed in numbers of standard deviations, i.e., column 3 divided by σ
5: per cent area (probability) between interval limit and the mean (obtained from standard normal, Table 8-5)
6: per cent area (probability) in each group, obtained by successively subtracting values of column 5
7: per cent frequencies f_t, obtained by multiplying column 6 values by the total number of observations (300 in the sample problem). Theoretical frequencies less than 5 must be combined here.
8: differences, $f_o - f_t$
9: squares of differences $(f_o - f_t)^2$
10: squares of differences divided by f_t, $(f_o - f_t)^2/f_t$

The summation of column 10 values is the value of χ^2.

The area method for calculating values of theoretical frequencies is illustrated in Figure 8-5.

Column 5 values represent the per cent area or probability between the class limit and the mean. In Figure

TABLE 8-8. χ^2 FIT TEST ON SPOT SPEED DATA FIT TO A NORMAL CURVE

1	2	3	4	5	6	7		8	9	10
SPEED GROUP (mph)	NUMBER OF VEHICLES OBSERVED f_o	DEVIATION OF CLASS LIMIT FROM MEAN	DEVIATION IN TERMS OF σ	PER CENT OF AREA BETWEEN CLASS LIMIT AND MEAN	PER CENT OF AREA IN CLASS INTERVAL	THEORETICAL FREQUENCY f_t		$f_o - f_t$	$(f_o - f_t)^2$	$\frac{(f_o - f_t)^2}{f_t}$
15–19.9	6	−23.7	−2.76	49.71	1.17	3.51	15.90	−1.90	3.61	0.23
20–24.9	8	−18.7	−2.18	48.54	4.13	12.39				
25–29.9	29	−13.7	−1.59	44.41	10.03	30.09		−1.09	1.19	0.04
30–34.9	60	−8.7	−1.01	34.38	17.74	53.22		6.78	45.97	0.86
35–39.9	63	−3.7 +1.2	−0.43 +0.14	16.64 5.57	22.21	66.63		−3.63	13.18	0.20
40–44.9	74	+6.2	0.72	26.42	20.85	62.55		11.45	131.10	2.10
45–49.9	29	+11.2	1.30	40.32	13.90	41.70		−12.70	161.29	3.86
50–54.9	19	+16.2	1.88	46.99	6.67	20.01		−1.01	1.02	0.05
55–59.9	10	+21.2	2.46	49.31	2.32	6.96	8.67	3.33	11.09	1.28
60–64.9	2	+26.2	3.04	49.88	0.57	1.71				

Known values of sample spot speed survey: x = 38.7 mph
σ = 86 mph
n = 300

χ = summation of column 10 = 8.62
degrees of freedom = 8 − 3 = 5

Probability level (Table 8–9) $= 0.10 + \frac{0.62}{2.61}(0.15) = 13.6\%$

As 13.6% > 5%, the data may be assumed to be normal

TABLE 8-9. UPPER PERCENTAGE POINTS OF THE χ^2 DISTRIBUTION

$$P(\chi^2) = \int_{\chi^2}^{\infty} \frac{1}{(f-2/2)!\,2^{1/2}} (\chi^2)^{(f-2)/2} e^{-\chi^2/2}\, d(\chi^2)$$

f \ $P(\chi^2)$	.995	.990	.975	.950	.900	.750	.500	.250	.100	.050	.025	.010	.005
1	3927×10^{-2}	1571×10^{-7}	9821×10^{-7}	3932×10^{-6}	0.01579	0.1015	0.4549	1.323	2.706	3.841	5.024	6.635	7.879
2	0.01003	0.02010	0.05064	0.1026	.2107	.5754	1.386	2.773	4.605	5.991	7.378	9.210	10.60
3	.07172	.1148	.2158	.3518	.5844	1.213	2.366	4.108	6.251	7.815	9.348	11.34	12.34
4	.2070	.2971	.4844	.7107	1.064	1.923	3.357	5.585	7.779	9.488	11.14	13.28	14.86
5	.4117	.5543	.8312	1.145	1.610	2.675	4.351	6.626	9.236	11.07	12.83	15.09	16.75
6	.6757	.8721	1.237	1.635	2.204	3.455	5.348	7.841	10.64	12.59	14.45	16.81	18.55
7	.9893	1.259	1.690	2.167	2.833	4.255	6.346	9.037	12.02	14.07	16.01	18.48	20.28
8	1.344	1.646	2.180	2.733	3.199	5.071	7.344	10.22	13.36	15.51	17.53	20.09	21.96
9	1.735	2.088	2.700	3.325	4.168	5.899	8.343	11.39	14.68	16.92	19.02	21.67	23.59
10	2.150	2.558	3.247	3.940	4.865	6.737	9.342	12.55	15.99	18.31	20.48	23.21	25.19
11	2.603	3.053	3.816	4.575	5.578	7.584	10.34	13.70	17.28	19.68	21.92	24.72	26.76
12	3.074	3.571	4.404	5.226	6.304	8.458	11.34	14.85	18.55	21.03	23.34	26.22	28.30
13	3.565	4.107	5.009	5.892	7.042	9.299	12.34	15.98	19.81	22.36	24.74	27.69	29.82
14	4.075	4.660	5.629	6.571	7.790	10.17	13.34	17.12	21.06	23.68	26.12	29.14	31.32
15	4.601	5.229	6.262	7.261	8.547	11.04	14.34	18.25	22.31	25.00	27.49	30.58	32.80
16	5.142	5.812	6.908	7.962	9.312	11.91	15.34	19.37	23.54	26.30	28.85	32.00	34.27
17	5.697	5.408	7.564	8.672	10.09	12.79	16.34	20.49	24.77	27.59	30.19	33.41	35.72
18	6.265	7.015	8.231	9.390	10.86	13.68	17.34	21.60	25.99	28.87	31.53	34.81	37.16
19	6.844	7.644	8.907	10.12	11.65	14.56	18.34	22.72	27.20	30.14	32.85	36.19	38.58
20	7.434	8.260	9.591	10.85	12.44	15.45	19.34	23.83	28.41	31.41	34.17	37.57	40.00
21	8.034	8.897	10.28	11.59	13.24	16.34	20.34	24.93	29.62	32.67	35.48	38.93	41.40
22	8.643	9.542	10.98	12.34	14.04	17.24	21.34	26.04	30.81	33.92	36.78	40.29	42.80
23	9.260	10.20	11.69	13.09	14.85	18.14	22.34	27.14	32.01	35.17	38.08	41.64	44.18
24	9.886	10.86	12.40	13.85	15.66	19.04	23.24	28.24	33.20	36.42	39.36	42.98	45.56
25	10.52	11.52	13.12	14.61	16.47	19.94	24.34	29.34	34.38	37.65	40.65	44.31	46.93
26	11.16	12.20	13.84	15.38	17.29	20.84	25.34	30.43	35.56	38.89	41.92	45.64	48.29
27	11.81	12.88	14.57	16.15	18.11	21.75	26.34	31.53	36.74	40.11	43.19	46.96	49.64
28	12.46	13.56	15.31	16.93	18.94	22.66	27.34	32.62	37.92	41.34	44.46	48.28	50.99
29	13.12	14.26	16.05	17.71	19.77	23.57	28.34	33.71	39.09	42.56	45.72	49.59	52.34
30	13.79	14.95	16.79	18.49	20.60	24.48	29.34	34.80	43.26	43.77	46.98	50.89	53.67
40	20.71	22.16	24.43	26.51	29.05	33.66	39.34	45.62	51.80	55.76	59.34	63.69	66.77
50	27.99	29.71	32.36	34.76	37.69	42.94	49.33	56.33	63.17	67.50	71.42	76.15	79.49
60	35.53	37.48	40.48	43.19	46.46	52.29	59.33	66.98	79.08	79.08	83.30	88.38	91.95
70	43.28	45.44	48.76	51.74	55.33	61.70	69.33	77.58	85.53	90.53	95.02	100.42	104.22
80	51.17	53.54	57.15	60.39	64.28	71.14	79.33	88.13	96.58	101.88	106.63	112.33	116.32
90	59.20	61.75	65.65	69.13	73.29	80.62	89.33	98.65	107.56	113.14	118.14	124.12	128.30
100	67.33	70.00	74.22	77.93	82.36	90.13	99.33	109.14	118.50	124.34	129.56	135.81	140.17
χr	−2.576	−2.326	−1.960	−1.645	−1.282	−0.6745	0.0000	+0.6745	+1.282	+1.645	+1.960	+2.326	+2.576

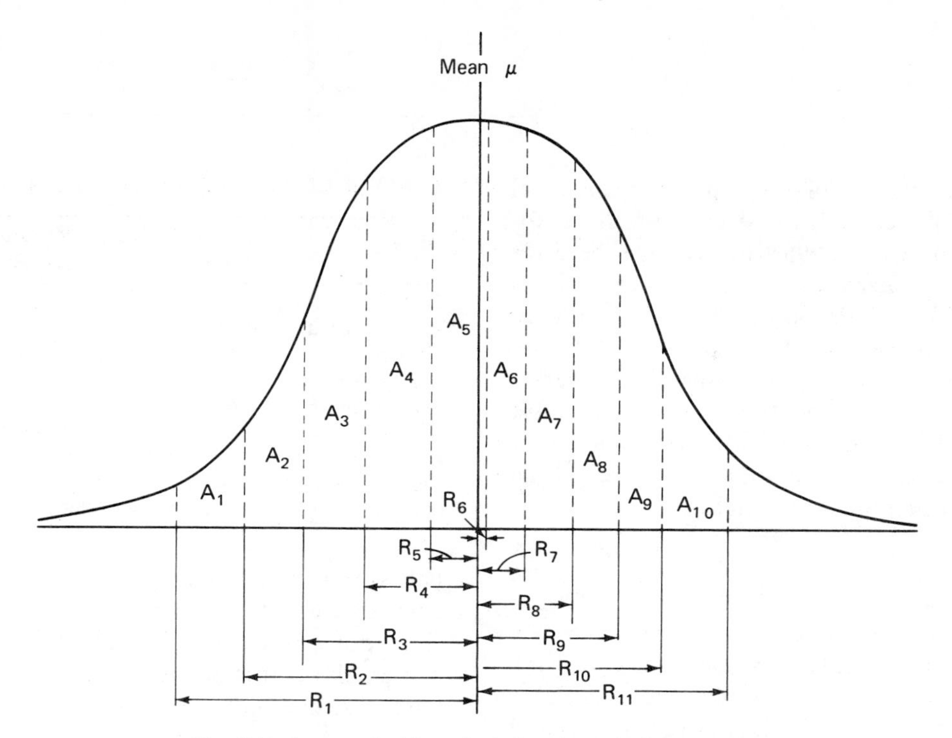

Fig. 8-5. Area method for calculating theoretical frequencies

8-5, these are illustrated as R_1, R_2, R_3, etc. These are found by taking the column 4 values and using Table 8-5, the standard normal distribution.

Column 6 values, A_1, A_2, etc., represent the area, or per cent occurrence in each group. Pictorially, these are computed as follows:

$$A_1 = R_1 - R_2$$
$$A_2 = R_2 - R_3$$
$$A_3 = R_3 - R_4$$
$$A_4 = R_4 - R_5$$
$$\boxed{A_5 = R_5 + R_6} \longleftarrow \text{Note the special case of the speed group which straddles the mean.}$$
$$A_6 = R_7 - R_6$$
$$A_1 = R_8 - R_1$$
$$A_8 = R_9 - R_8$$
$$A_9 = R_{10} - R_9$$
$$A_{10} = R_{11} - R_{10}$$

The χ^2 test is carried out in Table 8-8.

SPOT SPEED CHARACTERISTICS

Speed Trends

Changes in the operating characteristics of motor vehicles with respect to speed and braking ability, and improvements in highway design, have resulted in steady increases in operating speeds on all highways. National average (arithmetic mean) speeds on main rural highways have increased (based on the results of 638 speed studies compiled by the BPR for 26 states in 1957),[11] as is shown in the following values for the national average speeds of all vehicles: 1941, 47.1 miles per hour, 1948, 47.7, 1952, 49.5, 1957, 51.4, 1963, 55.8 (see Table 8-10.)

TABLE 8-10. NATIONAL AVERAGE SPEEDS–1963

VEHICLE TYPE	*AVERAGE SPEED (mph)*	*PERCENT OVER 50 mph*
Passenger Cars	57.1	77.5
Trucks	51.3	55.0
Buses	58.1	77.5

(Source: *Highway Statistics, Summary to 1965,* USGP, March, 1967)

TABLE 8-11. AVERAGE SPEEDS ON N.Y. STATE FACILITIES IN 1967

FACILITY	*THRUWAY*	*INTERSTATE*	*INTERSTATE*	*STATE HIGHWAY*
Speed Limit (mph)	65	65	60	50
Average speed (mph)				
PC	64.3	62.5	61.2	48.6
T	59.6	58.6	58.5	47.3
B	65.4	66.0	60.0	48.7
All	63.8	62.1	60.8	48.4
85th Percentile (mph)				
All	70.1	68.8	66.0	54.8
Pace, (mph)				
All	60–70	59–69	56–66	44–54

(Source: *1967 Annual Speed Survey* PR7-70200-01, NYSDOT December, 1967)

Although the national average speeds on main rural highways in 1964 were slightly less than those for passenger cars and buses in 1963, the average speed for trucks was slightly increased.[56] Speed characteristics on various types of facilities in the state of New York for 1967 are given in Table 8-11.

It is important to note that all of the above characteristics relate to free-flowing traffic under favorable conditions, and that they are not representative of various traffic and weather conditions. They exceed the legal speed limit for all highway systems.

Speed Distribution

Drivers are influenced in the selection of their speeds by many factors, such as traffic volume, traffic density, roadway conditions, vehicle conditions, and environmental conditions, as well as the restraints of speed regulations.

When traffic and roadway conditions are such that drivers can select whatever speed they desire, i.e., a free-flowing condition, there is always a wide range in the speeds at which the various individuals operate their vehicles. Under these circumstances the speed distribution follows a normal distribution. (The significance of different portions of the speed distribution curve was discussed previously.) As shown by the speed trends in the tables, it is uncommon for the average speed to be much in excess of 50 miles per hour. Comparatively few passenger-car drivers want to travel at speeds which approach the potential speeds of their vehicles, even under the most favorable traffic and roadway conditions. Maximum speeds attainable by production-model passenger cars increased from 75 miles per hour in 1930 to 115 miles per hour for 1958 models.

Effect of Volume on Speed

Investigations conducted on an extensive scale have definitely shown that there is a straight-line relationship between traffic volume and average speed when other conditions are identical and the critical traffic density is not exceeded.[64] The *critical traffic density* occurs at about 30 miles per hour, with a total of approximately 2,000 passenger cars per hour for a two-lane, two-direction highway, and with approximately 2,000 passenger cars per hour, per lane, for divided four-lane facilities. The average speed decreases, linearly as the traffic volume increases.[62-69]

The effect of lane width on speed was found for two bridges, one with 9-foot, $8\frac{3}{4}$-inch lanes, and the other with 11-foot, 3-inch lanes. Both traffic volume and average speed were higher for the wider lanes.[58] The wider lanes produced an average speed of 39 miles per hour compared with 27 miles per hour on the narrower lanes. Conditions at the terminal of the bridges might have had an effect. It is possible that one bridge had a better street distribution system than the other.

TABLE 8-12. SPEEDS ON ILLINOIS STATE HIGHWAYS

	PASSENGER CARS	*LIGHT TRUCKS*	*MEDIUM TRUCKS*	*HEAVY TRUCKS*	*BUSES*
Daytime Average Speed (mph)	54.9	46.5	44.7	46.8	52.5
Nighttime Average Speed (mph)	53.0	49.0	47.9	47.7	56.3

(Source: "Report of Speeds at High Speed Locations on Ilinois Rural State Highways," in *IDH*, February, 1956)

Variation in Speeds by Time of Day

It was found in a study of freeways[63] that the daytime and nighttime average speeds were essentially the same, but that there was considerable difference between daytime and nighttime speed distributions. These same results were reported in another study,[70] but this included urban locations ranging from controlled-access highways to a surface arterial with parking along commercially developed frontage. Relatively large fluctuations in speed occurred during the early morning hours, mainly because of low traffic volumes during these periods. For most of the locations the magnitude of the fluctuations during daytime periods was relatively small, except for periods of peak volume.

In an area of transition from rural to urban conditions, it was found that the hourly average speeds did not differ greatly throughout the daylight and evening hours at each location, although the speed did naturally vary from one location to another along the transition. At none of the locations was the traffic volume sufficiently close to the capacity of the roadway to cause significant reduction in speed due to volume alone.[15]

At selected high-speed locations on Illinois rural state highways in 1955, the variations noted in Table 8-12 were found.

Variation in Average Speeds for Different Types of Vehicles

For free-flowing traffic, the average speed for buses exceeds that of passenger cars, which exceeds that of trucks. In a comparison of different types of passenger cars,[17] there was no great difference in speed on urban expressways for compact or foreign cars as compared to standard American passenger cars.

Variation in Speeds with Road Conditions

In New York State, on rural highways, it was found that as drivers approached vertical curves with short sight distances (between 150 and 500 feet) they invariably reduced their speeds to some extent. However, the reduction in speed was far less than that required to provide safe operating conditions.[18] In other studies,[19] motorists reduced their speeds to a much greater extent when they approached horizontal curves with short sight distances, than for vertical curves. Substantial speed reductions occurred when sight distance was below 1,000 to 1,200 feet.[19]

As concerns the effects of roadway delineation on speed, no significant changes in speed patterns took place after improved delineation at hazardous locations, but nighttime passenger-car speeds were lower than daytime speeds by 1 to 5 miles per hour.[20] In measuring the effects of highway illumination and delineation,[21] similar results were obtained, in that there was no significant difference in average nighttime speeds due to illumination or delineation. However, it was found that center-line-marked two-lane roads had an average speed 4 miles per hour higher than those without.[22]

In observing acceleration and deceleration lanes at a number of locations, there were significant differences in speed between the acceleration-and deceleration-lane traffic at the time of merging or diverging and that of the through lanes, but the acceleration- or deceleration-lane traffic had little effect on the speed of through-lane traffic.[23]

In considering direct-taper type and parallel type off-ramps, it was found that the direct taper was definitely superior with respect to speed characteristics,[24] while left-hand exit ramps on freeways had no adverse effect on the speed of through traffic.[25]

The speed of traffic has been found to be significantly higher when the spacing between interchanges was increased. Interchange spacing was changed by closing intermediate ramps.[26-28] Closing off-ramps caused no improvement in overall operating conditions.[27]

In formulating models (multiple linear regression equations) which could be used to estimate mean spot speeds on two-lane rural highways, it has been found that, of eight variables considered, the most pronounced influence on speed characteristics was caused by curvature.[29,30] The other seven variables were out-of-state car, truck combination, gradient, sight distance, lane width, roadside development, and volume. It has also been found that stream friction (essentially geometric cross-section elements) and vehicle composition accounted for the major proportion of variation in average speed.[31]

The installation of a guardrail in a narrow median (4 feet wide) had no effect on speed characteristics, but accident frequency was increased.[32] Objects placed on the highway shoulder have a tendency to reduce speeds, but not significantly, regardless of how close the object is located to the pavement edge. The average speed reduction for passenger cars was 3 miles per hour on two-lane pavements 16 to 20 feet wide, and 1 miles per hour on pavements 22 to 26 feet wide.[33] Shoulder surface types of approximately the same width do not affect speed, but speed increased slightly (2 to 5 miles per hour) on sections of two-lane roads with 8-foot shoulders as compared with 4-foot shoulders.[34,35]

The Effects of Development on Speed

Continuous residential development causes a reduction in speeds.[15] The extent of this reduction is more dependent on the lateral restrictions it imposes than on the relative density of development. Extensive planting of trees and shrubbery near the roadway contributes to this result. Concentrated commercial development in transition areas causes substantial reductions in speeds. Also, it was found that the faster drivers in the rural areas are generally the faster drivers throughout the transition areas, and the speeds of vehicles making trips entirely within a developed area are consistently slower than vehicles traveling through the developed area.

Overall travel speeds are significantly influenced by four factors. Commercial development, urban development, and stream friction all adversely affect travel speed, and rural development affects speed in a positive manner. *Commercial development* means sections with a concentration of commercial establishments, access drives, and related conditions; *urban development* means those sections which pass through urban areas; *stream friction* describes those sections where conditions within the traffic stream cause congestion; and *rural development* includes sections with little roadside development. Commercial development has the greatest adverse affect on travel speed and accounts for a reduction of about 30 per cent.[59]

Effects of Weather on Speeds

Bad weather will lower speeds, the amount of reduction being dependent on the severity of the weather. The effect of inclement weather is to reduce the average speed of all traffic as much as 40 per cent, and congestion can occur at about 60 per cent of the critical traffic density.

These percentages are based on free-flowing conditions during favorable weather.[36]

Effects of the Driver on Speed

Eighty-four per cent of passenger-car drivers in one study felt that the 65-miles per hour speed limit was reasonable, and it is interesting to note that the 85th percentile speed is used to establish speed limits.[39] In one study drivers were guided to various speeds without the aid of a speedometer and were asked to estimate the speed, while in another study drivers were asked to produce various speeds. Estimated values were greater than the produced values in the high-speed range and lower in the low-speed range and the two regression lines intersected in the vicinity of 40 miles per hour.[40] Speed estimation is more accurate from peripheral vision than from acute vision, and deceleration is more effectively sensed than acceleration.[41]

Although drivers are unable to judge accurately the speed of an opposing vehicle, drivers are able to make good judgments of the distance to an oncoming vehicle. Experiments have shown that drivers can make effective use of oncoming vehicle speed information in executing passing maneuvers.[61]

Variability of Spot Speed Measurements

In determining the reliability of spot speed samples for predicting the speed of the traffic stream, and in evaluating changes in speed behavior produced by artificial means, it appears that there is an inherent variability in speeds.[42] It was found that samples taken at different time periods, even though environmental conditions were comparable, cannot be used to interpret the significance of a difference in the means of samples.

There are significant differences in hourly, daily and monthly mean speeds even though study locations were uncongested during the time period of sampling.[43,44]

9

Volume Studies

Traffic volume studies are conducted to obtain accurate information about the number and movement of vehicles and/or pedestrians within or through an area, or at selected points within the area.

DEFINITIONS[1]

Volume: The number of vehicles passing a given point during a specified period of time, or the number of vehicles that pass over a given section of a lane or a roadway during a specified period of time.

Average annual daily traffic: The total yearly volume divided by the number of days in the year, commonly abbreviated as AADT.

Average daily traffic: The total volume during a given time period, in whole days greater than one day and less than one year, divided by the number of days in that time period, commonly abbreviated as ADT.

Maximum annual hourly volume: The highest hourly volume that occurs on a roadway in a designated year.

Thirtieth highest annual hourly volume: The hourly volume that is exceeded by 29 hourly volumes during a designated year. Corresponding definitions apply to any other ordinal highest hourly volume, such as the tenth, twentieth, etc.

Density: The number of vehicles occupying a unit length of a moving lane or moving lanes of a roadway at a given instant, usually expressed in vehicles per mile.

Average density: The average number of vehicles per unit length of roadway over a specified period of time.

Volume and density should not be confused and are not interchangeable. The three basic factors of traffic flow are *volume* (V), *speed* (S), and *density* (D). The relationship between these factors, when volume is expressed in vehicles per hour, is $V = DS$. The density (in vehicles per mile) is equal to the volume (in vehicles per hour) divided by the space mean speed (in miles per hour). It is possible to have a very low traffic volume with a high traffic density. In fact, the highest traffic densities occur when vehicles are practically at a standstill, in which case the traffic volume would approach zero. Therefore, density may be a better measure of roadway service than volume, since it continues to increase as congestion increases. For any given volume, the efficiency of flow can be determined by observing the density. The smaller the density, the quicker the movement of traffic, and the greater the efficiency of flow.[2]

Critical density: The density of traffic when the volume is at capacity on a given roadway or lane. At a density either greater or smaller than the critical density, the volume of traffic will be decreased. Critical density occurs when all vehicles are moving at or about the same speed.

NEED FOR DATA ON TRAFFIC VOLUMES

Traffic volume information is extremely important in traffic planning, design, operation, and research. Types of volume information vary according to the application of the data.

Annual total traffic volumes are used for:

1. Measuring and establishing trends in traffic volume
2. Determining annual travel in vehicle miles as economic justification for proposed expenditures
3. Computing accident rates per 100 million vehicle-miles
4. Estimating highway user revenues

AADT or ADT volumes are used for:

1. Highway planning activities, such as developing freeway systems and major or arterial street systems, selecting through streets, and selecting the best route for a new facility or improvement
2. Highway programming to determine the need for and priority of street improvements
3. Measuring the present demand for service by the street or highway
4. Evaluating the present traffic flow with respect to the existing street or highway system

The direct use of AADT in geometric design is not appropriate because it does not indicate the significant variations in volume occurring during the various months of the year, days of the week, hours of a day, and intervals of an hour.

Peak hourly volumes are used for:

1. Geometric design with respect to number and width of lanes, channelization, intersection design, ramp design, shoulder design, and similar geometric features of the highway
2. Determining deficiencies in capacity
3. Justifying, planning, and locating traffic control devices such as traffic signals, traffic signs, and traffic markings
4. Developing traffic operation programs, such as designation of through streets, one-way streets, unbalanced or offset lane flow, and traffic routing
5. Justifying other regulatory measures, such as parking, turning, and stopping restrictions
6. Highway classification
7. Justifying and planning enforcement

Classified volumes giving the types of vehicles, number of axles, weight and dimensions are used for:

1. Geometric design with respect to minimum turning paths, clearances, grades, etc.
2. Structural design of the pavement, bridges, and other highway facilities
3. Analysis of capacity in determining the effect of commercial vehicles.
4. Estimating highway user revenues
5. Adjusting machine counts

Short-term volumes for intervals less than an hour (1-, 5-, 6-, 10- or 15-minute intervals) are used for:

1. Analyzing maximum rates of flow and variations within peak hours
2. Determining capacity limitations in urban areas, since hourly traffic flows are not sufficiently definitive of flow variations
3. Determining the characteristics of peak volumes
4. Providing an economical means of obtaining volume data

Intersectional volume counts are made to determine:

1. Total traffic entering the intersection for all legs
2. Total traffic executing each of the possible turning movements
3. Total traffic by time periods
4. Classification of vehicles by type

Information relative to volumes and classification of vehicles involved in turning movements are important for intersection design, interchange design, channelization, accident analysis (to help establish remedy measures), capacity analysis, congestion analysis, planning of effective controls (such as traffic signal timing and phasing), and turning and parking restrictions.

Mid-block volume counts are made to determine:

1. Total traffic from each direction
2. Total traffic by time periods
3. Classification of vehicle by type

Information about directional distribution of volumes and classification of vehicles are important for capacity analysis, planning one-way streets, unbalanced or offset lane flow, signal systems and timing, parking restrictions, and justification of various types of traffic control measures.

Cordon volume counts are made to determine the accumulations of vehicles and/or persons, during a typical time period, within a district called the *cordon area.* A *cordon line* defines the area, and each street crossing the cordon line is a *count station,* where all vehicles and/or persons entering and leaving the district are counted. Cordon count information is important for planning adequate parking facilities, long range planning of arterials and freeways, planning and adjusting transit service to needs, providing a basis for the evaluation and introduction of traffic operational techniques including traffic control devices and traffic regulatory measures, and planning selective enforcement measures.

Screen-line volume counts are made at crossings of natural or man-made barriers, such as waterways, railroad tracks, expressways, parks, and ridges or at crossings of an imaginary line. *Screen lines* are used to divide the area into large districts for the purpose of expanding origin and destination data and traffic assignments. This is accomplished by detecting long-range changes in volume and direction of traffic from significant changes in land use and travel patterns.

Pedestrian volume counts are made at problem locations (often where pedestrians are involved in accidents), at crosswalks, at bus and subway stations, and at mid-block locations. Pedestrian counts provide the basic data needed to evaluate the adequency of existing walkways, crosswalks, and protection and control facilities, such as pedestrian barriers, pedestrian signals and timing of traffic signals. They provide data for planning future facilities such as pedestrian overpasses and underpasses and for studying the causes of pedestrian accidents.[3]

METHODS FOR CONDUCTING VEHICULAR COUNTS

Mechanical Counters[4, 5]

Fixed or *permanent counters* are employed for continuous counts, recording the distribution of traffic by hours of the day, days of the week, months of the year, and from year to year. Such counts are extremely important in establishing traffic trends and characteristics, and for developing adjustment factors to convert short-term counts to estimated AADT. For detecting or sensing vehicles, permanent counters use any of the following devices:

Electric Contact Device. This consists of a subsurface detector which provides for a positive electrical contact for each vehicle axle crossing it. It consists of a steel base plate over which a molded and vulcanized rubber pad (which is flush with the road surface) holds suspended a strip of spring steel. Electric contacts are installed in each lane.

Photoelectric Device. Detection is accomplished by the vehicle passing between a source of light and a photocell which is capable of distinguishing between light and lack of light. The equipment is necessarily mounted above the road surface.

Radar Device. Detection is accomplished by continuously comparing the frequency of a transmitted radio signal with the frequency of the reflected signal. When a moving vehicle intercepts the signal, a frequency difference exists. The unit is normally mounted above the center of the lane or lanes for which detection is desired.

Magnetic Device. Detection is accomplished by a signal or impulse caused by a vehicle passing through and disturbing a magnetic field. The unit is installed in each lane immediately below the road surface.[6]

Ultrasonic Device. This is similar in its operation and installation to a radar unit.

Infrared Device. This type utilizes a pickup cell, which is similar to a photoelectric cell, but is sensitive to infrared (heat) radiation rather than to visible light. The unit containing the source and the pickup is mounted above the road surface on a bridge, sign structure, mast arm, etc.

The location of detectors is of great importance. Generally the location will be determined by purpose, type of detector, type of vehicle or pedestrian to be detected, and kind of installation.[66] Various types of recorders can be used together with the above detecting devices to obtain traffic volumes. These include:

1. *Printed tape,* recorders which receive the impulse from a detector, store it in a register having print-face wheels, and periodically print out the results on a continuous tape. The counts are accumulated for a specific time period, say 15 minutes or an hour.
2. *Circular graphical charts,* which are capable of recording traffic volumes from 0 to 1,000 vehicles for intervals of 5,- 10-, 15-, 20-, 30- or 60-minute counting periods, and for periods of 24 hours or 7 days. This operates on the principle that the distance the recording pen moves out from the center of the chart is a function of volume, and the rotation is a function of time.
3. *Special computer tapes,* which, when connected to a keypunch machine, produce punched cards to be used in conventional computers.

In 1965, the Georgia State Highway Department implemented an automatic volume data acquisition and processing procedure. Four continuous counting stations were connected via special telephone lines, provided by an independent utility, to an IBM 1710 real-time computer system, which was programmed to receive, format, edit, and analyze the data. At two stations, pneumatic road tubes were used; at one, an ultrasonic detector was employed; the remaining station utilized a magnetic detector. In all cases electrical counting devices were employed. After one year of exhaustive testing, the telemetry system and computer operation employed in this data reduction and collection proved successful, and produced greater efficiencies than other methods in use in Georgia.[62]

Portable counters are employed for short-term, periodic counts at various locations throughout an area. They are battery-operated and employ pneumatic detectors (road tube) placed across all or part of a roadway, and air-impulse switches to actuate the counting mechanism. When the counter is used in conjunction with printed tape, it is called a *recording counter*, and volumes are printed on tape every 15 minutes or every hour. When the counter is used in conjunction with a visual register, it is called a *nonrecording counter*, and cumulative volumes are shown on dials for the period of operation. These dials must be read by an observer at desired intervals.[8]

Advantages of mechanical counters are:

1. They have a relatively low cost per hour of counting.
2. They provide an extensive coverage of time, from which variations and trend data are obtained.
3. With some detecting devices a definite separation of vehicle volumes by lane is obtained, but with road tubes the probability of error from simultaneous wheel passings is not serious (for traffic volumes under 3,000 vehicle per hour).

Disadvantages of mechanical counters are:

1. They cannot obtain turning movements or vehicle classification data. Ultrasonic detectors have been used to obtain the latter by means of a relay closing a contact whenever a vehicle (whose top surface is above a preset height) is within the detection zone. Normally, the units are adjusted so that vehicles whose top surfaces are higher than 68 inches cause this relay to close its contact. Actuation of this relay therefore indicates the presence of a high (commercial) vehicle in the detection zone.[9]
2. Detectors and recorders are subject to vandalism and many traffic hazards, which accounts for the need for extensive maintenance.
3. With a visual register there is no way of knowing if the unit has been inoperative between readings.
4. With some detectors it is necessary to determine the percentage of vehicles having three or more axles to obtain true volumes.

Manual Counts

This kind of count uses field observers to obtain volume data which cannot be collected by mechanical counters. For light volumes, observations are recorded by tally marks on prepared field data sheets, and for heavier volumes mechanical hand counters are used. Manual counts are used to determine:

1. Turning movements
2. Vehicle classification
3. Occupancy studies (the number of occupants of vehicles)
4. Pedestrian counts

Manual counts result in relatively high costs and are subject to the limitations of human factors, generally precluding 24-hour continuous counts.

Moving Vehicles Method

The procedure and typical data analysis for this method of obtaining traffic volume data are described in Travel Time and Delay, Chapter 7.

Photographic Techniques

These are discussed in Spot Speed Studies, Chapter 8. The Port of New York Authority has developed a system for traffic data acquisition utilizing time-sequenced, vertical aerial photography which it calls *sky count*. It is designed to accomplish two basic kinds of aerial traffic studies. One uses a fixed-wing aircraft flying at relatively high altitudes, to cover large areas and identify problem locations. The other uses a helicopter which hovers over problem locations to obtain detailed data for analysis.

In addition to obtaining vehicular and pedestrian volume data, the techniques have been used to study stream-flow characteristics, and for a variety of transportation planning purposes.[10-14]

Correlation of Drivometer Events and Volume

Based on data collected with a drivometer-equipped test car operating on an urban arterial, a correlation between recorded drivometer events and traffic volume was formulated[97]. The *drivometer* is an electromechanical device which provides, on a time and/or distance basis, a visual measure of driver performance and vehicle motion. These characteristics include total travel time, change in speed, running time, small steering reversals ($\frac{3}{8}$-inch change in position of steering wheel), large steering reversals (1-inch change), brake applications, accelerator applications, and changes in direction of travel. The variables which correlated best with traffic volume were large steering reversals and brake applications. By combining these two significant variables and using a multiple regression analysis, the following equation was developed to predict the traffic volume on the facility:

$$Y = 0.18364 + 0.16027X_1 + 0.11922X_2$$

where Y = 15-minute traffic volume
X_1 = number of large steering reversals
X_2 = number of brake applications

Comparisons between measured and computed volumes showed that for 12 of the 30 test runs, computed values were within ± 10 per cent of actual values.

SCHEDULING COUNTING PERIODS

The counting period at a specific location depends upon the method used to obtain data and the purpose or planned use of the data. The counting period should avoid special event conditions such as holidays, sporting events, exhibitions or fairs, transit strikes, special sales, unusual weather, or temporary street closures, unless the count is for the specific purpose of studying the effects of the special event condition. Mechanical counts are usually made for periods of 24 hours or longer. Manual counts, including turning movements and vehicle classifications, and possibly pedestrians, are made for the following periods.

1. 12 hours; 7:00 A.M.–7:00 P.M.
2. 8 hours; 7:00 A.M.–11:00 A.M, 2:00 P.M.–6:00 P.M.
3. 4 hours; 7:00 A.M.–9:00 A.M, 4:00 P.M.–6:00 P.M.

Scheduling of Periodic Volume Counts in Rural Areas

The Bureau of Public Roads (BPR) method is outlined below:[15]

1. *Permanent stations:* An average of 30 to 40 permanent-count stations are distributed throughout the state to cover the various types of highways. These continuous-count stations should be carefully located to reflect traffic patterns on the various types of highways throughout the state, since the data are used to expand shorter counts at other locations.

2. *Control count stations:* These are required to establish seasonal and daily traffic volume characteristics, and factors for expanding data from single counts obtained at coverage stations. Two types of control counts are recommended.

(a) *Major control counts:* A monthly or bimonthly count should be conducted, with each count to consist of three weekdays, a Saturday, and a Sunday. Counts are to be made with recording counters. It is estimated that one major control station is required for each 90 coverage stations.

(b) *Minor control counts:* These should be conducted four to six times a year, with each count covering 48 hours on weekdays only. Counts are to be made with recording counters. It is estimated that two minor control stations are required for each 90 coverage stations.

3. *Coverage count stations:* These are required to provide sufficient data to permit reasonably accurate estimates of ADT on each road section. Counts are not usually made on roads having an ADT of less than 25. Counts are made for a 48-hour period (minimum of 24 hours) once a year on weekdays, normally with non-recording counters.

The number of coverage stations required may be approximated by dividing the total rural mileage by 2, to eliminate low-volume roads, and then dividing again by the average length of road section (considered as 2 miles). Therefore, the approximate number of coverage stations may be obtained by dividing the total rural mileage by 4.

In a state having 90,000 miles of rural road, approximately 22,500 coverage stations, 500 minor stations, and 250 major control stations would be required.

4. *Classification counts:* Manual counts are made at selected control stations. The suggested schedule of classification counts at continuous-count stations is:

(a) Where the volume is 2,000 vehicles per density or greater, make a 24-hour classification count on two weekdays, a Saturday, and a Sunday, six times a year, in alternate months, and also include a holiday count.

(b) At $\frac{1}{5}$ of the remaining stations make a 16-hour count on one weekday, a Saturday, and a Sunday, in each of the four seasons.

(c) At the remaining stations make an 8-hour count on a weekday in each of the four seasons.

Scheduling of Periodic Volume Counts in Urban Areas

The procedure recommended by the National Committee on Urban Transportation is outlined below:[16]

1. *Street classification:* The first step in developing the counting program is to determine street use classifications.[17]

(a) *Major streets* are designated as *expressways, major arterials*, and *collector* streets.

(b) *Minor* or *local streets* are designated as *residential, commercial*, and *industrial streets.*

2. *Selecting control stations:* Counts at these stations provide the control factors necessary to record volume counts on a common basis. The procedure recommended is a sampling process, and special counts are required to develop factors that may be applied to the sample counts for adjustment to ADT volumes.

(a) *Major control stations* are selected to sample representatively the traffic volume on the major street system. One major control station should be located on each major street. The minimum duration and frequency of counting is a 24-hour, directional, recording machine count every second year. The purpose of the major control count is to obtain a pattern of hourly and directional variations in traffic volume which is typical of the entire route.

(b) *Minor control stations* are selected to sample typical streets of each class in the minor street system. A minimum of nine minor control stations should be established on the local or minor street system in a very small city. Three stations would be located on each class of streets. The initial number of stations selected will vary with the size of the city and with the importance attached to the determination of local street volumes. A 24-hour, nondirectional, recording machine count is made at each station every second year.

(c) *Key counts* are made to obtain an index of daily and seasonal variations in volume for each of the various classes of streets. At least one key station should be selected from the control stations to represent each class of street in both the major and minor systems. The duration and frequency of counts for determining daily variations is a nondirectional, recording machine count for a seven-day period performed every year. For determining seasonal variations, a 24-hour, nondirectional, recording machine weekday count at three-month intervals is performed annually. In areas where the weather is subject to pronounced variations, a 24-hour count should be made at each key station during each month.

3. *Coverage counts on major street systems:* The major street system is divided into *control sections* that are homogeneous for the purposes of inventory and keeping records. Manual 1A[17] describes the method of classifying streets and designating control sections. One 24-hour, nondirectional machine, weekday count should be made within each control section. The counts should be repeated every four years. The purpose of these counts is to estimate ADT volumes throughout the system.

4. *Coverage counts on minor street system:* To obtain a broad coverage of the minor street system, one 24-hour, nondirectional, nonrecording machine count should be made for every mile of minor street. No specific frequency is recommended, but the counts should be made when local circumstances indicate a need. Instead of stationary counts, the moving vehicle method may be used to obtain minor street coverage counts.

5. *Central traffic district cordon count:* The cordon count is used to measure the transportation activity generated by the central business district (CBD). This is done by counting the vehicles and persons entering and leaving the CBD by all modes of travel during a typical time period. The procedure for obtaining information on persons arriving by all modes of travel except mass transit is available.[16] Information on persons arriving by transit is obtained by procedures described in Manual 4A.[18] The data obtained from the cordon count are particularly important in determining the transportation needs of the area with regard to transit service, arterial planning, and need for parking facilities.

Cordon counting stations are established at mid-block locations on all streets crossing the cordon line. Mid-block locations are selected to simplify the counting by eliminating turning movements. At stations where the peak-hour volume exceeds 600 vehicles, a directional count should be made. The counts are usually made from 7:00 A.M. to 6:00 P.M. on a normal weekday (Tuesday, Wednesday, or Thursday). The counts are made with recording machines, and vehicle occupancy and classification are sampled by manual counts. The cordon line should be counted every other year.

6. *Screen-line study:* Counts are taken at screen-line locations to detect long-range changes in volume and direction of traffic due to significant changes in land use and travel patterns. The screen line is counted in the same manner as the cordon line. An 18-hour count should be taken at each station from 6:00 A.M. to midnight. Screen lines should be counted every other year.

PRESENTATION OF TRAFFIC VOLUME DATA

Traffic Flow Maps

These maps show volume along various routes by using bands proportional to the traffic volume carried. This provides rapid visualization of the relative volumes carried by different streets and highways throughout the area. The numerical traffic volumes are often entered on the bands for more accurate interpretation. It is also advisable to show separate bands representing flows in opposite directions, since large differences may exist between opposing volumes. The volumes shown on a traffic flow map may be for the peak hour, ADT, or volumes for other time intervals.[3,4,16,46]

Intersection Flow Diagrams

These diagrams give the direction and volume for all movements, by bands, through an intersection. Generally, peak-hour counts are shown if it is a problem intersection or if volume data are taken for traffic control purposes. For general planning surveys it may be desirable to prepare additional flow diagrams for ADT volume or for any other time interval that may be of interest.[3,46]

Variation of Fluctuation Charts

These are charts which present the hourly, daily, or monthly changes in volume through an area, along a given route, or across a screen line. They readily disclose the peak periods of flow.[3,4,16,46]

Trend Charts

These show the volume changes over a period of years.

TABLE 9-1. SAMPLE INTERSECTION SUMMARY PEAK-HOUR FLOWS—POLY AVE. & CIVIL ST.

FROM SOUTH ON POLY AVE.						*FROM WEST ON CIVIL ST.*					
L *into Civil*		S *into Poly*		R *into Civil*		L *into Poly*		S *into Civil*		R *into Poly*	
PC	T	PC	T	PC	T	PC	T	PC	T	PC	T
20	9	593	73	37	20	44	6	400	40	58	32
FROM NORTH ON POLY AVE.						*FROM EAST ON CIVIL ST.*					
L *into Civil*		S *into Poly*		R *into Civil*		L *into Poly*		S *into Civil*		R *into Poly*	
PC	T	PC	T	PC	T	PC	T	PC	T	PC	T
19	1	516	14	65	5	58	2	299	41	70	5

Summary Tables

These are tables which summarize traffic volume data, such as ADT, peak-hour flows, classification counts, and counts for various time periods, in tabular form.

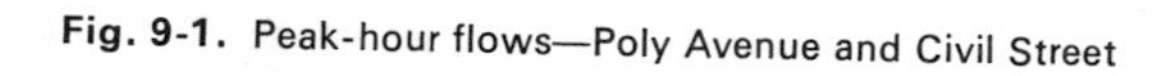

Fig. 9-1. Peak-hour flows—Poly Avenue and Civil Street

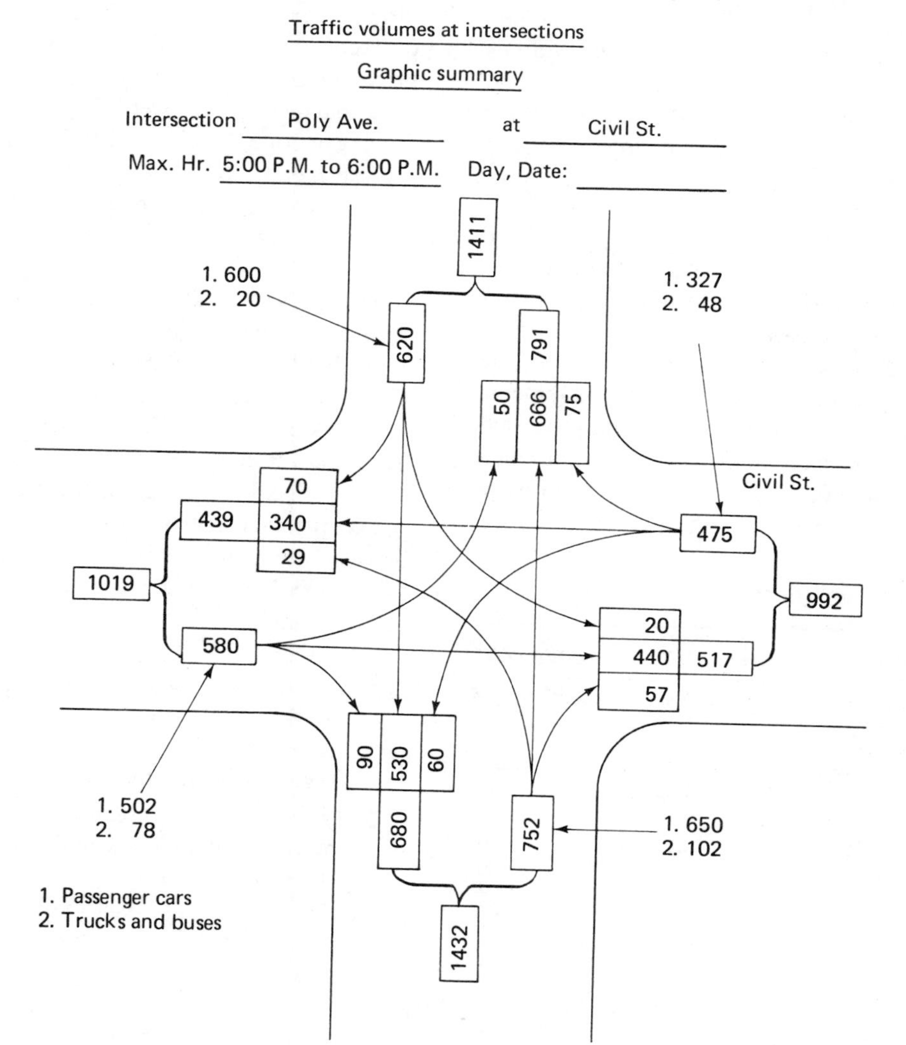

Graphic Summary Intersection Sheet

These sheets present a picture of how much traffic flows through the intersection and which way it flows for a peak hour, ADT, or full period of survey. This is extremely desirable in studies of accidents, signal timing, and other intersection studies. Normally a graphic summary sheet or intersection flow diagram is used.[4] For example, consider the following data summarized from a field sheet for a peak-hour flow. The graphic summary for the data shown in Table 9-1 is shown in Figure 9-1.

TRAFFIC VOLUME CHARACTERISTICS

Traffic volumes can never be considered to be static and, therefore, volume data are only accurate for the time of the counts. However, since variations in volumes are generally rhythmic and repetitive, knowledge of volume characteristics is important in scheduling traffic counts, in relating volumes at one time or place to volumes as some other time or place, and in determining the accuracy of counts.

Elements in Volume Variations[1,19]

An understanding of the variations in traffic volume within peak hours, by hours of the day, days of the week, months of the year, directional distribution, and lane distribution, is important in traffic engineering operations and in planning.

Traffic Pattern. A traffic pattern is a tabular or graphical representation of the fluctuation in traffic volume over a specified period of time. The volume, during the increments of time used in the pattern, may be expressed as numbers of vehicles or as the percentage which these numbers are of the average number for all the increments covered in the pattern. Different facilities may be said to have like traffic patterns if similar curves result when traffic volumes are shown as a percentage of the average volume for the period covered by the pattern.

TABLE 9-2. AVERAGE VOLUMES BY LANE[20-27]

TOTAL HOURLY VOLUME	*MEDIAN LANE*		*CENTER LANE*		*SHOULDER LANE*	
	Per cent	*Volume*	*Per cent*	*Volume*	*Per cent*	*Volume*
1,000	22	220	47	470	31	310
2,000	31	620	43	860	26	520
3,000	35	1.050	40	1,200	25	750
4,000	37	1,480	38	1,520	25	1,000
5,000	37	1,850	37	1,850	26	1,300
6,000	37	2,220	37	2,220	26	1,560

Hourly Traffic Pattern. Traffic volumes for consecutive time increments of less than one hour (such as 1-, 5-, 6-, or 15-minute increments) are shown for the entire hour, usually the peak hour.

Daily Traffic Pattern. Traffic volumes for each of 24 consecutive hours are shown. If the period of time is extended to include 168 consecutive hours, it is termed a *daily traffic pattern for one week.* In like manner, daily traffic patterns for one month, one year, or any other period of time may be obtained.

Weekly Traffic Pattern. Daily traffic volumes for each of 7 consecutive days are shown. If the period of time is extended to include 30 consecutive days, the pattern is termed a *weekly traffic pattern for one month*, and if volume for 365 consecutive days are shown it is termed a *weekly traffic pattern for one year.*

Monthly Traffic pattern. Monthly traffic volumes for each of 12 consecutive months are shown.

Directional Distribution. Distribution of movement shows marked variations during peak-hour flows. Conditions vary widely among different facilities and in different locations within urban and rural areas. During peak periods it is possible for the traffic volume to be so unbalanced that over 80 per cent of the traffic is flowing in one direction. Distribution by directions on a given facility usually does not change materially from year to year. Typical directional distributions for urban and rural highways have been developed.[1]

Lane Distribution. The distribution of total traffic volume among the various lanes of a multi-lane facility varies with the lane location and with changes in volume. These conclusions result from a number of studies of freeway operations[20-30]. The lane distributions observed from these studies were fairly consistent, and the average values for six-lane, urban facilities are given in Table 9-2.

The primary reasons for the rather large difference between the volume carried by the shoulder lane and the other two lanes, when the facility is approaching its capacity, include:

1. Points of conflict, primarily ramps, are usually adjacent to the shoulder lane.
2. Certain drivers prefer to drive at slower speeds, and they normally use the shoulder lane.
3. Trucks normally use the shoulder lane and usually travel at a speed less than the desired rate of speed of passenger cars.

Type and Location of Highway or Street

Traffic patterns depend on the type of route and its location. Rural highways have different patterns from urban streets and highways. In rural areas patterns have distinctive variations which are generally repetitive although differing by type of highway. Freeway, primary, secondary, and local roads as well as toll roads all have different patterns, as do recreational routes. Urban streets have distinctive patterns, and so do expressways, major arterials, collector streets, residential streets, commercial streets, and industrial streets. However, traffic shows great and reasonably predictable fluctuation by time period.

Hourly Variations. There is usually no pronounced early morning peak in rural areas, but in urban areas there is a pronounced peak which occurs in the 7:00 to 9:00 A.M. interval. For both rural and urban locations, the greatest peak throughout the day occurs in the 4:00 to 6:00 P.M. interval. Local rural roads exhibit greater variations than do primary state routes. Local rural roads also have distinctive morning peaks.

In urban areas there is a high directional peak in-bound in the morning on expressways and radial arterials leading into the CBD, and out-bound in the evening. However, on some circumferential routes there are no pronounced peaks, and the directional distribution is more evenly balanced.

For many urban locations the variation within one hour may be considerable. A Michigan study of seven different multi-lane urban highways of various types revealed that the average peak 1-minute, 5-minute, and 15-minute volumes, expressed as a percentage of the peak-hour volume, were 2.8, 10.7, and 28.5 per cent, respectively.[26] A study of 792 signalized intersection approaches on all types of streets revealed that the average peak 15-minute volume was 29.3 per cent of the total hourly volume.[31] Relationships of peak 5-minute volumes to peak-hour volumes for freeways located in metropolitan areas of different population are available.[29,30] In larger urban areas where peak periods exceed one hour in duration on freeways, the peak 5-minute percentages approach the average 5-minute percentages of $8\frac{1}{3}$ per cent.[20] Short-term peak periods are important in analyzing frequencies of gaps in traffic of sufficient length to permit pedestrians and vehicles to enter and cross the stream, and for measuring peaking characteristics in capacity analysis.

Duration of peak flows is very important in planning traffic controls, such as peak period parking and stopping restrictions, restrictions on turning movements, and signal timing. The duration of peak flows will indicate the appropriate times of the day such controls should be in effect.

Hourly variations are fairly consistent for one type of urban street, but may be significantly different from one type of street to another. A report on the traffic counting program in Cincinnati[33] showed good consistency in hourly patterns by months of the year for a group of different streets. About 80 per cent of 24-hour total traffic occurs in 12 hours of daylight on rural routes, and about 70 per cent on urban routes.

Daily Variations. The heaviest daily traffic volumes for main rural highways occur on the weekend, on Saturday or on Sunday. Typical patterns for rural freeways in California show that maximum flow occurs on Saturday, representing about 112 per cent of average daily flow. The variation throughout the weekdays is relatively small. Recreational rural routes are subject to high weekend travel near the metropolitan area. Some recreational routes in California carry about 190 per cent on Sunday and 126 per cent on Saturday of average daily flow. Weekday travel on such routes averages about 70 per cent, with the only significant variation occurring on Friday, with approximately 94 per cent of average daily flow. Although local rural roads, such as farm-to-market roads, exhibit large variations of hourly flow, there is very little variation of daily flow except during winter months in those regions where significant weather changes take place throughout the year.

The variation in daily urban traffic for weekdays is not very pronounced on urban arterials. However the heaviest daily traffic volumes for weekdays occur on Friday. For example, on Friday, flow on urban arterials in California represent 109 per cent, in Cincinnati[33] from 103 to 116 per cent and in New York City about 108 per cent of average daily flow. Recreational-type facilities within New York City are subject to significant seasonal variations, and maximum daily travel occurs on Sunday throughout the year.

Monthly Variations. On main rural routes, traffic volumes are generally greatest during the months of July and August, lowest during the months of January and February, and about equal to the average month during April, May and November. In California, the peak flow on rural freeways was about 145 per cent of the average month during August, and on recreational routes the peak flow was in July, representing about 180 per cent of average monthly flow. Local rural roads are subject to greater fluctuations than main rural roads. In Minnesota, seasonal travel ranged from about 37 per cent of ADT in January to about 242 per cent of ADT in July.[34]

Monthly variations in urban areas are usually much less than in rural areas, but on recreational-type facilities the seasonal fluctuations are quite significant. For example, in New York and California on commuter-type facilities, the peak month's travel is about 110 per cent of the average, while on recreational-type facilities the peak month's travel is about 190 per cent of the average.

Traffic Volume Trends

Traffic volumes in urban and rural areas show continual increases, with no apparent indication of leveling off. This is reflected also by total vehicle-miles of travel. Approximately one-half of the nation's total vehicle-miles of travel take place in urban areas, and this will increase in the future because of suburbanization, or city spread. With the exception of the war years and a few years thereafter, the trend of national travel follows closely the economic trend as represented by the Gross National Product (GNP). Although, historically, traffic growth has closely paralleled the growth in GNP, it has been noted that since 1950 there has been a definite divergence in the two trends with traffic growth increasing at a more rapid rate than GNP. The facts given in Table 9-3 illustrate the trend in growth of national travel.

Over the ten-year period 1956–1966, annual travel in New York State increased by almost 40 per cent. Estimates of AADT and design hour (based on 30th peak hour) volumes on all traffic control sections on state routes have been made.[58]

TABLE 9-3. TRENDS IN VEHICLE-MILES OF TRAVEL IN THE U.S.[35,36,37]

YEAR	*TOTAL VEHICLE-MILES OF TRAVEL*	*PER CENT OF TRAVEL IN URBAN AREAS*	*PER CENT INCREASE IN TOTAL TRAVEL BASED ON 1955*
1955	$603,434 \times 10^6$	44.4	—
1960	$718,845 \times 10^6$	46.1	19.2
1965	$887,640 \times 10^6$	47.7	47.1
1980	$1,277,500 \times 10^6$	60.6	112

Composition of Traffic Stream

This varies as different percentages of passenger cars, trucks of various types, and buses are represented in the stream. AASHO gives the distribution of traffic by type for main rural highways. The data are broken down into different regions of the country; in addition, the following United States averages are given: Passenger Cars, 87.4 per cent; Trucks, 12.1 per cent; Buses, 0.5 per cent. The distribution of vehicles registered, and travel by vehicle types is shown in Table 9-4.

PEAK PERIODS OF TRAFFIC VOLUME

Annual daily variations:[19]

Based on traffic data collected during 1936–1941 throughout the United States, a study of 48 selected rural highways in 45 states showed that the traffic volume representing the AADT was exceeded during 160 days for the year at the average location. At some locations, the AADT was exceeded on as few as 70 days a year, and at other locations it was exceeded on as many as 228 days during the year. It is evident that facilities planned to accommodate only the traffic of an average day will be definitely overtaxed for a considerable portion of the year. The minimum day's volume was from 30 to 50 per cent of the AADT, and the peak day's volume from 140 to 340 per cent (with an average of 230 per cent) of the AADT.

TABLE 9-4. REGISTRATIONS AND MILES TRAVELED

YEAR	*VEHICLE TYPE*	*REGISTRATIONS*	*VEHICLE-MILES*	*PER CENT INCREASE IN TRAVEL FROM 1950*
1950	PC	$40,339 \times 10^3$(82.0%)	$363,613 \times 10^6$(79.3%)	
	T	$8,599 \times 10^3$(17.5%)	$90,552 \times 10^6$(19.8%)	
	B	224×10^3(0.45%)	$4,081 \times 10^6$(0.9%)	
1960	PC	$61,682 \times 10^3$(83.4%)	$588,083 \times 10^6$(81.8%)	62
	T	$11,914 \times 10^3$(16.2%)	$126,409 \times 10^6$(17.6%)	40
	B	271×10^3(0.37%)	$4,353 \times 10^6$(0.6%)	6.7
1967	PC	$80,414 \times 10^3$(82.9%)	$770,971 \times 10^6$(80.5%)	111
	T	$16,193 \times 10^3$(16.7%)	181,445 % 10^6(19.0%)	190
	B	338×10^3(0.4%)	$4,979 \times 10^6$(0.5%)	22

(Source: *Automobile Facts and Figures,* AMA, 1969)

Annual hourly variations:

Traffic volume during an interval of time shorter than a day more appropriately reflects operating conditions which should be used for design, if traffic is to be properly served. The brief but frequently repeated rush-hour periods are significant in this regard. In nearly all cases a practical and adequate time period is one hour.

The traffic pattern on any road shows considerable variation in traffic volumes during the different hours of the day, and even a greater fluctuation in hourly volumes throughout the year. It must be decided which of these hourly volumes to use for design. It is not economical to provide for the extreme hourly volumes of traffic that may occur but a few times a year. On the other hand, the design hourly volume should not be exceeded too often. An excellent guide in deciding on the hourly traffic best fitted for use in design is a curve showing variations in hourly traffic volumes during the year. Examples of yearly traffic patterns are available for different types of roadways.[46] By considering the shape of the yearly traffic pattern, the most feasible hourly volume that should be used for design purposes can be determined. Examination of these curves tells us that at about the *thirtieth highest hour* the slope of the curve changes rapidly, and it is at this point that the ratio of benefit to expenditure is near the maximum. Providing for an hourly traffic volume that is not exceeded at least 30 times a year will show an extremely small return in terms of driver benefit, whereas little will be saved in the construction cost and a great deal lost in driver benefit if provision is not made for the 50th highest hourly volume of the year. It is for this reason that a design which will accommodate the 50th highest hourly traffic volume can usually be justified, whereas a design to accommodate a traffic volume greater than that occurring during the 30th highest hour is generally not warranted.

The results of an analysis of traffic count data for 171 stations in 48 states, on main rural roads, revealed that with the average fluctuation in traffic flow, the peak hourly volume was about 25 per cent of the AADT (modified by climate, about 28 per cent in the northern portion and 20 per cent in the southern portion of the country). At 15 per cent of the locations, the peak hourly volume exceeded 32 per cent of AADT (maximum was 70 per cent of AADT), and at 15 per cent of the locations, the peak hourly volume was less than 16 per cent of AADT (minimum was 10 per cent of AADT).

The 30th highest hourly volume ranged between 8 and 38 per cent of AADT with an average of 15.3 per cent of AADT. The 30th hour volume was less than 10 per cent in only 2 per cent of the locations, and greater than 20 per cent in only 16 per cent of the locations. However, for recreational routes the 30th hourly volume may be as high as 38 per cent of the AADT.[19,42] (See Table 9-5.) Remarkably similar volume characteristics to those above exist for different types of routes in Switzerland, as shown in Table 9-6. Past indications have suggested that the 30th highest hourly volume on a percentage basis changed little from year to year. This would add greatly to its worth as a design criterion. However, more recent studies[39] have indicated that the factor exhibits a tendency to decline slightly with the passing of time.

TABLE 9-5. ESTIMATES OF PEAK HOURS, PERCENT OF AADT VOLUME

TYPE OF HIGHWAY	*PEAK HOURS*		
	30th	*50th*	*100th*
Primarily recreational route	38	34	28
Partial recreational route	23	21	18
Rural local route	20	18	17
Rural through route	15	14	13
Surburban route	12	11	10
Urban through route	8	8	7.5

(Source: *Traffic Engineering Handbook*, 3rd ed., ITE, 1965, Figure V-8)

TABLE 9-6. ESTIMATES OF PEAK HOURS, SWITZERLAND

TYPE OF HIGHWAY	*PEAK HOURS, PER CENT OF AADT*		
	30th	*50th*	*100th*
Primarily recreation route	32	27	22
Partially recreation route	19	17	15
Main rural route	17	16	14

(Source: Muranyi, "Estimating Traffic Volumes by Systematic Sampling," in *HRBB 281*, 1961)

Analysis of traffic data collected during 1961 and 1962 throughout the United State revealed the relationships shown in Table 9-7 between peak hours and AADT for different types of facilities. Similar results were obtained from a study of 30th peak-hour trends over a 10-year period at 69 counting stations in New Jersey.[40] The following conclusions were reported:

1. The 30th peak-hour factor generally declines as the AADT increases.
2. The reduction rate for high 30th peak-hour factors is much greater than for low 30th peak-hour factors. For example, the average yearly rate of change was found to be −0.133 per cent when 30th peak hour was 10–15 per cent of AADT, and −0.59 per cent when 30th peak hour was 30–40 of per cent AADT.

TABLE 9-7. PEAK HOURS, PER CENT OF AADT

Per cent of AADT in Peak Hours

TYPE OF FACILITY	ONE DIRECTION			BOTH DIRECTIONS		
	Maximum	*30th*	*200th*	*Maximum*	*30th*	*200th*
Rural:						
Freeway	23.6	15.4	11.4	18.3	13.5	10.9
Expressway	21.5	14.1	10.6	19.2	12.7	9.7
Highway, more than two lanes	21.2	13.7	10.3	16.4	12.7	9.9
two-lane, two-way highway	—	—	—	19.7	13.6	11.2
Urban:						
Freeway	15.0	12.7	10.7	13.6	11.0	9.6
Expressway	14.6	11.4	8.9	11.6	9.5	8.3
Street, more than two lanes	13.8	11.1	9.6	12.0	10.0	8.7
two-lane, two-way street	—	—	—	13.4	10.6	9.0

(Source: *Highway Capacity Manual*, HRB SR 87 1965)

TABLE 9-8. RELATIONSHIPS DEVELOPED FOR ESTIMATING 30th PEAK HOURS

PTR GROUP	NUMBER OF STATIONS	1961 AADT	LINEAR REGRESSION EQUATION	30th HOUR, ESTIMATED	PER CENT AADT, ACTUAL
I	4	5,963	$Y_c = 56 + 0.11658X$	12.7	12.2
II	6	2,463	$Y_c = 46 + 0.11439X$	13.3	11.9
III	8	4,133	$Y_c = 60 + 0.13437X$	14.9	14.9
IV	5	3,377	$Y_c = 16 + 0.15912X$	16.4	16.2
V	3	2,694	$Y_c = 8 + 0.18747X$	19.0	19.3
VI	5	1,232	$Y_c = 26 + 0.13225X$	15.4	14.6

(Source: Heibl, "A Method for Estimating Design Hourly Traffic Volume", *HRR 72*, 1965)

3. Low population and sparsely developed areas have a high 30th peak-hour factor.
4. A method was also developed to estimate the change that will take place in the peak-hour factor over a period of time.

An exponential curve was developed to predict the trend rate of reduction in the 30th peak-hour factor. A reduction rate of 2.3 per cent compounded per year was found to best express the relationship.

$$Y = b^x(a - 4.2) + 4.2 \tag{9-1}$$

where Y = future 30th peak-hour factor
b = rate of reduction (constant 0.977 based on 2.3 per cent compounded)
x = number of future years
a = existing 30th hour factor

An additional four years of data were used to test the validity of the above formula, and it was found to be very satisfactory.[59]

An analysis of counting station data for Pennsylvania revealed that an exponential trend curve was also valid to predict reduction in the 30th peak hour factor, but the rate of reduction was found to be 1.4 per cent as compared to New Jersey's 2.3 per cent.[60] For the three-year period 1964–1967, there was no conclusive evidence of a similar decay in the 30th peak hour factor in New York State.[61]

A linear regression analysis of data from continuous count stations operated for a period of years on main state rural routes in Wiscosin produced the following relationships for estimating 30th peak hour volumes from AADT volumes. The stations were divided into groups which reflected similar traffic patterns. (See Table 9-8.)

Y_c = 30th peak hour volume
X = AADT volume
PTR = permanent traffic recorder

SHORT COUNT METHODS FOR VOLUME STUDIES

Reasons for Short Count Methods

In many cases, manpower, time, and financial limitations

may rule out the feasibility of taking long counts. Manual counts have an advantage over machine counts in that the manual counts give classification and turning movements. They would be used more frequently if the cost could be reduced by reducing the time required to obtain reliable data. Making a short count at a number of locations several times throughout the day would permit the counting of more locations with the same personnel normally used for duty at locations counted continuously; at the same time manual count data would be obtained.

Short Count Method Number 1[42,43,44]

This involves counting for short periods of 5, 6, 10, or 12 minutes at each location or station for every hour. If traffic at each station is counted for 5 minutes out of each hour of the count period, say 7:00 A.M. to 7:00 P.M., the observer can cover about six locations, assuming it takes an average of 5 minutes to travel between locations. The hourly flow at each station is obtained by multiplying the 5-minute count by 12.

Representative control stations, selected throughout the area of the counts, are counted for each day of the survey, but they need not be counted continuously. Volume counts are made for the same interval of time as at the other stations, say 5 minutes. These control station counts are needed to adjust counts made in different sections and on different days to a common basis, if all counts are not made simultaneously on one day.

The accuracy of this short-count method is very much dependent upon the magnitude of the traffic volume. The heavier the traffic volume, the greater the accuracy since a short count (say for 5 minutes) would approach the average 5-minute percentage of $8\frac{1}{3}$ per cent. When traffic is not light or unduly erratic, counts of 5 or 6 minutes duration are quite satisfactory.

Short Count Method Number 2[3]

This involves counting for one-half hour between 9:00 and 11:00 A.M. and one-half hour between 2:00 and 4:00 P.M. on one day at each station. The counts are confined to the four off-peak hours on the premise that traffic volumes during these periods of the day are generally more stable and subject to the least variation from day to day, thus providing more reliable samples. This hour count is then expanded by appropriate factors developed from counts taken at control stations. The basic theory of this method is that the per cent of daily traffic during any given period of the day is a constant value at all points on routes of the same classification.

A procedure using one-hour counts, but otherwise similar to the above method, is also described.[45]

EXPANDING AND ADJUSTING TRAFFIC VOLUME COUNTS

It is necessary to expand and adjust short counts and coverage counts to a common base, or they may not be comparable. For example, a 24-hour coverage count made on a Wednesday in January cannot be compared with a 24-hour coverage count made on a Friday in August unless they are adjusted to a common base. The AADT is the most desirable base because it represents the average for a year and it is the most frequently used value to indicate the level of service provided by a given facility. For less comprehensive studies, the average day of the count period can be used to compare stations in a limited area.

The most frequently encountered situation is one of expanding and adjusting a single count, taken at a coverage station, to an estimated ADT value. This is accomplished by applying factors obtained from control stations to the coverage count. Control stations are of two types: *Continuous-count* stations, where traffic is counted continuously for a year, and *seasonal control* stations, where several counts are taken in such a manner that the relation between the seasonal or monthly volume of traffic and the ADT can be established.

The use of expansion and adjustment factors from control stations to adjust coverage counts is predicated on the premise that similar traffic patterns exist for like facilities in a given area. Therefore, it is only necessary to identify the control station having a traffic pattern similar to that of the coverage station to obtain the appropriate adjustment factors. The total 24-hour volume at different locations in an area may vary considerably in total amount, but the percentage of the total volume that is recorded for any particular hour of the day is likely to be similar at all stations in the area. Although traffic volumes vary from one month to another, the relationship of each monthly volume to the yearly volume remains fairly consistent over a period of years.

Adjustment of Counts taken in a Limited Area[3]

In making basic traffic planning surveys of volume, a number of locations are included in one study. The relationship of volumes between different locations is important, so they must be adjusted to represent an average day for the period of the counts, or one particular day. To accomplish this, it is necessary to take control station counts on any day that the surrounding coverage stations are counted. It is necessary to select a control station in each of the different districts, that is, one in the CBD, one in the residential district, one in the industrial district, etc., if the area to be covered includes more

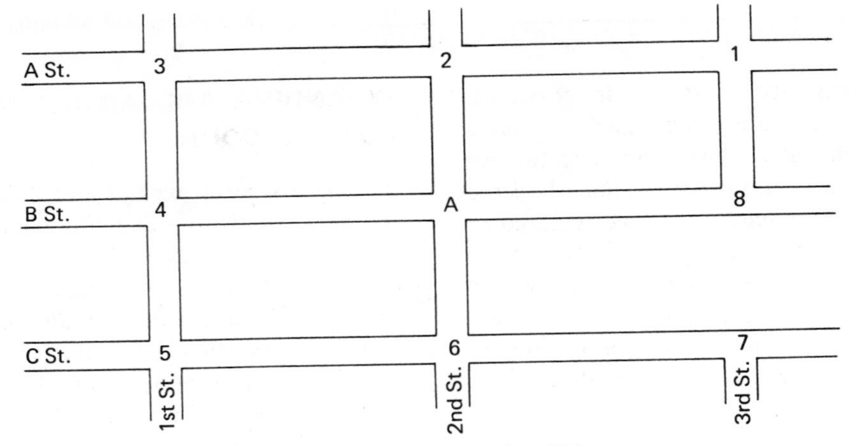

System of volume counts in a CBD

Adjustment of coverage station counts

Day	Total 12-hr flow at sta. A	Ratio to average day of area count	Coverage sta.	Total 12-hr. count	Adjusted vol. count
Monday	9,000	1.05	1	2,500	2,400
Tuesday	8,500	0.99	2	9,600	9.700
Wednesday	8,300	0.97	3	4,800	5,000
Thursday	8,600	1.00	4	9,200	9,200
Friday	9,100	1.06	5	5,200	4,900
Monday	8,800	1.03	6	7,900	7,700
Tuesday	8,300	0.97	7	2,100	2,200
Wednesday	8,100	0.95	8	3,500	3,700
Total	68,700				
Avg. Day	8,588				

Schedule of counts for ½ hr. between 9-11 A.M. and ½ hr. between 2-4 P.M. on one day

Coverage Station	Day	Hours of count
1	Monday	9:00- 9:30 A.M.
2	Monday	9:45-10:15 A.M.
3	Monday	10:30-11:00 A.M.
1	Monday	2:00- 2:30 P.M.
2	Monday	2:45- 3:15 P.M.
3	Monday	3:30- 4:00 P.M.
4	Tuesday	9:00- 9.30 A.M. 2:00-2:30 P.M.
5	Tuesday	9:45-10.15 A.M. 2:45-3:15 P.M.
6	Tuesday	10:30-11:00 A.M. 3:30-4:00 P.M.
7	Wednesday	9:00- 9:30 A.M. 2:00-2:30 P.M.
8	Wednesday	9:45-10:45 A.M. 2:45-3:15 P.M.

Fig. 9-2. Example of adjustments to cordon count

than one type of district. A minimum of one control station in each area should be chosen so that the traffic pattern will be representative of all traffic movement in each area or district. For example, consider a cordon area made up of streets in the CBD, as shown in Figure 9-2. Assume the counting schedule is set up for a 12-hour count at each coverage station, and that with the number of observers available it is only possible to count one coverage station per day and the control station. Then to complete the volume counts, eight days are necessary for the eight coverage stations. The schedule of counts is shown in Figure 9-2. The control station A must be counted for 12 hours for each of the eight days that it takes to count the coverage stations. The counts at the control station are averaged, and a factor is developed for each day. This factor indicates how much higher or lower that day was as compared to the average day of the control station count. It is assumed that the same ratios would be found for each coverage station if it were counted on the same days. This is the basic premise upon which the accuracy of the adjustment rests. In order to adjust each coverage count to the average day for the duration of the counts in the area, the coverage count is divided by the ratio found for that day at the control station. Thus, the adjusted counts at each coverage station will be comparable, as they do not reflect the traffic volume variation from day to day.

Expansion of Volume Counts Taken by Short-Count Method Number 2[3]

Considering the same CBD area as above, the same control station and coverage stations would be selected. However, the counting schedule is different. Instead of counting each coverage station for 12 hours, they are counted for one-half hour between 9:00 and 11:00 A.M. and for one-half hour between 2:00 and 4:00 P.M. on one day. The counting schedule is set up as shown in Figure 9-2. The control station A would therefore be counted for 12 hours on Monday, Tuesday, and Wednesday to provide the basic data for converting the coverage short-count stations to estimated 12-hour totals. The ½-hourly flow totals at the control station, for those increments of counts made at coverage stations, are expressed as percentages of the 12-hour total flow for each day. These percentages are used to expand each short count to the 12-hour total estimated for the day of its count. These estimated 12-hour flows are then treated exactly like the 12-hour counts made at each of the coverage stations as described above.

An alternate adjustment procedure can be used by finding the average ½-hourly flows at the control station (expressed as percentages) combining all daily control station counts. Using these ½-hourly percentages, it is not necessary to adjust with control station factors, since the estimated counts will have already been adjusted through the averaging of ½-hourly control station volumes.

Example

Assume that the short count at station 4 made on Tuesday showed a total of 358 vehicles entering from 9:00 to 9:30 A.M. and 410 vehicles entering between 2:00 and 2:30 P.M. The corresponding percentages at the control station were 3.82 per cent and 4.09 per cent for that same Tuesday.

Estimated 12-hour flow at station 4 based on morning sample $= \frac{358}{0.0382} = 9{,}400$

Estimated 12-hour flow at station 4 based on afternoon sample $= \frac{410}{0.0409} = 9{,}800$

Estimated 12-hour flow at station 4 for Tuesday $= \frac{9{,}400 + 9{,}800}{2} = 9{,}600$

Estimated 12-hour flow at station 4 for period of area count $= \frac{9{,}600}{0.99} = 9{,}700$ vehicles

Adjusting Coverage Station Counts to ADT

The following method is used in conjunction with the procedure recommended by the National Committee on Urban Transportation[67] for scheduling periodic volume counts in urban areas.

The coverage counts are adjusted to ADT by use of factors obtained from key station counts. The key station is counted for a seven-day period in each year to obtain daily variation factors, and a 24-hour count is made on the same weekday in each month during the year to obtain seasonal variation factors.

Daily factor =

$$DF = \frac{\text{7-day average for 24-hour count}}{\text{24-hour count for a specific day}}$$

Seasonal factor =

$$SF = \frac{\text{average 24-hour count for 12 months}}{\text{24-hour count for a specific month}}$$

The 24-hour count made at a coverage station on a specific day in a specific month is adjusted to ADT as follows:

$$ADT = \text{coverage count} \times DF \times SF \qquad (9\text{-}2)$$

The peak-hour directional volume for the coverage station is obtained by multiplying the estimated ADT by a factor obtained from the major control count for that street.

Adjustment and Expansion Factors from Continuous-Count Stations

Hourly Adjustments. The per cent of total daily volume for each hour and/or the per cent of the average hourly volume for each hour are determined. These are plotted or tabulated as shown in Table 9-9. The per cent of total daily volume is computed as

$$\text{Per cent of total daily volume} = \frac{\text{average volume for particular hour}}{\text{average volume for 24-hour period}} \qquad (9\text{-}3)$$

These values are used for hourly adjustments and for expanding counts of shorter duration than 24 hours into estimated 24-hour volume.

Daily Adjustments. The per cent of total weekly volume for each day and the per cent of the average daily volume for each day are determined. The latter results in a multiplicative factor which converts the volume on a particular day to a volume indicative of the average day in the week.

TABLE 9-9. TYPICAL HOURLY, DAILY, AND SEASONAL FACTORS FOR OBTAINING AVERAGE TRAFFIC VOLUMES

HOURLY VARIATION OF TRAFFIC VOLUME ON A STATE HIGHWAY SYSTEM

Hour	*Per Cent of Total 24 Hour Volume*	*Hour*	*Per Cent of total 24-Hour Volume*
6:00– 7:00 A.M.	2.53	6:00– 7:00 P.M.	6.12
7:00– 8:00 A.M.	3.69	7:00– 8:00 P.M.	5.72
8:00– 9:00 A.M.	4.42	8:00– 9:00 P.M.	4.74
9:00–10:00 A.M.	5.34	9:00–10:00 P.M.	3.85
10:00–11:00 A.M.	5.73	10:00–11:00 P.M.	3.18
11:00–12:00 A.M.	5.42	11:00–12:00 Midnight	2.61
12:00– 1:00 P.M.	5.34	12:00– 1:00 A.M.	1.89
1:00– 2:00 P.M.	6.18	1:00– 2:00 A.M.	1.32
2:00– 3:00 P.M.	6.56	2:00– 3:00 A.M.	0.90
3:00– 4:00 P.M.	6.88	3:00– 4:00 A.M.	0.76
4:00– 5:00 P.M.	7.71	4:00– 5:00 A.M.	0.76
5:00– 6:00 P.M.	7.30	5:00– 6:00 A.M.	1.05

Example: An 8-hour count (7:00–11:00 A.M. and 2:00–6:00 P.M.) constitutes 47.6 per cent (factor 2.09) of total 24-hour volume. Average 24-hour count = 8-hour count (7:00–11:00 A.M. and 2:00–6:00 P.M.) × 2.09.

DAILY VARIATION OF TRAFFIC VOLUME ON A STATE HIGHWAY SYSTEM

Day	*Per Cent of Total Weekly Volume*	*Per Cent of Average Day*	*Weekly Factor*
Sunday	18.10	126.73	0.789
Monday	13.32	93.25	1.072
Tuesday	12.75	89.14	1.121
Wednesday	12.89	90.22	1.108
Thursday	13.00	91.04	1.096
Friday	14.06	98.44	1.015
Saturday	15.88	111.18	0.899

Example: To convert a Tuesday count to average day for week, multiply by 1.121.

SEASONAL VARIATION BY MONTHS

Month	*Per Cent of Average Month*	*Monthly Factor*	*Month*	*Per Cent of Average Month*	*Monthly Factor*
January	82.24	1.215	July	109.51	0.913
February	83.94	1.191	August	113.38	0.882
March	90.89	1.100	September	113.10	0.884
April	100.79	0.992	October	107.46	0.931
May	105.29	0.949	November	97.38	1.026
June	108,89	0.918	December	87.13	1.114

(Source: *Traffic Engineering Handbook*, 3rd ed., ITE, 1965)
Example: To convert a count in May to average month for year, multiply by 0.949.

Per cent of total weekly volume

$$= \frac{\text{average volume for particular day}}{\text{average total volume for week}}$$

factor

$$= \frac{\text{average total for week/7}}{\text{average volume for particular day}}$$

Seasonal Adjustments. The per cent of total annual volume for each month and the per cent of the average monthly volume for each month are determined.

Per cent of total annual volume

$$= \frac{\text{total volume for particular month}}{\text{total yearly volume}}$$

factor

$$= \frac{\text{total yearly volume /12}}{\text{total volume for particular month}}$$

For example, assume that an 8-hour count (7–11 A.M. and 2–6 P.M.) is made on a Tuesday in the month of May. Estimate the ADT for this station if the 8-hour count was 6,300 vehicles. From the hourly variation table, the 8-hour count period constitutes 47.6 per cent of the total 24-hour volume; therefore, the factor is 100/47.6 = 2.09. Estimated 24-hour volume = 2.09 × 6,300 = 13,200 vehicles. From the daily variation table, the Tuesday factor = 1.121. Therefore, the estimated average daily volume = 1.121 × 13,200 = 14,800 vehicles. From the monthly variation table, the May factor = 0.949. Therefore, the estimated ADT = 0.949 × 14,800 = 14,100 vehicles.

Table 9-9 is an example of a variation factor table. Such tables should be established for each highway system district with homogeneous characteristics.

Adjustment and Expansion Factors from Control Stations

The following directive from the BPR explains a procedure for the expansion of rural coverage station counts by utilizing factors derived from control station counts that are not extended over the entire year. The procedure is similar to the method based on continuous count station data.

BUREAU OF PUBLIC ROADS DIRECTIVE

SUBJECT: Expansion of coverage traffic counts

Inquiries from the States have indicated a need for a practical method of computing factors for the expansion of coverage traffic counts on rural roads. To satisfy this need, the two tabulations attached

TABLE 9-10. COMPUTATION OF FACTORS FOR THE EXPANSION OF COVERAGE COUNTS
Weekdays and final factors

	CONTROL STATIONS														
	MAJOR				*MINOR*								*Total of Factors*	*Average factor N*	*Final factor F = N × C**
	A		*B*		*1*		*2*		*3*		*4*				
Month	*Vehicles*	*Factor*	*Vehicles*	*Factor*	*Vehicles*	*Factor*	*Vehicles*	*Factor*	*Vehicles*	*Factor*	*Vehicles*	*Factor*			
January	631	1.177	558	1.213			472	1.252			589	1.241	4.883	1.221	1.275
February	586	1.268	543	1.247	366	1.372			527	1.332			5.219	1.305	1.362
March	665	1.117	608	1.113			488	1.211			614	1.191	4.632	1.158	1.209
April	745	0.997	674	1.004	486	1.033			628	1.118			4.152	1.038	1.084
May	776	0.957	683	0.991			543	1.088			721	1.014	4.050	1.012	1.057
June	811	0.916	737	0.919	536	0.937			823	0.853			3.625	0.906	0.946
July	818	0.908	786	0.861			708	0.835			859	0.851	3.455	0.864	0.902
August	893	0.832	822	0.824	625	0.803			897	0.783			3.242	0.811	0.847
September	881	0.843	798	0.848			684	0.864			852	0.858	3.413	0.853	0.891
October	770	0.965	696	0.973	506	0.992			750	0.963			3.866	0.967	1.010
November	684	1.086	637	1.063			652	0.906			751	0.973	4.028	1.007	1.051
December	656	1.133	578	1.171	491	1.022			589	1.192			4.518	1.130	1.180
Total	8,916		8,122		3,010		3,547		4,214		4,386				
Avg *K*	743		677		502		591		702		731				

* From Table 9-11.

Example 1: 48-hour weekday count at a coverage station in March, 1,144 vehicles or 572 vehicles mean for 24 hours. Estimated volume for the annual average day = 572 × 1.209 = 792 vehicles.

Example 2: 48-hour weekday count, 718 vehicles in June at a coverage station similar to minor control station No. 3, or 359 vehicles mean for 24 hours. Estimated volume for the annual average weekday = 359 × 0.853 (from minor control station No. 3) = 306 vehicles. Estimated volume for the annual average day = 306 × 1.044 (constant *C* from Table 9-11) = 319 vehicles.

TABLE 9-11. COMPUTATION OF FACTORS FOR THE EXPANSION OF COVERAGE COUNTS (Saturdays and Sundays)

	SATURDAYS		*SUNDAYS*	
Month	*Vehicles, Station A*	*Vehicles, Station B*	*Vehicles, Station A*	*Vehicles, Station B*
January	640	562	651	609
February	595	548	615	531
March	685	636	697	685
April	810	750	875	873
May	835	786	979	898
June	876	813	1,096	984
July	892	881	1,158	1,003
August	968	931	1,203	1,112
September	954	856	1,182	971
October	840	742	950	806
November	743	675	810	715
December	701	603	764	794
Total	9,549	8,783	10,980	9,982
Average *L*	796	732	915	832
Average *K*	743	677	743	677
$L \div K = M$	1.071	1.081	1.231	1.229
Average *M*	1.076		1.230	

$$\text{Constant } C = \frac{5.000 + 1.076 + 1.20}{7} = 1.044$$

demonstrate the mechanics of the calculations of the factors.

For the purposes of illustration only, two major and four minor control stations were used, although in practice a large number of stations would normally be combined into a control system. Table 9-10 shows the treatment of weekdays and the computation of the final factors. Table 9-11 shows the computation of constant "C" which represents the relationship of the annual average day of the week to the annual average weekday.

In Table 9-10 under the heading "vehicles" are recorded the 24-hour means of 48-hour or longer weekday counts. Under the "factor" headings is shown the ratio that the annual average K is of each monthly mean. All other computations are believed to be self-explanatory.

In the computation of the ratio M, as shown in Table 9-11 the nature of the value K is unity. It is possible, therefore, to write the equation for constant C so that 5 in the numerator represents the five weekdays. By dividing the numerator by 7, the mean value of the day of the week is obtained.

If the pattern of traffic flow is known, then the knowledge of a single point in the pattern permits estimating of the rest of the pattern. In actual practice it is presumed that a number of coverage stations have similar traffic patterns. Then, from control stations which are expected to have patterns similar to coverage stations, the mean pattern is computed which is the most probable for that group of stations. The significant feature of the method here illustrated is that for each control station the pattern was computed in ratios. Consequently, the data under the head "average" represent the mean of the patterns of all stations or the most probable pattern applicable to coverage stations of similar characteristics.

The chief use of the data thus presented is in the appliation of final factors F to a bulk of suitable coverage counts as shown in Example 1. Another use is in the selection from Table 9-10 of factors from the individual control stations when it is believed that the traffic at the coverage station has a pattern most similar to that of a particular control station.

If the factors are computed by averaging the volumes at the different control stations instead of averaging the patterns, then the factors reflect the pattern of the total traffic volume at all stations used. The station having a traffic volume larger than the others will have the predominant effect on the final factors. A pattern so computed is biased in favor of stations having the greatest traffic volume and is therefore not the most probable pattern of a group of stations, except when the patterns are alike.

Statistical Accuracy of AADT Estimates

The only time the AADT is obtained with no error is when traffic is counted continuously for a complete year. When traffic volume is obtained for a period of less than one year, the count is considered to be a *sample*.

The application of statistical analysis to evaluate the magnitude of error associated with sample volume counts, which have been expanded and adjusted by factors derived from continuous count stations, was first introduced in Oregon.[47] It is not possible to evaluate the magnitude of the error for a particular station by statistical analysis, but it can be applied to the determination of the size of the error that may be expected at a number or group of stations.

Experience shows that the estimates of AADT, with errors measured by standard deviation (coefficient of variation) of about 10 to 12 per cent for high volume rural roads (over 500 vehicle per density), are satisfactory to the users of the data. The procedure developed to statistically analyze volume data is, briefly, as follows:

1. At each continuous-count station, monthly expansion factors are computed. These factors represent the ratios of the ADT to the average weekday volume of each month. It should be noted that adjustment factors are

in terms of average weekday traffic. Coverage counts are usually made on weekdays; when Saturday and Sunday are included, only the weekday counts should be used for estimating AADT. As a rule, the variations of Saturdays and Sundays within a month are greater than those of the weekdays, thus the AADT estimates based on counts which include weekends tend to be less accurate than those based on weekdays.

2. The continuous-count stations are then grouped so that the range of variation among the factors for each month does not deviate by more than ±10 per cent. Since the ultimate objective is the determination of expansion factors for coverage counts, only months during which coverage stations are operated are included in the grouping of continuous-count stations. If during these months the factors between two continuous-count stations do not differ by more than ±10 per cent they are put into the same group.
3. Mean group factors are then computed for each month for the group, and these factors are used to expand and adjust coverage counts taken within the area covered by the group.
4. It has been found that the standard deviation of the group is a significant measure of errors of estimates of AADT that can be expected at coverage stations. These errors distribute themselves in accordance with the normal curve. This conclusion is based upon the results of many chi-square tests which have indicated probability levels well in excess of 5 per cent.
5. Experience with seasonal control stations supports the assumption that the continuous-count stations are representative of the population as far as monthly expansion factors are concerned. Where it is substantiated that the expansion of coverage counts can be made on the basis of continuous-count data alone, it would no longer be necessary to operate seasonal control stations. Seasonal control stations may be grouped so that factors do not vary by more than ±15 per cent from the mean factor for each month.
6. Since the chi-square test has indicated that a good fit exists to a normal distribution, the standard deviations are a very reliable measurement of the magnitude of error and the frequency of its distribution.

An example of the goodness of fit to a normal distribution for a grouping of continuous-count stations is given in Tables 9-12 and 9-13. Here $d = (A - B)100$, where A is the ratio of the station's AADT to the average

TABLE 9-12. CONTINUOUS-COUNT STATION GROUPING

	APR		*MAY*		*JUN*		*JUL*		*AUG*		*SEP*		*OCT*		*NOV*	
Station	*Ratio*	*d*	*Ratio*	*d*	*Ratio*	*d*	*Ratio*	*d*	*Ratio*	*d*	*Ratio*	*d*	*Ratio*	*d*	*Ratio*	*d*
152	1.03	−5	1.02	−1	0.93	+1	0.90	+1	0.86	−1	0.92	0	1.02	−2	1.14	+3
189	1.18	+10	1.03	0	0.94	+2	0.92	+3	0.94	+7	0.94	+2	1.11	+7	1.20	+9
197	1.04	−4	1.02	−1	0.91	−1	0.90	+1	0.86	−1	0.93	+1	1.01	−3	1.07	−4
198	1.05	−3	1.01	−2	0.92	0	0.87	−2	0.88	+1	0.91	−1	1.00	−4	1.07	−4
205	1.09	+1	1.06	+3	0.91	−1	0.84	−4	0.82	−5	0.92	0	1.06	+2	1.09	−2
Mean	1.08		1.03		0.92		0.89		0.87		0.92		1.04		1.11	
$\sum d^2$		151		15		7		40		77		6		82		126

TABLE 9-13. CHI-SQUARE TEST FOR GOODNESS OF FIT

CLASS INTERVAL OF VALUE d	*CLASS LIMIT x*	$\frac{x}{\sigma}$	*PER CENT AREA BETWEEN* $+x/\sigma$ *AND MEAN*	*PER CENT AREA*	*THEORETICAL FREQUENCY* f_t		*OBSERVED FREQUENCY* f_o	$f_o - f_t$	$(f_o - f_t)^2$	$\frac{(f_o - f_t)^2}{f_t}$
0–1.99	2.00	0.56	42·46	17.0	17.0		17	0	0	0
2–3.99	4.00	1.13	74.36	29.7	12.7		12	−0.7	0.49	0.039
4–5.99	6.00	1.69	90.90	36.4	6.7		7	0.3	0.09	0.013
6–7.99	8.00	2.25	97.56	39.0	2.6	3.4	2	0.6	0.36	0.106
8–9.99	10.00	2.82	99.52	39.8	0.8		2			

DF = 4 − 3 = 1 Probability level = 71 per cent $\sum = 0.158 = \chi^2$

day of the month, and B is the mean monthly ratio for all stations.

$$\text{Total } \sum d^2 = 504, \qquad \sigma = \sqrt{\frac{504}{40}} = \pm 3.55$$
$$n = 40$$

Since the mean of all ratios is practically unity, the standard deviation is almost equivalent to the coefficient of variation. Therefore, when a 24-hour traffic volume count on a given day is compared with the average weekday for the month, it could be expected that about $\frac{2}{3}$ (actually 68.3 per cent) of such 24-hour weekday counts would not differ by more than ± 3.55 per cent from the respective monthly means, and about 95 per cent of such counts should not differ from their respective monthly means by more than ± 7.1 per cent.

The accuracy of counting low-volume rural roads has been studied as well as the accuracy of volume counts in urban areas.[35,49]

BPR Statistical Approach to Estimating AADT from Sample Counts

The BPR released a procedural guide[51,62] essentially based on other material[34,43,47,48,49] developed for estimating AADT volumes. The primary purpose of the guide is to provide efficient procedures for making statistically accurate estimates of AADT based on sample counts. Separate procedures are recommended for rural highways with AADT volumes exceeding 500, for rural highways with AADT volumes between 25 and 500, and for urban roads and streets.

Rural Highways with AADT Greater than 500. The procedure recommended for high-volume roads results in a coefficient of variation of less than ± 10 per cent in estimating AADT and can be divided into the three major steps:

1. The first step is to Group continuous-count stations into similar patterns of monthly traffic volume variation. This is accomplished by following the steps outlined in under Expanding and Adjusting Traffic Volume Counts, this chapter.
2. The next step is to assign road sections to groups of similar patterns of monthly variation. This is accomplished by assigning different colors to each group and marking each continuous-count station with the appropriate color on a location map. Stations of the same group usually fall along a continuous route or routes. Road sections on these continuous routes are connected and designated by the color assigned to the stations which fall upon them.

   The number of continuous-count stations is not ordinarily sufficient to assign all road sections, with an AADT exceeding 500, to pattern groups. Therefore, continuous-count stations are supplemented by *seasonal control stations*, at which traffic counts are made at equally spaced intervals during the year. The seasonal control stations are grouped by following a procedure similar to that for grouping continuous-count stations. The remaining high-volume road sections are then assigned to pattern groups utilizing the seasonal control station counts.

   Since monthly ratios persist over a period of years, it may be expected that the great majority of road sections will fall into the same monthly pattern gtoups year after year.
3. The last step involves locating and operating traffic counting stations.
   - **a.** Continuous-count stations: In general, a minimum of four stations should be located in each group of road sections with an independent set of monthly factors. For example, applying the BPR procedure in Wisconsin resulted in six groups of continuous-count stations. In 1961, 35 stations were operated.[59] Applying the BPR procedure in Georgia resulted in four groups from 26 continuous-count stations in 1963.[62] Using the BPR procedure in the Delaware Valley Region of Pennsylvania resulted in four groups from 24 continuous-count stations in 1966–67.[63]
   - **b.** Seasonal control stations: After all road sections have been grouped, most of the seasonal control stations can be eliminated, because the pattern of monthly variations of traffic volumes persists over long stretches of highway.
   - **c.** Coverage stations: The bulk of ADT data comes from coverage station counts. In general, approximately 25 coverage stations are required for each 100 miles of rural roads. In a comprehensive traffic volume survey, information is needed for each section of road between intersections. Therefore, it is advisable to locate coverage stations at alternate intersections. The coverage station counting program may be completed in one year or spread over a period of several years. If a five-year cycle is used, one-fifth of the coverage counts would be made each year.

An extensive analysis of rural coverage counts using the above procedure revealed that coverage counts taken for 48 hours on weekdays will have a mean annual coefficient of variation of ± 9.0 to ± 9.5 per cent when compared to the average weekday of the month. The study also showed that a coefficient of ± 10 per cent or less is not to be expected for coverage counts of 24 hours taken on a weekday.[51]

Rural Highways with AADT Volumes Between 25 and 500. The procedure recommended for low-volume roads results in a coefficient of variation within 20 to 25 per cent in estimating AADT. Past studies have indicated that the coefficient of variation increases at a much greater rate when traffic volume is less than 500, and therefore low-volume roads must be treated differently than high-volume roads.

All low-volume rural roads, regardless of the administrative system, can generally be represented by one group for the purpose of computing monthly adjustment factors to obtain estimates of AADT. Data from five or six continuous-count stations are normally sufficient to compute average adjustment factors. An equal number of seasonal control stations can replace continuous-count stations if control station counts are made in each month. These stations should not be located on roads carrying less than 100 AADT. The procedure for locating coverage stations is the same as that for high-volume roads.

Adjustment Factors. Coverage station counts made on road sections within a specific group are adjusted by multiplying the coverage count by the appropriate group mean factor. For example, if a 48-hour count of 4,000 vehicles is recorded in the month of July at a coverage station located within a group defined by the continuous-count station shown in the table of the previous section, the estimated AADT for this coverage-count station is $4{,}000/2 \times 0.89 = 1{,}780$ vehicles.

Mechanical Data Processing. Mechanical analysis of traffic count data, using computers, provides a practical means of obtaining more statistics and of improving the degree of accuracy in estimating AADT. Computers make possible the use of weekly or daily adjustment factors instead of monthly adjustment factors.[52,53,54] A method was presented[38] to estimate AADT and peak hours of traffic on different types of rural roads in Switzerland utilizing traffic pattern curves and application of statistical methods based on weekly adjustment factors. Additional applications of computers for the analysis of traffic count data have been made.[55,56,62,64,65]

Urban Roads and Streets. Based on detailed analysis of long-term counts on highvolume streets, it was found that the normal traffic volumes on weekdays could be considered the same as the AADT without the application of adjustment factors. The accuracy of such counts was measured to have a coefficient of variation of ± 10 per cent. This error could be reduced to about ± 7 per cent by application of monthly adjustment factors based on data obtained at control stations. If short count methods are used, the estimated AADT can be expected to be characterized by an error of ± 12 per cent.

10

Traffic Theory: Flow and Control

Traffic theory (*traffic flow theory*) is that body of knowledge which is concerned with the analytic formulation of traffic phenomena and mechanisms and the extension of these formulations to realize greater understanding or utility in traffic. This particular discipline is marked by the use of mathematical analysis and modeling, the techniques of system and control engineering, and computer simulation and process control. Studies in this discipline range from abstract mathematical treatises with little basis in data or direct physical modeling, to almost completely empiric model-building procedures. The place of many papers and branches of study will only be properly ascertained as the field is further defined and advanced.

This chapter will not give a comprehensive and detailed treatment of the discipline of traffic theory because of the diversity of both the work and its sources, and because of the state of much of the theory. This chapter is limited to a selection of some of the more relevant and illustrative work, with particular emphasis on utility in operations and control.

For broader treatments of traffic theory, see a survey article by Gazis and Edie,[1] texts by Drew[2] and Ashton,[3] and the bibliographies therein. Other prime references are HRB Special Report 79 (which is currently in revision), and the publications *Transportation Science* and *Transportation Research*, the issues of the *Journal of The Operations Research Society of America* predating *Transportation Science*, and the *Highway Research Records*. In addition, a bibliography of literature on unidirectional traffic flow has been published,[4] and the Highway Research Information Service (HRIS) is available for surveying the literature in specific topics.

CAR-FOLLOWING AND MACROSCOPIC FLOW

A car-following model is a mathematical expression relating the movement of a single vehicle to that of the vehicle it follows. Such expressions may be manipulated to yield descriptive expressions for the flow of an entire traffic stream.

Major work in car following was undertaken at the General Motors Research Laboratories by Herman and Rothery.[5,6] The basic tenet of this work is that the driving pattern of an individual driver can be effectively modeled by a differential equation relating him to other drivers in a (single) lane of traffic. Typical of the equations considered is:

$$\frac{dv_n(t)}{dt} = K\left\{\frac{v_n(t-T) - v_{n-1}(t-T)}{x_n(t-T) - x_{n-1}(t-T)}\right\} \tag{10-1}$$

where v_i = the speed of the ith driver
x_i = position of the ith driver
T = a delay or response lag

The equation states that the acceleration of a driver (number n) is determined by the difference between his speed and that of the vehicle in front of him (number $n-1$), with the sensitivity to speed discrepancies determined by some constant K, and the spacing between the two vehicles; the further away, the lower the sensitivity. Experiments have determined that the above relationship was an adequate descriptor, was more appropriate than a simple constant-sensitivity model, and did not need additional linkages to other downstream or upstream (following) vehicles.

Among the experiments conducted with car-following models, typical K and T values were computed for vehicles on a test track and in a tunnel in New York City. Experiments were also conducted with buses. On the analytic side, the stability of pairs and of strings of vehicles were studied for various K and T ranges. Studies by other researchers investigated the effects of road geometrics in observed traffic patterns (speed and volume fluctuations) via car-following[7] and the use of car following in simulation models.[8]

Other interesting results with car-following equations are: (1) The small-scale (microscopic) car-following relationships can be manipulated to yield large-scale (macroscopic) average relationships in terms of hourly flow Q and density D in vehicles per mile, and (2) these macroscopic relationships can have meaningful flow-density profiles, and the microscopically-determined values (K, for instance) yield meaningful macroscopic values (free speed, or speed at maximum flow, for instance).

To illustrate this relationship to the macroscopic expressions, consider Equation (10-1) for a stable equilibrium condition on a string of vehicles, that is, no acceleration; all vehicles are equally spaced and at same speed. Note that

$$\frac{dV}{dt} = K\frac{\Delta V}{\Delta x} \tag{10-2}$$

where V is speed, ΔV is speed difference, and ΔX is spacing. For a differential change in speed note that

$$dV = K\frac{\Delta V}{\Delta x}dt = K\frac{d(\Delta x)}{\Delta x}$$

so that from any known pair (V_0, ΔX_0), one may write

$$\int_{V_0}^{V} dV = K\int_{\Delta x}^{\Delta x_0}\frac{d(\Delta x)}{\Delta x}$$

or

$$V - V_0 = K\ln\frac{\Delta x}{\Delta x_0} = K\ln\frac{D_0}{D} \tag{10-3}$$

where D is density (vehicles per mile), $D = 1/\Delta x$. Denoting flow as Q, noting $Q = VD$, and observing $V = 0$ at jam density D_J, it is seen from Equation 10-3 that:

$$Q = KD\ln\frac{D_J}{D} \tag{10-4}$$

As shown in Figure 10-1, this is one form of the familiar humpback flow-density curve. It may be shown by differentiation that the speed at maximum flow is given by K. In the illustration, the knowledge that $V = 30$ miles per hour at a density of $D = 30$ vehicles per mile provides a $K = 20.6$ miles per hour and a maximum flow of 990 vehicles per hour (single lane).

OTHER FLOW-DENSITY MODELS

In addition to the car-following derivations of macroscopic flow relationships, such expressions have been arrived at by curve-fitting hypotheses,[9] by observation of safe headways,[10] by heat-flow analogies,[11] and by fluid-flow analogies.[11] The heat and fluid analogies center on equilibrium conditions for partial differential

Fig. 10-1. Macroscopic flow relationships

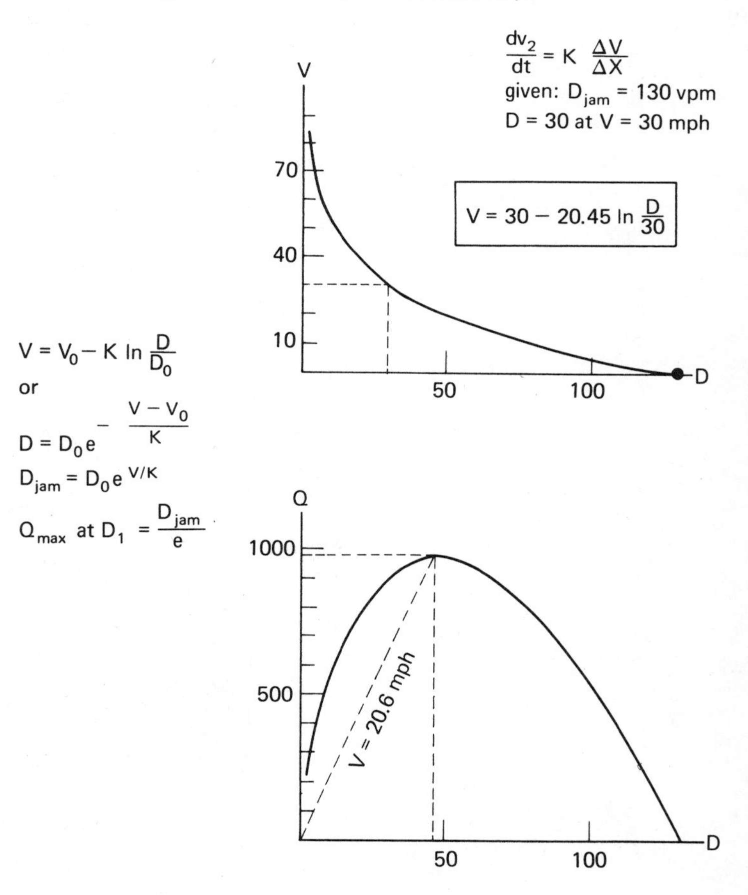

equations expressing a heat or mass balance (equation of continuity). The derivation on safe headways allows for a space headway which includes a vehicle length L, a reaction-time distance $C_1 V$ (where C_1 is the reaction or "dead" time), and a deceleration distance $C_2 V^2$ (where C_2 is determined by braking capability):

$$\Delta x = L + C_1 V + C_2 V^2$$

or, since $Q = VD = V/X$,

$$Q = \frac{V}{L + C_1 V + C_2 V^2} \qquad (10\cdot5)$$

and it may be shown that the maximum flow Q_M is attained at a speed $V = \sqrt{L/C_2}$ and spacing $\Delta x = 2L + C_1\sqrt{L/C_2}$.

It should be noted that it is generally taken that the density D is the basic independent variable, with speed V and thus flow Q being dependent on it. This is reflected in the above derivations.

Other notable models on flow-density relationships include the classic treatment by Lighthill and Whitham via kinematic waves of the propagation and growth of a disturbance by shock waves,[12] and the treatment by Prigogine,[13,4] based on a probabilistic description of individual vehicle speeds. The former also presents an interesting theory of bottlenecks, and some notes on traffic flow at junctions.

The mechanisms of the bottleneck formulation by Lighthill and Whitham are illustrated in Figure 10-2, in which the flow-density relationships of both the main road and its bottleneck sector are shown. Given an arriving flow (point A) in excess of the bottleneck capacity, a capacity operation (point B) is realized above the bottleneck area, and the same flow is realized beyond it (point F), but at considerably better level of service. This is because the chord (the line OF) which indicates the speed is significantly faster than that indicated by chord OB. The propagation of the "shock wave" is at a velocity given by the interface of the two flows (the chord AB) and, being negative, is "turned back" upstream. Its position may be computed by its duration and propagation speed. Vehicles between its position and the physical bottleneck are forced to a "crawl" at speed OB. This "crawl"

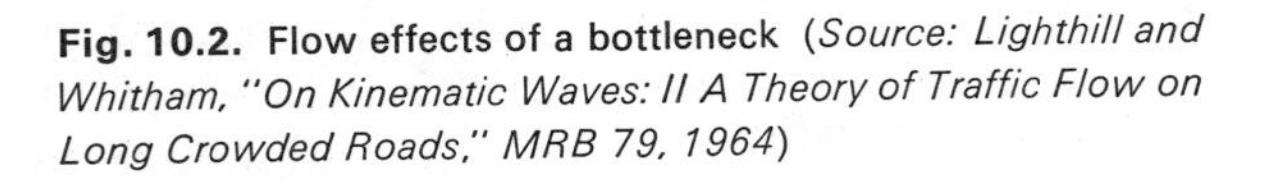

Fig. 10.2. Flow effects of a bottleneck *(Source: Lighthill and Whitham, "On Kinematic Waves: II A Theory of Traffic Flow on Long Crowded Roads," MRB 79, 1964)*

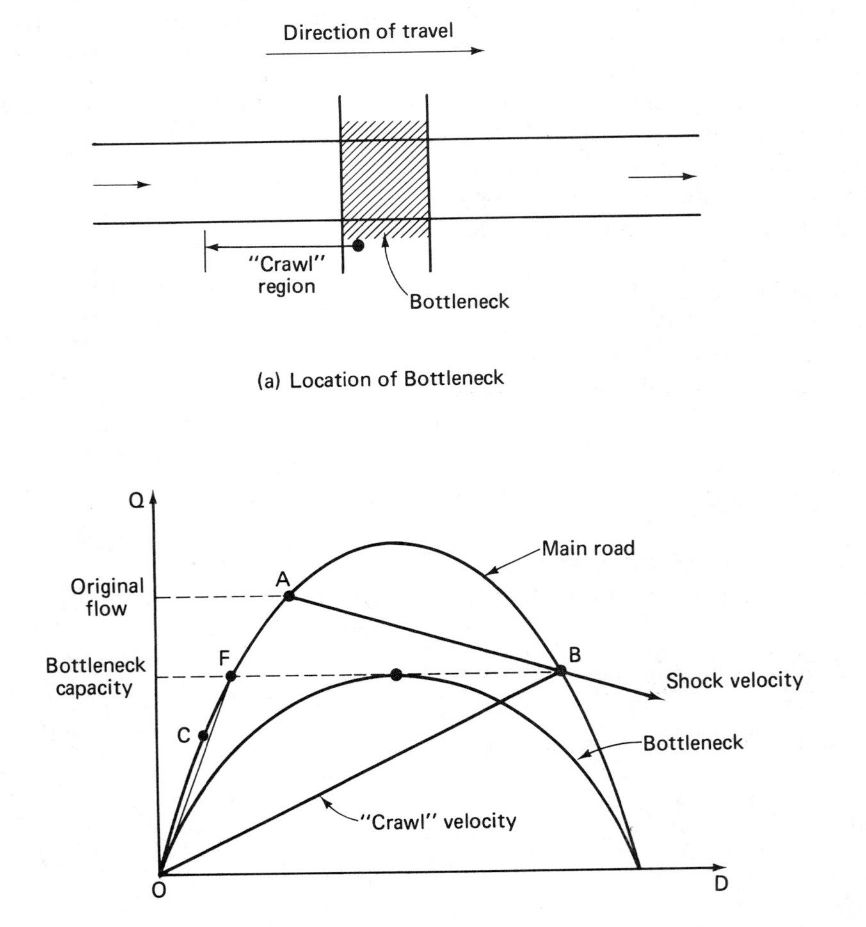

region propagates upstream, until the input flow originally given by point A falls below the bottleneck capacity (point F), say at point C. At this time, the region affected decreases at rate CB until it reaches the original bottleneck position. This information, with the upstream arrival pattern (with respect to time) known, can be used to compute the duration of the "crawl" or level of service F region both in time and in space (extent). These mechanisms, including dissipation, are further treated in the paper.

PROBABILISTIC MODELS

The use of probabilistic descriptions, both in simple form and in particular disciplines (queuing and renewal theory applications), underlies much of traffic flow theory. This section does not purport to represent all of these applications, but simply illustrates the utility and approach of some cases not primarily classified into other sections below.

Flow Out of a Freeway Section

The flow-density relationship has been discussed in the previous two sections of this chapter, and is frequently accepted by analysts for estimating the gross performance of a facility. It is concurrently recognized that: (1) in fact density and flow fluctuate even within approximately "trendless" time periods (2) excursions of the flow-density into the right-hand portion of the relationship (level of service F) are generally undesirable, since the facility tends to degenerate and to stay in this region.

One approach that has been illustrated[14] in the literature in considering these elements has been to accept a deterministic relationship for flow, $Q = Q(D)$, and to consider the density D to be a random variable with mean μ_D and variance σ^2. Given this, a mean flow μ_Q is computed, and other simple probabilistic measures are taken. An improvement on this might be to consider the flow Q to have a density-dependent component and an additive random component, thus allowing perturbations from the basic curve [Figure 10-3 (c)]:

$$Q = Q_{\text{FIX}}(D) + Q_{\text{RANDOM}} \tag{10-6}$$

To illustrate the application without the complication of Equation (10-6), consider $Q = KD(D_0 - D)$, and note that

$$\begin{aligned} \mu_D = E(Q) &= E[KD(D_0 - D)] \\ &= KD_0\, E(D) - KE(D^2) \end{aligned}$$

but that

$$\sigma_x^2 = E(X^2) - [E(X)]^2$$

in general, so that

$$\mu_Q = KD_0\mu_D - K(\sigma_D^2 + \mu_D^2)$$

or

$$\mu_Q = Q(\mu_D) - K\sigma_D^2 \tag{10-7}$$

where $Q(\mu_D)$ is obtained by substituting μ_D in the defining relationship $Q = KD(D_0 - D)$.

Note that for a given μ_D, flow is maximized by minimizing the fluctuation in the density. For a possible range of density, the maximum flow is obtained at $\mu_D = D_0/2$ or at the closest obtainable value. Refer to Figure 10-3 (a).

It must also be noted that maximum flow is not the only possible criterion, nor is it necessarily obtained so simply. Note that for $\mu_D = D_0/2$, excursions into level of service F are extremely common. If the density distribution were symmetric (normal, for instance), these excursions would occur 50 per cent of the time. Such behavior is certainly to be avoided.

Given that the distribution of D is normal (an approximation), and that excursions into the right portion of the basic curve are to be avoided at least 97.5 per cent of time, one would require

$$\mu_D \leq \frac{D_0}{2} - 1.96\sigma_D \tag{10-8}$$

as illustrated in Figure 10-3 (b).

Signalized Control

A significant number of models have been developed for investigation of delays at signalized intersections, and for other criteria in signal setting. Much of this work addresses *delays* at pre-timed signals,[15,16] the *timing* of traffic signals,[17] and rationalization of delays at *traffic-actuated* signals.[18,19] A summary of a number of such studies, including those referenced, is available.[20]

In addition to delay studies and related "optimal" cycle lengths, probabilistic models may be constructed for such objectives as undersaturation at a signalized intersection. Maintenance of undersaturation is treated in Chapter 17 under Signal Timing, on a peak-hour (peak 15-minute) average basis. The *Traffic Engineering Handbook*[21] contains a procedure based on Poisson random arrivals.

The derivation of an undersaturation procedure for Poisson arrivals is based on the fact that the probability $P_R(x)$ of x arrivals for service during any cycle of length

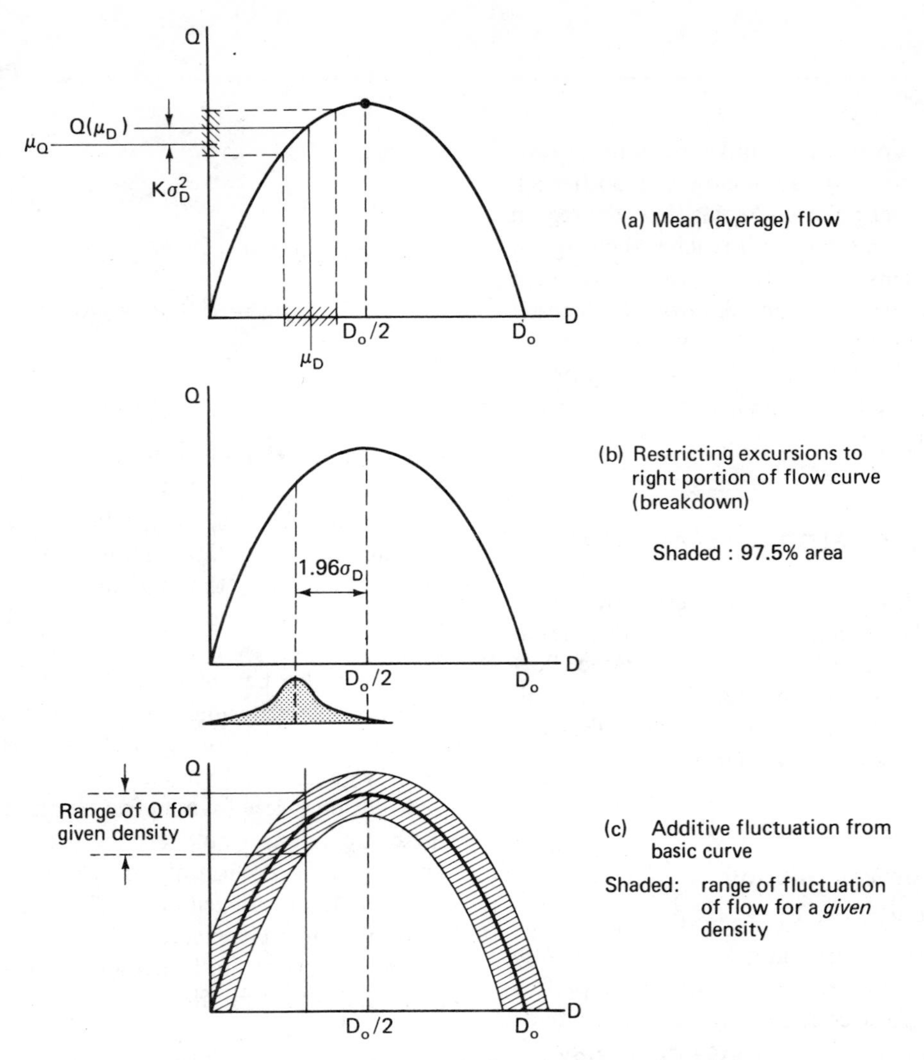

Fig. 10.3. Random elements in flow-density: one formulation

C is given by

$$P_R(x) = \frac{e^{-\lambda C}(\lambda C)^x}{x!} \tag{10-9}$$

where λ is the *average* arrival rate. For a given cycle length and statement on undersaturation (for instance, require that vehicles arriving during a cycle length C be served during the available green G at least 95 per cent of the time), the number of arrivals N that *must* be provided for may be computed from

$$P_R(x \leq N) = \sum_{x=0}^{N} \frac{e^{-\lambda C}(\lambda C)^x}{x!} \geq 0.95 \tag{10-10}$$

where $P_R(x \leq N)$ states the "probability that x is *less than or equal* to N," and N is the *smallest* number satisfying the relation. The green time G required may then be computed from $2.5N$, or $3.0N$, as appropriate (2.5 seconds per passenger car; 3.0 seconds for a typical mix of vehicles).

To illustrate, assume $C = 60$ seconds and $\lambda = 360$ vehicles per hour. Thus $\lambda C = (360/3{,}600)(60) = (1$ vehicle per 10 seconds)(60 seconds) $= 6$. From tables of the Poisson distribution, it may be found that

$$P_R(X \leq 9) = 0.916$$
$$P_R(X \leq 10) = 0.957$$

so that $N = 10$ vehicles must be provided for during every cycle, if 95 per cent undersaturation is to be achieved.

Merging Capacity

The ability of freeway lanes to absorb additional traffic is a critical factor in traffic engineering. This ability is treated from an operational view in Chapter 8 (Ramps) of the *Highway Capacity Manual*,[22] and to a degree in Chapter 7 (Weaving). The approaches of Chapter 8, particularly, may be classed as empiric model-building in the context of traffic theory. These procedures are currently under evaluation on an NCHRP contract.[23]

In a more limited view, Drew[2] has presented a probabilistic model of maximum ramp flow (ramp or merging capacity) which is explicitly in terms of shoulder-lane gap distribution and the gap acceptance pattern. In particular, it is assumed that if a time gap t in the shoulder lane is between 0 and some T $(0 \leq t \leq T)$, no one will accept it. If t is greater than T, but less than $T + T_1$, then one vehicle will use it; if it is between $T + T_1$ and $T + 2T_1$ two vehicles will, and so forth. For any gap, the probability of $(i + 1)$ vehicles entering it is given by

$$P_R(T + iT_1 < t \leq T + \{i + 1\}T_1)$$

The average number of vehicles entering each gap is given by

$$\sum_{i=0}^{\infty} (i + 1)P_R(T + iT_1 < t \leq T + \{i + 1\}T_1)$$

For a shoulder-lane flow of q vehicles per hour, there are q gaps per hour, so that the ramp (merging) capacity q_R is given by

$$q_R = q\sum_{i=0}^{\infty} (i + 1)P_R(T + iT_1 < t \leq T + \{i + 1\}T_1) \tag{10-11}$$

For a distribution of shoulder-lane gaps which is exponential (probability density function $f(t) = qe^{-qt}$), this equation may be written as

$$q_R = q\sum_{i=0}^{\infty} (i + 1)e^{-iqT_1}(1 - e^{-qT_1})e^{-qT}$$

or

$$q_R = q\frac{e^{-qT}}{1 - e^{-qT_1}} \tag{10-12}$$

This analysis may be done for any other gap distribution. It is cited[2] that the Erlang distribution $f(t) = (aq)^a t^{a-1} e^{-aqt}/(a - 1)!$ is a good descriptor of the gaps, with the integer a dependent on the shoulder-lane flow. Figure 10-4 shows the relationship of q, q_R, and T for the exponential case and $T_1 = T$. Note that the sensitivity of q_R to T for a given q is significant. For example, at $q = 1000$ vehicles per hour, a decrease of T from 3.5 to 3.0 seconds increases q_R by 150 vehicles per hour.

Other Applications

There are numerous other applications of probabilistic modeling, both in range of problems (parking, gap acceptance, other freeway and queueing situations) and in depth and sophistication of treatment beyond the above illustrations. Within the recent literature there are analyses of several modes of gap-acceptance injection,[24] and a Markov model of lane-changing on multilane highways.[25]

RAMPS AND FREEWAYS

Contributions to freeway studies have been made in three prime areas: (1) underlying mechanisms and measures, (2) individual ramp considerations and control, and (3) facility surveillance and control.

Underlying Mechanisms and Measures

There are numerous papers in the literature addressed to, or impinging on, freeway mechanisms. Significant among these are the applications treated in Chapters 9 and 14 of Drew.[2] These include work on the effect of ramp geometrics on critical gap size, work on gap distribution and acceptance patterns, and on an energy-momentum model of freeway performance. A regression equation was developed, for instance, which relates the critical gap size T to the geometrics as follows:

$$\begin{aligned} T = {} & 5.547 + 0.828\theta - 1.043L + 0.045L^2 \\ & - 0.042\theta^2 - 0.874S \end{aligned} \tag{10-13}$$

where θ = angle of approach (degrees)
L = length of acceleration lane (100-foot stations)
$$S = \begin{cases} 1 & \text{for taper-type ramp} \\ 0 & \text{for parallel-type ramp} \end{cases}$$

The importance of T may be noted from Figure 10-4.

In regard to the energy-momentum model, equations of continuity of the type cited above were developed, and the kinetic energy, total energy, and lost energy were identified. Two points are noteworthy: (1) lost energy is identifiable in terms of the acceleration noise (treated below) of the stream, and (2) the objective of maximizing efficiency or kinetic energy results in a flow allocation which is other than maximum flow.

The acceleration noise σ of a vehicle was defined by Jones and Potts[26] as the standard deviation of the acceleration record $a(t)$ over a given test section:

$$\sigma = \sqrt{\frac{1}{T}\int_0^T \{a(t) - a_{av}\}^2\,dt} \tag{10-14}$$

In addition to its identification with lost energy, acceleration noise has been independently related to individual

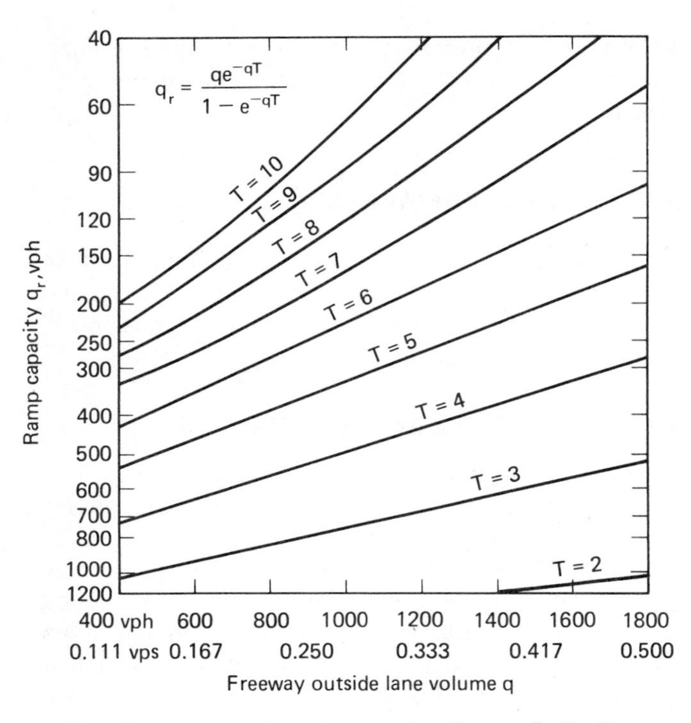

Fig. 10.4. A case of merging capacity (*Source:* Traffic Flow Theory & Control, *Drew, McGraw-Hill, 1968, Figure 9.22*)

driver satisfaction, and to safety, since more hazardous roads are observed to have a higher σ.

On the Gulf Freeway in Houston, the *natural* acceleration noise was observed to correlate with the rolling terrain of that facility (refer to Figure 10-5). This was done by dividing the facility into a number of short test sections, running test vehicles over the entire length, and dividing the obtained record appropriately. The *total* acceleration noise (including driver interaction) was shown to be favorably altered by freeway control. (See Figure 10-6.)

Individual Ramp Consideration

The types of ramp control policies in existence are: (1) simple metering, (2) demand-responsive metering, (3) gap-acceptance injection, and (4) pacer, or "follow-the-rabbit," injection. In regard to the injection policies (3) and (4), it is noteworthy that they can serve two prime functions: (1) access limitation as part of a facility plan; and (2) guidance of individual vehicles into the traffic stream with less turbulence. As noted in Equation (10-6), the latter can, in itself, be a direct aid to facility performance.

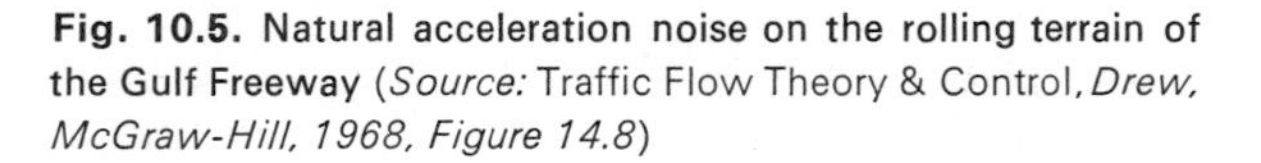

Fig. 10.5. Natural acceleration noise on the rolling terrain of the Gulf Freeway (*Source:* Traffic Flow Theory & Control, *Drew, McGraw-Hill, 1968, Figure 14.8*)

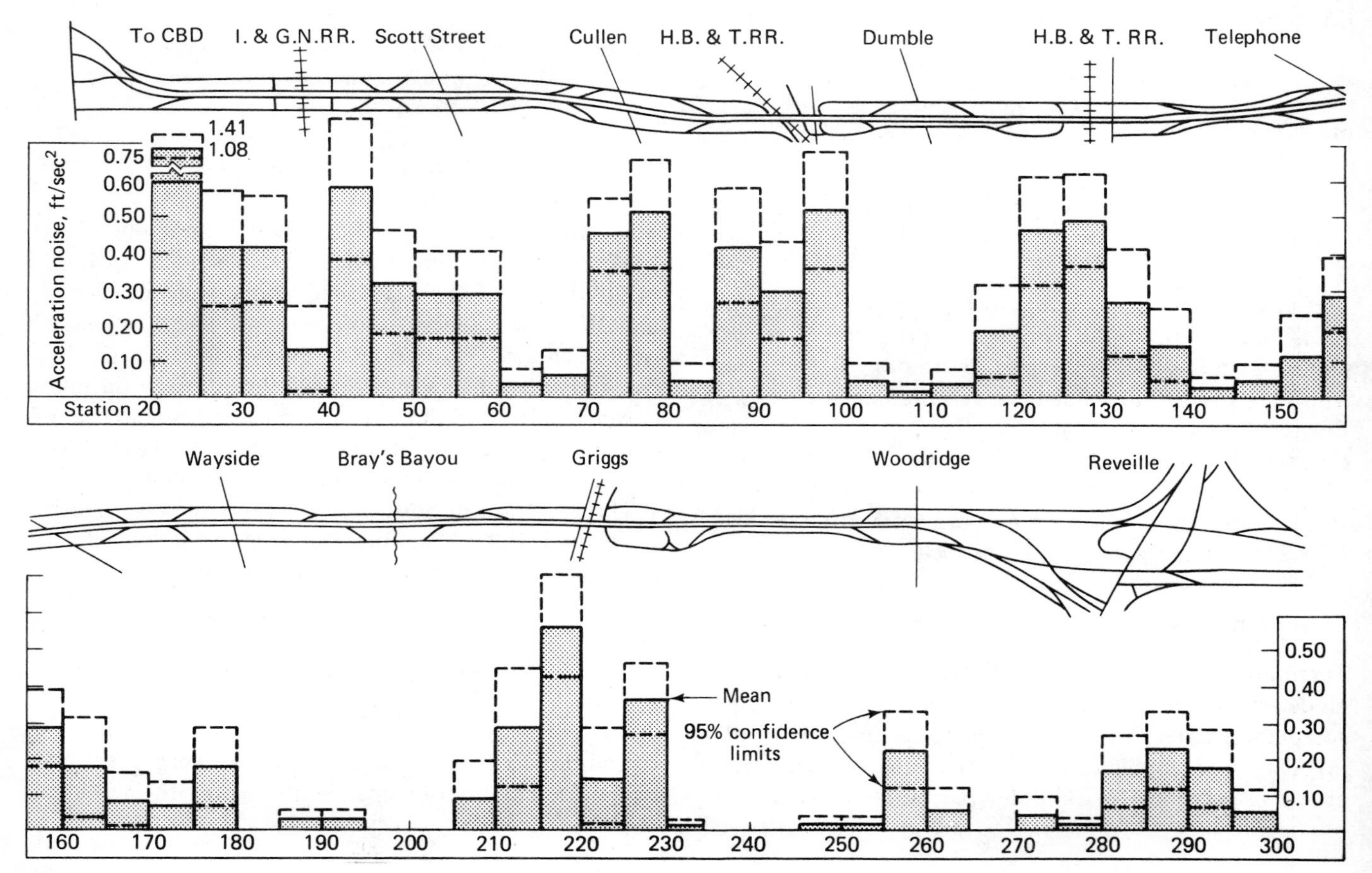

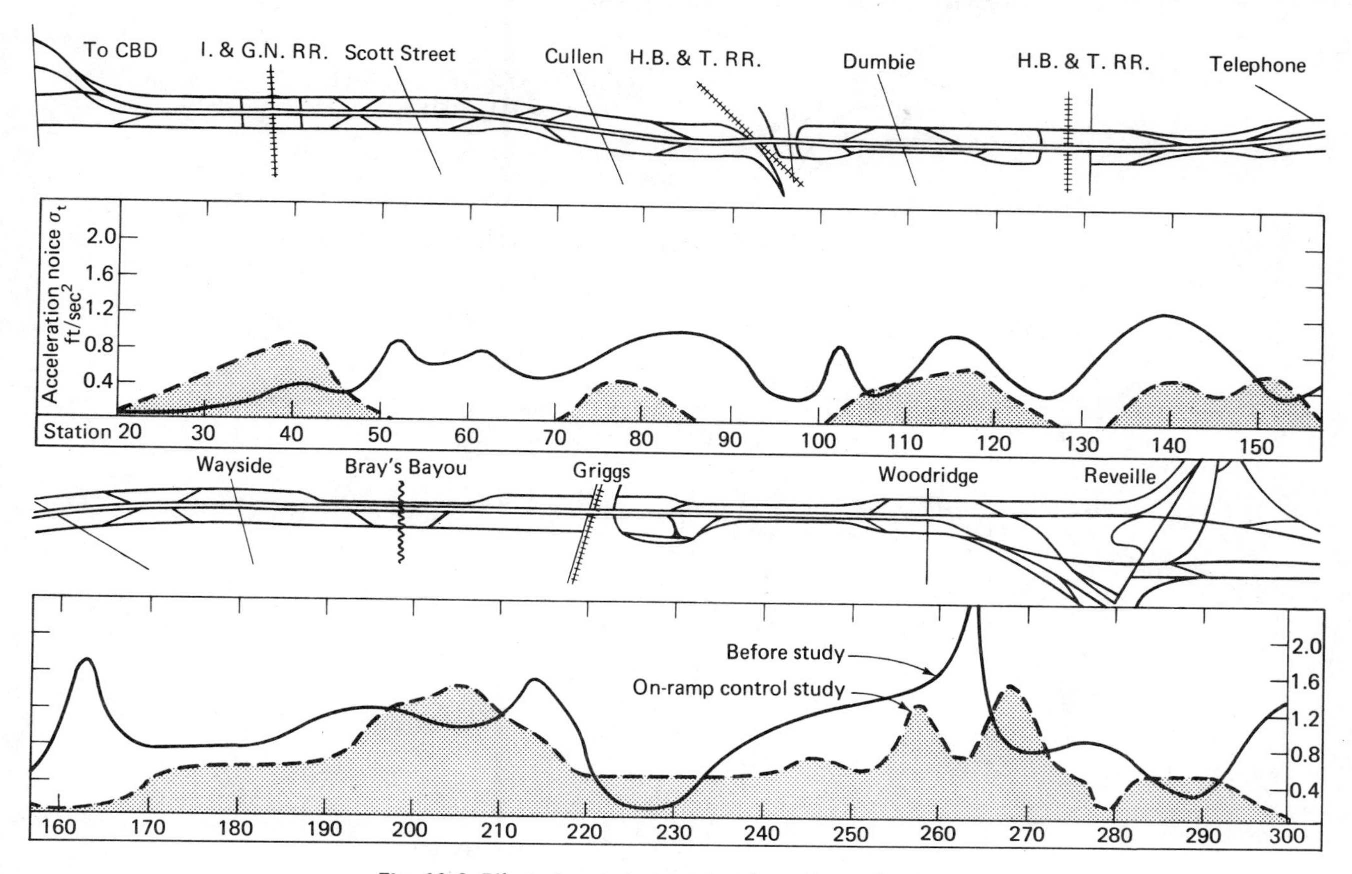

Fig. 10-6. Effect of control on acceleration noise profile, Gulf Freeway (*Source:* Traffic Flow Theory & Control, *Drew, McGraw-Hill, 1968, Figure 14.9*)

As illustrated in Figure 10-7, there are several complex functions in a gap-acceptance controller: (1) detection of ramp vehicles desiring service, (2) detection of gaps acceptable to the subject ramp vehicle (probabilistic), (3) projection of the subject gap forward to the merge area, and a statement on its survival probability (stability), and (4) timed release of the subject ramp vehicle so as to optimize the probability of insertion (acceptance). Each of these functions requires detectors. The second requires a string of detectors for complete treatment, since the ramp-position-to-merge travel time is relatively fixed, and the gap must be detected some T seconds upstream, with the distance depending upon freeway speed. The problem is not to use the entire string, but to determine which detector to use.

In addition to the above functions, detectors may be used to: (1) determine completion of a merge, (2) ascertain that a vehicle has in fact left the signal, and (3) determine some maximum extent, such as impending intersection blockage. Potential detector positions are illustrated in Figure 10-7(b). The first type is particularly important for certain gap-acceptance policies, there being alternates such as release-after-completion (no vehicle released until previous one out of merge area), release-before-completion, and multiple-vehicle release. The theoretic aspects of gap acceptance are treated in

Fig. 10-7. Gap acceptance policy and detection (*Source:* Gap Acceptance and Traffic Interaction in The Freeway Merging Process, *Final Report, TTI, BPR, Figure 3.1*)

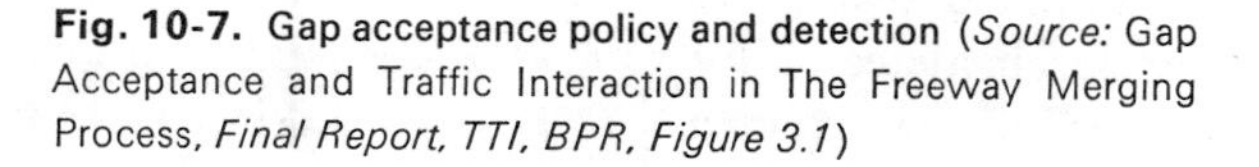

(a) Gap acceptance policy illustrated

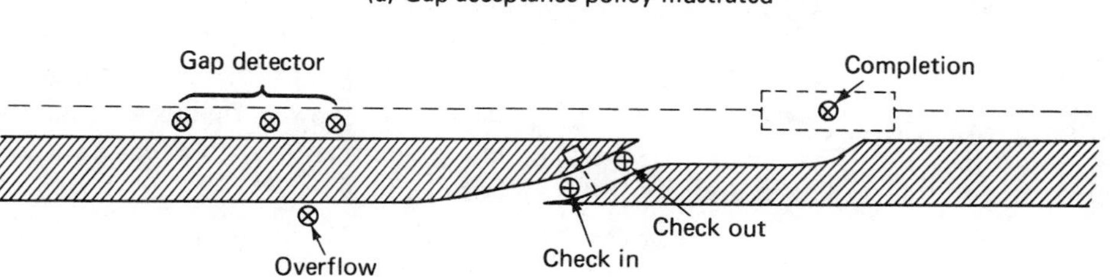

(b) Potential detector locations

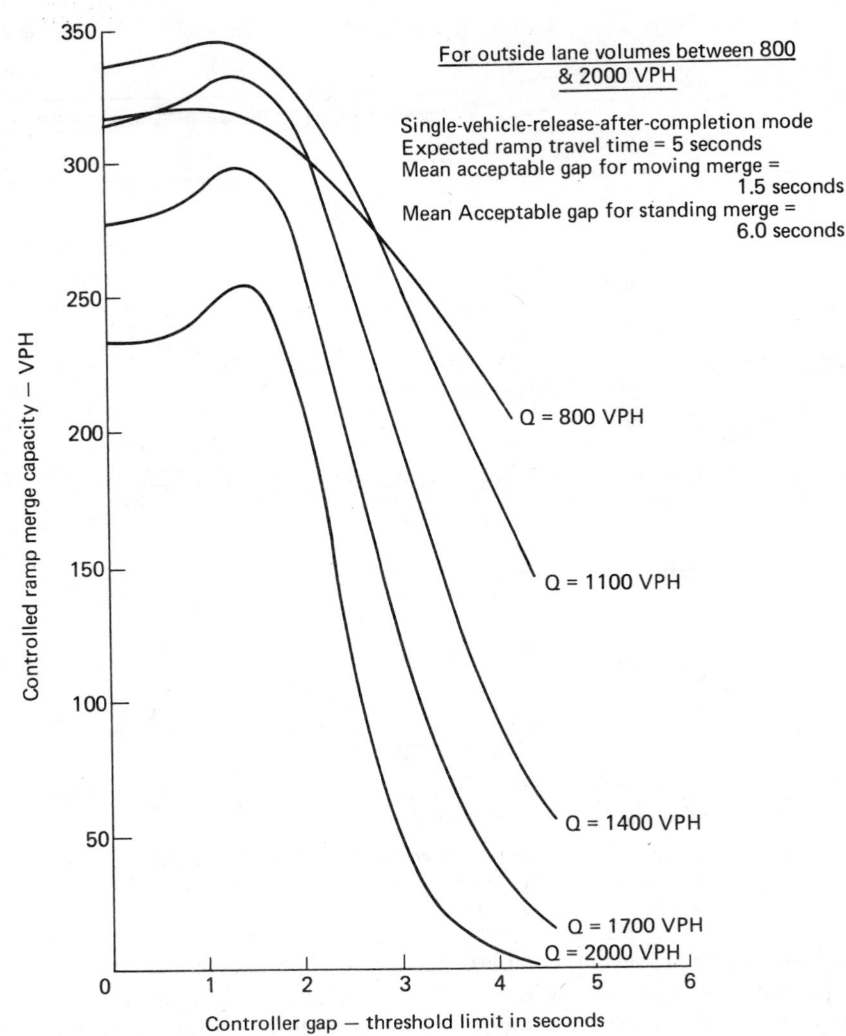

Fig. 10-8. Controller merge capacity vs. controller threshold setting (*Source: Yagoda and Pignataro, "The Analyses and Design of Freeway Entrance Ramp Control Systems,"* HRR303, *1970, Figure 6*)

the literature.[24,27,28] Figure 10-8 illustrates the application of the acceptance gap setting in effectively using gap acceptance for metering.

The demand-responsive ramp controller (type 2) is also an interesting study: The metering rate, which varies with time, may be pre-programmed to known demand profiles, or may use an upstream detector for on-line responsiveness. The former is illustrated in a study[29] which includes an analysis of the delay of the restricted vehicles and indicates the extent (queue) of the stored vehicles.

Freeway Surveillance and Control

A short discussion of freeway surveillance and control and a description of existing projects is contained in the section Freeway and Ramp Controls of Chapter 17. It is noted therein that despite the considerable advances made by these projects and by continuing research, there is, to date, no comprehensive management or allocation policy which can be accepted for application to other than constant-load averaged conditions (i.e., to varying peak-period input profiles, or to changed capacity due to incident occurrence). This area is one of high current interest and research. Among the best studies to date have been a formulation of the allocation of the flows as a linear program for constant demands,[27] and a limited use of a "power function" (comparable in result to energy-momentum) to effect an allocation between two parallel facilities.[14]

ARTERIAL AND NETWORK CONTROL

Although there have been some theoretical and systems-feasibility articles in the literature on routing control and advisory routing,[30-32] most traffic-theoretic material with current-applications utility has generally been related to aspects of signal timing, to signal pattern policies and algorithms, and to hierarchical structures for both control policies and control devices (computers). As such, much of this material has been treated in the section Control of Coordinated Signals, in Chapter 16. This includes: (1) presentation of certain criteria (maximum bandwidth, queue clearance, stops and delay), (2) signal timing algorithms, including SIGOP, and (3) enumeration of control projects and digital computer control.

This section will supplement the above with a treatment of a control policy for an oversaturated controller, and two measures of effectiveness: mean velocity gradient and energy ratio.

Oversaturated Intersection Control

There are many cases of "oversaturated" intersections, in which queues develop and exist for long periods within a peak period on all rights-of-way at an intersection. During this time it is not possible to clear a vehicle within the green time following his arrival. This problem was addressed by Gazis and Potts,[31] and it was found that the control strategy should not depend on instantaneous measurement of queues and arrival flows, but should be dependent on the historic demand during the entire saturation period.

The problem involved an oversaturated intersection with convex, cumulative, demand-for-service curves (smoothed from true sawtooth behavior) as illustrated in Figure 10-9), with constrained green times (maximum, minimum per cycle), and with fixed cycle. The criterion considered was minimization of total delay. The techniques of optimal control theory were employed, and it was found that the optimal controller was what is called a "bang-bang" controller in the control field literature: The maximum allocation (maximum green) is given to one direction for a certain time, at which point a switch is made to minimum allocation (maximum on other direction) until the end of the control period. This control policy is illustrated in Figure 10-9. The controller was constrained to achieve end-of-saturation at the common time T on both directions.

Some Measures of Effectiveness

A measure of effectiveness (MOE) is defined as an indicator of the performance of a system. Common MOE's are speed, travel time, and number of stops. A recent study[34] analyzed a list of MOE's known to traffic engineers, with the objective of obtaining a basic, non-redundant set. A set of three was recommended: travel time T, energy ratio η_E, and service rate R. R reflects the rapidity with which vehicles are serviced from an intersection.

Of the set of three, energy ratio η_E was newly defined. It is defined to be the ratio of effective kinetic energy E_{eff} to the measured free-movement kinetic energy E_{meas}. The former is defined in terms of travel time (a speed indicator) as $E_{\text{eff}} = D\{L/T\}^2$, where D is density and L is section length. The latter is defined in terms of the free-movement spot speed V_F as $E_{\text{meas}} = DV_F^2$. The energy ratio is thus

$$\eta_E = \frac{E_{\text{eff}}}{E_{\text{meas}}} = \left\{\frac{L/T}{V_F}\right\}^2 \tag{10-15}$$

Note that η_E ranges between 0 and 1; η_E is a measure of lost energy, as is acceleration noise [Equation (10-14)]; and lost energy is minimized by maximizing η_E. One marked advantage cited for η_E is that it is simpler to

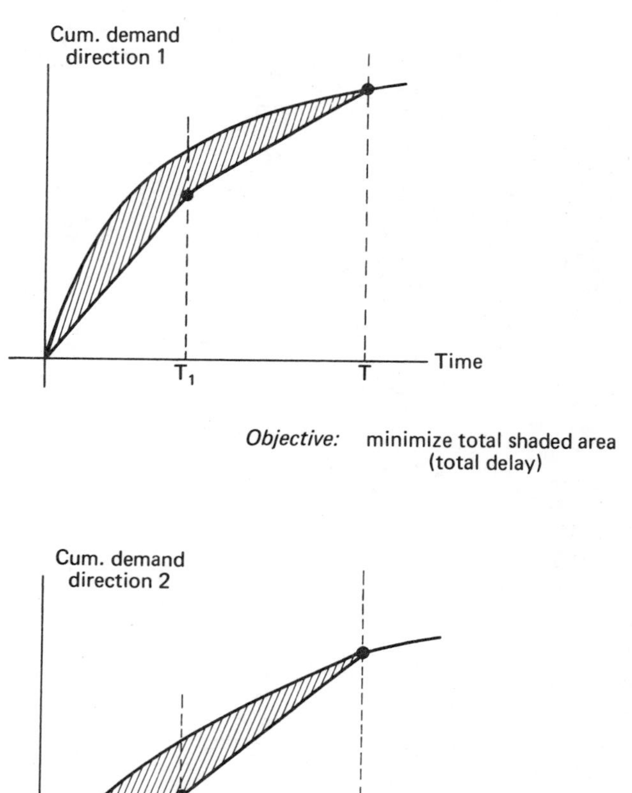

Fig. 10-9. A signalization policy for an oversaturated intersection

compute than acceleration noise, requiring only V_F and T.

The literature[34] also treats measurement error and their effects extensively, concluding among other things that queue measurement errors can cause significant errors in the MOE indications. The problems of measurement and detection are also treated extensively in the UTCS system description.[35]

Another MOE of note is the mean velocity gradient G,[36] defined by:

$$G = \frac{\sigma}{\bar{V}} \tag{10-16}$$

where σ is acceleration noise, and $\bar{V}$ is average speed. This measure was proposed for urban arterials since the acceleration noise, although useful for continually running MOE situations, could not effectively distinguish between such situations as stoppages on a virtually empty but poorly-timed arterial, and a well-timed arterial with normal interaction with dense traffic, although the driver satisfaction would be markedly different.

OTHER ASPECTS

As noted in the introduction, this presentation of traffic-theoretic studies has been limited in both scope and depth. Particularly lacking are treatments of queuing theory applications, probabilistic modeling in depth on signal problems, and pedestrian and safety analysis models. Reference is made to the general literature, and to bibliographies cited in the introduction, for this and other material.

11

Highway Capacity: Introduction and Background

The subject of highway capacity not only concerns itself with the ultimate carrying ability of various facilities, but also with the relative service characteristics of facilities operating at some fraction of their capacity volume. Thus, a study of highway capacity is at once a quantitative and a qualitative study, which permits evaluation of both the adequacy and the quality of vehicle service being provided by the facility under study.

The *Highway Capacity Manual*[1] is the most complete and authoritative source on capacity and, like the original edition,[2] is largely based on empirical data rather than theoretical development.

Capacity analyses are necessary inputs to many traffic engineering evaluations:

1. Deficiencies in the existing highway system may be evaluated by comparing measured volumes to the capacity of existing facilities.
2. Proposed changes in the existing street system, such as changes in geometrics, signalization, parking regulations, converting to one-way operation, turn restrictions, etc., must all be evaluated for their effect on capacity.
3. The design of new facilities must always be based upon capacity analyses coupled with projected demands.
4. The comparison of the relative effectiveness of alternate modes of transportation in serving a particular demand is often partially based on capacity analyses.

DEFINITIONS

Capacity terms

The original Capacity Manual utilized three capacity terms: *basic capacity*, *possible capacity*, and *practical capacity*.

Basic capacity was the maximum number of passenger cars that could pass a given point on a lane or roadway under ideal traffic and roadway conditions in one hour.

Possible capacity was similarly defined for prevailing roadway and traffic conditions.

Practical capacity referred to the maximum number of passenger cars that could pass a point in one hour without causing unreasonable delay, hazard, or restriction.

The new Manual replaces these with the terms *capacity* and *level of service*.

Capacity is defined as the maximum number of vehicles that can pass over a given section of a lane or roadway, in one direction (or in both directions for a two-lane or three-lane highway), during a given time period (one hour unless otherwise specified), under prevailing roadway and traffic conditions. The current term of *capacity* is synonymous with the earlier term of *possible capacity*. As previously, it is the volume of traffic that cannot be exceeded in actuality without changing one or more of the conditions that prevail. Roadway conditions refer to the physical features of the roadway, which do not change unless some construction or reconstruction is performed. Traffic conditions refer to the characteristics of traffic using the roadway, which may change hourly.

The former *basic capacity* has been replaced with the term *capacity under ideal conditions*. Ideal prevailing roadway and traffic conditions are characterized by:

1. Uninterrupted flow, free from side interferences of vehicles and pedestrians
2. Passenger cars only in the traffic stream
3. Twelve-foot traffic lanes, with adequate shoulders and no lateral obstructions within 6 feet of the edge of pavement
4. For rural highways, horizontal and vertical alignment satisfactory for average highway speeds of 70 miles per hour or greater, and no restricted (less than 1,500 feet) passing sight distances on two-lane and three-lane highways

The concept of *level of service* has replaced *practical capacity*. Level of service is associated with different operating conditions that occur on a facility when it accommodates various traffic volumes. It is a qualitative measure of the effect of a number of factors which include:

1. Speed and travel time
2. Traffic interruptions
3. Freedom to maneuver
4. Driver comfort and convenience
5. Safety
6. Vehicle operating costs

The concept of level of service is carried throughout the new Manual, and it is applied to all highway elements. Six levels of service have been established, designated by the letters A through F, providing for best to worst service in terms of driver satisfaction. For a given highway facility, different levels of service will be selected to provide for appropriate operating characteristics on the various components of the facility. However, these operating conditions should be in harmony with each other; that is, they should be of approximately equal acceptability to average drivers. Different highway elements and types of facilities include: intersection, ramp, weaving section, ramp terminal, speed-change lane, freeway, uncontrolled-access multi-lane highway, two-lane or three-lane highway, arterial street, downtown street, etc. There is an important distinction between capacity and level of service, and it should be clearly understood. A given lane or roadway may provide a wide range of levels of service (depending essentially on speed and volume), but the lane or roadway has only one capacity. In practice, any given highway, or component thereof, may operate at a wide range of levels of service, depending upon the time of day, day of week, and period of the year.

Service volume is the maximum number of vehicles that can pass over a given section of a lane or roadway, in one direction on multi-lane highways (or in both directions on a two- or three-lane highway), during a specified time period, while operating conditions are maintained corresponding to the selected or specified level of service. In the absence of a time modifier, service volume is an hourly volume.

Roadway terms

Control of Access.

Full control of access: The authority to control access is exercised to give preference to through traffic, by providing access connections with selected public roads only, and by prohibiting crossings at grade or direct private driveway connections.

Partial control of access: The authority to control access is exercised to give preference to through traffic to a degree that, in addition to access connections with selected public roads, there may be some crossings at grade and some private driveway connections.

Uncontrolled access: The authority having jurisdiction over a highway, street, or road does not limit the number of points of ingress or egress, except through the exercise of control over the placement and the geometrics of connections as necessary for the safety of the traveling public.

Functional Types.

Arterial highway: A highway primarily for through traffic, usually on a continuous route, not having access control.

Expressway: A divided arterial highway for through traffic, with full or partial control of access, and generally with grade separations at major intersections.

Freeway: An expressway with full control of access.

Parkway: An arterial highway for noncommercial traffic with full or partial control of access.

Major street or major highway: An arterial highway with intersections at grade and direct access to abutting property, and on which geometric design and traffic control measures are used to expedite the safe movement of through traffic.

Through street or *through highway:* Every highway, or portion thereof, at the entrance to which vehicular traffic from intersecting highways is required by law to stop or yield before entering or crossing.

Local street or *local road:* A street or road primarily for access to residence, business, or other abutting property.

Frontage road: A road contiguous to, and generally parallelling, an expressway, freeway, parkway, or through street. It is designed so as to intercept, collect, and distribute traffic desiring to cross, enter, or leave such a highway and to furnish access to property that otherwise would be isolated as a result of the controlled-access feature. It is sometimes called a *service road.*

Terrain.

Level terrain: Any combination of gradients, length of grade, or horizontal or vertical alignment that permits trucks to maintain speeds that are equal to, or approach the speeds of, passenger cars.

Rolling terrain: Any combination of gradients, length of grade, or horizontal or vertical alinement that causes trucks to reduce their speeds substantially below that of passenger cars on some sections of the highway, but which does not involve a sustained crawl speed by trucks for any substantial distance.

Mountainous terrain: Any combination of gradients, length of grade, or horizontal or vertical alinement that will cause trucks to operate at crawl speed for considerable distances or at frequent intervals.

Sustained grade: A continuous highway grade of appreciable length and consistent or nearly consistent gradient.

Traffic Operations Terms

Peak-hour traffic: The highest number of vehicles found to be passing over a section of a lane or a roadway during 60 consecutive minutes. This term may be applied to a daily peak hour or a yearly peak hour.

Rate of flow: The hourly representation of the number of vehicles that pass over a given section of a lane or a roadway for some period less than one hour. It is obtained by expanding the number of vehicles to an hourly rate by multiplying the number of vehicles during a specified time period by the ratio of 60 minutes to the number of minutes during which the flow occurred.

Interrupted flow: A condition in which a vehicle traversing a section of a lane or a roadway is required to stop by a cause outside the traffic stream, such as signs or signals at an intersection or a junction. Stoppage of vehicles by causes internal to the traffic stream does not constitute interrupted flow.

Uninterrupted flow: A condition in which a vehicle traversing a section of a lane or a roadway is not required to stop by any cause external to the traffic stream, although vehicles may be stopped by causes internal to the traffic stream.

Peak hour factor: A ratio of the volume occurring during the peak hour to the maximum rate of flow during a given time period within the peak hour. It is a measure of peaking characteristics, whose maximum attainable value is unity. The term must be qualified by a specified short period within the hour; this is usually 5 or 6 minutes for freeway operation, and 15 minutes for intersection operation.

Land Use and Development Terms

Central Business District (CBD): That portion of a municipality in which the dominant land use is for intense business activity. This district is characterized by large numbers of pedestrians, commercial vehicle loadings of goods and people, a heavy demand for parking space, and high parking turnover.

Fringe area: That portion of a municipality immediately outside the CBD in which there is a wide range in type of business activity. It generally includes small businesses, light industry, warehousing, automobile service activities, and intermediate strip development, as well as some concentrated residential areas. This area is characterized by moderate pedestrian traffic and a lower parking turnover than is found in the CBD, but it may include large parking areas serving the CBD.

Outlying business district: That portion of a municipality, or an area within the influence of a municipality, normally separated geographically by some distance from the CBD and its fringe area, in which the principal land use is for business activity. This district is characterized by relatively high parking demand and turnover and moderate pedestrian traffic.

Residential area: That portion of a municipality, or an area within the influence of a municipality, in which the dominant land use is residential development, but where small business areas may be included. This area is characterized by few pedestrians and a low parking turnover.

STREAM CHARACTERISTICS

Stream characteristics of a roadway is a measure of its ability to accommodate traffic. Although this ability depends to a large extent on the physical features of the roadway, there are other factors not directly related to roadway features which are of major importance in determining the capacity of any roadway. Many of these factors relate to variations in the traffic demand and the interaction of vehicles in the traffic stream. An understanding of stream characteristics is basic to achieving a thorough insight into capacity analysis techniques.

Spacing and Headway Characteristics

Spacing is defined as the interval in distance from head to head of successive vehicles, and *headway* as the interval in time between individual vehicles measured from head to head as they pass a given point. These two measures describe the longitudinal arrangement of vehicles in a traffic stream.

The relationship between spacing and headway is dependent on speed, with:

$$\text{Headway (second)} = \frac{\text{spacing (feet)}}{\text{speed (feet per second)}} \qquad (11\text{-}1)$$

The relationship between average spacing and density is as follows:

$$\text{Density (vehicles per mile)} = \frac{\text{5,280 (feet per mile)}}{\text{average spacing (feet per vehicle)}} \qquad (11\text{-}2)$$

The relationship between average headway and volume may be expressed as follows:

$$\text{Volume (vehicles per hour)} = \frac{\text{3,600 (second per hour)}}{\text{average headway (second per vehicle)}} \qquad (11\text{-}3)$$

Spacing as a Measure of Capacity. Few drivers, if any, operate their vehicles in identically the same manner, or react in the same manner when exposed to similar conditions. It is impossible, therefore, to predict the effect of various roadway and traffic conditions on an individual driver. It has been found, however, that the combined effect on traffic as a whole can be predicted with reasonable accuracy.

All drivers do not maintain the same spacing to the vehicle ahead when traveling at a given speed. Figure 11-1 shows the minimum distance-spacings allowed by the average driver at different speeds for several conditions. Similar curves for other conditions may also be presented, but the curves in Figure 11-1 are sufficient to show that the average driver increases the distance-spacing between vehicles as his speed increases, and that the spacing is also influenced by the characteristics of the highway.

Using the data shown in Figure 11-1, it is possible to determine the maximum number of passenger cars, one behind the other, that can pass a point in one hour at any given speed, if this given speed is maintained by all vehicles. The following relationship could be used to compute the maximum volume of the traffic lane at the

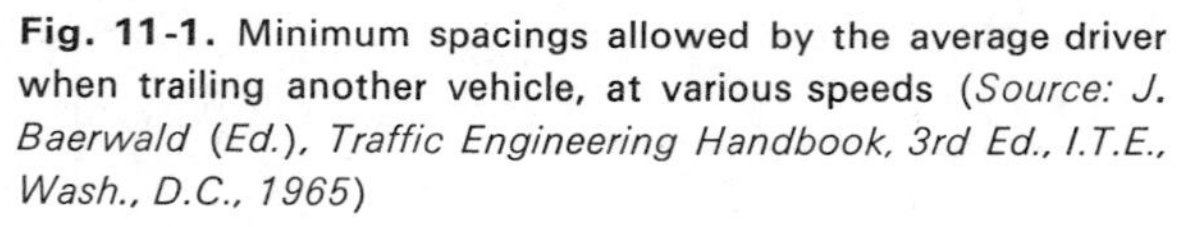
Fig. 11-1. Minimum spacings allowed by the average driver when trailing another vehicle, at various speeds (*Source: J. Baerwald (Ed.), Traffic Engineering Handbook, 3rd Ed., I.T.E., Wash., D.C., 1965*)

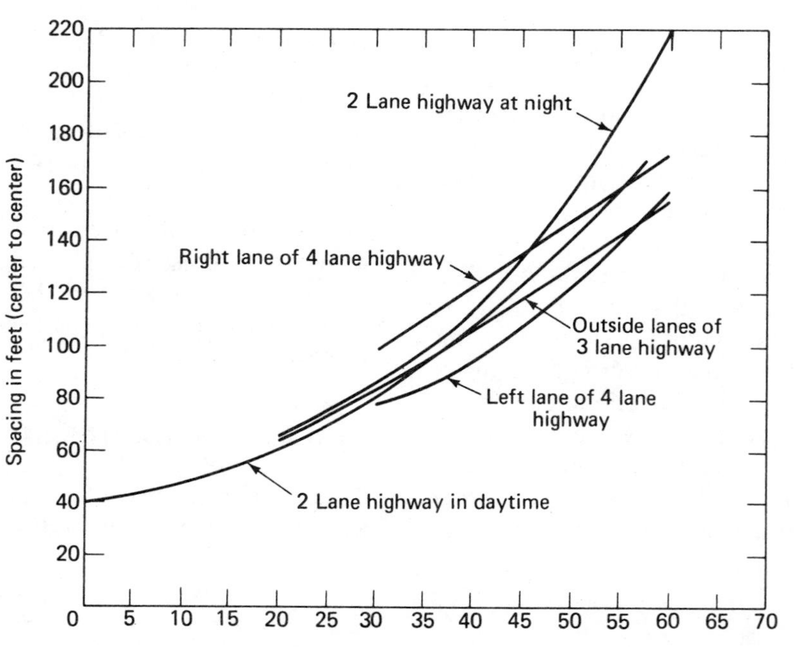

given speed:

$$\text{Volume (vehicles per hour)} = \frac{5{,}280 \text{ (feet per mile)} \times \text{speed (miles per hour)}}{\text{spacing (feet per vehicle)}} \quad (11\text{-}4)$$

If volume is plotted against speed using the above relationship, it will be found that the highest volume of a traffic lane is approximately 2,000 passenger vehicles per hour, when vehicles travel at about 30 miles per hour. This is the maximum capacity of a traffic lane operating under conditions of uninterrupted flow. Any traffic variable, or any roadway condition that prevents vehicles from moving safely at a speed of 30 miles per hour, lowers the capacity of the traffic lane.

A traffic lane can attain this maximum capacity only if the following conditions exist:

1. There must be at least two lanes for the exclusive use of traffic traveling in one direction.
2. There must be no differentials in speeds of vehicles. All vehicles must travel at approximately the same speed.
3. There must be practically no commercial vehicles.
4. The widths of traffic lanes, shoulders, and clearances to vertical obstructions beyond the edge of the traffic lane must be adequate.
5. There must be no merging, weaving, or turning movements.
6. There must be no parking, loading, unloading, or stopping of vehicles.
7. There must be no restrictive sight distances, grades, improperly superelevated curves, signalized intersections, or interference by pedestrians.

Under actual operating conditions, if the above requirements are satisfied, it is possible to attain the following basic capacities for the different types of highways, assuming the capacity of a traffic lane is 2,000 passenger cars per hour.

1. *Two-lane, two-way highways:* On two-lane roads, with few opposing vehicles, traffic can fill one lane by immediately passing into gaps that form. This single lane might reach the capacity of 2,000 passenger cars per hour. However, as passing is restricted by vehicles from the opposite direction, the spaces that develop in the lane cannot be filled by passing maneuvers. Instead, breaks form in the traffic stream in each direction. Capacity in one direction, then, is affected by volume in the opposite direction. It has been established that the ideal capacity of a two-lane road in both directions is 2,000 passenger cars per hour, regardless of the distribution between lanes.
2. *Three-lane, two-way highways:* The center lane of a three-lane road serves vehicles performing passing maneuvers in either direction. Vehicles can, therefore, completely fill the outside lanes by utilizing the center lane for passing. Thus, the ideal capacity of a three-lane, two-way road is 4,000 passenger cars per hour total in both directions. The ideal capacity is limited to 2,000 passenger cars for one direction.
3. *Multi-lane highways:* Multi-lane roads can be designed to meet all the conditions listed above. Therefore, the ideal capacity for a multi-lane highway is 2,000 passenger cars per hour per lane, regardless of the number of lanes.

On two- and three-lane roads, two lanes for the exclusive use of traffic traveling in the one direction are not available. The character of operation is therefore entirely different on these roads from that on multi-lane highways, since vehicles performing passing maneuvers are forced to use a traffic lane that is provided for vehicles traveling in the opposite direction. Consequently, the ideal capacities of two- and three-lane roads are much lower than for multi-lane roads.

Headway Distribution. If all the vehicles using a highway were equally spaced, determination of maximum volumes or levels of congestion would be fairly simple. However, vehicles tend to form groups or "platoons," even at low volumes. Individual headways show a wide degree of variation, with many vehicles queuing at short headways and others separated by large time gaps.

Figures 3.29 and 3.30 of the Capacity Manual show headway distributions for vehicles traveling in the same direction on typical two- and four-lane rural highways for various volumes with uninterrupted flow. Under nearly all volume conditions, these curves show that approximately two-thirds of all vehicles are spaced at, or less than, the average headway between vehicles.

On multi-lane facilities, some drivers will accept smaller headways than others, and these tend to gravitate to the median lane. Thus, volumes for median freeway lanes on many facilities have consistently carried volumes of 2,200 vehicles per hour. However, drivers accepting larger headways tend to remain in the right-hand lane, and capacity for this lane may be lower than 2,000 passenger cars per hour. Therefore, the indicated maximum capacity of 2,000 passenger cars per hour is reasonable when one takes into account the headway variation from lane to lane.[3]

Further description of vehicular spacing characteristics can be made in mathematical terms. Under certain circumstances, vehicle spacing at a point will follow a random distribution. Such a distribution can be described mathematically by the *Poisson distribution.* Complete discussions of the application of the Poisson distribution to vehicular spacing characteristics are available.[4-23]

Relationships of Speed, Volume, and Density

Principles of physics, dynamics, hydraulics, and the laws of various sciences are being applied to the theory of

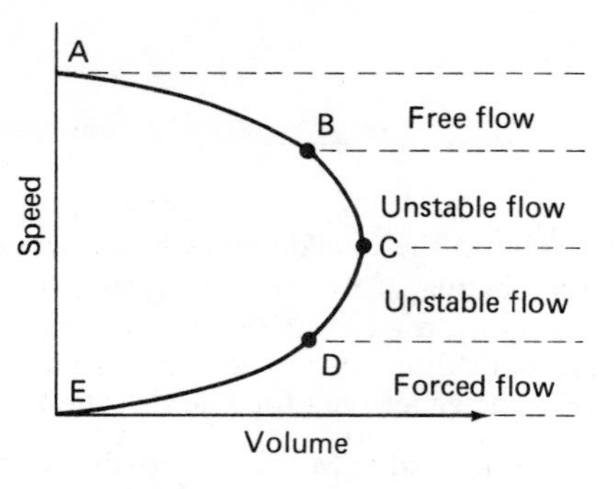

Fig. 11-2. Speed vs. volume

movement of traffic with varying degrees of success. Analytical models of traffic flow are extremely useful, even though they have not, as yet, been developed to completely describe the complex interrelationships affecting the traffic stream. These models permit a more thorough approach to operational problems and can lead to the application of optimizing techniques for the control of traffic.

Speed-Volume Relationships. The basic speed-volume relationship for uninterrupted flow is that as volume increases, the space-mean-speed of traffic decreases until the critical density is reached. Thereafter both volume and speed decrease. The general form of this relationship is shown in Figure 11-2. The *AB* portion of the relationship is normally classified as the *free flow* condition, the *BCD* portion as the *unstable* flow condition (with volume at point *C* representing critical density), and the *DE* portion as the *forced* or *breakdown flow* condition. Numerous studies have confirmed the characteristic shape of the speed-volume relationship, and have indicated that the zone of free flow is essentially a linear relationship.[1,2,24,25]

An extensive study to establish speed-volume relationships on many urban signalized streets (interrupted flow) in Chicago revealed that speed remained essentially constant when volume did not exceed about 70 per cent of capacity. Thereafter, speed decreased almost to less than 5 miles per hour, linearly, as volume approached capacity.[25] Somewhat similar results were obtained in another study.[27]

Speed-Density Relationships. The speed-density relationship for uninterrupted flow is similar to the speed-volume relationship in that in the upper range, speed decreases with increasing volume and density. Density, however, continues to increase past the point of critical density, whereas volume decreases. Early studies found a linear relationship between average density and space-mean-speed. More recent studies have mathematically fitted best curves to observed data, resulting in exponential curves and bell-shapted curves.[1] A statistical analysis of these various speed-density relationships was undertaken, and the linear relationship exhibited excellent correlation.[37]

The speed-density relationship for interrupted flow is very similar to that under uninterrupted flow conditions, and most of the studies found a linear relationship.[1]

Fig. 11-3. Volume vs. density

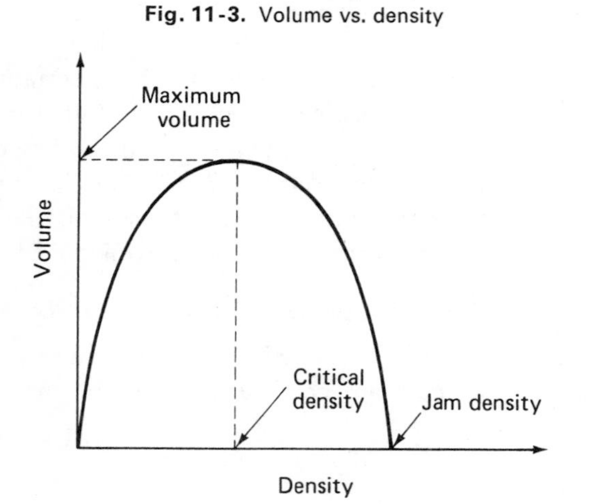

Volume-Density Relationships. The basic volume-density relationship for both uninterrupted and interrupted flow conditions is that as density increases from zero (when there are no vehicles), volume increases to the point of critical density. Thereafter volume decreases as density continues to increase to a maximum value, known as *jam density* (when all vehicles are stopped). The general form of this relationship is shown in Figure 11-3. Numerous studies have confirmed the characteristic shape of the volume-density relationship.[28-32,8] A study[33] measured stopped vehicle spacings to determine the jam density. Jam density was found to be 210 vehicles per lane-mile.

CAPACITY AND LEVEL OF SERVICE

Capacity for Uninterrupted Flow Conditions

Under ideal roadway and traffic conditions, the fundamental capacities for uninterrupted flow conditions for different types of highways are shown in Table 11-1. Justification for the use of these values is essentially based on the reasoning outlined under Spacing as a Measure of Capacity.

Capacity for Interrupted Flow Conditions

Unlike uninterrupted flow, it is not feasible to define fundamental capacities under ideal conditions because of the large number of variables involved. In general, the following two basic limitations can be established.

1. Rarely does a traffic lane on an urban arterial carry

TABLE 11-1. CAPACITY UNDER IDEAL CONDITIONS

TYPE OF FACILITY	*CAPACITY (pcph)*
Two-lane, two-way roadways	2,000, total both directions
Three-lane, two-way roadways	4,000, total both directions
Multi-lane roadways	2,000, each lane, average

volumes at a rate greater than 2,000 passenger cars per hour of green signal, even with ideal signal progression.

2. A line of vehicles, all of which are stopped by an interruption, will rarely move away from the interruption at a rate greater than 1,500 passenger cars per hour per lane. This is based on an average departure headway of 2.4 seconds.

Levels of Service

This concept is best defined in terms of units of measure familiar to drivers using the given highway. For uninterrupted flow on freeways, rural highways, and some suburban highways, operating speeds are used to define the levels. On urban arterials, where interrupted flow exists and the degree of delay varies widely, average overall travel speed is a better measure. On downtown streets, approximate speed measures are used. At individual intersections, it is not possible to use speeds as a measure. Instead, the degree of loading is used, that is, the percentage of green signal phases fully utilized by moving traffic.

A second factor used to evaluate level of service is either the ratio of demand volume to capacity, or the ratio of service volume to capacity. Demand volume is used instead of service volume when the level of service is evaluated for an existing facility.

The following criteria have been established for determining capacity and level of service relationships:

1. Volume and capacity are expressed in *numbers of passenger cars per hour* for subsections of each section of roadway.
2. Level of service, strictly defined, applies to a section of roadway of significant length.
3. Analysis of volume and speed is made for each point or subsection of the highway having relatively uniform conditions. The weighted operating speed, or average overall travel speed, is then determined for the entire section, and a corresponding level of service is identified.
4. Variables used to measure capacity include roadway type, geometrics, average highway speed, traffic composition, and time variations in volume. For level of service, additional elements include speed and volume-to-capacity ratios.
5. Values of speed and volume-to-capacity ratio which define levels of service are established for each of the following types of facilities:
   a. Freeways and other expressways
   b. Other multi-lane highways
   c. Two- and three-lane highways
   d. Urban arterial streets
   e. Downtown streets

   Related levels of service are established for different point elements, including intersections, ramp junctions, and weaving sections.

Attempts have been made to model the level of service concept.[34,35,36]

Operating Conditions for Levels of Service

Parameters for Measuring Levels of Service. The six levels of service each represent a range, the extreme of which is defined by the upper volume limit and the lower speed limit. Traffic-operational freedom on a highway is considered equal to or greater than Level of Service A, B, C, D, or E as the case may be, when two conditions are met. For largely uninterrupted flow, these conditions are:

1. Operating speeds or average overall speeds are equal to or greater than a standard value for the level considered.
2. The ratio of the demand volume or service volume to the capacity of any subsection does not exceed a standard value for that level.

Levels of Service: Definitions. The six levels of service are generally described as follows for simple uninterrupted flows. More specific descriptions for each highway element, including interrupted as well as uninterrupted flow, are presented later under appropriate headings.

Level of Service A: This is a condition of free flow, accompanied by low volumes and high speeds. Traffic density will be low, with uninterrupted flow speeds controlled by driver desires, speed limits, and physical roadway conditions. There is little or no restriction in maneuverability due to the presence of other vehicles, and drivers can maintain their desired speeds with little or no delay.

Level of Service B: This occurs in the zone of stable flow, with operating speeds beginning to be restricted somewhat by traffic conditions. Drivers still have reasonable freedom to select their speed and lane of operation. Reductions in speed are not unreasonable, with a low probability of traffic flow being restricted. The lower limit (lowest speed, highest volume) of this level of service has been used in the design of rural highways.

Level of Service C: This is still in the zone of stable flow, but speeds and maneuverability are more closely controlled by the higher volumes. Most of the drivers are restricted in their freedom to select their own speed, change lanes, or pass. A relatively satisfactory operating speed is still obtained, with service volumes suitable for urban design practice.

Level of Service D: This level of service approaches unstable flow, with tolerable operating speeds being maintained, though considerably affected by changes in operating conditions. Fluctuations in volume and temporary restrictions to flow may cause substantial drops in operating speeds. Drivers have little freedom to maneuver, and comfort and convenience are low. These conditions can be tolerated, however, for short periods of time.

Level of Service E: This cannot be described by speed alone, but represents operations at lower operating speeds, typically, but not always, in the neighborhood of 30 miles per hour, with volumes at or near the capacity of the highway. Flow is unstable, and there may be stoppages of momentary duration. This level of service is associated with operation of a facility at capacity flows.

Level of Service F: This describes a forced-flow operation at low speeds, where volumes are below capacity. In the extreme, both speed and volume can drop to zero. These conditions usually result from queues of vehicles backing up from a restriction downstream. The section under study will be serving as a storage area during parts or all of the peak hour. Speeds are reduced substantially and stoppages may occur for short or long periods of time because of the downstream congestion.

FACTORS AFFECTING CAPACITY AND SERVICE VOLUMES

It is seldom that traffic and roadway conditions are ideal, and therefore fundamental capacities must be decreased to take into consideration the many factors that adversely affect traffic flow. Service volumes are affected in a similar way.

The various factors affecting capacity and service volumes are divided into two categories: roadway factors and traffic factors.

Roadway Factors

Lane Width. Twelve-foot lanes are considered ideal for heavy volumes of mixed traffic, and a lane of narrower width will restrict capacity.

Lateral Clearance. Objects closer than 6 feet from the edge of the pavement reduce the effective width of the roadway. The magnitude of the effect depends upon the closeness of the objects to the pavement and their frequency. Adjustments for lane width and lateral clearance are combined into one correction factor which is applied to the capacity under ideal conditions.

Shoulders. Adequate shoulders must be provided as a refuge for stopped vehicles if capacities are to be maintained on the through lanes.

Auxiliary Lanes. These include parking lanes, speed-change lanes, turning and storage lanes, weaving lanes, and truck-climbing lanes. Each of these lanes provides additional pavement width to accommodate special uses, helping to maintain the capacity of the through roadway.

Surface Conditions. Poor pavement surface conditions may influence the attainment of high speed, thereby affecting the better levels of service, but capacity may be very little affected.

Alinement. Poor alinement prevents the attainment of high speed, thereby affecting the better levels of service. It also affects capacity on two- and three-lane roads when passing sight distance is restricted to less than 1,500 feet.

Grades. Grades affect service volumes and capacity in three ways:

1. Vehicle braking distance is less on upgrades and greater on downgrades than on level grades. This permits shorter spacings between vehicles that are climbing grades, and requires longer spacings between vehicles going downgrades, in order to maintain a safe headway.
2. The presence of a grade generally causes a restriction in the sight distance, thereby affecting the percentage of highway on which passing maneuvers can be performed safely. This would only apply to two- and three-lane highways.
3. Commercial vehicles, with their normal loads, travel at slower speeds on upgrades than on level grades, especially if the grade is long and steep. This is also true to some extent with passenger cars. Most passenger cars, however, can negotiate long 6 and 7 per cent grades at speeds above 30 miles per hour. The effect that grades up to 7 per cent have on capacity as related to the performance of passenger cars is therefore generally negligible. Capacity may be little influenced until speeds of heavy vehicles are reduced to about 30 miles per hour, but service volumes for specific levels of service are significantly affected.

Providing climbing lanes can greatly reduce the adverse effect of grades by almost entirely removing the influence of commercial vehicles. This is as true for multilane highways as it is for two-lane highways.

Traffic Factors

Highways which have identical roadway factors may have different capacities depending upon the composition, habits, and desires of the traffic using them, and the controls which must be exercised over that traffic.

These considerations are taken into account by means of traffic factors.

Trucks. Commercial vehicles (vehicles with dual tires on one or more axles) under all conditions take up more space than passenger cars, and their presence is taken into consideration by determining the "passenger car equivalent" which represents the number of passenger cars that each truck is equivalent to under specific conditions.

If E_T = passenger car equivalent, and P_T = percentage of trucks, a service volume in passenger cars can be converted to mixed traffic through multiplication by the truck adjustment factor, $100/(100 - P_T + E_T P_T)$. Similarly, a volume of mixed traffic can be converted to equivalent passenger cars through multiplication by the reciprocal of the truck adjustment factor.

Buses. Intercity buses have better performance characteristics than do trucks, and one bus is assumed to be equivalent to 1.6 passenger cars for a wide variety of level and rolling conditions on multi-lane highways and streets.[38] Local transit buses operating on city streets affect capacity in quite a different manner, and special procedures and charts have been developed to determine their influence.

Lane Distribution. The distribution of total traffic volume among the various lanes of a multi-lane facility varies with the lane location and with changes in volume. Even though the lane adjacent to the shoulder carries a smaller volume than the other lanes, no special correction is made because the fundamental capacity is stated as an average of 2,000 passenger cars per hour per lane, regardless of distribution.

Variations of Traffic Flow. Traffic volume variations within the peak hour can seriously affect flow conditions. Capacity analyses are based on traffic volume over a full hour, since capacity is defined in terms of vehicles per hour. However, the rate of traffic flow for intervals of less than one hour can substantially exceed the peak-hour rate. Therefore, it is necessary to provide excess capacity over the full hour to accommodate the peak intervals of flow, because when demand exceeds capacity, congestion will extend over a much longer time than just for the duration of the peak interval. For such purposes, a period of 5 to 15 minutes is employed for peak intervals. The relationship between the peak interval and the peak-hour volume is expressed by a peak-hour factor as follows:

For intersections:

$$\text{Peak-hour factor} = \frac{\text{peak-hour volume}}{4(\text{15-minutes peak volume})} \quad (11\text{-}5)$$

For freeways:

$$\text{Peak-hour factor} = \frac{\text{peak-hour volume}}{12(\text{5-minutes peak volume})} \quad (11\text{-}6)$$

The maximum value of the peak-hour factor in unity.

Traffic Interruptions. Any feature or device installed on a street or highway which requires some or all traffic to stop will reduce that highway's ability to carry traffic. Once stopped, traffic can depart at a rate of about 1,500 passenger cars per hour per lane, based on an average headway of 2.4 seconds. Since uninterrupted flow may reach 2,000 passenger cars per hour per lane, any stops imposed on heavy uninterrupted flows are likely to cause back-up of traffic.

Typical interruptions include at-grade intersections, toll gates, drawbridges, and railroad grade crossings. As long as rates of flow are below 1,500 passenger cars per hour per lane, probably only level of service will be affected in the latter three cases; but at higher volumes, queuing will develop. At-grade intersections cause significant restrictions to flow, and special procedures have been developed to analyze them.

12

Highway Capacity: Freeways and Expressways

Freeways and expressways generally provide a high level of service by eliminating free access to abutting lands in favor of exclusive service to moving traffic. These facilities come closest to providing "ideal" conditions, i.e., 12-foot lanes, no lateral obstructions, at least two lanes in each direction, and 70-miles per hour geometrics.

The freeway is made up of several elements, which together make up the facility. These include: basic freeway sections, weaving sections, and ramp junctions. In the case of expressways, at-grade intersections may also be present.

In order to properly evaluate the capacity of a given facility, each of its component sections must be separately evaluated before the composite effect can be considered. The *Highway Capacity Manual*[1] presents methods of analysis for each of the components of freeways, as well as techniques for evaluating overall performance.

Given any existing freeway with known traffic characteristics, the Manual enables the traffic engineer to evaluate both the capacity of the facility and the level of service it is providing. More importantly, it enables the engineer to take an expected traffic demand, select a desired level of service, and design a new facility to provide that level of service. Of course, the success of such designs depends heavily on the accuracy of the traffic forecast.

In design, it is important that a balanced level of service be maintained throughout the facility. This is not to say that if a freeway is being designed for Level of Service B that every element must provide Level of Service B. However, levels of service should be of equal acceptability to drivers on all elements. For example, if the basic design level of service is B, most drivers would accept Level of Service C operation at major ramp junctions or weaving sections. In general, levels of service for various elements should not differ by more than one.

In general, capacity or service volume computations involve the determination of an unadjusted value for a given set of standard conditions. This value is then corrected via a series of multiplicative factors to reflect prevailing conditions.

Standard conditions for freeways include:

1. No commercial vehicles in the traffic stream
2. Geometrics suitable for 70-mile-per-hour operating speeds
3. 12-foot lane widths
4. No lateral obstructions within 6 feet of the pavement edge

BASIC FREEWAY SECTIONS—NO RAMPS OR WEAVING

Unadjusted Capacity and Service Volumes

The unadjusted capacity and service volumes for basic freeway sections are given in Table 12-1. These volumes are unadjusted for traffic conditions and nonideal roadway conditions. Volumes for prevailing conditions are obtained by multiplying the values in Table 12-1 by appropriate adjustment factors.

The volumes shown in the right side of Table 12-1 are for ideal, 70-mile-per-hour highway alinement. These service volumes were obtained by multiplying the ideal capacity of 2,000 passenger cars per hour per lane, by the volume-to-capacity ratios shown on the left side of Table 12-1. Volume-to-capacity ratios for less than 70-mile-per-hour geometrics are also shown here, but no corresponding service volumes are included on the right side of the Table, and the volume-to-capacity ratio must be used directly.

For Levels of Service C and D, peak-hour variations must be considered. The average factors (PHF's) for metropolitan areas are shown in Table 12-2. These should be used only when the exact PHF for the location under study is not known or cannot be easily measured.

Adjustment for Lane Width and Lateral Obstructions

Ideal conditions include a 12-foot lane width and no lateral obstructions closer than 6 feet to the pavement edge. Any conditions which are less than the "ideal" standard will cause capacity or service volume to decrease from the ideal value. The adjustment factors for lane width and lateral clearance are found in Table 12-3. All adjustment factors are applied as multipliers to the unadjusted capacity or service volume.

Adjustment for Prevailing Percentage of Trucks and Buses

Another of the ideal conditions specified is the absence of trucks and buses in the traffic stream. The presence of such vehicles, which occupy more freeway space and have operating characteristics inferior to those of passenger cars, has an adverse effect on capacity or service volume.

The *Highway Capacity Manual* presents two procedures for determining the proper adjustment factor. One method is used for specific subsections, such as a 2-mile grade, where the subsection effect may be crucial. The other is used for longer, general sections, which may contain some upgrades, downgrades, and level subsections, no one of which is severe enough to warrant separate consideration.

Adjustments for Specific Subsections of Critical Grade. The Manual procedure consists of selecting the appropriate *passenger-car equivalent* for a truck or bus under given conditions of grade and traffic makeup. The bus equivalents are given in Table 12-4, the truck equivalents in Table 12-5. These values are then used to determine the appropriate adjustment factor in Table 12-6.

Adjustments for General Freeway Sections. The Manual procedure consists of selecting a truck or bus passenger-car equivalent from Table 12-7, and then selecting the appropriate adjustment factor. Truck adjustment factors are selected from Table 12-8, and the bus factors from Table 12-6 of the previous section. Table 12-8 gives truck adjustment factors directly, not requiring a prior determination of the passenger-car equivalent in Table 12-7.

Where the percentage of buses is insignificant in comparison to the percentage of trucks, they may be included as trucks, and only one factor need be used. A detailed study of truck equivalency is available.[61]

Formulas for Computation of Capacity and/or Service Volume

The formulas presented below are used in conjunction with the factors given in Tables 12-1 through 12-8 to determine capacity and service volumes.

Capacity Under Prevailing Conditions. Capacity is determined directly by multiplying the Level of Service E service volume from Table 12-1, 2,000 passenger cars per hour per lane, by appropriate adjustment factors.

$$C = 2{,}000 N W T_c \qquad (12\text{-}1)$$

where
C = capacity (mixed vehicle per hour, total for one direction)
N = number of lanes (in one direction)
W = adjustment for lane width and lateral clearance, from Table 12-3
T_c = truck factor at capacity, from Table 12-6 or Table 12-8

Service Volumes. The Manual presents four procedures for obtaining service volume for a given level of service, but the first method has the widest applicability. Regardless of which method is employed, it is necessary to

TABLE 12-1. LEVELS OF SERVICE AND MAXIMUM SERVICE VOLUMES FOR FREEWAYS AND EXPRESSWAYS UNDER UNINTERRUPTED FLOW CONDITIONS

LEVEL OF SERVICE	TRAFFIC FLOW CONDITIONS		SERVICE VOLUME/CAPACITY (v/c) RATIO[a]					MAXIMUM SERVICE VOLUME UNDER IDEAL CONDITIONS, INCLUDING 70-MPH AVERAGE HIGHWAY SPEED (Total Passenger Cars per Hour, One Direction)			
	DESCRIPTION	OPERATING SPEED[a] (MPH)	BASIC LIMITING VALUE FOR AVERAGE HIGHWAY SPEED (AHS) OF 70 MPH, FOR:			APPROXIMATE WORKING VALUE FOR ANY NUMBER OF LANES FOR RESTRICTED AVERAGE HIGHWAY SPEED OF		4-Lane Freeway (2 Lanes One Direction)	6-Lane Freeway (3 Lanes One Direction)	8-Lane Freeway (4 Lanes One Direction)	Each Additional Lane Above Four In One Direction
			4-Lane Freeway (2 Lanes/Direction)	6-Lane Freeway (3 Lanes/Direction)	8-Lane Freeway (4 Lanes/Direction)	60 Mph	50 Mph				
A	Free flow	$\gtrsim$60	$\lesssim$0.35	$\lesssim$0.40	$\lesssim$0.43	—[b]	—[b]	1400	2400	3400	1000
B	Stable flow (upper speed range)	$\gtrsim$55	$\lesssim$0.50	$\lesssim$0.58	$\lesssim$0.63	$\lesssim$0.25	—[b]	2000	3500	5000	1500
	Peak-Hour Factor (PHF)[c]							0.77 0.83 0.91 1.00[d]	0.77 0.83 0.91 1.00[d]	0.77 0.83 0.91 1.00[d]	0.77 0.83 0.91 1.00[d]
C	Stable flow	$\gtrsim$50	$\lesssim$0.75 (PHF)	$\lesssim$0.80 (PHF)	$\lesssim$0.83 (PHF)	$\lesssim$0.45 (PHF)	—[b]	2300 2500 2750 3000	3700 4000 4350 4800	5100 5500 6000 6600	1400 1500 1650 1800
D	Approaching unstable flow	$\gtrsim$40		$\lesssim$0.90 (PHF)		$\lesssim$0.80 (PHF)	$\lesssim$0.45 (PHF)	2800 3000 3300 3600	4150 4500 4900 5400	5600 6000 6600 7200	1400 1500 1650 1800
E[f]	Unstable flow	30–35[e]			$\lesssim$1.00			4000[e]	6000[e]	8000[e]	2000[e]
F	Forced flow	<30[e]	←——Not meaningful——→					←——Widely variable (0 to capacity)——→			

(Source: *Highway Capacity Manual,* HRB SR 87 1965, Table 9-1)

[a] Operating speed and basic *v/c* ratio are independent measures of level of service; both limits must be satisfied in any determination of level.

[b] Operating speed required for this level is not attainable even at low volumes.

[c] Peak-hour factor for freeways is the ratio of the whole-hour volume to the highest rate of flow occurring during a 5-minute interval within the peak hour.

[d] A peak-hour factor of 1.00 is seldom attained; the values listed here should be considered as maximum average flow rates likely to be obtained during the peak 5-minute interval within the peak hour.

[e] Approximately.

[f] Capacity.

TABLE 12-2. AVERAGE VALUES OF THE PEAK-HOUR FACTOR

POPULATION	*PEAK-HOUR FACTOR (PHF)*
under 500,000	0.77
500,000–1,000,000	0.83
over 1,000,000	0.91

TABLE 12-3. COMBINED EFFECT OF LANE WIDTH AND RESTRICTED LATERAL CLEARANCE ON CAPACITY AND SERVICE VOLUMES OF DIVIDED FREEWAYS AND EXPRESSWAYS WITH UNINTERRUPTED FLOW

DISTANCE FROM TRAFFIC LANE EDGE TO OBSTRUCTION	*ADJUSTMENT FACTOR,* FOR LANE WIDTH AND LATERAL CLEARANCE*							
	Obstruction on One Side of One-Direction Roadway				*Obstruction on Both Sides of One-Direction Roadway*			
	12-ft Lanes	11-ft Lanes	10-ft Lanes	9-ft Lanes	12-ft Lanes	11-ft Lanes	10-ft Lanes	9-ft Lanes
	(*a*) 4-Lane Divided Freeway, One Direction of Travel							
6	1.00	0.97	0.91	0.81	1.00	0.97	0.91	0.81
4	0.99	0.96	0.90	0.89	0.98	0.95	0.89	0.79
2	0.97	0.94	0.88	0.79	0.94	0.91	0.86	0.76
0	0.90	0.87	0.82	0.73	0.84	0.79	0.74	0.66
	(*b*) 6- and 8-Lane Divided Freeway, One Direction of Travel							
6	1.00	0.96	0.89	0.78	1.00	0.96	0.89	0.78
4	0.99	0.95	0.88	0.77	0.98	0.94	0.87	0.77
2	0.97	0.93	0.87	0.76	0.96	0.92	0.85	0.75
0	0.94	0.91	0.85	0.74	0.91	0.87	0.81	0.70

(Source: *Highway Capacity Manual,* HRB SR 87, 1965, Table 9.2)
* Same adjustments for capacity and all levels of service.

TABLE 12-4. PASSENGER-CAR EQUIVALENTS OF INTERCITY BUSES ON FREEWAYS AND EXPRESSWAYS ON SPECIFIC SUBSECTIONS OR GRADES

GRADE (%)*	*PASSENGER-CAR EQUIVALENT,*** E_B — *Levels of Service A Through C*	*Levels of Service D and E (Capacity)*
0–4	1.6	1.6
5†	4	2
6†	7	4
7†	12	10

(Source: *Highway Capacity Manual,* HRB SR 87, 1965, Table 9-5)
* All lengths.
** For all percentages of buses.
† Use generally restricted to grades over ½-mile long.

TABLE 12-5. PASSENGER-CAR EQUIVALENTS OF TRUCKS ON FREEWAYS AND EXPRESSWAYS ON SPECIFIC SUBSECTIONS OR GRADES

GRADE (%)	LENGTH OF GRADE (mi)	*PASSENGER CAR EQUIVALENT, E_T* Levels of Service A through C for:					Levels of Service D and E (Capacity) for:				
		3% Trucks	*5% Trucks*	*10% Trucks*	*15% Trucks*	*20% Trucks*	*3% Trucks*	*5% Trucks*	*10% Trucks*	*15% Trucks*	*20% Trucks*
0–1	All	2	2	2	2	2	2	2	2	2	2
2	¼–½	5	4	4	3	3	5	4	4	3	3
	¾–1	7	5	5	4	4	7	5	5	4	4
	1½–2	7	6	6	6	6	7	6	6	6	6
	3–4	7	7	8	8	8	7	7	8	8	8
3	¼	10	8	5	4	3	10	8	5	4	3
	½	10	8	5	4	4	10	8	5	4	4
	¾	10	8	6	5	5	10	8	5	4	5
	1	10	8	6	5	6	10	8	6	5	6
	1½	10	9	7	7	7	10	9	7	7	7
	2	10	9	8	8	8	10	9	8	8	8
	3	10	10	10	10	10	10	10	10	10	10
	4	10	10	11	11	11	10	10	11	11	11
4	¼	12	9	5	4	3	13	9	5	4	3
	½	12	9	5	5	5	13	9	5	5	5
	¾	12	9	7	7	7	13	9	7	7	7
	1	12	10	8	8	8	13	10	8	8	8
	1½	12	11	10	10	10	13	11	10	10	10
	2	12	11	11	11	11	13	12	11	11	11
	3	12	12	13	13	13	13	13	14	14	14
	4	12	13	15	15	14	13	14	16	16	15
5	¼	13	10	6	4	3	14	10	6	4	3
	½	13	11	7	7	7	14	11	7	7	7
	¾	13	11	9	8	8	14	11	9	8	8
	1	13	12	10	10	10	14	13	10	10	10
	1½	13	13	12	12	12	14	14	13	13	13
	2	13	14	14	14	14	14	15	15	15	15
	3	13	15	16	16	15	14	17	17	17	17
	4	15	17	19	19	17	16	19	22	21	19
6	¼	14	10	6	4	3	15	10	6	4	3
	½	14	11	8	8	8	15	11	8	8	8
	¾	14	12	10	10	10	15	12	10	10	10
	1	14	13	12	12	11	15	14	13	13	11
	1½	14	14	14	14	13	15	16	15	15	14
	2	14	15	16	16	15	15	18	18	18	16
	3	14	16	18	18	17	15	20	20	20	19
	4	19	19	20	20	20	20	23	23	23	23

(Source: *Highway Capacity Manual,* HRB SR 87 1965, Table 9-4)

TABLE 12-6. ADJUSTMENT FACTORS* FOR INTERCITY BUSES AND TRUCKS ON SPECIFIC SUBJECTIONS OR GRADES OF FREEWAYS AND EXPRESSWAYS**

PASSENGER-CAR EQUIVALENT, E_T OR E_B	TRUCK ADJUSTMENT FACTOR T_r OR T_1, (B_r OR B_1, FOR BUSES)†														
	Percentage of Trucks, P_T (or of Buses, P_B) of:														
	1	2	3	4	5	6	7	8	9	10	12	14	16	18	20
2	0.99	0.98	0.97	0.96	0.95	0.94	0.93	0.93	0.92	0.91	0.89	0.88	0.86	0.85	0.83
3	0.98	0.96	0.94	0.93	0.91	0.89	0.88	0.86	0.85	0.83	0.81	0.78	0.76	0.74	0.71
4	0.97	0.94	0.92	0.89	0.87	0.85	0.83	0.81	0.79	0.77	0.74	0.70	0.68	0.65	0.63
5	0.96	0.93	0.89	0.86	0.83	0.81	0.78	0.76	0.74	0.71	0.68	0.64	0.61	0.58	0.56
6	0.95	0.91	0.87	0.83	0.80	0.77	0.74	0.71	0.69	0.67	0.63	0.59	0.56	0.53	0.50
7	0.94	0.89	0.85	0.81	0.77	0.74	0.70	0.68	0.65	0.63	0.58	0.54	0.51	0.48	0.45
8	0.93	0. 88	0.83	0.78	0.74	0.70	0.67	0.64	0.61	0.59	0.54	0.51	0.47	0.44	0.42
9	0.93	0.86	0.81	0.76	0.71	0.68	0.64	0.61	0.58	0.56	0.51	0.47	0.44	0.41	0.38
10	0.92	0.85	0.79	0.74	0.69	0.65	0.61	0.58	0.55	0.53	0.48	0.44	0.41	0.38	0.36
11	0.91	0.83	0.77	0.71	0.67	0.63	0.59	0.56	0.53	0.50	0.45	0.42	0.38	0.36	0.33
12	0.90	0.82	0.75	0.69	0.65	0.60	0.57	0.53	0.50	0.48	0.43	0.39	0.36	0.34	0.31
13	0.89	0.81	0.74	0.68	0.63	0.58	0.54	0.51	0.48	0.45	0.41	0.37	0.34	0.32	0.29
14	0.88	0.79	0.72	0.66	0.61	0.56	0.52	0.49	0.46	0.43	0.39	0.35	0.32	0.30	0.28
15	0.88	0.78	0.70	0.64	0.59	0.54	0.51	0.47	0.44	0.42	0.37	0.34	0.31	0.28	0.26
16	0.87	0.77	0.69	0.63	0.57	0.53	0.49	0.45	0.43	0.40	0.36	0.32	0.29	0.27	0.25
17	0.86	0.76	0.68	0.61	0.56	0.51	0.47	0.44	0.41	0.38	0.34	0.31	0.28	0.26	0.24
18	0.85	0.75	0.66	0.60	0.54	0.49	0.46	0.42	0.40	0.37	0.33	0.30	0.27	0.25	0.23
19	0.85	0.74	0.65	0.58	0.53	0.48	0.44	0.41	0.38	0.36	0.32	0.28	0.26	0.24	0.22
20	0.84	0.72	0.64	0.57	0.51	0.47	0.42	0.40	0.37	0.34	0.30	0.27	0.25	0.23	0.21
21	0.83	0.71	0.63	0.56	0.50	0.45	0.41	0.38	0.36	0.33	0.29	0.26	0.24	0.22	0.20
22	0.83	0.70	0.61	0.54	0.49	0.44	0.40	0.37	0.35	0.32	0.28	0.25	0.23	0.21	0.19
23	0.82	0.69	0.60	0.53	0.48	0.43	0.39	0.36	0.34	0.31	0.27	0.25	0.22	0.20	0.19
24	0.81	0.68	0.59	0.52	0.47	0.42	0.38	0.35	0.33	0.30	0.27	0.24	0.21	0.19	0.18
25	0.80	0.67	0.58	0.51	0.46	0.41	0.37	0.34	0.32	0.29	0.26	0.23	0.20	0.18	0.17

(Source: *Highway Capacity Manual,* HRB SR 87, 1965, Table 9-6)

* Computed by $100/(100 - P_T + E_T P_T)$, or $100/(100 - P_B + E_B P_B)$, as presented in Chapter Five of the HCM. Use this formula for larger percentages.

** Used to convert equivalent passenger-car volumes to actual mixed traffic; use reciprocal of these values to convert mixed traffic to equivalent passenger cars.

† Trucks and buses should not be combined in entering this table where separate consideration of buses has been established as required, because passenger-car equivalents differ.

TABLE 12-7. PASSENGER-CAR EQUIVALENTS OF TRUCKS AND INTERCITY BUSES ON GENERAL FREEWAY SECTIONS

LEVEL OF SERVICE		EQUIVALENT, E, FOR: Level Terrain	Rolling Terrain	Mountainous Terrain
A		Widely variable; one or more trucks have same total effect, causing other traffic to shift to other lanes. Use equivalent for remaining levels in problems.		
B through E	E_T, for trucks	2	4	8
	E_B, for buses*	1.6	3	5

(Source: *Highway Capacity Manual,* HRB SR 87, 1965, Table 9-3a)

* Separate consideration not warranted in most problems; use only where bus volumes are significant.

TABLE 12-8. ADJUSTMENT FACTORS FOR TRUCKS ON EXTENDED GENERAL FREEWAY SECTIONS*

PERCENTAGE OF TRUCKS, P_T	FACTOR, T, FOR ALL LEVELS OF SERVICE: Level Terrain	Rolling Terrain	Mountainous Terrain
1	0.99	0.97	0.93
2	0.98	0.94	0.88
3	0.97	0.92	0.83
4	0.96	0.89	0.78
5	0.95	0.87	0.74
6	0.94	0.85	0.70
7	0.93	0.83	0.67
8	0.93	0.81	0.64
9	0.92	0.79	0.61
10	0.91	0.77	0.59
12	0.89	0.74	0.54
14	0.88	0.70	0.51
16	0.86	0.68	0.47
18	0.85	0.65	0.44
20	0.83	0.63	0.42

(Source: *Highway Capacity Manual,* HRB SR 87, 1965, Table 9-3b)

* Not applicable to buses where they are given separate consideration.

check the result by means of Table 12-1 to confirm that both the operating speed and the volume criteria for the desired level of service are met, with consideration for the prevailing average highway speed.

1. *Computed directly from capacity under ideal conditions:*

$$SV = 2{,}000N\frac{v}{c}WT_L \qquad (12\text{-}2)$$

where SV = service volume (mixed vehicles per hour, total for one direction)
N = number of lanes (in one direction)
v/c = volume-to-capacity ratio, from Table 12-1; for levels C and D, selection of the v/c ratio also involves the use of the PHF as a multiplier
W = adjustment for lane width and lateral clearance, from Table 12-3
T_L = truck factor at given level of service, from Table 12-6 or Table 12-8

2. *Computed from maximum service volume for ideal conditions.* This procedure is only applicable when alinement is ideal, that is, average highway speed is 70 miles per hour.

$$SV = (MSV)WT_L \qquad (12\text{-}3)$$

where MSV = maximum service volume in passenger cars per hour for the appropriate number of lanes and PHF, from Table 12-1

The attainment of desired level of service is confirmed by using Table 12-1 to check the resulting operating speed.

3. *Computed from capacity under prevailing conditions:*

$$SV = C\frac{v}{c}\frac{T_L}{T_c} \qquad (12\text{-}4)$$

where C = capacity as computed under prevailing conditions in (1) above.

4. *Determined from level of service limits.* This method is applicable for the design of new freeways, where the level of service is established in advance; service volumes can be obtained directly from Table 12-1, provided conditions are ideal.

Level of Service. The determination of the level of service for an existing or proposed freeway, while accommodating a given demand volume, may be accomplished approximately by inspection of Table 12-1, if pertinent parameters are known and trucks are neglected. Refined analysis, considering trucks and peaking characteristics, involves a "trial-and-error" approach, since the level of service must be known to choose an appropriate truck factor and PHF (if necessary). Therefore, a level must be assumed, and new computations be carried out if the results prove the assumption incorrect. The procedure is illustrated in Example 9.3 of the *Capacity Manual.*

Sample Problems

Problem 1:

The following conditions are given:

Four-lane freeway in a city of 700,000 population
Twelve-foot lanes
Level terrain
Ten-foot shoulder on right
Frequent bridge abutments 2 feet from left lane
Ideal alinement (AHS = 70 miles per hour)
Traffic: 5 per cent trucks; 1 per cent buses

Determine service volume for Level of Service C.

Solution:

PHF = 0.83 (Table 12-2)
MSV = 2,500 passenger cars per hour (Table 12-1)
E_T = 2 (consider 1 per cent buses as trucks) (Table 12-7)
T_L = 0.94 (Table 12-8)
W = 0.97 (Table 12-3: obstruction on one side, 2 feet)

$SV = (MSV)WT_L = 2{,}500 \times 0.94 \times 0.97 = 2{,}280$ vehicles per hour

Problem 2:

The same conditions as in Problem 1 are given, but the highway is on a subsection of 5 per cent grade, 2 miles long. Determine service volume for Level of Service.

Solution:

The following values are the same as in Problem 1:
MSV = 2,500 passenger cars per hour
W = 0.97

However, now

E_T = 14 (Table 12-5)
T_L = 0.61 (Table 12-6)
E_B = 4 (Table 12-4)
B_L = 0.97 (Table 12-6)
$SV = (MSV)WT_LB_L = 2{,}500 \times 0.94 \times 0.61 \times 0.97$
1,368 vehicles per hour

WEAVING SECTIONS

Weaving is the crossing of traffic streams moving in the same general direction, accomplished by merging and diverging movements. Knowledge of the capacities of weaving sections is an extremely important requirement in highway design and traffic operation, especially for

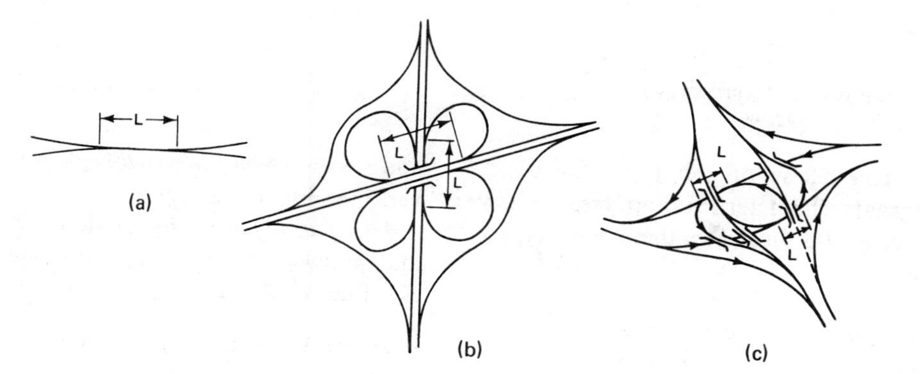

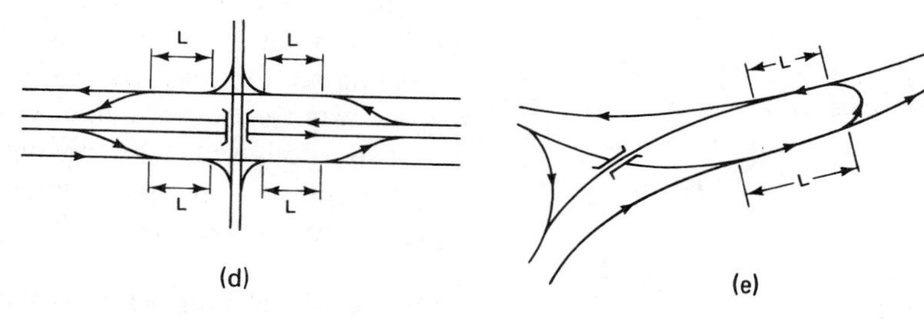

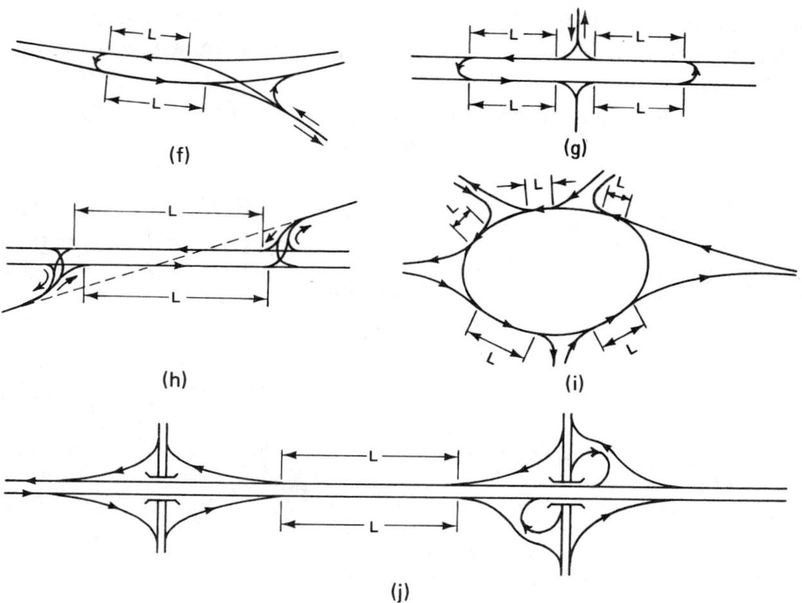

Fig. 12-1. Formation of weaving sections (*Source:* Highway Capacity Manual, *HRB SR 87, 1965, Figure 7.1*)

freeways, channelized intersections, and other designs where the crossing of two or more traffic streams is not controlled by traffic signals.

Weaving sections are provided for the purpose of permitting the crossing at grade of vehicle pathways with the least possible interference between vehicles. Weaving sections are usually selected as a compromise between the conventional intersection at grade and the grade separation. In fact, weaving sections are often provided as adjuncts to grade separations. The traffic rotary is, in actuality, a series of weaving sections, and there are many other applications of the principle in the layout of freeways. Some weaving sections are not readily recognized. One case is the roadway between the two inner loops of a cloverleaf, where vehicles entering a freeway on one loop must cross the path of vehicles exiting on the other inside loop. Another case is the section of freeway between an on-ramp and an off-ramp. These, as well as other examples illustrating the formation of weaving sections, are shown in Figure 12-1.

Unless consideration is given to that volume of traffic which must cross the path of other vehicles in reaching its destination, the capacity of high-type facilities may very easily be overestimated. If the capacity of such sections is not given special consideration, they could become perpetual "bottleneck" areas which limit the capacity and effectiveness of the entire facility.

Types of Weaving Sections

Weaving sections may be considered as *simple* or *multiple*, and they are further subdivided into *one-sided* or *two-sided* sections. Regardless of the type, however, the same operational characteristics and analyses generally apply. Figure 12-3 illustrates these basic types of weaving sections.

Simple Weaving Sections. *Simple weaving* involves a single entrance junction followed by a single exit junction. Simple weaving sections are classified as follows:

1. *Single-purpose* weaving sections, where all traffic weaves. That is, all the vehicles entering the weaving section from one approach are destined to cross the path of all vehicles entering from the other approach. See Figure 12-2(a).

2. *Dual-purpose* weaving sections, which serve both weaving and nonweaving traffic. See Figure 12-2(b).

3. *Compound* weaving sections, where traffic enters the section on approaches having more than one lane, and some vehicles are involved in two weaving maneuvers. See Figure 12-2(c).

4. *Separated* weaving sections where weaving traffic and nonweaving traffic are separated from each other. See Figure 12-2(d).

Multiple Weaving Sections. A section of highway consisting of two or more overlapping weaving sections is referred to as a *multiple weaving section.* A multiple weaving section may also be defined as that portion of a one-way roadway which has two consecutive entrance junctions followed closely by one or more exit junctions, or one entrance junction followed closely by two or more exit junctions. See Figure 12-3(b).

One-Sided and Two-Sided Weaving Sections. Both simple and multiple weaving sections may be further subdivided into those in which weaving takes place only on one

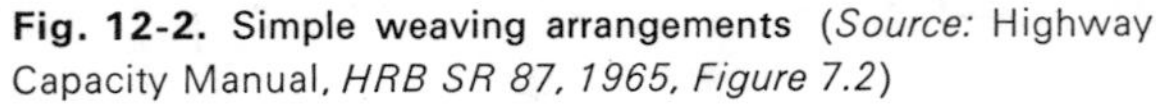

Fig. 12-2. Simple weaving arrangements (*Source:* Highway Capacity Manual, *HRB SR 87, 1965, Figure 7.2*)

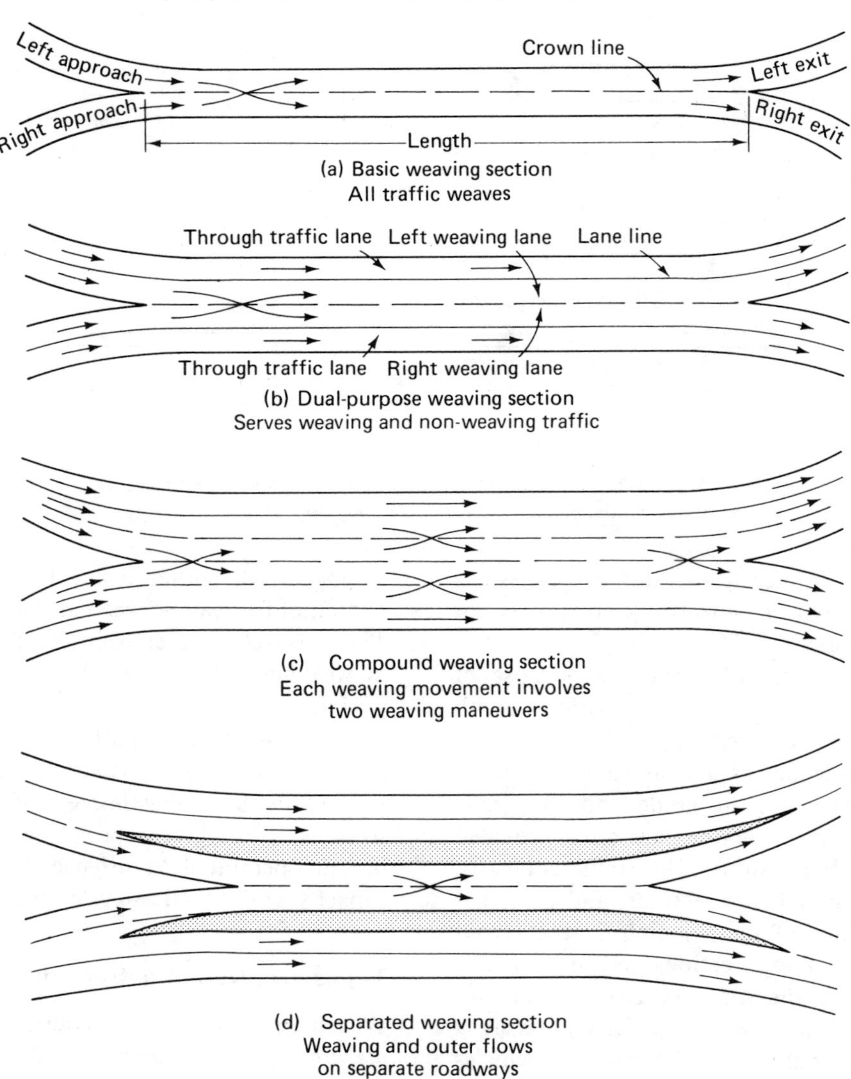

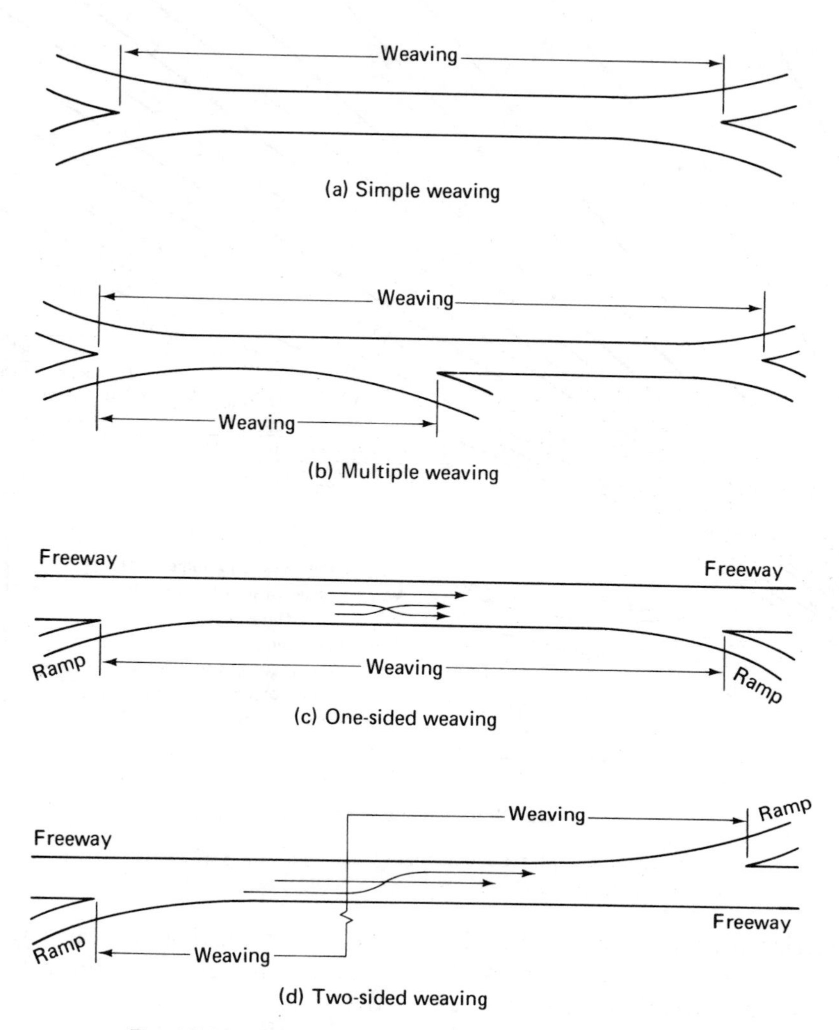

Fig. 12-3. Basic types of weaving sections (*Source:* Highway Capacity Manual, *HRB SR 87, 1965, Figure 7.3*)

side of the roadway and those which require maneuvers on both sides, causing weaving to occur across the roadway. The two types are shown in Figures 12-3(c) and 12-3 (d), respectively.

One-sided weaving is typical on most freeways where the entrances and exits of interchange ramps are on the same side of the roadway. *Two-sided* weaving occurs where the roadways of two major routes, crossing each other, are combined through a weaving section, as illustrated in Figures 12-1 (a), (e), (h), and (i).

Weaving problems represented by Figures 12-3 (a), (b) and (d) are handled by methods described in this section; those represented by Figure 12-3(c) may be handled either by methods described in the following section, ramp junctions, or by those presented here.

Operating Characteristics of Weaving Sections

Classification of Vehicles. The vehicles using a weaving section fall into two classes:

1. The first type describes those entering, passing through, and leaving the section without crossing the normal path of other vehicles. These are the *nonweaving vehicles*, and are designated as V_{o1} and V_{o2} in Figure 12-4, the o subscript signifying *outer flow*.
2. The second type are those that must cross the paths of other vehicles after entering the section. These are the *weaving vehicles*, which make the facility necessary, and they are designated as V_{w1} and V_{w2} in Figure 12-4, the w subscript signifying *weaving flow*.

Figure 12-4 illustrates the basic relationship between the length of a weaving section and the number of vehicles which may be expected to weave through the section, and it is based on nationwide studies conducted by the BPR.

Weaving Movements. If all the vehicles entering a weaving section from either approach are destined to cross the

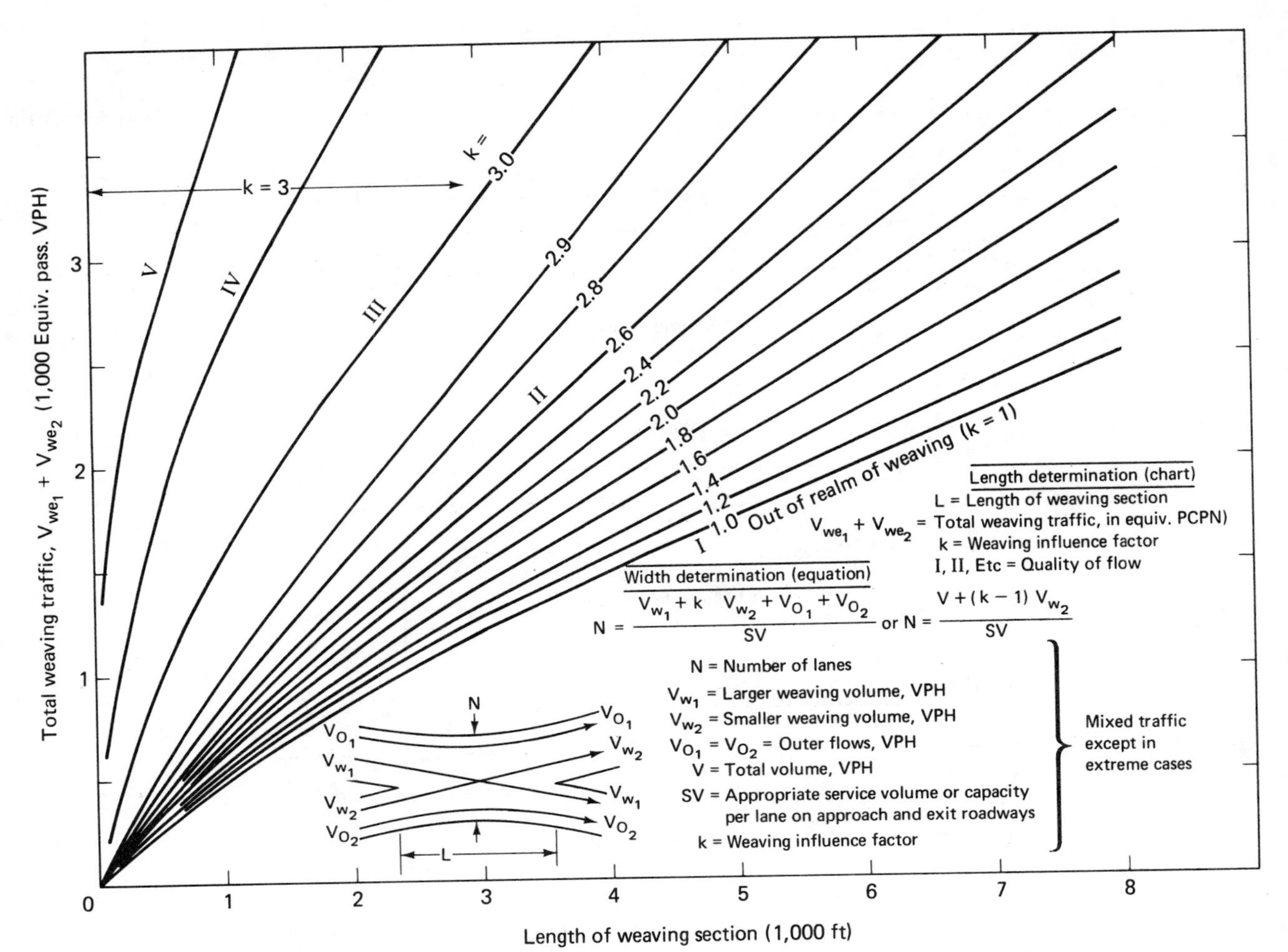

Fig. 12-4. Length requirements of weaving sections (*Source:* Highway Capacity Manual, *HRB SR 87, 1965, Figure 7.4*)

path of all vehicles entering from the other approach, that is, if all traffic is *weaving traffic*, it is necessary that every vehicle must cross the *crown line* somewhere between its extremities. At no instant can the number of vehicles in the act of crossing this crown line exceed the number that can crowd into a single lane, provided that the facility is operating as it should, without vehicles being required to come to a stop before merging with the stream of traffic from the other approach. Therefore, the total number of vehicles entering the weaving section (if all are weaving vehicles) cannot exceed the capacity of a single lane. This is one of the requirements governing the capacity of a weaving section.

When the number of weaving vehicles exceeds the capacity of a traffic lane, some of the vehicles are involved in *two* weaving maneuvers, and *compound weaving* exists, which requires greater width and length of weaving section. The effect of compound weaving is included in Figure 12-4.

Nonweaving Movements (Outer Flows). The reason for the existence of a weaving section is to accommodate weaving traffic. However, it serves a secondary purpose in that it must accommodate the nonweaving traffic by means of added lanes adjoining either side of the weaving lanes. Determining the capacity of these nonweaving lanes involves no new principle, since the procedure is the same as that for any traffic lane on a multi-lane facility. If any weaving section is to function properly and efficiently, it is important that these added lanes have adequate capacity to serve the nonweaving vehicles. To discourage or eliminate the use of weaving lanes by nonweaving vehicles, signs should be used to direct the drivers to the proper side of the approach road.

Service Criteria: Quality of Flow. Weaving section service is measured in terms of *quality of flow*. Figure 12-4 contains a family of curves designated I through V, representing best to worst service, respectively. Level of service is not used directly for two reasons:

1. Weaving may occur on a variety of facility types, each having different level of service standards. Quality of flow provides an intermediate service measure which is, in turn, related to levels of service for the various facility types in Table 12-9.
2. It is theorized that for any given level of service, drivers will reduce their speeds 5 to 10 miles per hour, regardless of traffic conditions, in response to the more restrictive nature of the weaving maneuver. Quality of flow standards reflect this reduction.

TABLE 12-9. RELATIONSHIP BETWEEN LEVEL OF SERVICE AND QUALITY OF FLOW

	QUALITY OF FLOW*			
	FREEWAYS AND MULTI-LANE RURAL HIGHWAYS			
Level of Service	*Highway Proper*	*Connecting Collector-Distributor Roads and Other Interchange Roadways*	*Two-Lane Rural Highways*	*Urban and Suburban Arterials*
A	I-II	II-III	II	III-IV
B	II	III	II-III	III-IV
C	II-III	III-IV	III	IV
D	III-IV	IV	IV	IV
E**	IV-V	V	V	V
F	Unsatisfactory†			

(Source: *Highway Capacity Manual,* HRB SR 87, 1965, Table 7-3)
* As represented by curves of Figure 12.4. Relationships below heavy line not normally considered in design. Where two entries are given, that on the left is desirable, that on the right is minimum.
** Capacity operation.
† Maximum volume equivalent to V, but may be much lower.

In applying the procedure, both a quality of flow and a level of service are used. Quality of flow describes the operation of weaving vehicles only, and is used in designing the length of the weaving section. Level of service describes the average behavior of all vehicles, weaving and nonweaving, in the section, and is used in designing the width of the weaving section.

Length Requirements. The length of a weaving section is a significant factor in its operation, and the quality of flow can be improved by increasing the length of the section. The length of a weaving section is measured from a point at the merge end to a point at the diverge end, as shown in Figure 12-5. The length of the weaving section to provide a desired quality of flow is obtained from Figure 12-4 where total weaving traffic, expressed in equivalent passenger vehicles, is plotted against length of weaving section.

Width Requirements. In a complete solution of a weaving section problem, both length and width requirements must be met. The width (number of lanes) of the weaving section is dependent on the weaving volume, the outer flow volumes, and the lane service volume or capacity on approach and exit roadways.

The number of lanes required for the outer flows is calculated as for any uninterrupted flow facility, i.e., by dividing the volume by the appropriate lane service volume (SV) or capacity. Therefore, the number of lanes required would be $(V_{o1} + V_{o2})/SV$. For freeway facilities, SV is taken from Table 12-1 and expressed as a per-lane value.

The additional lanes required for the weaving volume are found in a similar manner, but it has been demonstrated that for equivalent volumes more width is required for weaving than for uninterrupted flow. Based on observed data, a formula reflecting this fact has been developed, and the number of lanes required is obtained by $(V_{w1} + kV_{w2})/SV$, where V_{w1} is the larger weaving volume in vehicles per hour, V_{w2} is the smaller weaving volume in vehicles per hour, SV is the lane service volume, or capacity, in vehicles per hour (same value as used in determining the number of lanes required for the outer flows), and k is a weaving influence factor,

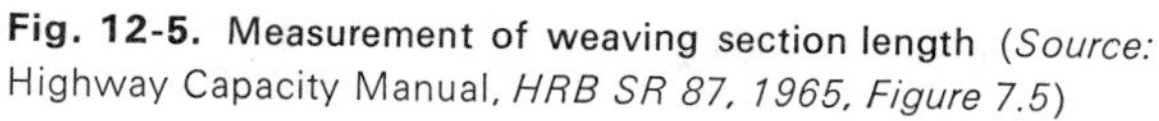
Fig. 12-5. Measurement of weaving section length (*Source:* Highway Capacity Manual, *HRB SR 87, 1965, Figure 7.5*)

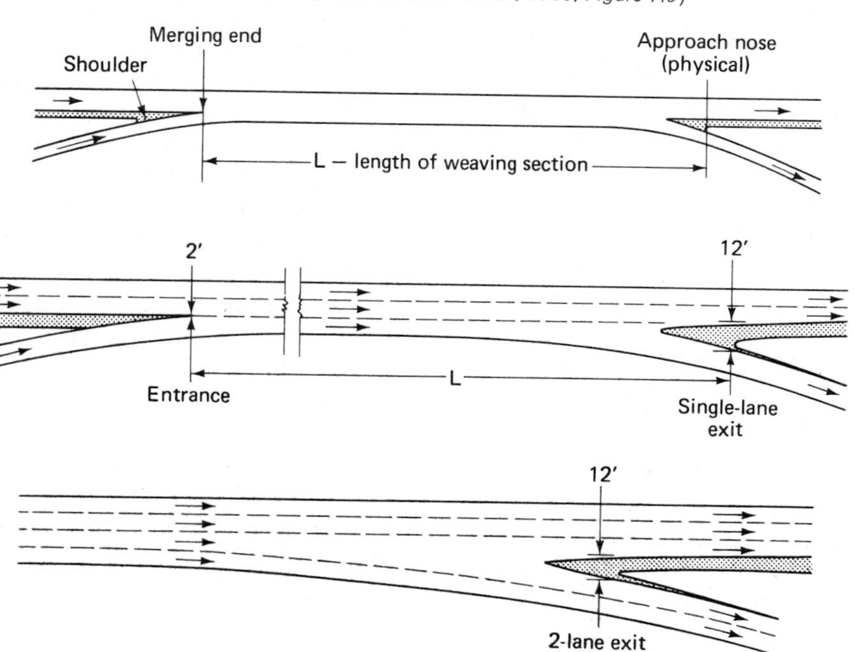

which ranges between 1.0 and 3.0. The k factor reflects the need for expanding the effect of the weaving traffic and is given as part of Figure 12-4.

In Figure 12-4 a series of curves is provided for various k values. The influence of weaving traffic is more pronounced on shorter weaving sections whose operation is represented by curves III, IV, and V, and therefore the maximum value ($k = 3.0$) is used in association with these qualities of flow. When the weaving section length is greater than the minimum required, as in the case for the better qualities of flow represented by the conditions to the right of curve III, the adverse effect of weaving becomes progressively less, with the k factor gradually reduced to a value of 1.0 for curve I.

The total number of lanes required to accommodate outer flow and weaving volumes becomes

$$N = \frac{V_{w1} + kV_{w2} + V_{o1} + V_{o2}}{SV}$$

$$= \frac{V + (k - 1)V_{w2}}{SV} \qquad (12\text{-}5)$$

where $\quad V = V_{w1} + V_{w2} + V_{o1} + V_{o2}$

The value of SV used in the equation should normally be the average service volume per lane, for the level of service selected for the approach and exit roadways, and, in the case of Levels C and D, for the appropriate peak-hour factor. For freeways, these values are given in Table 12-1 and they must be adjusted downward as necessary to reflect the prevailing conditions, such as percentage of trucks, grades, lane widths, etc.

Speed Relationships. The approximate speeds which are associated with the various qualities of flow are shown in Table 12-10. These speeds refer to the average speeds of weaving vehicles only. In general, on freeways, operating speeds through weaving sections for a given level of service will be from 5 to 10 miles per hour below those for the same level on the roadway sections forming the entrances and exists. Expected average speeds for all vehicles (on freeways) correspond to those given in Table 12-1 for the level of service used in determining SV, minus 5 to 10 miles per hour.

TABLE 12-10. RELATIONSHIP BETWEEN SPEED AND QUALITY OF FLOW

QUALITY OF FLOW	*SPEED OF WEAVING TRAFFIC (mph)*
I	50 or more
II	45–50
III	40–45
IV	30–35
V	20–30

TABLE 12-11. VOLUME-LENGTH COMBINATIONS CONSIDERED OUT OF THE REALM OF WEAVING

WEAVING VOLUMES (pcph)	*MINIMUM LENGTH OF SECTION (ft)*
500	1,000
1,000	2,400
1,500	4,000
2,000	6,000

(Source: *Highway Capacity Manual,* HRB SR 87, 1965, Table 7-2)

Sections Out of the Realm of Weaving. The adverse effect of weaving on traffic flow is reduced as the length of weaving section is increased. Eventually, the distance between an entrance and an exit along a highway can become sufficiently long that the effect of weaving maneuvers is no more than that of normal lane changing. Table 12-11 contains minimum lengths for various weaving volumes in equivalent passenger vehicles. If these lengths are exceeded, it is not necessary to consider the operation of the roadway section on the basis of weaving section criteria, and procedures for uninterrupted flow apply.

Application of Procedures

Simple Weaving Sections. Direct analysis is relatively straightforward, involving use of the basic weaving chart and equation in Figure 12-4, with reference to Table 12-9, to determine length and width of section for given demand volumes. The traffic flows through the weaving section must be separated into *outer flows, larger weaving volume,* and *smaller weaving volume.* The procedure can be used in reverse when geometrics are given and operating characteristics are required.

It is important to note that weaving volumes in Figure 12-4 are expressed in terms of *equivalent passenger cars per hour.* In the typical problem, *level of service* rather than *quality of flow* will be designated, and Table 12-9 is used to correlate these two measures.

The level of service desired controls the value of SV to use in the width determination of the weaving section, and the appropriate SV is obtained from Table 12-1 for freeways. Similar tables for other facilities appear later under the appropriate headings. Normally, the average of the service volumes for the approach and exit roadways is used for SV.

Complex Weaving Sections. It is possible that a weaving section may have more than two entrances or exits adjacent to one another, or so closely spaced as to form

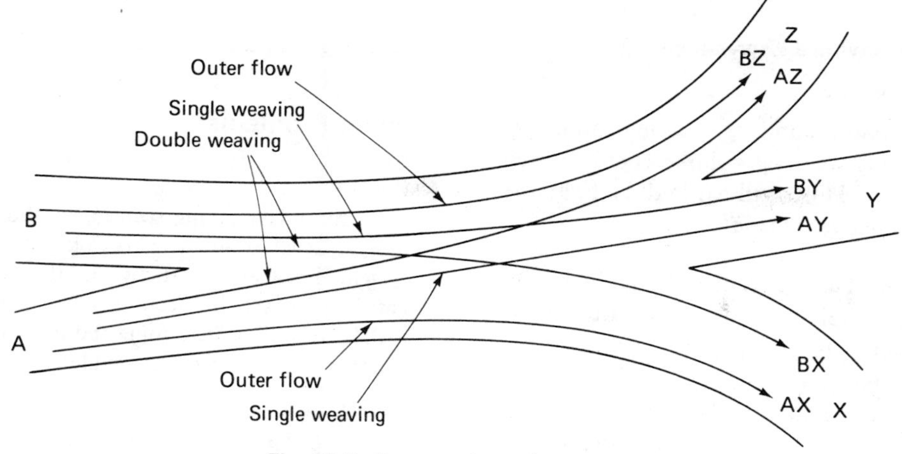

Fig. 12-6. Compound weaving analysis

a compound weaving section, or two weaving sections that overlap. When there are two entrances and three exits, or three entrances and two exits, there may be six different movements rather than four, as in a weaving section with two entrances and two exits. Figure 12-4 may be used to analyze these conditions by employing a somewhat modified treatment.[2]

The required length of a section with three entrances or exits adjacent to one another may be determined by using the total number of weaving maneuvers that must be performed. In calculating the total number of lanes required, the number of vehicles involved in the minor weaving movements, wherever two movements cross, are multiplied by the appropriate k factor (as with a weaving section with two entrances and two exits) and added to the major crossing movements. For example, if the two entrances from left to right are A and B and the exits from left to right are X, Y, and Z, the two outer movements not involved in weaving would be AX and BZ, as shown in Figure 12-6. The movements involved in weaving would be AY, AZ, BY, and BX. AY and BY may be thought of as *single* weaving movements because they cross only one weaving stream; AZ and BX may be thought of as *double* weaving movements because they cross two weaving streams. In calculating the required length of the weaving section, the number of weaving maneuvers would become $AY + 2AZ + 2BX + BY$. In calculating the number of lanes required (N) when $AY > AZ > BX < BY$, the numerator of the equation becomes $AY + 3(2BX) + 2AZ + 3BY + AX + BZ$, assuming a k factor $= 3.0$; therefore,

$$N = \frac{AY + 3(2BX) + 2AZ + 3BY + AX + BZ}{SV}$$

For determining the required length of the weaving section, the number of weaving vehicles $= AY + 2AZ + 2BX + BY$.

Multiple Weaving Sections. Since multiple weaving sections are made up of overlapping simple weaving sections, each analyzed separately, the basic relationships in Figure 12-4 and the related tables apply to the analysis of multiple weaving sections as well. Some of the varieties of multiple weaving sections include a single entrance followed by (1) two exits and (2) three exits; two successive entrances followed by (1) one exit and (2) two exits; or three successive entrances followed by one exit.

The manner in which weaving traffic divides itself between the various segments of a multiple weaving section can only be estimated. Considerable variation occurs, depending on geometrics, truck traffic, signing, and other factors. For purposes of analysis, it is considered reasonable to assume that weaving along the longer sections is proportional to the lengths of segments within these sections, and thus to allocate the weaving on that basis. Figure 12-4, then, is used in a manner similar to that for simple sections to establish the length requirements for each segment, using the summation of all the weaving movements within the segment (still on an equivalent passenger car basis). The required number of lanes for each segment is also determined individually in accordance with basic formula, modified as follows:

$$N = \frac{V + \sum (k - 1)V_{w2}}{SV} \tag{12-6}$$

The summation sign $\sum$ is introduced ahead of the second term in the numerator to account for the more than one set of weaving movements which may have a different value of k.

In setting up the weaving diagram, caution must be taken to identify and separate the various weaving movements (primary and secondary weaves) to avoid double counting. This principle is discussed in one of the illustrative problems, in which a movement already taken into consideration in one step is omitted in subsequent calculations.

In applying the above formula for determining the number of lanes, when the omitted volume happens to be the smaller of the two secondary weaving moments, the unduplicated (larger) weaving volume is substituted for V_{w2} in the formula. This is illustrated in Example 7.6 of the *Highway Capacity Manual.*

Weaving Under Other Than Freeway Conditions. On ordinary highways other than freeways in rural areas the procedures described for freeways remain generally valid as long as reasonably free flow is achieved. This applies to weaving along the highway between reasonably closely spaced junctions, as well as to weaving within well-designed rotary intersections of adequate dimensions. On the other hand, where certain design features of low standard are prevalent within the weaving section, care should be exercised to account for these inadequacies. Usually this is accomplished by appropriate reduction in the lane service volume or capacity (SV) in determining the number of lanes.

On major streets in urban areas, disturbing elements within the weaving section (such as signals, driveways, exits and entrances to business establishments, pedestrians, parking, or vehicles stopping to pick up or discharge passengers) can place severe limitations on the use of weaving sections. Although the individual effects of these factors cannot be evaluated directly, their influence should be recognized in design and in analysis of operation. First, the value of SV which determines the width should be appropriately reduced to reflect these influences, as on any highway. Second, the effect of adverse conditions should be further recognized through use of judgment in choosing the proper curve of Figure 12-4 for determining the length; that is, for any intended operating level, a length greater than that represented in the chart should be selected.

A computer program for the solution of weaving capacities has been developed.[22]

Sample Problems

Problem 1:

A multiple weaving section, as shown below, is under design. Level of Service C is to be provided on the freeway system. There are no trucks, and the basic number of lanes entering the weaving section is 3. (See Figure 12-7; assume a PHF of 0.83.) Determine the number and arrangement of lanes and the adequacy and length of weaving segments.

Solution:

Two of the weaving movements will occur over both segments of the weaving section; that is, movements AD and BE will take place in both L_1 and L_2. As each section will be considered separately, it is necessary to divide these weaving volumes into two groups, one which will be assumed to occur in L_1, the other in L_2. This is accomplished by distributing the volumes by the ratio of segment lengths:

$$AD(L_1) = 1{,}800 \times \frac{1{,}000}{4{,}000} = 450 \text{ vehicles per hour}$$

$$AD(L_2) = 1{,}800 - 450 = 1{,}350 \text{ vehicles per hour}$$

$$BE(L_1) = 800 \times \frac{1{,}000}{4{,}000} = 200 \text{ vehicles per hour}$$

$$BE(L_2) - 800 - 200 = 600 \text{ vehicles per hour}$$

The weaving diagram is constructed in Figure 12-8. Note that the vehicles of weaving flows AD and BE, which will execute their weaving movement in segment L_2, are outer flows in segment L_1. They are noted separately in the weaving diagram to act as a reminder that these become the secondary weave, shown below the primary diagram, in segment L_2. The 1,350 vehicles of the secondary weave in segment L_2 are denoted by a dashed line, since these vehicles have already been included as part of the 1,800 vehicles of the primary weave. These 1,350 vehicles cross the path of 250 vehicles of movement CE, and 600 vehicles of movement BE, which did not weave in segment L_1.

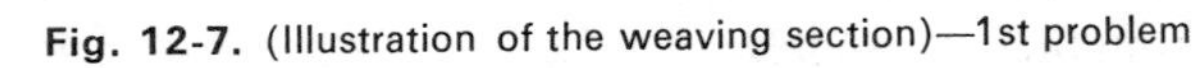

Fig. 12-7. (Illustration of the weaving section)—1st problem

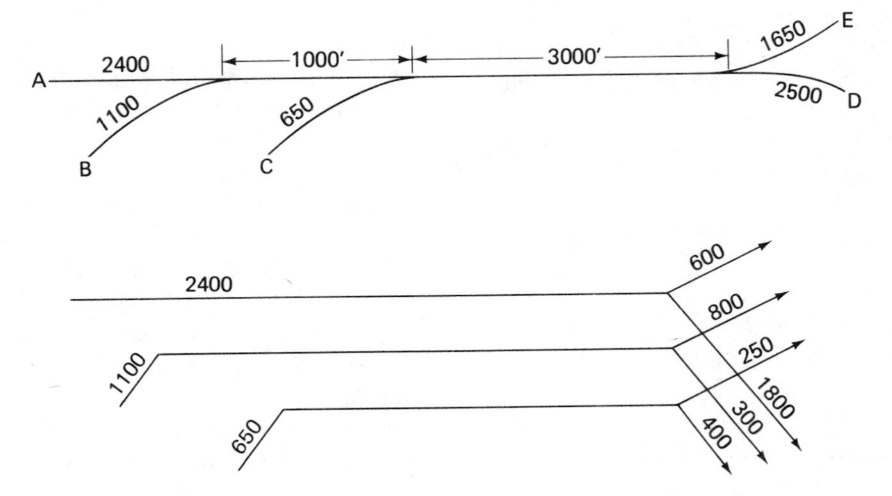

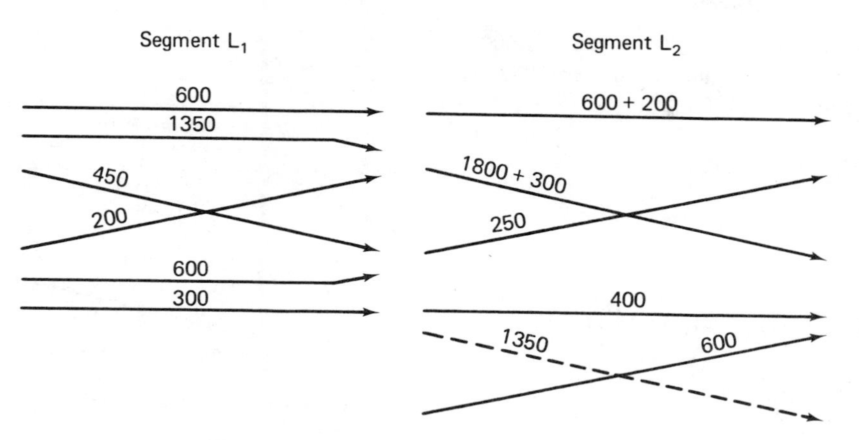

Fig. 12-8. (The weaving diagram)—Problem 1

1. Considering segment L_1:

$V_{we} = 450 + 200 = 650$ passenger cars per hour (volumes in passenger cars, no correction necessary)

For Level of Service C, Quality of Flow II or III is acceptable. Checking the length of segment,

L required $= 800$ feet (Figure 12-4, Quality of Flow II, $V_{we} = 650$)

Therefore, length is adequate.

Number of lanes

$$N = \frac{V + \sum (k-1)V_{w2}}{SV}$$

where $k = 2.0$ (Figure 12-4, $V_{we} = 650$, $L = 1{,}000$)

$V_{w2} = 200$

$SV = 4{,}000/3 = 1{,}350$ passenger cars per hour (Figure 12-1, Level of Service C, $PHF = 0.83$)

2. Considering L_2:

$V_{we(\text{TOTAL})} = 1{,}800 + 300 + 250 + 600 = 2{,}950$ passenger cars per hour

$V = 1{,}800 + 300 + 250 + 600 + 600 + 200 + 400 = 4{,}150$ passenger cars per hour

$SV = 1{,}350$ (as above)

Checking the length of segment,

L required $= 2{,}600$ feet (Figure 12-4, Quality of Flow III, $V_{we} = 2{,}950$ vehicles per hour)

Therefore length is sufficient.

Number of lanes:

$$N = \frac{V + \sum (k-1)V_{w2}}{SV}$$

Primary weave:

$V_{we} = 1{,}800 + 300 + 250 = 2{,}350$ passenger cars per hour

$k = 2.85$ (Figure 12-4)

Secondary weave:

$V_{we} = 1{,}350 + 600 = 1{,}950$ passenger cars per hour

$k = 2{,}60$ (Figure 12-4)

$$N = \frac{4{,}150 + 1.85(250) + 1.60(600)}{1{,}350} = 4.20$$

Use Five lanes.

It should be pointed out that the indicated section design, that is, three lanes in the first section and five lanes in the section following, is not feasible. A one-lane ramp and a three-lane mainline cannot merge into a five-lane facility without destroying all lane balance and continuity.

The final design would have to weigh the relative merits of providing only four lanes in the second section, versus adding a fourth lane in the first section (this implies a third lane prior to the first ramp, where only two lanes are necessary), which would permit a geometrically feasible five-lane second section.

Problem 2:

Design a weaving section to accommodate the following flows at Level of Service B. The freeway mainline consists of two lanes; on- and off-ramps consist of one lane each, (See Figure 12-9.)

Solution:

1. A check with Table 12-1 reveals that the two-lane mainline is sufficient to handle the incoming volume of 1,700 pcph and the outgoing volume of 1,900 pcph at Level of Service B. The standard is 2,000 pcph maximum.

2. From Table 12-9, for Level of Service B on a freeway, Quality of Flow II should be used.

Fig. 12-9. (The weaving diagram)—Problem 2

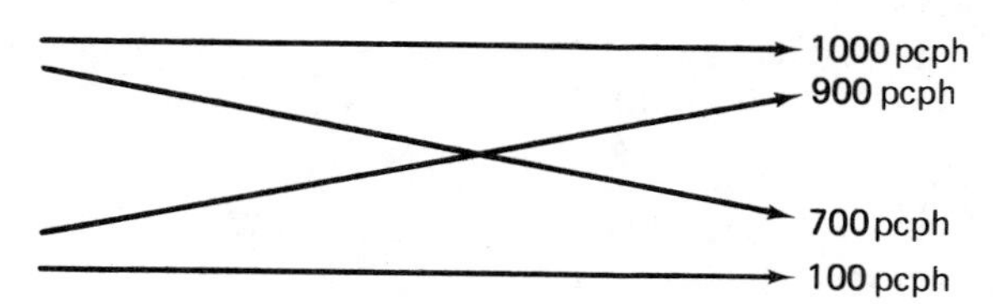

3. From Figure 12-4, for $V_{we} = 900 + 700 = 1{,}600$, and Quality of Flow II, the minimum length necessary is 2,300 feet. For Quality of Flow II, $k = 2.6$.

4.

$$N = \frac{V_t + (k - 1)V_{w2}}{SV}$$

where

$$V_t = 2700$$
$$k = 2.6$$
$$V_{w2} = 700$$
$$SV = \frac{2{,}000}{2} = 1{,}000 \qquad \text{(Table 12-1)}$$

$$N = \frac{2{,}700 + (1.6)700}{1{,}000}$$

$$N = 3.82$$

Use four lanes.

5. It is not feasible to provide a geometric design consisting of single-lane ramps, a two-lane mainline, and a four-lane weaving section. This would require two auxiliary lanes, an undesirable condition due to the increased lane-changing it induces, and lack of lane balance.

6. Investigate the length necessary if $N = 3$.

$$3 = \frac{2{,}700 + (k - 1)700}{1{,}000}$$

$$3{,}000 = 2{,}700 + 700(k - 1)$$

$$k - 1 = \tfrac{300}{700}$$

$$k = (\tfrac{300}{700}) + 1$$

$$k = 1.43$$

7. From Figure 12-4 for $V_{we} = 1{,}600$, $k = 1.43$, the necessary $L = 4{,}250$ feet.

8. If space permits, a design of $L = 4{,}250$ feet and $N = 3$ would be adopted, which requires the construction of an auxiliary lane.

Note that length was substituted for width. This is possible due to the decreased weaving friction, and a decreased k factor which reduces the total "equivalent" nonweaving volume $[V + (k - 1)V_{w2}]$. This is only possible between curves I and III in Figure 12-4, where k varies from 1.0 to 3.0. An increase in length from curve V to curve III, for example, will not permit a reduction in width, as k is a constant (3. 0) in this range.

The previous two problems were of the *design* variety. Frequently, the traffic engineer will be confronted with an existing situation in which the predicted level of service is to be determined.

Problem:

At what level of service is the segment as shown in Figure 12-10 operating?

Solution:

1. From Figure 12-4, with V_{we} 1,800, and $L = 800$ feet, $k = 3 - 0$, Quality of Flow k = III-IV.

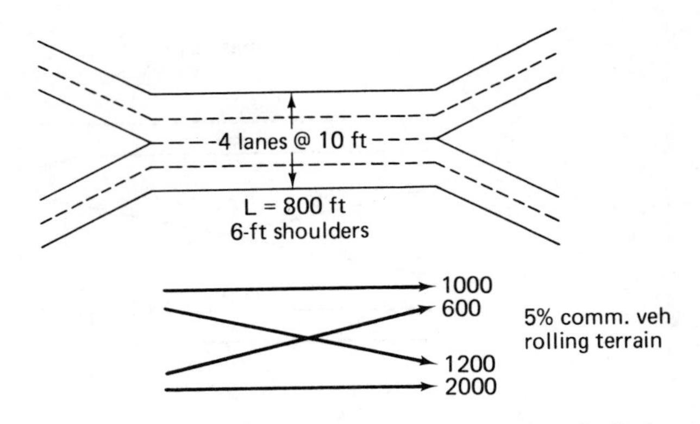

Fig. 12-10. (The weaving arrangement and diagram)—Problem 3

2. To determine SV, consider the equation

$$N = \frac{V + (k - 1)V_{w2}}{SV}$$

Solving for SV,

$$SV = \frac{V + (k - 1)V_{w2}}{N}$$

$$SV = \frac{4{,}800 + (2)600}{4}$$

$$SV = 1{,}500 \text{ vehicles per hour}$$

3. Before entering Table 12-1 to determine the level of service, the volume of 1,500 vehicles per hour must be expanded to passenger cars per hour under ideal geometric and traffic conditions. Since a volume in vehicles per hour under prevailing conditions is being converted to an appropriate volume in passenger cars per hour, the volume must be *divided* by appropriate factors W and T.

$$W = 0.89 \qquad \text{(Table 12-3)}$$
$$T = 0.87 \qquad \text{(Table 12-3)}$$
$$SV = \frac{1{,}500}{(0.89 \times 0.87)} = 1{,}937 \text{ passenger cars per hour}$$

Since the mainlines of the weaving section are two-lane freeways, Table 12-1 is entered for a two-lane freeway of total $SV = 2 \times 1{,}937 = 3{,}874$ passenger cars per hour. From Table 12-1, level of service is E.

4. Quality of Flow III-IV corresponds to Level of Service D (Table 12-9). However, the computed level of service is E. In cases where the quality of flow and level of service do not agree, the worse of the two (in this case Level of Service E) should be taken as the expected condition.

DEFINITIONS PERTAINING TO RAMP FUNCTIONS

Ramp: An interconnecting roadway of a traffic interchange, or any connection between highways at different levels, or between parallel highways, on which vehicles may enter or leave a designated roadway.

Inner loop: A ramp used by traffic executing a left turn from one highway to another, when such movement is accomplished by turning right through about 270 degrees.

Outer connection: A ramp used by traffic executing a right turn from one highway to another, when such movement is accomplished by turning right through about 90 degrees.

Direct connection: A ramp which provides for direct paths of travel for left and right turns.

Ramp terminal: the general area where a ramp connects with a roadway. Ramps have both entrance and exit terminals. The entrance terminal relates to a *merging* condition; the exit terminal relates to a *diverging* condition.

Merging: The process by which two separate traffic streams, moving in the same general direction, combine or unite to form a single stream.

Diverging: The dividing of a single stream of traffic into separate streams.

Acceleration lane: A speed-change lane, for the purpose of enabling a vehicle entering a roadway to increase its speed to a rate at which it can more safely merge wtih through traffic.

Deceleration lane: A speed-change lane, for the purpose of enabling a vehicle exiting from a roadway to slow to the safe turning speed of the curve it is approaching.

Auxiliary lane: An additional width of pavement, adjoining the traveled way, connecting ramps.

Upstream: The direction along the roadway from which the vehicle flow under consideration has come.

Downstream: The direction along the roadway toward which the vehicle flow under consideration is moving.

RAMP JUNCTIONS

The efficiency of traffic movement on the through lanes of a freeway depends directly upon the adequacy of the facilities that are provided for entering and leaving the freeway. Merging and diverging sections are often the controlling elements in establishing service volumes on the freeway and inadequate ramp design and placement can seriously affect freeway operating conditions. The original Manual contained broad generalities on ramp operation, but the new edition contains detailed recommendations on this subject. The procedures presented in the Manual are based on the analysis of data obtained from a great many studies of ramp operations conducted throughout the country, and they are applicable to both operational and design problems.

General Considerations

The following are some of the more important factors influencing ramp operation and design:

Weaving Between Ramps. Studies have shown that *volume* distribution by lane at certain critical points in the merging and diverging sections between on-ramp and off-ramp junctions (separated by a short distance) as well as the *rate* at which weaving occurs, are significant factors in providing for appropriate level of service.

Peak Period Volumes. The peak flow on the ramp may occur at a different time within the hour than the peak flow on the freeway served. Therefore it is essential to determine the peaking characteristics within the peak period of ramp flow in order to evaluate the level of service provided by existing ramps, or for design considerations.

Influence of Design. The geometric design of the ramp and its junction affect overall operations. Sharp ramp curvature, inadequate sight distance, inadequate length for accomplishing merging, diverging, or speed-change maneuvers result in poor operating conditions. Proper terminal design should provide smooth transitional travel paths between ramp and freeway.

The computation procedures in the Manual are based on adequate designs. Analysis of the available field data did not reveal specific adverse effects of various design deficiencies. Improper driver actions and unfamiliarity often appear to overshadow the effects of design variations.

Factors Controlling Ramp Capacity. The overall capacity of a ramp is the *least* of the following three values:

1. Capacity of the ramp proper
2. Capacity of the ramp-freeway junction
3. Capacity of the ramp-street junction

The ramp roadway or ramp-street junction are seldom the limiting element in the amount of traffic carried. In the great majority of cases, merging and diverging conditions at the ramp-freeway junction are the controlling factors. Therefore, the procedures presented in the Manual provide for the determination of *ramp junction* capacity and service volumes. Operation of traffic at the ramp junction is largely dependent on the traffic volume in freeway lane 1 (lane adjacent to ramp). The ability to estimate lane 1 volume at merging and diverging sections is the essential factor in ramp capacity and service volume determinations.

Single-lane entrance and exit ramps are normally adequate for most locations, but conditions which may require additional width include:

1. Where ramp volumes are too great to provide the desired level of service, two-lane entrance and exit ramps may be required.
2. To avoid a back-up onto the freeway, it may be necessary to widen a one-lane exit ramp to two or more lanes at its junction with a cross street in order to provide sufficient approach width to a signalized intersection.
3. If a high-volume entrance ramp roadway exceeds 1,000 feet in length, or if it is on an upgrade and carries appreciable truck volume, two lanes are necessary to allow for passing and smoother arrival rates at the freeway. A two-lane entrance ramp may be funneled to one lane at the merge, if the volume does not exceed 1,500 vehicles per hour, or 30 vehicles per minute over any 5-minute peak period. If the volume exceeds this amount, the two lanes should be retained, but a parallel auxiliary lane should be provided.

Computational Procedures for Ramp Junctions

The basic procedures for determining acceptable volumes at ramp junctions provide for comparison of demand volume with a given service volume at certain critical points. If demand volume does not exceed service volume at the checkpoints, good operating conditions will result on both the freeway and the ramp, and no further detailed analysis is required. If the demand volume exceeds the service volume for the desired level of service, but remains below capacity, the ramp will operate at a lower level of service. These critical checkpoints are shown in Figure 12-11 and include the following:

1. Diverge volume
2. Volume across all freeway lanes
3. Weaving volume
4. Merge volume

Table 12-12 contains volume criteria for critical checkpoints for the different levels of service. The demand volumes for the various traffic movements should be expressed as full-hour volumes, because peaking characteristics have been taken into consideration in Table 12-12. The service volumes shown in the table correspond to full-hour volumes for the different peak-hour factors, and demand volumes should be compared directly with these service volumes. It is necessary to determine the peak-hour factor, however, so that the appropriate service volumes can be used for evaluation.

The Manual gives two procedures for determining the acceptable merge or diverge volume at the junction. One procedure applies to Levels of Service A, B, and C, covering a wide variety of geometric combinations of on-ramp and off-ramp (with one or two lanes) junctions with four-, six- and eight-lane freeways. The other procedure applies for heavy volume conditions covering Levels of Service D and E. Both procedures have been developed for an average base condition, assuming 5 per cent or less of trucks, on relatively level terrain, with grades not exceeding 3 per cent, for conditions on both lane 1 and ramp. For any combinations of trucks and grades *less* than these limits, no adjustments are recommended, but when either one or both of these percentages is exceeded, it is necessary to apply a correction as a final step before comparison is made with the volumes given in Table 12-12. To determine the adjustment for trucks to be applied to lane 1 and ramp volumes, Table 12-6 or 12-8 is used. After the adjustment factors have been evaluated, the lane 1 volume, or the ramp volume, is multiplied by 0.91/(adjustment factor), which converts the volume to the base condition of 5 per cent trucks. The 0.91 factor applies to the base condition.

Procedure for Calculation of Service Volumes, Levels A Through C. In the normal case, whether it deals with the analysis of an existing interchange or a proposed new design, the traffic demand for each of the movements is established, and it is necessary to determine if the interchange will operate satisfactorily at the level of service under consideration.

In all, 18 equations and corresponding nomographs

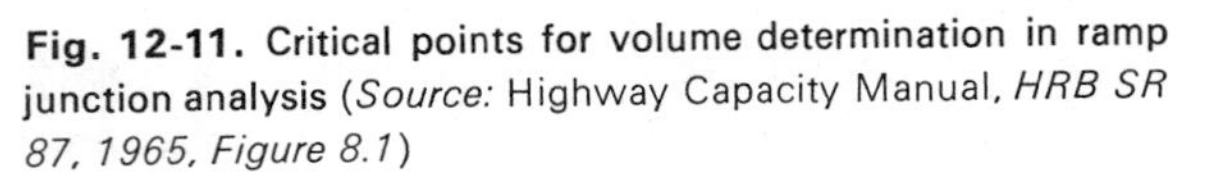
Fig. 12-11. Critical points for volume determination in ramp junction analysis (*Source:* Highway Capacity Manual, *HRB SR 87, 1965, Figure 8.1*)

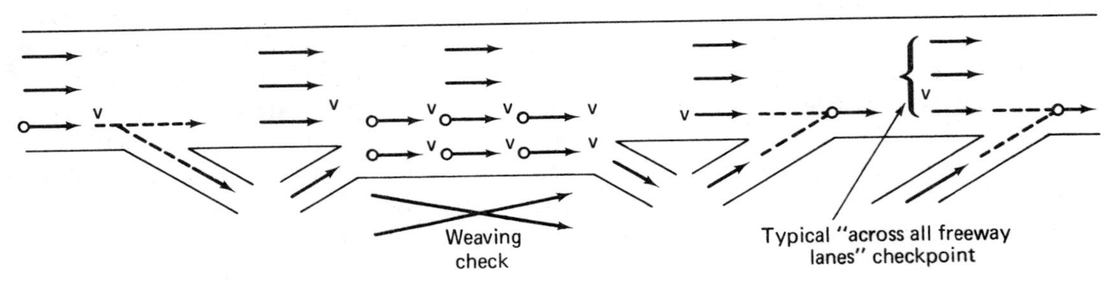

TABLE 12-12. SERVICE VOLUMES AND CAPACITY IN VICINITY OF RAMP TERMINALS
(VPH of mixed traffic in one direction, assuming level terrain and not over 5% trucks)

LEVEL OF SERVICE	*FREEWAY VOLUME, ONE DIRECTION (vph)*[b]												*CHECKPOINT VOLUME (vph)*								*WEAVING VOLUMES*[e]			
	4-Lane				*6-Lane*				*8-Lane*				*Merge*[c]				*Diverge*[d]				*vph*			
A	1400				2400				3400				1000				1100				800			
B	2000				3500				5000				1200				1300				1000			
Peak-hour factor[f]	0.77	0.83	0.91	1.00[g]	0.77	0.83	0.91	1.00[g]	0.77	0.83	0.91	1.00	0.77	0.83	0.91	1.00[g]	0.77	0.83	0.91	1.00[g]	0.77	0.83	0.91	1.00[g]
C	2300	2500	2750	3000	3700	4000	4350	4800	5100	5500	6000	6600	1300	1400	1550	1700	1400	1500	1650	1800	1100	1200	1350	1450
D	2800	3000	3300	3600	4150	4500	4900	5400	5600	6000	6600	7200	1400	1500	1650	1800	1500	1600	1750	1900	1400	1500	1650	1800
E[h]	$\lesssim$4000				$\lesssim$6000				$\lesssim$8000				$\lesssim$2000				$\lesssim$2000				$\lesssim$2000			
F	← Widely variable →																							

(Source: *Highway Capacity Manual*, HRB SR 87, 1965, Table 8-1)

a. Upper limit volume for each level of service.
b. To be used in making "across all through freeway lanes" service volume checks between ramp-freeway terminal and/or interchanges.
c. Represents the merge taking place, which is determined by the computed lane 1 volume plus the on-ramp volume.
d. Represents the volume in lane 1 immediately upstream from an exit ramp; includes both through vehicles and prospective exit-ramp vehicles.
e. For weaving between on-ramps and off-ramps per 500 ft. of roadway segment.
f. For freeways, the ratio of the whole-hour volume to the highest hourly rate of flow occurring during a 5-min. interval within the peak hour.
g. A peak-hour factor of 1.00 is rarely attained; the values given should be considered as maximum average flow rates likely to be obtained during the peak 5-min. interval within the peak hour.
h. Capacity.

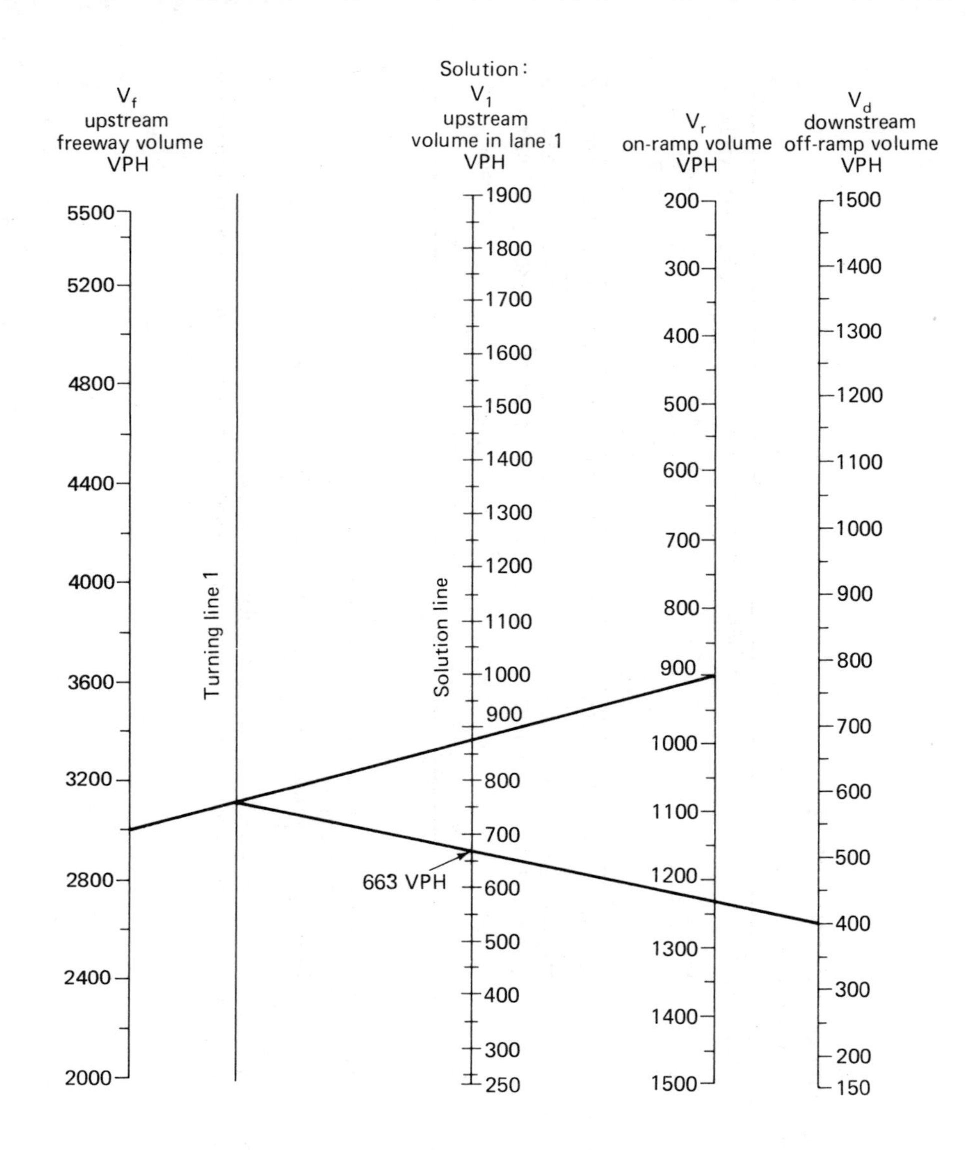

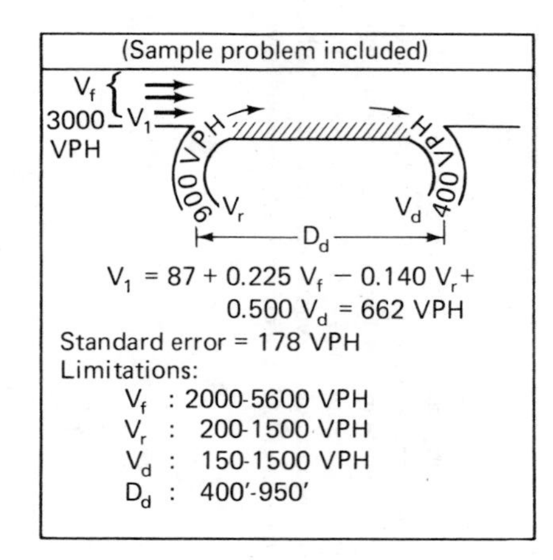

$V_1 = 87 + 0.225\ V_f - 0.140\ V_r + 0.500\ V_d = 662$ VPH

Standard error = 178 VPH

Limitations:

V_f : 2000-5600 VPH
V_r : 200-1500 VPH
V_d : 150-1500 VPH
D_d : 400′-950′

CONDITIONS FOR USE

To determine the lane 1 volume at the inner loop on-ramp nose at a cloverleaf interchange containing an auxiliary lane between the loops. The interchange may or may not have enter connections. The equation does not consider their effects and applies only to inner loop on-ramps. The inner loop off-ramp should be located between 400 and 850 ft downstream.

STEPS IN SOLUTION

(1) Draw a line from V_f value to V_r value intersecting Turning line 1.
(2) Draw a line from Step 1 intersection of Turning line 1 to V_d value. The intersection point of this line with Solution line is V_1.

Fig. 12-12. Nomograph for determination of lane 1 volume upstream of on-ramp junction, 6-lane freeway, at cloverleaf inner loop with auxiliary lane. *(Source:* Highway Capacity Manual, *HRB SR 87, 1965, Figure 8.11)*

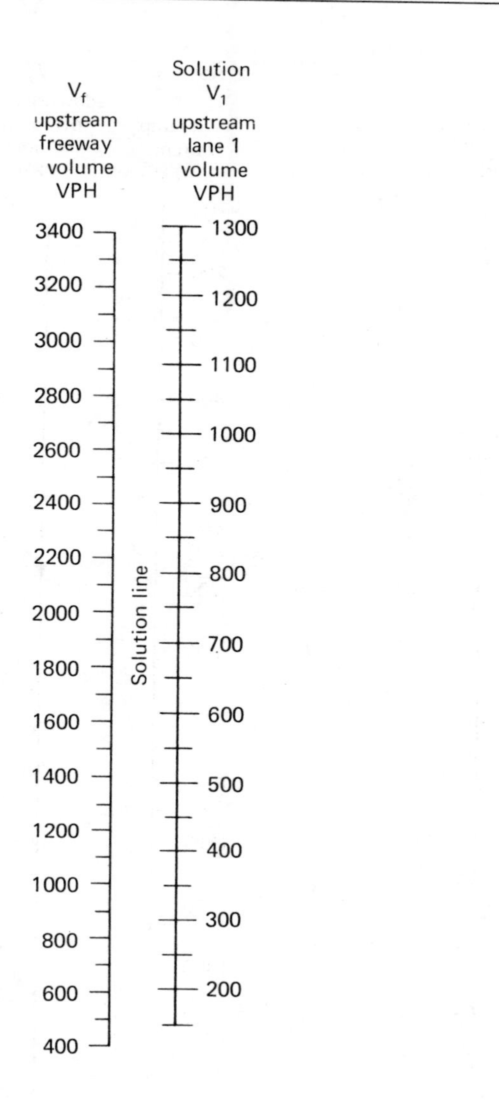

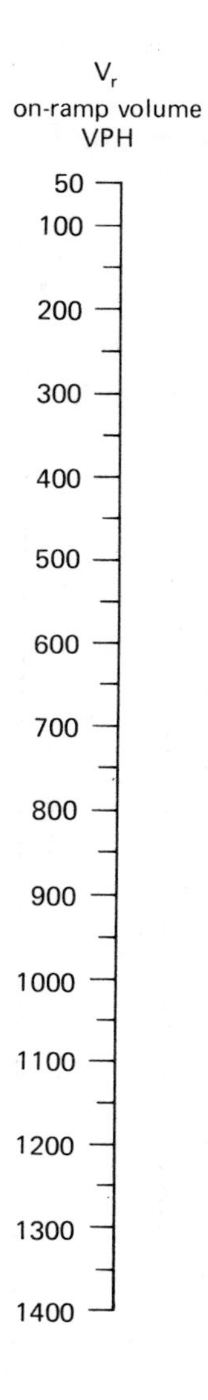

$V_f \{ V_1$

V_r

$V_1 = 136 + 0.345\ V_f - 0.115\ V_r$

Standard error = 113 VPH

Limitations:

V_f : 400-3400 VPH

V_r : 50-1400 VPH

No upstream on-ramp within 2,000 feet.

CONDITIONS FOR USE

To determine the lane 1 volume at an on-ramp nose before merging takes place. The ramp can be of any single-lane type except cloverleaf loop. An acceleration lane may or may not be present.

If there is an adjacent upstream on-ramp within 2000 ft, use Figure 8.8 of *Highway Capacity Manual.*

STEPS IN SOLUTION

Draw a line from V_f value to V_r value intersecting V_1 on Solution line.

Fig. 12-13. Nomograph for determination of lane 1 volume upstream of on-ramp junction, 4-lane freeway, not applicable to cloverleaf inner loop *(Source:* Highway Capacity Manual, *HRB SR 87, 1965, Figure 8.2)*

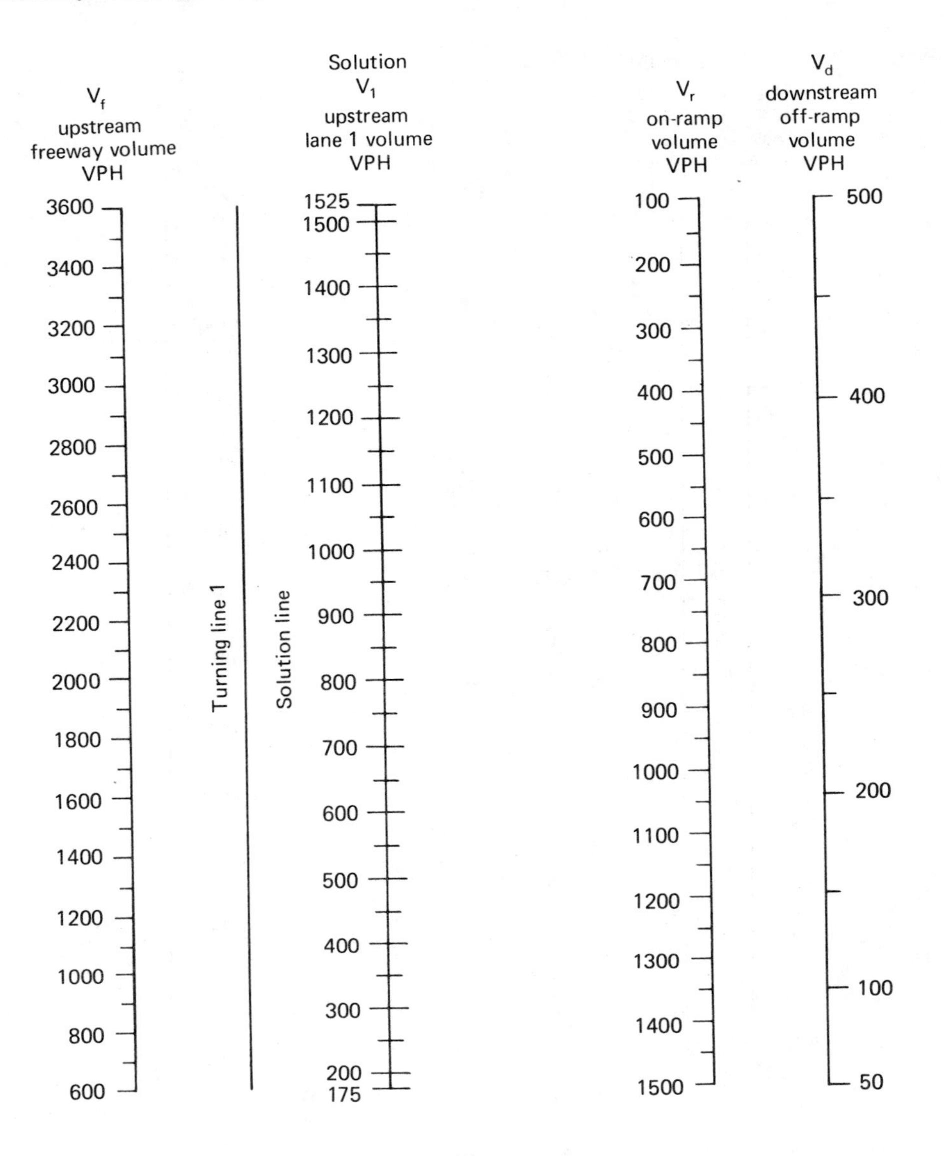

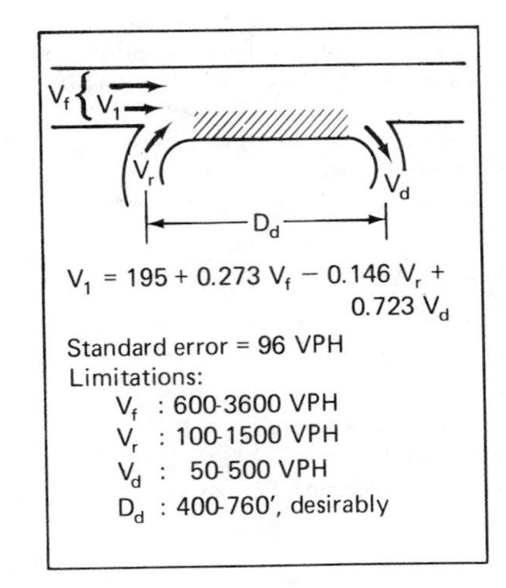

$$V_1 = 195 + 0.273\,V_f - 0.146\,V_r + 0.723\,V_d$$

Standard error = 96 VPH

Limitations:

V_f : 600-3600 VPH

V_r : 100-1500 VPH

V_d : 50-500 VPH

D_d : 400-760′, desirably

CONDITIONS FOR USE

To determine the lane 1 volume at an inner loop on-ramp at a cloverleaf interchange containing an auxiliary lane between the loops. The interchange may or may not have outer connections. Desirably, the distance between ramps should be 400 to 750 ft.

Assuming all prospective off-ramp vehicles are in lane 1 at the on-ramp nose, it is possible to determine the number of through vehicles in lane 1 by subtracting the off-ramp volume from the computer lane 1 volume.

STEPS IN SOLUTION

(1) Draw a line from V_f value to V_r value intersecting Turning line 1.

(2) Draw a line from Step 1 intersection of Turning line 1 to V_d value. The intersection point of this line with Solution line is V_1.

Fig. 12-14. Nomograph for determination of lane 1 volume upstream of on-ramp junction, 4-lane freeway, at cloverleaf inner loop with auxiliary lane (*Source:* Highway Capacity Manual, *HRB SR 87, 1965, Figure 8.6*)

are given in the Manual for solution of lane 1 volumes. Three of these nomographs are shown in Figures 12-12, 12-13, and 12-14.

Once lane 1 volume is determined upstream from an on-ramp, the allowable ramp volume can be found. This, in combination with the lane 1 volume, will provide a specified level of service. At off-ramps, lane 1 volume is determined upstream from the exit point, before divergence takes place. Therefore, with given freeway and exit-ramp demands, it is possible to establish *when* overloading will occur upstream from an exit ramp. The basic premise is that if the lane 1 volume can be kept at an acceptable level of service, the freeway as a whole will be operating at an acceptable level.

Computational steps

1. Establish the geometrics of the location under study, including number of freeway lanes, grades, location and type of adjacent ramps upstream and downstream from the junction under consideration, and existence of auxiliary lanes.
2. Establish demand volumes for traffic movements involved.
3. Select appropriate equation or nomograph, and determine volume in lane 1 at checkpoints.
4. Determine acceptable ramp volumes based on criteria given in Table 12-12.

Factors affecting computations

The equations take into consideration freeway volume, ramp volume, and traffic action of adjacent ramps which have an effect on the lane volume at the ramp being investigated.

Lane designations

1. *Lane 1*: The right-hand through lane of the freeway.
2. *Ramp Lane A:* For a two-lane ramp, the ramp lane closest to the freeway; for a major fork, the merge or diverge lane closest to the adjacent roadway.
3. *Ramp Lane B:* For a two-lane ramp, the ramp lane farthest from the freeway; for a major fork, the merge or diverge lane farthest from the adjacent roadway.

Dependent variables (all volumes expressed in vehicles per hour)

1. V_1 = the lane 1 volume immediately upstream from the ramp, or the lane 1 volume immediately downstream from a two-lane off-ramp or major fork.
2. V_{1+A} = for two-lane ramps, the initial merge volume, consisting of lane 1 plus ramp lane A, or combined lane 1 and lane A volumes just before the divergence takes place.
3. V_c = for a major fork on a six-lane freeway, the center lane volume before the freeway splits into lane 1 and lane A of the two fork legs.

Independent variables (all volumes in vehicle per hour)

1. V_f = the freeway approach volume immediately upstream of the on-ramp.
2. V_t = the total freeway approach volume immediately upstream of the off-ramp.
3. V_r = the on-ramp volume, or the off-ramp volume, or the volume using the right-hand roadway of a major fork.
4. D_u = the distance, in feet, from the ramp under consideration to an adjacent upstream on-ramp or off-ramp.
5. V_u = the volume on an adjacent upstream on-ramp or off-ramp.
6. D_d = the distance, in feet, from the ramp under consideration to an adjacent downstream on-ramp or off-ramp, as measured in Figure 7-5 of the Manual. Where an auxiliary lane is provided, this distance coincides with the length of the auxiliary lane.
7. V_d = the volume on an adjacent downstream on-ramp or off-ramp.

Computations involving auxiliary lanes

The use of an auxiliary lane somewhat modifies the computational procedures from those used in conventional merging and diverging problems. Conventionally, the computed lane 1 volume is added to the ramp volume to yield a merge volume. Also, the lane 1 volume upstream from the ramp is the expected diverge. At auxiliary lane locations, the opportunity to weave or change lanes between lane 1 and the auxiliary lane makes it necessary to compute the volume in each of these lanes at selected points between ramp noses. Usually a volume check at the 0.5 point will suffice when the sum of the volumes in lane 1 and the on-ramp does not exceed the merge service volume given in Table 12-12. If the off-ramp volume is high, an additional check should be made at the 0.2 point; if the on-ramp volume is high, an additional check should be made at the 0.8 point. The 0.2, 0.5, and 0.8 points refer to the fractional parts of the distance from the on-ramp to the off-ramp.

The computed lane 1 and auxiliary lane volumes should be checked separately against the service volume in Table 12-12. If the check is made at the 0.5 point or closer to the on-ramp, the *merge* service volume is used; for other points the *diverge* service volume is used.

Checks of weaving volume per 500 feet of roadway should also be made. In making a volume check across

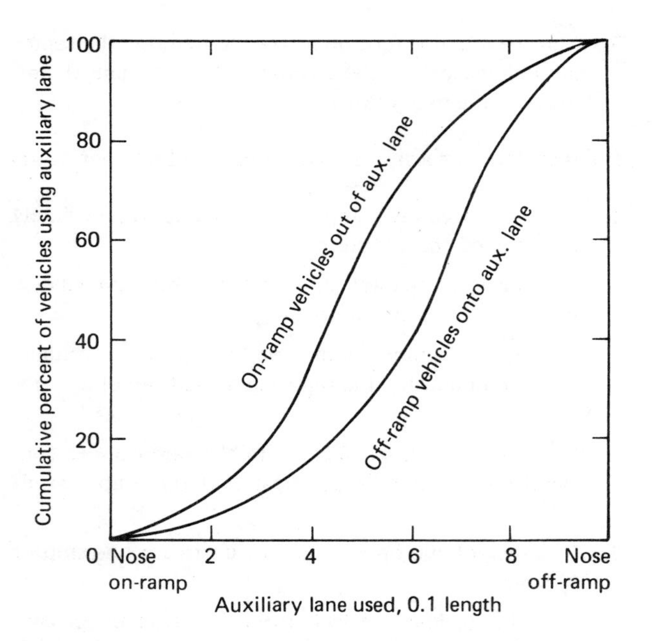

Fig. 12-15. Use of auxiliary lane between adjacent on- and off-ramps (*Source:* Highway Capacity Manual, *HRB SR 87, 1965, Figure 8.20*)

all freeway lanes, the auxiliary lane and its volume are not to be included in the check.

Figure 12-15 is used in conjunction with the appropriate equation or nomograph. Figure 12-15 shows the cumulative per cent of on-ramp vehicles out of the auxiliary lane, and the cumulative per cent of off-ramp vehicles in the auxiliary lane at various points along the auxiliary lane. The computational procedure consists of the following:

1. Lane 1 volume at the on-ramp nose is determined, using the appropriate equation or nomograph. For simplicity in determining the lane 1 through volume, 100 per cent of the off-ramp vehicles which intend to exit are considered to be in lane 1 at the on-ramp nose (in actuality this percentage is closer to 95).
2. Compute the through volume in lane 1 by subtracting the off-ramp volume from the lane 1 volume at the on-ramp nose.
3. Make checks of the lane 1 and auxiliary lane volumes at points between ramps. The volumes will consist of:

Lane 1 volume = lane 1 through + on-ramp vehicles out of auxiliary lane + off-ramp vehicles still in lane 1

Auxiliary lane volume = on-ramp vehicles still in auxiliary lane + off-ramp vehicles which have moved onto auxiliary lane

Equations and nomographs

The equations and the nomographs presented in the Manual and herein do not fit all geometric conditions which may be encountered, and suggestions are given in the Manual for the adaptation of equations to other geometric conditions for which there was insufficient data to develop specific equations. These suggestions are contained in Table 8.2 of the Manual which provides an excellent index for selecting the appropriate equation for various geometric combinations. Table entries which are shown in parentheses give suggested adaptations of criteria not developed specifically for the geometrics shown and/or references to discussion of these cases.

Procedure for Calculation of Service Volumes, Levels D and E. Table 12-13 and Figure 12-16 are provided to cover all geometric combinations of ramp entrances and exits. Table 12-13 contains the percentage of through traffic (traffic not involved in a ramp movement) likely to remain in lane 1 through the ramp junction area on four-, six- and eight-lane freeways. Figure 12-16 shows the percentage distribution, at various points, of on-ramp and off-ramp traffic likely to be in lane 1, and in the auxiliary lane if one is provided.

The footnote to Figure 12-16, which states that the minimum amount of traffic in lane 1 cannot be less than that determined using the percentages given in Table 12-13, should be carefully considered. In essence, this requirement means that lane 1 volume at any point

TABLE 12-13. APPROXIMATE PERCENTAGE OF THROUGH* TRAFFIC REMAINING IN LANE 1 IN THE VICINITY OF RAMP TERMINALS AT LEVEL OF SERVICE D

TOTAL VOLUME OF THROUGH TRAFFIC, ONE DIRECTION (vph)	*THROUGH TRAFFIC REMAINING IN LANE 1 (%)*		
	*8-Lane** Freeway*	*6-Lane† Freeway*	*4-Lane‡ Freeway*
6,500 and over	10	—	—
6,000–6,499	10	—	—
5,500–5,999	10	—	—
5,000–5,499	9	—	—
4,500–4,999	9	18	—
4,000–4,499	8	14	—
3,500–3,999	8	10	—
3,000–3,499	8	6	40
2,500–2,999	8	6	35
2,000–2,499	8	6	30
1,500–1,999	8	6	25
Up to 1,499	8	6	20

(Source: *Highway Capacity Manual,* HRB SR 87, 1965, Table 8-3)
* Traffic not involved in a ramp movement within 4,000 ft. in either direction.
** 4 lanes one way.
† 3 lanes one way.
‡ 2 lanes one way.

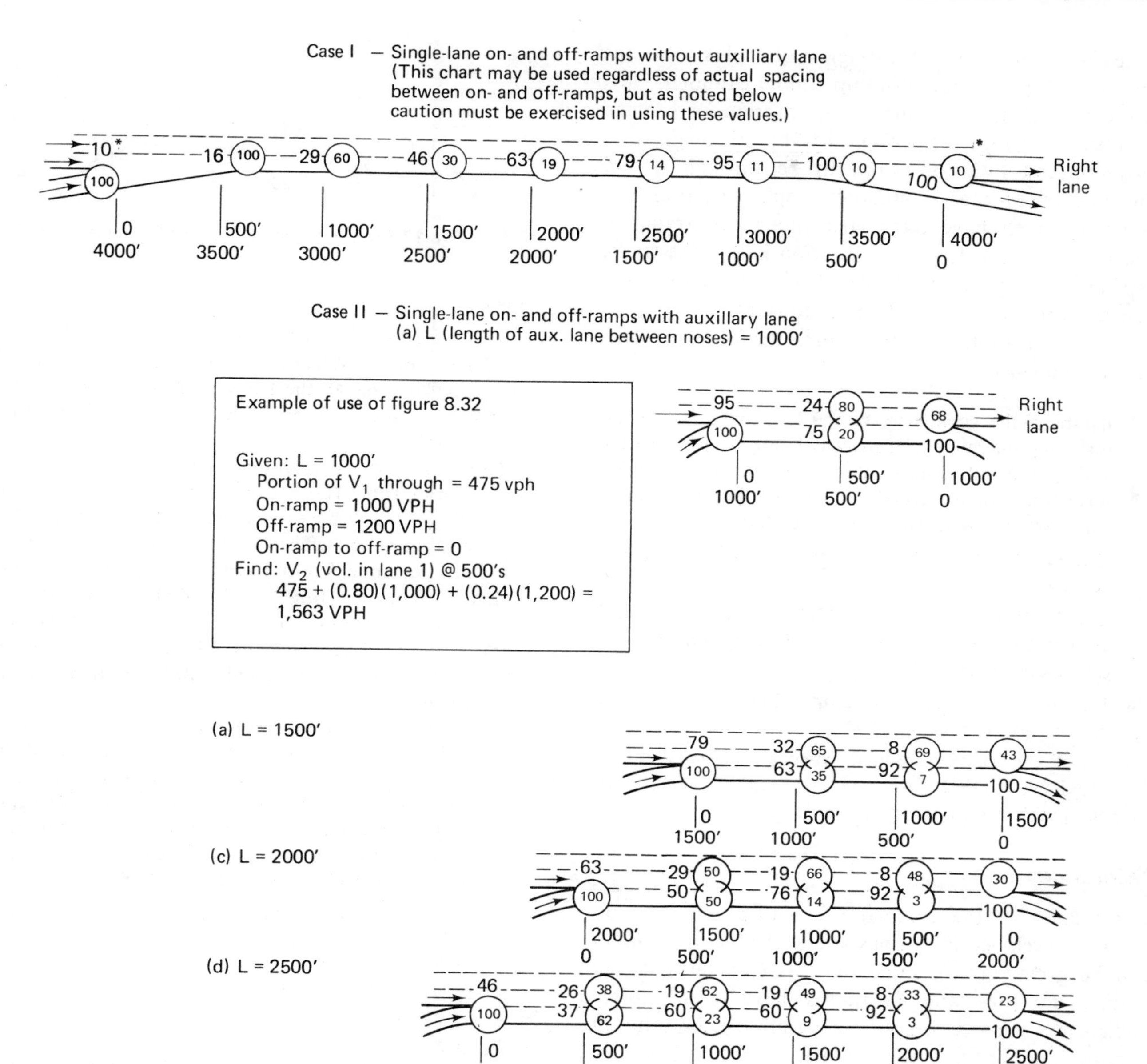

Circle values (0) indicate percentage of on-ramp traffic in lane shown. Uncircled values indicate percentage of off-ramp traffic in lane shown. (Remaining portion of traffic is in lane(s) to left of lane 1.)

These percentages are not necessarily the distributions under free flow of light ramp traffic, but under pressure of high volumes in the right lanes at the point being considered and with room available in other lanes.

*Minimum % in right lane cannot be less than % of through traffic in right lane as determined from Table 12.13.

Fig. 12-16. Distribution in lane 1 and auxiliary lane of on- and off-ramps (*Source:* Highway Capacity Manual, *HRB SR 87, 1965, Figure 8.23*)

(which may contain some through traffic, plus some on-ramp traffic, plus some off-ramp traffic) cannot be less than that obtained by using the appropriate percentage from Table 12-13, considering all traffic (through, total on-ramp, and total off-ramp) to be through traffic. The adjustment most commonly applies to four-lane freeways, which have larger percentages of traffic in lane 1. The reason for this restriction is that the traffic distributions depicted in Table 12-13 and Figure 12-16 are predicated on heavy volume conditions.

In the application of the procedure, the following steps are recommended:

1. Establish the geometrics of the location under study, including the number of freeway lanes, grades, location and type of all ramps within a distance of 4,000 feet upstream and downstream from the junction under consideration, and existence of auxiliary lanes.
2. Establish the demand volumes for the traffic movements involved.
3. Use Table 12-13 and Figure 12-16 to determine lane 1 volume and auxiliary lane (if provided) volume at various checkpoints.
4. Determine the acceptable ramp volumes based on the criteria given in Table 12-12.
5. Check the weaving volume, when an auxiliary lane is provided, at 500 foot intervals, and compare with acceptable values given in Table 12-12.

References

This section of the Manual was based primarily on material developed in various studies.[3,4,5] Other studies have been made in the same area.[6-21] Geometric design features for on- and off-ramps have been closely studied.[23-24] Also, a computer program for the solution of ramp-junction problems has been developed.[45]

Sample Problems

Problem 1:

On an eight-lane freeway, where a PHF = 0.83 is applicable, one section has an on-ramp, off-ramp combination. Average grades and numbers of trucks prevail. Geometrics are adequate, and demand volumes are as follows (see Figure 12-17):

A to B = 4,200 vehicles per hour
A to Y = 500
X to B = 1,200
X to Y = 0
Total = 5,900 vehicles per hour between ramps

Determine if requirements for Level of Service D are met. Redesign to provide for Level of Service D if criteria are not met.

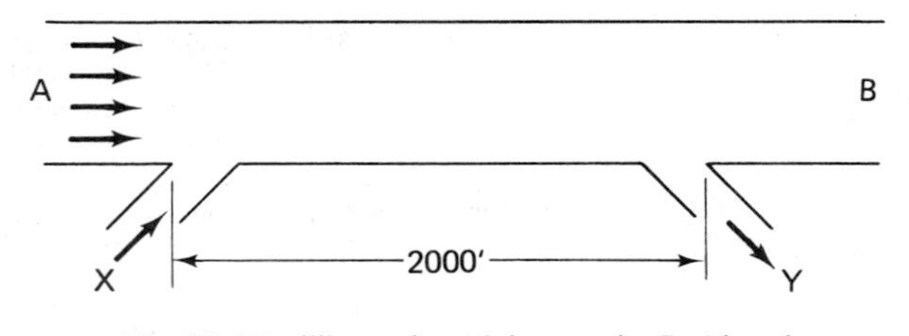

Fig. 12-17. (Illustration of the ramp)—Problem 1

Solution:

Volume across all freeway lanes check:

Total volume between ramps = 5,900 vehicles per hour. 5,900 < 6,000, the Level D freeway service volume for a PHF = 0.83 from Table 12-12; therefore requirement is met.

500 feet downstream of on-ramp:

1. From Table 12-13, through traffic in lane 1 = 0.08 × 4,200 = 340 vehicles per hour. From Figure 12-16 (Case I), on-ramp traffic in lane 1 = 1.00 × 1,200 = 1,200 vehicles per hour. From Figure 12-16 (Case I), off-ramp traffic in lane 1 = 0.79 × 500 = 400 vehicles per hour. Total volume = 340 + 1,200 + 400 = 1,940 vehicles per hour. 1,940 > 1,500, the Level D merge service volume for a PHF = 0.83 from Table 12-12; therefore requirement is *not* met.

2. One modification in the design is to consider the feasibility of splitting the high-volume on-ramp into two on-ramps, so that some of the first on-ramp traffic will have an opportunity to distribute into available gaps in lanes adjacent to Lane 1 before the second on-ramp traffic enters. Geometrics are as shown. Assume that demand volumes are as follows (see Figure 12-18):

A to B = 4,200 vehicles per hour
A to Y = 500
X_1 to B = 500
X_2 to B = 700
X_1 to Y = 0
X_2 to Y = 0
Total = 5,900 vehicles per hour between on-ramp and off-ramp

Total volume across all freeway lanes = 5,900 vehicles per hour. 5,900 < 6,000; therefore requirement for Level D is met.

500 feet downstream of first on-ramp:

From Table 12-13, through traffic in lane 1 = 0.08 × 4,200 = 340 vehicles per hour. From Figure 12-16 (Case I), on-ramp traffic in lane 1 = 1.00 × 700 = 700 vehicles per hour. From Figure 12-16 (Case I), off-ramp traffic in lane 1 = 0.46 × 500 = 230 vehicles per hour. Total volume = 340 + 700 + 230 = 1,270 vehicles per hour. 1,270 < 1,500; therefore requirement for Level D is met.

500 feet downstream of second on-ramp:

From Table 12-12, through traffic in lane 1 = 0.08 × 4,200 = 340 vehicles per hour. From Figure 12-16, first on-ramp traffic in lane 1 = 0.30 × 700 = 210 vehicles per hour. Through traffic plus first on-ramp = 340 ×

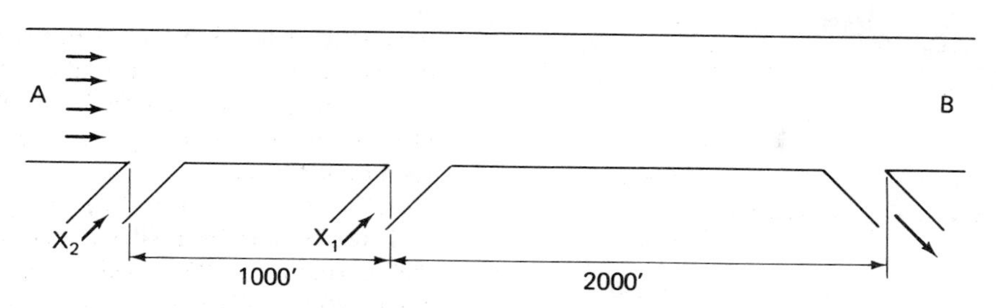

Fig. 12-18. (Illustration of the ramp)—Problem 16

210 = 550 vehicles per hour. From Figure 12-16, second on-ramp traffic in lane 1 = 1.00 × 500 = 500 vehicles per hour. From Figure 12-16, off-ramp traffic in lane 1 = 0.79 × 500 = 400 vehicles per hour. Total volume = 550 + 500 + 400 = 1,450 vehicles per hour. 1,450 < 1,500; therefore requirement for Level D is met.

Weaving in first 500 feet downstream of second on-ramp:

From Figure 12-16 (Case I), the percentages of first on-ramp traffic in lane 1 at second on-ramp nose, and at 500 feet downstream of second on-ramp nose, are equal to 60 per cent and 30 per cent, respectively. Therefore, the difference between these percentages must be involved in weaving maneuvers, or first on-ramp weaving = (0.60 − 0.30)700 = 210 vehicles per hour. Similarly, second on-ramp weaving = (1.00 − 1.00)500 = 0 and off-ramp weaving = (0.79 − 0.63)500 = 80 vehicles per hour. Total weave = 210 + 80 = 290 vehicles per hour. 290 < 1,500, the Level D weave service volume for a PHF = 0.83 from Table 12-12; therefore requirement is met.

At 1,000 feet downstream of second on-ramp:

From Table 12-13, through traffic in lane 1 = 0.08 × 4,200 = 340 vehicles per hour. From Figure 12-16, first on-ramp traffic in lane 1 = 0.19 × 700 = 130 vehicles per hour. Through traffic plus first on-ramp = 340 + 310 = 470 vehicles per hour. From Figure 12-16, second on-ramp traffic in lane 1 = 0.60 × 500 = 300 vehicles per hour. From Figure 12-16, off-ramp traffic in lane 1 = 0.95 × 500 = 480 vehicles per hour. Total volume = 470 + 300 + 480 = 1,250 vehicles per hour. 1,250 < 1,500; therefore requirement for Level D is met.

Weaving in second 500-foot segment downstream of second on-ramp:

First on-ramp weaving = (0.30 − 0.19)700 = 80 vehicles per hour. Second on-ramp weaving = (1.00 − 0.60)500 = 200 vehicles per hour. Off-ramp weaving = (0.95 − 0.79)500 = 80 vehicles per hour. Then the total weave = 80 + 200 + 80 = 360 vehicles per hour. 360 < 1,500; therefore requirement for Level D is met.

Therefore provision of the second on-ramp has resulted in meeting all requirements for Level of Service D.

3. Another modification in the design that should be investigated is the provision of an auxiliary lane between the two ramps. Geometrics are as shown in Figure 12-19, and demand volumes are as follows:

A to B = 4,200 vehicles per hour
A to Y = 500
X to B = 1,200
X to Y = 0
Total = 5,900 vehicles per hour between ramps

At 500 feet downstream of on-ramp:

Lane 1 volume: From Table 12-13, through traffic = 0.08 × 4,200 = 340 vehicles per hour. From Figure 12-16 (Case II), on-ramp traffic = 0.50 × 1,200 = 600 vehicles per hour. From Figure 12-16 (Case II), off-ramp traffic = 0.29 × 500 = 150 vehicles per hour. Total volume = 340 + 600 + 150 = 1,090 vehicles per hour. 1,090 < 1,500; therefore requirement for Level D is met.

Auxiliary lane: From Figure 12-16 (Case II), on-ramp traffic 0.50 × 1,200 = 600 vehicles per hour. From Figure 12-16 (Case II), off-ramp traffic = 0.50 × 500 = 250 vehicles per hour. Total volume = 600 + 250 = 850 vehicles per hour. 850 < 1,500; therefore requirement for Level D is met.

Volume across all freeway lanes: Total volume on through lanes = 5,900 − 850 = 5,050 vehicles per hour. 5,050 < 6,000; therefore requirement for Level D is met.

Weaving in first 500-foot segment: From Figure 12-16 (Case II), 50 per cent of the on-ramp vehicles have entered lane 1, on-ramp weaving = 0.50 × 1,200 = 600 vehicles per hour. From Figure 12-16 (Case II), 50 per cent of the off-ramp vehicles have entered auxiliary lane, off-ramp weaving = 0.50 × 500 = 250 vehicles per hour. Total weave = 600 + 250 = 850 vehicles per hour. 850 < 1,500; therefore requirement for Level D is met.

At 1,000 feet downstream of on-ramp:

Lane 1 volume: From Table 12-13, through traffic = 0.08 × 4,200 = 340 vehicles per hour. From Figure 12-16, on-ramp traffic = 0.66 × 1,200 = 790 vehicles per hour. From Figure 12-16, off-ramp traffic = 0.19 × 500 = 100 vehicles per hour. Total volume = 340 + 790 + 100 = 1,230 vehicles per hour. 1,230 < 1,500; therefore requirement for Level D is met.

Auxiliary lane: From Figure 12-16, on-ramp traffic

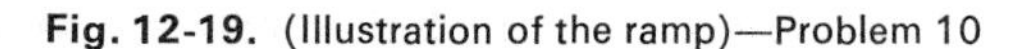

Fig. 12-19. (Illustration of the ramp)—Problem 10

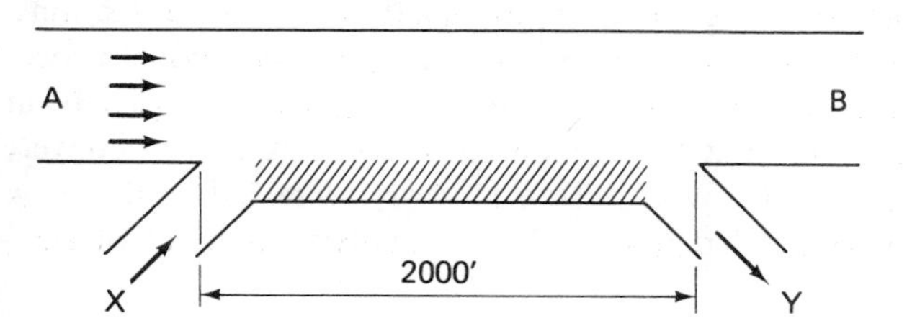

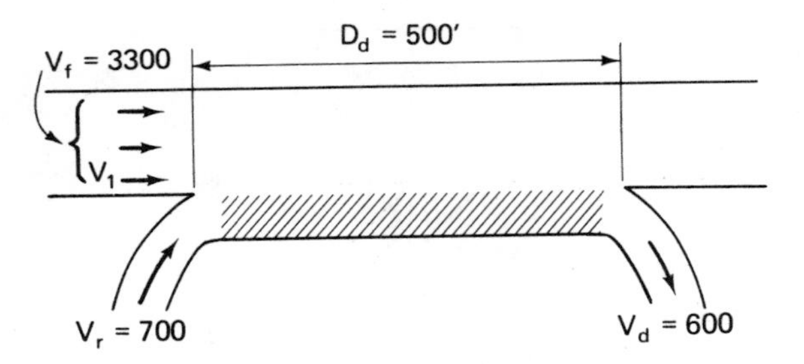

Fig. 12-20. (Illustration of the ramp)—Problem 2

= 0.14 × 1,200 = 170 vehicles per hour. From Figure 12-16, off-ramp traffic = 0.76 × 500 = 380 vehicles per hour. Total volume = 170 + 380 = 550 vehicles per hour. 550 < 1,500; therefore requirement for Level D is met.

Volume across all freeway lanes: Total volume on through lanes = 5,900 − 550 = 5,350 vehicles per hour. 5,350 < 6,000; therefore requirement for Level D is met.

Weaving in second 500-foot segment: Maximum weave occurs in first 500 feet; therefore the computation is unnecessary. To illustrate the procedure, however, it will be carried out: From Figure 12-16, on-ramp weaving = (0.50 − 0.14)1,200 = 430 vehicles per hour. From Figure 12-16, off-ramp weaving = (0.76 − 0.50)500 = 130 vehicles per hour. Total weave = 430 + 130 = 560 vehicles per hour.

Therefore, provision of the auxiliary lane has resulted in meeting all requirements for Level of Service D.

Problem 2:

A cloverleaf interchange on a six-lane freeway has an auxiliary lane between inner loops. A PHF = 0.91 has been found applicable. Geometrics and demand volumes are as indicated. Truck volumes are less than 5 per cent and grades are under 3 per cent. (See Figure 12-20.) Determine the level of service provided.

Solution:

Figure 12-12 should be used in conjunction with Figure 12-15 to check on operations. At on-ramp (Figure 12-12):

$$V_1 = -87 + 0.225V_f - 0.140V_r + 0.500V_d$$

$$V_1 = -87 + 0.225 \times 3{,}300 - 0.140 \times 700 + 0.500 \times 600 = 860 \text{ vehicles per hour}$$

at the nose of the on-ramp.

$$V_1 + V_r = 860 + 700 = 1{,}560 < 150 \text{ per cent of } 1{,}200$$

the Level B merge service volume from Table 12-12; therefore, check at 0.5 point is sufficient.

Lane 1 volume at 0.5 point between ramps:

Lane 1 through volume = 860 − 600 = 260 vehicles per hour. From Figure 12-15 the cumulative per cent of on-ramp vehicles in lane 1 at 0.5 point = 58 per cent, and the cumulative per cent of off-ramp vehicles onto auxiliary lane at 0.5 point = 25 per cent. Total in lane 1 at 0.5 point = 260 through traffic + 0.58 × 700 from on-ramp + 0.75 × 600 destined to off-ramp = 1,120 vehicles per hour. 1,120 < 1,200; therefore the requirement for Level B is met.

Auxiliary lane volume at 0.5 point:

Total volume in auxiliary lane at 0.5 point = 0.42 × 700 on-ramp + 0.25 × 600 off-ramp = 440 vehicles per hour. 440 < 1,200; therefore requirement for Level B is met.

Volume across all freeway lanes check at 0.5 point:

Freeway volume at 0.5 point > 3,300 upstream of on-ramp + 700 on-ramp − 440 in auxiliary lane = 3,560 vehicles per hour. 3,560 is approximately = 3,500, the Level B freeway service volume from Table 12-12.

Weaving volume check:

700 vehicles per hour from on-ramp + 600 to off-ramp = 1,300 vehicles per hour weaving in 500 feet. This condition meets the Level C criterion of not over 1,350 vehicles per hour weaving per 500 feet for a PHF = 0.91 from Table 12-12. In this case, the weaving volume over the distance between ramps is the same with and without an auxiliary lane. The auxiliary lane, however, provides for a reduction of 440 vehicles per hour across all freeway lanes at the critical section.

INTERSECTIONS AT GRADE

Expressways may contain some intersections at grade. These are covered later in the discussion of city street and urban arterial capacity (Chapter 13). The analysis of expressway intersections is identical to that described for urban signalized intersections.

THE FREEWAY AS A TOTAL FACILITY

The freeway is a system comprised of the component parts discussed herein: through sections, weaving sections, ramps, and possibly some intersections at grade. To understand the operation of the freeway, it is necessary to understand how these components interact.

Combined Analysis of Freeway Elements

Once each point or subsection has been evaluated on a freeway section, it is necessary to investigate the overall operation of the facility.

Balanced design requires the relating of the through roadway subsection operation to the operation of other elements, including ramp junctions, weaving sections, grades, lane reductions, etc. Since computational procedures for weaving sections and ramps are basically in terms of mixed traffic, and it is relatively easy to convert through freeway section analysis to mixed traffic, it is recommended that problems involving a series of freeway elements be carried out in terms of mixed traffic.

Problems usually involve analyzing an existing freeway to disclose the location of "weak links," or designing a new freeway for fully balanced operation. The procedures are similar, since the design of a new freeway is conveniently carried out by analyzing trial designs.

Some general observations can be made regarding weighting computations. Weighting for fully balanced design is meaningless, since the level of service is identical at all points. However, on existing freeways or on new designs where perfect balance is not achieved, it is desirable to develop a general measure of performance.

Weighting of operating speeds and v/c ratios, the indicators of level of service, is done in terms of the relative lengths of the sections involved, or their relative influence areas. For some freeway elements, lengths of section are obvious; for others, such as ramps or weaving sections, the area of influence for use in weighting is not readily apparent. As a rule of thumb, a ramp junction influence distance of 3,000 feet can be assumed. In the case of an on-ramp, this is composed of about 500 feet upstream and 2,500 feet downstream; in the case of an off-ramp it consists of 2,500 feet upstream and 500 feet downstream. Where an overlap occurs, the poorer of the two sets of operating conditions is considered as applicable to the overlap area. For weaving sections, an influence distance of about 1,000 feet (500 feet upstream and 500 feet downstream) in addition to the section length may be assumed.

Determination of a weighted level by precise methods is usually limited to cases where operating speeds, v/c ratios, and lengths are readily available. The procedure involves the computation of the weighted average of the operating speeds obtained for each subsection by multiplying the length of each section by its operating speed, summing the several results, and dividing by the overall length. Similarly, the weighted average of the several v/c ratios is computed. With the use of these weighted averages, the overall level of service is obtained from Table 12-1.

In other cases, where ramp junctions, weaving sections, and at-grade intersections are involved, an approximate weighting by inspection of the letter designations of the various levels is adequate for most purposes. A graphical plot is a convenient aid in carrying out an approximate weighting.

Sample Problems

The problems below illustrate both numerical and approximate procedures (HCM Problem 9.8).

Problem 1:

Determine the weighted level of service for the four-lane freeway section shown in Figure 12-21, which has already determined level of service measures as shown (PHF = 0.77).

Solution:

v/c weighting:

$$\begin{array}{rcl} 0.45 \times 2 &=& 0.90 \\ 0.50 \times 3 &=& 2.00 \\ 0.66 \times 2 &=& 1.32 \\ 0.45 \times 3 &=& 1.35 \\ 0.55 \times \underline{2} &=& \underline{1.10} \\ 13 & & 6.67 \end{array}$$

6.67/13 = 0.51, weighted v/c ratio; in Level C.

Operating speed weighting:

$$\begin{array}{rcl} 57 \times 2 &=& 14 \\ 58 \times 4 &=& 232 \\ 43 \times 2 &=& 86 \\ 61 \times 3 &=& 183 \\ 53 \times \underline{2} &=& \underline{106} \\ 13 & & 721 \end{array}$$

721/13 = 55.5 miles per hour, weighted operating speed; in Level B

Overall level of service is C, governed by v/c ratio. (This is only 0.01 over the dividing line, while operating speed is slightly better than limit of level B.)

Fig. 12-21. (Illustration of the sample freeway)

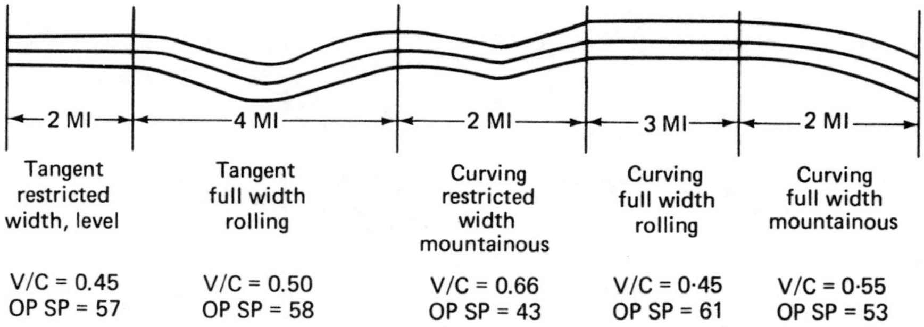

2 MI — Tangent restricted width, level — V/C = 0.45, OP SP = 57

4 MI — Tangent full width rolling — V/C = 0.50, OP SP = 58

2 MI — Curving restricted width mountainous — V/C = 0.66, OP SP = 43

3 MI — Curving full width rolling — V/C = 0·45, OP SP = 61

2 MI — Curving full width mountainous — V/C = 0·55, OP SP = 53

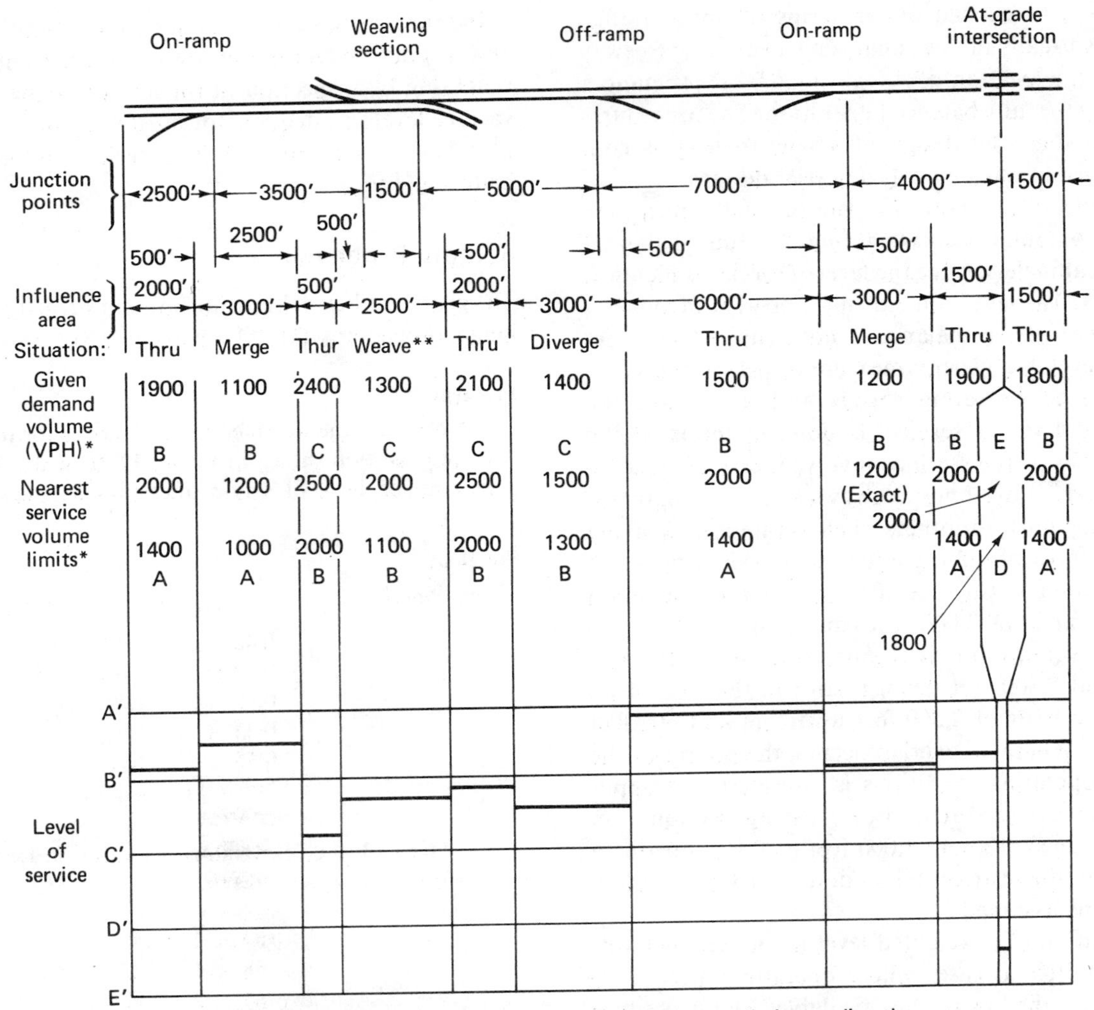

*Demand volumes and surrounding service volume limits are directly available from subsection - by - subsection analysis of elements

**Weave limits taken from chap. seven for major weaves, in figure 7.4, number of weaving vehicles rather than number of lanes found to control.

Fig. 12-22. (The determination of level of service)

Fig. 12-23. Delay characteristics of a bottleneck (*Source: Newman,* Introduction to Capacity, *California Division of Highways*)

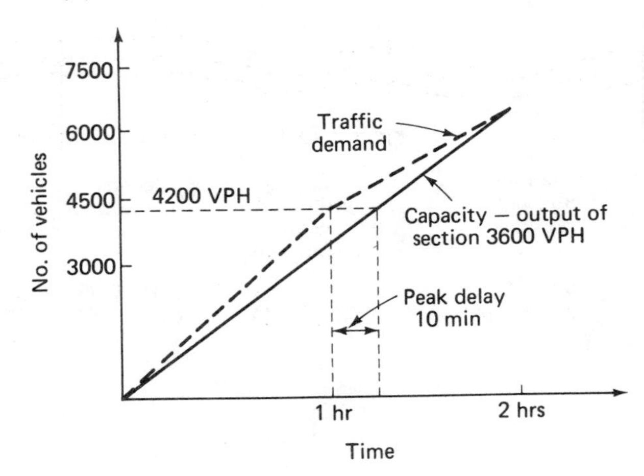

Problem 2:

Determine the approximate overall Level of service for the freeway and expressway section shown, which has a variety of elements without readily available numerical level of service measures. (PHF = 0.83, geometrics are ideal, and there are no trucks.)

Solution:

See Figure 12-22. Average overall level of service is approximately the limit of B, with the exception of the at-grade intersection, which is very close to capacity. This point restriction is so far from agreement with the remaining operation that it should be reported separately, rather than averaged.

Capacity and Delay—The Bottleneck[46]

A *bottleneck* occurs on a freeway when demand for service on a particular point or subsection exceeds the capacity of that section. This will occur where one point or subsection of the facility has a lower capacity than adjoining sections. When the capacity of that point is reached and demand surpasses it, vehicles will queue behind the section waiting for service, blocking adjoining sections, even though their capacity has not been reached.

For example, assume that a two-mile section of roadway has a capacity of 3,600 vehicles per hour, and for two hours cars arrive at exactly this rate. The average speed would be on the order of 30 miles per hour. In terms of expected urban peak conditions, this speed would be acceptable to most drivers, and there would be no delay associated with the operation of the facility at that speed. If, however, 30-miles per hour operation were compared to expected rural conditions or driver desires (60 miles per hour +), some delay would be assigned to the operation through the two-mile section. This delay would be 2 miles/30 mile-per-hour — 2 miles/60 miles per hour, or $\frac{1}{30}$ hour (2 minutes) per vehicle. However, cars rarely arrive at a uniform rate, matching capacity exactly for two hours. Now assume that traffic demand for the first hour is 4,200 vehicles, and 3,000 vehicles for the second hour. Cars will immediately begin to back up. At the end of the first hour, 600 cars will be stored waiting to get through the section. The 4,200th car will be delayed for 10 minutes, since there are 600 cars ahead of this vehicle getting through at a rate of 3,600 vehicle per hour. It will take until the end of the second hour to dissipate the congestion. Average delay through the section is 5 minutes. This is graphically shown in Figure 12-23. Bottlenecks on freeways are to a certain extent self-perpetuating. As noted in Chapter 11, while the capacity of a lane is 2,000 passenger cars per hour under ideal conditions with moving traffic, vehicles do not depart from standing queues at rates greater than 1,500 passenger cars per hour. Thus, given that a bottleneck has caused a breakdown, the queue will not begin to dissipate until demand at the bottleneck drops below 1,500 passenger cars per hour. Whereas prior to the breakdown the section may have accommodated up to 2,000 passenger cars per hour, once breakdown occurs only 1,500 passenger cars per hour may be accommodated until the standing queue has dissipated.

TABLE 12-14. CAPACITIES UNDER LANE CLOSURE*

Normal Number of Lanes	2	3–4	4
Lanes Open	1	2	3
Median Barrier or Guard Rail Repair	1,500	3,200	4,800
Pavement Repair	1,400	3,000	4,500
Striping	1,200	2,600	4,000
Pavement Markers	1,100	2,400	3,600
Middle Lane Closure for any reason		2,200	3,400

(Source: Kenrede and Myra "Freeway Lane Closures," in *TE*, February, 1970)

* Capacities are given in terms of total passenger cars per hour.

Freeway Lane Closures

Often it become necessary to close a portion of a freeway lane because of road repairs or accidents. A study[47] was conducted to measure capacity effects under conditions of lane closure. Results are given in Table 12-14.

Freeway Traffic Surveillance and Control Systems

Many limited-access facilities experience breakdown, or Level of Service F, operation during urban peak periods of travel. In many cases, these situations cannot be corrected by new or remedial construction. In such cases, freeway surveillance and ramp control may be used to maintain capacity operation of the facility.

It has been noted previously that capacity, or Level of Service E, operation is highly unstable, and that any interruption to flow will cause an "accordion effect" breakdown into Level of Service F. On congested urban freeways, such interruptions are common as vehicles enter jammed traffic streams from ramps. Freeway surveillance and control is a mechanism whereby the gaps in the traffic stream are observed using different types of equipment. Gap observance information is translated to ramp metering signals, which permit vehicles to enter the stream only when a gap is available. By preventing disruptions to the freeway stream, it is hoped that the degeneration of Level of Service E into Level of Service F operation may be prevented or forestalled. In a complete discussion of control aspects, equipment, and an evaluation of existing projects in freeway surveillance and control, see Chapter 25 and the literature.[48-59]

CONTINUING RESEARCH

The increasing importance of limited-access facilities in the highway networks of the United States was a principal factor in the need for revising the 1950 *Highway Capacity Manual.* Since the publication of the 1965 Manual, the sections regarding limited-access facilities have undergone heavy use and close scrutiny. Recognizing the need for further study the National Cooperative Highway Research Program (NCHRP) has sponsored a project involving an evaluation and revision of the weaving and ramp capacity procedures with regard to weaving operations. A final report on the initial phase of that project is available.[60]

Two principal findings of that study are of interest in the application of current capacity procedures:

1. Despite the fact that the Manual recommends the use of the *ramp* procedure for situations involving weaving between adjacent on- and off-ramps, the use of *weaving* procedures tends to produce more accurate estimates of resultant levels of service.
2. For cases of an on-ramp followed by an off-ramp with an auxiliary lane, the Levels A–C ramp procedure produced more accurate lane 1 volume estimates than the Levels D–E procedure for *all* levels of service.

The study also disclosed major internal difficulties in the weaving procedures of the 1965 Manual; principally, there were inconsistencies in the specifications of quality of flow and level of service, and difficulties in the concept and calibration of the k expansion factor. These difficulties led to the formulation of a new weaving algorithm in the form of a series of multidimensional equations involving V_w, L, and W (W = number of lanes utilized by weaving traffic).

Work is continuing on the calibration and use of the new algorithm, and development of a new procedure is expected to be completed in late 1973.

Application of procedures of the *Highway Capacity Manual* has been slightly revised herein as a result of the above research.

13

Highway Capacity: Urban Streets and Arterials

The operation of city streets and urban arterials differs greatly from that of freeways and expressways. These facilities are not designed exclusively to favor the movement of through traffic, but to provide access to abutting lands as well. All or most of the intersections are at-grade, causing interruptions to flow. Parking, driveway entrances, city buses, turning movements, pedestrians, and similar factors further inhibit flow.

As was the case in freeway analysis, the overall operation of a street or arterial cannot be adequately evaluated until each subsection or point of interest is investigated. For city streets and urban arterials, these points of interest include intersections and mid-block interferences. No specific method of analysis has been developed for mid-block interferences, since these are often singular cases which must be evaluated. The *Highway Capacity Manual*[1] presents a methodology for the analysis of signalized intersections, as well as the combined analysis of downtown streets and urban and suburban arterials. There has been a negligible amount of work done in the area of unsignalized intersections. It can be generally assumed that any intersection with a significant capacity problem would be signalized but more research in the area of unsignalized intersections is certainly needed.

CAPACITY OF SIGNALIZED INTERSECTIONS

Highway Capacity Manual Method

The capacity of an intersection is important only in its effect on the streets or arterials it serves. For this reason, the capacity of an intersection is discussed in terms of the capacity of each intersection approach. Thus, for one intersection, there are generally four capacities to consider.

Unadjusted Capacity and Service Volumes for Standard Conditions. The *Highway Capacity Manual* presents six figures for unadjusted service volumes which reflect several important factors:

1. Unadjusted service volumes are given in terms of vehicles per hour of green time. The values given would be the capacity or service volume if the approach leg had a green signal indication for the entire hour.
2. Street width in feet is used as a determinant, rather than the number of lanes. The *Highway Capacity Manual* data suggest that an approach of 40 feet would have the same capacity whether it was striped for four ten-foot lanes or five eight-foot lanes. This point has been disputed in several studies[2,3] and other capacity methods, some of which are discussed below, have used the number of lanes rather than the width in feet.

TABLE 13-1. LEVELS OF SERVICE VERSUS LOAD FACTORS

LEVELS OF SERVICE	*TRAFFIC FLOW DESCRIPTION*	*LOAD FACTOR*
A	Free Flow	0.0
B	Stable Flow	0.1
C	Stable Flow	0.3
D	Approaching Unstable Flow	0.7
E*	Unstable Flow	1.0
F	Forced Flow	**

(Source: *Highway Capacity Manual,* HRB SR 87, 1965, Table 6-3)
* Capacity.
** Not applicable.

3. Unadjusted service volumes are given in terms of mixed vehicles per hour of green for *standard conditions,* which include 5 per cent trucks and intercity buses, 10 per cent left-turning vehicles, 10 per cent right turning vehicles, and no local transit buses.

   For conditions other than these, correction factors must be applied as multipliers.

4. Level of service is correlated to the *load factor.* The load factor is defined as the percentage of fully utilized green phases within the peak hour. The correlation between level of service and load factor is shown in Table 13-1. Arguments have been raised as to whether or not the load factor is a good indicator of level of service. Delay has been proposed as an alternative. Studies have shown[4,5] that vehicle delay and load factor do not correlate well. Load factor is used since it is easier to measure than delay. A load factor of unity is almost impossible to achieve, making a factor of 0.85 more representative of Level of Service E, or capacity.

   Based on a simulation study, vehicle delay was correlated with load factor.[6] The results given in Table 13-2 were obtained.

   Because of the small difference in delay for Levels B and C on the one hand, and the large disparity in delay for Levels C and D on the other hand, it was suggested that the load factor limits be revised to provide for more consistent results. The preliminary recommendations are shown in Table 13-3. Further studies involving the relationships between load factor, cycle failure rate, and delay have been conducted.[4,32]

TABLE 13-2. CORRELATION OF LOAD FACTOR AND DELAY

LEVEL OF SERVICE	*LOAD FACTOR*	*AVERAGE INDIVIDUAL DELAY (sec/veh)*
A	0.0	≤ 11
B	≤ 0.1	≤ 15
C	≤ 0.3	≤ 19
D	≤ 0.7	≤ 32–55
E	≤ 1.0	≤ 32–55

(Source: May and Pratt, "A Simulation Study of Load Factor at Signalized Intersections," in *TE,* February, 1968)

TABLE 13-3. REVISED LEVEL OF SERVICE STANDARDS-PRELIMINARY RECOMMENDATIONS

LEVEL OF SERVICE	*REVISED LOAD FACTOR LIMITS*	*AVERAGE INDIVIDUAL DELAY (sec/veh)*
A	≤ 0.1	≤ 15
B	≤ 0.58–0.66	≤ 30
C	≤ 0.66–0.82	≤ 45
D	≤ 0.72–0.91	≤ 60
E	≤ 1.0	≤ 60

(Source: May and Pratt, "A Simulation Study of Load Factor at Signalized Intersections," in *TE,* February, 1968)

5. The six basic figures for service volume, Figures 13-1 through 13-6, represent different combinations of parking regulations and one- and two-way operation. Both of these have a marked effect on the capacity of the approach leg. Table 13-6 is used for rural intersections.

Adjustments to Capacity or Service Volume for Standard Conditions.

Adjustment for PHF and metropolitan area size

Figures 13-1 through 13-6 also contain tables which give adjustment factors for the combined effect of the PHF and city size. Data revealed that larger cities exhibited markedly better capacities at intersections, due probably to better traffic and pedestrian controls and driver familiarity with congested conditions in the larger cities. Peaking characteristics (PHF) within the peak hour must be accounted for if the approach is not to be overloaded for a portion of the peak hour.

Adjustment for location within city

Figures 13-1 through 13-6 also contain tables which give adjustment factors reflecting the effect of location within the city. Four classifications are used:

1. Central business district (CBD)
2. Fringe of CBD
3. Outlying business district
4. Residential area

Service volumes in the CBD are generally lowest due to the increase in pedestrian conflicts and greater numbers of turning movements.

Adjustment for green time

Since unadjusted service volumes are given in terms of vehicles per hour of green signal time, a factor must be applied to convert this to a total in vehicles per hour. This is done by multiplying the service volume in vehicles

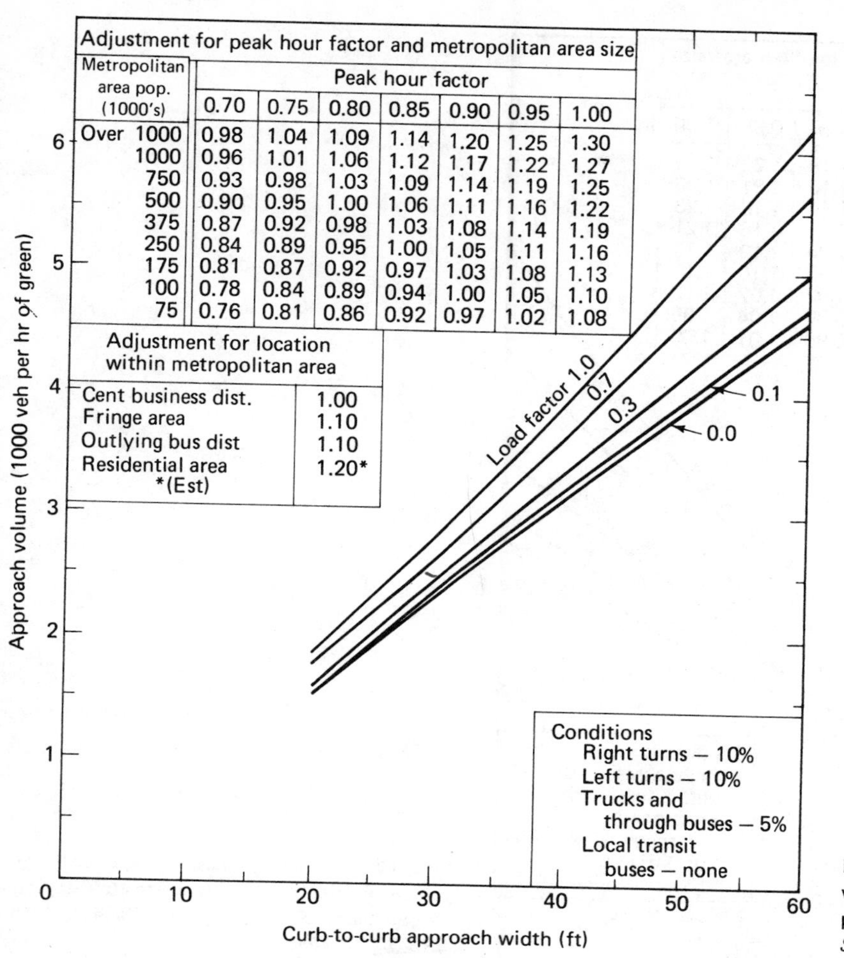

Adjustment for peak hour factor and metropolitan area size

Metropolitan area pop. (1000's)	Peak hour factor 0.70	0.75	0.80	0.85	0.90	0.95	1.00
Over 1000	0.98	1.04	1.09	1.14	1.20	1.25	1.30
1000	0.96	1.01	1.06	1.12	1.17	1.22	1.27
750	0.93	0.98	1.03	1.09	1.14	1.19	1.25
500	0.90	0.95	1.00	1.06	1.11	1.16	1.22
375	0.87	0.92	0.98	1.03	1.08	1.14	1.19
250	0.84	0.89	0.95	1.00	1.05	1.11	1.16
175	0.81	0.87	0.92	0.97	1.03	1.08	1.13
100	0.78	0.84	0.89	0.94	1.00	1.05	1.10
75	0.76	0.81	0.86	0.92	0.97	1.02	1.08

Adjustment for location within metropolitan area

Location	Adjustment
Cent business dist.	1.00
Fringe area	1.10
Outlying bus dist	1.10
Residential area	1.20*

*(Est)

Fig. 13-1. Urban intersection approach service volume, in VPHG, for one-way streets with no parking (*Source:* Highway Capacity Manual, *HRB SR 87, 1965, Figure 6.5*)

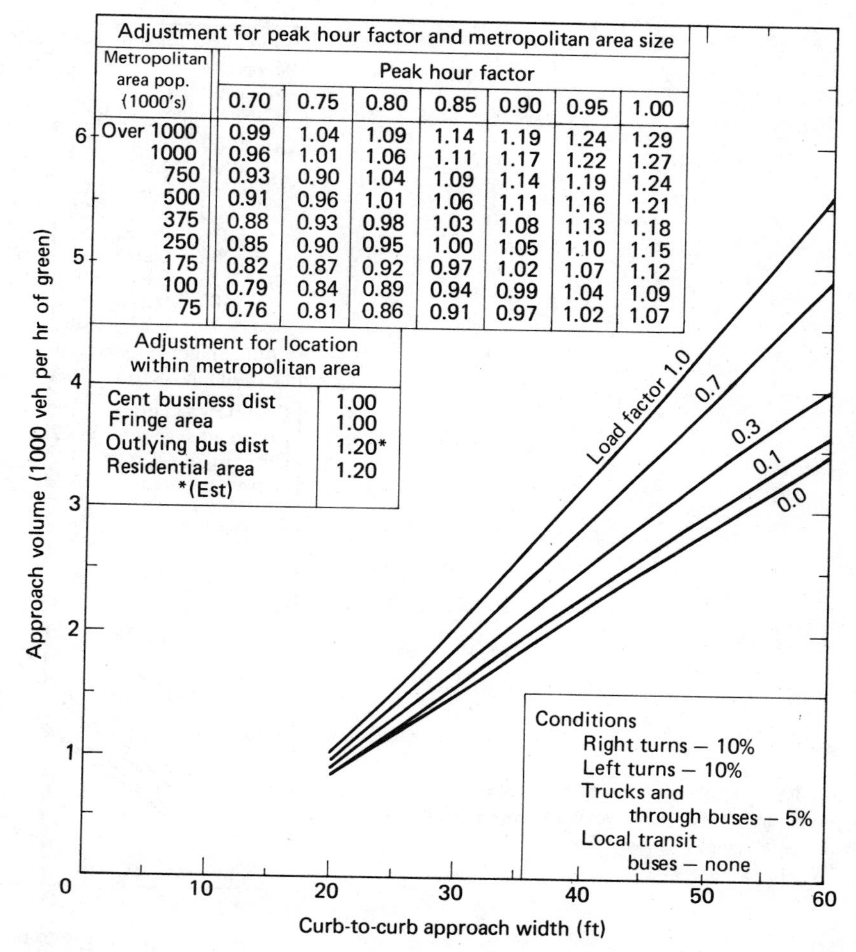

Adjustment for peak hour factor and metropolitan area size

Metropolitan area pop. (1000's)	Peak hour factor 0.70	0.75	0.80	0.85	0.90	0.95	1.00
Over 1000	0.99	1.04	1.09	1.14	1.19	1.24	1.29
1000	0.96	1.01	1.06	1.11	1.17	1.22	1.27
750	0.93	0.90	1.04	1.09	1.14	1.19	1.24
500	0.91	0.96	1.01	1.06	1.11	1.16	1.21
375	0.88	0.93	0.98	1.03	1.08	1.13	1.18
250	0.85	0.90	0.95	1.00	1.05	1.10	1.15
175	0.82	0.87	0.92	0.97	1.02	1.07	1.12
100	0.79	0.84	0.89	0.94	0.99	1.04	1.09
75	0.76	0.81	0.86	0.91	0.97	1.02	1.07

Adjustment for location within metropolitan area

Location	Adjustment
Cent business dist	1.00
Fringe area	1.00
Outlying bus dist	1.20*
Residential area	1.20

*(Est)

Fig. 13-2. Urban intersection approach service volume, in VPHG for one-way streets with parking on one side (*Source:* Highway Capacity Manual, *HRB SR 87, 1965, Figure 6.6*)

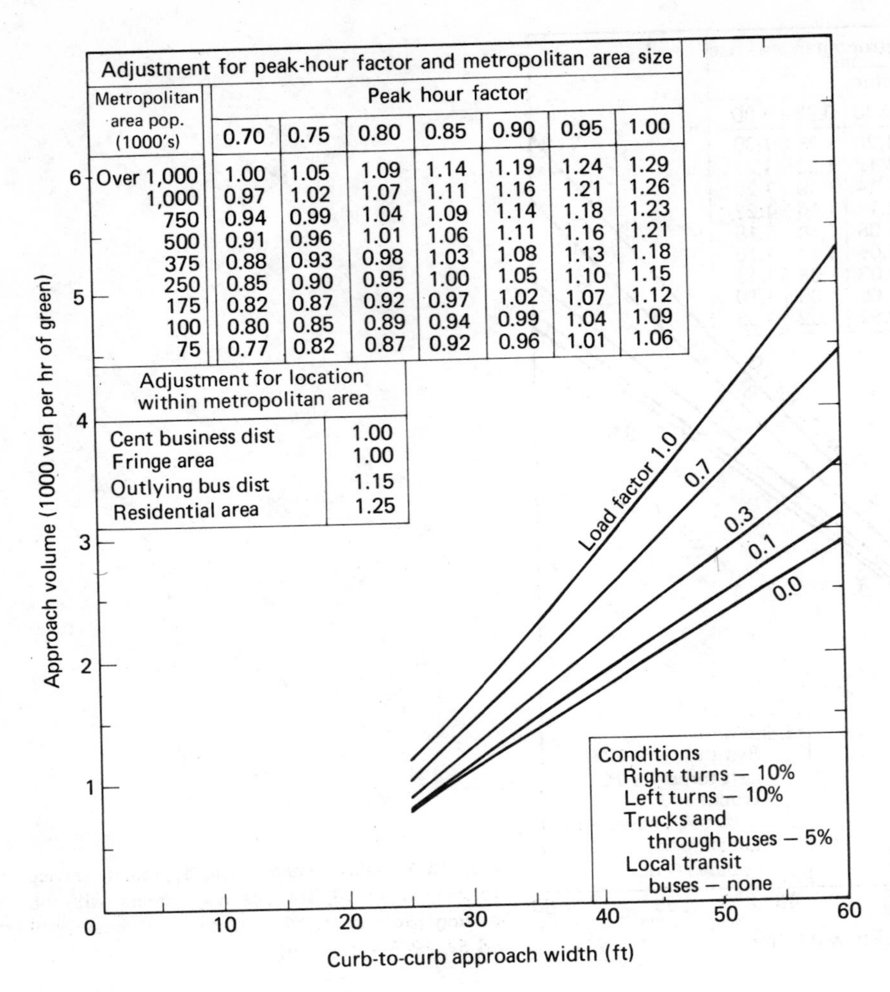

Metropolitan area pop. (1000's)	Peak hour factor 0.70	0.75	0.80	0.85	0.90	0.95	1.00
Over 1,000	1.00	1.05	1.09	1.14	1.19	1.24	1.29
1,000	0.97	1.02	1.07	1.11	1.16	1.21	1.26
750	0.94	0.99	1.04	1.09	1.14	1.18	1.23
500	0.91	0.96	1.01	1.06	1.11	1.16	1.21
375	0.88	0.93	0.98	1.03	1.08	1.13	1.18
250	0.85	0.90	0.95	1.00	1.05	1.10	1.15
175	0.82	0.87	0.92	0.97	1.02	1.07	1.12
100	0.80	0.85	0.89	0.94	0.99	1.04	1.09
75	0.77	0.82	0.87	0.92	0.96	1.01	1.06

Adjustment for location within metropolitan area	
Cent business dist	1.00
Fringe area	1.00
Outlying bus dist	1.15
Residential area	1.25

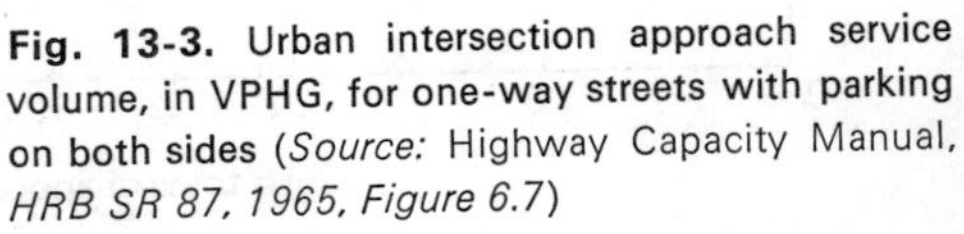

Fig. 13-3. Urban intersection approach service volume, in VPHG, for one-way streets with parking on both sides (*Source:* Highway Capacity Manual, *HRB SR 87, 1965, Figure 6.7*)

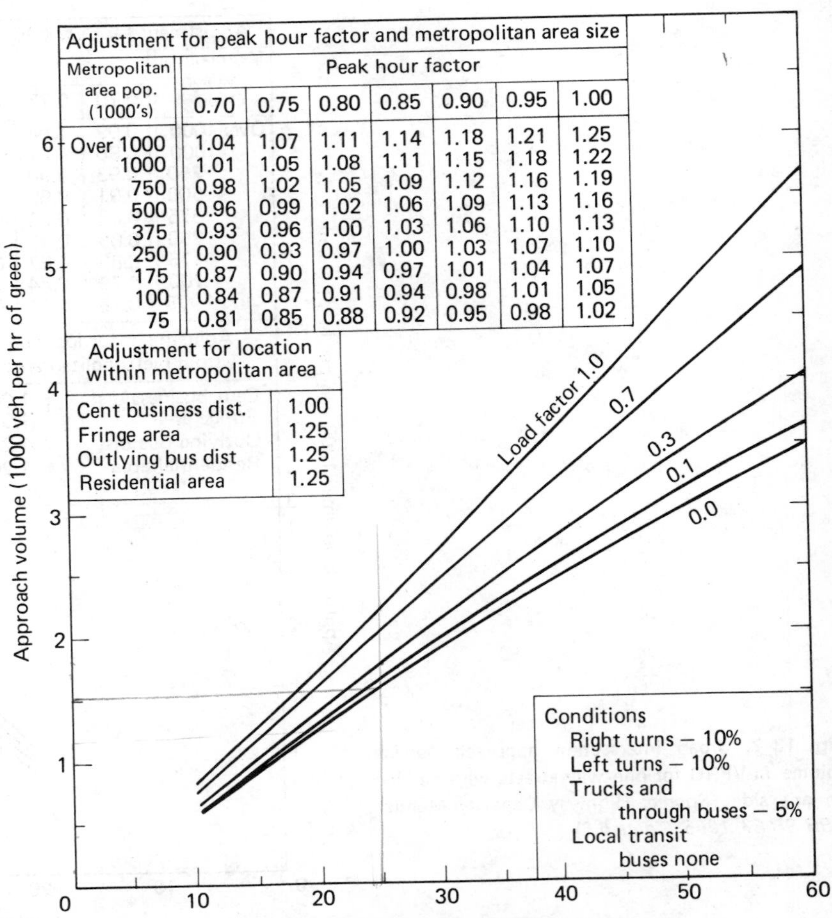

Metropolitan area pop. (1000's)	Peak hour factor 0.70	0.75	0.80	0.85	0.90	0.95	1.00
Over 1000	1.04	1.07	1.11	1.14	1.18	1.21	1.25
1000	1.01	1.05	1.08	1.11	1.15	1.18	1.22
750	0.98	1.02	1.05	1.09	1.12	1.16	1.19
500	0.96	0.99	1.02	1.06	1.09	1.13	1.16
375	0.93	0.96	1.00	1.03	1.06	1.10	1.13
250	0.90	0.93	0.97	1.00	1.03	1.07	1.10
175	0.87	0.90	0.94	0.97	1.01	1.04	1.07
100	0.84	0.87	0.91	0.94	0.98	1.01	1.05
75	0.81	0.85	0.88	0.92	0.95	0.98	1.02

Adjustment for location within metropolitan area	
Cent business dist.	1.00
Fringe area	1.25
Outlying bus dist	1.25
Residential area	1.25

Fig. 13-4. Urban intersection approach service volume, in VPHG, for two-way streets with no parking (*Source:* Highway Capacity Manual, *HRB SR 87, 1965, Figure 6.8*)

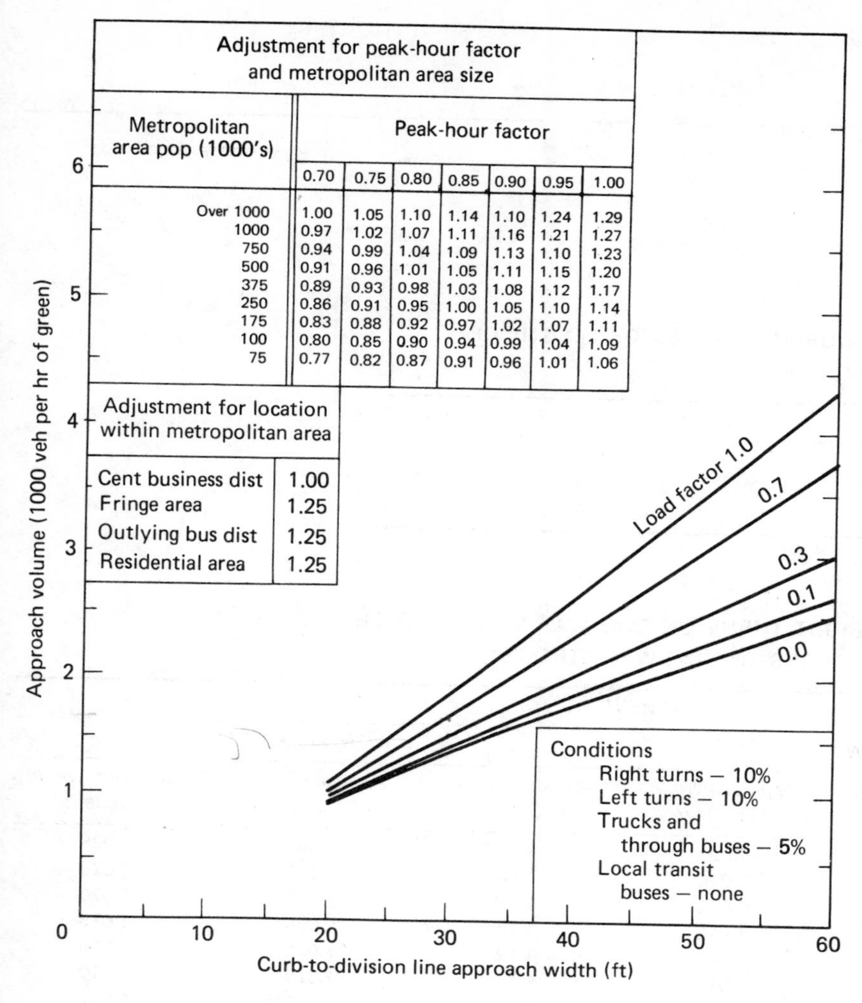

Metropolitan area pop (1000's)	Peak-hour factor						
	0.70	0.75	0.80	0.85	0.90	0.95	1.00
Over 1000	1.00	1.05	1.10	1.14	1.10	1.24	1.29
1000	0.97	1.02	1.07	1.11	1.16	1.21	1.27
750	0.94	0.99	1.04	1.09	1.13	1.10	1.23
500	0.91	0.96	1.01	1.05	1.11	1.15	1.20
375	0.89	0.93	0.98	1.03	1.08	1.12	1.17
250	0.86	0.91	0.95	1.00	1.05	1.10	1.14
175	0.83	0.88	0.92	0.97	1.02	1.07	1.11
100	0.80	0.85	0.90	0.94	0.99	1.04	1.09
75	0.77	0.82	0.87	0.91	0.96	1.01	1.06

Adjustment for location within metropolitan area	
Cent business dist	1.00
Fringe area	1.25
Outlying bus dist	1.25
Residential area	1.25

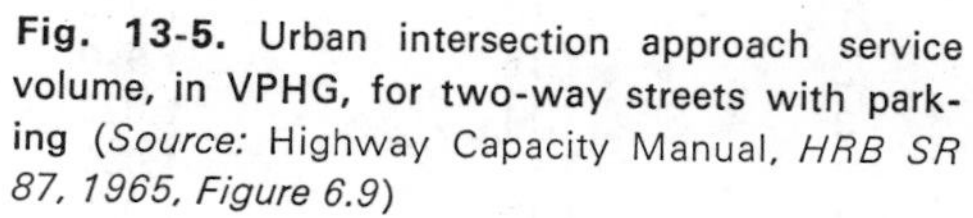
Fig. 13-5. Urban intersection approach service volume, in VPHG, for two-way streets with parking (*Source:* Highway Capacity Manual, *HRB SR 87, 1965, Figure 6.9*)

Adjustment for peak hour factor	
Peak-hour factor = 0.70 (average rural conditions)	Peak-hour factor = 1.00 (recreation route during peak flow)
Adjustment = 1.00	Adjustment = 1.40

Approach volume (1000 veh per hr of green)

Load factor 1.0

0.7

0.3

0.1

0.0

Conditions
Right turns — 10%
Left turns — 10%
Trucks and through buses — 5%

0 10 20 30 40 50 60

Curb-to-division-line approach width (ft)

Fig. 13-6. Rural intersection approach service volume, in VPHG, for two-way highways with no parking on the traveled way (*Source:* Highway Capacity Manual, *HRB SR 87, 1965, Figure 6.10*)

TABLE 13-4. NORMAL RANGES OF LENGTH OF GREEN

LF	*LENGTH OF GREEN PHASE (sec)*
0.0–0.1	15–20
0.1–0.3	25–35
0.3–1.0	40–60

TABLE 13-5. ADJUSTMENT FACTORS FOR RIGHT TURNS ON TWO-WAY STREETS*, RIGHT TURNS ON ONE-WAY STREETS, AND LEFT TURNS ON ONE-WAY STREETS*

*TURNS*** (%)	*ADJUSTMENT FACTOR*					
	WITH NO PARKING†			*WITH PARKING*‡		
	Approach Width ≤ *15* ft	*Approach Width 16 to 24 ft*	*Approach Width 25 to 34 ft*	*Approach Width* ≤ *20 ft*	*Approach Width 21 to 29 ft*	*Approach Width 30 to 39 ft*
0	1.20	1.050	1.025	1.20	1.050	1.025
1	1.18	1.045	1.020	1.18	1.045	1.020
2	1.16	1.040	1.020	1.16	1.040	1.020
3	1.14	1.035	1.015	1.14	1.035	1.015
4	1.12	1.030	1.015	1.12	1.030	1.015
5	1.10	1.025	1.010	1.10	1.025	1.010
6	1.08	1.020	1.010	1.08	1.020	1.010
7	1.06	1.015	1.005	1.06	1.015	1.005
8	1.04	1.010	1.005	1.04	1.010	1.005
9	1.02	1.005	1.000	1.02	1.005	1.000
10	1.00	1.000	1.000	1.00	1.000	1.000
11	0.99	0.995	1.000	0.99	0.995	1.000
12	0.98	0.990	0.995	0.98	0.990	0.995
13	0.97	0.985	0.995	0.97	0.985	0.995
14	0.96	0.980	0.990	0.96	0.980	0.990
15	0.95	0.975	0.990	0.95	0.975	0.990
16	0.94	0.970	0.985	0.94	0.970	0.985
17	0.93	0.965	0.985	0.93	0.965	0.985
18	0.92	0.960	0.980	0.92	0.960	0.980
19	0.91	0.955	0.980	0.91	0.955	0.980
20	0.90	0.950	0.975	0.90	0.950	0.975
22	0.89	0.940	0.980	0.89	0.940	0.980
24	0.88	0.930	0.985	0.88	0.930	0.985
26	0.87	0.920	0.990	0.87	0.920	0.990
28	0.86	0.910	0.995	0.86	0.910	0.995
30+	0.85	0.900	1.000	0.85	0.900	1.000

(Source: *Highway Capacity Manual*, HRB SR 87, 1965, Table 6.4)

* No separate turning lanes or separate signal indications.

** Handle right turns and left turns separately in all computations: do not sum.

† No adjustment necessary for approach width of 35 ft or more; that is, use factor of 1.000.

‡ No adjustment necessary for approach width of 40 ft or more; that is, use factor of 1.000.

per hour of green by G/C, where G is the length of green phase for the approach under consideration and C is the cycle length.

The cycle length has a marked effect on capacity. The shorter the cycle length, the greater the number of cycles per hour. For each cycle, a fixed amount of time is lost to the amber or clearance phase during which the intersection is not used. Thus, the larger the number of cycles per hour, the more time is lost to amber phases. For this reason, longer cycle lengths will result in higher capacities, assuming the percentage phase split remains unchanged. However, as cycle lengths are increased, delay to some vehicles also increases. Guide relationships between load factor and length of green phase are shown in Table 13-4.

Adjustment for turning movements

Basic Case—No separate turning lanes or signal phases

Unadjusted service volumes are given in terms of mixed vehicles for a basic case of 10 per cent left and 10 per cent right turns. Separate adjustment factors are necessary for any percentage of left or right turns other than 10 per cent.

Table 13-5 gives adjustment factors for right turns on two-way streets, and left turns or right turns on one-way streets. Table 13-6 gives factors for left turns on two-way streets. On intermediate-width streets (25–34 feet with no parking or 30–39 feet with parking) the effect of right turns is greatest at the 20 per cent level. Thereafter, the effect reduces as the right lane is used almost as an exclusive turning lane. The effect of turns on wide streets is less marked, and right turns on streets wider than 34 feet without parking, or 39 feet with parking have no effect.

All left turns against opposing traffic should be checked versus a limiting value of $V_L = (1{,}200 - V_o)$ passenger cars per hour of green time, where V_L is the maximum allowable left-turn capacity, and V_o is the volume of opposing traffic, total for all lanes, both expressed in terms of passenger cars per hour of green. V_L can never be lower than two passenger cars per cycle, which will make their turn on the amber phase. This check takes into account the fact that heavy opposing volume may prevent left turns, a factor ignored by the *Highway Capacity Manual.*

Separate turning lanes with signal control

Tables 13-5 and 13-6 do not apply to this case. The turning lane should be about twice as long as that length necessary to accommodate the average turning volume per cycle.

Through volume V_T (that volume not utilizing separate

TABLE 13-6. ADJUSTMENT FACTORS FOR LEFT TURNS ON TWO-WAY STREETS*

TURNS (%)	ADJUSTMENT FACTOR					
	WITH NO PARKING			WITH PARKING		
	Approach Width ≤ 15 ft	Approach Width 16 to 34 ft	Approach Width ≥ 35 ft	Approach Width ≤ 20 ft	Approach Width 21 to 39 ft	Approach Width ≥ 40 ft
0	1.30	1.10	1.050	1.30	1.10	1.050
1	1.27	1.09	1.045	1.27	1.09	1.045
2	1.24	1.08	1.040	1.24	1.08	1.040
3	1.21	1.07	1.035	1.21	1.07	1.035
4	1.18	1.06	1.030	1.18	1.06	1.030
5	1.15	1.05	1.025	1.15	1.05	1.025
6	1.12	1.04	1.020	1.12	1.04	1.020
7	1.09	1.03	1.015	1.09	1.03	1.015
8	1.06	1.02	1.010	1.06	1.02	1.010
9	1.03	1.01	1.005	1.03	1.01	1.005
10	1.00	1.00	1.000	1.00	1.00	1.000
11	0.98	0.99	0.995	0.98	0.99	0.995
12	0.96	0.98	0.990	0.96	0.98	0.990
13	0.94	0.97	0.985	0.94	0.97	0.985
14	0.92	0.96	0.980	0.92	0.96	0.980
15	0.90	0.95	0.975	0.90	0.95	0.975
16	0.89	0.94	0.970	0.89	0.94	0.970
17	0.88	0.93	0.965	0.88	0.93	0.965
18	0.87	0.92	0.960	0.87	0.92	0.960
19	0.86	0.91	0.955	0.86	0.91	0.955
20	0.85	0.90	0.950	0.85	0.90	0.950
22	0.84	0.89	0.940	0.84	0.89	0.940
24	0.83	0.88	0.930	0.83	0.88	0.930
26	0.82	0.87	0.920	0.82	0.87	0.920
28	0.81	0.86	0.910	0.81	0.86	0.910
30+	0.80	0.85	0.900	0.80	0.85	0.900

(Source: *Highway Capacity Manual,* HRB SR 87 1965, Table 6-5)

* No separate turning lanes or separate signal indications.

TABLE 13-7. TURNING LANE VOLUMES PER 10 FEET OF LANE WIDTH IN VEHICLES PER HOUR OF GREEN

LEVEL OF SERVICE	*L*	*ASSUMED PERCENTAGE OF TRUCKS*
A, B, C	800	5
D	1,000	5
E	1,200	5

(Source: *Highway Capacity Manual,* HRB SR 87, 1965)

turning lanes) is calculated as in the Basic Case above, using an approach width not including the width of the turning lane and a turning percentage of zero for the movement accommodated by the separate lane.

Left turn V_L or right turn V_R volume is evaluated separately, using the following expression:

$$V_L \text{ or } V_R = L \times \frac{G}{C} \times \frac{W}{10}[1 + 0.8(N - 1)][1 - 0.01(T - 5)] \quad (13\text{-}1)$$

where L = turn lane volume in vehicles per hour of green time per 10 feet of width. Values of L are tabulated in Table 13-7.
G/C = factor for separate signal phase to convert to vehicles per hour.
W = width of turning lane, in feet.
N = number of turning lanes
T = percentage of trucks using turning lane

Separate turning lanes without signal control

Through volume V_T is calculated as described in above. Right-turn volume V_R is determined as described above if pedestrians are not present. Where turns must be made simultaneously with pedestrian crossings, use $L = 600$ and follow the procedure above.

Left-turn volume V_L, for any level of service, is calculated as the difference between 1,200 and the total opposing traffic volume in terms of passenger cars per hour of green, but not less than two passenger cars per signal cycle. That is,

$V_L = (1{,}200 - V_o)$ passenger cars per hour of green time
V_o = opposing volume in passenger cars per hour of green time
$V_L \geq 2$ passenger cars per signal cycle
Total approach volume $= V_T + V_L + V_R$

Conversion of units from vehicles per hour of green time to passenger cars per hour of green time may be accomplished by multiplying the volume in vehicles per hour of green time by $1.00 + 0.01T$, where T is the percentage of trucks.

Separate signal control, no separate lanes

This condition exists in special cases. One example is where certain turning movements are permitted for times different from the phase length for through traffic by means of green arrow indications. Another example is the provision of leading or lagging green which permits turns free of opposing traffic for part, but not all, of the time.

Where there is opposing traffic, apply the basic intersection capacity computation procedures to the entire approach width, incrementally for each different combination of signal indications. Where left turns are unopposed, each such increment is computed by the basic procedure, but left turns are considered as left turns from one-way streets.

Adjustment for trucks and through buses

Table 13-8 presents adjustment factors which will convert the base condition of 5 per cent trucks to any existing percentage. Through buses are those which do not pick up or discharge passengers in the local area, and they are considered as trucks. The factor can be calculated as follows:

$$\text{Truck factor} = 1 - 0.01(T - 5)$$

TABLE 13-8. TRUCK AND THROUGH BUS ADJUSTMENT FACTORS

TRUCKS AND THROUGH BUSES (%)	*CORRECTION FACTOR*	*TRUCKS AND THROUGH BUSES* (%)	*CORRECTION FACTOR*	*TRUCKS AND THROUGH BUSES* (%)	*CORRECTION FACTOR*
0	1.05	7	0.98	14	0.91
1	1.04	8	0.97	15	0.90
2	1.03	9	0.96	16	0.89
3	1.02	10	0.95	17	0.88
4	1.01	11	0.94	18	0.87
5	1.00	12	0.93	19	0.86
6	0.99	13	0.92	20	0.85

(Source: *Highway Capacity Manual,* HRB SR 87, 1965, Table 6-6)

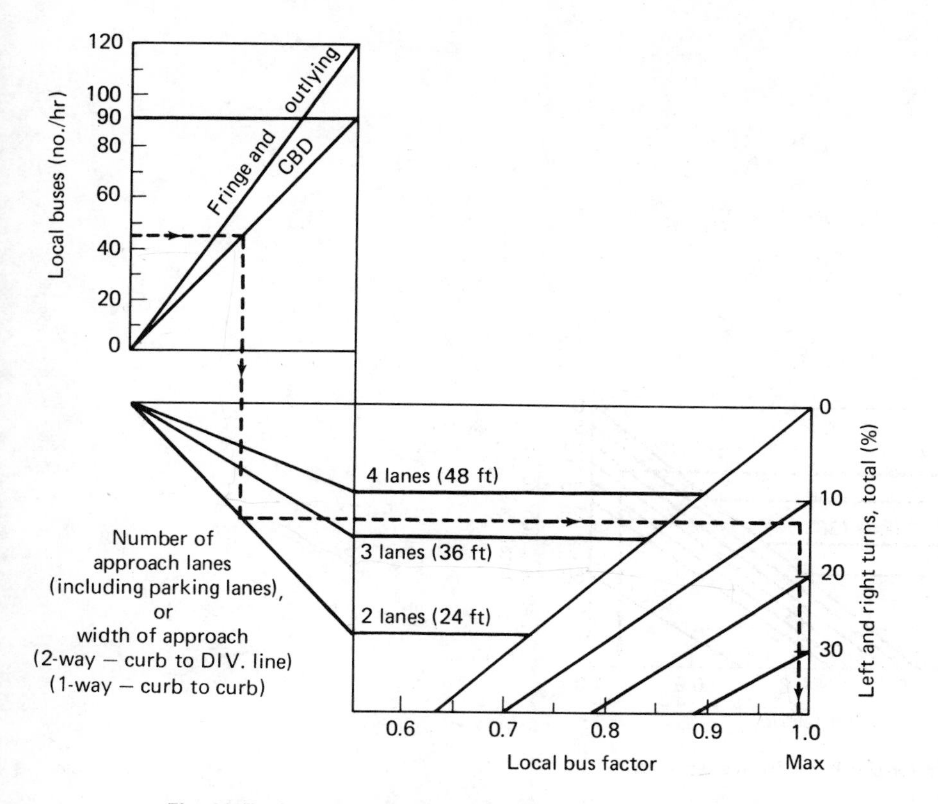

Fig. 13-7a. Local bus factor for far-side bus stop on street with parking

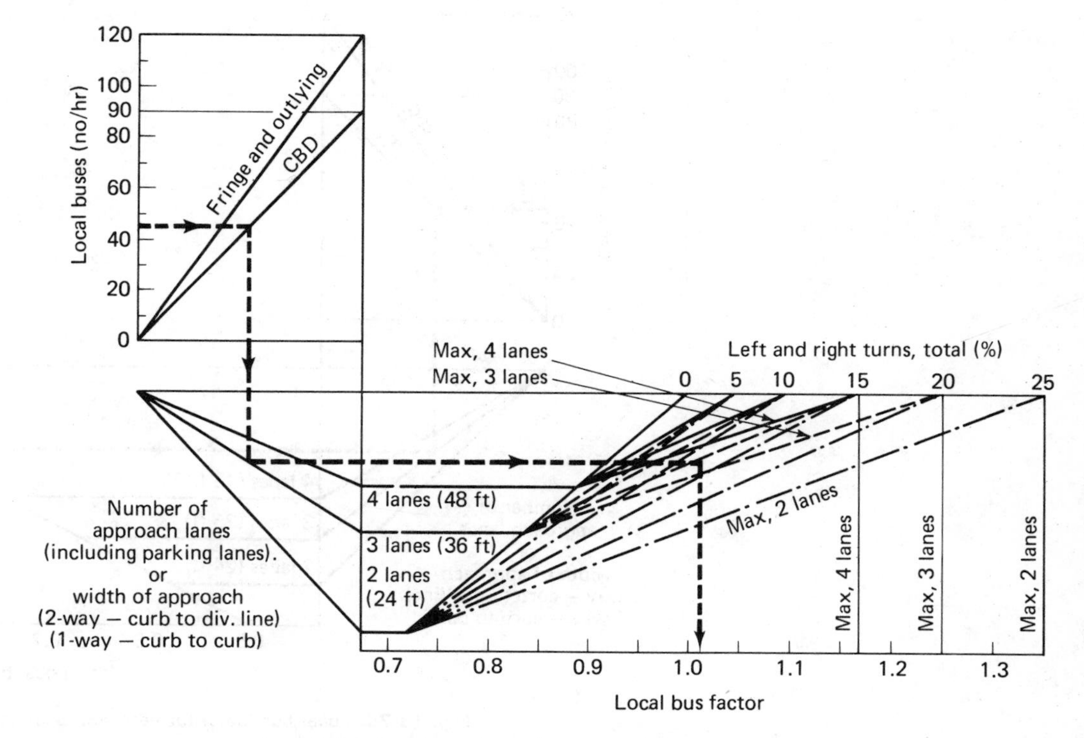

Fig. 13-7b. Local bus factor for near-side bus stop on street with parking

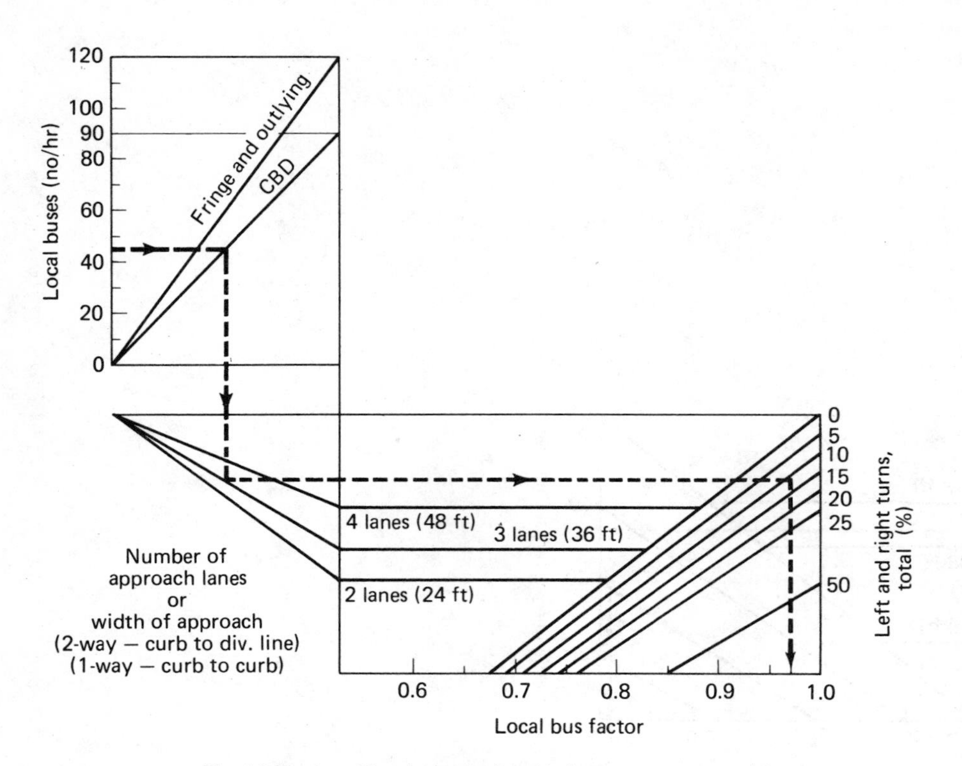

Fig. 13-7c. Local bus factor for far-side bus stop on street with no parking

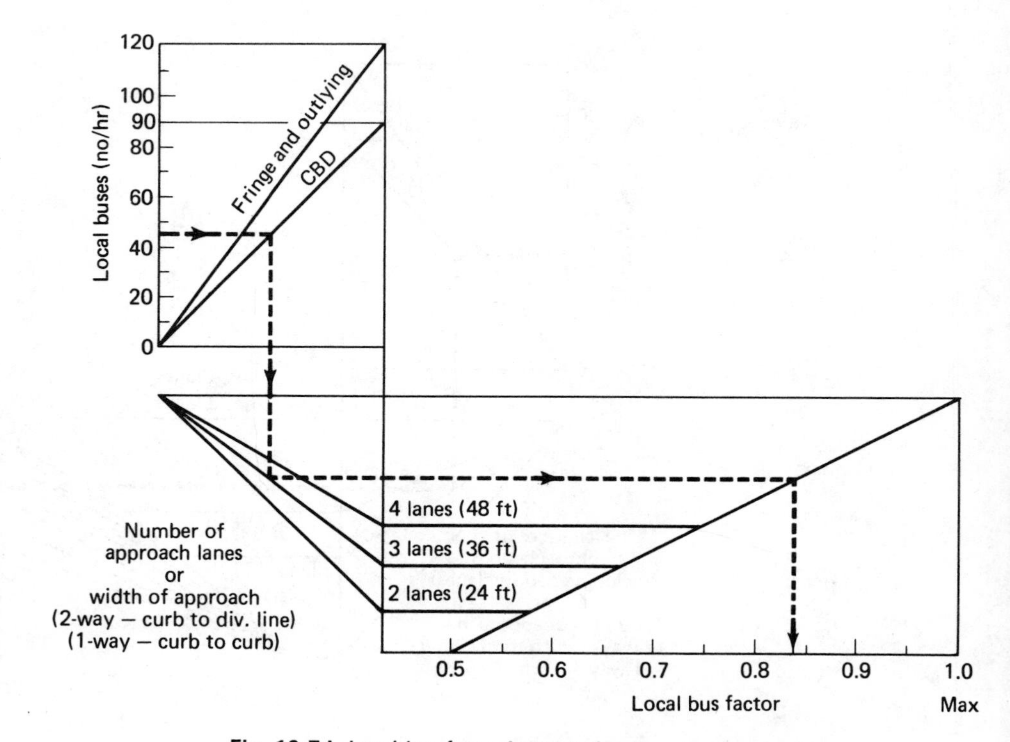

Fig. 13-7d. Local bus factor for near-side bus stop on street with no parking (*Source:* Highway Capacity Manual, *HRB SR 87, 1965, Figure 6.11–6.14*)

Adjustment for local buses

Figure 13-7 contains four nomographs for the selection of an appropriate factor. The nomographs represent combinations of parking regulations and near- or far-side of corner stops. Local buses are those that make regular stops along the street of interest.

Total Capacity.

No special turning cases

Where there are no special turning considerations (separate lanes, signals, etc.), the capacity or service volume of an intersection approach becomes

$$V_i = (USV_i) \times P \times L \times \frac{G}{C} \times LT \times RT \times T \times LB \tag{13-2}$$

where
- V_i = service volume in vehicles per hour for a specified level of service i (capacity = Level of Service E)
- USV_i = unadjusted service volume for level of service i
- P = correction factor for population and PHF
- L = correction factor for location within city
- G/C = ratio of green time to cycle length for approach in question
- LT = correction factor for left-turn percentage
- RT = correction factor for right-turn percentage
- T = correction factor for trucks and through buses
- LB = correction factor for local buses stopping within intersection

Special turning considerations

As discussed previously, where separate turning lanes and/or signals exist, the service volume or capacity of the separate lane is considered separately from that of the through lanes.

A through lane service volume, V_T, is computed as above, where the width of the turning lane is not included in the approach width, and the percentage of turns for the movement handled by the separate lane is zero. The capacity of a service volume of the separate lane (V_L if a left-turn lane, V_R if a right-turn lane) is computed as previously discussed).

The total service volume may be taken to be the sum: $V_{\text{Total}} = V_T + V_{L \text{ or } R}$. If however, the expected percentage of turning vehicles is known, this becomes unrealistic, as V_T and $V_{L \text{ or } R}$ need not be in the expected proportion. It can then be said that the total service volume is either

$$V_{\text{Total}} = \frac{V_T}{P_T} \tag{13-3}$$

where
- V_T = through service volume, including turns not made from separate lane
- P_T = percentage of vehicles not using separate turning lane

or

$$V_{\text{Total}} = \frac{V_{L \text{ or } R}}{P_{L \text{ or } R}} \tag{13-4}$$

where
- $V_{L \text{ or } R}$ = service volume of left- or right-turn lane
- $P_{L \text{ or } R}$ = percentage of vehicles utilizing left- or right-turn lane

The *minimum* of the two values is the most realistic indication of the total service volume which the approach may handle without breaking down below the level of service in question.

Applications of Procedures. The procedures outlined above provide for the solution of a variety of problems, including:

1. Operational studies, involving the determination of service volumes for various levels of service.
2. New designs, involving the determination of required approach width to provide a given level of service for a known demand volume.
3. System evaluations, involving determinations of level of service for existing conditions. This technique is used to identify deficient locations.
4. Design of new signalization schemes and the determination of cycle splits.

Nomographs for Intersection Capacity. The Bureau of Public Roads has made available a special issue of *Public Roads*, which contains a series of nomographs for the direct solution of intersection capacity as well as many illustrative examples. The special issue, called "Capacity Analysis Techniques for Design of Signalized Intersections" is a reprint of previously published nomographs.[7,8] The charts are prepared for design capacity, that is, Level of Service C in urban areas and B in rural areas. Conversion to other levels is provided for by a set of convenient factors. Similar nomographs were prepared for the first edition of the Manual.[9] Computer programs for intersection capacity have been developed.[10,11]

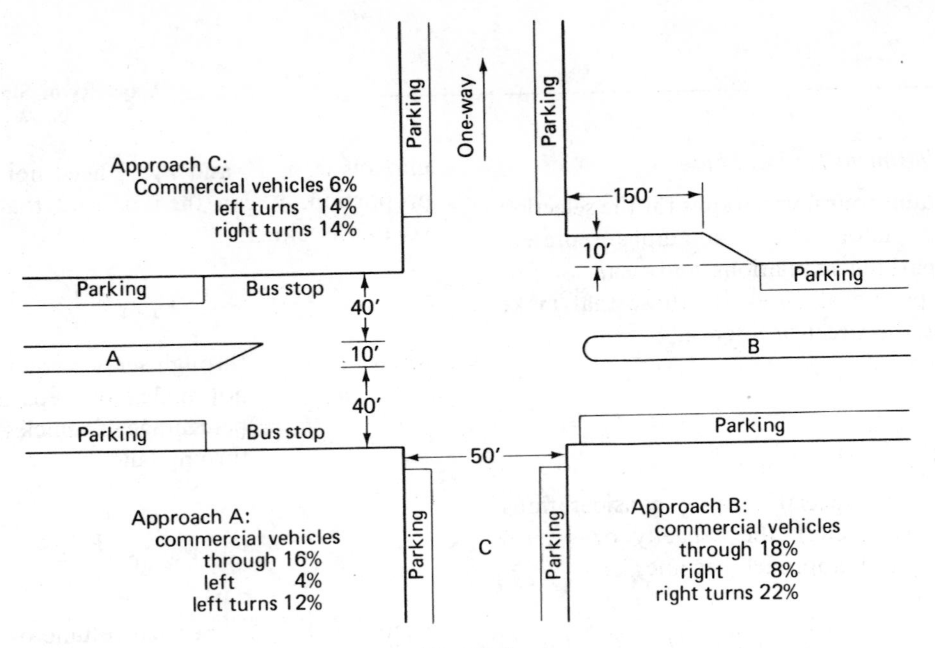

Fig. 13-8.

Sample Problems

Problem 1:

Determine the service volumes for Level of Service C for all approaches of the intersection shown in Figure 13-8. The intersection is located in the CBD of a metropolitan area having a population of 1,000,000. A directional distribution of traffic is 70 per cent-30 per cent (70 per cent eastbound). The PHF is 0.85. There are 60 local buses per hour on approach A and 20 per hour on approach B. The signalization is in three phases and is shown below in Figure 13-9. Cycle length is 75 seconds, and pedestrian interference is negligible.

Solution:

The basic method will be to find the basic unadjusted capacity for each approach leg, as predicted by the appropriate tables, and then add the necessary adjustment factors for such things as turns, trucks, buses, separate lanes or signal phases, etc. For each approach, it is advisable to isolate the pertinent factors before beginning.

1. *Approach A:*
   Separate left turn lane: 10 feet
   Left turn signal phase: 15 seconds of 75-second cycle
   Through signal phase: 45 seconds of 75-second cycle
   Through street width: 40 feet
   Turns: 12 per cent left
   0 per cent right
   Commercial vehicles: 16 per cent through
   4 per cent left
   Bus stop on near-side of corner: 60 per hour
   CBD: population 1,000,000
   PHF: 0.85
   No pedestrian interference
   Parking one side

The basic unadjusted service volume of the leg is obtained from Figure 13-5. Since left turns will be handled separately, we are now dealing with through service volume (V_T), left turns being taken as 0 per cent for this purpose. From Figure 13-5, for Level of Service C (load factor = 0.3), width is 40 feet, and

$$V_T = 2{,}000 \text{ vphg}$$

From Figure 13-5, the following correction factors are obtained:

For population: 1.11
CBD: 1.00

From Table 13-5 the correction for right-turning vehicles is obtained. Since Figure 13-5 was prepared for a base percentage of 10 per cent right-turning vehicles, it is expected that 0 per cent will result in a correction greater than, or equal to, unity:

Right-turn correction = 1.00

since the approach width of 40 feet requires no correction. From Table 13-6, the correction for left-turning vehicles is obtained. Using 0 per cent left-turning vehicles, it is again expected that a correction greater than, or equal to, unity will result:

Left-turn correction = 1.05

The correction for trucks is obtained from Table 13-8.
Truck correction factor (for 16 per cent trucks) = 0.89
The correction for local buses is obtained from Figure 13-7(b) (buses with near-side stop on street with parking).
Bus factor (for 60 buses per hour and total turns) = 0.88
Applying these,

$$V_T = 2{,}000 \text{ vehicles per hour of green time} \times \tfrac{45}{75} \times 1.11 \times 1.00 \times 1.05 \times 0.89 \times 0.88$$

$$V_T = 1{,}095 \text{ vehicles per hour}$$

Now, the left-turning movement must be considered, using the formula

$$V_L = L \times \frac{G}{C} \times \frac{W}{10}[1 + 0.8(N - 1)][1 - 0.01(T - 5)]$$

where $L = 800$ for Level of Service C (Table 13-7)
$W = 10$ feet, width of turning lane
$N = 1$, number of turning lanes
$T = 4$ per cent, percentage of left-turning trucks
$G/C = \frac{15}{75}$
$V_L = 800 \times \frac{15}{75} \times 10[1 + 0.8(1 - 1)][1 - 0.01(4 - 5)]$
$V_L = 800 \times 0.20 \times 1[1][1.01] = 162$ vehicles per hour

The total service volume might be taken as $V_T + V_L$, or $1{,}095 + 162 = 1{,}257$ vehicles per hour. However, it is known that the percentage of vehicles executing a left turn is 12 per cent. Therefore, through traffic makes up 88 per cent of the total, and service volume can be said to equal

$$\text{Service volume} = \frac{1{,}095}{0.88} = 1{,}244 \text{ vehicles per hour}$$

$$\text{or } \frac{162}{0.12} = 1{,}350 \text{ vehicles per hour}$$

Since 1,244 is the smaller of the two values, it is taken as the most reasonable estimate of service volume. The through capacity exerts the controlling influence.

2. *Approach B:*

Separate right turn lane: 10 feet
Right-turn signal phase: 45 seconds of 75-second cycle
Through signal phase: 27 seconds of 75-second cycle
Through street width: 40 feet
Turns: 22 per cent right
0 per cent left
Commerical vehicles: 18 per cent through
8 per cent right
Bus stop on far-side of corner: 20 per hour
CBD: population 1,000,000
PHF: 0.85
Parking one side
No pedestrian interference

Once again, the right-turn traffic is handled separately. Therefore, the service volume for through traffic is calculated first, taking right turns to be 0 per cent.

Unadjusted service volume = 2,000 vehicles per hour of green time (Figure 13-5). Adjustment factors from appropriate tables and figures are as follows:

Population:	1.11	(Figure 13-5)
CBD:	1.00	(Figure 13-5)
Left-turn adjustment:	1.05	(Table 13-6)
Right-turn adjustment:	1.00	(Table 13-5; no correction is needed for approach widths greater than 39 feet)
Truck adjustment (for 18 per cent trucks):	0.87	(Table 13-8)
Local bus adjustment (for far-side stop with parking, 20 buses):	0.99	[Figure 13-7(a)]

Through service volume,

$$V_T = 2{,}000(\tfrac{27}{75})(1.11)(1.05)(0.87)(0.97)$$
$$V_T = 708 \text{ vehicles per hour}$$

Now, the right-turning volume is considered:

$$V_R = L \times \frac{G}{C} \times \frac{W}{10}[1 + 0.8(N - 1)][1 - 0.01(T - 5)]$$

where $L = 800$
$G/C = \frac{45}{75}$
$W = 10$
$N = 1$
$T = 8$ per cent

Thus,

$$V_R = 800(\tfrac{45}{75})(\tfrac{10}{10})(1)(0.97) = 465 \text{ vehicles per hour}$$

Fig. 13-9.

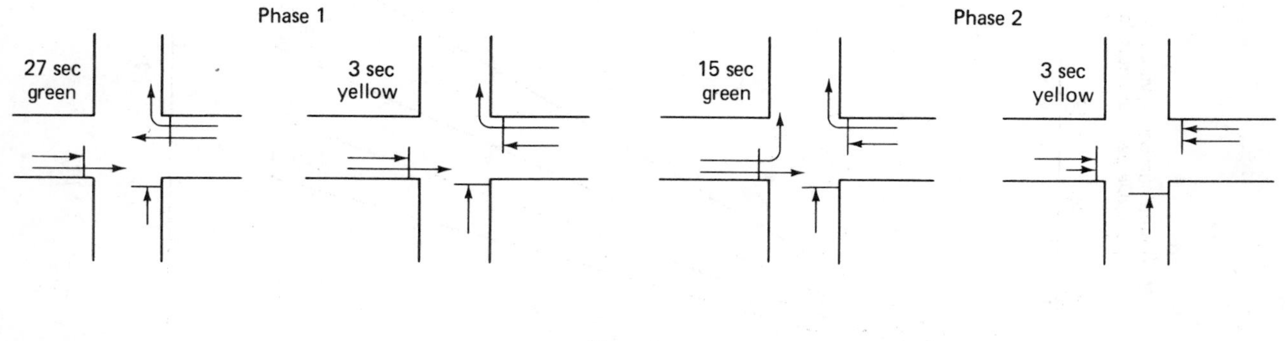

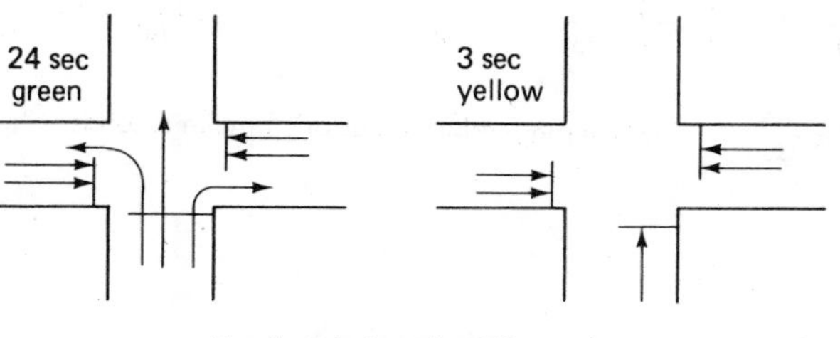

However, we know that only 22 per cent of the total traffic will execute the turn. Therefore, the total service volume will be

$$\text{Service volume} = \frac{708}{0.78} = 908 \text{ vehicles per hour}$$

$$\text{or } \frac{465}{0.22} = 2{,}113 \text{ vehicles per hour}$$

This means that 908 − 708 = 200 vehicles per hour will execute the turn. As this is below the maximum allowable values of 465 vehicles per hour, this is acceptable, and again excess turning capacity is evident.

3. *Approach C:*

One way
Width: 50 feet
No turning lanes or signals
No buses
Turns: 14 per cent left
Commercial vehicles: 6 per cent left
14 per cent right
Parking both sides
Signal phase: 24 seconds of 75-second cycle
CBD: population 1,000,000
PHF: 0.85
No pedestrian interference

Since there are no separate turning lanes or signals, these need not be considered separately. Correction factors may be obtained directly from the appropriate tables and figures.

Unadjusted service volume:	2,800 passenger cars per hour of green time (Figure 13-3)	
Population:	1.11	(Figure 13-3)
CBD:	1.00	(Figure 13-3)
Left-turn adjustment:	1.00	(Table 13-5)
Right-turn adjustment:	1.00	(Table 13-5; no adjustment is necessary for approach width greater than 39 feet)
Truck adjustment:	0.99	(Table 13-8)
Service volume:	$2{,}800(1.11)(0.99)(\frac{24}{75}) =$ 985 vehicles per hour	

Consider the service volume of Approach A if the signalization were as shown in Figure 13-10.

The through service volume is calculated as previously, except that the new G/C used. In this case, the signalization is such this is also the same as previously, so that $V_T = 1{,}095$ vehicles per hour, as before. Since the turning percentages are 12 per cent left and 0 per cent right, the total service volume, as before, is expressed as 1,095/0.88 = 1,244 vehicles per hour. This means that 149 vehicles per hour execute the left turn (1,244 − 1,095). However, left turn capacity must be taken as V_o − 1,200 passenger

Fig. 13-10. Bellis method capacity chart (*Source: Bellis, "Capacity of Traffic Signals and Traffic Signal Timing," HRBB 271, 1960*)

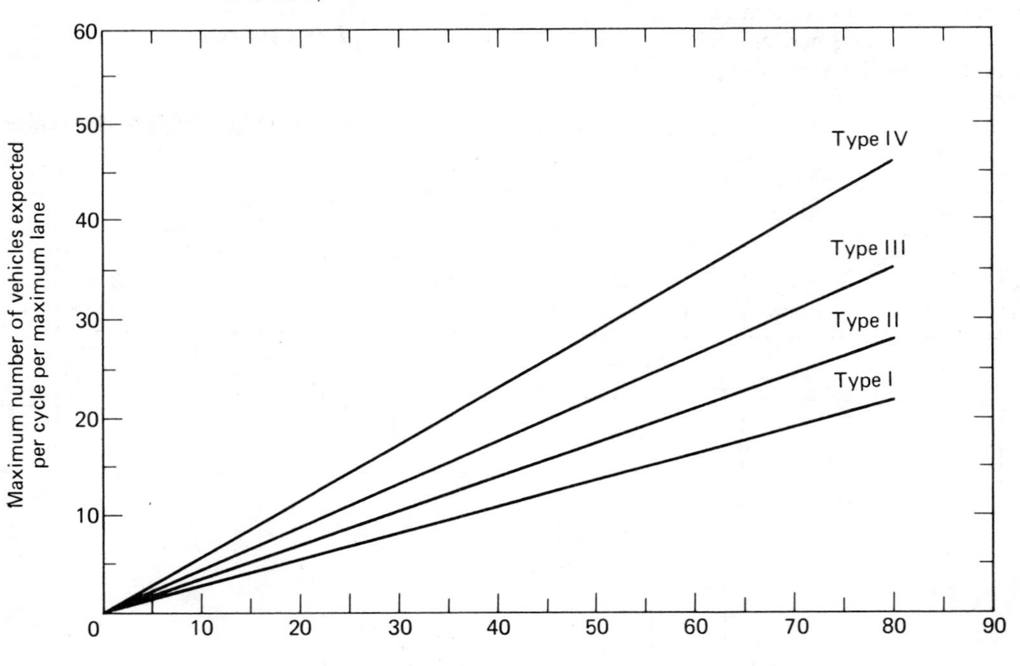

cars per hour of green time. The opposing volume is $\frac{3}{7}(1{,}244) = 533$ vehicles per hour. Of these, 22 per cent execute right turns and, therefore, do not oppose left turns from Approach A. Therefore, $V_o = 533(0.88) = 469$ vehicles per hour. Converting to passenger cars per hour of green time, $V_o = 469(\frac{75}{45})(1.18) = 922$ passenger cars per hour of green time. V_L may now be computed:

$V_L = 1{,}200 - 922 = 278$ passenger cars per hour of green time

$$V_L = \frac{278(\frac{45}{75})}{(1.18)} = 141 \text{ vehicles per hour}$$

Since 149, the actual number of turns at Level of Service C, is greater than 141, the capacity for left turns, a slight constriction, which could cause the intersection to break down, is evident. While the constriction is only 8 vehicles per hour, and the problem would not be expected to be critical, it is one which requires consideration of remedial measures.

Problem 2:

Consider the following one-way street approach:

Signal phase: 30 seconds green of 60-second cycle
One way
No parking
Population: 500,000
PHF: 0.85
CBD location
Street width: 36 feet
Turns: 10 per cent left
15 per cent right
Commercial vehicles: 8 per cent
Bus stop on near-side of corner: 30 local buses per hour
Volume now serviced: 1,500 vehicles per hour

Determine the level of service provided.

Solution:

In this problem, the volume is given. To determine the level of service, the unadjusted service volume (actual) in vehicles per hour of green time for standard conditions must be computed and compared with standards of the HCM. The various factors are computed as previously.

For 8 per cent commercial vehicles,

$T = 0.97$ (Table 13-8)

For 10 per cent left turns,

$LT = 1.00$ (Table 13-5)

For 15 per cent right turns

$RT = 1.00$ (Table 13-5)

For 30 local buses per hour,

$LB = 0.87$ [Figure 13-7(d)]

For 500,000 population and PHF 0.85,

$P = 1.06$ (Figure 13-1)

For CBD,

$L = 1.00$ (Figure 13-1)

$G/C = 0.50$ (given)

Consider the equation:

$$V = (USV) \times P \times L \times \frac{G}{C} \times LT \times RT \times T \times LB \qquad (13\text{-}5)$$

Since V is known, and USV is to be computed,

$$USV = \frac{V}{P \times L \times (G/C) \times LT \times RT \times T \times LB}$$

$$USV = \frac{1{,}500}{(1.06 \times 1.00 \times 0.50 \times 1.00 \times 1.00 \times 0.97 \times 0.87)}$$

$USV = 3{,}354$ passenger cars per hour of green time

With $USV = 3{,}354$ passenger cars per hour of green time and street width $= 36$ feet, load factor is determined from Figure 13-1 as 1.0. From Table 13-1, the level of service is E, or at capacity.

Comparison of Predicted Service Volumes with Actual Intersection Data[3]. The Missouri State Highway Department conducted an extensive survey which compared service volumes predicted by the *Highway Capacity Manual* with data from actual intersections. The study reached the following conclusions:

1. Estimates of intersection approach service volumes for levels of service A, B and C by the 1965 *Highway Capacity Manual* are unreliable for Missouri intersections, and tend to be high, on the average. At Level of Service D, estimates were more accurate, and were most accurate for capacity (Level E).
2. Traffic does not appear to operate in the manner prescribed by the *Capacity Manual* until forced to by the pressure of heavy traffic conditions.
3. On intersections where the exit lanes are not sufficient to carry the volume indicated by calculations, on very wide streets, on one-way streets, and where left turns are made against traffic from a through lane, actual service volumes may be less than computed.
4. On intersections where the exit legs are much wider than the approach leg, or where traffic tends to be heavily loaded or under "pressure" conditions for part of the peak hour, the actual service volume may be higher than the computed value.
5. It appears that service volume is not as closely related to approach width as indicated in the Manual. When service volumes for all approaches in the 0.3 to 1.0 load factor group were recomputed using a standard lane width of 10 feet (parking lanes, 8 feet), the range of errors (STD) was reduced from 26.1 to 21.1 per cent.
6. City size appears to have less influence on service volume than indicated in the *Capacity Manual.*

Sources. This section of the *Capacity Manual* is based primarily on previously published material.[12] Other interesting studies are available.[13-22]

Bellis Method of Intersection Capacity[3]

This method differs from the *Highway Capacity Manual* (HCM) method in that it uses the number of lanes as the basic unit of street width. Four classifications of streets are used:

TYPE I: All central business district (CBD) streets
TYPE II: All streets, outside the CBD, which do not fall into the following categories
TYPE III: Expressways, arterials, major highways, major streets, and through streets with only right turns at intersections
TYPE IV: Expressways, arterials, major highways, major streets, and through streets with no turns at intersections or with separate phases and turn lanes (including jughandles) provided

Figure 13-10 shows the basic capacity relationship for the four roadway types. This figure plots maximum number of vehicles expected per cycle per maximum lane versus effective green time, in seconds. Effective green time is taken to be the green phase + amber phase − 1, in seconds, to allow for possible use of the amber phase by some vehicles.

The Bellis method uses a volume per maximum lane. To compute full-street capacities, lane distributions need to be known. The lane distribution assumed by the Bellis method is shown in Table 13-9 and should be used for analysis unless local data for this is available. Bellis does not specify which lane is assumed to be maximum, allowing for variability on this point. However, whichever lane is maximum, the percentage split among lanes is assumed to remain the same.

Capacity is adjusted for turns, trucks, and local buses, where applicable, using the HCM procedure. The Bellis method is good only for capacity determination, and cannot be used for determinations of service volumes at levels of service other than E.

A study[23] was undertaken to compare capacity predicted by both the HCM and Bellis methods with actual observed capacities. The results of the study were:

1. HCM estimates for capacity were high for sites with parking.
2. HCM estimates for capacity were low for sites without parking.
3. Bellis method capacities were at least 10 per cent low at all CBD locations.

TABLE 13-9. LANE DISTRIBUTION FOR BELLIS METHOD

NUMBER OF LANES	*PERCENTAGE DISTRIBUTION OF TRAFFIC AMONG LANES*
2	55%-45%
3	40%-35%-25%

(Source: Bellis, "Capacity of Traffic Signals and Traffic Signal Timing," HRBB 271, 1960)

TABLE 13-10. LANE SATURATION FLOWS BY LANE TYPE AND ENVIRONMENT (tcu's/hr)

	LANE TYPE		
ENVIRONMENT	*R*	*T*	*L*
CBD	1,270	1,580	1,550
Industrial	1,570	1,700	1,670
Suburban shopping	1,670	1,810	1,770
Residential	1,700	1,850	810

(Source: *Australian Road Capacity Guide,* ARRBB, No. 4, June, 1968)

TABLE 13-11. AVERAGE UTILIZED PROPORTION OF CURB LANES WITH A BAN ON PARKING OR STANDING

WIDTH OF APPROACH	*% OF CURB LANE CAPACITY UTILIZED*
Two-Lane	100
Three-Lane with no parking	40*
Three-Lane with no standing	60*

(Source: *Australian Road Capacity Guide,* ARRBB, No. 4, June, 1968)
* This figure is higher where there is a high percentage of right-turning vehicles. If the percentage of right-turners times the number of approach lanes exceeds the 40 or 60 per cent given, use that figure; if it exceeds 100 per cent, use 100 per cent.

TABLE 13-12. ADJUSTMENTS TO SATURATION FLOW DUE TO LANE WIDTH (%)

Lane width (ft)	8	9	10–12	13	14	15
Adjustment (%)	−12	−7	0	+3	$+4\frac{1}{2}$	+6

(Source: *Australian Road Capacity Guide,* ARRBB, No. 4, June, 1968)

Australian Road Capacity Guide Method

The Australian Guide method differs from the HCM in two major areas. The Australian Guide method, like the Bellis method, makes use of the lane as the unit of width, rather than width in feet. It also differs in its approach to corrections for turning vehicles and commerical vehicles. Rather than applying correction factors to calculated capacities, volumes are expressed in terms of equivalent "through car units" (tcu's).

The capacity of an intersection approach is determined by the formula

$$C = Sg/c$$

where C = capacity
S = saturation flow, in vehicles per hour of green time
g = effective green time
c = cycle length

Saturation flows for various types of lanes are given in Tables 13-10 and 13-11. Adjustments for lane width are given in Table 13-12.

Four types of lanes are considered:

1. Type R = lanes containing both right-turning and through vehicles, exclusive right turn lanes, lanes containing left-turning vehicles from one-way streets
2. Type T = lanes containing only through vehicles
3. Type L = lanes containing left-turning vehicles, either exclusively or with through vehicles, with or without separate signal control
4. Curb lanes, where vehicles are prohibited from parking in curb lanes. It is assumed that some illegal standing, parking, or stopping will reduce the effectiveness of that lane to through traffic. Thus, Table 13-11 gives the effective percentage of capacity which may be realized in the curb lane. Curb lanes are still characterized as L, T, or R lanes.

Note: Since Australians drive to the left all references have been adjusted from the Australian Manual to fit American Practice.

Tables 13-13 and 13-14 give the values for through car equivalents for turning and commercial vehicles. Equivalents for commercial vehicles and right-turning vehicles are assumed constant. Left-turning vehicle equivalents are tabulated versus s, the saturation flow in vehicles per hour, the g/c rate, q, the arrival rate of vehicles on the approach in vehicles per hour, and cycle length c.

Saturation flows in tcu's per hour may be converted to vehicles per hour by the following formula:

$$S(\text{vehicles per hour}) = \frac{S(\text{tcu's per hour})}{1 + \frac{1}{4}P_{RT} + 1\frac{1}{2}P_{RTCV} + P_{CV}(E_{LT} - 1) + P_{LTCV}E_{LT}}$$

where P_{RT} = percentage right-turning cars
P_{RTCV} = percentage right-turning commercial vehicles
P_{CV} = percentage through commercial vehicles
P_{LT} = percentage left-turning cars
P_{LTCV} = percentage left-turning commercial vehicles
E_{LT} = left-turn equivalent

The Australian Guide method considers service in a totally different manner from the HCM. A value y is defined as follows:

$$y = \frac{q}{s}$$

where q = arrival rate of vehicles on approach in vehicles per hour
s = saturation flow on approach in vehicles per hour of green

It then follows that:

$$Y = \sum_{i=1}^{n} y_i = \sum_{i=1}^{n} \frac{q_i}{s_i} \leq \sum_{i=1}^{n} \frac{g_i}{c} = 1 - \frac{L}{c}$$

where Y = the sum of the demands for green time, as a fraction of one available green hour
L = lost time per cycle due to amber phase, delay in start of queue movement
$1 - L/c$ = available effective green time in one hour, sum for all legs

The roadway is operating at capacity when $Y = 1 - (L/c)$. The Australian Guide recommends that Y never be more than 0.70 times $1 - (L/c)$ for "acceptable service." The Guide does not attempt to relate the $Y/[1 - (L/c)]$ ratio to the levels of service in the HCM.

The Australian Road Capacity Guide is based primarily on material presented in previous literature.[2]

A regression analysis was performed on intersection flow data to develop predictive equations for demand volume at intersections under varying conditions.[25] An interesting study relating level of service to delay, rather than load factor, has been made.[5]

COMBINED ANALYSIS OF URBAN AND SUBURBAN ARTERIALS

This category of facility includes highways and streets which have sufficient interruptions to flow that these interruptions, rather than geometrics, tend to govern capacity. Urban and suburban arterials are defined as major streets and highways outside the CBD, having either signalized intersections at average spacings of 1 mile or less, or speed limits of 35 miles per hour or less.

TABLE 13-13. THROUGH CAR EQUIVALENTS FOR DIFFERENT CLASSES OF VEHICLE AND DIFFERENT MOVEMENTS

RIGHT TURN		*THROUGH*		*UNOPPOSED LEFT TURN*		*OPPOSED LEFT TURN*	
Car	CV	Car	CV	Car	CV	Car	CV
1.25	2.50	1.00	2.00	1.00	2.00	E_{LT}*	$E_{LT}+1$

(Source: *Australian Road Capacity Guide,* ARRBB, No. 4, June, 1968)
* See Table 13-14.

TABLE 13-14. LEFT-TURNING VEHICLE EQUIVALENTS, E_{LT}

c = 40 sec

q (veh/hour)	*s = 1,300 veh/hour (g/c ratio)* 0.2	0.4	0.6	*s = 3,600 veh/hour (g/c ratio)* 0.2	0.4	0.6	*s = 5,400 veh/hour (g/c ratio)* 0.2	0.4	0.6
200	1.6	1.5	1.6	1.3	1.5	1.6	1.2	1.4	1.5
400	*	2.3	2.1	1.7	1.9	1.9	1.5	1.8	1.9
600	*	3.6	2.7	2.2	2.3	2.3	1.8	2.1	2.2
800	*	*	3.8	*	2.8	2.7	2.1	2.4	2.6
1000	*	*	6.0	*	3.4	3.2	2.5	2.8	2.9

c = 60 sec

q (veh/hour)	*s = 1,800 veh/hour (g/c ratio)* 0.2	0.4	0.6	*s = 3,600 veh/hour (g/c ratio)* 0.2	0.4	0.6	*s = 5,400 veh/hour (g/c ratio)* 0.2	0.4	0.6
200	1.9	1.8	1.7	1.5	1.6	1.7	1.4	1.6	1.6
400	*	2.7	2.3	2.1	2.1	2.1	1.8	2.0	2.0
600	*	4.7	3.1	3.1	2.7	2.5	2.3	2.4	2.4
800	*	*	4.5	*	3.4	3.1	2.9	2.9	2.9
1000	*	*	8.0	*	4.3	3.7	3.7	3.4	3.4

c = 90 sec

q (veh/hour)	*s = 1,800 veh/hour (g/c ratio)* 0.2	0.4	0.6	*s = 3,600 veh hour (g/c ratio)* 0.2	0.4	0.6	*s = 5,400 veh/hour (g/c ratio)* 0.2	0.4	0.6
200	2.3	1.9	1.8	1.7	1.7	1.7	1.6	1.7	1.7
400	*	3.0	2.5	2.6	2.3	2.2	2.2	2.2	2.1
600	*	5.2	3.4	4.2	3.0	2.7	2.9	2.7	2.6
800	*	*	5.4	*	3.9	3.4	3.9	3.3	3.1
1000	*	*	10.2	*	5.2	4.1	5.3	4.0	3.7

c = 120 sec

q (veh/hour)	*s = 1,800 veh/hour (g/c ratio)* 0.2	0.4	0.6	*s = 3,600 veh/hour (g/c ratio)* 0.2	0.4	0.6	*s = 5,400 veh/hour (g/c ratio)* 0.2	0.4	0.6
200	2.5	2.0	1.9	1.9	1.8	1.3	1.7	1.8	1.3
400	*	3.2	2.5	2.9	2.4	2.3	2.4	2.3	2.2
600	*	6.5	3.6	5.1	3.2	2.3	3.3	2.8	2.7
800	*	*	5.5	*	4.3	3.5	4.6	3.5	3.3
1000	*	*	11.9	*	5.9	4.3	6.8	4.4	3.9

(Source: *Australian Road Capacity Guide,* ARRBB, No. 4, June, 1968)
* In these cases, the capacity is exceeded for the opposing traffic.

Levels of Service

In the past, arterials were evaluated primarily in terms of the capacity of individual intersection approaches. Though each intersection on an arterial may be operating with reasonable efficiency, as a group they may produce rather poor (stop and go) operation. Therefore, when general service to traffic over the arterial as a whole is considered, it is not realistic to analyze the arterial by means of a series of separate intersection studies alone, and the determination of levels of service for arterials involves consideration of relatively long sections.

Since interrupted flow exists and the degree of delay varies widely, the speed measure used in urban arterial analyses is *average overall travel speed.*

Capacity represents the maximum utilization of that portion of the hour when the facility has a green signal indication, that is, the number of vehicles per hour of green. Under perfect signal progression, the arterial may carry traffic at flow rates approaching uninterrupted flow values while traffic is moving on a green signal indication. But there are many periods when traffic does not move, or when long gaps exist between platoons of vehicles. Therefore, the capacity in actual vehicles passing per hour is far less than with uninterrupted flow.

Levels of service are based on excellent, but not perfect, signal progression, and a 35-miles per hour speed limit. They are defined in Table 13-15.

Elements to Consider in Arterial Capacity

Signalization. Signalized intersections are important restrictions to flow, and therefore they must be considered as fundamental elements in any determination of the capacity of an arterial section. The methods described under Capacity of Signalized Intersections are used for this purpose.

Signalization with Progression. Perfect progression is a special case which provides largely nonstop operation, with flow rates approaching 2,000 passenger cars per hour of green per lane. This type of operation can only be attained where there are relatively few turning movements, with all signal cycles being very close to fully loaded, and with no mid-block frictional elements. It is susceptible to abrupt breakdown whenever any abnormality in the traffic flow develops. On two-way streets, near perfect progression usually can be obtained only in one direction.

One-Way Versus Two-Way Operation. One-way operation for the same width of street is generally more efficient than two-way operation, but the degree of one-way superiority varies considerably. A valid comparison of the traffic-carrying capabilities of streets under one-way and two-way operation requires consideration of an adjacent pair of one-way streets versus the same pair which was formerly operated as two two-way streets.

It should be stressed that one-way operation will provide a better level of service even if capacity is not improved substantially in all cases. This is due to the fact that fewer conflicts will exist within the traffic stream, to the relative ease with which signal progression may be provided, and to the minimizing of delays because two-phase signals are normally adequate.

TABLE 13-15. LEVELS OF SERVICE FOR URBAN AND SUBURBAN ARTERIAL STREETS

LEVEL OF SERVICE	*TRAFFIC FLOW CONDITIONS (TYPICAL APPROXIMATIONS, NOT RIGID CRITERIA)*				
	Description	*Average Overall Travel Speed (mph)*	*Load Factor*	*Likely Peak-Hour Factor*	*Service Volume/Capacity Ratio*
A	Free flow (relatively)	≳30	0.0	≲0.70	≲0.60 (0.80)
B	Stable flow (slight delay)	≳25	≲0.1	≲0.80	≲0.70 (0.85)
C	Stable flow (acceptable delay)	≳20	≲0.3	≲0.85	≲0.80 (0.90)
D	Approaching unstable flow (tolerable delay)	≳15	≲0.7	≲0.90	≲0.90 (0.95)
E	Unstable flow (congestion: intolerable delay)	Approx. 15	≲1.0 (0.85 typical)	≲0.095	≲1.00
F	Forced flow (jammed)	<15	(Not meaningful)	(Not meaningful)	(Not meaningful)

(Source: *Highway Capacity Manual,* HRB SR 87, 1965, Table 10-13)

Other Interruptions and Interferences. Along most arterials there exists a variety of other factors which sometimes impede the smooth flow of traffic. Problem elements likely to be encountered include:

1. Unsignalized intersections
2. Mid-block driveways and related turning movements
3. Curb parking in mid-block
4. Off-street parking in mid-block
5. Inadequate signs and markings
6. Lack of channelization
7. Restricted lateral clearances
8. Pedestrian interferences
9. Transit operations
10. Nonenforcement of regulations

Insufficient data are available to develop individual correction factors for these various elements, but their overall effects should be considered.

Methods of Increasing Arterial Capacity

There are several methods available to increase the capacity of an arterial and its intersections. These include prohibition of parking, prohibition of turns, channelization to create separate turn lanes, utilization of lane markings, signal timing improvements, and one-way operation.

Prohibition of parking is one of the most effective measures because it adds to the total street width available for moving traffic. Converting two parallel streets to one-way operation may increase capacity significantly, particularly where the curb-to-curb width of the streets is such that two-way operation permits only two lanes of moving traffic, while three lanes may be striped for one-way flow. Under other circumstances, there would be somewhat smaller increases in capacity due to reduction of conflicts at intersections and better opportunity to time signals progressively.

A step-by-step procedure is available[26] that could be employed to increase the capacity of an arterial. The steps range from practically no expenditure of funds to the construction of grade separations at major intersections.[27-30]

Methods have been developed[31] in which police power, access control provisions, and land use and planning controls, can be utilized to prevent the deterioration of arterials.

Computations for Arterials

Procedures for determining overall levels of service of arterials are less specific, and general procedures are recommended which must be coupled with good judgment.

Basic Components. The primary components available for the evaluation of an arterial include:

1. Capabilities of each important intersection
2. All significant mid-block restrictions
3. Average overall travel speeds

Overall Analysis of Urban Arterial Streets. The following general steps are recommended for the overall analysis of arterials.

Capacity determinations

1. Make an overall review of the arterial section under study, and select potential elements that may influence the capacity. Usually these elements will include all signalized intersections, mid-block locations restricted by geometrics, traffic interferences such as entrances or exits, or special traffic controls.
2. Compute the capacities of significant intersection approaches, of significant mid-block restrictions, and of any uninterrupted flow sections (section between two signals spaced more than 1 mile apart).
3. Interpret the results of the above analyses to identify obvious bottleneck locations having capacities considerably less than the arterial as a whole, and a capacity level for the remainder of the section, exclusive of bottlenecks, governed by the minimum capacity of remaining elements.
4. Attempt to increase capacity of bottlenecks. If this is not possible, their specific adverse effects should be analyzed in detail to determine actual influence on overall capacity.

Service volume determinations

Generally the same steps apply as for capacity determinations, except that the inability of a bottleneck to meet a service volume criterion is not as critical as it would be to meet a capacity criterion. To a degree it will still be a bottleneck, providing poorer service than the remainder of the section, but usually accommodating the demand.

Level of service determinations

The desirable goal is usually the determination of an overall level of service for an entire arterial, or a major section of an arterial, the analysis generally requires subdivision, and the procedure is as follows:

1. Subdivide the arterial into sections having relatively uniform geometrics and traffic conditions.
2. Determine the capacities of all significant points. Sepa-

rate the abnormal points (bottlenecks) and establish the general capacity level of the remainder.

3. Determine whether the demand volume exceeds either the general capacity level or the capacity of any bottleneck.
4. Where capacity is not controlling, establish a single average v/c ratio where possible, and establish a level of service. If various sections have differing levels, establish them accordingly.

The procedure can be used in reverse to establish service volumes provided by the facility for a desired level of service or average overall speed.

Sample Problem

(*Highway Capacity Manual* Example 10.6)

Problem:

The following conditions are given (see Figure 13-11):

Urban signalized two-way arterial street segments
Widths as shown in sketch
Curbed (6-inch curbs)
Level
No parking
3 per cent trucks throughout
Bus stop location as shown in sketch: 30 local buses per hour
Outlying business district
City size: population 500,000
PHF: 0.85
Pedestrian interference negligible
Intersection and turning movement characteristics as shown
Timed runs give an average overall travel speed of 19 miles per hour
Eastbound flow under consideration
Demand volumes as shown

Determine the following:

1. Through level of service indicated by average overall travel speed.
2. Level of service indicated by intersection and mid-block restriction performance.
3. Evaluate results.

Solution:

1. From Table 13-15, for 19 miles per hour, level of service is D, but not far from the limit of Level of Service C.
2. Review indicates that intersection 1, driveway entrance area 2, and intersection 3 are the main controlling elements.

Intersection 1:

Figure 13-4 applies. Determination of chart volume. From Figure 13-4, for 500,000 population and PHF = 0.85, factor = 1.06. From Figure 13-4 for outlying business district, factor = 1.25. From Table 13-5, for 10 per cent right turns, factor = 1.00. From Table 13-6, for 5 per cent left turns, factor = 1.05. Because there is no bus stop, local buses can be considered as trucks. Therefore, $30/1{,}100 = 2.7$ per cent buses (say 3 per cent). From Table 13-8, for 3 per cent +3 per cent = 6 per cent trucks, factor is 0.99. G/C ratio $= 35/70 = 0.50$. Given volume = 1,100. Chart volume $= 1{,}100/(1.06 \times 1.25 \times 1.00 \times 1.05 \times 0.99 \times 0.50) = 1{,}597$ vehicles per hour of green time under base conditions.

The intercept of 1,597 vehicles per hour of green time and the 24-foot width shows LF = 0.15, intersection Level of Service C indicated.

Capacity from Figure 13-4 = 2,100 vehicles per hour of green time, $v/c = 1{,}597/2{,}100 = 0.76$, in Level C, from Table 13-15.

Driveway entrance area 2:

Approximate method of handling must be developed. Demand volume $= 1{,}100 - 1{,}100(0.10 + 0.05) + 100 + 100 + 110 = 1{,}145$ vehicles per hour. Because opposing turns obstruct the through flow, flow through the block is stop and go, as it would be through a signalized inter-

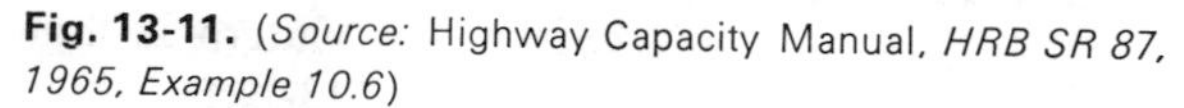

Fig. 13-11. (*Source:* Highway Capacity Manual, *HRB SR 87, 1965, Example 10.6*)

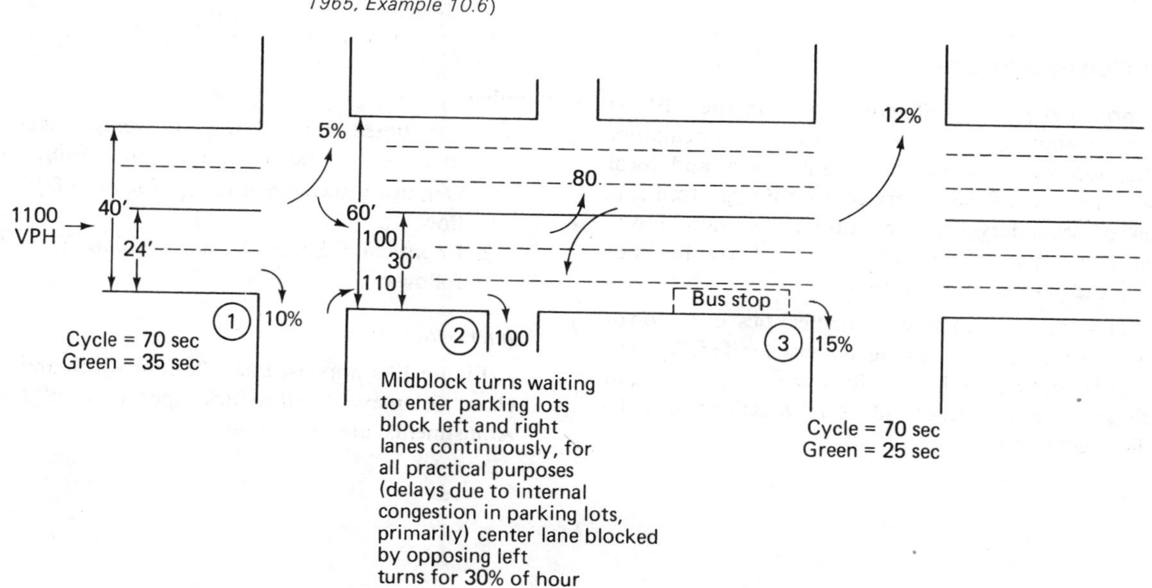

section. Assume signalized intersection with no turns, 10-foot approach width, no parking, and 70 per cent green time (100 − 30). Given volume = 1,145. 1,145/(1.06 × 1.25 × 1.20 × 1.30 × 0.99 × 0.70) = 800 vehicles per hour of green time under base conditions.

From Figure 13-4, the intercept of 800 vehicles per hour of green time and the 10-foot width shows LF = 0.9, within Level of Service E, and at or near capacity.

Intersection 3:

Demand volume is 1,145 − 100 − 80 = 965 vehicles per hour. Factors are as follows:

Population and PHF:	1.06	(Figure 13-4)
Outlying business district:	1.25	(Figure 13-4)
15 per cent right turns:	0.99	(Table 13-5)
12 per cent left turns:	0.98	(Table 13-6)
3 per cent trucks:	1.02	(Table 13-8)
30 buses per hour near-side stop:	0.91	[Figure 13-7(d)]
G/C ratio:	25/70 = 0.36	
Given volume:	965	

965/(1.06 × 1.25 × 0.99 × 0.98 × 1.02 × 0.91 × 0.36) 2,246 vehicles per hour of green time under base conditions.

The intercept of 2,246 vehicles per hour of green time and the 30-foot width shows LF = 0.5; center of intersection Level of Service D is indicated. The capacity, from Figure 13-4 = 2,700 vehicles per hour of green time; v/c = 2,246/2,700 = 0.83 in Level D, from Table 13-15.

3. The street is moderately heavily utilized. Overall, it is near the start of Level of Service D, and the individual intersections are in Levels C and D, respectively. (For urban arterials, through and intersection levels of service would be expected to be nearly alike, by definition.) However, the mid-block pair of driveways is a serious "spot" interference. This location, rather than the signalized intersections, controls the capacity of this portion of the street.

COMBINED ANALYSIS OF DOWNTOWN STREETS

General Considerations

1. The principal purpose of most streets in the CBD is service to abutting property. Consequently, circulatory traffic flow exists, with many local buses and local service trucks. Efficient service to through traffic is often of secondary concern, although certain downtown streets may be operated as arterial-type facilities during the peak commuter hours.
2. Present knowledge of flow characteristics over downtown street grids as a whole is very limited; specific criteria cannot be established for overall capacities and levels of service. Instead, individual locations must be studied separately.
3. Downtown operation would be considered poor, based upon the rating scale for any other highway type.
4. A rudimentary level of service scale is presented, based on average overall travel speeds, in Table 13-16.

Procedure for Determining Capacity, Service Volume, or Level of Service

1. Analyze the operation of each signalized intersection individually by methods presented under Capacity of Signalized Intersections.
2. Determine travel times through the section to establish the average overall travel speeds.
3. Select a general level of service from Table 13-16.

Sample Problem (*Highway Capacity Manual* Example 10.8)

Problem:

The following conditions are given (see Figure 13-12):

Downtown street segment, four blocks long
All intersections signalized
Two way
Parking as shown
Width: 56 feet, curb to curb
City size: population 175,000
PHF: 0.85
Intersection and traffic characteristics as shown
Bus stop as shown: 40 local buses per hour
No separate turning lanes or signal phases
Eastbound flow under consideration

Determine the following:

1. The approximate through level of service being provided, if timed runs show an average overall travel speed of 14 miles per hour.
2. Intersection service volumes, at intersection level of service equivalent to the through level determined above.
3. The controlling intersection, for a traffic demand pattern as shown in part (2) of solution. (See Figure 13-12.)
4. Evaluate results.

Solution:

1. The timed runs show an average overall travel speed of 14 miles per hour. From Table 13-16, this is in downtown street Level of Service D, for through flow.
2. Intersection Level D service volumes are found as follows:

Intersection 1:

Figure 13-5 applies. For a 28-foot width and $LF = 0.7$, chart volume = 1,550 vehicles per hour of green time. Adjustments are as follows:

TABLE 13-16. LEVELS OF SERVICE FOR DOWNTOWN STREETS

LEVEL OF SERVICE	TRAFFIC FLOW CONDITIONS (APPROXIMATIONS, NOT RIGID CRITERIA) DESCRIPTION	AVERAGE OVERALL SPEED (mph)
A	Free flow (relatively; some stops will occur)	≂25
B	Stable flow (delays not unreasonable)	≂20
C	Stable flow (delays significant but acceptable)	≂15
D	Approaching unstable flow (delays tolerable)	≂10
E*	Unstable flow (congestion not due to backups ahead)	Below 10 but moving
F	Forced flow (jammed)	Stop-and-go

(Source: *Highway Capacity Manual,* HRB SR 87, 1965, Table 10-14)

* Level E for the downtown street as a whole cannot be considered as capacity; capacity is governed by that of controlling intersections or other interruptions.

Population and PFH: 0.97 (Figure 13-5)
Downtown: 1.00 (Figure 13-5)
G/C ratio: $\frac{30}{60} = 0.50$

In this problem where all trucks turn, special consideration is justified when considering right turns. Three-eighths of the turns are trucks, with a passenger-car equivalent of at least 2. Thus, $\frac{5}{8} \times 1 + \frac{3}{8} \times 2 = \frac{11}{8}$, and we assume turns to be 11 per cent equivalent passenger cars. Therefore,

11 per cent right turns: 0.995 (Table 13-5)
5 per cent left turns: 1.05 (Table 13-6)

In this problem, we have 40 buses per hour, no stop. We consider them as trucks, with an approximate percentage, based on inspection of factors thus far, of 5 per cent. Therefore,

Trucks: 3 per cent actual + 5 per cent buses
= 8 per cent: 0.97 (Table 13-8)

$$SV_D = 1.550 \times 0.97 \times 1.00 \times 0.50 \times 0.995 \times 1.05 \times 0.97 = 760 \text{ vehicles per hour}$$

Intersection 2:

Figure 13-5 again applies. Adjustments are:

Population and PHF: 0.97
Downtown: 1.00
G/C ratio: $\frac{35}{60} = 0.58$
13 per cent right turns: 0.985
0 per cent left turns: 1.10
0 per cent trucks: 1.05
40 buses, far-side stop: 1.00(maximum) [Figure 13-7(a)]

$$SV_D = 1.550 \times 0.97 \times 1.00 \times 0.58 \times 0.985 \times 1.10 \times 1.05 \times 1.00 = 992 \text{ vehicles per hour}$$

Intersection 3:

Figure 13-4 applies. For 28-foot width and $LF = 0.7$, chart volume = 2,250 vehicles per hour. Adjustments are:

Population and PHF: 0.97
Downtown: 1.00
G/C ratio: $\frac{45}{80} = 0.56$
10 per cent right turns: 1.00
8 per cent left turns: 1.02
~~40 buses, no stop:~~ 1.05

Inspection shows 40 buses to be probably about 3 per cent; to be considered as trucks.

0 per cent trucks + 3 per cent through buses: 1.02

$$SV_D = 2{,}250 \times 0.97 \times 1.00 \times 0.56 \times 1.00 \times 1.02 \times 1.05 \times 1.02 = 1{,}335 \text{ vehicles per hour}$$

Intersection 4:

Figure 13-4 again applies. Adjustments are:

Population and PHF: 0.97

Fig. 13-12. (*Source:* Highway Capacity Manual, *HRB SR 87, 1965, Example 10.8*)

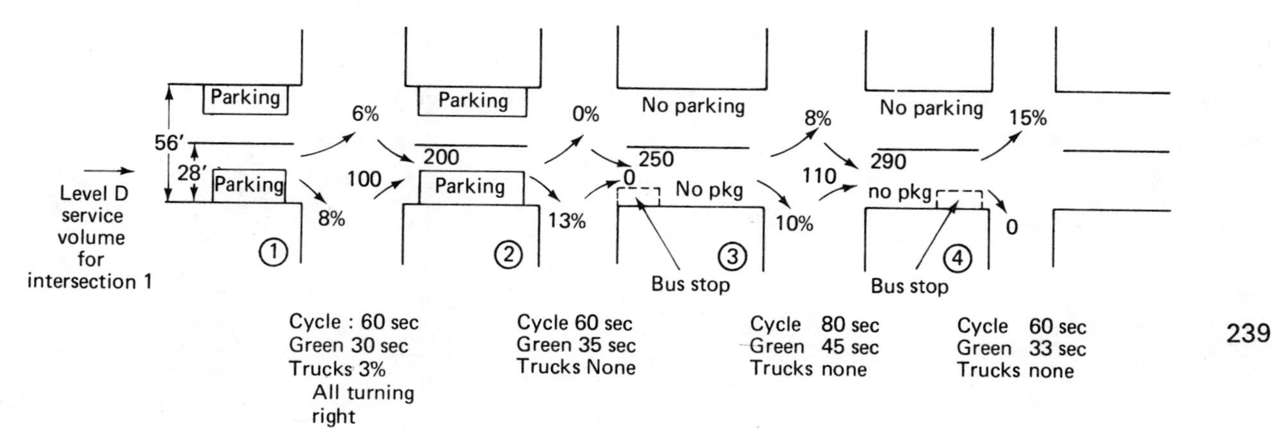

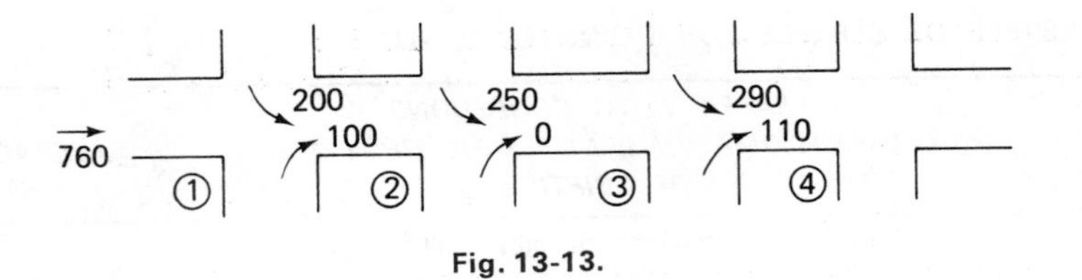

Fig. 13-13.

Downtown: 1.00
G/C ratio: $\frac{33}{60} = 0.55$
0 per cent right turns: 1.025
15 per cent left turns: 0.95
0 per cent trucks: 1.05
40 buses, near-side stop: 0.82 [Figure 13-7(d)]

$$SV_D = 2{,}250 \times 0.97 \times 1.00 \times 0.55 \times 1{,}025 \times 0.95 \times 1.05 \times 0.82 = 1{,}006 \text{ vehicles per hour}$$

Therefore, we have the following service volumes for Level D:

Intersection 1 = 760 vehicles per hour
Intersection 2 = 992 vehicles per hour
Intersection 3 = 1,335 vehicles per hour
Intersection 4 = 1,006 vehicles per hour

Intersection 1 appears to be the controlling intersection.

3. Particularly in a downtown area, where turns into the street under consideration from cross streets are likely to be quite heavy, a review of the actual demand volume pattern is necessary before final conclusions are drawn regarding the controlling intersection.

Assume the following entering traffic volumes (Figure 13-13):

Intersection 1:

Approaching = 760
Leaving to side 760(0.05 + 0.08) = − 99
Entering from side 200 + 100 = +300

Intersection 2:

Approaching = 961 < 992, SV_D, Intersection 2; satisfactory.
Leaving to side 961(0.13) = −125
Entering from side = 250

Intersection 3:

Approaching = 1,086 < 1,335, SV_D, intersection 3; satisfactory.
Leaving to side 1,086(0.18) = −195
Entering from side = 400

Intersection 4:

Approaching = 1,291 < 1,006; SV_D, intersection 4; not satisfactory.

4. Although intersection 1 appeared at first to control, based on service volumes alone, further analysis shows that under the given traffic circulation pattern in this downtown area, intersection 4 is actually more critical. It will reach capacity first, producing back-ups extending to other locations.

This problem demonstrates the fact that, where turning movements are significant along either a downtown street or an arterial, an apparent restriction may well be less controlling than it at first appears.

14

Highway Capacity: Rural Highways Without Access Control

Most rural highway are built without control of access because they must provide land use service in addition to traffic service, and therefore procedures for determining capacity, service volumes, and levels of service are somewhat different from those for highways with access control. Separate consideration is given to multi-lane and two-lane highways under essentially uninterrupted flow conditions. As for freeways, operating speed and v/c ratio are used to define levels of service. The basic v/c ratio limits apply to ideal alinement, permitting 70 miles per hour average highway speeds. To provide for the situations where lower design standards exist, working v/c ratios more appropriate for lower average highway speeds are included.

MULTI-LANE RURAL HIGHWAYS

This category includes multi-lane highways that cannot be classified as freeways or expressways because they are undivided, or because they lack significant control of access features, or both. The treatment of these highways does not differ greatly from that of freeways. However, an important difference is that the presence of more than two lanes in one direction on an uncontrolled multi-lane highway does not necessarily produce the same predictable increase in efficiency that occurs at intermediate volume levels on freeways. Therefore, ordinary multi-lane highways are considered to carry the same service volumes per lane in one direction, regardless of the number of lanes in that direction, and operating speeds reflect the average conditions in all lanes.

Ideal conditions for ordinary multi-lane highways are the same as for freeways, including 12-foot lanes, adequate lateral clearance and shoulders, alinement for 70 miles per hour average highway speed, and no commercial vehicles. The average per lane capacity of 2,000 passenger cars per hour, under ideal conditions, is also the same.

Levels of Service

For uninterrupted flow conditions, the characteristics of the several levels of service are very similar to those for freeways. Table 14.1 summarizes the relationships between levels of service, operating speeds, and v/c ratios for both ideal and restricted alinement.

Critical Elements Affecting Capacity and/or Service Volume

It is seldom that ideal conditions exist. Generally, a highway will have restricting elements which adversely affect traffic operations, and many highways are unable to provide service at Levels B or A. Factors are provided,

TABLE 14-1. LEVELS OF SERVICE AND MAXIMUM SERVICE VOLUMES FOR MULTI-LANE HIGHWAYS, UNDIVIDED AND/OR WITHOUT ACCESS CONTROL, UNDER UNINTERRUPTED FLOW CONDITIONS (NORMALLY REPRESENTATIVE OF RURAL OPERATION)

LEVEL OF SERVICE	*TRAFFIC FLOW CONDITIONS*		*SERVICE VOLUME CAPACITY (v/c) RATIO*			*MAXIMUM SERVICE VOLUME UNDER IDEAL CONDITIONS, INCLUDING 70-MPH AHS (TOTAL PASSENGER CARS PER HOUR, ONE DIRECTION)*		
	Description	*Operating Speed* (mph)*	*Basic Limiting Value* for ahs of 70 mph*	*Approximate Working Value for Restricted ahs of*		*Four-Lane hwy. (two-lanes One Direction)*	*Six-Lane hwy. (three-lanes One Direction)*	*Each Additional Lane*
				60 mph	*50 mph*			
A	Free flow	≥60	≤0.30	–**	–**	1200	1800	600
B	Stable flow (upper speed range)	≥55	≤0.50	≤0.20	–**	2000	3000	1000
C	Stable flow	≥45	≤0.75	≤0.50	≤0.25	3000	4500	1500
D	Approaching unstable flow	≥35	≤0.90	≤0.85	≤0.70	3600	5400	1800
E†	Unstable flow	30‡		≤1.00		4000	6000	2000
F	Forced flow	< 30‡	Not Meaningful §				Widely variable (0 to capacity)	

(Source: *Highway Capacity Manual,* HRB SR 87, 1965, Table 10.1)

* Operating speed and basic ratio are independent measures of level of service: both limits must be satisfied in any determination of level.

** Operating speed required for this level is not attainable even at low volumes.

† Capacity.

‡ Approximately.

§ Demand volume capacity ratio may well exceed 1.00 indicating overloading.

TABLE 14-2. COMBINED EFFECT OF LANE WIDTH AND RESTRICTED LATERAL CLEARANCE ON CAPACITY AND SERVICE VOLUME OF UNDIVIDED MULTI-LANE HIGHWAYS WITH UNINTERRUPTED FLOW

Distance from traffic lane edge to obstruction (ft)	*ADJUSTMENT FACTOR,* W, FOR LATERAL CLEARANCE AND LANE WIDTH*							
	Obstruction on Right Side Only, of One-Direction Traveled Way (Includes Allowance for Opposing Traffic on Left)				*Obstructions on Both Sides of One-Direction Traveled Way*			
	12-ft Lanes	11-ft Lanes	10-ft Lanes	9-ft Lanes	12-ft Lanes	11-ft Lanes	10-ft Lanes	9-ft Lanes
	(*a*) 4-Lane Undivided Highway, One Direction of Travel							
6	1.00	0.95	0.89	0.77	N.A.	N.A.	N.A.	N.A.
4	0.98	0.94	0.88	0.76	N.A.	N.A.	N.A.	N.A.
2	0.95	0.92	0.86	0.75	0.94	0.91	0.86	N.A.
0	0.88	0.85	0.80	0.70	0.81	0.79	0.74	0.66
	(*b*) 6-Lane Undivided Highway, One Direction of Travel							
6	1.00	0.95	0.89	0.77	N.A.	N.A.	N.A.	N.A.
4	0.99	0.94	0.88	0.76	N.A.	N.A.	N.A.	N.A.
2	0.97	0.93	0.86	0.75	0.96	0.92	0.85	N.A.
0	0.94	0.90	0.83	0.72	0.91	0.87	0.81	0.70
	(*c*) Divided Highways, One Direction of Travel							
	Use adjustment factors from Table 12-3							

(Source: *Highway Capacity Manual,* HRB SR 87, 1965, Table 10-2)

* Same adjustments for capacity and all levels of service.

N.A. = Not applicable—use adjustment for obstruction on right side only.

as for freeways, for specific adjustments for some restricting elements, and elements requiring consideration are as follows:

Lane Width and Lateral Clearance. The combined effects on capacity and service volumes of pavement width and lateral clearance are given in Table 14-2. The adjustment factors (W) should be applied as multipliers to correct for any such limitations. The factors in Table 14-2 take into consideration the effect of opposing traffic, and therefore the adjustment for "obstruction on both sides" should only be used when some physical obstruction encroaches on the center of the roadway that would be closer than the opposing flow of traffic.

Trucks, Buses, and Grades. The adjustment factors for these effects are identical to those for freeways, and reference should be made to the procedures and tables described under Capacity of Freeways and Expressways. Specifically, these tables include Table 12-4, Table 12-5, Table 12-6, Table 12-7, and Table 12-8.

Weaving Areas and Ramp Junctions. The procedures described under Weaving Sections and Ramp Junctions generally apply to highways having no control of access, even though they are commonly associated with freeways.

Alinement. The influence of adverse alinement, as reflected by average highway speeds less than 70 miles per hour, is incorporated directly into the computational criteria as indicated in Table 14-1.

Traffic Interruptions and Interferences. Where fixed interruptions (traffic signals, stop signs, railroad grade crossings, etc.) average *more* than 1 mile apart, and/or where speeds of 40 miles per hour or greater are attainable between interruptions, the procedures for uninterrupted flow apply directly, without detailed consideration of the interruptions. Of course, the capacity of the interruption is reduced, in the case of a traffic signal, in proportion to the amount of red time in the total cycle assigned to the multi-lane approach; however, there is normally a very low percentage of green time assigned to the cross road.

Where fixed interruptions are *less* than 1 mile apart, or if speeds are restricted to less than 35 miles per hour, the highway should be treated as an urban arterial, as described under Combined Analysis of Urban and Suburban Arterials.

Computational Procedures

The generalized procedures are similar to those described for freeways involving, first, determination of capacities, service volumes, and levels of service of near-uniform highway subsections, followed by development of overall measures for sections of substantial length formed by several subsections combined.

Basic uniform multi-lane subsections. Operating speed and service or demand volume/capacity ratio (v/c ratio), are the basic measures used in making level-of-service determinations. The limitations defining the several levels of service are summarized in Table 14-1.

Capacity under prevailing conditions

This is determined directly, by multiplying the Level E volume from Table 14-1, 2,000 passenger cars per hour per lane, by appropriate adjustment factors.

$$C = 2{,}000NWT_c \tag{14-1}$$

where C = capacity (mixed vehicles per hour, total for one direction)
N = number of lanes (in one direction)
W = adjustment for lane width and lateral clearance, from Table 14-2
T_c = truck factor at capacity, from Table 12-6 or 12-8

Service volumes

The Manual presents four procedures for obtaining service volume, for a given level of service, which are very similar to the procedures for freeways in their application. In all cases, the attainment of the desired level of service is confirmed by checking the resulting operating speed from Table 14-1 or the corresponding figures in the HCM for the given average highway speed.

1. Computed directly from capacity under ideal conditions.

$$SV = 2{,}000N\frac{v}{c}WT_L \tag{14-2}$$

where SV = service volume (mixed vehicles per hour, total for one direction)
v/c = volume-to-capacity ratio, from Table 14-1
T_L = truck factor at given level of service, from Table 12-6 or Table 12-8

2. Computed from maximum service volume for ideal conditions.

This procedure is only applicable when alinement is ideal, that is, average highway speed is 70 miles per hour.

$$SV = (MSV)WT_L \tag{14-3}$$

where MSV = maximum service volume in passenger cars per hour from Table 14-1.

Level of service.

The determination of the level of service for an existing or proposed multi-lane highway, while accommodating a given demand volume, may be accomplished approximately by inspection of Table 14-1, if pertinent parameters are known, and trucks are neglected. Refined analysis considering trucks involves a "trial-and-error" approach, since the level of service must be known to choose the appropriate truck factor. Therefore, a level must be assumed, and recomputations are carried out if the results prove the assumption incorrect.

Combined Analysis of Subsections. In overall concept, the procedures described under Capacity of Freeways and Expressways apply to the determination of weighted average levels of service for multi-lane highways.

In the case of point or short-distance lateral clearance restrictions (such as at narrow bridges), in the absence of more definitive information, an influence distance equivalent to 5 seconds travel time plus the actual length of the restriction can be assumed. For example, a narrow bridge, 100 feet long, on a highway with a 50-mile-per-hour operating speed would have an approximate advance influence distance of $50 \times 1.47 \times 5 + 100 = 468$ feet.

Sample Problem (*Highway Capacity Manual,* Example 10.2)

Problem:

The following conditions are given (see Figure 14-1):

Rural four-lane highway, undivided, without access control
11-foot lanes
No shoulders
Obstructions at pavement edge
Individual grade, 6 pr cent, 1 mile long
Alinement provides 50-miles per hour AHS
7 per cent trucks
3 per cent intercity buses
Demand volume = 2,100 vehicles per hour, in upgrade direction
Determine the level of service being provided on this upgrade.

Fig. 14-1. (*Source:* Highway Capacity Manual, *HRB SR 87, 1965, Example 10.2*)

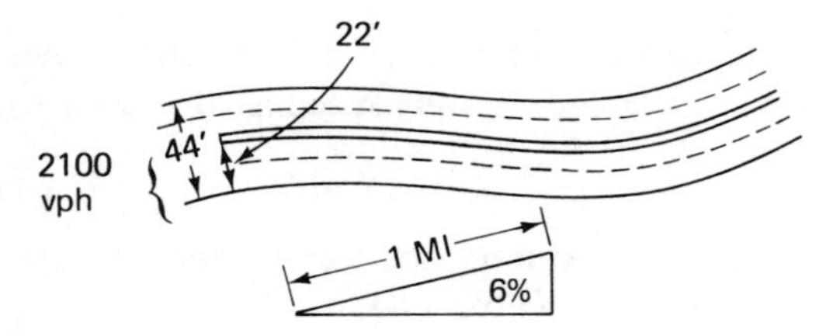

Solution:

Inspection of Table 14-1, given an ordinary four-lane highway on heavy grade, indicates that operation probably is in Level D or poorer. Assume Level D for use in selecting adjustments, dependent upon a known level. On this heavy grade, buses should be considered separately, using bus adjustment factor B_L. The formula is

$$\text{Capacity} = 2{,}000NWT_LB_L$$

where $N = 2$
$W = 0.85$ (from Table 14-2 for 11-foot lanes with 0-foot clearance, one side = 0.85)
$T_L = 0.52$ (from Table 12-5 for Level D, given 7 per cent trucks on 6 per cent, 1-mile grade, $E_T = 14$; from Table 12-6 for $E_T = 14$ and $P_T = 7$, $T_L = 0.52$)
$B_L = 0.92$ (from Table 12-4 for Level D, given 3 per cent buses on 6 per cent, 1-mile grade, $E_b = 4$; from Table 12-6 for $E_b = 4$ and $P_b = 3$, B_L 0.92)

Therefore,

Capacity $= 2{,}000 \times 2 \times 0.85 \times 0.52 \times 0.92 = 1{,}627$ vehicles per hour

$$\frac{v}{c}\text{ratio} = \frac{2{,}100}{1{,}627} = 1.29$$

That is, the highway would be overloaded; it could not operate at Level D. A computation assuming Level E would make no change, because all factors used for Level D would be the same at E; overloading would remain. Hence, level of service is F, and the demand volume of 2,100 vehicles per hour cannot be accommodated. Correction would involve the provision of a climbing lane for the heavy truck volume.

TWO-LANE HIGHWAYS

In terms of mileage, two-lane highways constitute the majority of main rural highways. Two basic characteristics differentiate traffic operations on two-lane highways from multi-lane facilities. First, distribution of traffic by direction has little effect on operating conditions. Therefore, the capacity and service volumes of two-lane highways are expressed in total vehicles per hour, regardless of distribution of traffic by direction. Second, overtaking and passing maneuvers must be made in the lane normally occupied by opposing traffic. Whenever service volumes are considered for two-lane roads, the corresponding range in available passing sight distance (1,500 feet or greater) must also be taken into account. An excellent literature review of reported studies of overtaking and passing on two-lane rural highways has been published.[1]

TABLE 14-3. LEVELS OF SERVICE AND MAXIMUM SERVICE VOLUMES ON TWO-LANE HIGHWAYS UNDER UNINTERRUPTED FLOW CONDITIONS (NORMALLY REPRESENTATIVE OF RURAL OPERATION)

LEVEL OF SERVICE	*TRAFFIC FLOW CONDITIONS*		*PASSING SIGHT DISTANCE >1,500 FT (%)*	*SERVICE VOLUME/CAPACITY (v/c) RATIO*						*MAXIMUM SERVICE VOLUME UNDER IDEAL CONDITIONS INCLUDING 70-MPH AHS (PASSENGER CARS, TOTAL, BOTH DIRECTIONS, PER HOUR)*
				Basic Limiting Value for ahs of*	*Working Value for Restricted Average Highway Speed** of*					
	Description	*Operating Speed (mph)*		*70 mph*	*60 mph*	*50 mph*	*45 mph*	*40 mph*	*35 mph*	
A	Free flow	≥60	100	≤ 0.20	—	—	—	—	—	400
			80	0.18	—	—	—	—	—	
			60	0.15	—	—	—	—	—	
			40	0.12	—	—	—	—	—	
			20	0.08	—	—	—	—	—	
			0	0.04	—	—	—	—	—	
B	Stable flow (upper speed range)	≥50	100	≤ 0.45	≤ 0.40	—	—	—	—	900
			80	0.42	0.35	—	—	—	—	
			60	0.38	0.30	—	—	—	—	
			40	0.34	0.24	—	—	—	—	
			20	0.30	0.18	—	—	—	—	
			0	0.24	0.12	—	—	—	—	
C	Stable flow	≥40	100	≤ 0.70	≤ 0.66	≤ 0.56	≤ 0.51	—	—	1400
			80	0.68	0.61	0.53	0.46	—	—	
			60	0.65	0.56	0.47	0.41	—	—	
			40	0.62	0.51	0.38	0.32	—	—	
			20	0.59	0.45	0.28	0.22	—	—	
			0	0.54	0.38	0.18	0.12	—	—	
D	Approaching unstable flow	≥35	100	≤ 0.85	≤ 0.83	≤ 0.75	≤ 0.67	≤ 0.58	—	1700
			80	0.84	0.81	0.72	0.62	0.55	—	
			60	0.83	0.79	0.69	0.57	0.51	—	
			40	0.82	0.76	0.66	0.52	0.45	—	
			20	0.81	0.71	0.61	0.44	0.35	—	
			0	0.80	0.66	0.51	0.30	0.19	—	
E†	Unstable flow	30‡	Not applicable §	≤1.00						2000
F	Forced flow	< 30‡	Not applicable §	Not Meaningful						Widely variable (0 to capacity)

(Source: *Highway Capacity Manual,* HRB SR 87, 1965, Table 10-7)

* Operating speed and basic v/c ratio are independent measures of level of service: both limits must be satisfied in any determination of level.

** Where no entry apprears, operating speed required for this level is unattainable even at low volumes.

† Capacity.

‡ Approximately.

§ No passing

‖ Demand volume capacity ratio may well exceed 1.00, indicating overloading.

TABLE 14-4. COMBINED EFFECT OF LANE WIDTH AND RESTRICTED LATERAL CLEARANCE ON CAPACITY AND SERVICE VOLUMES OF TWO-LANE HIGHWAYS WITH UNINTERRUPTED FLOW

DISTANCE FROM TRAFFIC LANE EDGE TO OBSTRUCTION (FT)	ADJUSTMENT FACTORS W_L AND W_e FOR LATERAL CLEARANCE AND LANE WIDTH*															
	Obstruction on One Side Only**								Obstructions on Both Sides**							
	12-ft Lanes		11-ft Lanes		10-ft Lanes		9-ft Lanes		12-ft Lanes		11-ft Lanes		10-ft Lanes		9-ft Lanes	
	Level B	Level E†	Level B	Level E†	Level B	Level E†	Level B	Level E†	Level B	Level E†	Level B	Level E†	Level B	Level E†	Level B	Level E†
6	1.00	1.00	0.86	0.88	0.77	0.81	0.70	0.76	1.00	1.00	0.86	0.88	0.77	0.81	0.70	0.76
4	0.96	0.97	0.83	0.85	0.74	0.79	0.68	0.74	0.92	0.94	0.79	0.83	0.71	0.76	0.65	0.71
2	0.91	0.93	0.78	0.81	0.70	0.75	0.64	0.70	0.81	0.85	0.70	0.75	0.63	0.69	0.57	0.65
0	0.85	0.88	0.73	0.77	0.66	0.71	0.60	0.66	0.70	0.76	0.60	0.67	0.54	0.62	0.49	0.58

(Source: *Highway Capacity Manual*, HRB SR 87, 1965, Table 10-8)
* Adjustment W_e given for level *E*, capacity, and W_L for level B, interpolate for others.
** Includes allowance for opposing traffic.
† Capacity.

Levels of Service

Table 14-3 summarizes, for uninterrupted flow conditions, the relationships between levels of service, operating speeds, and v/c ratios, for both ideal and restricted alinement, in terms of the percentage of the highway with passing sight distances in excess of 1,500 feet.

Critical Elements Affecting Capacity and/or Service Volumes

Lane Width and Lateral Clearance. The combined effects of pavement width and lateral clearance on capacity and service volume for Level B (interpolation is recommended for other levels), are given in Table 14-4, and the adjustment factors (W) should be applied as multipliers to correct for any such limitations. The factors take into consideration the effect of opposing traffic.

Trucks, Buses, and Grades. Tables 14-5, 14-6, 14-7, 14-8, and 14-9 are provided to determine the effect of trucks, buses, and grades on two-lane highways, and procedures for application of factors are essentially the same as for freeways and uncontrolled multi-lane highways.

Sometimes a special analysis must be carried out to determine a passenger car equivalency factor for use in Table 14-9. This situation normally arises on intermediate grades of 2 to 4 per cent, to properly take into consideration downgrades, or where the percentages of trucks in the two directions differ significantly. An upgrade for one direction of flow is a downgrade for the other direction. Since capacities and service volumes are expressed as total volumes for both directions, the effects of the opposing downgrade are included in any consideration of a particular upgrade.

Ramp Junctions. The procedures described under Signalized Intersections and Ramp Junctions, depending on whether the ramp junction with the two-lane highway is signalized or not, generally apply.

Alinement. The influence of adverse alinement on operating speed and volume is incorporated directly into the computational criteria as indicated in Table 14-3.

Traffic Interruptions and Interferences. Where fixed interruptions and interference average more than 1 mile apart, and/or where speeds of 35 miles per hour or greater are attainable between interruptions, the procedures for uninterrupted flow apply directly. Where these limits are not met, the highway normally should be treated as an urban arterial.

TABLE 14-5. AVERAGE GENERALIZED PASSENGER CAR EQUIVALENTS OF TRUCKS AND BUSES ON TWO-LANE HIGHWAYS, OVER EXTENDED SECTION LENGTHS (INCLUDING UPGRADES, DOWNGRADES, AND LEVEL SUBSECTIONS)

EQUIVALENT	LEVEL OF SERVICE	EQUIVALENT FOR: Level Terrain	Rolling Terrain	Mountainous Terrain
E_T, for trucks	A	3	4	7
	B and C	2.5	5	10
	D and E	2	5	12
E_B, for buses*	All levels	2	4	6

(Source: *Highway Capacity Manual*, HRB SR 87, 1965, Table 10-9a)
* Separate consideration not warranted in most problems; use only where bus volumes are significant.

TABLE 14-6. AVERAGE GENERALIZED ADJUSTMENT FACTORS FOR TRUCKS ON TWO-LANE HIGHWAYS, OVER EXTENDED SECTION LENGTHS

PERCENTAGE OF TRUCKS, P_T	TRUCK ADJUSTMENT FACTOR, T								
	LEVEL TERRAIN			ROLLING TERRAIN			MOUNTAINOUS TERRAIN		
	Level of Service A	Levels of Service B and C	Levels of Service D and E*	Level of Service A	Levels of Service B and C	Levels of Service D and E*	Level of Service A	Levels of Service B and C	Levels of Service D and E*
1	0.98	0.99	0.99	0.97	0.96	0.96	0.94	0.92	0.90
2	0.96	0.97	0.98	0.94	0.93	0.93	0.89	0.85	0.82
3	0.94	0.96	0.97	0.92	0.89	0.89	0.85	0.79	0.75
4	0.93	0.95	0.96	0.89	0.86	0.86	0.81	0.74	0.69
5	0.91	0.93	0.95	0.87	0.83	0.83	0.77	0.69	0.65
6	0.89	0.92	0.94	0.85	0.81	0.81	0.74	0.65	0.60
7	0.88	0.91	0.93	0.83	0.78	0.78	0.70	0.61	0.57
8	0.86	0.90	0.93	0.81	0.76	0.76	0.68	0.58	0.53
9	0.85	0.89	0.92	0.79	0.74	0.74	0.65	0.55	0.50
10	0.83	0.87	0.91	0.77	0.71	0.71	0.63	0.53	0.48
12	0.81	0.85	0.89	0.74	0.68	0.68	0.58	0.48	0.43
14	0.78	0.83	0.88	0.70	0.64	0.64	0.54	0.44	0.39
16	0.76	0.81	0.86	0.68	0.61	0.61	0.51	0.41	0.36
18	0.74	0.80	0.85	0.65	0.58	0.58	0.48	0.38	0.34
20	0.71	0.77	0.83	0.63	0.56	0.56	0.45	0.46	0.31

(Source: *Highway Capacity Manual*, HRB SR 87, 1965, Table 10-9b)
* Not applicable to buses where they are given separate specific consideration; use instead Table 10-9a in conjunction with Table 10-12.

Computational Procedures

The generalized procedures for freeways and multi-lane highways are also applicable to two-lane highways.

Basic Uniform Two-Lane Subsections. Operating speed, service or demand volume/capacity ratio, and percentage of available passing sight distance are the basic measures used in making level-of-service determinations. The limitations defining the several levels of service are summarized in Table 14-3.

Capacity under prevailing conditions

This is determined directly by multiplying Level E volume from Table 14-3, 2,000 passenger cars per hour total for both directions, by appropriate adjustment factors.

$$C = 2{,}000 W_c T_c \qquad (14\text{-}4)$$

where C = capacity (mixed vehicles per hour, total in both directions)
W_c = adjustment for lane width and lateral clearance, from Table 14-4
T_c = truck factor at capacity from Table 14-6 or Table 14-8

Service volumes

The Manual[2] presents four procedures for obtaining service volumes, for a given level of service, which are similar to freeways and multi-lane highways in their application. In all cases, the attainment of the desired level of service is confirmed by checking the resulting operating speed from the appropriate chart in Table 14-3 or the corresponding figures in the HCM for the given average highway speed. The two most commonly used procedures are detailed below:

1. Computed directly from capacity under ideal conditions.

$$SV = 2{,}000 \frac{v}{c} W_L T_L \qquad (14\text{-}5)$$

where SV = service volume (mixed vehicles per hour, total in both directions)
v/c = volume-to-capacity ratio, from Table 14-3
W_L = adjustment for lane width and lateral clearance at given level of service, from Table 14-4
T_L = truck factor at given level of service, from Table 14-6 or Table 14-8

2. Computed from maximum service volume for ideal conditions

This procedure is only applicable when alinement is ideal, that is, average highway speed is 70 miles per hour and there is 100 per cent passing sight distance.

$$SV = (MSV) W_L T_L \qquad (14\text{-}6)$$

TABLE 14-7. PASSENGER CAR EQUIVALENTS OF TRUCKS ON TWO-LANE HIGHWAYS, ON SPECIFIC INDIVIDUAL SUBSECTIONS OR GRADES

GRADE (%)	LENGTH OF GRADE (Mi)	PASSENGER CAR EQUIVALENT, E_T (FOR ALL PERCENTAGES OF TRUCKS)		
		Levels of Service A and B	Level of Service C	Levels of Service D and E (Capacity)
0–2	All	2	2	2
3	$\frac{1}{4}$	5	3	2
	$\frac{1}{2}$	10	10	7
	$\frac{3}{4}$	14	16	14
	1	17	21	20
	$1\frac{1}{2}$	19	25	26
	2	21	27	29
	3	22	29	31
	4	23	31	32
4	$\frac{1}{4}$	7	6	3
	$\frac{1}{2}$	16	20	20
	$\frac{3}{4}$	22	30	32
	1	26	35	39
	$1\frac{1}{2}$	28	39	44
	2	30	42	47
	3	31	44	50
	4	32	46	52
5	$\frac{1}{4}$	10	10	7
	$\frac{1}{2}$	24	33	37
	$\frac{3}{4}$	29	42	47
	1	33	47	54
	$1\frac{1}{2}$	35	51	59
	2	37	54	63
	3	39	56	66
	4	40	57	68
6	$\frac{1}{4}$	14	17	16
	$\frac{1}{2}$	33	47	54
	$\frac{3}{4}$	39	56	65
	1	41	59	70
	$1\frac{1}{2}$	44	62	75
	2	46	65	80
	3	48	68	84
	4	50	71	87
7	$\frac{1}{4}$	24	32	35
	$\frac{1}{2}$	44	63	75
	$\frac{3}{4}$	50	71	84
	1	53	74	90
	$1\frac{1}{2}$	56	79	95
	2	58	82	100
	3	60	85	104
	4	62	87	108

(Source: *Highway Capacity Manual*, HRB SR 87, 1965, Table 10-10)

TABLE 14-8. ADJUSTMENT FACTORS FOR TRUCKS AND BUSES ON INDIVIDUAL ROADWAY SUBSECTIONS OR GRADES ON TWO-LANE HIGHWAYS (INCORPORATING PASSENGER-CAR EQUIVALENT AND PERCENTAGE OF TRUCKS OR BUSES)

PASSENGER CAR EQUIVALENT, E_T OR E_B	*TRUCK ADJUSTMENT FACTOR T_e OR T_B, (B_e OR B_B, FOR BUSES)*														
	PERCENTAGE OF TRUCKS, P_T (OR OF BUSES, P_B) OF:														
	1	*2*	*3*	*4*	*5*	*6*	*7*	*8*	*9*	*10*	*12*	*14*	*16*	*18*	*20*
2	0.99	0.98	0.97	0.96	0.95	0.94	0.93	0.93	0.92	0.91	0.89	0.88	0.86	0.85	0.83
3	0.98	0.96	0.94	0.93	0.91	0.89	0.88	0.86	0.85	0.83	0.81	0.78	0.76	0.74	0.71
4	0.97	0.94	0.92	0.89	0.87	0.83	0.83	0.81	0.79	0.77	0.74	0.70	0.68	0.65	0.63
5	0.96	0.93	0.89	0.86	0.83	0.81	0.78	0.76	0.74	0.71	0.68	0.64	0.61	0.58	0.56
6	0.95	0.91	0.87	0.83	0.80	0.77	0.74	0.71	0.69	0.67	0.63	0.59	0.56	0.53	0.50
7	0.94	0.89	0.85	0.81	0.77	0.74	0.70	0.68	0.65	0.63	0.58	0.54	0.51	0.48	0.45
8	0.93	0.88	0.83	0.78	0.74	0.70	0.67	0.64	0.61	0.59	0.54	0.51	0.47	0.44	0.42
9	0.93	0.86	0.81	0.76	0.71	0.68	0.64	0.61	0.58	0.56	0.51	0.47	0.44	0.41	0.38
10	0.92	0.85	0.79	0.74	0.69	0.65	0.61	0.58	0.55	0.53	0.48	0.44	0.41	0.38	0.36
11	0.91	0.83	0.77	0.71	0.67	0.63	0.59	0.56	0.53	0.50	0.45	0.42	0.38	0.36	0.33
12	0.90	0.82	0.75	0.69	0.65	0.60	0.57	0.53	0.50	0.48	0.43	0.39	0.36	0.34	0.31
13	0.89	0.81	0.74	0.68	0.63	0.58	0.54	0.51	0.48	0.45	0.41	0.37	0.34	0.32	0.29
14	0.88	0.79	0.72	0.66	0.61	0.56	0.52	0.49	0.46	0.43	0.39	0.35	0.32	0.30	0.28
15	0.88	0.78	0.70	0.64	0.59	0.54	0.51	0.47	0.44	0.42	0.37	0.34	0.31	0.28	0.26
16	0.87	0.77	0.69	0.63	0.57	0.53	0.49	0.45	0.43	0.40	0.36	0.32	0.29	0.27	0.25
17	0.86	0.76	0.68	0.61	0.56	0.51	0.47	0.44	0.41	0.38	0.34	0.31	0.28	0.26	0.24
18	0.85	0.75	0.66	0.60	0.54	0.49	0.46	0.42	0.40	0.37	0.33	0.30	0.27	0.25	0.23
19	0.85	0.74	0.65	0.58	0.53	0.48	0.44	0.41	0.38	0.36	0.32	0.28	0.26	0.24	0.22
20	0.84	0.72	0.64	0.57	0.51	0.47	0.42	0.40	0.37	0.34	0.30	0.27	0.25	0.23	0.21
22	0.83	0.70	0.61	0.54	0.49	0.44	0.40	0.37	0.35	0.32	0.28	0.25	0.23	0.21	0.19
24	0.81	0.68	0.59	0.52	0.47	0.42	0.38	0.35	0.33	0.30	0.27	0.24	0.21	0.19	0.18
26	0.80	0.67	0.57	0.50	0.44	0.40	0.36	0.33	0.31	0.29	0.25	0.22	0.20	0.18	0.17
28	0.79	0.65	0.55	0.48	0.43	0.38	0.35	0.32	0.29	0.27	0.24	0.21	0.19	0.17	0.16
30	0.78	0.63	0.53	0.46	0.41	0.36	0.33	0.30	0.28	0.26	0.22	0.20	0.18	0.16	0.15
35	0.75	0.60	0.49	0.42	0.37	0.33	0.30	0.27	0.25	0.23	0.20	0.17	0.16	0.14	0.13
40	0.72	0.56	0.46	0.39	0.34	0.30	0.27	0.24	0.22	0.20	0.18	0.15	0.14	0.12	0.11
45	0.69	0.53	0.43	0.36	0.31	0.27	0.25	0.22	0.20	0.19	0.16	0.14	0.12	0.11	0.10
50	0.67	0.51	0.40	0.34	0.29	0.25	0.23	0.20	0.18	0.17	0.15	0.13	0.11	0.10	0.09
55	0.65	0.48	0.38	0.32	0.27	0.24	0.21	0.19	0.17	0.16	0.13	0.12	0.10	0.09	0.08
60	0.63	0.46	0.36	0.30	0.25	0.22	0.19	0.17	0.16	0.15	0.12	0.11	0.10	0.09	0.08
65	0.61	0.44	0.34	0.28	0.24	0.21	0.18	0.16	0.15	0.14	0.12	0.10	0.09	0.08	0.07
70	0.59	0.42	0.33	0.27	0.22	0.19	0.17	0.15	0.14	0.13	0.11	0.09	0.08	0.07	0.07
75	0.57	0.40	0.31	0.25	0.21	0.18	0.16	0.14	0.13	0.12	0.10	0.09	0.08	0.07	0.06
80	0.56	0.39	0.30	0.24	0.20	0.17	0.15	0.14	0.12	0.11	0.10	0.08	0.07	0.07	0.06
90	0.53	0.36	0.27	0.22	0.18	0.16	0.14	0.12	0.11	0.10	0.09	0.07	0.07	0.06	0.05
100	0.50	0.34	0.25	0.20	0.17	0.14	0.13	0.11	0.10	0.09	0.08	0.07	0.06	0.06	0.05

(Source: *Highway Capacity Manual*, HRB SR 87, 1965, Table 10-12)

TABLE 14-9. PASSENGER-CAR EQUIVALENTS OF INTERCITY BUSES ON TWO-LANE HIGHWAYS, ON SPECIFIC INDIVIDUAL SUBSECTIONS OR GRADES

GRADE (%)	*PASSENGER CAR EQUIVALENT**, E_B*		
	Levels of Service A and B	*Level of Service C*	*Levels of Service D and E (Capacity)*
0–4	2	2	2
5*	4	3	2
6†	7	6	4
7†	12	12	10

(Source: *Highway Capacity Manual*, HRB SR 87, 1965, Table 10-11)

* All lengths.

** For all percentages of buses.

† Use generally restricted to grades over $\frac{1}{2}$ mile long.

where MSV = maximum service volume in passenger cars per hour, from Table 14-3.

Level of Service. The procedure for determining the level of service for an existing or proposed two-lane highway, while accommodating a given demand volume, is the same as that described for freeways and multi-lane highways.

Combined Analysis of Subsections. The combined analysis of subsections, to obtain a weighted average level of service, is the same as that described for freeways and multi-lane highways.

Three-Lane Highways. Although three-lane highways are seldom, if ever, built in the United States today, a number are still in operation. Basic operational characteristics of three-lane highways are similar to those of two-lane highways.

The capacity of a three-lane highway under ideal conditions is about 4,000 passenger cars per hour, total for both directions, with an operating speed of about 30 miles per hour. For Level C, involving operating speeds of about 40 miles per hour, 2,000 passenger cars per hour total for both directions can be carried under ideal conditions; and for Level B, involving operating speeds of about 50 miles per hour, 1,500 passenger cars per hour total for both directions can be carried under ideal conditions.

The adjustment factors and procedures described for two-lane highways are generally applicable to three-lane highways, including the consideration of the criteria for percentage of available passing sight distance.

Sample Problem (*Highway Capacity Manual* Example 10.5)

Problem:

The following conditions are given (see Figure 14-2):

Rural two-lane highway of intermediate design
10-foot lanes
6-foot paved shoulders
No nearby obstructions
Individual 3 per cent grade, upgrade EB, 7,700 feet long (1.5 miles), starting at left end of sketch
Alinement as shown
Demand volumes: 300 vehicles per hour, with 5 per cent trucks, EB
200 vehicles per hour, with 12 per cent trucks, WB
Negligible number of buses

Determine the following:

1. Average highway speed.
2. Percentage of passing sight distance
3. Level of service being provided.

Solution:

1. Use an approximate method, in which each curve and related transitions are assigned an average influence distance of 800 feet, at the design speed of the curve, regardless of the actual length and degree of curvature. The resulting influence distances are shown in the sketch.

SUBSECTION	*INFLUENCE DESIGN*		*DESIGN SPEED*
1	1,050	×	(70) = 73,500
2	800	×	(50) = 40,000
3	3,600	×	(70) = 252,000
4	800	×	(60) = 48,000
5	1,450	×	(70) = 101,500
	7,700		515,000

$$\frac{515,000}{7,700} = 67 \text{ miles per hour, approximate, ahs.}$$

2. For the purposes of this part of the problem, zones where no passing is possible in either direction, as shown by the passing distance scale above the sketch, are taken as those portions of the roadway in the sketch which have a double solid centerline. In actual problems, they would be established as those portions not having 1,500-foot passing sight distance in either direction, regardless of markings.

Available 1,500-feet passing sight distance, eastbound:

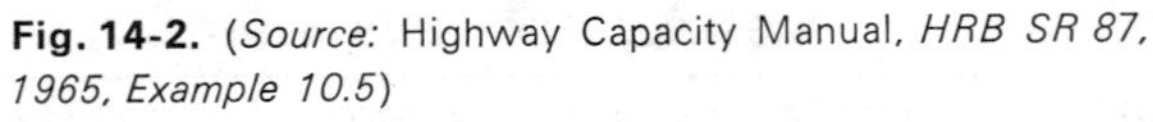

Fig. 14-2. (*Source:* Highway Capacity Manual, *HRB SR 87, 1965, Example 10.5*)

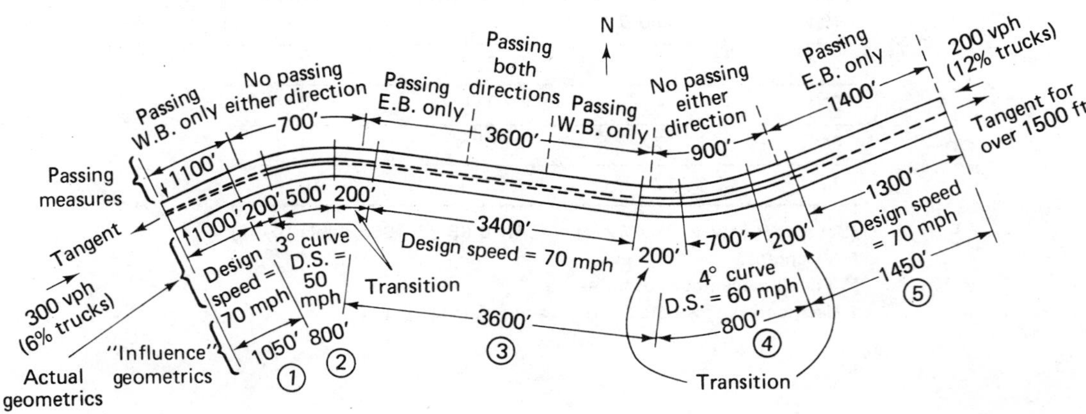

Subsection 1		none
Subsection 2		none
Subsection 3	3,600 − 1,500 =	2,100
Subsection 4		none
Subsection 5		1,400
		3,500

The available passing sight distance in the opposing direction is nearly the same, as is true in most typical problems; seldom will it justify separate consideration or averaging.

Percentage of passing sight distance

$$= \frac{3{,}500}{7{,}000} = 0.45 \quad \text{or} \quad 45 \text{ per cent}$$

3. Assumption of level must be made before certain adjustment factors can be selected. Given 500 vehicles, with a significant number of trucks, on a relatively long grade, experience with other problems indicates that service probably will be in Level C or D. Assume factors for Level D for use in trial computations.

Lanes can be considered as 11 feet wide. Lanes narrower than 12 feet can be considered 1 foot wider if a paved shoulder 4 feet or more in width is present.

Here, an intermediate grade exists, as do unbalanced percentages of trucks. Data permit refinement of the procedure to consider upgrade and downgrade effects separately.

$$5 \text{ per cent of } 300 = 15 \text{ trucks upgrade}$$
$$12 \text{ per cent of } 200 = 24 \text{ trucks downgrade}$$
$$\text{Total vehicles} = 500$$
$$\tfrac{15}{500} = 3 \text{ per cent of total volume, upgrade trucks}$$
$$\tfrac{24}{500} = 4.8 = 5 \text{ per cent of total volume, downgrade trucks}$$

Local observations of downgrade truck speeds, applied to HCM Chapter 5 procedures, have determined passenger-car equivalent E_T of 10 downgrade.

Capacity under prevailing conditions

$$= 2{,}000\left(\frac{T_{L(up)}P_{T(up)} + T_{L(dn)}P_{T(dn)}}{P_{T(up)} + P_{T(dn)}}\right) \times W_L$$

where $W_L = 0.87$ (from Table 14-4, for 11-foot equivalent lanes and adequate clearance, at Level D)

$T_{L(up)} = 0.57$ (from Table 14-7 for Level D, given 3 per cent trucks on 3 per cent grade $1\frac{1}{2}$ miles long, $E_T = 26$; from Table 14-8, for $E_T = 26$ and 3 per cent trucks, $T_L = 0.57$)

$T_{L(dn)} = 0.69$ (from Table 14-8 for $E_T = 10$ and 5 per cent trucks, $T_L = 0.69$)

Capacity under prevailing conditions

$$= 2000 \times \left(\frac{0.57 \times 3 + 0.69 \times 5}{8}\right) \times 0.87$$
$$= 1{,}122 \text{ vehicles per hour}$$

$$v/c \text{ ratio} = \frac{500}{1{,}122} = 0.45$$

From Table 14-3, for AHS = 67 miles per hour and 45 per cent passing sight distance, service is in Level C. Assumption of level was incorrect. Recompute for Level C for final check.

CAPACITY RESEARCH NEEDS

The Highway Capacity Committee of the Highway Research Board has set up guidelines for further research on capacity, identifying areas of deficient data and study.[3,4]

15

Pedestrian Studies

One of the elements requiring the attention of the traffic engineer is the pedestrian. Particularly in urban and CBD locations, the pedestrian presents an element of sharp conflict with vehicular traffic, resulting in high accident rates and traffic delays. Pedestrian movements and characteristics must be studied for the purpose of providing a design which minimizes pedestrian-vehicle conflicts, increases pedestrian safety, and minimizes vehicle delays.

The pedestrian must be accommodated and considered in a variety of situations. At any intersection where pedestrians must cross the street at grade, they must be able to do so safely and conveniently. Where high volumes of pedestrians exist, these volumes must be adequately provided for by crosswalks. In such cases, it will also be vital to determine peak crossing volumes and peak queues of waiting pedestrians,[30] so that adequate space and crossing time can be provided. In business districts, sidewalks must be wide enough to accommodate prevalent pedestrian volumes. Pedestrian walking speeds also need to be studied, as this will affect the timing of signals at intersections and the amount of space required. Where pedestrians board mass-transit vehicles, it is important that boarding times and other characteristics be studied in order that the bus loading zone, or transit station, may be properly designed as to size and location.

In these cases, the pedestrian and his characteristics are important inputs into traffic engineering decisions, and as such they merit a thorough analysis.

PEDESTRIAN MEASUREMENTS

Because of the complexity of pedestrian movement, measurements are best taken manually.

Volume counts may be taken by stationing an observer with a hand counter at any convenient vantage point. It is generally inconvenient to have one observer count more than one movement at a time. Elevated vantage points are useful for intersection studies.

Speed measurements are taken by timing (with a stop watch) persons walking a measured course. Pedestrians may be observed from elevated vantage points, assuming that parallax does not preclude the accurate observation of entrance and exit from the measured course. One study[3] had observers, following closely behind pedestrians in the Port of New York Authority Bus Terminal, timing their walk across a measured section. This method has the disadvantage that persons when followed will often alter their natural movement pattern because of an awareness of being followed.

Photographic methods are also useful where elevated

Pedestrian group size study

Date ________ Time; From ________ to ________ Location ________

Crosswalk across: ________ Curb to curb distance ________

Divided roadway? Yes No Width of island ________

Group size	Number of rows	Number of groups		Cumulative	Computations
		Tally	Total		
46–50	10				
41–45	9				
36–40	8	1	1	1	
31–35	7	111	3	4	
26–30	6	~~1111~~ 11	7	11 ←	This figures included "9", the cutoff for the 85th percentile group size, therefore N = 6
21–25	5	~~1111~~ ~~1111~~ 111	13		
16–20	4	~~1111~~ ~~1111~~ ~~1111~~ 111	18		
11–15	3	~~1111~~ ~~1111~~ 11	12		
6–10	2	~~1111~~	5		
5 or less	1	1	1		
		Total no of groups	60	x 15 = 9	N = 6

vantage points are available. The study above[3] made use of photography to study pedestrian densities, volumes, and speeds. Data reduction by this method is time-consuming, but much information may be gained from a single set of films.

One method, developed[1,2] to study school crossing locations, may be adapted to any crossing location. The method involves counting group sizes of crossing school children. In the adaptation to other locations, these could be groups of any pedestrian population. The 85th percentile group size is then determined, in terms of the number of rows of five pedestrians in each row. It is assumed that crosswalks 10 feet wide will allow crossings of rows of five pedestrians across. A sample field sheet is shown in Figure 15-1.

Having determined the 85th percentile group size, it is necessary to determine the size of the time gap in the traffic stream which will permit groups of this size to safely cross the street.

The *adequate time gap* is given by the equation

$$G = \frac{W}{3.5} + 3 + (N - 1)2 \qquad (15\text{-}1)$$

where
G = adequate time gap, in seconds
W = width of pavement to be crossed, in feet
3.5 = assumed walking speed, in feet per second
3 = pedestrian reaction time, in seconds
N = number of rows of 5 across to be crossed
2 = spacing between rows, in seconds

A study of actual gaps in the traffic stream will enable the determination of the adequacy of existing control measures, or the necessity of initiating control at an uncontrolled location.

The actual delay to pedestrians is also a useful value in analyzing the control situation. At unsignalized locations, this may be estimated by measuring gaps in the traffic stream, and considering the percentage of time during which a gap of at least G is available. This percentage is given as

$$D = \frac{T - t}{T} \qquad (15\text{-}2)$$

where
D = percentage of time that 85th percentile pedestrian group cannot safely cross
T = total study time, in seconds
t = total of all gaps $\geq G$, in seconds

It is important to note that gaps must be measured considering all lanes and directions of traffic.

For school crossing locations, control is generally considered necessary when gaps arrive at an average rate of less than one per minute.[2]

PEDESTRIAN CHARACTERISTICS

Walking Speeds

A study[4] reported in the *Traffic Engineering Handbook* gives representative walking speeds for persons crossing streets as indicated in Table 15-1.

TABLE 15-1. REPRESENTATIVE PEDESTRIAN WALKING SPEEDS WHILE CROSSING STREETS (fps)

GROUP	*15th PERCENTILE*	*MEAN*	*85th PERCENTILE*
Men	3.75	4.45	5.25
Women	3.40	4.00	4.75
All	3.55	4.20	5.00

(Source: *Traffic Engineering Handbook*, ITE, 1965)

TABLE 15-2. PEDESTRIAN WALKING SPEEDS AT MID-BLOCK AND INTERSECTION LOCATIONS (fps)

	MID-BLOCK	*INTERSECTION*
Men	4.93	4.93
Women	4.63	4.53
All	4.80	4.72

(Source: Hoel, "Pedestrian Travel Rates in Central Business Districts," in *TE*, January, 1968)

TABLE 15-3. LEVEL OF SERVICE STANDARDS FOR QUEUED PEDESTRIANS

LEVEL OF SERVICE	*PEDESTRIAN MODULE* (ft^2)	*INTERPERSON SPACING* (*ft*)	*CIRCULATION THROUGH QUEUE*
A	More Than 13	4	Unrestricted
B	10–13	$3\frac{1}{2}$–4	Slightly Restricted
C	7–10	3–$3\frac{1}{2}$	Restricted, but Possible By Disturbing Others
D	3–7	2–3	Severely Restricted
E	2–3	2	Not Possible
F	Less Than 2	—	Not Possible

(Source: Fruin, *Designing for Pedestrians, A Level of Service Concept.* Dissertation, Polytechnic Institute of Brooklyn, January, 1970, Table 9-1)

TABLE 15-4. LEVEL OF SERVICE STANDARDS FOR PEDESTRIAN WALKWAYS

LEVEL OF SERVICE	*PEDESTRIAN MODULE* (ft^2)	*PEDESTRIAN VOLUME* (*ppm/ft*)	*NORMAL FLOW**	*REVERSE FLOW***	*CROSS FLOW*†
A	More Than 35	7	F	F	F
B	25–35	7–10	F	F	R
C	15–25	10–15	F	R	R
D	10–15	15–20	R	R	S
E	5–10	20–25	R	S	S
F	Less Than 5	Variable up to 25	S	S	S

PPM/FT = Persons per minute per foot width of walkway
F = relatively free, minimum of restrictions or inconvenience
R = restricted, higher probabilities of conflict and inconvenience
S = severely restricted
(Source: Fruin, *Designing for Pedestrians, A Level of Service Concept.* Dissertation, Polytechnic Institute of Brooklyn, January, 1970, Table 5-4)
* Direction of major flow.
** Opposite direction to major flow.
† Direction at right angles to major flow.

TABLE 15-5. APPROXIMATE TIME PER PASSENGER, LOADING HEADWAY BASED ON FARE COLLECTION

FARE	*SECONDS*
Single Coin, or Token Fare Box, or Pass	2–3
Odd-Penny Cash Fares	3–4
Multiple-Zone Fares	
Prepurchased Tickets	4–6
Cash	6–8

(Source: *Traffic Engineering Handbook*, ITE, 1965, Table 13-7)

TABLE 15-6. REQUIRED NUMBER OF BUS LOADING STALLS

PASSENGER LOADING TIME (*sec*)		3			5			7		
Passengers Per Bus		30	45	60	30	45	60	30	45	60
Scheduled Bus Headway (min)	2	1	2	2	2	2	3	2	3	3
	5	1	1	1	1	1	1	1	2	2
	10	1	1	1	1	1	1	1	1	1

(Source: *Traffic Engineering Handbook*, ITE, 1965, Table 13-8)

It has also been observed[1,4] that pedestrians alter their speed depending upon the closeness of vehicles when crossing the street. Mean walking speeds varied from 6.4 feet per second when the gap before the arrival of the next car was 2 seconds, to 3.8 feet per second when the gap was 9 or more seconds.

A study of walking speeds[3] at the Port Authority Bus Terminal and Pennsylvania Railroad Station in New York City yielded mean walking speeds of 4.3 and 4.5 feet per second respectively, under free flow conditions. At both locations walking speeds for women were lower than for men.

Another study[5] of pedestrian movement in the CBD produced the results shown in Table 15-2.

There are other studies[6,8] concerning pedestrian speed characteristics to be found in the literature.

Pedestrian Volumes and Capacity Considerations

A set of capacity and level of service standards for pedestrians have been produced.[3] Levels of Service A through F are defined in a manner similar to the levels of service for vehicular flow in the *Highway Capacity Manual.*[7] As in the Capacity Manual, levels of service range from A, representing free movement, to F, representing breakdown flow. Level of Service *E* refers to the capacity or maximum volume condition.

Criteria are given in terms of *pedestrian modules* (*M*), or the number of square feet per pedestrian necessary to give the desired level of service. Unlike the HCM, where vehicle levels are defined by speed and volume, pedestrian levels are defined in terms of the inverse of density (number of pedestrians per unit area).

Criteria are given for queued pedestrians and for walkways. These are shown in Tables 15-3 and 15-4[3] on p. 254.

Transit Loading Characteristics

Also of concern to the traffic engineer is the way in which pedestrians board mass-transit vehicles. Space must be provided for loading areas, both for pedestrians and vehicles. Inadequate provisions for either vehicles or pedestrians may cause extended peak-hour delays for entire intersections or downtown streets.

Pedestrian boarding times are shown below in Table 15-5.

Table 15-6 gives the necessary number of bus loading positions, depending on the scheduled headway and pedestrian boarding times.

If different bus routes are involved, each having separate stalls, it is possible that a bus from one route may block the stall of another. Under these circumstances additional space may be required.

Table 15-7 shows the required length of stall required for one and two bus loading zones at curbside.

CLOSING COMMENTS

The foregoing has discussed the most common engineering measurements which one uses in the analysis and design of pedestrian movements at crosswalks, sidewalks, or in boarding a transit vehicle. Some of the criteria illustrated were taken from sources which were primarily concerned with traffic engineering studies of pedestrian-vehicular conflicts occurring at grade crossings. Others dealt primarily with the concept of levels of service for the pedestrians in the traffic stream. None, however, touched upon the concepts of desirable pedestrian circulation.

In the highly developed parts of urban land, it is not uncommon to witness complete traffic stoppages at times where the peaks of pedestrian movements and those of

TABLE 15-7. MINIMUM DESIRABLE LENGTH FOR CURBSIDE BUS STOPS

NOMINAL BUS LENGTH (ft)	*LENGTH IN FEET FOR*					
	One Bus Stop			*Two Bus Stop*		
	*Near-Side**	*Far-Side***	*Mid-Block*	*Near-Side**	*Far-Side***	*Mid-Block*
25	90	56	125	120	90	150
30	95	70	130	130	100	160
35	100	75	135	140	110	170
40	105	80	140	150	120	180

(Source: *Traffic Engineering Handbook,* ITE, 1965, Table 13-5)

* The length of near-side stops should be increased by 15 feet if buses are required to make a right turn. Increase by 30 feet if there is a heavy flow of other right-turning vehicles.

** Lengths for far-side stops are based on roadways 40 feet wide or wider. To enable buses to leave without crossing the centerline, increase the length by 15 feet if the road is 36 feet wide, 30 feet if it is 32 feet wide

Fig. 15-2. The Penn. Center complex *(Source: Philadelphia City Planning Commission, 1964.)*

vehicular movements coincide. Typically these occur in the morning, at noon, and in the evening peak periods. Little, if anything, can be done in these instances to eliminate the pedestrian-vehicular conflict through traffic engineering control and design. This is so because we are faced with an already existing network of streets and sidewalks which cannot be significantly altered without large capital expenditures.

Nevertheless it is possible to take advantage of design concepts which are directed to eliminating at-grade pedestrian-vehicle conflicts at the time of project-conception. It is both feasible and desirable to do so, especially in large-scale urban renewal projects located in downtown business districts. This involves physically separating vehicle movement from pedestrian movements by means of elevated or depressed pedestrian ways, so as not to interfere with crossing vehicular flows. A good example of this efficient design is in the Penn Center Complex planned for Philadelphia, Pennsylvania as shown in Figure 15-2. A proper term with which to describe this traffic planning concept is "traffic architecture."[11] The Penn Center plan conveys the desirable design concept of the building-circulatory road relationships to achieve efficient handling of traffic, both vehicular and pedestrian.

Another excellent example of a desirable traffic circulation design for a suburban community is illustrated in Figure 15-3. This is taken from the Radburn, New Jersey, plan. It will be noted that pedestrian walkways are completely separated from vehicular routes, thus entirely eliminating the dangerous vehicle-pedestrian conflicts which are the causes of most pedestrian traffic fatalities or injuries.

Too often have traffic engineers applied their talent and technical know-how to analyzing problems which are attributable to poor urban planning and design. In these cases the traffic engineer has limited hope of successfully coping with the problems created by the

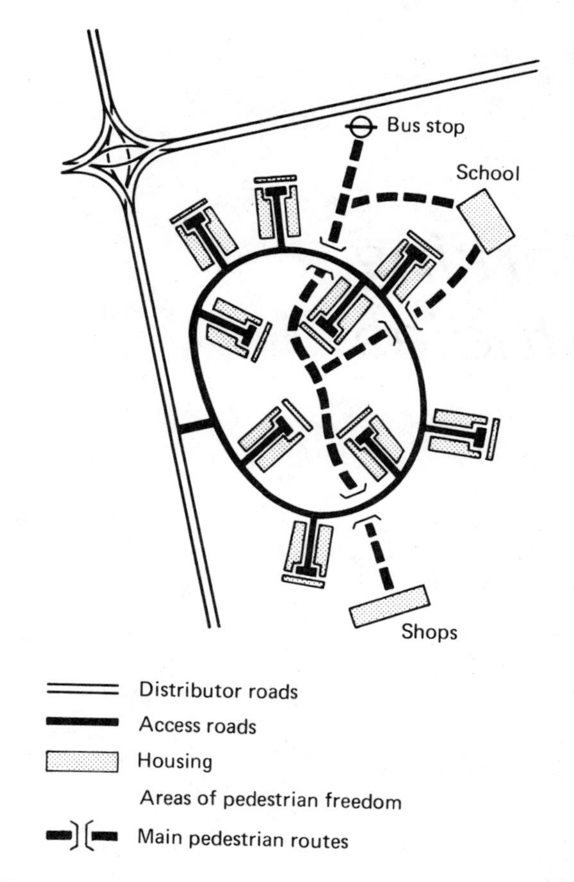

Fig. 15-3. An example of the Radburn circulation plan (*Source:* Traffic In Towns, *The Buchanan Report, Penguin Books, Inc., Baltimore, Md., 1963, Figure 15*)

vehicle and pedestrian crossing. The most effective traffic engineering contribution can and must be registered during the plan-conception phase of any significant civic project. The traffic engineer must become involved with city and regional planners, and his office must work together with the planning counterpart. Through this approach alone can one expect significant traffic improvements in the urban areas of the future.

16

Parking Studies

Parking is a major urban land use. Anyone who drives a car needs no introduction to the difficulties of finding a parking space in areas which are intensively used for business, commercial, or residential purposes. An area containing a central business district (CBD), a regional or community shopping center, an industrial park, an airport, a civic center, or a stadium is usually an area where extensive parking problems are found.

The parking problem becomes more pronounced as the size of the city increases. Figure 16-1 shows that the number of parking spaces per 1,000 population in the CBD declines with increasing city size, even though the *total number* of spaces in the CBD increases with increasing city size.

Historically, expenditures for parking facilities have not kept pace with those for highways, and this lack of balance in investments has compounded the traffic congestion which is witnessed daily in major urban centers.

THE NATURE OF THE PROBLEM

The terminal, the roadway, and the vehicle are the basic elements of the highway transportation system. The arrangement and design of each element influences the performance of the total system. For this reason, the problem of terminal design, its regulation and control, must be tied to problems of traffic flow requirements and vehicle characteristics.

The land space allocated for motor vehicles may be divided between space for movement and space for vehicle storage. Like any economic good, this space resource is scarce in high-density areas such as the CBD. In these areas, therefore, the vehicle is in competition with itself for its space needs. A classic example of this competition is the curb parking problem, a problem of how to allocate the total street space resources and apportion them into space for vehicles in motion and space for vehicles at rest. However simple this may sound, there is no straightforward solution, since the criteria to be followed in providing a solution are dependent on the goals of the community, and these tend to vary from one community to another.

However philosophies may differ in the approaches taken to solve the dilemma between the competing needs of the vehicle in motion and the vehicle which is to park, an agreement can be developed as to what principles should be followed in proper planning for vehicles in motion and vehicles at rest. For example, the highway network could be viewed as a composite of three subsystems, arterial, collector, and local, each performing different functions:

1. *The Arterial Subsystem.* Arterial streets exist for the purpose of moving traffic. The traffic engineer has the responsibility of insuring that each arterial highway be used to maximize flow under safe conditions. The conflict between the moving vehicle and the stationary vehicle on the arterial should be resolved in favor of the vehicle in motion.
2. *The Collector Subsystem.* Collector streets are for the purpose of channeling traffic between the arterial subsystem and the local street subsystem. In some cases they serve abutting land uses. Conflicts between movement of traffic and parking may not always be solved in favor of the vehicles in motion.
3. *The Local Subsystem.* The primary function of the local streets is to serve abutting land use. Here the conflict between the vehicle in motion and the vehicle which is to park should be resolved in favor of the vehicles at rest.

If these goals and objectives of each subsystem are clearly defined *prior* to tackling any area requiring assistance in the improvement of traffic flow and traffic parking, street circulation systems could be planned in accordance with, and respecting, the basic principles of each subsystem. Thus, the job of reconciling the conflicting needs of the vehicle in motion and the vehicle parked would be greatly minimized. What is implied is the necessity for developing criteria on land use controls along each street subsystem. If, for example, the arterial street subsystem is to function efficiently, land use control

Fig. 16-1. Central business district parking spaces and urbanized area population (*Source: Wilbur Smith and Associates,* Parking In The City Center, *Automobile Manufacturer's Association, 1965*)

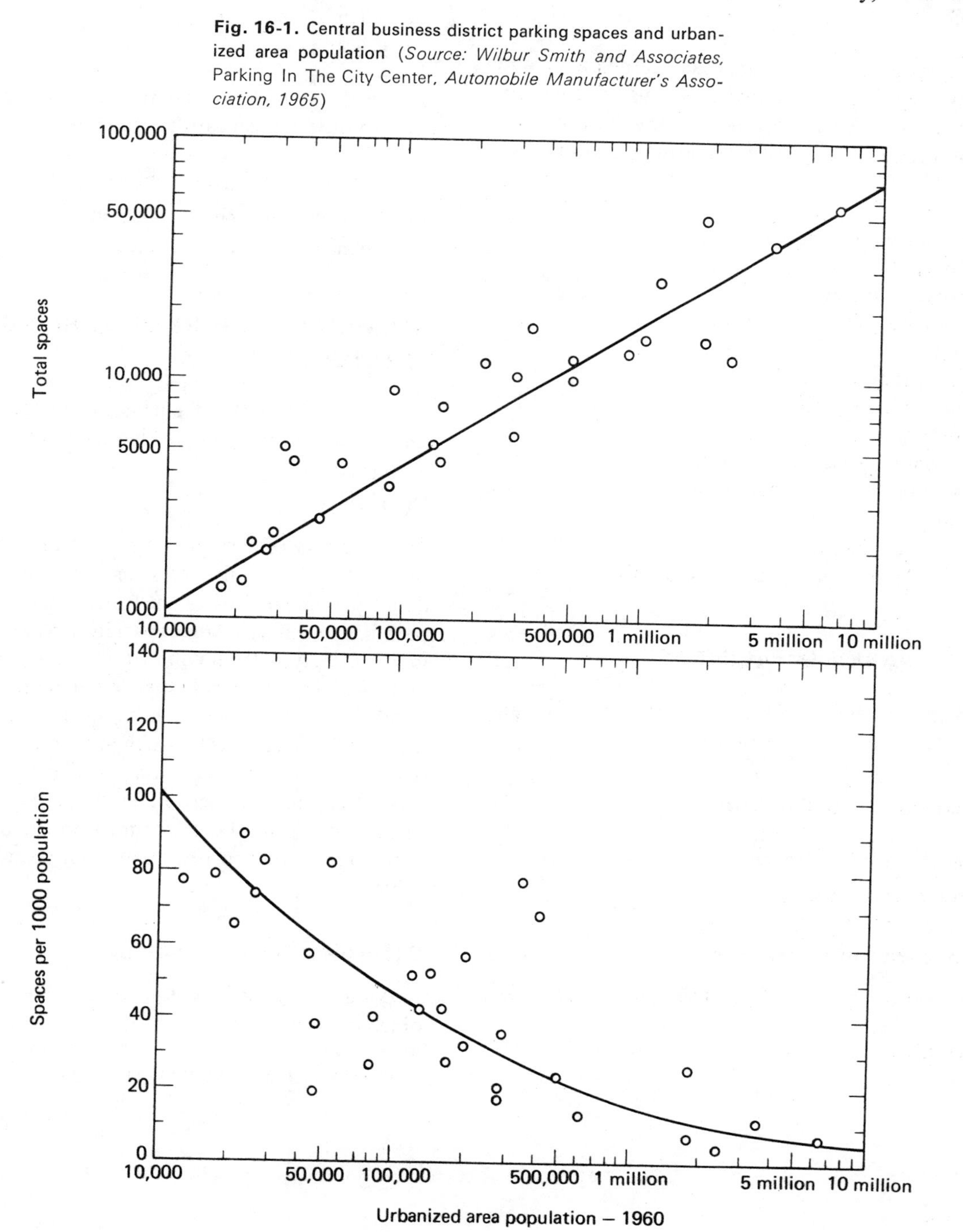

along the land abutting the arterials should insure that the use of this land will be compatible with the function of the road.

Lack of land use controls along the land abutting major traffic facilities contributes substantially to the problem of traffic flow on these roads. Limited choices are found in existing, built-up CBD's, but great opportunities exist for achieving functional balance between the requirements of vehicles in motion and vehicles at rest in newly developing suburban areas, or in other urban areas undergoing large-scale redevelopment.

GROUPS AND INTERESTS AFFECTED BY PARKING

There are many groups and interests concerned with, and affected by, parking. Accompanying this diversity of interests is a wide variety of views on how best to solve the parking problem. Some of these groups and interests are:

Downtown businesses
Motorists
Property owners
Commercial fleet operators
Taxis
Emergency vehicles
Mass transportation
Commuters
Economic losses due to parking
Parking lot and garage owners
Police
Traffic engineers
City planners and planning commissions

TYPES OF PARKING FACILITIES

There are several ways in which parking supply is made available. These include the following:

On-Street or Curb Facilities

1. Unrestricted curb parking
2. Restricted curb parking

Off-Street Parking Facilities

There are two basic types of off-street parking facilities:

1. Surface lots
2. Garages

Garages, classified by general type, are considered:

1. Above ground
2. Underground
3. Integral

If garages are classified by means of interfloor travel, they are considered:

1. Mechanical
2. Ramps

Off-street facilities, classified by method of operation, are:

1. Attendant parking
2. Self-parking

Off-street facilities, considered in terms of ownership, develop other classifications, such as:

1. Privately owned and operated
2. Publicly owned and privately operated
3. Publicly owned and operated

GEOMETRIC DESIGN OF PARKING TERMINALS

The two basic types of parking terminals are the curb space and the off-street parking facilities.

Curb Parking

The geometry of curb parking, illustrated in Figure 16-2, shows that the number of spaces that can be accommodated per linear foot of curb increases as the angle of parking increases. Angle parking, however, interferes more severely with moving traffic than parallel parking, and accident rates are higher for angle parking than for parallel parking. Angle parking is therefore recommended for local streets which are wide, have good sight distance, and carry low traffic volumes.

The required geometric layout for buses, trucks, and taxi stands should be determined on the basis of vehicle dimensions and operational characteristics of the vehicles.

Off-Street Surface Parking

Parking lots must be designed to achieve the following objectives:

1. Provide maximum number of spaces

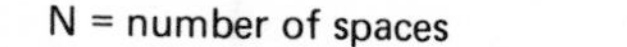

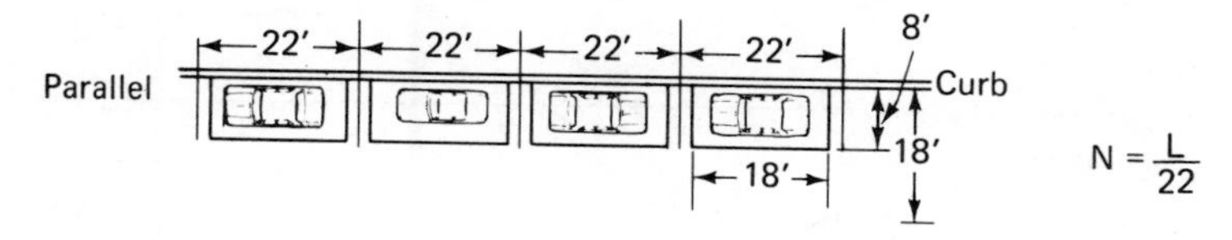

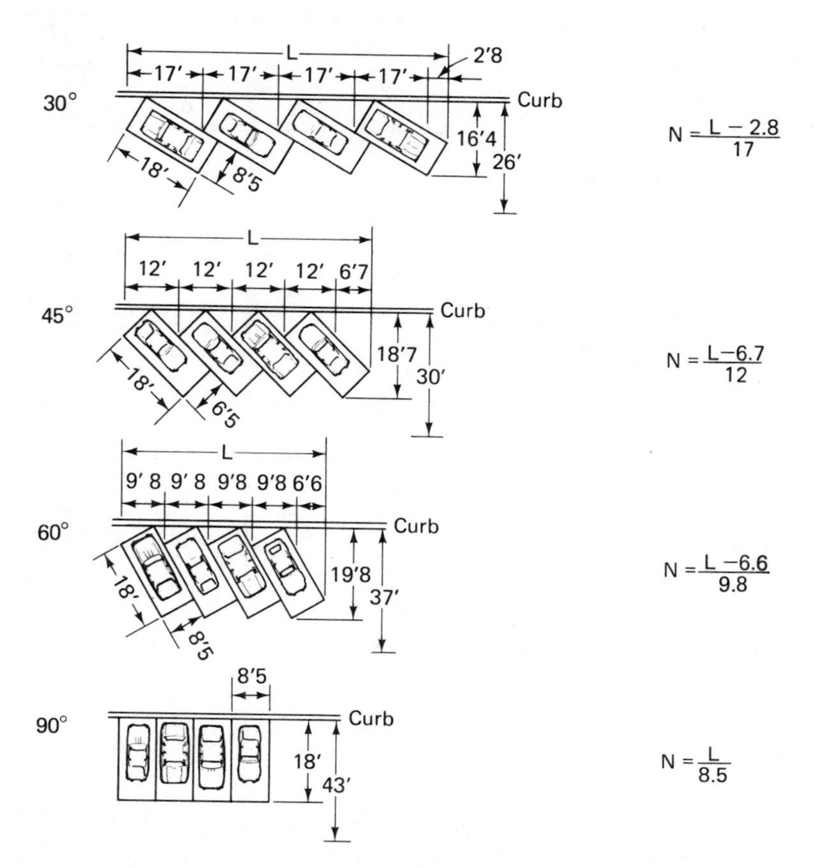

Fig. 16-2. Street space used for various parking (*Source:* Parking, *ENO Foundation, 1957, Figure IV.1*)

2. Minimize travel discomfort while parking, unparking, and driving within the lot
3. Minimize interference of entrance and exit lanes with pedestrian and vehicle movements external to the lot

Various forms of parking-stall layouts are available for use in a parking lot. Selection of the best parking angle, however, depends primarily on the size and shape of the parking lot. More than one parking angle layout may be used in a parking lot to maximize space utilization.

Typical parking lot capacity figures as a function of the parking angle are shown in Table 16-1. The headings of Table 16-1 are illustrated in Figure 16-3 for a 90-degree parking angle.

Layouts for self-parking should use stall widths no less than 8.5 feet. A stall width of 9.0 feet would be highly desirable. To allow the maneuver into and out of a stall with comfort and convenience, the minimum width of an aisle should be 12 feet for one-way movement, and 24 feet for two-way travel. The "unit parking depth" is a convenient unit of measure for the purpose of selecting the best parking layout for a lot.

90-degree Parking. This layout uses the space most efficiently. Cars can use the aisle in either direction, and travel distances are reduced. It permits use of dead-end aisles, thereby minimizing wasted space.

Other Angle Parking. If the parking angle is less than 90 degrees, the travel aisles must be made one-way. One-way circulation is desirable (but not essential) for a busy lot, since the 30-degree and 45-degree angles are more easily accessible by self-parkers. These layouts are generally used for customer parking, and where space is available.

Regardless of the parking angle used, however, the traffic engineer should insure that the circulating system of the lot should permit easy and efficient movement of both the car and the pedestrian. Entrances and exits should be placed with the objective of minimizing potential conflicts within the parking lot, and between lot traffic and access street traffic. Entrances and exits

TABLE 16-1. TYPICAL PARKING LOT CAPACITY FIGURES

	SINGLE PARKING LINES						INTERMESHED MULTIPLE PARKING LINES			
α	*L*	*D*	*W*	*A*	*upd*	*N*	D_i	A_i	*upd*	N_i
Parking Angle (deg)	*Curb Length per Car (ft)*	*Depth of Stall (ft)*	*Width of Aisle (ft)*	*Gross Area per Car (sq ft)*	*Unit Parking Depth (ft)*	*Approximate Number of Cars per Acre*	*Depth of Stall (ft)*	*Gross Area per Car (sq ft)*	*Unit Parking Depth (ft)*	*Approximate Number of Cars per Acre*
0	22	8	12	308	28	141	8	308	28	141
20	24.9	14.2	12	502.9	40.4	87	10.1	400.9	32.2	109
25	20.1	15.4	12	430.1	42.8	101	11.4	349.7	34.8	125
30	17.0	16.4	12	380.8	44.8	114	12.7	309.8	37.4	141
35	14.8	17.3	12	344.8	47.6	126	13.7	291.6	39.4	149
40	13.2	18.1	12	318.1	48.2	137	14.8	274.6	41.6	159
45	12.0	18.7	12	296.4	49.4	147	15.8	261.6	43.6	167
50	11.1	19.2	12	279.7	50.4	156	16.6	250.9	45.2	174
55	10.4	19.6	12	266.2	51.2	164	17.2	241.3	46.4	181
60	9.8	19.8	14.5	265.1	54.1	164	17.8	245.5	49.6	177
65	9.4	19.9	17	267.0	56.8	163	18.2	250.9	53.4	174
70	9.0	19.8	20	268.2	59.6	162	18.4	255.6	56.8	170
75	8.8	19.6	23	273.7	62.2	159	18.6	264.9	60.2	164
80	8.6	19.2	24	268.3	62.4	162	18.4	261.4	60.8	167
85	8.5	18.7	24	260.9	61.4	167	18.3	257.6	60.6	169
90	8.5	18.0	24	255.0	60.0	171	18.0	255.0	60.0	171

(Source: Burrage and Mogren, *Parking,* ENO, 1957, Fig. VIII-4)

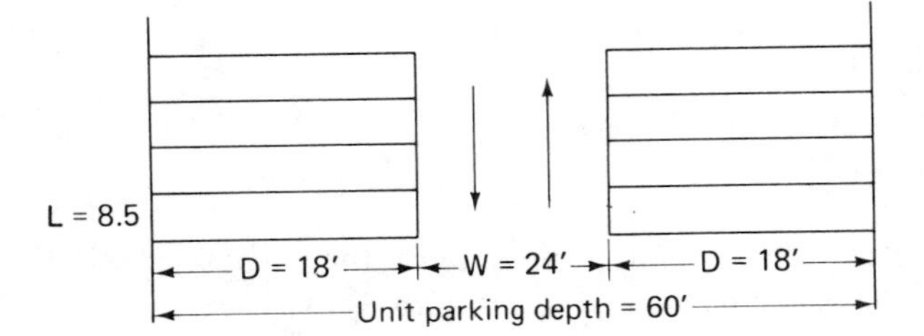

Fig. 16-3. Typical lot: 90-degree angle parking

of parking lots should be located at mid-block to provide the least interference with street intersection areas where pedestrians and vehicles are in greatest competition for street space.

Parking areas should be located, preferably, on the local street subsystem, and should be avoided along major arterial streets, whose primary function is to move traffic and not to give land use service.

Storage Area. Parking facilities using attendant parking should be designed with enough reservoir space to avoid encroachment of the access street by cars queued to gain access to the lot. The amount of reservoir space is determined from the arrival rate of cars during the peak period, the number of attendants, and their rate of vehicle storage. Figure 16-4 aids in the determination of reservoir space on the basis of the above variables.

Parking Garages. Many of the concepts described for surface parking apply to the layout of garages. The gross area per car space required for garages is, however, larger than that required for surface parking. This stems from the extra space used by ramps, elevators, stairs, columns, etc. For self-parking garages, the gross area per car space varies from 350 to 400 square feet.

USAGE OF PARKING FACILITIES

Table 16-2 shows the distribution of the available supply of parking spaces, and distribution of parkers by type of parkings. It is seen that in the largest cities, curb spaces are only 16 per cent of the total, but nevertheless accommodate 50 per cent of all the parkers. In the largest cities, garages provide 24 per cent of the space supply, but handle only 12 per cent of the parkers. In the smallest cities, 93 per cent of the parking is at the curb and there are no garages.

Table 16-3 indicates that as cities grow, the average length of time a vehicle is parked increases, no matter what the trip purposes of the parker may be. The shortest durations result from parkers on sales and service calls, stops for shopping and business calls average slightly under one hour, and the parking time of workers is the longest. In small towns, a trip downtown often

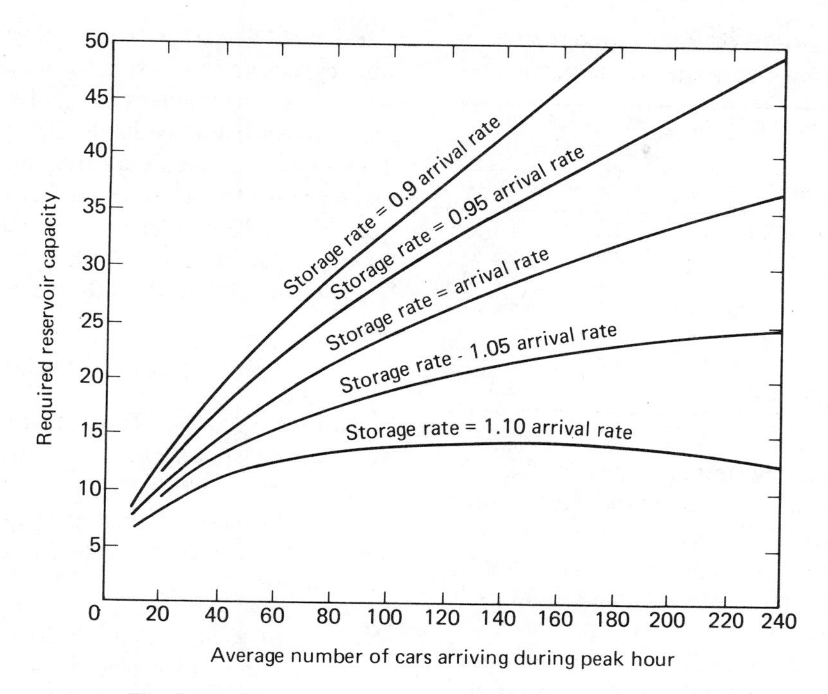

Fig. 16-4. Reservoir space (*Source: Burrage and Mogren,* Parking, *ENO, 1972, Figure VI.11*)

TABLE 16-2. PARKING SUPPLY AND USAGE

	PERCENTAGE DISTRIBUTION OF AVAILABLE SUPPLY OF PARKING SPACES				*PERCENTAGE DISTRIBUTION OF PARKERS BY TYPE OF PARKING*			
Population Group	*Curb*	*Lot*	*Garage*	*Off-Street*	*Curb*	*Lot*	*Garage*	*Off-Street*
5–10	88	12	0	12	93	7	7	7
10–25	64	32	4	36	85	14	1	15
25–50	61	35	4	39	84	15	1	16
50–100	38	38	7	45	79	19	2	21
100–250	44	42	14	56	76	20	4	24
250–500	30	54	16	70	66	28	6	34
500–1000	23	51	26	77	63	26	11	37
Over 1,000	16	60	24	84	50	38	12	50

TABLE 16-3. AVERAGE LENGTH OF TIME PARKED FOR VARIOUS TRIP PURPOSES

POPULATION GROUP	*NUMBER OF CITIES*	*SHOPPING*	*BUSINESS*	*WORK*	*SALES AND SERVICE*	*OTHER*	*ALL*
5–10	2	0.5	0.5	2.8	0.5	0.7	1.0
10–25	14	0.6	0.6	3.1	0.6	0.9	1.1
25–50	16	0.6	0.7	3.4	0.6	1.0	1.3
50–100	5	0.7	0.7	3.8	0.6	1.1	1.4
100–250	13	1.0	0.9	3.8	0.5	1.3	1.6
250–500	6	1.3	1.1	4.8	0.7	1.4	1.9
500–1,000	4	1.3	1.3	4.8	1.0	1.4	2.2
Over 1,000	3	1.8	1.5	5.6	1.0	1.9	3.0

TABLE 16-4. AVERAGE LENGTH OF TIME PARKED AT VARIOUS TYPES OF PARKING FACILITIES

POPULATION GROUP	*NUMBER OF CITIES*	*LENGTH OF TIME PARKED IN HOURS*				
		Curb	*Lot*	*Garage*	*Off-Street*	*All Spaces*
Under 25	17	0.9	2.1	5.2	2.2	1.1
25–50	15	1.0	2.5	4.3	2.6	1.2
50–100	5	1.1	2.5	5.1	2.7	1.4
100–250	12	1.0	3.2	4.9	3.5	1.6
250–500	8	1.1	3.7	4.5	3.9	2.1
500–1000	5	1.0	4.0	4.5	4.4	2.1
Over 1,000	3	1.3	4.5	4.6	4.5	2.9

requires no more than a few minutes, and parking is simple. Many trips are made on impulse, perhaps several trips a day. As cities increase in size, the drive becomes longer, and parking more of a problem. As a result, fewer trips are made, and more is accomplished on each trip.

The influence of the type of facility on parking durations is shown in Table 16-4. The average stay at the curb is much shorter than the duration off the street, both because the curb is more attractive to short-time parkers, and because a large proportion of the curb space is restricted to short-time parking.

Population has little influence on length of time parked at curbs or in garages. However, at lots, the average durations definitely increase as the city size increases. Curb durations are held down by restrictions, and it appears that as cities grow (and greater numbers of parkers require more than an hour), those needing from one to three hours prefer lots, leaving the garages for the long-time parkers.

DEFINITIONS

To conduct a parking analysis, it is necessary to establish common units of measurement which are useful in describing various states of parking activity. The most commonly used terms are:

Space Hour: One parking space for one hour.

Parking Accumulation: The total number of vehicles parked at a given time.

Parking Volume: The total number of vehicles parking in a particular area over a given period of time. It is usually expressed in vehicles per day.

Parking Load: The total number of space hours used during a given period of time. It corresponds to the area under the accumulation curve. Its peak is reached at peak accumulation, when capacity is used to its fullest extent.

Practical Capacity: This is always less than the available capacity. This is due to the time wasted in parking and unparking maneuvers, and the lack of knowledge that a free space is available. Practical capacity is usually 5 to 15 per cent less than theoretical capacity.

Turnover: The average number of times that a parking space is used by different vehicles during a given period of time. For example, if 100 spaces were used by 1,000 vehicles in a 10-hour study period, the turnover would be:

$$1000 \div (100) = 10 \text{ vehicles per space}$$

Parking Duration: The average amount of time spent in a parking space. This is a measure of the frequency of availability of parking spaces. The expected number of vehicles that can park in a given area can be calculated in terms of average parking duration by the following relationship:

$$\text{Number of vehicles that can park in a given area} = \frac{\text{(number of spaces) [period covered (in hours)]}}{\text{average parking duration (hours per vehicle)}} \times (0.85 \text{ to } 0.95) \qquad (16\text{-}1)$$

Parking duration is also dependent on the type of parking facility. For example, 70–80 per cent of curb parking is less than 1 hour in duration, while corresponding figures for off-street parking are 10–20 per cent of less than 1 hour in duration.

RELATIONSHIPS OF PARKING SUPPLY, USAGE, AND DEMAND

Comprehensive parking studies include a record of where each driver parked, how long he parked, and his destination after parking. It is therefore possible to determine the demand for every block in the CBD, both in number of vehicles and in space-hours needed.

TABLE 16-5. DEMAND FOR AND SUPPLY OF SPACES

POPULATION GROUP	*NUMBER OF CITIES*	*SUPPLY, SPACE-HOURS*	*DEMAND, SPACE-HOURS*	
			Number	*Per 1000 Population*
5–10	2	6,380	3,600	426
10–25	17	13,200	7,970	451
25–50	16	21,400	13,430	372
50–100	5	39,970	31,800	397
100–250	13	44,200	30,520	189
250–500	7	94,960	63,340	175
500–1000	5	104,740	79,670	145
Over 1,000	3	185,510	146,660	112

Table 16-5 compares, in space-hours, the supply of parking space in the CBD with the demand. The total demand increases as the population increases, but at a very much slower rate, as shown by the decline in per capita demand. The larger cities have mass transit available, and attraction outside the CBD. These account for the decline in per capita demand, in addition to the delay, congestion, and high cost of off-street parking.

Most business districts have enough parking space, but not where it is needed. The presence of retail stores, banks and office buildings in the core causes the high land values which prevent the establishment of parking facilities, and at the same time generates the demand for them. These factors are illustrated in Table 16-6.

TABLE 16-6. SUPPLY AND DEMAND IN THE CORE

POPULATION GROUP	*NUMBER OF CITIES*	*DEMAND, SPACE-HOURS*	*SUPPLY, SPACE-HOURS*	*RATIO DEMAND TO SUPPLY*
Under 25	17	3,930	3,050	1.4
25–50	15	7,040	4,280	1.6
50–100	4	22,670	11,730	2.0
100–250	13	14,460	8,850	1.8
250–500	7	37,400	19,950	2.2
500–1,000	4	45,500	11,040	4.5
Over 1,000	3	83,380	22,940	3.8

WALKING DISTANCES IN PARKING

In any decision to locate additional parking spaces, the factor of walking distance is introduced. If walking distance were not a factor, then there would be no parking problem in any city, for space can always be found by walking far enough. Table 16-7 gives data on the walking habits of parkers. It is clear that the distances people walk after parking their vehicles become larger as the population increases. In cities under 25,000 population, the average distance was only 223 feet, and 77 per cent of the parkers walked less than one block. In the largest cities the average distance was 549 feet, and only 45 per cent walked less than a block.

Based upon parking duration, parkers tend to accept greater walking distances as their parking durations increase. Parking facilities intended primarily to serve short-time parkers, such as shoppers, should be located within one or two blocks, depending on the size of the city.

TABLE 16-7. AVERAGE WALKING DISTANCE BY TRIP PURPOSE (ft)

	TRIP PURPOSE		
POPULATION	*Work*	*Shopping*	*Sales and Service*
25,000–50,000	408	295	216
100,000–250,000	539	539	221
500,999–1,000,000	698	656	419

PARKING STUDIES

The purpose of a parking study is to provide recommendations for the development of a parking program which meets the requirements of an area. To carry out a parking study, it is necessary to have information on the following points:

1. The supply and type of parking facilities
2. How and for what purpose parking facilities are used
3. The demand for parking space
4. The characteristics of parking demand
5. The location of parking generators
6. The legal, financial, and administrative factors associated with the parking situation

The scope of a parking study which will provide all the information necessary for developing a program of parking facility improvements must be comprehensive in nature. However, due to limitations of funds, manpower, or time, it may be necessary to conduct limited parking studies to obtain the essential, basic information on parking conditions in a city or an area.

Traditionally, parking studies were designed to analyze the needs of the central business district (CBD). Several procedures have been developed for this purpose, ranging from the very comprehensive to the more limited studies. Some of these procedures, for various purposes, are illustrated below.

CBD Parking Generators

The average requirements of some typical CBD establishments are shown in Table 16-8. Expected changes are given in Table 16-9 for various United States cities. Relationships between CBD parking space requirements and urban population are shown in Figure 16-5.[4] Parking studies for the CBD follow.

Comprehensive Parking Study[2]. The procedure for conducting a comprehensive type of study has been developed by the Bureau of Public Roads (BPR). Federal aid funds that have been allocated to the states for engineering and economic investigations may be used to finance parking studies.

General procedures and preliminary steps

Some preliminary or organizational work of a general

TABLE 16-8. AVERAGE PARKING SPACE REQUIREMENTS FOR SELECTED DOWNTOWN ESTABLISHMENTS

TYPE OF ESTABLISHMENT	*SPACE PER 1,000 SQUARE FEET*	
	Average	*Range*
Banks	5.4	1.8–10.8
Bus Depots	4.8	1.7– 7.9
Libraries	4.1	3.9– 4.3
Medical Buildings	3.8	1.1– 8.6
Grocery Stores	3.7	1.4– 7.5
City-County Offices	3.6	1.2– 6.0
Post Offices	3.4	2.0– 4.9
Utility Company Offices	2.9	0.4–10.7
Drug Stores	2.9	1.4– 5.5
Department Stores	2.8	1.4– 5.1
Clothing Stores	2.5	1.1– 6.3
Restaurants	2.1	0.9– 3.3
YMCA-YWCA	1.6	1.2– 2.2
Offices	1.5	0.4– 2.9
Auto Sales	1.2	0.9– 1.5
Variety Stores	1.1	0.6– 1.9
Hotels	0.6	0.4– 1.0
Furniture Stores	0.6	0.3– 1.2

(Source: Wilbur Smith and Associates, *Parking in the City Center,* Automobile Manufacturers Association, 1965, Table 3)

nature is necessary to prepare for the various phases of the study. This includes the following:

1. Select overhead personnel. The necessary overhead personnel for planning the parking study, supervising its operations, and analyzing its results should comprise the following, as a minimum: director, assistant director, traffic studies manager, office manager, one or two draftsmen, stenographer.
2. Provide for office space.
3. Delineate area to be studied. The area will generally cover the central parking district, which includes the CBD, broadly defined as streets with business frontage plus the fringe area, where vehicles are parked by persons destined within the business district. This fringe area will usually extend two or three blocks outside the CBD.
4. Prepare a map of the central parking district. A large-scale map (about 1 inch = 200 feet) of the central parking district should be prepared for use in the inventory and for assigning interview and cordon stations.
5. Establish a system for designating blocks. If a previous origin and destination study has established a code system for numbering blocks, it should be followed. A preferred system for numbering blocks is to consider a block as both sides of a street between two intersections, the block area including both curbs and any lots, garages, buildings, or alleys entered from those curbs.
6. Establish scheduling of operations.
7. Prepare land use and assessed valuation maps for study area.
8. Study existing parking and zoning ordinances.
9. Prepare interview forms.
10. Prepare manpower requirements, equipment, and forms for the cordon count.
11. Prepare manuals for field personnel.
12. Enlist cooperation of appropriate agencies and organizations. The success of the parking study, as well as the development program to result from it, will largely depend upon the cooperation of the police department, the city government, the retail merchants, the parking lot and garage operator, the chamber of commerce, the board of trade, the planning board, and other municipal and civic bodies.
13. Prepare publicity and campaign.
14. Select and train field personnel.
15. Equip field office.

Inventory of parking facilities

A physical inventory of existing facilities in the central parking district, at the curb and off-street, lists their location and capacity (in numbers and space-hours), physical features, operating features, regulations, and fee schedules. Included also are data on land use and assessed valuations, as well as areas suitable for additional facilities. The purposes of the inventory are to determine (1) the existing supply of space and the necessary auxiliary information to permit its evaluation, and (2) areas or locations for potential development of additional parking spaces.

The data collected are summarized in tables and on a map of the area under study. The data permit the determination of the theoretical available space-hours of parking capacity for each facility. Usually the space-hour capacities are expressed for an 8-hour (10 A.M. to 6 P.M.) or 10-hour (8 A.M. to 6 P.M.) period. Thus, any space available for the entire 8- or 10-hour period, regardless of the time limits on parking, will provide 8 or 10 space-hours of parking.

Parking interviews

The purposes of this phase of the study are to determine (1) the extent to which the inventoried facilities are being used, and (2) the extent and location of the demand for parking facilities. The information on demand, and its location, can come only from the drivers who create it. It is obtained by interviewers assigned to the various parking facilities (the curbs, parking lots, and garages).

The driver of practically every car parking in the area

TABLE 16-9. EXISTING AND PROJECTED PARKING SPACE SURPLUSES AND DEFICIENCIES
Selected Central Business Districts

URBAN AREA	TOTAL PARKING SPACES	SURPLUS (+) OR DEFICIENCY (−) IN STUDY YEAR (SPACES)			SURPLUS (+) OR DEFICIENCY (−) IN FUTURE YEAR (SPACES)		
		Study Year	CBD	Core	Future Year	CBD	Core
Chattanooga, Tenn.	6,908	1960	+1,070	− 800	1970	−2,325	−1,560
Hartford, Conn.	10,423	1961	− 773	−1,294	1971	+4,461	−1,645
Nashville, Tenn.	15,089	1959	+1,990	− 720	1970	+ 252	−1,152
Charlotte, N. C.	12,117	1961	+ 763	− 580	1966	− 991	−1,940
Pittsburgh, Pa.	14,830	1955	−2,358	−3,464	1965	−7,033	− 865
New Orleans, La.	13,634	1960	+ 771	− 920	1970	−1,573	−2,600
Philadelphia, Pa.	39,024	1957	− 999	−2,997	1970	−8,274	−6,177

(Source: Comprehensive parking studies in each urban area, Wilbur Smith and Associates, *Parking in the City Center,* Automobile Manufacturers Association, 1965, Table 6)

Fig. 16-5. Parking space demand factor (*Source: Wilbur Smith and Associates,* Parking In The City Center, *Automobile Manufacturers Association, Figure 14*)

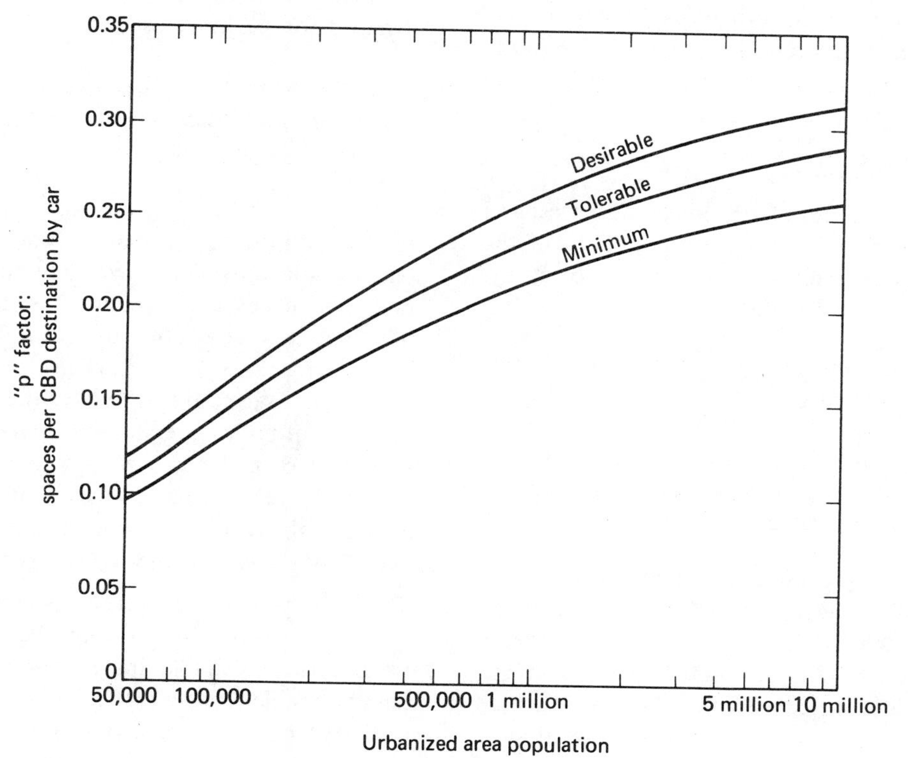

Application of curve

1 — Estimate CBD person trip destinations

2 — Estimate percentage of downtown person trips by car

3 — Calculate the daily CBD person trip destinations by car (multiply 1 by 2)

4 — Read the appropriate "p" from above curves

5 — Calculate the CBD parking space requirements (multiply 3 by 4)

6 — Calculate additional parking spaces required by comparing space demands with available spaces

is interviewed. An attempt is made to interview 100 per cent of parkers over an 8- or 10-hour period, usually from 8 or 10 A.M. to 6 P.M., on a typical weekday, to determine (1) trip origin, home address, and purpose of stop, and (2) destination to which driver is going to walk, which, together with the observed location of parking, will determine the distance walked by parker.

Observed data by the interviewer, in addition to parking location, include (1) times of arrival and departure, to obtain duration of parking, (2) type of vehicle, and (3) type of parking space, i.e., whether parking is at an unrestricted curb, a metered space, an illegal place, or a fee lot or garage. Normally, the interview requires about 30 seconds, and one interviewer for every 15 spaces is required. Answers to questions, and observed data, permit determination of turnover, violation of space and time regulations, and use of loading zones and other special-privilege facilities.

Turnover is a measure of the rate of usage of a facility, and it is determined by dividing the number of vehicles parked in that facility, during the study period, by the number of spaces available.

Cordon count

A cordon count is a count of all vehicles, inbound and outbound, on each street crossing the boundary of the CBD or any other congested area designated for study. The purposes of the cordon count are:

1. To determine the volume and classification of traffic, inbound and outbound, on each street crossing the boundary of the area under study.
2. To estimate the total parking load imposed on the area during the business day, or other selected period, and during half-hour periods throughout the day.

Data are obtained by conducting manual counts and classification of traffic at locations where streets cross the cordon for at least that period of the day when parking interviews are made. A 12-hour count (7 A.M. to 7 P.M.) is preferable. In addition, continuous, 24-hour machine counts are made at control stations for the duration of the study. Volumes should be recorded at least hourly, and preferably at 15-minute intervals. The continuous counts are used to expand and adjust the cordon counts to average-day volumes.

Cordon count data are summarized in tabular form, showing the accumulation of vehicles (by classification), by half-hour periods within the study area, which permits an estimate of the total volume of vehicles inside the cordon at any particular time. To evaluate the accumulation of vehicles, it is necessary to determine the number of vehicles in motion and the number of vehicles parked within the area at the beginning of the cordon count.

Data from the cordon count and parking interviews are combined, and normally plotted, to indicate total accumulation of vehicles, total parking accumulation (on and off-street), accumulation of commercial vehicles, accumulation of vehicles parked off-street, and accumulation of commercial vehicles parked.

Analysis of data

This includes the routine coding and tabulation necessary to summarize the data, as well as the interpretation of the data for making recommendations and for the development of an improvement program.

Demand for parking space

Analysis of data will give (block by block within the CBD) the supply of available parking facilities, the use being made of those facilities, and the space-hours of demand for parking, as measured by:

1. Number of drivers legally parked and duration
2. Number of drivers illegally parked and duration
3. Number of drivers parked just outside the CBD and duration

The total of these demands, when compared with the parking space available, will demonstrate the priorities and urgency of requirement for additional space. The data will disclose whether the need of these parkers is for long-time or short-time parking.

Parking usage data will reveal the location and type of violation of parking regulations, parking duration habits, the effect of metered areas on turnover as against free zones, the relative patronage of off-street facilities, the extent of the enforcement of current regulations, and the extent of all-day parking by workers.

Determination of practical space-hours of supply

The inventory of parking facilities will yield the theoretically available space-hours, but it is practically never possible to obtain 100 per cent efficiency in the use of this space. This is due to the time lost in each turnover and, more importantly, to the fact that demand and supply are not parallel throughout the day. The available space-hours must, therefore, be discounted by application of factors based upon the best performance which can be expected from each type of facility (curb, lot, or garage).

The efficiency factor (a factor for converting theoretically available space-hours into practical space-hours) for curbs may be determined by comparing the supply in space-hours with the legal curb usage in space-hours for the blocks where the demand is heaviest. In these blocks the usage will usually be found to fall between 80 and 95 per cent (say an average of 90 per cent), and this figure is adopted as the factor for all curbs in the district. Similar procedures will permit the determination of factors for lots and garages, which will be found to range between 75 and 90 per cent (say 85 per cent for lots and 80 per cent for garages). The practical space-hours are determined by multiplying the available space-hours by the efficiency factor.

Deficiency in Parking Space in the Core of the CBD. Within the CBD there may be a section or sections in which the deficiency of parking spaces is large. This will usually be the core, where the tallest buildings are located, where land value is the highest, and where few if any parking lots and garages have been developed.

The core can be identified by comparing statistics, for each block, which show usage (space-hour used by parked vehicles), demand (space-hours needed by parkers, as determined by parker's destinations in interviews), and practical supply (available space-hours multiplied by efficiency factors). This will result in the determination of net deficiency or net surplus, in space-hours, for each block.

Some blocks, which show a net deficiency in space, may be adjacent or close to other blocks having a surplus of space. The extent to which surpluses in one block may be applied to make up deficiencies in another block depends upon the distance between the two. There is no standard as to what constitutes a reasonable or acceptable walking distance. However, the larger the city, the greater the distance that will be tolerated. Table 16-10 contains suggested criteria for acceptable walking distances.

The deficiencies determined in space-hours in the core may be minimized by providing parking facility improvements located as close to the core as possible, to serve first the short-time parkers and, if additional space is available, to serve the longer-time parkers.

TABLE 16-10. SUGGESTED CRITERIA FOR ACCEPTABLE WALKING DISTANCES

POPULATION GROUP (urbanized area)	*WALKING DISTANCE (feet)*
Under 25,000	300
25,000–50,000	345
50,000–100,000	490
100,000–250,000	530
250,000–500,000	740
Over 500,000	750

Origins of Parking Volume. Analysis of interview data and cordon-station counts will give the volume of vehicles which enter each side of the CBD and park in the district. Parking facilities should be planned and located with consideration of entering volumes of parked vehicles, as well as of destinations, so that it should not be necessary to cross the business districts to find a parking space.

Special Analysis of Truck Parking. Analysis should be made of the use of loading zones by trucks and of the extent of trucks parking elsewhere. The extent of double parking by trucks is often significant. This analysis will reveal the location and amount of demand for truck loading spaces.

Potential Improvements. The analysis of inventory data should include a study and appraisal of the various lots, buildings, areas, developments, or improvements selected and recommended as a means toward relief of parking congestion. Any recommendation involving the acquisition or improvements of real property should include data supporting its selection and data on its ownership, condition, and assessed valuation, and calculation of financing costs, improvements, and operation. Study of the city's pertinent zoning ordinances and state statutes will provide information relative to the possibility and desirability of condemnation of property for parking facilities.

Check on Accuracy of Survey. A simple check will reveal the thoroughness of the field work of the study. The number of vehicles which enter the cordon area and do not stop to park, plus the number of vehicles which leave parking spaces, should very nearly equal the number of outbound vehicles at the cordon in any given time interval. Exact agreement will not be obtained, because it is impractical to determine the few vehicles parking for services and repairs, or parking at individual private spaces. In addition, there are lags involving traffic moving on the streets at the beginning and end of a given time interval.

Legal Aspects, Administration, and Financing. This phase of the parking study involves two general steps. The first involves an inventory of the existing laws and ordinances, present administrative machinery, and existing means of financing parking facilities within the city. The adequacy of these features can be measured in light of the determined need for parking facilities. Deficiencies in legislation, in administrative implementation, and in finance will then become apparent. The second step involves recommendations as to the need for additional legal authority to deal adequately with the parking problem, proper administrative machinery, and possibilities for financing. For a complete discussion of the legal administration and financing aspect of parking, see the literature.[10]

Peak-Moment Type of Parking Study

The comprehensive type of parking study described above involves the interviewing of practically every person parking a vehicle in the CBD during the business day, and thus is relatively expensive.

An alternate interview method is the *peak-moment* study procedure, which confines its concern to those parked in the area at the time of its greatest accumulation of parked cars, usually around 2 P.M. Interviewers, placed at that time, make a complete list of all parked vehicles. They then remain only as long as is necessary to interview those particular parkers, disregarding later arrivals, but including all vehicles then parked, whether legally or not. The interviews are conducted as in the comprehensive study, recording the same data. The inventory is also conducted in the same manner as in the comprehensive study.

This procedure, although much less complex and expensive, will not produce all the valuable data of the comprehensive study. It should be satisfactory in determining the extent of the need for space, and where spaces should be provided—the primary function of a parking study. However, it does not give representative data on illegal parking, overtime parking, trip purposes, or truck requirements for loading zones.

Simplified Study for Small Cities

In larger cities, where walking distance is an important factor, it is necessary to determine the destinations of parkers. This is obtainable only through interviewing, which is expensive. In smaller cities (under 25,000 population) the CBD, with relatively few blocks, involves only short distances, and it is possible to obtain reasonably satisfactory information on parking habits without the expense of conducting interviews.

The recommended procedure involves the following phases:

Inventory of Parking Facilities. The inventory is conducted in the same manner as in the comprehensive study.

Usage of Parking Facilities. Each curb face and each lot or garage is toured every 15 minutes, from 10 A.M. to 6 P.M. On the first tour, the observer makes a record of every vehicle parked (type of vehicle and license number are recorded), whether legally or illegally. On subsequent tours, a check is made next to each vehicle which was present on the preceding tour, and each new vehicle is recorded.

Cordon Count. The cordon count is conducted in the same manner as in the comprehensive study.

Limitations. Since no data are obtained on destinations, it is not possible to determine block-by-block demand for space and to determine space deficiencies. However, the study will provide data on block-by-block supply and usage, on parking durations, and on illegal parking.

Limited Parking Study[3]

The National Committee on Urban Transportation has developed a procedure for conducting a limited parking study, composed of the following phases:

Inventory of Parking Facilities. The inventory is conducted in the same manner as in the comprehensive study.

Parking Usage and Duration Studies. Data on usage of parking spaces and parking duration are obtained by hourly tours of each curb face and each lot or garage from 7 A.M. to 6 P.M. Tours are made with an automobile, and data are recorded by two observers.

Parking Meter Revenue Study. An analysis of meter revenues is made to determine the average daily collection per meter in various sections of the study area. The collection is an indication of the percentage of occupancy for the metered space.

Non-CBD Parking Generators

To plan for the parking requirements of non-CBD traffic generators, it is necessary to know the characteristic functions of such facilities, their size, and the type of traffic demand they generate. A significant source which analyzes travel patterns for airports, shopping centers, and industrial plants is NCHRP Report No. 24.[5]

Shopping Centers. The parking requirements of shopping centers were studied by Alan M. Voorhees in his report for the Urban Land Institute.[6] The average shopping center peak hour (7 to 8 P.M.) is about 16 per cent of the daily trip-making total. During the normal highway peak period (4 to 6 P.M.), shopping center traffic amounts to approximately 7 per cent of the daily trips.

The automobile is the only significant form of transport to the typical shopping center and, where there is virtually no walk-in trade nor public-transit usage, the provision of 5.5 car parking spaces for every thousand

TABLE 16-11. USAGE BY MODE OF DEPARTING DOMESTIC AIR PASSENGERS (per cent of total)

MODE	*LA GUARDIA AIRPORT (per cent)*	*NEWARK AIRPORT (per cent)*	*J. F. KENNEDY INTERNATIONAL (per cent)*	*TOTAL (per cent)*
Auto	37	60	44	45
Taxi	46	12	31	32
Airport Bus	9	13	12	11
Bus	2	10	3	4
Suburban Limousine	5	4	8	6
Other	1	1	2	2
Total	100	100	100	100

(Source: Bender, "Airport Surface Traffic Demands", *TQ*, July, 1970)

square feet of gross leasable area is adequate as a standard to meet the demand for parking.

Industrial Plants. About 95 per cent of the daily trip-making total to a typical industrial plant consists of work trips. A recent study by a committee of the Institute of Traffic Engineers[7] recommends at least two parking spaces for every three employees, to accommodate typical parking demand. It may be necessary, however, to provide additional spaces for visitor parking. In any event, for a more precise estimate of the daily maximum parking requirements at any given plant, the committee suggests the following procedure:

1. Obtain employment by category: executive, office, and plant workers, by shift.
2. Estimate parking requirements of each employment category, considering transit usage and car occupancy.
3. Examine shift "start" and "end" times to determine maximum parking demand.
4. Allow a contingency for seasonal fluctuations, inefficiency in space usage, overtime schedules, etc., as appropriate variables.

Airports. Planning parking facilities at airports entails a thorough knowledge of user characteristics. Users should be classified in such a way that the needs of each class can be determined and evaluated individually and in aggregate.

Each class of airport users has different needs for airport parking. The need depends on the purpose for going to the airport and the availability of alternate modes of transportation to the airport from various points in the region.

Table 16-11 illustrates the modes used by air passengers at three New York area airports. It may be noted that the modes used by departing passengers in reaching each airport are not equally utilized.

In a recent survey of 13 airports,[8] wide ranges in mode usage were found:

Auto or rental car:	38–88 per cent
Taxi:	5–47 per cent
Public bus:	1–10 per cent
Limousine or airport bus:	5–25 per cent

The reasons for going to the airport are also not consistent from airport to airport, as illustrated in Table 16-12. This points out the need for treating each airport separately, in evaluating its parking requirements, because of the differences in parkers at different airports. Table 16-13 might, however, serve as a general guide to provide some insight into the characteristic parking needs of various classes of airport population.

Public parking facilities at major airports typically accommodate the largest proportion of users who park three hours or less. The distribution of parking duration for public parking facilities at four major airports is illustrated in Figure 16-6.

TABLE 16-12. PURPOSE DISTRIBUTION OF PERSON TRIPS TO AND FROM SELECTED AIRPORTS, ALL TRAVEL MODES*

	TRIPS TO AIRPORT (%)			*TRIPS FROM AIRPORT (%)*		
AIRPORT	*To Work*	*To Social Recreation*	*To Air Travel*	*To Home*	*To Personal Business*	*To Other*
Atlanta	67.8	5.8	26.4	—	—	—
Buffalo	23.3	33.7	43.0	55.7	14.1	30.2
Chicago (Midway)	34.7	25.7	39.5	82.6	6.0	11.4
Minneapolis–St. Paul	46.8	19.7	33.6	80.3	7.1	12.6
Philadelphia	24.2	32.8	43.1	70.0	9.7	20.3
Pittsburgh	43.0	20.6	36.5	85.9	4.7	9.3
Providence	39.8	37.7	22.5	—	—	—
San Diego	45.9	21.6	32.4	—	—	—
Seattle–Tacoma	35.0	24.2	40.8	81.3	12.4	6.3
Washington (National)	69.8	15.8	14.4	80.1	9.9	10.0

(Source: *Urban Travel Patterns for Airports, Shopping Centers, and Industrial Plants,* NCHRP 24, Table 3)
* From transportation study data (home interviews) for the various cities.

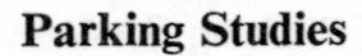

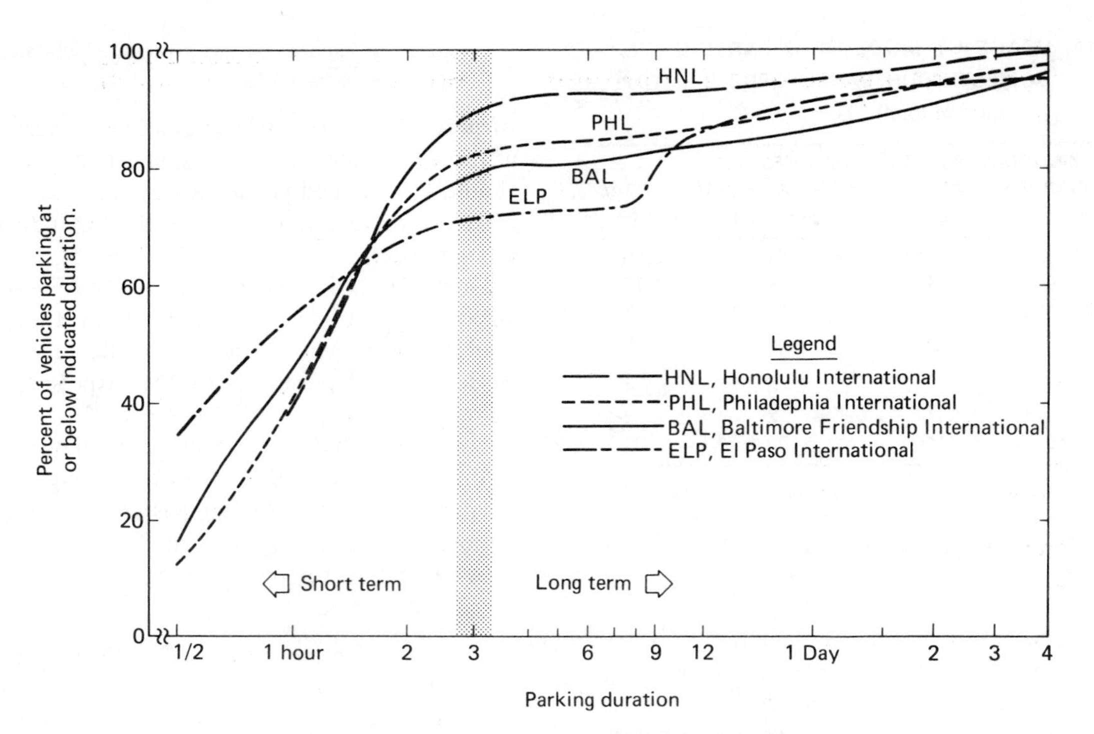

Fig. 16-6. Cumulative parking duration distributions at four airports (*Source: J. C. Orman, "Parking's Place in Airport Planning," in TE, February, 1971*)

OTHER METHODS FOR ESTIMATING PARKING NEEDS

As illustrated by the foregoing approaches, parking demand and parking usage are usually measured by interviewing the parker and by measuring the amount of time he parks his car. There are other means, however, which are available to obtain similar information.

TABLE 16-13. GENERAL AIRPORT PARKING GUIDE

AIRPORT POPULATION CLASSES USING AUTOS	*PARKING REQUIREMENTS*
1. (a) Air passengers (b) Air commuters	One day or more Less than one day
2. Visitors accompanying air passengers, sightseers, persons doing business at the airport, etc.	Three hours or less
3. Employees	Depends on length of shifts
4. Operators of ground transportation services	Normally park in reserved areas. Requirements depend on requirements for their services.
5. General aviation	24 hours or less. Peak demands occur on weekends.

(Source: Adapted from Orman, "Parking's Place in Airport Planning," *TE*, February, 1971)

Origin-Destination Studies

The Federal Highway Administration (formerly the Bureau of Public Roads) has developed computer programs which produce parking data for the CBD with the use of origin-destination data.

Indices of Parking Demand

In cases where it is not feasible to conduct a parking study, where no origin-destination studies are available, or where facilities which are not yet in operation require estimates of parking demand, it is customary to use the results from other studies to provide estimates of parking requirements. In their book *Parking in The City Center*,[4] Wilbur Smith and associates have developed a useful guide for estimating the parking needs of CBD establishments. Tables 16-8 and 16-9, and Figure 16-5, discussed earlier, were reproduced from that publication.

17

Accident Studies

Analysis of traffic accidents is of the utmost importance to traffic engineers. The tremendous toll of motor vehicle accidents not only causes much suffering and misery, but, in addition, large economic losses. In 1967, the economic loss to the United States resulting from motor vehicle accidents was estimated at 12.4 billion dollars, and the 1968 figure was 14.2 billion, a 14 per cent increase in one year.

A few statistics can serve to focus attention on the staggering toll of deaths resulting from motor vehicle accidents in the United States. Traffic fatalities on the nation's highways were 53,100 in 1967, and 55,200 in 1968. More than twice as many Americans have been killed by motor vehicle accidents than by all the wars the United States has been engaged in since 1776. From 1900 through 1967 there were approximately 1,700,000 motor vehicle fatalities, and based on current trends, this figure will reach about 2 million by 1975. There were 3,081 deaths in New York State and 868 in New York City in 1968.

Accurate accident analysis is largely dependent on thorough knowledge of the characteristics of drivers, vehicles, and roadways, and their interrelationships, and upon uniform and accurate reporting of accidents.

Since passage of the National Traffic and Motor Vehicle Safety Act of 1966 there has been a great deal of ferment in the field of traffic safety. A study of the current "state of the art" was made.[1] In June, 1967, a National Highway Safety Bureau was established within the Department of Transportation. By 1971, the Bureau was raised to the status of Administration within the Department.

The Bureau issued 13 Highway Safety Program Standards[2] concerned with motor vehicle inspection and registration, motorcycle safety, driver education and licensing, codes and laws, traffic courts, alcohol, identification and surveillance of accident locations, traffic records, medical services, highway design, and traffic control devices.

Since highway safety programs need to cover all these areas, a publication of the Automotive Safety Foundation[3] describes the requirements of setting up the total program.

The literature on the broad subject of traffic safety is now burgeoning.[4] The legislative hearings before and since the Highway Safety Act of 1966 have brought out material of great interest to the traffic engineer.[5,6,7] The Highway Safety Act has spawned a massive research effort in the field.[8]

The international community has been interested in the subject for some time, however.[9] In the United States, interest is translated into legislation and conferences concerning manpower needs,[10] strategies for action,[11] guidelines for programs,[12] improvements of geometric design,[13] and roadside hazards.[14]

CAUSES OF TRAFFIC ACCIDENTS

Most accidents result from a combination of several contributing factors: violations or unsafe acts by drivers or pedestrians; roadway, vehicular, driver, or pedestrian defects; bad weather; or poor visibility.

Human Factors

Drivers. In 1968 approximately 90.6 per cent of the nationwide total of accidents were attributed to improper driving. The principal kinds of improper driving were: *fatal accidents*—excessive speed, failure to yield right-of-way, and driving left of centerline; *nonfatal injury accidents*—failure to yield right-of-way, excessive speed, and following too closely; *total accidents* (predominately property damage)—failure to yield right-of-way, excessive speed, and following too closely.[15]

Alcohol was reported to be a factor in half the fatal accidents in the nation. A study by the Washington State Department of Motor Vehicles in 1968 reported that drug users were involved in more reckless driving, hit-and-run accidents, negligent driving, and had more defective equipment violations.[15]

In 1968, some 33 per cent of total accidents in New York State were attributed to speeding, 17 per cent to reckless driving, and 10 per cent to driving left of center.[16]

Based upon reports from 23 states, 67 per 100,000 drivers of all ages were involved in fatal accidents. The involvement rate for drivers between 20 and 24 years of age was 118, which was the highest rate. Drivers under 25 constituted about 21 per cent of all drivers, and they were involved in over 34 per cent of the fatal accidents and over 34 per cent of the total accidents.[15] Based upon a study of the distribution of accidents by age groups for the year 1968 in New York State, it was found that although drivers under 29 constituted only 28 per cent of all drivers, they were involved in 43 per cent of the fatal accidents, 35 per cent of the injury accidents, and 31 per cent of the property damage accidents.[16] The above involvement rates and percentages, however, do not consider the amount of driving done by drivers of different age groups, or the environment of their driving. Both of these factors could affect the results. A detailed analysis, by the BPR, of accident characteristics on 600 miles of main rural highways in 11 states revealed that drivers under 25 had the highest total accident involvement rate (per 100 million vehicle-miles of travel) of any age group.[17]

Although approximately 59 per cent of the drivers in 1968 were males, they constituted about 75 per cent of all drivers involved in accidents. This is not surprising since males do nearly three-fourths of all driving, and much of the driving by females is done during daylight hours in off-peak traffic volumes. The total accident involvement rate, based on vehicle-miles of travel, indicated a male rate about 1.3 times that of the female rate.

The Bureau's study of main rural highways[17] revealed that male drivers did about 87 per cent of all driving during the day, and 93 per cent during the night. Considering the drivers of all types of vehicles, the total accident involvement rate was 210 for males and 247 for females during the day, and at night the rate of males was 419 and of females 579. However, at night the predominately male truckdrivers have an exceptionally low involvement rate. Therefore, to obtain a more meaningful comparison, rates were computed for passenger-car drivers only. This comparison gave an involvement rate of 200 for males and 246 for females during the day; at night the rate for males was 578 and for females 593.

Pedestrians. In 1968, pedestrians constituted 31 per cent of the total killed in traffic accidents in New York State and 18 per cent in the nation, and 8 per cent of the total injured in both New York State and the nation. Older people are more frequently involved. More than 83 per cent of the deaths associated with crossing at the intersection involve persons 45 years of age and older in both New York State and New York City. Pedestrians of 14 years of age or younger accounted for over 45 per cent of persons injured while standing or playing in the roadway and about 68 per cent of those coming from behind parked cars.[15,16]

Vehicular Factors

The percentages of vehicles involved in accidents by type are, in general, nearly the same as their share of total registration. However, several studies found that the involvement rates are varying. In 1968 the involve-

TABLE 17-1. TYPES OF VEHICLES INVOLVED IN ACCIDENTS—NEW YORK STATE 1968

VEHICLE TYPE	*PER CENT OF TOTAL REGISTRATION*	*PER CENT VEHICLES INVOLVED IN ACCIDENTS*
Passenger cars	87.0	88.9
Trucks and trailer-trucks	10.1	7.5
Taxis and buses	1.7	2.8
All other	1.2	0.8

(Source: *Accident Facts,* NY State Dept. of Motor Vehicles, 1969)

TABLE 17-2. TYPE OF VEHICLE INVOLVED IN ACCIDENTS BY DAY AND NIGHT

VEHICLE TYPE	DAY		NIGHT	
	Per cent Vehicles Involved in Accidents	*Involvement Rate*	*Per cent Vehicles Involved in Accidents*	*Involvement Rate*
Passenger cars	75.7	207	80.2	580
Truck, four tires	9.4	281	6.2	398
Truck, six or more tires	13.1	208	12.6	164
Bus	0.8	266	0.3	119
Other	1.0		0.7	

(Source: Solomon, *Accidents on Main Rural Highways Related to Speed, Driver and Vehicle,* BPR, USSPC, July, 1964.)

ment total accident rates by type of vehicle in New York State were as shown in Table 17-1.[14] The Bureau's study of main rural highways[17] revealed that the total accident involvement rate, per 100 million vehicle-miles, by vehicle type, was nearly the same for day driving. However, the situation changed radically for night driving as shown in Table 17-2.

The proportion of accidents attributed to vehicle defects is well under 2 per cent, and the majority of these accidents is due to defective brakes and improper lights. In the last few years the federal government has specified design changes to vehicles. Organizations are studying the need, for example, of additional lights on vehicles. There is some indication that vehicle "running lights" can reduce accident occurrence.[18,19]

Environmental Factors

Though unfavorable weather or road conditions do have an influence on traffic accidents, the extent of such influence is yet to be determined. However, the driver and the pedestrian are the major factors in traffic accidents. Thus, the traffic engineer must strive to change driver and pedestrian behavior, by reasonable regulations and enforcement, so as to reduce their dangerous acts. The highway designer has the responsibility to incorporate as many safety features as possible in his design to minimize the number of accidents due to poor geometry. Total accident occurrence under various weather and road conditions in New York State for 1968 is given in Table 17-3.

ACCIDENT REPORTS AND REPORTING SYSTEMS

Accident Records

Complete accident records are essential to analyzing accidents and planning accident prevention measures through engineering, education and enforcement.[20,21,22] The users of accident reports and their general interests or purposes can be summarized as shown in Figure 17-1.

Reporting Procedures

Accident reports should be made for all accidents involving a fatality or bodily injury of any person, or total property damage in excess of $100. The minimum damage reportable in a motor vehicle accident is established by state law in most states. In New York State, it is required to report property damage in excess of $150. Regulations throughout the 50 states vary considerably, ranging from reporting any damage to reporting property damage in excess of $200.

It is essential to adopt uniform accident reporting procedures among all levels of government. In this regard, the concepts and procedures recommended by the Presidential Committee for Traffic Safety[23] should be fully implemented.

Accidents are, in general, reported by the drivers

TABLE 17-3. ACCIDENT OCCURRENCE BY ROAD AND WEATHER CONDITIONS—1968

PER CENT TOTAL ACCIDENT OCCURRENCE	
Weather Conditions	*New York State*
Clear	78.2
Rain, snow, sleet	20.6
Fog	1.2
PER CENT TOTAL ACCIDENT OCCURRENCE	
Road Conditions	*New York State*
Dry	64.8
Wet, snowy	35.2
PER CENT TOTAL ACCIDENT OCCURRENCE	
Road Character	*New York State*
Straight	88.6
Curve	11.4

(Source: *Accident Facts,* NY State Department of Motor Vehicles, 1969)

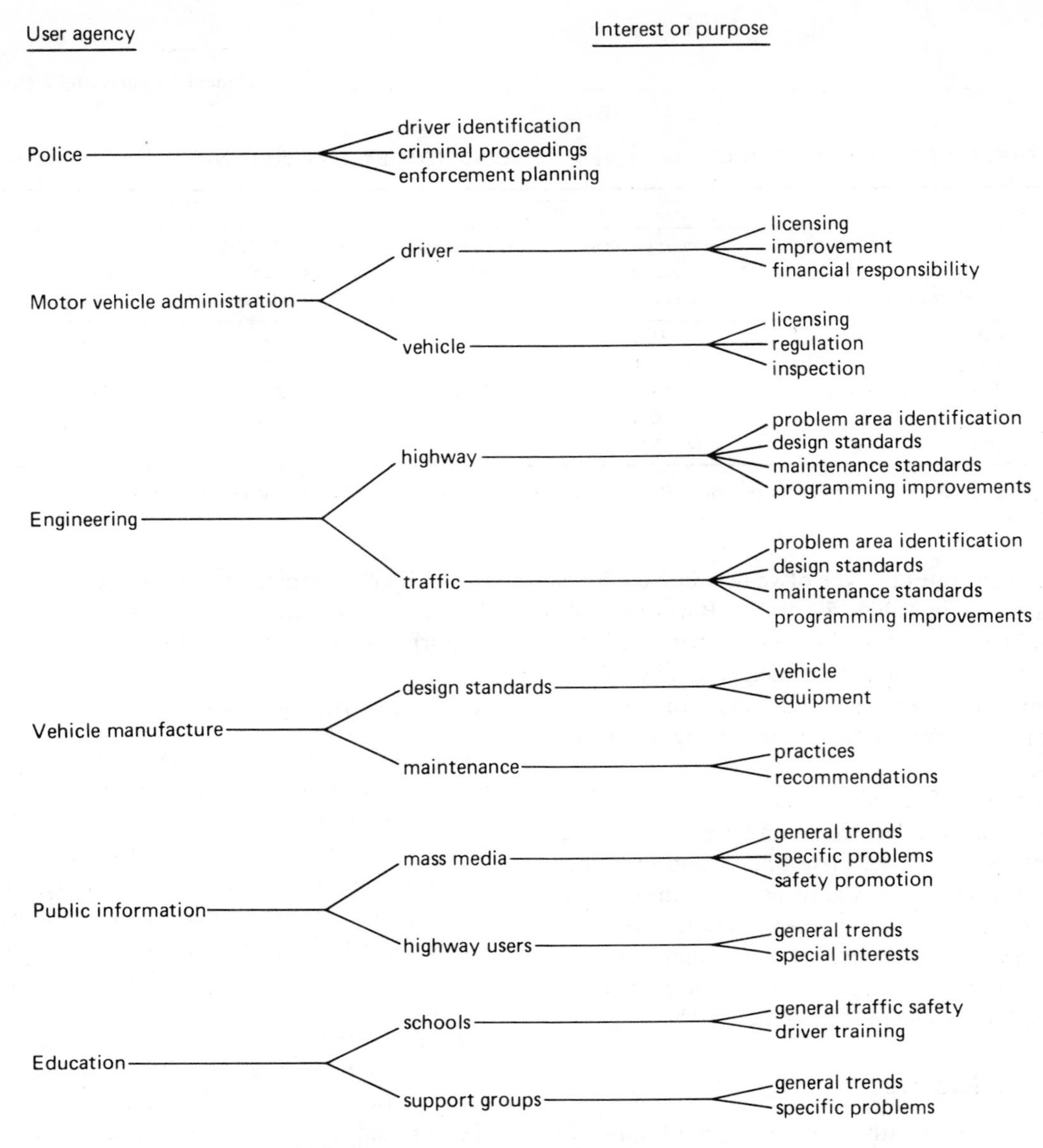

Fig. 17-1. The summary of the users of accident reports and their general interests or purposes

involved and by investigating police officers. Standard forms for use by drivers in reporting to state motor vehicle departments have been developed by the National Safety Council, for example. The form should be filled out after police have carefully investigated the accident, including questioning participants and witnesses, making physical measurements, examining for evidence of intoxication, and reconstructing the accident scientifically. Forms made out by drivers and police should include detailed information on:

- Time of accident
- Location of accident
- Driver
- Vehicles
- Persons injured
- Extent of damage of vehicles
- Location and description of traffic control devices
- Regulations in force
- Roadway and weather conditions
- Possible violations
- Probable causes
- Diagram of accident

Location Systems

Reporting accurate locations of accidents is essential for later studies of high accident locations and the implementation of remedial action. Various reference marker systems are in use for this purpose. Reference markers are especially useful in rural areas where no distinctive features are available for reference in locating accidents.

The interstate system has instituted a reference marker system using $\frac{1}{10}$-mile markers. New York City numbers all lamp posts on expressways, parkways, and other

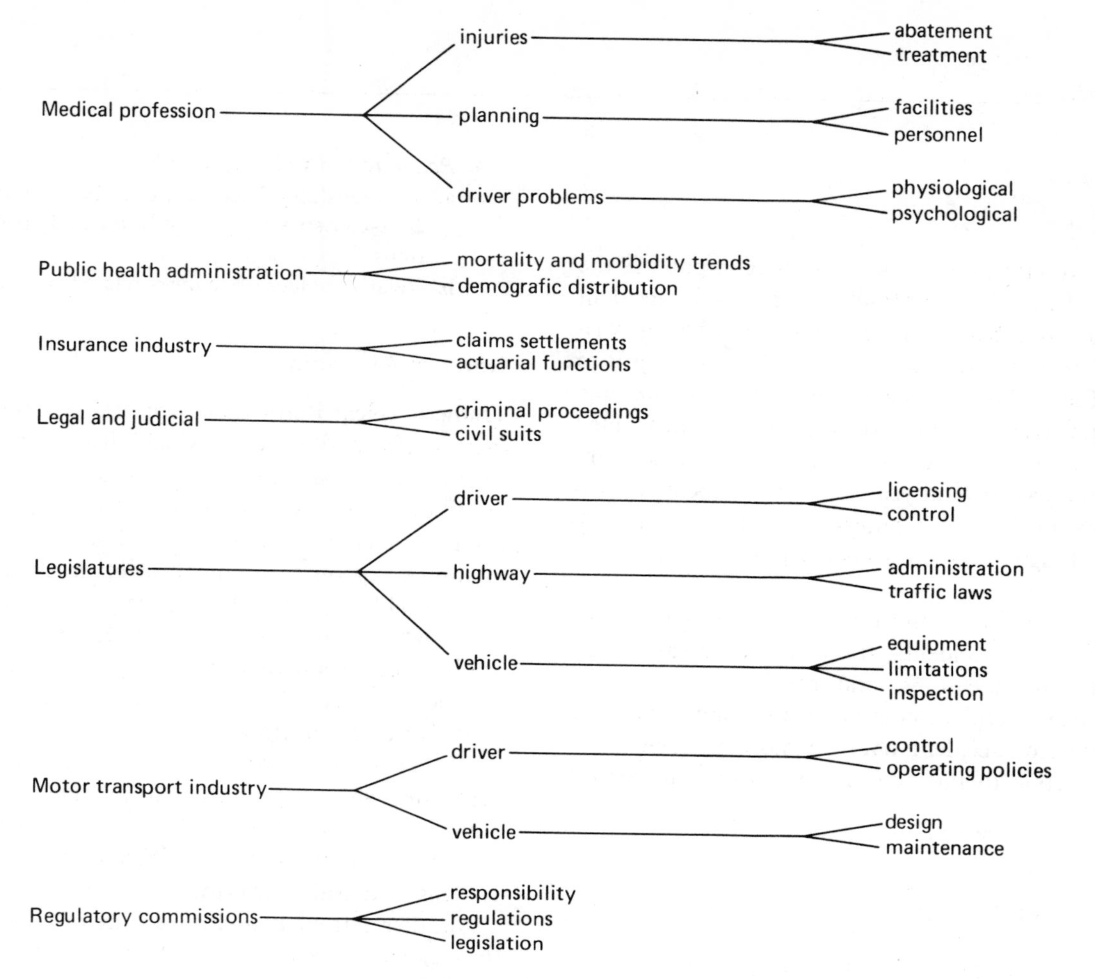

Fig. 17-1. continued

limited-access roadways. The Garden State Parkway in New Jersey has numbered overpasses to aid in this identification process. The Port of New York Authority has zoned all of its facilities (bridges, tunnels, airports, seaports, terminals) by specific location, each ranging in length from $\frac{1}{20}$ to 1 mile, determined by type (merging area, diverging area, straightaway) and coded for electronic data processing.

Accident Recording System

It is extremely important that traffic accident data be compiled on a uniform basis by all levels of government. Uniform classification of traffic accidents and close cooperation among all agencies concerned with investigating and reporting traffic accidents will insure the attainment of uniform accident statistics. In this regard, the definitions and recommendations of the National Safety Council should be adopted.

The following classifications are used to characterize the manner of occurrence of motor vehicle traffic accidents.

1. Running off road
2. Noncollision on road
   a. Overturning on road
   b. Other noncollision on road
3. Collision on road
   a. With pedestrian
   b. With another motor vehicle in traffic
   c. With parked motor vehicle
   d. With railroad train
   e. With bicyclist
   f. With animal
   g. With fixed object
   h. With other object

Furthermore, collisions between motor vehicles should be further classified on a uniform basis such as:

1. *Angle:* Collision between vehicles moving in different directions, *not opposing directions*, usually at a right angle.
2. *Rear-End:* A vehicle being hit from behind by another vehicle moving in the same direction, usually in the same lane.
3. *Sideswipe:* A vehicle being hit by another *from the side* while traveling in the same direction, or in opposite directions, usually in different lanes.
4. *Head-on:* Collision between vehicles traveling in opposite directions (not sideswipe).

5. *Backing*
6. *Others*

Automated data processing using the computer is the only hope of an effective attack on the traffic accident problem of our urban areas. The huge number of variables involved in the causation of every accident leads to the need for high-speed means of sorting, tabulating, computing, correlating, and printing out the transformed data collected from hundreds of thousands of accident forms. In the past several years, several accident recording systems utilizing computers were developed.[25] A functional design for such a system is shown in Figure 17-2.

The computer is very useful in storing a large amount of data related to accidents, traffic and roadway conditions at various locations, and other pertinent data. It is also very useful in compiling and summarizing a great amount of accident data on more frequent time intervals suitable to the needs of engineers for further analysis.

ACCIDENT ANALYSIS

The purpose of accident analysis is to find the possible causes of accidents, as related to drivers, vehicles, and roadways, and to plan measures to protect the motoring public by reducing the frequency and severity of accidents. The enormous number of traffic accidents and the consequently large amount of data make it impossible today to analyze accidents manually. Electronic computers provide not only accurate and efficient accident recording systems, but make accident analysis easier by readily providing information such as accident rates, accident summaries, and tabulations by various related variables.[25]

Accident analyses are generally made to develop information such as the following:

1. *Driver and pedestrian*
   - **a.** Identification of drivers with high infraction-arrest-accident records
   - **b.** Accident occurrence by age groups
   - **c.** Relationships of accidents to physical capacities and to psychological test results
   - **d.** Driver performance by areas of residence
2. *Vehicles*
   - **a.** Accident occurrence related to characteristics of vehicles
   - **b.** Severity and extent of damage related to vehicles
   - **c.** Severity, extent, and location of injury related to vehicles
3. *Roadway and Roadway Conditions*
   - **a.** Relationship of accident occurrence and severity to characteristics of the roadways and roadway conditions
   - **b.** Relative values of changes related to roadways

General Analysis

Traffic Accident Rates. Since accident reporting laws and their interpretations differ widely, it is usually necessary to confine comparisons between cities or states to traffic fatalities. Within an area where reporting is uniform, a traffic accident rate may be based separately on deaths or injuries or on a combination of deaths and injuries. Total accidents may also be used, but care must be taken when injuries or property damage accidents are used to be certain that reporting is comparable. Computation of any of these rates will reflect the accident exposure of an entire area.

Of increasing importance is the computation of accident rates which reflect accident involvement by type of highway or street. These rates provide a means of comparing the relative safety of different highway and street systems and traffic controls.

Another extremely useful measure of relative exposure to accidents is accident involvement by characteristics of drivers and vehicles associated with accidents. This accident involvement rate is an effective measure of the chance of a driver or vehicle being involved in an accident because of any roadway or traffic conditions.

Accident rate per mile

On this basis, the total accident hazard is expressed as the number of accidents of all types per mile of each highway and street classification.

$$R = \frac{A}{L} \tag{17-1}$$

where R = total accident rate per mile for one year
A = total number of accidents occurring in one year
L = length of control section in miles

This is useful in comparing accident rates of a series of sections of a route which has a relatively uniform traffic flow.

Accident involvement rates

Accident involvement is expressed as the number of drivers of vehicles with certain characteristics who were involved in accidents per 100 million vehicle-miles of travel (of all drivers of vehicles observed with those

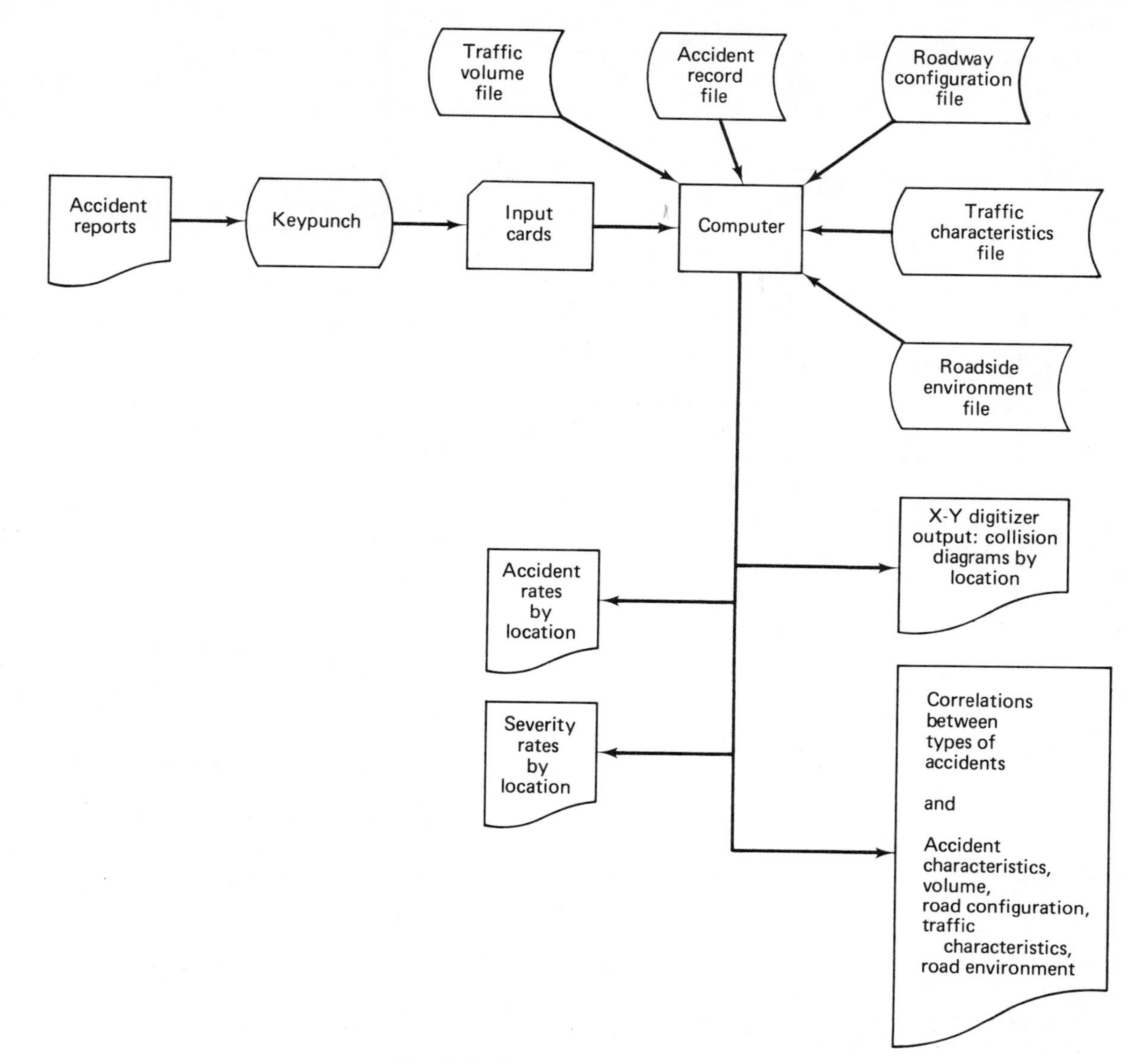

Fig. 17-2. Computer accident system

particular characteristics). For example, the accident involvement rate could indicate the number of drivers involved in accidents at any particular speed related to the amount of travel at that speed. Traffic volume counts and speed studies over the length of road section under investigation would provide the necessary data for computation of vehicle-miles of travel at a particular speed.

$$R = \frac{N \times 100{,}000{,}000}{V} \tag{17-2}$$

where R = accident involvement per 100 million vehicle-miles
N = total number of drivers of vehicles involved in accidents during the period of investigation
V = vehicle-miles of travel on road section during the period of investigation

Death rate based on population

The traffic hazard to life in a community is expressed as the number of traffic fatalities per 100,000 population. This rate reflects the accident exposure for the entire area.

$$R = \frac{B \times 100{,}000}{P} \tag{17-3}$$

where R = death rate per 100,000 population
B = total number of traffic deaths in one year
P = population of area

Death rate based on registration

The traffic hazard to life in a community can also be expressed as the number of traffic fatalities per 10,000 vehicles registered. The rate reflects the accident exposure for the entire area, and has a use similar to death rate based on population.

$$R = \frac{B \times 10{,}000}{M} \tag{17-4}$$

where R = death rate per 10,000 vehicles registered

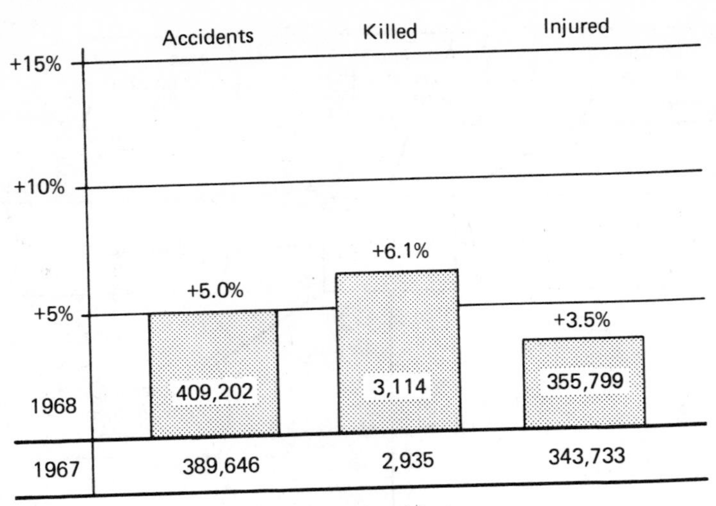

Fig. 17-3. Number of accidents (*Source:* Accident Facts, *NY State Department of Motor Vehicles, 1969, p. 6*)

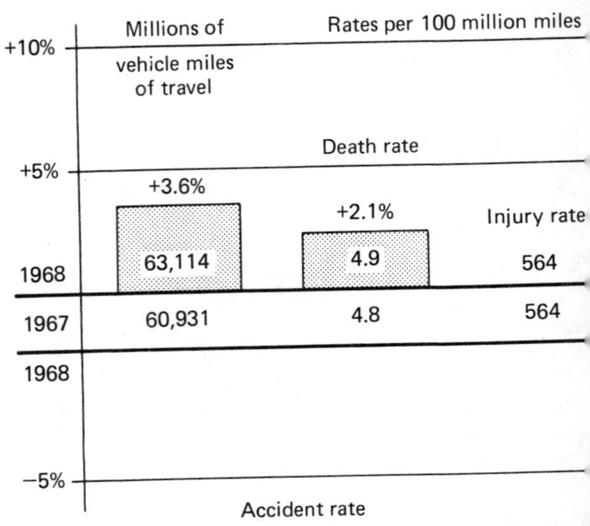

Fig. 17-4. Accident rate (*Source:* Accident Facts, *NY State Department of Motor Vehicles, 1969, p. 6*)

Fig. 17-5. Accident rate trend (*Source:* Accident Facts, *NY State Department of Motor Vehicles, 1969, p. 11*)

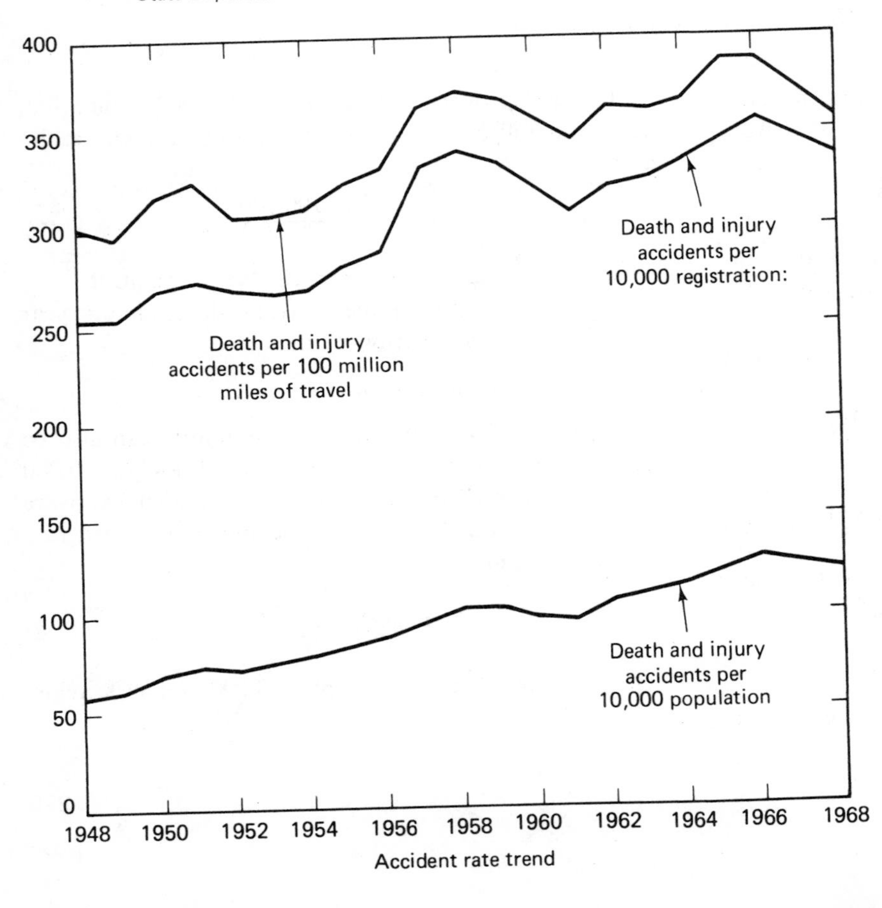

B = total number of traffic deaths in one year
M = number of motor vehicles registered in the area

Accident rate based on vehicle-miles of travel

The accident hazard in this case is expressed as the number of accidents per 100 million vehicle-miles of travel. The true exposure to accidents is probably more nearly approximated by the miles of travel by motor vehicles than by either the population or registration. The vehicle-miles rate may be expressed in terms of deaths, injuries, or total accidents per 100 million vehicle-miles.

$$R = \frac{C \times 100{,}000{,}000}{V} \qquad (17\text{-}5)$$

where R = accident rate per 100 million vehicle-miles
C = number of accidents (deaths or injuries or total accidents) in one year
V = vehicle-miles of travel in one year

For cities, the total motor-vehicle-miles is usually difficult to determine, but in states it is possible to approximate this figure with fair accuracy from motor fuel consumption rates, obtained from tax records, for the state as a whole. Vehicle-miles of travel may then be calculated by multiplying the fuel consumption by an average figure for miles of travel per gallon of fuel. This conversion factor is determined on a national basis by the BPR, and it has gradually decreased from 13.60 in 1946 to 12.47 miles per gallon in 1966. This value is a weighted average for all types of motor vehicles (14.00 for passenger cars and 8.52 for trucks).[26]

For example, in 1968 the motor fuel consumption in New York State was 5.082 million gallons, there were 3,114 motor vehicle fatalities, 355,799 motor vehicle injuries, 6,721,049 motor vehicle registrations, and an estimated population of 18,190,238.

Approximate vehicle-miles of travel

$$= \frac{5.08 \times 10^9}{63.1 \times 10^9} \times 12.42$$

Registration death rate

$$= \frac{3.114 \times 10{,}000}{6.72 \times 10^6} = 4.63$$

Population death rate

$$= \frac{3{,}114 \times 100{,}000}{18.2 \times 10^6} = 17.1$$

Vehicle-miles death rate

$$= \frac{3{,}114 \times 100 \times 10^6}{63.1 \times 10^9} = 4.93$$

The various accident rates presented above are very useful in describing the general picture of accident problems such as those shown in Figures 17-3, 17-4, and 17-5. However, they do not present any clue to the problems at individual locations, which require microscopic studies.

There is a great need for the numeralization of accident severity. Current methods merely divide severity into the three categories of *fatal*, *personal injury*, and *property damage*. But there are great variations in amount of injury and damage that are not shown in statistics.[23,27]

Accident Patterns and Characteristics. Most current accident reports do not provide adequate information to determine basic causes of accidents, but accident reports executed by police officers are normally more revealing. Accident reporting was initially developed for the purpose of providing statistics upon which the traffic engineer or the enforcement agency could base remedial measures. However, present methods of accident analysis are too time-consuming to be of significant value. Accident statistics are normally compiled on a yearly basis, but analysis is required more often if remedial measures are to be effective. With computers it is possible to make accident analyses on shorter time intervals, such as monthly or quarterly.

Time patterns of traffic accidents

Characteristic patterns of traffic accidents exhibit only slight changes from year to year. Traffic accident fatalities vary significantly for the different months of the year and different days of the week. The yearly pattern of the nation's fatality rates shows very slight variations, with a small decline since 1965. National rates of deaths per 100,000 population, per 10,000 motor vehicles, and per 100 million vehicle-miles, respectively, are shown in Table 17-4.[15] Statistics for New York State for the same period of time are given in Table 17-5.[16]

TABLE 17-4. NATIONAL MOTOR VEHICLE DEATH RATES

YEAR	*1960*	*1961*	*1962*	*1963*	*1964*	*1965*	*1966*	*1967*
Population rate	21.2	20.8	22.1	23.1	24.9	25.4	27.1	26.7
Registration rate	5.1	5.0	5.1	5.2	5.5	5.4	5.5	5.0
Vehicle-Miles rate	5.3	5.2	5.3	5.4	5.7	5.5	5.7	5.5

(Source: *Accident Facts,* 1969 Edition, National Safety Council, 1969, Chicago, Illinois)

TABLE 17-5. NEW YORK STATE FATALITY AND INJURY RATES

	FATALITY RATES		*INJURY RATES*	
YEAR	*Per 10,000 Registrations*	*Per 100 Million Vehicle-Miles*	*Per 10,000 Registrations*	*Per 100 Million Vehicle-Miles*
1960	4.02	4.4	526	578
1961	4.12	4.6	490	549
1962	4.31	4.8	514	572
1963	4.15	4.6	501	556
1964	4.70	5.2	494	543
1965	4.40	4.9	531	589
1966	4.47	4.9	548	597
1967	4.56	4.8	535	564
1968	4.63	4.9	529	564

(Source: *Accident Facts,* NY State Department of Motor Vehicles, 1969)

On a national basis, traffic accident deaths from January through May are below the annual monthly average, while from June through December they are above the average. In recent years, the lowest level occurs in January and February. Then they increase steadily to a peak during the summer months, and generally this high level prevails through the latter months of the year. The highest monthly total normally occurs in August. Significantly greater travel and school vacations are the primary factors contributing to the high level during the summer months. Longer hours of darkness and increased pedestrian activity occasioned by the holiday season are factors contributing to the high level later in the year.

Weekly patterns of traffic accidents indicate that the numbers of traffic deaths from Monday through Thursday are less than average, while from Friday through Sunday they are above average. From January through May, daily averages are below the annual daily average, while from June through December they are above the average. The weekends exhibit the highest level when the traffic volume is the greatest. The highest daily total normally occurs on Saturday.

Daily patterns indicate that traffic accidents by hour of day vary for different days of the week. The most dangerous time to drive on weekdays is during the evening peak period from 4:00 to 6:00 P.M. During this period, total accidents (fatal, personal injury, property damage) are at their highest level. However, weekend frequencies indicate that the highest level is from 3:00 to 6:00 P.M. On weekdays traffic fatalities are highest during the hours from 4:00 to 8:00 P.M. Although the frequency on weekends is extremely high during these same hours, the highest levels occur from 11:00 P.M. to 2:00 A.M. on Saturday, and from midnight to 3:00 A.M. on Sunday. The highest weekend hourly total occurs from 1:00 to 2:00 A.M. During the period from dusk to dawn (6:00 P.M. to 6:00 A.M.) about 54 per cent of fatal accidents and about 36 per cent of all accidents occur, even though traffic volumes are considerably lower than during the hours of daylight. Although the percentage of fatal accidents during the night is only slightly greater than during the day, the night death rate on a vehicle-miles basis is significantly higher than the day death rate. The death rate during the day is about 4, and the death rate during the night is about 10. The death rate is much more indicative of the greater hazard associated with driving during the hours of darkness.

Urban-Rural Differences. The manner of occurrence of traffic accidents varies greatly in urban and rural areas. Approximately one-half of total *urban* accidents take place at intersections versus about one-quarter for *rural* accidents. In addition, there is a significantly higher proportion of pedestrian accidents in urban areas. The total number of accidents in urban areas exceeds that in rural areas; however, the severity of accidents is greatest in rural areas. These differences are primarily due to greater population density and motor vehicle registrations as well as more restricted regulation of traffic in urban areas as compared to higher speeds associated with rural areas.

Nationally, the distribution of traffic deaths and nonfatal injuries between urban and rural areas for 1968 was as follows:

	DEATHS		*NONFATAL INJURIES*	
AREA	*Number*	*Per cent*	*Number*	*Per cent*
Urban	17,500	32	1,150,000	58
Rural	37,700	68	850,000	42
Total	55,200	100	2,000,000	100

Type and Location of Traffic Accidents. The distribution of predominate types of accidents on a national basis for 1968 in urban and rural areas is shown in Table 17-6.

Accidents and Characteristics of Traffic Operation and Highway Design Features. To plan any kind of corrective measures related to traffic regulations and operation and physical features of roadways, it is essential for engineers to understand the relationship between highway design features, operating characteristics, and accident occurrence. Such understanding will provide an insight into many contributing factors associated with highway accidents and will thereby reveal what possible corrective measures should be instituted to minimize accident occurrence. An excellent summary of research related to highway elements and highway safety and an extensive pertinent bibliography are available in the literature.[28,29]

TABLE 17-6. TYPE AND LOCATION OF TRAFFIC ACCIDENTS

TYPE OF ACCIDENT	FATALITIES		NONFATAL INJURIES	
	Urban	Rural	Urban	Rural
Vehicle collision with:				
Pedestrian	6,400	3,400	125,000	25,000
Other vehicle	5,500	17,000	820,000	520,000
Fixed object	1,200	1,400	40,000	20,000
Bicycle	420	380	32,000	6,000
All other	480	1,220	3,000	9,000
Noncollision	3,500	14,300	130,000	270,000

(Source: *Accident Facts,* 1969 Edition, National Safety Council, Chicago, Illinois, 1969)

Access Control. Studies comparing the accident rates on highways with controlled access to those without access control illustrate the importance of location and design to traffic safety. The BPR has for several years been analyzing data on this subject from 30 states.[23] The conclusions of this study indicate that full control of access is the most important single factor in accident reduction ever developed. Accident and fatality rates on fully controlled-access highways are one-third to one-half as great as those on highways with no control of access as shown below. The results of a study[30] of sections of interstate highways are also shown in Table 17-7 for comparison purposes. Although when interstate highways are fully improved to interstate standards they will have full control of access, some sections analyzed in the study had less than full control of access. These sections were opened to traffic, but they were not fully improved for the period of study. While the number of accidents per 100 million vehicle-miles is higher in urban areas than in rural areas, the rural accidents are more severe in terms of resulting fatalities. In comparing the results of the two studies, it is seen that, regardless of the degree of access control, interstate highways had lower total accident rates than other highways. This finding indicates that other design features such as wide medians, easy curvature and gradient, and long sight distances may have also contributed to the improved safety record of the interstate system.

Various estimates have been made of the number of lives that will be saved because of the interstate system.[31,32] The most recent estimate of the number of lives that would be saved for the first full year (1972–1973) of operation on the entire 41,000-mile interstate system is approximately 8,000.[30]

In a study[33] of accident experience on freeways as compared to other types of highway facilities in California, rates for different accidents were determined. The results, which are shown in Table 17-8, indicate that the really significant differences in rates occur in head-on accidents and accidents at intersections. It is seen that the total accident rate for freeways is less than one-fourth the rate for undivided four-lane highways and less than one-half the rates for two-lane and three-lane highways.

Traffic Volumes. Traffic volume has a decided effect on the accident rate for nearly all types of highways. A study[34] of accident rates on tangent, two-lane, rural highways, in relation to ADT, indicated that the accident rate increases steadily with increasing volume, reaching a maximum for roads carrying 8,000 to 9,000 vehicles per day. Heavier traffic reduces the accident rate primarily because the extreme congestion at higher volumes makes it difficult for drivers to execute passing maneuvers. Good agreement with the above results was obtained from a study[35] of traffic accidents relative to hourly volume on two-lane, tangent sections of California highways. It was found that accidents increase with increasing volume, reaching a maximum at about 650 vehicles per hour.

TABLE 17-7. ACCESS CONTROL AND ACCIDENT RATES

AREA AND ACCESS CONTROL	ACCIDENT RATE*			
	All Accidents		Fatalities	
	Reference 23	Reference 30	Reference 23	Reference 30
Urban				
Full control	186	161	2.0	2.5
Partial	496	264	4.6	3.3
None	526	380	4.0	5.5
Rural				
Full control	151	122	3.3	2.3
Partial	211	94	6.1	6.6
None	332	169	8.7	8.4

(Source: *Accident Facts,* 1969 Edition, National Safety Council, Chicago, Illinois, 1969)
* Per 100 million vehicle-miles.

TABLE 17-8. ACCIDENT RATES ON RURAL HIGHWAYS RELATED TO DESIGN STANDARDS

	TWO-LANE	THREE-LANE	FOUR-LANE UNDIVIDED	FOUR-LANE* DIVIDED	DIVIDED** CONTROLLED ACCESS	FREEWAY
Miles	10,450	45	167	210	794	430
ADT	2,191	14,239	15,997	16,130	12,224	19,449
Million vehicle-miles	8,358	232	976	1,234	3,543	3,052
	Rate†	Rate	Rate	Rate	Rate	Rate
Total reported accidents	2.38	2.57	4.09	2.91	1.69	1.00
Single vehicle accidents	0.84	0.49	0.38	0.40	0.43	0.28
Collisions between two or more vehicles	1.54	2.08	3.71	2.51	1.26	0.72
A. Between intersections:	1.70	1.65	1.57	1.22	1.08	0.95
1. Head-on	0.42	0.29	0.20	0.06	0.06	0.045
2. Non head-on	0.44	0.87	0.99	0.77	0.59	0.63
B. At intersections	0.68	0.92	2.52	1.69	0.61	0.045‡

(Source: Solomon, *Accidents on Main Rural Highways Related to Speed, Driver and Vehicle,* BPR, USSPC, July, 1964)

* Four-lane divided roads have a median separating opposing traffic but roadside access is not controlled.

** Divided, controlled-access roads are primarily four-lane roads, with no access except at intersections. Intersections at grade are frequent and traffic enters and exits at large angles, approximating 90 degrees.

† Rate is number of accidents per million vehicle-miles.

‡ Accidents at ramps.

Cross-Section Elements.

Lane width. Studies relating roadway width to traffic accidents have indicated that accident rates decrease with increasing lane width. In one study[34] the total accident rate per million vehicle miles decreased from 5.5 to 2.4 as the pavement width increased from 16 feet to 25 feet. Another study,[36] covering about 240 miles of highway which had been widened from 18 to 22 feet, indicated that accident rate reduction ranged from 21.5 per cent for low-volume roads to 46.6 per cent for higher-volume roads.

Highway shoulders. Studies relating shoulder width to traffic accidents have indicated that accident rates decrease with increasing shoulder width. Based on a thorough study of accidents in relation to highway shoulders in New York State,[37] it was found that roadway sections with wide shoulders had much lower accident rates than those with narrow shoulders. A study[38] of the entire system of rural two-lane roads in New York State revealed that, in general, the wider the shoulder the lower the accident rate. This was especially true for property damage accidents.

The results of most studies conducted to determine the effect of pavement edge markings on traffic accidents showed a reduction in accidents, but some studies indicated that no change in accident experience took place.[39-44]

Medians. In studying the effect of medians on accident experience, medians are divided into three basic types:

1. The *traversable type* includes those medians which have no obstructions and are easily crossed by vehicles.
2. The *deterring type* discourages deliberate entrance or crossing of the median by means of a physical obstruction. Medians with mountable curbs and most of the earth medians with flat cross slopes are in this group.
3. The *nontraversable type* is physically designed to prevent crossing from one roadway to another. Medians with barrier curbs, earth medians with steep cross slopes, medians with continuous obstructions such as posts and guard rails, separate roadways, and medians with widths in excess of 100 feet, are included in this group.

In addition, medians are also classified according to width. *Narrow* medians are those which are less than 16 or 20 feet in width. *Intermediate* medians are those which are more than 20 feet but less than 50 or 60 feet in width, and *wide* medians are those which are greater than 60 feet in width.

In studies of narrow medians, results have generally indicated that deterring-type medians are safer than non-traversable or traversable-type medians. However, non-traversable medians utilizing rigid barriers or dividers appear to be safer under high-volume conditions.[45-53]

Most studies of intermediate-width medians have indicated little or no correlation between median width and accident experience.

The cross-median accident rate decreases rapidly as the median width is increased up to 50 feet. Between 50 and 100 feet there is only slight reduction in the reportable accidents due to crossing the median, based on the limited data available. However, care must be taken to remove potentially dangerous objects from the median area, and cross slopes should be 1 on 6 or flatter.[46,48,49,51,54,55]

Alinement.

Horizontal alinement. Horizontal curves introduce an element of hazard on all types of highways. Sharp curves on grades are far more hazardous than curves on level alinement. Curves are also likely to be the scene of skidding accidents. There are a number of possibilities for reducing the accident potential of curves, including the improvement of superelevation, of visibility at the curve, and of signs and markings, particularly by the addition of indicated speed signs and delineators.

Vertical alinement. Long, steep grades and steep grades in combination with horizontal curves are extremely hazardous. Grades over 5 per cent in combination with curves over 5 degrees are associated with higher accident rates and skidding accidents. Higher accident rates are also associated with crest and sag vertical curves and with sight distance restrictions.[34,37,56,57]

Interchanges. Study of interchange accidents has shown that certain design arrangements are safer than others. With respect to ramp location, the following conclusions have been reached:

1. Ramp terminals located where the freeway through lanes are on a downgrade are safer than those located where the through lanes are on an upgrade, or at the crest or sag of vertical curves.
2. Long auxiliary lanes result in safer ramp terminals than do short tapers.
3. On-ramps which are located so that the merging point is less than 700 feet to a structure are considerably more dangerous than those located farther away.

Data on ramp spacing tend to be contradictory, with the evidence being nearly equally divided as to whether a single large-volume ramp is or is not safer than two smaller-volume ramps.

With respect to the relative safety of left-hand versus right-hand ramps, all available data indicate that the right-hand ramp is the safer of the two[57-61].

A detailed study of accidents on an urban expressway revealed that the most predominant type of accident on entrance ramps was *sideswipe*, and *rear-end* collisions represented the second highest in frequency[62]. On exit ramps, the most predominant type of accident was *rear-end* collision, and *sideswipe* was the second highest in frequency.

Intersections. During 1968 approximately 36 per cent of all reported accidents, nearly 37 per cent of all fatal accidents in urban areas, and about 15 per cent of rural fatal accidents, occurred at intersections.[16] The number and type of accidents at intersections are significantly influenced by the type of intersection, the individual details of design, the volume of traffic, and the control devices used.

The number of approaches has an effect on the accident rate of the particular intersection. Three-way intersections are inherently safer than four-way.[63-65] This is undoubtedly due to fewer points of possible conflict in three-way intersections.

The study of main, rural, two- and four-lane highways showed that the accident involvement rate increased as the number of intersections per mile increased.[17] This indicates the importance of controlling access to the highway to provide greater safety.

The type of traffic control used at intersections has a decided influence on the frequency of traffic accidents. Traffic control devices such as YIELD and STOP signs and traffic signals are used primarily at relatively high-volume intersections.

Four-way and two-way stops.[66] The accident rate for four-way stops was found to increase significantly when the major street volume is greater than 12,000 ADT, and the accident rate at two-way stops indicated a reverse trend of decreasing rate with increasing major street volume.

Flashing beacons.[67,68] This catalog includes traffic signals which flash yellow to the main street and red to the side street. In theory the red indication means *stop and then proceed when it is safe to do so*, while the yellow indication simply means *caution*. Flashing beacons have no value in expediting traffic, but they have definite safety values.

YIELD signs.[69-71] This sign is more recently used as a traffic control device in this country, but it has definitely shown its value in reducing accident frequency, particularly *right-angle* collisions.

Traffic signals.[66,67,72-75] Traffic signals by no means result in fewer total accidents, but they reduce certain types of accidents. Generally, after signalization, *right-angle* collisions and others involving vehicles on intersecting approaches, decrease, while *rear-end* and *turning* collisions between vehicles on the same street increase.

Such turning collisions may be reduced by special turning intervals, channelization, and turn prohibitions. In general, traffic signals reduce accidents at high-volume intersections.

Railroad Crossings.[76,77] There are several types of railroad crossing traffic controls in use, ranging from flagmen, wigwag signals, flashing light signals, and signs (particularly the crossbuck), to gates which drop across the roadway. The effectiveness of different types of control devices, based on several studies, may be indicated by assigning relative rating values. For this purpose, the crossbuck sign is usually assigned an index of 1.0.

RELATIVE EFFECTIVENESS OF DIFFERENT TYPES OF CONTROL

TYPE OF CONTROL	RELATIVE INDEX
Crossbuck signs	1.0
Wigwag signals	0.6–0.8
Flashing lights	0.3–0.6
Automatic gates	0.1–0.2

Roadside Objects.[6,14,78] There is no doubt that roads built in former years can be improved, from the standpoint of reducing the effects of collision between a car and a roadside object, or even completely avoiding a collision. The single-vehicle accident, or collision between a car and a bridge abutment, a lamppost, or some other roadside appurtenance, can account for as much as 32.3 per cent of all accidents on the open road. In addition, such single-vehicle accidents are generally more severe than other accident types: a National Safety Council survey indicates that the 7.5 per cent of urban accidents accounted for by such one-car accidents caused 21.3 per cent of the deaths. In rural areas they increase to 32.3 per cent of accidents and account for 39.6 per cent of fatalities.[15]

Guardrails, prows, walls, and any other structures in or near the road are to be used to minimize injury to motorists and damage to vehicles when vehicles leave the roadway. Their basic purpose is not to protect a lamppost, sign support, or bridge abutment *from* a colliding car, but the exact opposite.

In the past such devices have been very formidably effective in protecting the roadside appurtenances from the errant vehicle—at the expense of vehicle, driver, and passengers. These devices should never be installed so that they are ultimately more dangerous than the original condition. Of course, roadways should be designed to minimize the need for protective devices.

Pedestrian and School Child Safety. Approximately 18 per cent of all traffic fatalities are pedestrians, and a great deal more effort and research should be applied to alleviate this condition. The specific control devices and pedestrian regulations presently in use include the following:

Crosswalks. Crosswalk marking has significant safety value. Drivers tend to yield right-of-way to pedestrians more often at marked crosswalks.

Sidewalks. These are an essential part of urban development and are becoming increasingly more important for many sections of rural highways.

Exclusive pedestrian interval. At signalized intersections where pedestrian crossings and vehicle turning movements are heavy, exclusive pedestrian intervals are being used to increase pedestrian safety. However, an exclusive pedestrian interval is not suitable for use under all circumstances, since pedestrians are required to wait longer and good pedestrian observance of the signals is essential.

Pedestrian barriers. These are used to prohibit pedestrian crossing of major roadways at points where such crossing would cause exceptional hazard or would impede and delay vehicle movement.

School child pedestrian protection. Studies have shown that of the different types of protection at school crossings, the most effective measures are those that utilize school crossing guards or police officers.

Speed. One national study[23] presents a comprehensive analysis of speed and its relation to accidents on main rural highways. The study indicated that moderately high speeds are safer than slow speeds or excessively high speeds. It was found that the daytime accident involvement rate (number of vehicles involved in accidents per 100 million vehicle-miles) is at its minimum in the speed range of 55 to 70 miles per hour. However, accident severity increases rapidly at higher driving speeds, particularly beyond 60 miles per hour. These conclusions cannot be interpreted to mean that any given speed is safer than another under all conditions. Speed must be related to the highway and traffic conditions, and the lowest speeds can be unsafe under particularly adverse conditions, as can relatively high speeds under the most favorable conditions.

The BPR extended the above study to provide much greater detail on the relationship between accident involvement and speed on two- and four-lane main rural highways.[17]

Over the years traffic engineers have developed the following guiding principles for the application of speed zoning:

1. Motorists govern their speed more by traffic and roadway conditions than by indicated speed regulations.

2. Speed limits must be enforceable if they are to be effective. This means that a speed limit must be such that a majority of motorists will observe it voluntarily, and enforcement can be directed to the minority.
3. Any speed limit is reasonable only for the roadway and traffic conditions for which it was set.
4. Speed limits based on studies of the prevailing speeds and the roadway and traffic conditions tend to reduce the spread in speeds.
5. Accidents are not as much related to average speed as to the spread in speeds from the highest to the lowest.

Illumination. The nighttime fatal accident rate is more than 2.5 times the daytime rate. Night accidents can be significantly reduced by street and highway lighting. Experience to date suggests that the greatest benefit comes from the provision of some minimum level of illumination, that is, a level of illumination which brings significant reductions in night accidents though further step-up of lighting intensity produces only minor or negligible reductions.[79-83]

Human Variables in Traffic Accidents.[84-106] Psychological tests properly developed and administered have much to offer in reducing accidents and licensing of drivers. Knowledge of why drivers commit dangerous acts can be extremely helpful in developing corrective measures and regulations. Increased knowledge of driver characteristics and more comprehensive driver education are also required.

Accident Costs

Accurate data on costs of motor vehicle accidents is essential for analysis and evaluation of highway systems from the standpoint of losses through traffic accidents. Accident costs are one of the most useful measures of the importance of accident prevention efforts. However, the costs of accidents are very difficult to determine, because of the variety of factors whose monetary values are hard to measure. There are also a great many differences of opinion in the treatment of indirect costs of accidents. As a result, accident costs have been of relatively little use in practical problems of highway planning and traffic control, where detail and precise measurements are required. The high loss of lives and property from highway traffic accidents has increased the need for accurate and uniform determination of costs.

A rough approximation of the total economic loss resulting from traffic accidents may be obtained by methods recommended by the National Safety Council. These methods are based on the calculable costs of motor vehicle accidents which include wage loss, medical expenses, overhead cost of insurance, and property damage.

Wage loss includes loss of wages (or the value of services) due to temporary inability to work, lower wages when returned to work due to permanent partial disability, and the present value of anticipated future earnings for permanent total disability or death. *Medical expenses* include doctors' and hospitals' fees. *Insurance administrative costs* include all administrative, selling and claim settlement expenses for insurance companies and self-insurers.

In 1968, motor vehicle accident costs reached a total of $11.3 billion in the United States, according to National Safety Council (NSC) estimates. Costs averaged $204,000 per death for all accidents (fatal, injury, and property damage). This figure would be $460,000 for urban areas and $90,000 for rural areas.

In recent years several states have conducted relatively comprehensive studies of accident costs. Most accident costs are classified by severity (fatal, injury, and property damage only), by areas (urban and rural), and by facility types (expressway and nonexpressway). Principal studies have been conducted in Massachusetts,[107] Utah,[108] Illinois,[109] Washington, and California.[110] The Illinois study is significant in that it considers the costs of unreported accidents. The California study is largely based upon the results of the Illinois study. A great deal of difference in average accident costs is seen among these studies. The principal reasons are the methods of treating individual cost rather than any geographical differences.

Approximate unit cost by severity, as used by the NSC, is:

Death	$36,000
Personal injury	2,000
Property Damage	340

The Washington study assigns a cost of $70,000 to a fatal injury, $40,000 for a nonfatal injury.[111-113]

A correlation and synthesis of these various studies into a consistent format has been undertaken under NCHRP projects. This work has resulted in an important single document presentation of available accident cost data.[114]

Specific Analysis

Accident Location File. Detailed records of motor vehicle accidents by location should be maintained through the use of accident location files. The accident reports are filed alphabetically by intersection or street. Some authorities prefer to file the original reports chronologically, and provide a location file made up of cross-

TABLE 17-9

ACCIDENT TYPE	SYMBOLS	
	Fatal Accidents	Non-Fatal Accidents
Motor vehicle versus pedestrian	◎	◍
Motor vehicle versus motor vehicle	◍	◍

reference cards. The grouping of reports by location is of considerable benefit to the engineer and the enforcement officer.

Reports of accidents occurring at intersections are filed directly under the name of the intersecting street standing second in alphabetical order, behind the index card indicating the street standing first. For example, a report for an accident at Alpha Ave. and River Rd. would be filed in the section under primary card Alpha Ave. and behind River Rd., secondary index card.[115] Reports for accidents occurring between intersections are filed behind a primary card bearing the name of the street on which the accident took place. If there are several between-intersection accidents occurring on the same street, and the between-intersection portion of the road is long, secondary cards are used to divide the street into sections.

For rural roads and controlled-access highways, accident reports are usually filed by routes and sections, which generally include one or more interchanges, depending on the frequency of accidents.

Some of the uses of the accident location file are the following.

1. It furnishes complete and quick information about accidents at any location.
2. It reveals the most hazardous intersections and detailed accident facts about each.
3. It is an important aid in constructing collision diagrams.
4. It aids in the preparation of selective law enforcement programs.

Accident Spot Maps (Accident location spot map). This map, through the use of variously shaped or colored pins or marks, presents a visual record of the location and number of traffic accidents. It shows at a glance the locations of high accident frequency. It is important to limit the different types of pins or symbols to as few as possible, to permit quick and easy visual interpretation of the map. The commonly used symbols and classification are given in Table 17-9.

For spotting urban accidents a map drawn to a scale of 400 to 600 feet to the inch, showing street names is recommended. In rural areas, a scale of one mile to the inch, or $\frac{1}{2}$ mile to the inch, is generally used for heavily traveled areas.

Normally, a one-year accident record is kept on the map. It is advisable to maintain two spot maps simultaneously, one for the current year and one for the preceding year, allowing instant comparison. Before removing the pins (if pins are used), maps for each year should be photographed to maintain a record.

(Pedestrian accident spot map.) This map is used to show where the day and night pedestrian accidents occurred, using two types of pins or marks to distinguish day from night accidents.

(Fatal accident location spot map.) This is used to dramatize traffic fatalities for publicity and educational purposes. A distinction should be made between pedestrian and vehicle occupant fatalities.

In a similar manner, accidents involving drinking drivers can also be dramatized.

(Driver and pedestrian residence spot map.) This is a map used to indicate where drivers and pedestrians reside who are involved in traffic accidents. Two types of pins or marks are necessary to distinguish drivers and pedestrians. This map will show sections of the community where special safety education activities can be most effectively applied.

(Night accident location spot map.) This map is used to indicate night accidents separately from others. It is helpful in revealing a need for improved street lighting, interference of illuminated advertising signs with traffic control devices, or lack of visibility of traffic signs and markings.

(Child accident location spot map.) This map is used to indicate the relationship of child accidents to location of schools, school districts, and playgrounds.

(Uses of spot maps.)

1. Spot maps are used as a guide to traffic control and engineering in indicating the most hazardous locations and streets where condition, collision, and traffic volume diagrams should be maintained up-to-date to aid in the application of the most effective engineering and control measures.
2. Another use is as an aid to safety educational efforts and to public speakers.
3. Spot maps are also used as an aid in planning selective enforcement to direct proper emphasis as to locations, time of day, and character of accidents.
4. Lastly, they are used as an aid to determining street lighting needs on the basis of night accidents.

Collision Diagram. A collision diagram is a sketch showing the nature of accidents and, by means of arrows, the approximate paths of vehicles and pedestrians involved in accidents. Collision diagrams are seldom drawn to scale; they are schematic, showing each individual accident. The path of each vehicle involved

TABLE 17-10. COLLISION DIAGRAM SYMBOLS

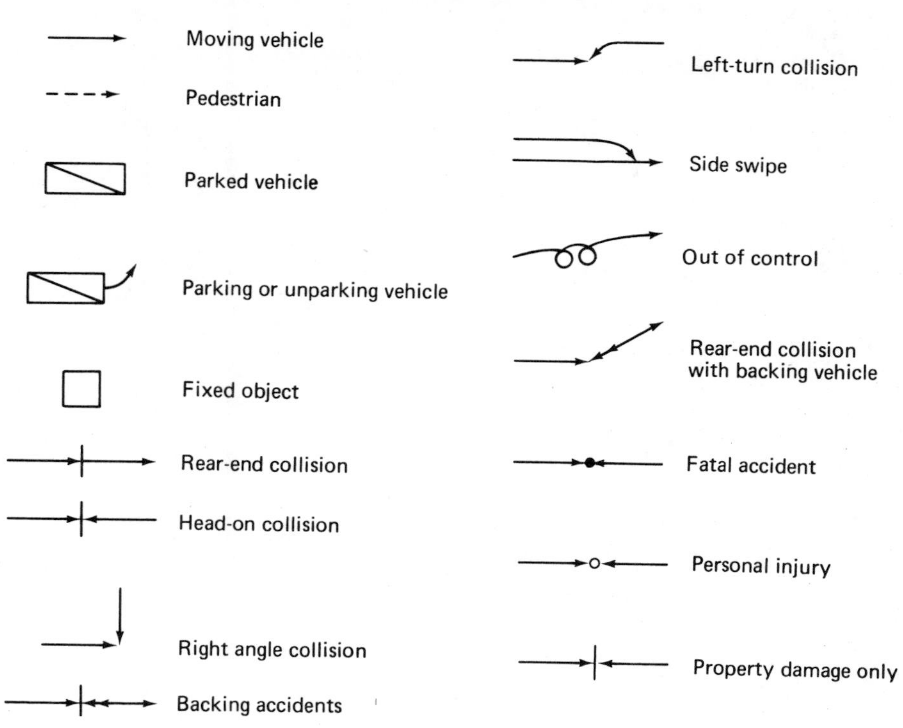

is represented by a solid line and of each pedestrian by a broken line. Sometimes the paths of vehicles before and after collision are shown by two different types of line. It is useful to indicate secondary collision. The exact path of vehicles and spot of accident need not be indicated because of possible overlap.[114]

On one of the arrows representing each accident the date and time of day can be shown. Weather, road condition, and unusual driver or pedestrian conditions can be indicated if and only if they are out of the ordinary, or if they have some important bearing on the accident. The nature of accidents is indicated by various symbols which reveal, to some degree, the type and cause of the accident. The various collision diagram symbols are shown in Table 17-10.

Collision diagrams are used to study accident patterns, to determine what kind of remedial measures are required (in conjunction with traffic flow and condition diagrams), and the results from their application. When diagrams for equal periods before and after application are compared, they show the types of accidents that have been eliminated, those that have continued, and new ones that have developed.

Condition Diagram. A condition diagram is a drawing to scale, showing the important physical conditions at a location to be studied and used as an aid in the interpretation of accident patterns. Important features that can affect traffic movement are shown. The observations and measurements which should be noted at the location to prepare a condition diagram are:

1. Street or roadway widths (curb to curb), shoulder width, if any, and corner radii
2. Property lines, sidewalks, and driveways
3. View obstructions at corners and physical obstructions in the roadway
4. Location, size, type, and legends of all signs, street markings, and traffic islands at or near the location
5. Location, type, visibility, and timing of traffic signals
6. Type, grade, and surface condition of roadway
7. Position and type of street lighting
8. Types of occupancy of adjacent property
9. Parking regulations

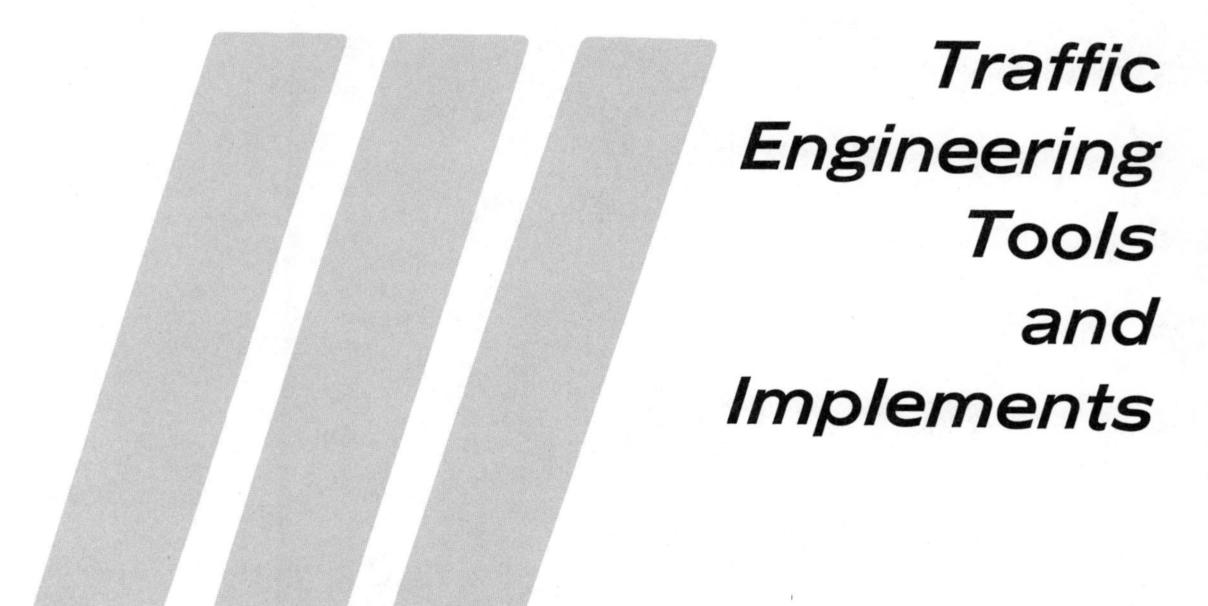

Traffic Engineering Tools and Implements

18

Traffic Laws and Ordinances

The need for traffic regulations becomes abundantly clear when one considers that there are over 100 million motor vehicles registered in the United States, driven by more than 103 million drivers, covering over 1,000 billion vehicle-miles per year, over 3.7 million miles of roads and streets! In addition, motor vehicle accidents in 1968 resulted in 55,200 deaths and 2.0 million injuries.[2] All of these facts completely justify the need for traffic, driver, and pedestrian regulations. Each intersection and every mile of highway is a potential point of accident or congestion.

The question of adequate traffic regulations is intimately related to the problem of moving people and goods safely and efficiently. The general public resents regulations, and from the public's standpoint, the observance of traffic rules is largely dependent on the clarity, reasonableness, and uniformity of the regulations. The proper purpose of all traffic legislation is not to impose unnecessary or unreasonable restrictions on highway traffic, but to insure, as far as this can be done by law and its enforcement, that traffic should move smoothly, expeditiously, and safely and that no legitimate users of the highway, whether in a vehicle or on foot, shall be killed, injured or frustrated in such use by improper behavior of others. On the official's side, traffic administration can be effective only to the degree that essential functions are properly integrated, with authority and responsibility fixed by law.

Both traffic laws and ordinances are legislative actions. They differ only in the legal terminology applied. Traffic *laws* are generally enacted by state legislatures. *Ordinances* are classified as "legislative acts," and are generally passed by local municipal legislative bodies. As such, the latter can usually be effected or rescinded more quickly than the former, which may or may not be a benefit.

REGULATORY AUTHORITY

The authority to regulate comes from several basic sources:

1. Some phases of highway transportation involve federal rules and regulations. The federal government is authorized to regulate interstate commerce and thereby controls many factors of commercial highway transportation. In addition, the Federal Aid Highway Acts and Highway Safety Acts involve the government in highway traffic matters, since the Bureau of Public Roads must pass on construction standards and the use of traffic control devices on federally aided highways.
2. Primary authority comes directly from state legislative bodies (traffic laws).

3. Some powers come from local authorities (traffic ordinances).
4. Some powers are designated in the general police powers. Police powers vary greatly in different states. They depend upon state constitution, city charter, and court opinions and interpretations.

Ideally, basic traffic regulations should be uniform throughout the state. Local authorities should, however, be given powers to adopt additional traffic regulations which provide for public safety and welfare, and which are not in conflict with the basic state traffic regulations.

Traffic laws, ordinances, and regulations cover three primary phases:

1. Control of the driver as to age, ability to operate a motor vehicle, and financial responsibility
2. Control of the vehicle from the standpoint of ownership and mechanical fitness
3. Control of operation of the vehicle in the traffic stream

Together these define the right of individuals and specify the responsibility of individuals.

UNIFORMITY OF REGULATIONS

Need for Uniformity

The need for basic uniformity of regulations and regulatory devices is clear. Obvious factors, such as driving on the right-hand side of the road, must clearly be uniform in all regions. However, due to the high, continuous volume of vehicular flow between municipalities and states, the need for uniformity in all areas of control becomes vital. Thus follows the need for standard devices, as to colors, shapes, and messages, so that a driver from New York will be easily able to understand and interpret the laws and ordinances, as relayed through control devices, in California or any other state. Uniformity in licensing and registration procedures is also desirable, to a less urgent degree. Confusion due to inconsistencies may be a major problem to travelers where recognition of out-of-state licenses is unclear, especially as pertains to younger drivers, where age requirements for the various states differ. Also, the inconvenience of changing licenses and registrations for drivers who change their state of residency could be minimized by the introduction of more uniform regulations.

Historical Development of Uniformity

The motivating force behind the development of traffic laws was the First National Conference on Street and Highway Safety held in Washington in 1924. This conference was sponsored almost exclusively by industries and associations and made several important recommendations, which included that:

1. The states adopt driver license laws
2. The states establish motor vehicle departments
3. Certificates of title be issued for vehicles
4. Accident reporting be made compulsory
5. Basic road construction standards be established
6. Uniform colors for signs and signals be designated

The second conference was held in 1926, and an effort was made to have public officials present from the 48 states. There were 43 states represented, and the most important event was the adoption of the Uniform Vehicle Code and the Model Traffic Ordinance. The Uniform Vehicle Code reflected the need, which even then was clearly evident, for uniformity in traffic regulations throughout the United States. To achieve this end, the code presented a reliable guide for the use of the legislatures of the several states. Since that time the Code and Ordinance have been reviewed periodically and additions and changes have been made where warranted by new developments and by practical experience.

In 1947 the National Conference was disbanded, and since then the work has been continued by the National Committee of Uniform Traffic Laws and Ordinances. This committee is made up of more than 100 representatives of federal, state, and local governmental units (legislators, police officers, highway and traffic engineers, motor vehicle administrators, judges, prosecutors, educators, mayors, county officials, and attorneys general), motor vehicle equipment manufacturers and dealers, insurance companies, motor clubs, safety councils, trade associations, national transportation associations, and other civic, legal, engineering, business, and labor organizations.

The Code and Ordinance are not based upon theory but on actual experience under the various state laws throughout the nation. The members of the National Committee are people in daily contact, such as officials or otherwise, with the complex problems of highway transportation.

Since 1926, the Code has been used by the legislators of all the states as a guide in drafting new statutes or amendments. In the same way, the Model Traffic Ordinance has been used as a guide to all municipalities. Several states and municipalities have adopted entire sections from the Code and Ordinance, practically verbatim. Both publications[3,4] are available from the National Committee on Uniform Traffic Laws and

Ordinances, 1319 18th Street, N.W., Washington 6, D.C. The latest revision in the Code and Ordinance took place in 1968. Prior to that date, they were revised in 1962 and 1956. Discussions of changes in the code included in the 1962 edition and proposed changes to account for traffic operation on the interstate system are available.[5,6]

Documents for Uniformity

Uniform Vehicle Code. The Code serves as a guide in establishing state traffic laws. The sections of particular concern to traffic engineers deal with:

1. Creation and operation of the department of motor vehicles
2. Certificates of title and registration of vehicles
3. Licensing of motor vehicle drivers and limitations imposed on licenses
4. Financial responsibility of motor vehicle operators
5. Method of treatment of accidents and accident reports
6. Rules of the road
7. Standards for vehicle equipment
8. Inspection of vehicles
9. Standards for vehicle size, weight, and load
10. Respective powers of state and local authorities

The Technical Committee on Safety of the Pan American Highway Congress is concentrating on uniform traffic signs, rules of the road and comprehensive vehicle code, uniform statistics, and recognizing outstanding achievements for the American continents.[11]

Model Traffic Ordinance. The Ordinance supplements the Code and serves as a guide in establishing local ordinances that may be necessary for public safety and welfare. The sections of particular concern to traffic engineers deal with:

1. Traffic administration
2. Enforcement of traffic regulations
3. Traffic control devices
4. Rules of the road
5. Designation of one-way streets

Observations on The Code and Ordinance. The Code and Ordinance do not necessarily recommend verbatim uniformity in all matters. However, the portions of the Code dealing with the rules of the road—the things that drivers and pedestrians shall and shall not do when on the public highways—should have complete uniformity so that the public may understand, remember, and observe these rules. In the portions dealing with motor vehicle equipment, substantial but not necessarily verbatim uniformity is a clear necessity, because vehicles are made and sold for use anywhere in the country. On other portions dealing with driver's license laws, financial responsibility, and vehicle registration, the need is not for *exact* uniformity but should embody the essential principles and substance of the Code as they have stood the test of time.

Since the publication of the revised editions of the Manual on Uniform Traffic Control Devices (1961) and of the Uniform Vehicle Code (1968), some inconsistencies in counterpart sections of the Manual and the Code were discovered. These inconsistencies should be eliminated to provide for more uniform interpretation of the laws and traffic control devices by police, courts, engineers, and drivers.[7,8]

The first in a series of annual reviews of state motor vehicles and traffic legislation, containing information on the status of traffic laws throughout the nation, has been published.[9] It is designed to be a supplement of the Uniform Vehicle Code (UVC) to promote the development of uniform traffic laws. Of particular interest to engineers are the reviews of new legislation and comparisons of all state laws on traffic control devices and traffic regulations.

A conference was held to explore the development of new ideas, new concepts, and new appreciations of the relationships between the motor vehicle and traffic law.[10] The conference accentuated the fact that considerable effort and research is required to fill in the gaps to bring about a compatibility between motor vehicle and traffic law.

ELEMENTS FOR REGULATION

Driver Controls

Control Techniques. Solution of the accident problem involves complex interrelationships. Among these are street and highway design and use, vehicle construction and operation, enforcement policies and their application, vehicle use legislation and driver licensing, public education, driver selection and education, and the nature and the limitations of the driver in terms of his physical, mental, and emotional traits and qualifications. While each of these factors is important in the accident reduction problem, it is upon the driver that paramount attention must be focused. Very few accidents would occur if all drivers handled themselves correctly in relation to the design and condition of the vehicle and the design and condition of the roadway.

Driver controls deal primarily with:

1. Licensing of drivers
2. Financial responsibility of drivers
3. Civil liability

Models for the granting of the privilege to drive a motor vehicle, for defining and establishing requirements for financial responsibility, for defining and establishing requirements for civil liability with respect to the liability of government agencies, the liability of vehicle owner relative to the operation of his vehicle by others with his consent and without his consent, liability for bodily injury to, or death of, a guest riding in a motor vehicle, and the liability of nonresident motor vehicle operators are all found in the Uniform Vehicle Code. The Code outlines, defines, and establishes proper procedures for driver controls. If all states adopted the Code recommendations, particularly with respect to driver licensing, there would be far fewer motor vehicle accidents. The more critical reasons for improper licensing are recognized as:

1. Many states still have motor vehicle laws well below the standard established by the Code.
2. Many states that have enacted laws which closely approximate or equal the standard have failed to appropriate the funds needed to administer the laws properly.
3. Strict administration of rigid driver licensing laws, even when adequate funds are available, is quite unpopular and all too often politically hazardous.
4. Driver license examiners are, at times, selected on a political basis rather than on a basis of fitness.
5. The volume of applicants in relation to the number of examiners is often so great that it becomes almost impossible to give a complete or thorough examination.
6. Many road tests given under the licensing laws are grossly inadequate as measures of driving skill.
7. Many of the facilities provided for driver licensing are definitely inadequate.

The problem of granting the privilege to drive is a very difficult and serious one. It is obvious that much still must be done before there can be even reasonable assurance that only the relatively competent will be granted that privilege.[12]

A positive step which has been given considerable national attention is the control of suspended drivers. There is an increasing trend to suspend licenses of drivers involved in multiple violations, arrests, or accidents. More than one million drivers forfeited their driving privileges in 1960, but conservatively it is estimated that half of these drivers continue to operate their vehicles while under suspension.[13]

To help prevent a driver who lost a license in one state from obtaining one in another state, Congress authorized, in 1959, the establishment of a National Driver Register. The Register, which is maintained by the BPR, serves as a clearing house to identify, for the states, those drivers whose licenses have been revoked primarily because of driving while intoxicated or conviction of a traffic violation resulting in fatal accident. The Register is operated as a voluntary state-federal enterprise. Those states that participate will furnish the BPR with identifying information on drivers whose licenses are being revoked for the specified causes, and may request the BPR to check new license applicants against the Register for a record of previous revocation.[14] The National Driver Register became operational on July 1, 1961.[15,16] The various states are now checking their driver license applications against the Register at a rate of 300,000 a month. As of mid-1965, the Register received almost 739,000 revocation reports at its Washington headquarters since it went into operation in mid-1961. During the same period, about 15.3 million requests from the states were processed, and over 96,000 identifications were returned.

Control of drivers will be considerably advanced when psychological tests can be practicably applied in the licensing of drivers. Although psychological testing alone is not a cure-all, it is a good way to determine certain personal characteristics proven by long observation to be important clues to unsafe driving tendencies. Most unsafe drivers exhibit traits including aggressiveness, impulsiveness, job drifting, poorly controlled drinking habits, resistance to regulation, lack of concern for the future, and a problem childhood. The safe driver, on the other hand, usually exhibits opposite characteristics.

The forward step taken by New York State requiring drivers to pass periodic eye tests should also be adopted by other states. Under this law, drivers are required to take vision tests every 9 years. If the driver fails to pass such a test, the state motor vehicle commissioner is empowered to cancel his old driver's license or impose restrictions on the use of any new license issued. Simple vision testing is necessary but still not sufficient for safe driving of a motor vehicle. A Snellen acuity of 20/40 and a visual field of 130 degrees or better appear to be reasonable levels for driver licensing. Retesting of accident repeaters should be required, and, based on age changes in vision, an appropriate time to retest the vision of drivers would be between the ages of 50 to 55.[17]

A dashboard-mounted device has been developed which provides fast, accurate measurement of driving skills.[18,19] The device measures driver skills by recording his actions and establishing impulses and responses, and it could be usefully applied in driver licensing road tests and driver education.

Impediments to Regulation. The reasons for regulation of drivers are clear. The unsafe, incapable driver is a physical hazard to those both in his vehicle and in other vehicles, as well as to pedestrians. However, strict regulation and restriction of driver licensing is not a particularly popular task. The man who is deprived of a license often feels that he has been deprived of a basic "right." Therefore, strict driver licensing procedures, which would undoubtedly improve the safety records on highways, are politically unpopular. Particularly in rural areas, the deprivation of a license severely restricts the mobility of those affected. Where mass transit is not available, nondrivers must rely on friends and relatives or hired cars (taxis, limousines, etc.) for transportation.

Another impediment to strict driver regulation is the economic infeasibility of setting up programs to adequately examine and reexamine drivers and prospective drivers in a wide variety of circumstances. Also, the technology of psychiatric testing is not advanced to the point where drivers can be evaluated on psychological grounds.

Because of these impediments, driver licensing is perhaps the least stringent item of control, although the effect of this item may well be of greatest importance.

Vehicle Controls

The physical characteristics of vehicles and the numbers of various-size vehicles using the highways are positive controls in geometric design. Some vehicle controls are more significant to the traffic engineer than others. Registration facts and figures provide the engineer with basic data which are necessary for correlating survey data. Limitations placed on vehicles regarding weight and dimensions and minimum requirements, relative to motor vehicle equipment and accessories, are extremely important in the planning, control, and operation of the traffic stream. Vehicle controls have a definite tendency to upgrade the vehicle. Vehicle controls deal with:

1. Certificates of title and registration of vehicles
2. Anti-theft laws
3. Licensing of dealers, wreckers, and rebuilders
4. Equipment of vehicles
5. Inspection of vehicles
6. Size, weight, and loads of vehicles

A reasonable, up-to-date model that outlines and defines in detail the requirements of adequate vehicle controls is found in the Uniform Vehicle Code.

A law in New York State makes it unlawful to drive with treadless tires. The state motor vehicle commissioner determines what amount of tread will be acceptable. His rules enable law enforcement officers to inspect tires visually and to measure the tread. In 1963, a law was also passed by the New York State legislature increasing the gross weight limit of highway vehicles to 71,000 pounds from the former limit of 65,000 pounds.

A proposed standard equipment measure that is receiving considerable support is the use of seat belts. The National Safety Council estimates that more than 8,000 lives would be saved annually and serious injuries would be reduced by one-third, if seat belts were installed and used in all motor vehicles. The basis for this statement is that persons ejected from accident vehicles are more likely to be seriously injured or killed than those not ejected, and seat belts reduce the possibility of ejection.[2] As of 1968, safety belts were available to nearly three-fourths of all passenger-car occupants.[2] According to the National Safety Council (NSC), the use of seat belts decreased 1964 traffic fatalities by about 750, based on estimates which indicated that approximately 30 per cent of the passenger cars had belts installed and that these were used about 50 per cent of the time.[2]

In New York State compulsory motor vehicle inspection consists of the checking of seven safety points: brakes, steering mechanism, wheel alignment, lights, tires, windshield wipers, and glass. Latest revisions in the inspection law require that all vehicles be inspected once a year, regardless of age, and when the vehicle is sold or resold.

Recent efforts by federal and state governments have been significantly intensified to coerce motor vehicle manufacturers to build more safety into motor vehicles. A series of articles have appeared in *The Highway User* magazine which report on what efforts automobile, truck and bus, and tire manufacturers take to make vehicles as safe as possible. The articles mainly express the views and comments of officials of the various manufacturers.[21,22,23]

Another area of growing governmental concern is automotive air pollution. It appears that national legislation will be enacted requiring that all new vehicles be equipped with emission control systems. However, meaningful legislation should provide for a sound periodic maintenance and inspection program, at the state level, to insure proper continued functioning of vehicle exhaust control systems.[24-29]

Control of Vehicle Operation

The basic rules of the road define the rights of individuals and specify the responsibility of individuals in their use of public traffic ways. These rules, which should be applied on a state-wide basis, generally cover all vehicles and persons using public ways. The basic rules of the road are essentially the same in both the Code and the Ordinance. However, the basic rules are modified when they become more urban in character.

The basic rules of particular concern to the traffic engineer include the following:

Traffic Control Devices. The Code defines the authority to install, obedience to, the meaning of, and unauthorized display of traffic signs, signals, and markings.

Use of Roadway.

1. Requires drivers to use right side of roadway except when:
   a. Overtaking and passing another vehicle
   b. An obstruction exists making it necessary to drive to the left of center
   c. A roadway is divided into three marked lanes
   d. A roadway is designated for one-way traffic
   e. Preparing for a left turn
   f. Authorized by official traffic control devices to use a portion of the roadway that is normally restricted in use to vehicles traveling in the opposite direction
2. Requires overtaking and passing on the left. Overtaking on the right is permitted only under the following conditions:
   a. When a vehicle overtaken is about to make a left turn
   b. Upon a roadway with unobstructed width for two or more moving lanes
   c. Upon a roadway designated for one-way traffic
3. In the use of the roadway, limitations are imposed on overtaking on the left when there is insufficient sight distance available when traveling on level grades or on vertical and horizontal curves. The driver is also prohibited in following another vehicle too closely. The proper use of limited access highways is also defined.

Right of Way.

1. Vehicles approaching or entering an intersection are restricted in the following ways:
   a. The driver of a vehicle approaching an intersection shall yield the right of way to a vehicle which has entered the intersection from another approach.
   b. When two vehicles enter an intersection from different approaches at approximately the same time, the driver of the vehicle on the left shall yield the right of way to the vehicle on the right.
   c. Rules (a) and (b) apply to uncontrolled intersections.
   d. The driver of a vehicle intending to turn to the left shall yield the right of way to any approaching vehicle from the opposite direction.
   e. Every driver approaching a STOP sign is required to stop and yield the right of way to any vehicle which has entered the intersection or which is approaching so closely as to constitute an immediate hazard.
2. Upon the immediate approach of an authorized emergency vehicle, the driver of every other vehicle shall yield the right of way and drive to a position as near as possible to the right-hand edge of the roadway.

Pedestrians' Rights and Duties.

1. At unsignalized intersections the pedestrian has the right of way after entering a marked or unmarked crosswalk.
2. Every pedestrian crossing a roadway at any point other than within a marked or unmarked crosswalk at an intersection shall yield the way to vehicles.
3. Any pedestrian crossing a roadway at a point where a pedestrian tunnel or overpass has been provided shall yield the right of way to vehicles.
4. Between adjacent intersections at which traffic control signals are in operation, pedestrians shall not cross at any place in the crosswalk.
5. Where sidewalks are provided, it shall be unlawful for any pedestrian to walk along and upon an adjacent roadway. Where sidewalks are not provided, the pedestrian shall walk on the left side of the roadway facing traffic approaching from the opposite direction.

Speed Restrictions. The rules dealing with speed controls will be discussed in a later chapter.

Stopping, Standing, and Parking. The rules dealing with stopping, standing and parking controls will be discussed under separate headings in later chapters.

19

An Introduction to Traffic Control Devices

Traffic control devices are all signs, signals, markings, and devices placed on or adjacent to a street or highway, by authority of a public body or official having jurisdiction to regulate, warn, or guide traffic.[5]

APPLICATION OF TRAFFIC CONTROL DEVICES[6]

Safety upon a highway and the ability of that highway to carry an adequate number of vehicles with a minimum of delay and inconvenience is dependent to a large extent on the orderly flow of traffic.

The majority of motorists drive in an orderly and safe manner provided they are given intelligent and reliable regulation and information which will enable them to do so. On the other hand, the public will not react favorably to unintelligent direction, and the disrespect and lack of obedience which some traffic control devices now receive is undoubtedly due to the fact that sufficient attention has not been given to their design or use. An essential factor in driver obedience is standardization. A motorist has a right to expect that any given device for the control of traffic will always have the same meaning and require the same action on his part regardless of where it is encountered. Application of sound principles in the establishment of regulations and in the selection, installation, and operation of traffic control devices is of the utmost importance. Their misapplication, besides wasting public funds, has in numerous cases accomplished the reverse of the purpose intended, causing delay, confusion, and disrespect for, and disregard of, all traffic control devices.

The applicability of traffic control devices in any specific case cannot be determined by guess work. It should be based upon sound engineering principles supplemented by factual studies of types and flow of traffic, accidents, speeds, delays, and physical conditions which will show the exact nature of the difficulty and indicate what particular devices or methods of control are needed. After such determination has been made, the devices or controls should conform to, and be applied as prescribed by, the principles established in the *Manual on Uniform Traffic Controls Devices* (MUTCD).[5,41]

By careful study and analysis of the factual data, the field in which judgment must be exercised will be limited, but in the last analysis the element of personal judgment must necessarily enter. Every solution of a traffic problem should therefore be considered as tentative only and, after being placed in effect, should be studied in the field to determine the reaction of traffic and to ascertain what changes may be necessary to produce the desired results.

In the application of traffic control devices, as in the case of any proposed traffic regulation, it is essential to obtain both official (head of local governmental unit) and public support.[7] Prior to the application of controversial traffic regulations and controls, it is necessary to obtain support from appropriate public officials. It is also vitally important that a program of public education be established to inform the community and interested public organizations of the meanings and objectives of the proposed program.

FUNCTION OF TRAFFIC CONTROL DEVICES

The function of traffic control devices is to provide the road user with information regarding performance requirements along his route of travel. Traffic control devices may either supplement or modify the basic rules of the road, and this information must be transmitted by appropriate devices at the specific time or location. Devices are classified into three functional groups as follows:

Regulatory Devices

These give the road user notice of traffic laws or regulations that apply at a given place or on a given roadway. Disregard of such devices is punishable as an infraction, violation, or misdemeanor.

Warning Devices

These call attention of the road user to conditions, on or adjacent to the roadway, that are potentially hazardous to traffic operations.

Guiding Devices

These provide directions and information to the road user regarding route designations, destinations, distances, roadway delineation, points of interest, and other geographical or cultural information.

REQUIREMENTS OF TRAFFIC CONTROL DEVICES[5, 41]

Any traffic control device should meet the following basic requirements:

1. It should be capable of fullfiling an important need.
2. It should command attention.
3. It should convey a clear, simple message.
4. It should command the respect of road users.
5. It should be located to give adequate time for response to the message.

Four basic considerations are employed to insure that the requirements are met. These include the following:

Design of Devices

1. Attention is drawn to the device by such features as size, contrast, color, shape, composition, lighting or reflectorization when required, and in some cases by movement and sound.
2. Clear meaning is produced by simplicity of legends and messages used, legibility, shape, size, and colors.
3. Command of respect is dependent upon uniformity, reasonableness, size, legibility, and brightness.
4. Adequate time for response is dependent upon legibility, size, and proper placement in relation to the traffic stream and speed of traffic.

Placement of Devices

1. To compel attention, the device must be positioned so that it is within the cone of clear vision of the normal road user (10 to 12 degrees) since vision beyond this cone becomes blurred.
2. To aid in conveying the proper meaning, the device should be located with respect to the point, object, lane, curb, or situation to which it applies.
3. To command respect, the device must be logically placed to carry its message to the road user for whom it is intended, and in some cases it may require advance warning.
4. To permit time for response, consideration must be given to location with respect to legibility, speed, and distance to point of required response.

Maintenance of Devices

1. To compel attention, the device must be clean, legible, and free from obstructions.
2. To make its meaning clear at a glance, and to permit time for response, the device must be maintained to high standards to assure that it is clean and bright so that legibility and visibility are retained.
3. To command respect, the device must be clean, legible, properly mounted, and in good working condition. The device should convey the impression that it is official and enforced. When a device is no longer needed it should be removed, so that the road user will not lose respect for all devices.

Uniformity of Devices

1. Attention is partly dependent upon uniformity of application of devices.
2. Clear meaning is dependent upon using the same device for the same purpose at the same relative position. Uniformity in the use of devices aids in instant recognition and understanding.
3. Command of respect is dependent upon standardized official devices, logically used for warranted traffic conditions.
4. Adequate time for response is dependent upon standards of design, placement, and application of devices.

UNIFORMITY OF TRAFFIC CONTROL DEVICES

Benefits of Uniformity

The efficiency and clarity with which messages are conveyed to the driver depends upon the motorist's ability to interpret automatically, rather than requiring specific decisions on the meaning of each individual message. Uniformity of devices helps produce this effect, as well as other efficiencies. Uniformity of devices will also serve to aid in the ways listed below:[9]

1. Aid in instant recognition and intent. With present-day high volumes and speeds of traffic, it is essential that the driver get the intended message of a device as quickly as possible. The road user must have sufficient time to react properly before the point of needed reaction is reached. This requires simple and clear messages which are large enough to read and understand. Uniformity achieves these objectives.
2. *Increased safety.* Properly applied, uniform devices provide the road user with required information, guidance, or control at the right place, at the right time, and in the right way. These all lead to increased traffic safety.
3. Aid to road user in unfamiliar areas. Many road users cross numerous municipal and county lines in daily or frequent trips, and they often cross state lines on longer trips. The road user expects to find control devices that are uniform and readily understandable.
4. Aid to traffic operations. Reasonable, warranted application of uniform devices provides for more orderly and efficient traffic movement and makes driving more convenient and pleasant.
5. *Avoid confusion.* Uniform devices help greatly to avoid driving uncertainty and confusion, which often result in accidents.
6. *Earn public support.* Properly applied, uniform devices will significantly help to obtain public support for traffic controls. Public support is extremely valuable in obtaining political support and in helping to overcome the objections of opponents.
7. *Produce economies.* Uniformity results in economy in manufacture, installation, replacement, maintenance, and administration.
8. *Eliminate questions on interpretation.* Adherence to standardized devices aids enforcement by giving police, traffic courts, and road users all the same interpretation.
9. *Help avoid governmental liability damages.* Proper application of uniform devices will help eliminate liability damage suits against the local governmental unit responsible for a device, as in cases where devices do not meet standards, whereas the state law requires local compliance with standards.
10. *Provide guidance.* Uniform standards are valuable in aiding engineering decisions, and in dealing with legislators and with citizen requests. Uniformity protects against unwarranted devices, since many local communities have no trained traffic engineer, and local officials can thereby avoid the misuse of devices.
11. *Promote local control.* Unless states, counties, and municipalities adopt relatively uniform standards, congressional legislation may be enacted for the development of federal standards for traffic control devices. If uniformity is not obtained, there would be a real need for federal government intervention.

The advantages of uniformity far outweigh the disadvantages or undesirable features of uniformity. However, the disadvantages of complete uniformity should be realized, and attempts made to avoid them. Principal disadvantages include the following:

1. *Possible lowering of a higher standard.* Uniformity for the sake of uniformity should not be carried to the extreme, which may result in not adopting an improved device or procedure simply because it is not in common use.
2. *Possible retarding of research.* Some degree of non-uniformity is desirable as it helps to stimulate research and experiments which could result in the development of superior equipment and new materials.
3. *Possible elimination of engineering judgment.* Uniformity may tend to replace or discourage good engineering judgment and imaginative application of devices. Standards and handbooks such as the MUTCD should be used as tools and guides to the engineer, but should not replace or eliminate the exercise of the engineer's judgment.
4. *Possible increase in expenditure.* Uniformity may tend

to make obsolete much equipment already in use when standards are changed or updated.

For a complete discussion of the issues surrounding uniformity, consult the literature.[18-23]

Development of National Uniformity and Standards

Uniformity and standardization of traffic control devices is extremely desirable on a national basis. National uniformity is a basic requirement in meeting future highway transportation needs. Driving safety under present-day traffic conditions is sufficiently difficult without the added handicap of nonuniform traffic control devices that confuse drivers and cause accidents. These conditions can only become more acute in the future, unless positive steps are taken to achieve national uniformity. However, prerequisite to the development of national standardization of traffic control devices is the necessity of providing reasonably uniform laws and ordinances which establish the legal authority for the various governmental units to effectively apply traffic control devices.[1-3] National uniformity in laws and ordinances provides the foundation for uniformity in the application of devices.

Effective and efficient application of traffic control devices and measures can be assured with greater certainty when they are applied simultaneously with, and become an integral part of, any street or highway improvement project. It would be desirable to require specific planning and location of traffic control devices on the plans of the improvement project prior to approval for construction.

Several constructive steps have been taken to promote national uniformity; however, to achieve this goal it is essential that cooperation be established among the various states and numerous local governmental units. Even with this cooperation, realistic progress is strongly dependent upon developing public support for the program. The significant steps taken to promote uniformity include the following:

The National Manual on Uniform Traffic Control Devices.[5,13,41] The first edition of the MUTCD was published in 1935 by a joint committee of the AASHO and the National Conference on Street and Highway Safety. The Institute of Traffic Engineers (ITE) joined the committee in 1942 and significant changes were made in the original edition to necessitate a revised edition, which was published in 1948 under the sponsorship of the three organizations. A special pamphlet of revisions was issued in 1954. A later edition of the MUTCD was published in 1961 by a joint committee of the AASHO, the ITE, the National Committee on Uniform Traffic Laws and Ordinances, the National Association of County Officials, and the American Municipal Association. The significant changes between that edition and the previous edition, with its pamphlet of revisions, are documented.[10-12]

Some references in the MUTCD to standards for the interstate system have been invalidated by a new revision of the *Manual for Signing and Pavement Marking of the National System of Interstate and Defense Highways*.[14,42] The interstate manual should be consulted for all signing and marking of the interstate system.

For additional comments about the 1961 edition of the MUTCD, consult the literature.[3,4,9,15-17] A new Manual[41] was published in May, 1971. Principal features of the new Manual include:

1. The definition of terms *shall*, *should*, and *may*. Where standards are specified by a *shall* stipulation, the standard is considered *mandatory*. Where *should* is used, the standard is *advisory;* and where *may* is stipulated, the standard is entirely *permissive*, with no requirement intended.
2. Specification of an overall color code for all devices:
   Red: Stop or prohibition
   Green: Indicated movements permitted, direction guidance
   Blue: Motorists' services guidance
   Yellow: General warning
   Black: Regulation
   White: Regulation
   Orange: Construction and maintenance warning
   Brown: Public recreation and scenic guidance
   Purple, *Bright yellow-green*, *Light blue*, *Coral*: Unassigned
3. The establishment of separate standards for treatment of school areas.
4. The adoption of several symbol signs in an effort to bring United States standards into closer concert with efforts at international uniformity.
5. The updating of all standards, including signs, signals, and markings.

Standards in the new Manual, particularly where changes have been made, should be implemented gradually, where possible, in conjunction with routine maintenance operations. New standards should be fully implemented by the following dates, as specified in the MUTCD: markings, December 31, 1974; signs, December 31, 1975; signals, December 31, 1976. Where new symbol signs are used, supplementary word messages should be used to aid the driving public in adjusting to symbols.

The Uniform Vehicle Code.[1] A MUTCD is ineffective unless the proper legal foundation is provided for its application by state laws. A very significant step towards uniformity of traffic control devices would be the adoption, by the various state governments, of a vehicle code that is in substantive conformity with the Uniform Vehicle Code. The pertinent portions of the Code relating to traffic control devices are Sections 11-201 through 11-206, and Sections 15-104 through 15-108. These sections provide the following:

1. Prescribe required traffic control devices and define meaning and obedience to various devices.
2. Provide for only official agencies to apply traffic control devices.
3. Require the state highway commission (department) to adopt a state manual of uniform traffic control devices. It further requires that this manual should correlate with and so far as possible conform to the national manual.
4. Require that traffic control devices installed by all local authorities within the state conform to the standards in the state manual.

The Model Traffic Ordinance.[2] To complete the legal step towards uniformity of traffic control devices, it is essential that every state develop its own traffic ordinance for its various local governmental units. The ordinance for municipalities should be developed so that it is in substantive conformity with the model traffic ordinance. The pertinent portions of the Ordinance relating to traffic control devices are Sections 4-1 through 4-13. These sections provide the following:

1. Prescribe required traffic control devices and define their meaning and obedience to various devices.
2. Provide for only local officials to apply traffic control devices.
3. Require conformance to state vehicle code and state manual.

State Manuals on Uniform Traffic Control Devices. Realistic uniformity of traffic control devices cannot be achieved unless the various state governments adopt state manuals that are in substantive conformity with the national manual. It is significant to keep in mind that the national manual represents the collective thinking of experienced specialists in the field of traffic control. The committee includes practicing highway traffic engineers, planning and maintenance engineers, state and local governmental officials, BPR officials, police officials, and safety and educational specialists from all over the country.

There are still considerable efforts that must be made by the state governments to take the initiative in promoting uniformity. Appropriation of sufficient funds should be made to replace obsolete devices and to conduct studies to make state manuals conform to the national manual and adopt adequate legislation to provide for proper application of devices.

For additional comments on the progress of uniformity, consult the literature.[19-23]

The AASHO Interstate Manual for Signing and Marking.[14,42] The Interstate Manual was first adopted by AASHO and approved by BPR in 1958. A second edition was issued in 1961 and was revised in 1962. The Manual provides for uniform signing and marking of the Interstate System, even though it is the responsibility of the various state highway departments.

It is required by law that the states conform to the Interstate Manual, for it implements Section 12 of the Federal Aid Highway Act of 1944, which requires concurrence of the BPR in the signing of federal aid projects as a means of achieving uniformity.

The current version of the Interstate Manual[42] was published in 1970. In keeping with the objective of uniformity, the new Interstate Manual and the 1971 MUTCD were designed to be in agreement with one another. For example, many of the sign colors initiated in the 1962 Interstate Manual were adopted for the 1971 MUTCD. Principal changes in the 1970 Interstate Manual include:

1. The requirement for mileposting the interstate system.
2. The requirement for interchange numbering on the basis of the mileposting system, rather than consecutive numbering, though supplemental consecutive numbers may be retained.
3. The adoption of white-on-brown sign colors for recreational and scenic guidance (in keeping with the similar standard in the 1971 MUTCD).
4. New standards concerning lettering, mounting, and placement of signs.

Federal Aid Projects. State programs for modernizing traffic control devices can be substantially aided, provided the modernization conforms with the standards in the National Manual. The BPR had provided strong financial inducements to achieve uniformity on all federal aid projects. The BPR had made the following announcements upon publication of the National MUTCD (1961 version):

1. Traffic control devices on all federal aid highway projects must conform with the National Manual, and a reasonable period was allowed to replace nonstandard devices. (January 1967 was the target date.) The use of

the National Manual was a further implementation of Section 12 of the Federal Aid Highway Act of 1944 for all roads on the primary and secondary systems. This provided a significant impetus for state manuals to conform with the National Manual, as it was highly unlikely that the states would apply two sets of standards on their highways.

2. Federal aid funds were made available towards the cost of standardizing and/or modernizing traffic control devices so as to conform with the National Manual.
3. Federal aid funds were made available for studies for existing traffic control devices on the federal aid systems to assure the orderly development of a state-wide program for improvement of traffic control devices in accordance with the National Manual.
4. Under the Highway Safety Act of 1966, each state had to have a program relating to the use of traffic control devices in conformance with standards issued or endorsed by the Federal Highway Administrator.[8]

Conferences. Updating and implementing standards for uniformity is an ongoing process. In 1962, thirteen regional conferences were sponsored by The National Joint Committee on Uniform Traffic Control Devices to introduce the 1961 edition of the MUTCD, and to present and discuss the need for uniformity to local officials from all parts of the nation. In November of 1962, a National Conference on Uniform Traffic Control Devices was sponsored by the President's Committee on Traffic Safety,[18] again to arouse support for uniformity. In November of 1967, the National Joint Committee met to complete several sections of the revised MUTCD, expected to be published shortly.[34] During May and June of 1969, ITE sponsored nine regional conferences for ITE members to discuss some of the proposed revisions to the MUTCD. All comments received as a result of the conferences were forwarded to the joint committee for consideration.[37,38] Discussions at the regional conferences and at the national conference brought out the necessity for the program to modernize traffic control devices.

To provide background information for developing presentations on the need for uniformity, six resumés were distributed to delegates at the national conference. These resumés are of extreme value to civic leaders and helpful to traffic engineers.[35,36]

Public Knowledge. Some results of a study relating the driver and traffic control devices are as follows:[24]

1. Drivers have a very limited knowledge of the traffic engineering intent of many traffic signs and markings.
2. Driver education programs should include substantial amounts of information on traffic control devices, and the certification of driver education instructors should include at least one basic course in the fundamentals of traffic engineering.
3. Some type of pamphlet should be distributed at the driver licensing bureaus. This pamphlet should present to the driver the basic concepts of design, placement, and intent of traffic control devices.
4. Research into the area of "driver understanding of traffic engineering controls and devices" is needed to provide a basis for eliminating poor devices and for improving controls and laws to provide for safer, more orderly traffic flow.

International Uniformity. Serious discussions have been held to promote an active program for developing an international system of uniform traffic control devices that would be intelligible to all road users regardless of where they may drive. Formidable obstacles block the path to international uniformity, including a reluctance to replace existing devices because of local familiarity and current large investments in local systems.[25-28] In one such case, 132 countries and 60 international organizations were invited by the Secretary-General of the United Nations to send representatives to the United Nations Conference on Road Traffic at Vienna to conclude new international agreements in road traffic and road signs and signals.[27]

To achieve any semblance of international uniformity, more emphasis must necessarily be placed on symbolic messages, rather than word messages, to overcome the difficulties presented by multi-lingualism. To supplement symbology, greater use must be made of shapes, pictures, and color. American standards extensively use word messages, while those of most European countries rely heavily on symbology, since they have to contend with the problems associated with multi-lingualism.[26,29,30]

Canada has made substantial advances in achieving national uniformity. It published its first manual on traffic control devices in 1960.[31] Although several provinces and local authorities have carried out major conversion programs based on the new standards, provincial government legislation embodying the standards as set forth in the manual is lagging.[32] A significant difference between the American and Canadian manuals is a much greater emphasis on symbolic messages in the Canadian manual, due, in part, to the bilingual characteristics of Canada.

In Great Britain, a great changeover of road traffic signs took place, involving the replacement of $1\frac{1}{2}$ million signs, costing 20 million pounds, and taking 10 years to complete. This modernization aims at matching the greater speeds and volumes of traffic for the present and the future, and to bringing the British signing system in line with the European system.[26]

A detailed presentation of the results of the 1968

TABLE 19-1. HIGHWAY SAFETY PROGRAM STANDARD 13 ON TRAFFIC CONTROL DEVICES

EACH STATE, IN COOPERATION WITH ITS COUNTY AND LOCAL GOVERNMENT, SHALL HAVE A PROGRAM RELATING TO THE USE OF TRAFFIC CONTROL DEVICES (SIGNS, MARKINGS, SIGNALS, ETC.) AND OTHER TRAFFIC ENGINEERING MEASURES TO REDUCE TRAFFIC ACCIDENTS.

I. The program shall provide, as a minimum, that:
  A. There is a method:
    1. To identify needs and deficiencies of traffic control devices.
    2. To assist in developing current and projected programs for maintaining, upgrading, and installing traffic control devices.
  B. Existing traffic control devices on all streets and highways are upgraded to conform with standards issued or endorsed by the Federal Highway Administrator.
  C. New traffic control devices are installed on all streets and highways, based on engineering studies to determine where devices are needed for safety. Such devices conform with standards issued or endorsed by the Federal Highway Administrator.
  D. There are programs for preventive maintenance, repair, and daytime and nighttime inspection of all traffic control devices.
  E. Fixed or variable speed zones are established, at least on expressways, major streets and highways, and through streets and highways, based on engineering and traffic investigations.
II. This program shall be periodically evaluated by the state, and the National Highway Safety Bureau shall be provided with an evaluation summary.

(Source: "The Relation of the National Highway Safety Program to Traffic Engineers," in SR TE, May, 1968)

United Nations Conference on Road Traffic and Draft Convention is available in the literature.[39,40]

LEGAL AUTHORITY

In addition to the above requirements of traffic control devices, it is essential that devices which control or regulate traffic be sanctioned by law. The meaning of each regulatory device and the actions required by it of motorists and pedestrians should be specified by state law or local ordinance.

Both the Uniform Vehicle Code and the Model Traffic Ordinance provide the content and language of legislation needed to apply traffic control devices and define the basic rules of the road. The Code and the Ordinance require the installations of signs or other traffic control devices to make some of their provisions effective, and both define the legal meaning of certain devices.

The Code directs the states to adopt manuals and specifications for a uniform system of traffic control devices (Section 15-104). The Ordinance requires that all traffic control devices, adopted by municipalities, conform to the state manual (Section 4-2).

In the Highway Safety Act of 1966, there is a Highway Safety Program for each activity for which federal assistance is provided. A set of desirable goals or performance standards has been established. Standard 13 is for traffic control devices and is shown in Table 19-1.

EFFECTS OF MISUSE OF TRAFFIC CONTROL DEVICES[32,33]

The injudicious use of traffic control devices can lead to several detrimental effects, including:

1. Wasted funds
2. Inordinate delay
3. Confusion
4. Accidents
5. Disregard for all devices

The misuse or over-use of traffic control devices, which often causes these effects, is generally a result of political pressures, lack of definitive warrants for each type of device, and for a desire on the part of police departments to save manpower. Proper consideration of the following factors may avoid these problems:

1. Traffic conditions, i.e., volume, speed, directional distributions, delays, and accidents
2. Physical conditions, i.e., sight distance, curvature, grades, snow and ice
3. Inability of the road user to follow directions of too many or poorly located devices
4. Vehicle characteristics, such as the limitation of the visibility of overhead signals caused by windshield height of commercial vehicles, speed, and braking distances.

20

Traffic Signs and Markings

Traffic signs and markings are inactive control devices. Except for variable message signs, these devices present a fixed, permanent message to the driver, informing him of a regulation, warning him of a hazard, or guiding him in the operation of his vehicle. Even with variable message signs, the given message generally remains constant over extended periods. Signs and markings, therefore, are permanent devices, presenting permanent messages until taken down or replaced. Unlike signals, whose operation may be adjusted to suit changing conditions, signs and markings are inflexible once installed.

TRAFFIC MARKINGS

Traffic markings are all lines, patterns, words, colors, or other devices, except signs, set into the surface of, applied upon, or attached to the pavement's surface or curbing, or to objects within or adjacent to the roadway, officially placed for the purpose of regulating, warning, or guiding traffic.

Functions and Limitations of Markings[1-3]

Markings have definite functions in a proper traffic control system. They are applied for the purposes of regulating and guiding the movement of traffic, and of promoting safety on the highway. In some cases they are used to supplement the regulations or warnings of other traffic control devices. In other cases they can solely provide very effective means of conveying certain regulations, warnings, and information in clearly understandable terms, without diverting the driver's attention from the roadway. In addition, the capacity of a highway is often increased by the orderly and properly regulated traffic flow which results from correct application of pavement markings.

The application of markings is more than a matter of painting lines. It is, in effect, the installation of a traffic regulating system on a highway. As with all other traffic control devices, marking must be readily recognized and understood, and this goal can only be achieved by using a uniform system of markings, when they are desirable or warranted. A motorist should be confronted with the same type of markings wherever he may travel, and these markings should convey exactly the same meaning wherever they are encountered.

Pavement markings have definite limitations. They may be entirely obliterated by snow, are not clearly visible when wet or partially covered by dirt or oil, and not very durable when painted on a surface subjected to traffic wear. In addition, pavement markings must be replaced at relatively frequent intervals, particularly where volumes are high, thus compounding the problem of restriping.

Markings are primarily used to control the lateral position of vehicles on the roadway. This function includes lane markings, marked channelization, and passing prohibitions on two-lane, two-way highways. Markings are also used to delineate the pavement edge and obstructions in, near, or adjacent to the pavement.

Marking Design[1,3]

Materials. The most common method of applying pavement, curb and object markings is by means of *paint.* Continuous improvement and development in paints, and in the equipment and methods of application, have resulted in increased use of pavement markings. Equipment in general use is versatile, in that it is capable of placing single, double, or triple lines on a road surface, solid or broken, in different colors, while operating at speeds of 6 to 10 miles per hour. Small, self-propelled stripers capable of laying a triple line are available, but their operating speed is about $1\frac{1}{2}$ miles per hour. Hand equipment is also available to place transverse or special markings at intersections or at other locations

A true measure of the relative value of a paint should consider its purchase price, application cost, durability, night visibility, appearance, and chemical formulation. Using this rationale, one agency developed a formula for determining the cost of material per mile per day of useful life. It was found, not unexpectedly, that a more expensive paint of higher quality is the most economical to use.[9,10]

Thermoplastic marking materials have shown extremely desirable characteristics, particularly for applications that are subject to extreme traffic wear, such as transverse markings at intersections. Thermoplastic marking compounds are applied at elevated temperatures, and they harden upon cooling, leaving a relatively permanent line on the pavement surface. Comparison with standard paint stripes has shown a relative service life of more than *three times* that of paint stripes, but costs may run as high as *six times* as much. Ease of application and reduction of maintenance may make the use of thermoplastics economical in the long run.

Flat units on or in the pavement surface are sometimes used to create pavement marking. They should be placed so that their upper surfaces are reasonably flush with the pavement surface, and they may be applied continuously or separated by small spaces approximately equal to the length of a single unit. To insure good appearance, particular care should be taken to see that alinement and spacing are accurate. Units such as those made of impregnated canvas, plastic, or other flexible material create relatively permanent lines.

Pavement inserts are also used to form markings. They come in a variety of shapes and sizes, and are usually made of metal or plastic. They should not be less than 4 inches in diameter if round, or equivalent minimum area if of another shape, and should be spaced not more than 16 inches on center. They should have rounded surfaces and should not project more than $\frac{3}{4}$ inch above the pavement surface. They are permanently fixed in place by bolting or cementing. Epoxy resin adhesives have proven to be extremely effective in attaching inserts to portland cement and asphaltic surfaces.

Metal and plastic inserts and flat marker units are principally used in urban areas where heavy traffic rapidly destroys painted markings, and frequent repainting is not only costly but causes undue traffic delays. However, the ease and speed of application makes painting the most widely used form of marking.

For a further discussion of pavement marking materials and their installation, consult the literature.[4-8,11-13]

Reflectorization. The night visibility of pavement markings is significantly increased by using *glass beads* with traffic paint. In addition, the use of beads lengthens the service of pavement marking. Reflectorization of paint may be obtained by distributing beads on the surface of the wet paint immediately after its application, or premixed with the paint prior to its application, or by using both methods simultaneously where a high degree of reflectivity is desired. These methods produce a retrodirective reflecting surface, which causes the markings to appear luminous at night under the light from vehicle headlamps.

Bead particle sizes normally range between 0.0059 and 0.0328 inches in diameter. When bead on paint application is used, particle size is predominantly < 0.0117 inches. However, with premixed application, particle size is predominantly < 0.0117 inches, so as to prevent clogging of the spray nozzle. Smaller-size beads can be more readily mixed with paint, and they help to reduce drying time and loss by rebound when mixed into paint and sprayed on the pavement. When beads are applied to the paint there is a high initial reflectance, with subsequent loss of reflectance. Loss of reflectance is primarily due to beads being dislodged from the paint. When beads are premixed with the paint, initial reflectance is low, and reflectance increases as the paint film initially covering the beads wears away. It may take a month before full reflectance is achieved. For additional comments on reflectorization, consult the literature.[16-21]

All pavement markings having application at night should be reflectorized. Reflectorization is not ordinarily required where high-level illumination is present; however, even on well-lighted streets in urban areas, it is normally desirable that markings which must be visible at night be reflectorized. Although reflectorization adds to the initial cost, the additional service life that is obtained and the significantly improved night visibility of reflectorized markings more than compensate for the difference in cost.

Marking Colors. The MUTCD specifies three colors for use as pavement markings: *white*, *yellow*, and *red*. The use of red pavement markings is a new addition made in the 1971 MUTCD. Previously, only white and yellow had been permissible. Black may be used on light pavements to provide a contrast for other markings, but it does not itself constitute a permissible marking color. Where contrast is not adequate, it is extremely important to the effectiveness of markings that some method be used to improve contrast.[22]

The MUTCD specifies the use of *yellow* to separate opposing traffic flows, *white* to separate flows in the same direction, and *red* to delineate a roadway that shall not be entered by or used by the viewer of the line.

A study comparing the visibility of fluorescent and conventional pigments under natural illumination revealed that target visibility for fluorescent pigments was significantly greater than that of conventional pigments.[22] Therefore, when high object visibility is essential, consideration should be given to fluorescent pigments.

Markings on vertical surfaces of objects within the roadway, or dangerously close to it, should be marked with a yellow or black diamond-shaped background with yellow button reflectors, a rectangular yellow reflector, or a rectangular reflector of alternating black and yellow or white 45-degree stripes. Roadway delineators should be yellow if used on the left side of a roadway, white if on the right. Red delineators may be used to mark the edge of a roadway that shall not be used by those viewing the delineators.

Types of Lines. In general, a *broken* line is permissive in character, and may be crossed at the discretion of the driver. *Solid* lines are restrictive in character, and may not be crossed. Center lines on two-lane highways, for example, consist of a *broken yellow* line where passing is permitted, a *solid yellow* line where it is prohibited. Where passing in only one direction is permitted, a *solid yellow* line and a *broken yellow* line (on the side where passing is permitted) is used. Yellow is used rather than white (as specified in the 1961 MUTCD), because opposing flows are being separated. A *solid white* line represents a line separating flows of traffic in the same direction which may not be crossed. An example of where such a line might be used is to demarcate a separate turning lane.

Width and Pattern of Lines. In general, the width of a line emphasizes its message. The greater the width, the greater its emphasis. A *normal* width line is from 4 to 6 inches wide. A *wide* line is usually twice this width. *Double* lines consist of two normal width lines separated by a discernible space. *Broken* lines are formed by line segments and gaps, usually in the ratio of 3: 5. On rural highways, 15 feet and 25 feet are generally used.

Marking Maintenance

To achieve driver observance of markings, it is essential that markings be properly maintained and repainted whenever necessary so as to provide necessary visibility and contrast. It is also necessary that markings be properly located and free from irregular lines. The frequency of repainting depends on the type of surface, composition and rate of application of paint, climate, and volume of traffic. Care should be taken to paint over the old markings as exactly as possible. Normally markings should be repainted at intervals of not more than one year. However, markings that are subjected to heavy traffic volumes may have to be renewed two or three times annually, or more often.

An excellent guide to marking maintenance operations is the AASHO *Maintenance Policy*.[23] It deals with all phases of maintenance for all types of markings as well as the equipment and methods necessary for proper application of markings.

Consult the literature for information on marking materials and practice in Europe.[24,24]

Applications of Pavement and Traffic Markings

Centerlines. A *centerline* divides a roadway between opposing flows. As such, all centerlines should be *yellow*, as specified in the 1971 MUTCD. Centerlines are desirable on all paved highways and, as a minimum, should be placed throughout the length of the following:

1. Two-lane pavements with an ADT in excess of 1,000 vehicles per day.
2. Two-lane pavements narrower than 20 feet, with an ADT in excess of 500 vehicles per day.
3. Two-lane pavements with widths between 16 and 18 feet, carrying an ADT in excess of 300 vehicles per day. Centerlines should not be used on pavements narrower than 16 feet.

4. All four-, six-, and eight-lane undivided pavements.
5. At other locations where the accident record indicates a need for them, and in areas where driver visibility is likely to be reduced frequently, as by fog.

Studies of driver behavior have revealed that vehicular speed, volume, and transverse placement of vehicles are affected by centerlines and lane lines.[26-29]

Lane Lines. A *lane line* separates two adjacent traffic lanes in the same direction. Lane lines, therefore, should always be *white*, in accordance with the MUTCD. Lane lines are helpful in producing orderly traffic flow and can increase the efficiency of existing pavement widths.

Lane lines should be used wherever traffic conditions of volume, safety, and ease of operation would be improved by their use, and always under the following conditions:

1. On all rural highways with an odd number of traffic lanes.
2. On all undivided rural highways of four or more lanes.
3. At approaches to important intersections and crosswalks, and in dangerous locations on both rural highways and city streets.
4. At congested locations, particularly on city streets, where the roadway will accommodate more lanes of traffic than would be the case without the use of lane lines. These include: locations between loading islands and sidewalk curbs, other locations where the normal lane width is decreased, and approaches to widened intersections.
5. On one-way streets and on roadways where maximum efficiency in utilizing existing pavement width is desired.

No-Passing Zones. No-passing zones are established at vertical and horizontal curves and at other locations on two- and three-lane highways where passing sight distance is restricted or other hazardous conditions exist. A no-passing zone is marked by a *solid yellow* barrier line, placed as the right-hand element of a combination line along the center line or lane line. The *combination line* consists of a *solid yellow* line along a *broken yellow* line.

On a two-lane highway, the combination line is marked along the center of the pavement throughout the no-passing zone. On a three-lane highway, the combination line starts in advance of the no-passing zone at the left-hand lane line of the center lane, and extends diagonally across the center lane to the right-hand line at the beginning of the no-passing zone, and then extends along the lane line to the end of the zone.

A no-passing zone at a horizontal or vertical curve is warranted where the sight distance is equal to or less than that listed below for the prevailing off-peak 85th percentile speed.

85th PERCENTILE SPEED (mph)	*MINIMUM SIGHT DISTANCE (ft)*
30	500
40	600
50	800
60	1,000
70	1,200

In no case should the no-passing zone marking be less than 500 feet long. The criterion used to measure passing sight distance on a vertical curve is the distance equal to the line of sight measured from the driver's eyes, which are assumed to be 3.75 feet above the pavement surface, to an object 4.5 feet high. On a horizontal curve, the available sight distance is measured along the centerline between points 4 feet above the pavement on a line tangent to the view obstruction.

It is standard practice to locate the barrier line at no-passing zones adjacent to the centerline or center lane line. A study was conducted to compare two types of no-passing zone markings.[30] One type was in accord with the recommendations in the 1961 MUTCD, that is, placing the barrier line adjacent to the center line; the other type placed the barrier line in the center of the lane from which passing was prohibited. The findings of the study revealed that speed characteristics differed to a minor degree on the two types of markings. However, for the critical condition of transverse vehicle placement, the advantage was consistently with the national standard. The national standard also resulted in better performance in a comparison of daylight and night driving, and in driver compliance with the no-passing restriction.

A practical method of determining proper no-passing zone locations is described in the literature.[31] The method utilizes two vehicles equipped with a foot survey meter and a two-way radio. The two vehicles maintain a distance apart equal to the design sight distance, and the driver of the lag vehicle observes the lead vehicle at all times. When the lead vehicle disappears from view, it marks the beginning of the no-passing zone, and the end of the no-passing zone is established when the lead vehicle again comes within view.

Some states utilize an 800-foot minimum sight distance for a warrant to establish no-passing zones for 85th percentile speeds of 50 miles per hour and above.[32]

Pavement Edge Lines. Pavement edge lines delineate roadway edges and should be *white* if used on the right-hand side of the roadway, and *yellow* if on the left. The use of edge markings may help to accomplish the following:

1. Reduction of travel on shoulders of lower structural capacity than adjacent pavement
2. More comfortable driving conditions, especially during night and inclement weather conditions, by providing the driver with a continuous guide
3. Reduction in accidents (particularly off-the-road accidents)

The need for right-hand edge markings varies with the adequacy of the shoulder. Absence of an adequate shoulder (either none at all, one less than 8 feet wide, or one which is unsurfaced) suggests the need for an edge line. Paved shoulders should be contrasted to the roadway edge by use of colored pavements or a white edge line.

A left-hand pavement edge marking is desirable to delineate the median area of a divided highway. Where special emphasis is required adjacent to curbs, in areas of poor visibility or on medians less than 16 feet in width, the *double yellow* barrier line should be used. Where the median is paved flush with the traffic lanes, the double barrier line should be used.

There is a strong demand by motorists for the use of pavement edge markings. A survey showed that during 1961 there was an increase of almost 45 per cent in the painting of white, reflective edge lines in the United States. The survey also indicated that as of January 1962, there were almost 89,000 miles of state highways and toll expressways which were marked with edge lines, and that preliminary plans for 1962 called for adding over 17,000 miles more highway to the program.[33]

A study was made to determine the effect of pavement edge marking on the lateral placement of vehicles.[34] On two-lane tangent sections of 24- and 20-foot pavement widths, a continuous edge line resulted in moving traffic closer to the centerline of the pavement, and this effect was much more significant at night. Another phase of the study was conducted on a 20-foot wide, 4 degree horizontal curve, which was marked as a no-passing zone. The effect of a continuous edge line on the curve was also to move traffic closer to the centerline; however, in the high-speed lane it resulted in vehicles crossing the barrier no-passing line to a greater degree than prior to the edge marking. The final phase of the investigation was conducted on a tangent section of a four-lane divided highway. Prior to the application of edge markings along both the shoulder and median edges, lane lines existed to subdivide the pavement. Two, 12-foot lanes were available for travel in one direction of flow, and the effect of the continuous edge markings was to move vehicles traveling in the right-hand lane closer to the lane line; but in the left lane, vehicles moved toward the median. Two, 10-foot lanes were available for travel in the opposing direction of flow, and the effect of the edge markings was to move vehicles traveling in both lanes closer to the lane line. Based on the results of this study, the Louisiana Department of Highways has adopted a policy to apply pavement edge markings on all 24-foot, two-lane highways, but not on two-lane highways narrower than 24 feet. Consult the literature for futher studies concerning the effect of edge markings.[35-38]

Lane-Width Transitions. Lane markings should be used to guide traffic at points where the pavement width changes to a lesser number of lanes. Converging barrier line treatment should have a length of not less than that determined by the formula $L = S \times W$, where L equals the length in feet, S the (off-peak) 85th percentile speed (or design speed) in miles per hour, and W the offset distance in feet.

Obstruction Approach Markings. Pavement markings are used to supplement signs and markings on obstructions and to guide traffic on the approach around fixed obstructions within a paved roadway. The diagonal approach marking should be given barrier line treatment for a length of not less than that determined by the formula $L = S \times W$, where the symbols have the same meaning as defined above. In no case should the diagonal line be less than 200 feet in length in rural areas, or 100 feet in urban areas. If all vehicles must pass to the right of the obstruction, *yellow* markings are used. If vehicles may pass to either side, *white* is used.

Channelizing Lines. The channelizing line is a *solid white* line which directs traffic, and indicates that traffic should not cross but may proceed on either side. Channelizing lines are useful for outlining traffic islands and to separate turn lanes from the main traffic lanes, where a more restrictive carrier is impractical or would create a hazard. It may also be used in lieu of the *broken* lane line to accentuate the lane marking in critical areas and more clearly define the traffic lanes where it is advisable to discourage lane-changing.

The application of a channelizing line to separate the two turning lanes at the intersection of two one-way streets resulted in the following beneficial effects:[39]

1. Dual turning movements better utilized available space by causing vehicles to drive nearer to each other in the turn.
2. There was a substantial reduction of interlane weaving movements.
3. There was a reduction in time necessary for vehicles to clear the intersection.

Consult the literature for further studies of the effectiveness of channelizing lines.[40-42]

Freeway Ramp Markings. Channelizing lines are utilized to demarcate a neutral area at the gore, which reduces the probability of collision with the curb nose. They direct exiting and entering traffic into the proper angle for smooth movements of divergence and convergence.

To provide for proper and safe use of entrance ramps, it is essential to inform the driver of that area which is set aside for his exclusive use when entering the expressway. Markings help the driver to adequately distinguish between through traffic lanes and the portion of the entrance ramp specifically designed for his use. This helps to eliminate much driver indecision, and greatly facilitates safe and efficient ramp operation.[43] Consult the literature for further studies on the effectiveness of ramp markings.[44-52]

Transverse Lines. Transverse lines are utilized for a variety of important purposes, particularly in intersection areas. Kinds of transverse lines are indicated below:

Stop Lines. Solid white lines should be used in both rural and urban areas where delineations of the stopping point is required. Stop lines should be at least 4 feet behind the nearest crosswalk or curb line of the intersecting street. They should not be more than 30 feet from the intersecting roadway edge.

Crosswalk Lines. Crosswalks are used to provide a legal and defined path for pedestrians, and to provide a safer crossing point for them. Crosswalks should be at least 6 feet wide and delineated by *white* crosswalk lines. They should be used wherever conflicts between vehicles and pedestrians are significant, or where pedestrians cross at an unexpected location (mid-block, for example).

Railroad Crossing Approach. Approaches to railroad crossings should be marked with transverse lines, symbols (X and RR, as per MUTCD), and appropriate word messages to supplement signs, signals, and/or gates.

Parking Space Limits. In downtown areas, it is often desirable to delineate the limits of individual parking spaces, particularly where parking meters are used. White lines should be used.

Transverse Lines at Toll Booth Approaches. Wide transverse lines are often used in toll plazas and approaches to alert drivers and reduce approach speeds.

Curb Markings for Parking Restrictions. Since *yellow* and *white* curb markings are used for standard edge markings, these should not be used to mark parking restrictions. It is generally desirable to establish parking regulations through the use of standard signs, but special curb colors may be used to supplement them.

Word and Symbol Messages. Word and symbol markings on pavements may be used whenever it is desirable to supplement signs or other traffic control devices for the purpose of guiding, warning or regulating traffic. They should be limited to as few words as possible, with never more than three; on high-speed roads messages of more than one line should be avoided. The letters and symbols should be greatly elongated in the direction of traffic movement, because of the low angle at which they are viewed by an approaching driver. It is necessary that all transverse lines making up letters or symbols be proportionately widened to give visibility equal to that of longitudinal lines.

Consult the MUTCD for accepted symbols, word messages, proportions, and colors. Arrow templates used to mark one-way streets in Long Beach, California are illustrated in the literature.[53]

Lane-use control markings are very desirable on the approach to an intersection to supplement lane-use control signs, indicating the types of movement that are permitted from specific lanes.

A study in England indicated that accidents were reduced 50 per cent at intersections where the word STOP had been marked on the pavement.[54]

Object Markings. Objects within or close enough to the roadway edge to constitute a hazard should be marked. Types of markings used have been discussed previously in this chapter, and are detailed in the MUTCD. Such objects as bridge abutments, underpass piers, and culvert headwalls should be marked. All objects in the roadway itself should receive extensive ground- and object-making treatment.

The use of reflectorized white paint on steel supporting columns located in the street at four underpasses in Washington, D.C., significantly helped to reduce accidents. During the first year after the application of markings, there was a 71 per cent reduction in nighttime accidents and a 44 per cent reduction in daytime accidents.[55]

Delineators. A delineator is a light-reflecting device mounted at the side of the roadway, in series with others, to indicate the alinement of the roadway. Delineators may be used on long, continuous sections of highway, or through short stretches where there are changes in horizontal alinement.

Single, *white* reflector units should mark the right side of roadways, and *yellow* reflectors should mark the left side of roadways. Delineators should be mounted on suitable supports at a height of 4 feet above the near pavement edge, and the lateral distance from the pavement edge should be between 2 and 6 feet, depending upon adjacent cross-section design. An important advantage of delineators is that they remain visible when there is snow on the ground, but if they are too close to the roadway, snowplows may rip them up. Normally, delineators should be spaced 200 to 528 feet apart, although shorter spacing may be required on approaches to, and throughout, horizontal curves. Table 1 of the MUTCD shows the recommended spacing for delineators on horizontal curves. For further discussion of delineators, consult the literature.[56-60]

TRAFFIC SIGNS

A traffic sign is a device mounted on a fixed or portable support, whereby a specific message is conveyed by means of words or symbols officially erected for the purpose of regulating, warning or guiding traffic.

Use of Traffic Signs[1,3]

The traffic sign is the most commonly used traffic control device, and it is the oldest device for controlling, safeguarding, or expediting traffic. Signs are not ordinarily needed to confirm basic rules of the road, but they are essential where special regulations apply at specific places or at specific times only, where hazards are not self-evident, and to furnish information. Many accidents occur because the driver has been suddenly confronted by the unexpected. He therefore should be warned, insofar as is possible, if abnormal driving conditions exist ahead.

The value of signs giving information as to highway routes, directions, and points of interest should not be overlooked as a safety factor. A motorist who enters an intersection, or highway interchange area, uncertain as to his direction, is confused, and is a potential cause of accidents. If he is properly informed before entering the intersection, this confusion can be minimized or eliminated.

Signs should only be used where they are warranted and justified by the application of engineering principles and factual studies. An adequate number of signs must be used to properly inform the motorist. Regulatory and warning signs should be used sparingly, since the excessive use of these signs tends to lead to a disregard of all signs. Therefore, a conservative use of regulatory and warning signs is recommended. However, a liberal use of guide signs, judiciously located, will be enthusiastically accepted by the motoring public, and will not lessen the value of such signs.

After a sign has been erected, its effect upon traffic should be observed and necessary changes made if the desired results are not being obtained. Upon erection or removal of a sign or signal, the effect of this action upon signs in the vicinity should be studied, and changes made as may be required by the changed situation.

Design of Signs[3]

Uniformity in design includes shape, color, dimensions, symbols, wording, lettering, and illumination or reflectorization. Detailed drawings of the standard signs illustrated in the MUTCD are available, upon request, from the FHWA.

Shape of Signs. In a study of the relative effectiveness of sign shape, size, color, and message, *shape* and *color* were found to be the most significant in producing recognition. In New York State, 87 out of 100 drivers failed to notice that a standard stop sign had the garbled message TOPS on it.[61]

The use of the following shapes is specified in the MUTCD:

1. *Octagon:* Exclusively for the STOP sign.
2. *Equilateral triangle* (one point downward): Exclusively for the YIELD sign.
3. *Round:* Advance warning of a railroad crossing.
4. *Pennant shape* (isosceles triangle with long dimension horizontal): Warning of no-passing zones.
5. *Diamond:* Warning signs for hazards or possible hazards existing on the roadway or adjacent thereto.
6. *Rectangle* (long dimension vertical): All regulatory signs *except* STOP and YIELD signs.
7. *Rectangle* (long dimension horizontal): All guide signs, with the exception of certain route markers and recreational area guide signs.
8. *Pentagon* (point up): All school advance warning and school crossing signs.
9. *Trapezoid:* Recreational guide signs.
10. *Special shapes* may be used for special purposes, such as the *crossbuck* for railroad crossings, and the various *shield* shapes for state and interstate route markings.

Of these, the *pentagon*, *trapezoid*, and *pennant* are new additions in the 1971 MUTCD. The standard for *round* signs was altered as well.

Sign Colors. Color standards in the 1971 MUTCD were coordinated with and are now in concert with standards in the 1970 Interstate Manual. This is due to (1) a desire for increased uniformity and (2) the demonstrated superiority of the interstate colors in visibility and legibility, particularly at night.[62]

In the MUTCD the following colors are specified for use:

1. *Red:* The background color for STOP signs, DO NOT ENTER messages, and WRONG WAY signs; the legend color for parking prohibition signs, route markers, and YIELD signs.
2. *Black:* The background color for ONE-WAY signs, weigh station signs, and night speed limit signs; the legend color on white, yellow, and orange backgrounds.
3. *White:* The background color for route markers, guide signs, fallout shelter directional signs, and regulatory signs (except for STOP signs); the legend color on brown, green, blue, black, and red backgrounds.
4. *Orange:* Only as a background for construction and maintenance signs.
5. *Yellow:* The background color for warning signs, except for construction, maintenance, and school areas.
6. *Brown:* The background color for recreational, scenic, or cultural guide signs.
7. *Blue:* The background color for motorist service guide or informational signs.
8. *Green:* The background color for guide signs, other than those using brown, white, or blue, and for mileposts; the legend color for permissive parking regulation signs.

Four other colors (*purple*, *light blue*, *coral*, and *strong yellow-green*) were identified as useful for highway purposes, but were reserved for future needs. The permitting of white-background guide signs represents the only remaining difference between interstate standards and the 1971 MUTCD.

Sign Dimensions. The size of a sign will depend primarily on the length of its message and the size and spacing of the letters that form the message, or on the size of any required symbol. The complete legend should be designed for adequate legibility.

The standard dimensions given in the MUTCD have been designed to provide clear legibility of the signs under normal highway conditions. Sizes larger than standard are desirable where greater legibility or emphasis is needed; for expressways, special categories of large signs are prescribed. In determining whether the standard size is adequate, consideration should be given to such factors as prevailing speeds and volumes, the width of roadway or number of lanes, the degree of hazard, and the competition offered by other signs, lighting, displays, or background.

Sign Symbols. Symbols that are instantly recognized are far superior to word messages. Symbols provide for greater safety and facilitation of traffic flow, and their use should be encouraged to the greatest degree. These views are strengthened by the general use and success of symbols in other countries.[63-65] Because of this, the 1971 MUTCD encouraged the use of symbol signs and standardized several such signs. It is recognized that public acceptance of symbology requires a long time span in terms of the necessary education and transition from word signs. To aid this process, where new symbology is used, supplemental word signs should be placed in accordance with MUTCD standards.

Word Messages. Where applicable, the standard wordings given in the MUTCD should be used for sign legends. Most messages, particularly those of regulatory and informational signs, cannot adequately be conveyed by symbols alone. Word messages should be as brief as possible, and the lettering should be large enough to provide the necessary legibility distance. Abbreviations should be kept to a minimum, and should include only those that are commonly recognized and understood.

A study was conducted to determine the effects of underlining (subdividing) on the readability of highway destination signs.[66] The investigation led to the following conclusions:

1. Underlining reduces the probability of a destination being associated with the wrong arrow in the unequal word-length guide signs. However, for place names of equal length on the same sign, underlining had no apparent beneficial effect.
2. The probability of a guide sign being misread is greater when the place names are of equal length than when the lengths are unequal.
3. It was found that misreading errors were proportional to the number of destinations that appear on the same sign. It is recommended, partly as a result of the study, that the maximum number of destinations be limited to *three*. This simply indicates the importance of brevity in sign messages.

This last point was illustrated in the following two studies:

1. A driving simulator was used to determine general driver ability to steer a vehicle while simultaneously searching for a specific sign. Results of the study showed the following:[67]
   a. Where time-sharing was required between two tasks, each type of performance was poorer than when only one task was required.
   b. An increase in the number of message units that appeared on the sign in the recognition task did not affect the simulated steering task, but increased the time required for recognition of the key word.
   c. As the speed of the simulated steering task increased, recognition time decreased.

2. In a field and laboratory study, it was found that a decrease in legibility distance was obtained when the number of legends on the sign was increased from four to six, but no differences were obtained when the number of legends was increased from two to four.[68]

Lettering. With the exception of destination names on guide signs, sign lettering should be in clear, open, capital letters. Destination names may be in lower-case lettering, with initial capitals.

Standard capital and lower-case alphabets are available from the Federal Highway Administration (abbreviated to FHWA to distinguish from FHA). The standard alphabets provide a graded series of widths, ranging from a narrowly condensed Series A to a broad, round Series F. The height-stroke ratio varies from 11: 1 for Series A to 5: 1 for Series F. The requirements of most signs are met by Series C and D. As a guide to choice of alphabets, the results of the following studies are extremely helpful:

1. Legibility distance (the distance at which a message can be read or recognized) at which signs could be read by stationary observers with 20/20 vision in the daytime varied as follows for different alphabet series.[69]

   Series B, 33 feet per inch of height (over 6 inches)
   Series C, 42 feet per inch of height (over 6 inches)
   Series D, 50 feet per inch of height (over 6 inches)

   For night conditions with the signs floodlighted, these distances were reduced by approximately 15 per cent.

2. Legibility distance at which signs could be read by observers riding in vehicles approaching the signs at 25 miles per hour yielded the following results for 4-inch letter height:[70]

	Day	*Night*
Series B	151 feet	—
Series C	186	117 feet
Series D	211	171
Series E	240	179

3. A study was conducted to determine the relative legibility of different types of letters.[71] It was found that rounded block letters (width equal to height) were superior to square block letters.

4. A study was conducted to compare lower-case and capital letters.[72] A white, series E alphabet was used with a black background. Observations were made by individuals (with about 20/20 vision) on foot, and nighttime observations were made with the signs artificially illuminated. Letter heights ranged from 5 inches to 18 inches, and messages were divided into three categories: familiar words with knowledge, familiar words without knowledge, and scrambled combinations. The following results were obtained:

   a. For both kinds of alphabet, legibility distances increased with size of letters and with degree of familiarity. The increase due to increasing familiarity was greater for lower-case letters than for capitals.
   b. Probably because lower-case letters were narrower, their legibility distance expressed in terms of letter height was lower than capitals.
   c. On the basis of width, the lower-case words could be seen farther than the capital words, probably because they were higher. Therefore, where length of sign is the controlling factor, lower-case letters could be used to greater advantage.
   d. Daytime legibility distance for scrambled capitals and place names without knowledge were 55 and 75 feet per inch of letter height, respectively. For lower-case letters of equal height, these distances were about 10 per cent greater. For night conditions, these distances were reduced by about 20 per cent.

5. A study was conducted to determine the effect of letter width and spacing on visibility of signs.[73] All observations were made at night by individuals in vehicles approaching the signs at about 30 miles per hour. The area was without illumination, and sign illumination was accomplished with the low beam of the headlamps. White, reflectorized letters, 10 inches high, were displayed on a black, nonreflectorized background. Three different capital alphabets were used: Series C, Series E, and an alphabet, designated as Series ED, for which letters were formed by plastic reflectors, with a width similar to Series E. The spacings between letters were increased so that the lengths of the test words were extended from normal to 20, 40, and 60 per cent above normal. The following results were obtained:

   a. As inter-letter spacing was increased, the legibility distance also increased for all three alphabets until word lengths were 40 per cent above normal. The resulting gain in legibility at this point was 15 per cent for Series C, 16 per cent for Series E, and 7 per cent for Series ED. Beyond the 40 per cent increase, legibility leveled off or declined.
   b. When word lengths were normal, or no more than 10 per cent above normal, test signs with the Series ED alphabet were found to have greater legibility. At wider spacings, the Series E alphabet was superior.
   c. Legibility distances for words formed with the Series E alphabet were found to be 23 to 27 per cent greater than those formed with the Series C alphabet. However, the Series C alphabet occupies less word length for a given spacing, and on the basis of legibility distance per inch of word length, it is somewhat superior to Series E. The two alphabets were equally legible at a point of equivalent legend area and spacing.

d. The study findings point to the importance of sign proportions and provide an improved means for efficient determination of legend design. Where vertical dimensions restrict sign letter heights to something less than desirable, increased spacing between letters can help to compensate for the loss of legibility distance that would otherwise occur.

For further studies concerning lettering and legend size, consult the literature.[74-76]

Illuminations and Reflectorization. All signs that are to convey their messages during hours of darkness, other than urban parking signs, should be reflectorized or illuminated. The majority of signs are of sufficient importance to be visible and legible to a motorist by night as well as by day. However, street lighting is normally sufficient to provide visibility for parking signs.

Overhead signs, 15 feet or more above the roadway, may not receive enough illumination from motor vehicle headlamps for adequate brightness. Therefore, it is necessary to illuminate overhead signs when reflectorization will not result in effective performance.

Methods of Providing Illumination. Illumination may be provided by the following means:

1. *Internal illumination*, which utilizes a light behind the sign face to illuminate the main message or symbol, or sign background, or both, through a translucent material
2. *External illumination*, which utilizes an attached or independently mounted light source designed to direct essentially uniform illumination over the entire face of the sign
3. *Other effective devices*, such as luminous tubing shaped to the desired lettering or symbol, patterns of incandescent light bulbs, or luminescent panels that will make the sign clearly visible at night

Ordinary street or highway lighting is not regarded as meeting the requirements for sign illumination. All illumination should provide for exhibiting the same sign colors by night and by day. A flashing light incorporated in a sign installation should be *red* when displayed with a stop sign, or *yellow* when displayed with a warning sign or other regulatory sign.

As a guide to choice of illumination, the following comments and results of studies are extremely helpful:

1. Experience with incandescent lamp sources for sign illumination has not been altogether satisfactory. Short life, sensitivity to voltage surges, and vibrations are among the difficulties.
2. Tube lighting, especially the rapid-start type of fluorescent lighting, is effective. This type is recommended and will serve most needs, either for exterior or interior illumination of the sign.
3. With the fixtures now commercially available, exterior lighting is feasible from a position in front of the bottom parallel to the sign face, at a distance of $\frac{1}{3}$ to $\frac{1}{2}$ the vertical dimension of the sign from the sign surface, and so aimed that from the driver's position the sign surface appears to be uniformly illuminated. Illumination from below avoids undesirable daytime shadows on the face of the sign.

Internally illuminated signs using translucent materials are very effective. They are particularly suited to overhead locations, where the fragile sign face is not subject to traffic hazards.

1. The city of Cincinnati found internal illumination of overhead traffic signs, displaying a constant message, to be very effective in providing the necessary target value and legibility.[77,78] Reflectorization of such signs was not found to be adequate, because of vehicle headlamps being directed downward, and because of competing background illumination from inadequate street lights, advertising signs, and approaching headlamps. A number of other applications of internally illuminated, constant-mesage signs are described.
2. Internal illumination has also been very successfully applied in Cincinnati for variable-message signs.
3. Internal illumination was recommended for use in areas of high-traffic volumes, wide roadways, locations where headlamps and street lights do not reflect the message, locations where better target value is needed to compete against a high back ground illumination, and locations where a variable message is required.
4. The characteristics of acrylic plastic (Plexiglas) panels are discussed in the literature.[80] This material is extensively specified for sign surface panels of internally illuminated signs. It possesses many desirable characteristics, such as its long-term resistance to weathering, its light weight, stability, shatter resistance, workability, and ease of installation. It also has excellent light-transmitting and diffusing properties which provide good visibility and legibility that are glare free.
5. General Motors Research Laboratories has developed a variable-speed limit sign that is instantaneously changeable to produce any number from 1 through 99.[83] The sign utilizes a matrix of incandescent bulbs to portray the desired numerical combinations.
6. Considerable success has been reported with the use of flashing lights on warning and STOP signs.[85] This practice was adopted to increase the ability of the signs to compel attention at hazardous locations, where motorists were subjected to many distracting features. Two alternately flashing beacons were installed in the signs; *red* was used with STOP signs and *yellow* with all other types of signs.

7. A study of flashes at very slow frequencies, of about 5 per second, on a reflectorized background, indicated that the attention-getting characteristics were significantly increased over higher-frequency displays. It was suggested that a group of such devices, preceding directive signing, could serve as an effective warning system at nighttime.[86]
8. California is experimenting with a new sight-and-sound warning system, designed to alert wrong-way motorists on freeway off-ramps. The instant the motorist's wheels touch a detector buried in the pavement, a 12-inch red light goes on, a horn emits a blaring sound, and the message "Go Back-You Are Going Wrong Way" is externally illuminated.[87] Another article[88] presents a design utilizing jughandle wrong-way lanes and signing to prevent wrong-way entries.

For further studies involving sign illumination, consult the literature.[79,81,82,84]

Methods of Providing Reflectorization. In a reflectorized sign, either the legend and border, or the background, or both, may be reflectorized, depending on the sign design and on the local conditions. On any particular highway a uniform policy should be followed for signs of the same type; however, differences in rural and urban requirements must be taken into consideration. As a general rule, both the legend and the background of urban signs, if other than black, should be reflectorized. Reflectorization may be effected by either of the following means:

1. Reflector buttons or similar units may be set into the symbol or message.
2. Reflective coatings may be used either on the sign background or, where a white legend is used on a black or colored background, in the symbol or message and border.

The combination of powerful vehicle headlamps and improved reflecting materials has made reflectorized signs almost as efficient as illuminated signs, provided the headlamp beam effectively strikes the reflecting material.

Reflecting materials used on signs should reflect white light or, if a reflective coating is used as the background of a colored sign, the color of the background. An effective reflecting material reflects a large portion of an incident beam of light directly back towards its source in a narrow cone having only enough divergence to reach the driver's eyes in his normal position above the headlamps; this is called *retrodirective reflection*.

Reflector buttons are small, individual reflecting units, arranged in rows or patterns on the face of the sign to form letters, symbols, or borders. In suitable sizes and spacings, they give the visual effect of continuous lines or areas of light. They are made of glass or transparent plastic, with lenses or prisms designed for retrodirective reflection.

A *reflective coating* is a coating or sheeting applied to the desired sign area to provide a bright reflection. The coating or sheeting is of a retrodirective reflecting character usually having minute glass spheres (beads) closely distributed and embedded in a painted surface or in a flexible plastic sheeting, or minute lenses molded in the surface of a plastic sheeting. Each bend or lens acts as an individual reflecting unit, but the effect is that of a uniformly brilliant area when viewed in the headlamp beam. A suitable incorporation of pigment in a reflecting coating causes it to reflect colored light. A method for calculating the brightness of different reflective coatings is presented in the literature.[89] Sign brightness was evaluated in relation to sign position and distance. It was found that reflective coatings have low brightness at near distance, maximum luminance at distances from 150 to 400 feet, depending upon the type of reflective coating, and decreasing luminance at greater distances. Letter size is a significant factor in the selection of reflective coatings. For signs with small letter size, little is gained by using more expensive high-brightness materials as small letters cannot be read at great distances regardless of their brightness, and adequate brightness is achieved by using any material at near distances. However, for large signs to be read at great distances the more expensive materials may be justified.

A study was conducted to evaluate the relative legibility of reflective coatings and externally illuminated signs.[90] Observations were made from vehicles approaching the test sign at 15 miles per hour. The study was conducted under rather ideal conditions, since observations were made in a dark rural area without glare from headlamps of opposing vehicles. This should be kept in mind when attempting to apply the results to urban areas in which the surrounding illumination is much higher and where bright glare sources are in the field of view. A much higher level of illumination would be required under these conditions. The alphabet used to form the messages was essentially a Series E, and the average daytime legibility distance was found to be about 88 feet per inch of letter height (letter height range was 8 to 18 inches). Optimum external illumination was found to be 10 foot-lamberts, which resulted in a legibility distance of about 15 per cent less than that for daytime. In the comparison of reflective coatings with illumination, it was found that with the upper headlamp beams, the overhead sign could almost be read at the same distance; however, for lower beams, the distance was significantly less than that for illumination. The reduction in distance ranged from 25 to 30 per cent. Therefore, it appears feasible that reflectorized overhead signs with high-brightness materials could be considered adequate in a dark rural area where no unfavorable alinement conditions exist. Another study showed that the average legibility distances for 12-inch letters was not significantly affected for reflective coatings under conditions of moderate illumination (as may be found in a suburban area) and total darkness (as may be found in a rural area). Nearly identical legibility distances were found for the two conditions.[91]

A study to determine whether reflectorized backgrounds contribute to the overall nighttime effectiveness of highway signing revealed that the differences between three degrees of background reflectorization were not statistically significant.[92] All test signs had reflectorized white legends and green backgrounds. The three degrees of background reflectorization were:

1. A nonreflectorized green background
2. A moderately bright, reflectorized green background
3. A relatively high-brightness, reflectorized green background

For further studies concerning sign reflectorization, consult the literature.[93-96]

Sign Borders. With few exceptions, all signs in the MUTCD have a narrow border, invariably of the same color as the legend, at or just inside the edge. The border improves the appearance of the sign and makes it more conspicuous.

Location of Signs

Standards for location of signs are detailed in the MUTCD. Circumstances, however, often preclude complete standardization of sign location. In general, signs are mounted on the right-hand side of the roadway, where drivers can see them more easily. They are located to be within the driver's cone of clear vision (12 degrees), and are placed at an angle to reflect headlight beams back to the driver. Guide and other directional signs must be placed sufficiently in advance of decision points to permit the driver to react properly to the sign message.

Overhead signs require expensive structures spanning the roadway to support them. Their use is justified under a variety of circumstances; for example:

1. Where the message applies to a particular lane of the roadway
2. In urban areas, where closely spaced interchanges and the multiplicity of other signs and competing distractions makes the positioning of all signs on the right-hand side impractical
3. Where sight distances for roadside signs are inadequate
4. Where high speeds or high volumes make the viewing of roadside signs difficult or hazardous
5. At gore areas, to avoid placing dangerous structures in the gore (Break-away sign posts may be used in lieu of overhead signs for this purpose.)
6. For consistency, where conditions necessitate large-scale use of overhead signs

On wide highways, it is often wise to supplement right-side signs with signs located in the median. Signs are often placed on channelizing islands, where they apply to specific movements, such as a left-turning lane.

For detailed specifications regarding the placement, positioning, and erection of roadside and overhead signs, consult the MUTCD.

The advent of the break-away sign support has enhanced the relative safety of many sign installations.[96] These supports, when hit by a vehicle, fracture at a predesigned weakness plane. The sign and supports are thrown over the colliding vehicle, thus avoiding major damage and injury. The collapsed sign is easily and inexpensively reerected.

For studies relating sign legibility to location, consult the literature.[89,97]

Sign Maintenance[8]

All traffic signs should be kept in proper position, clean, and legible at all times. Damaged signs should be replaced immediately. Special care should be taken to see that weeds, shrubbery, construction materials, mud, and snow are not allowed to obscure the face on any sign. Damaged, defaced, obscured, dirty, and missing signs are ineffective, and discredit the agency responsible for them. In addition, poorly maintained signs lose their authority as traffic control devices, may be contributory factors to vehicle accidents, and in some cases may be a liability to the agency responsible for their maintenance. Signs that are well-maintained add significantly to the general appearance of the highway. To insure adequate maintenance, a suitable schedule should be established for inspection, cleaning, and replacement of signs. All signs should be inspected at least twice a year, and those that are defective should be cleaned, touched up, replaced, or taken in for repair and refinishing or scrapping. In the case of illuminated signs, a regular schedule should be established for replacement of lighting elements such that they will be renewed before they would normally be expected to burn out.

An excellent guide to sign maintenance operations is the AASHO *Maintenance Policy*. It deals with field repairs, shop reconditioning, and erection methods. For further detailed discussion of sign maintenance procedures and erection techniques, consult the literature.[98-106]

Functional Classification of Signs

The MUTCD classifies signs by their function. Specifications as to size, shape, color, legend, and placement, as well as warrants for their use, are given for each sign standardized in the MUTCD. The use and application of signs in traffic control is discussed in other chapters of this book

Regulatory Signs. Regulatory signs are used to inform road users of traffic laws and regulations that apply at definite locations and specific times, or where statutory regulations are not obvious or are being disobeyed. However, regulatory signs cannot be expected to command respect and obedience unless the regulations are adequately enforced. Regulatory signs should normally be erected at those locations where the regulations apply, and they should be mounted so as to be easily visible and legible to the motorist whose actions they are to govern. The message on the sign should clearly indicate the requirements imposed by the regulation.

Specific details dealing with design and location for each of the regulatory signs are completely covered in the MUTCD. All regulatory signs, unless definitely excepted in pertinent sign specifications, should be illuminated or reflectorized. Unless otherwise specified, they are rectangular in shape (long dimension vertical) and have *black* legends on a *white* background. Regulatory signs are classified and include the following:

1. *Right-of way series:* STOP sign, YIELD sign
2. *Speed series:* Speed limit signs, night speed signs, minimum speed signs, REDUCED SPEED AHEAD signs
3. *Movement series:* Turn prohibition or restriction signs, lane-use control signs, DO NOT PASS signs, PASS WITH CARE signs, SLOW TRAFFIC KEEP RIGHT signs, KEEP RIGHT signs, DO NOT ENTER signs, WRONG WAY signs, selective exclusion signs, ONE WAY signs
4. *Parking series:* Parking prohibition signs, NO STANDING signs, NO PARKING signs, EMERGENCY PARKING ONLY signs
5. *Pedestrian series:* NO HITCHHIKING signs, WALK ON LEFT signs, PEDESTRIAN CROSSING signs
6. *Miscellaneous series:* Traffic signal supplementary signs, KEEP OFF MEDIAN signs, ROAD CLOSED signs, LOCAL TRAFFIC ONLY signs, WEIGHT LIMIT signs, TRUCK ROUTE signs, railroad crossbuck signs, other regulatory signs

Warning Signs. Warning signs are placed wherever it is considered prudent to warn traffic of an approaching hazard or potentially hazardous situation. Warning signs are diamond-shaped, with black symbols or messages on a yellow background. Warning signs should not be excessively or indiscriminately used, lest their effectiveness be weakened.

Warning signs are used to inform drivers of approaching hazards, such as abrupt changes in horizontal and vertical alinement, intersections, blind driveways, approaching control devices where high speeds and poor sight distance make advance warning necessary, railroad crossings, animal crossings, changes in roadway geometry, and others. Specifications for standard signs are given in the MUTCD.

Guide Signs. Guide signs are used for the purpose of directing motorists over specified routes, informing them of interesting routes, guiding them to their destination, and eliminating the confusion and potential danger caused by motorists who are uncertain of their direction. It is of extreme importance that a motorist be informed as to his future course before he enters an intersection and that he receive periodic confirmation of the route that he is following. All guide signs should therefore be located so as to meet these objectives. Unlike most other types of signs, guide signs do not lose their effectiveness by frequent use. When there is any doubt as to the advisability of any such sign, it should be erected. The amount of information on any one sign, or group of signs in close proximity to each other, should not be more than can be comprehended quickly by the motorists for whom they are intended. The legend on guide signs should be limited to three lines.

Guide signs fall into three general categories: *route markers*, *destination signs*, and *informational signs*. Route markers should be placed as frequently as is necessary to keep drivers following such markers certain of their route and location. Route markers should be located at all major intersections. Confirming route markers should be located between intersections and just beyond any built-up areas. Junction markers should be erected in advance of every intersection where a marked route is intersected or joined by another marked route. Directional assemblies should normally be located on the near right-hand corner of the intersection, and at major intersections it is desirable to erect additional assemblies on the far right-hand or left-hand corner. Destination signs should be located far enough in advance of the intersection to permit the motorist to reach necessary decisions before he reaches the intersection. Distance signs should be erected on important routes nearing municipalities and just beyond intersections of numbered routes in rural districts, or at intervals of about 10 miles along the route. Street names should be installed at all street intersections. Information signs should be erected in advance of, and at, the points of interest.

The MUTCD contains detailed specifications for placement and design of guide signs on conventional roads, expressways, and freeways. The Interstate Manual[107] should be consulted for guide signs on the interstate system.

School Area Controls

Because of the sensitive nature of controls and safety considerations in school areas, the MUTCD, in the 1971 edition, included a separate section on School Areas. Demands of the public and parents' groups for controls in these areas are often not justified under engineering considerations of normal pedestrian control.

In the interest of providing a specific standard for school areas which could be used as a guide in studying and controlling school areas, the MUTCD chapter gave detailed descriptions and standards for:

1. Designating school routes
2. Marking school crossings
3. Required engineering studies
4. Special signs and signing standards
5. Signal applications
6. School speed limits
7. Crossing guard programs

New Signs

As has been discussed, the new MUTCD made increased use of symbology in signing. Where new symbol signs were utilized, supplementary word signs were suggested. The new Manual also introduced several new signs, a selection of which are shown in Figure 20-1.

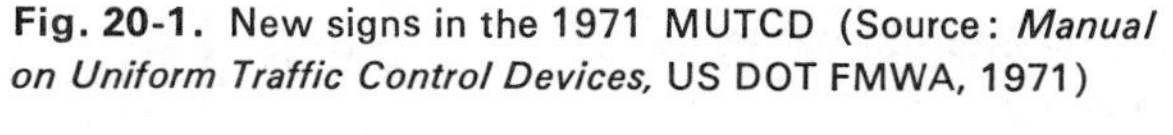

Fig. 20-1. New signs in the 1971 MUTCD (Source: *Manual on Uniform Traffic Control Devices,* US DOT FMWA, 1971)

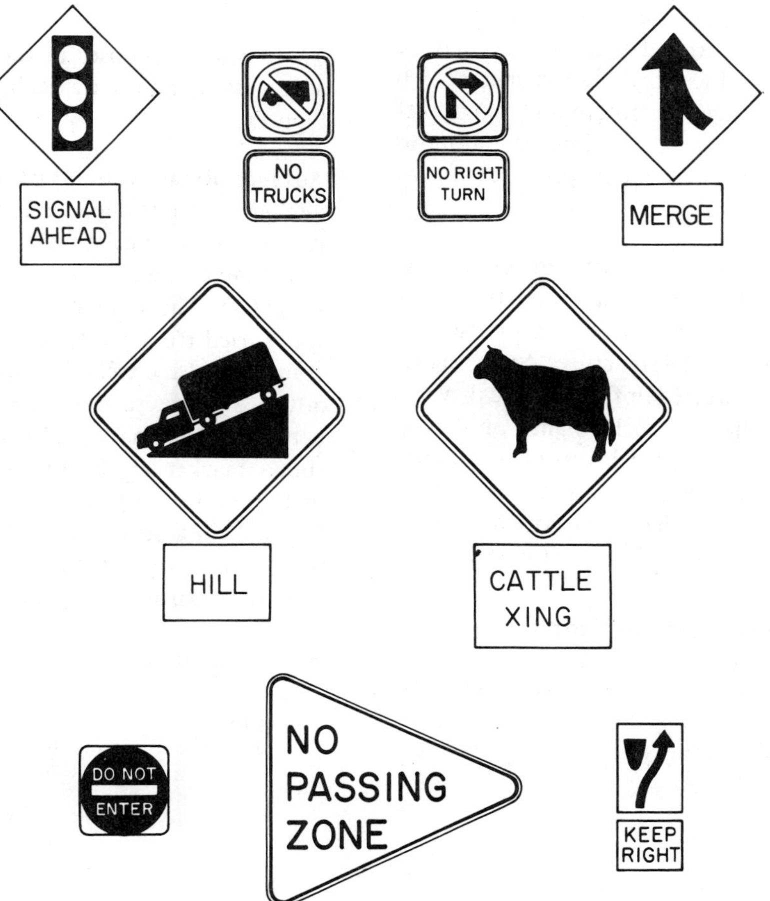

Type of Sign Equipment

Fixed-Message Signs. Fixed-message signs are those that present a fixed or permanent message to the road user. These are by far the most commonly used, and are the cheapest to produce and maintain.

Variable-Message Signs.[3] Variable signs, which at special times display emergency warnings, as of fog or icy surfaces ahead, or impose special regulations as to speed, turn prohibitions, or lane movements, are becoming increasingly more important on high-speed expressways. Such signs may be changed manually, by remote control, or for some purposes by automatic controls utilizing sensing elements for the conditions that require special sign messages. Since May, 1962, 67 signs on the New Jersey Turnpike are being remotely operated by radio to provide speed and reliability in alerting motorists of vital road conditions.

The variable-message sign may have a single message that is made legible only when required, or several messages of which one or more can be made legible at any time. The various methods of accomplishing this include patterns of incandescent lamps or of neon or fluorescent tubing that can be illuminated when desired, translucent panels having the legend on the back side legible only when lighted from behind, or mechanical devices for masking parts of the sign or for the interchanging of panels. Signs containing continuous lighting elements, which may include small light sources or small reflective beads illuminated by a controlled light source, may be remotely operated to produce any desired message. It is essential that such messages be totally "blanked out" when they are not intended to be on, since partial information transmitted to the driver can prove confusing and dangerous.

Holosigns. Experiments have been conducted with devices which are able to project lettered images on the atmosphere itself. These devices permit the projection of signs without the use of physical structures. Such "signs" could even be projected directly in front of the driver if desired. Of course, the effect of such "signs" on drivers must be evaluated. The equipment involved is complex and costly and not yet developed for practical use. It must be tested under many weather conditions and maintenance may prove extremely difficult and costly.[108]

Studies on the Effectiveness of Signs

Predicting the Effectiveness of Signs.[109] To properly evaluate the effectiveness of traffic control devices, it may take a period of several months to compile pertinent data. A study concerned itself with a film technique method to predict the comparative effectiveness of signs within a relatively short period of time. The method involves the proper placing on the highway of a series of different signs designed to accomplish the same purpose, and after each installation the highway is filmed with a 16 mm motion picture camera mounted on a moving vehicle. The filmed drives thus obtained are presented to groups of drivers who respond, after each showing, to questions designed to determine how accurately the intended sign message is perceived. Analysis of the response results permits an evaluation of the relative effectiveness of the different series of signs. Specifically, the study concerned a comparative evaluation of a series of warning signs to indicate the narrowing from four lanes in one direction of flow to three lanes. Lane signing configurations were studied as follows:

1. Sign A: Two diamond-shaped signs mounted on posts 400 feet apart: The first sign reads PAVEMENT NARROWS; the second sign reads SQUEEZE LEFT.
2. Sign B: Same as Sign A, except second sign reads MERGE LEFT.
3. Sign C: A single, diamond-shaped sign reading 3 LANES, underneath which are four vertical arrows with the one on the right having a slash drawn through it.
4. Sign D: A single rectangular sign reading LANE ENDS MERGE LEFT.

The written responses of the groups of drivers revealed a consistent and statistically significant superiority for Sign D.

Advance Route Turn Markers.[110] This study was conducted to determine the effectiveness of advance route turn markers on urban streets. The design of the advance directional assembly was in accordance with the 1961 MUTCD. The number of advance route turn markers was varied from none, one, and two. As a result of the study, it was concluded that there is a need for one advance route turn marker placed in advance of an urban intersection where a route turns. The lone directional marker at the intersection gives the driver too little time for response and, therefore, may create confusion, congestion, and potential hazard. However, there was little evidence to substantiate the use of a second advance turn marker.

Signs and Accidents.

1. A laboratory investigation was carried out to determine the effect of outdoor advertising signs on driving

efficiency. The study showed that numerous signs in the driver's field of vision did not adversely influence driving efficiency. A careful analysis of accident statistics along a 100-mile stretch of highway also revealed that advertising signs make no significant contribution to accidents.[114]

2. A change in the design of an assembly containing a curve sign and advisory speed sign for the curve helped to produce a significant (43 per cent) reduction in traffic fatalities[15]. The size of both signs was increased, and the background color of the advisory speed sign was changed from yellow to white. In addition, the advisory speed sign was placed above the curve sign. These last two changes were in conflict with the MUTCD; however, the significant reduction in traffic fatalities would appear to justify them, thus underscoring the need for revisions in the MUTCD. Neither of these two changes was included in the 1971 MUTCD.

For accounts of interesting studies regarding overhead signs,[111-113] accidents and signs,[116] and signing uniformity,[117-124] consult the literature.

21

Traffic Signals

All power-operated devices (except signs) for regulating, directing, or warning motorists or pedestrians, are classed as *traffic signals*. There is a widespread belief among laymen that traffic signals offer the solution to all traffic control and accident problems at intersections. In many instances signals have been installed where they were not warranted. The consequence has often been excessive delay, disobedience of signals, diversions to inadequate alternate routes, and increased accident frequency. As a guide to the proper use of signals, the MUTCD[1,2] contains minimum warrants for signal installations based on vehicular and pedestrian volumes, accident hazard, coordinated movement, and interruption to continuous flow. Separate recommendations are given for rural locations, where traffic is normally light and speeds high, and for urban locations, where these conditions are usually reversed.

When installed under conditions which justify its use, a signal is a valuable device for the control of traffic and for its safe and efficient movement. Therefore, it is important that the selection and use of this traffic control device be preceded by a thorough study of traffic and roadway conditions, and that the determination of the type of control and method of operation be based on the traffic factors developed by such study. Equally significant is the need for proper maintenance standards[3] and for checking the efficiency of the traffic signal, after installation, to determine the degree to which the type of installation, operation, and timing program meet the requirements of traffic and to permit intelligent decisions on operational adjustments.

MERITS OF SIGNALS

Signals offer the most positive form of control of all devices (excepting railroad grade crossing gates). A signal gives positive as well as negative control; it relays a message of what to do, and what not to do. Signals are useful in providing relief to congested situations where no other control device is adequate. The alternating assignment of right-of-way to intersection legs can eliminate most or all conflicting movements in the intersection area. However, since each traffic movement uses the intersection only for a portion of the time, delay may be increased and capacity reduced. Signals offer the maximum possible degree of control (short of physical barriers). A good general guide to the application of control devices is to utilize the *minimum* degree of control necessary to provide for safe and efficient movement of vehicles and pedestrians.

By alternately assigning right-of-way to various traffic movements, signals provide for orderly movement of heavy and/or conflicting flows. They may interrupt extremely heavy flows to permit the crossing of minor movements which otherwise could not have moved safely through the predominant flow. By coordinating systems of adjacent signals, smooth, continuous flow of traffic platoons may be developed. Signal progression timing may also be used to regulate speed on major streets and arterials.[4,5]

The installation of signals can significantly reduce the frequency of certain types of accidents, particularly right-angle collisions.[6,7] However, other types of accidents, notably rear-end collisions, may increase as a result of signal installation.

Signals should be used only where lesser forms of control will not suffice. As previously noted, signals generally result in increased total intersection delay.[8-10] In addition, excessive delay to minor movements may occur, particularly if the signal is improperly timed. Delay may cause diversion to inadequate alternate routes. Also, as is the case with all devices, overuse or misapplication will breed disrespect for all devices. Where all lesser forms of control prove inadequate, signals should be given careful consideration. The potential of signals for alleviating congestion, providing safe access for conflicting movements, and encouraging smooth, efficient flow, should be understood. Specifications for signals and placement, as well as warrants for their use, are detailed in the MUTCD. Warrants are discussed further in Chapter 23.

SIGNAL CLASSIFICATION

Signals may be classified by their operating function, rather than by the equipment characteristics of the signal, which vary from manufacturer to manufacturer. They are generally classified as follows:[1]

Traffic Control Signals (Stop and Go)

1. *Pre-timed signal.* This type of signal directs traffic to stop and permits it to proceed in accordance with a single, predetermined time schedule or a series of such schedules. The traffic signal is set to repeat a given sequence of signal indications regularly.
2. *Traffic-actuated signal.* The operation of this type of signal is varied in accordance with the demands of traffic as registered by the actuation of vehicle or pedestrian detectors on one or more approaches.
3. *Traffic-adjusted signals.* These are centrally controlled, as, for example, by a digital computer, and have settings which are updated from measurements of the system through detectors.

Pedestrian Signals

These are erected for the exclusive purpose of directing pedestrian traffic at signalized locations. They consist of WALK-DON'T WALK signals.

Special Traffic Signals

1. *Flashing beacon:* A *yellow* or *red* signal lens, illuminated by a rapid, intermittent, flashing light to warn traffic at hazardous locations. A recent study showed that a *flashing* device consisting of a horizontal or vertical "bouncing ball" (alternate illumination of two red or yellow signals placed vertically or horizontally) resulted in greater accident reductions at a hazardous location than other flasher types tested.
2. *Lane-use control signal:* Indicates and controls direction of traffic movement on reversible lanes.[11] These are often used on bridges.
3. *Drawbridge signal:* Traffic signal at drawbridge to warn of opening.
4. *Railroad crossing signal:* Train-approach signals at railroad grade crossings.

DESIGN AND LOCATION OF TRAFFIC SIGNALS

The MUTCD presents detailed specifications for design and location of signals with respect to color, position, diameter, illumination, visibility, shielding, number of lenses, and the meanings of standard indications. Among the standard indications specified are:

1. Solid red
2. Solid yellow
3. Solid green
4. Arrow indications
5. Flashing indications
6. Permissible combinations

Consult the MUTCD for specific standards. For supplementary information on design and location of signals, consult the literature.[12,13]

MAINTENANCE OF TRAFFIC SIGNALS

Maintenance of all control devices is vital, especially for signals, because the state or municipality which has jurisdiction can be held legally responsible for accidents which may occur while the signal is inoperative. An inoperative or improperly operating signal is more dangerous than other device malfunctions, since signals offer *positive* as well as *negative* messages. For example, simultaneous green indications on two legs of an intersection will be a *positive* cause for serious accidents. A broken or masked stop sign is not so serious, as drivers on the minor leg would be inclined to stop anyway. A driver approaching a green signal has no reason to suspect malfunction, and proceeds as though he had the right-of-way.

Of extreme importance is the periodic cleaning of lenses and the replacement of lamps before they malfunction. Consult the MUTCD and the literature for maintenance specifications and supplementary material.[13-17] A recent study[16] of signal maintenance practice showed the following results:

1. A maintenance program was formulated for traffic signals and flashers using systems analysis techniques. It determines optional lamp replacement intervals, calculates the shortest route for preventive maintenance, and staffs the crew to insure proper signal operation.
2. The preventive maintenance program affords certain economic advantages and improves the safety of intersections by reducing signal failure probability.
3. Lamps with long-rated lives are recommended because their operation is less costly and the anticipated failures per unit time are smaller.
4. An adequate maintenance record system is mandatory for the economic and efficient scheduling of realistic traffic signal and flasher maintenance.

PRE-TIMED SIGNALS

As previously defined, a pre-timed signal directs traffic to stop and permits it to proceed in accordance with a single, predetermined time schedule or a series of such schedules. Each schedule provides for a regularly repeated sequence of signal indications.

A comprehensive investigation of traffic and roadway conditions should be made to determine the necessity for a signal installation, and to supply necessary data for the proper design and operation of a signal that is found to be warranted. Such data should, ideally, include the following, as specified in the MUTCD:

1. *Hourly traffic volumes* from each approach, for 16 consecutive hours of an average day
2. *Classified volumes* for each movement during each 15-minute period of the morning and evening peak period
3. *Pedestrian volumes* on each crosswalk during peak vehicular and peak pedestrian periods
4. *The 85th percentile* speed on uncontrolled approaches to the intersection
5. A *condition diagram* showing the physical layout of the intersection
6. A *collision diagram* showing accident experience by type, location, direction, severity, date, and time of day for at least one year
7. *Average vehicular delay* for each approach during peak periods
8. *Gap distribution* in major flows where minor flows have difficulty crossing

While all of the above information is useful in determining the need for a signal installation, it is often not possible or practical to obtain data on all items. Data on peak-hour volumes (classified by movement and type), geometrics, and accident experience, are important for adequate consideration of warrants. If pedestrian volumes are unavailable, a general knowledge of pedestrian flow levels (light or heavy) during peak periods will suffice. Delay and gap distribution data is needed only in cases where detailed analytic studies are being conducted.

Warrants for Pre-Timed Signals

The MUTCD presents minimum warrants which justify the installation of signals. These warrants include:

1. Minimum vehicular volumes
2. Interruption of continuous traffic (to permit minor flows to cross safely)
3. Minimum pedestrian volumes
4. School crossings
5. Progressive movement (to provide for progressive flow)
6. Accident experience
7. Systems
8. Combinations of warrants

The specification of each warrant is given in the MUTCD. Satisfaction of one or more of these generally justifies the installation of signals. Warrants, as presented in the MUTCD, act as excellent guidelines for the application of devices, but should not be used to replace or negate sound engineering judgment.

Consult the literature for studies regarding the application, evaluation, and suggested modification of signal warrants.[18-26]

Pre-timed Controllers

A controller is an electrical mechanism for controlling the operation of traffic control signals, including the timer and all necessary auxiliary apparatus mounted in a cabinet. The type of controller selected to operate a pre-timed signal is primarily dependent upon whether or not coordination of signals is to be provided. However, the operating principles of the various types of pre-timed controllers are basically the same. The literature contains specifications and a complete list of definitions for pre-timed controllers.[27]

The timing unit of a pre-timed controller consists of an electric induction or synchronous motor which drives a timing dial through a gear. The size of the gear determines the cycle length. Standard gears, giving cycle lengths of from 30 to 120 seconds, are readily available. Gear sizes provide for 5-second incremental increases; however, commencing with 90 seconds, the gears usually provide for 10-second incremental increases. One complete revolution of the timing dial represents one complete cycle. The timing dial is a cylinder which has 100 equally spaced longitudinal slots on its periphery. This provides for a division of the cycle length by 1 per cent increments. By placing keys or pins in appropriate dial slots, the cycle can be proportioned into phases, and each dial can be set for a different cycle length by use of appropriate gears, and for different cycle divisions by the proper placing of keys or pins.

The keys on the timing dial actuate a microswitch, which causes a camshaft to be advanced one position at a time as each key, in turn, actuates the switch. Each cam has a maximum of 16 cam lobes, which can provide for 16 intervals; any lesser number of intervals may be used by simply removing the cam lobes. Switches in contact with the cams close when the cam shaft reaches a position where the lobe has been removed. A separate cam controls each signal indication, that is, one cam controls main street green, another main street red, another main street yellow, another cross street green, etc. Appropriate cam lobes are removed from each cam so that in any phase of the cycle the proper lights are illuminated.

Some controllers are provided with a means of discontinuing automatic control and substituting manual control. This permits the operator to vary cycle lengths and phasing as he desires. However, manual control should be used as sparingly as possible.

Figure 21-1 shows the standard dial arrangement(s) for pre-timed controllers.

The various timing mechanisms used are discussed below.

Nonsynchronous Pre-timed Controllers, for Isolated Intersections. This type of controller, which is usually timed by an induction motor, is little used because the timing is subject to changes in line voltages and temperature fluctuations. If such equipment must still be used, it should only be placed at minor, isolated intersections warranting signalization where it is unlikely that there will be any need for coordination with other intersections.

Fig. 21-1. Standard pre-timed signal controller dial

Synchronous Pre-timed Controller, for Isolated Intersections. This type of controller uses a synchronous timing motor which is only frequency dependent, and thus maintains a constant speed. All signal installations supplied by the same line voltage can be coordinated in the future or supervised by a master controller. These controllers are always to be preferred over the nonsynchronous, even at fully isolated locations, since they maintain accurate cycle lengths and phasing.

Program devices are available, for synchronous operation, which provide for changing cycle lengths and proportions a limited number of times during the day.

Controllers Providing for Coordination. Coordination of signal installations may be accomplished by using noninterconnected, or nonsupervised, synchronous controllers. Each individual controller is set to provide for the coordination of cycle phasing. However, noninterconnected coordination is not recommended for heavy traffic conditions, since it does not provide sufficient flexibility to accommodate the variations in traffic flow, and there is no assurance that the desired coordination will continue indefinitely. This latter quality requires frequent, periodic checking of the system.

Ideally, coordination of pre-timed signal installations is accomplished by using interconnected synchronous controllers. Coordination is provided by supervising secondary controllers with a *master controller*. Interconnections between the master controller and secondary controllers may be accomplished by the use of cables or telephone wires, or by radio. Underground or overhead cable or wire installations are used, depending upon available funds. Of course, radio interconnection requires no cable installation. Interconnection allows for complete flexibility of the system to accommodate traffic fluctuation.

Program changes of cycle lengths and proportions are easily accomplished with the interconnected system by installing time clocks in the master controller; however, noninterconnected systems cannot be programmed for changes in the entire system.

The *master controller* provides for supervising a system of secondary controllers, maintaining definite time interrelationships, or accomplishing other supervisory functions such as manually controlling the system, turning off the system, placing the system on flashing operation, etc. It can automatically provide for changing the length of the cycle of each secondary controller and if secondary controllers are equipped with more than one timing dial, it can also provide for automatically shifting to various timing dials. The master controller is the zero time reference base from which the beginning of the green indication at all secondary controllers in the system is offset in the range of 0 to 100 per cent, in 1 per cent increments, of the cycle length.

Traffic conditions may significantly change not only throughout the day, but may also change from day to day, particularly when comparing weekdays with Saturdays and Sundays. A *program controller* is used to provide for this additional flexibility. The program controller modifies control of the master controller on a day-by-day basis, and during each day, several changes of cycle length, division, offsets, and other supervisory functions can be provided.

TRAFFIC-RESPONSIVE SIGNALS

Traffic-responsive signals include all traffic-actuated or traffic-adjusted controllers and systems. The MUTCD applies the same warrants to responsive signals as to pre-timed signals. However, because responsive signals do not cause undue delay, consideration to their installation may be given where pre-timed signals might not be justified. Several considerations to include are:[1]

1. *Vehicular volumes.* At intersections where the volume of vehicular traffic is not great enough to warrant pre-timed signals, traffic-actuated signals may be applied if other conditions show a need for traffic control signals and justify the cost of the installation.
2. *Cross traffic.* When the volume of traffic on the major street is so great as to restrict and jeopardize the occasional movement of vehicular or pedestrian cross traffic on the minor street, semi-actuated signals may be installed for assignment of right-of-way to the minor movement.
3. *Peak-hour volumes.* When traffic control signals are required at an intersection during only a small portion of the day, such as during peak periods of traffic flow, traffic-actuated signals may be installed, if economically justified, since they do not unduly delay traffic at other times. Economic justification may be illustrated by carrying out *road user benefit analysis.*
4. *Pedestrian traffic.* When only minimum pedestrian volume warrants for pre-timed signals are met, traffic-actuated signals should be considered. They will delay vehicular movements only when there is pedestrian actuation of the signal.
5. *Accident hazard.* When only the minimum accident experience warrant for pre-timed signals is met, traffic-actuated signals should be considered, since they may lessen the stops and delays which are commonly associated with accidents after signalization.
6. *Wide traffic fluctuations between approaches.* When the relative volumes of traffic on the entering streets fluctuate widely at an intersection where pre-timed signals are warranted, full traffic-actuated control will usually provide for greater efficiency in intersection operation.
7. *Complicated intersections.* When traffic signals are warranted at complicated intersections requiring multiple traffic phases, traffic-actuated signals should be considered because, in addition to the usual advantage of traffic actuation, a phase can be skipped when there is no traffic demand to use it.
8. *Progressive signal systems.* When the spacing or character of an intersection in a pretimed progressive signal system is such that satisfactory progressive timing cannot be achieved, traffic-actuated control can be employed to advantage.

Actuated signals may also be justified at mid-block pedestrian crossings and at narrow crossings permitting only one direction of travel at a time.

Semi-Actuated Controllers

These devices provide means for traffic activation on one or more, but not all, intersection approaches. This type of control is applicable primarily to an intersection of a heavy-volume, urban or suburban traffic artery with a relatively lightly traveled minor road or street. The essential operating features of the controller are:

1. Detectors are on minor approaches only.
2. Major street receives minimum green period each cycle.
3. Major street retains green indefinitely after minimum period, until interrupted by minor phase detector actuation.
4. Minor phase receives green after actuation, provided major phase has completed minimum green period.

5. Minor phase receives initial portion (green).
6. Minor phase green is extended by additional actuations until preset extension limit is reached.
7. Memory feature remembers additional actuations if minor phase green is terminated by extension limit, and the green indication is returned to the minor phase after the major phase minimum green period.
8. Dials pre-set clearance (amber) intervals for both phases.

This kind of control is excellent for use where light side-street volumes cannot safely cross major flows without signalization. If side-street flows are sporadic, the regular interruption of the major flow with pre-timed control cannot be justified. Where both street volumes fluctuate widely, semi-actuated control should not be used, since there are no detectors on one or more legs.

Full-Actuated Controllers

This device provides for actuation by vehicles on all legs of the intersection. It is applicable primarily to an isolated intersection of streets or roads that carry approximately equal traffic volumes, but where distribution between approaches varies and fluctuates. It then becomes necessary to take into consideration the demands on all approaches. The essential operating features of the controller are:

1. Detectors are on all approaches.
2. Each phase has a pre-set initial portion which allows accumulated vehicles to start and enter the intersection.
3. Green interval is extended by a pre-set unit extension for each actuation after the expiration of the initial portion.
4. Green extension is limited by pre-set extension limit.
5. Clearance interval is pre-set for each phase.
6. Each phase has a recall switch.
   a. With all recall switches off, the green indication remains on the phase to which it was last called, provided there is no actuation on other approaches.
   b. With a recall switch on, the green indication will revert to that selected phase at every opportunity.
   c. With the recall switch on both phases of a two-phase control, the controller operates on a pre-timed basis, provided there is no demand on either phase.

Because of their actuated nature, full-actuated controllers cannot be coordinated with other signals without losing the flexibility for which they were designed. Demand patterns for which they are applicable, as well as the inability to coordinate, make the requirement of isolated locations (about a mile between adjacent signals) a fairly strong one.

Early models of traffic-actuated controllers were electromechanical. Modern equipment of this type is generally of solid-state design. A solid-state, full-actuated controller is shown in Figure 21-2.

Volume-Density Controllers

This class of device offers additional responsiveness in signalization for isolated intersections. Green time is allotted on the basis of volumes on approach legs. Unlike simple actuated signals, the volume-density signal does not merely react in a predetermined fashion to an actuation, but is able to record and retain information regarding volume, queue length, and delay times. In addition, a phase will lose the green by any one of three mechanisms:

1. There are no vehicles producing any further demand on the approach.
2. The maximum green phase is reached.
3. The time gap between vehicles on the approach exceeds the maximum standard.

The last mechanism is the "density" function of the signal. At the beginning of a green phase, the maximum time gap might be, for example, 5 seconds. As the green phase continues, the maximum time gap decreases. The phase is lost when the maximum gap is exceeded, or when the maximum length of phase is reached, whichever comes first. This type of control provides the greatest flexibility in traffic-actuated controllers, in that it is capable of taking into consideration the number of vehicles waiting on an approach, as well as the volume on the approach with the green indication. Its use is primarily applicable to an isolated intersection with wide

Fig. 21-2. A solid state, two-phase, full actuated controller (*Source:* Product Notes, *No. III, Model T-807R, Solid-State, Two-Phase, Full-Actuated Controller, Automatic Signal Division, Laboratory for Electronics, Inc.*)

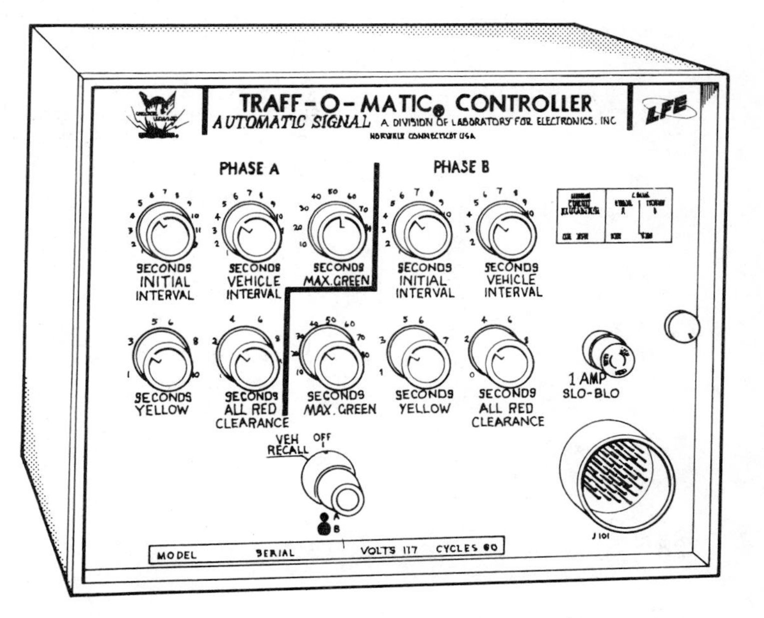

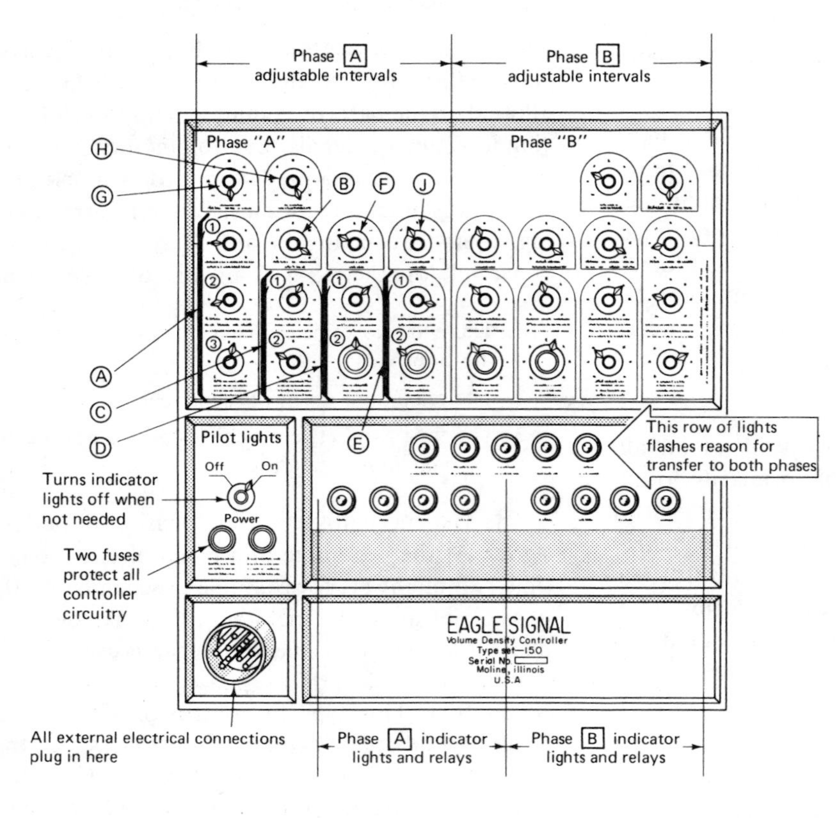

Ⓐ <u>Minimum green with density initial = assured green time</u>

<u>Dial 1</u>
Minimum green time
This amount of green time is guaranteed each phase, once each cycle.
<u>Dial 2</u>
<u>Density initial</u>
Each additional car (after first 10) receives an additional 1 second allowance, set on dial 3.

Ⓑ <u>Vehicle or unit extension</u>
<u>Passage time</u>
A fixed time interval set at the time required for a vehicle to travel from the detector to the intersection stop line, which could be called the "MAXIMUM ALLOWABLE GAP." This time serves as a reference to see that the vehicle will either get its required time or that the Controller, through its memory feature, will return to the phase at the first opportunity.

Ⓒ <u>Time waiting</u>
Reduction of the "ALLOWED GAP" for vehicles moving on green, initiated by a vehicle waiting on red.
<u>Dial 2</u>: Total time car waits on red.
<u>Dial 1</u>: Reduced gap time.
Results: Car waiting beyond 30 seconds reduces "ALLOWED GAP" on green to 5 seconds.

Ⓓ <u>Cars waiting</u>
Reduction of "ALLOWED GAP" for vehicles moving on green, initiated by number of cars waiting on red.
<u>Dial 2</u>: Maximum number of cars waiting on red.
<u>Dial 1</u>: Reduced gap time.
Results: Number of cars waiting beyond a total of 35, reduces "ALLOWED GAP" on green to 4 seconds.

Ⓔ <u>Density</u>
Reduction of "ALLOWED GAP" by movement of vehicles on green, automatically regulated by the pattern of the average vehicle spacing.
<u>Dial 2</u>: Minimum number of actuations required every 10 seconds.
<u>Dial 1</u>: Reduced gap time.
Results: 20 cars, failing to maintain a 10 second or less interval gap, reduces "ALLOWED GAP" to 2 seconds

Ⓕ <u>Platoon carryover</u>
Calibrated in percentage, this dial promotes coordinated movement of traffic through a group of similarly controlled intersections.
Result: Continuing "regulated platoons," approaching the green, but stopped when reaching the intersection, tend to continually lessen their waiting time until ultimately getting the green without stopping.

Ⓖ <u>Maximum green extension</u>
Determines maximum time green may be extended by successive actuations after a detector on the opposing street.
When properly set and applied, the Controller seldom reaches this setting. The resulting condition at the intersection would indicate virtually saturated "bumper to bumper" traffic in both phases.

Ⓗ <u>Traffic change</u>
Length of the yellow (caution) interval.

Ⓙ <u>Recall switch</u>
When this switch is on it insures that "dwell" occurs in that selected phase, giving the preferred street the green until a call is again received from the opposing street. This switch places an automatic call in its phase, but does not extend or reduce its timing. Street detectors continue to operate in the regular manner.

Fig. 21-3. Volume-density controller (*Source: Eagle Catalogue, 1969, ET 150 Series, Two-Phase, Full-Actuated Volume Density Traffic Control, Section VA, page 2e, Eagle Signal Division*)

traffic fluctuations between streets or roads. The essential operating features of the controller are:

1. Detectors on all approaches are placed sufficiently back from the intersection to enable the counting of relatively large numbers of queued vehicles.
2. Each phase has an assured green time, as set by three dials.
   a. Minimum green period
   b. Number of actuations before minimum green period is to increase
   c. Amount of increase of minimum green period for each additional actuation during the red phase.
3. *Passage time* is the unit extension, created by each additional actuation, *after the assured green time has elapsed.* Time is set for a vehicle to travel from the detector to the stop line. This value also becomes the maximum allowable gap between vehicles which will retain the green. This maximum gap, or passage time, may be reduced in several ways. The green will be lost when:
   a. A pre-determined low limit of passage time is reached when red-phase vehicles have waited a pre-set time.
   b. A pre-determined low limit of passage time is reached when the number of vehicles waiting on the red phase exceeds a pre-set value.
   c. A pre-determined low limit of passage time is reached when the number of green-phase vehicles per 10 seconds is less than a pre-set value.

   These constitute the density function of the controller. Specific dial settings differ from model to model.
4. *Platoon carryover* effect enables the controller to remember a pre-set percentage of the previous green-period traffic and synthetically applies that number of vehicles waiting on the red phase to insure a more prompt return to the green phase, when the next platoon of vehicles hits the detector.
5. *Green extension* is limited by pre-set *extension limit.* However, this feature seldom operates, because of the effect of the reduction factors on passage time.
6. Clearance intervals are pre-set for each phase.
7. Each phase has a recall switch that operates in the same manner as described for the full-actuated controller.

For additional comments on the operation of volume-density controller and auxiliary equipment, consult the literature.[28-30] A standard volume-density controller is shown in Figure 21-3.

Traffic-Adjusted Controllers

A combination of the advantages of a pre-timed, flexible, progressive signal system and traffic actuation can be realized by a system in which a traffic-actuated master controller is used to supervise either pre-timed or semi-actuated local controllers. Detectors are placed at representative locations in the progressive system if an artery is to be controlled, or at typical locations throughout an area if a grid system is to be controlled. The detectors transmit information about traffic volume and direction to a computer in the master controller. The information is analyzed, and the master controller selects the cycle length and offset combination predetermined to best serve the directional distribution and volume characteristics existing at that time. The local controllers are interconnected to the master controller and operate at any given instant upon the cycle and offset selected by the master controller.[31] Such installations are operating in numerous cities; for some examples, consult the literature.[31-49]

Auxiliary Actuated Controller Equipment

Pedestrian Interval Timer. This piece of equipment is used in conjunction with actuated controllers to provide sufficient green time to enable pedestrians to cross safely. However, the green indication is extended only upon push-button actuation and does not have to be extended for each green phase.

Minor Movement Controller. The most common use of this piece of equipment, in conjunction with actuated controllers, is for the addition of a phase to accommodate an intermittent traffic flow which has become too significant to operate without interference with major movements. Minor phase is skipped unless activation is received.

Advance Green Timer. This unit is used to provide a fixed advance green for one approach. The unit can be actuated or nonactuated. With actuated control, the advance green indication is displayed only after a vehicle actuation calls for it; with nonactuated control, the advance green indication is displayed in each cycle.

Detectors

All traffic-responsive equipment depends upon accurate receipt of demand information through vehicle and pedestrian detectors. Demand-responsive equipment is, therefore, entirely dependent upon the proper operation of detectors under all conditions. The most common types of detectors currently available are discussed below.[1,2,50]

Pressure-Sensitive Detector. The pressure-sensitive detector is installed in the roadway, and is actuated by the pressure of a vehicle passing over its surface. It is actuated by vehicles traveling at any speed up to 60 miles per hour, but it becomes inoperative if a vehicle is stopped upon it.

The detector may be nondirectional, that is, it is operated by traffic moving over its surface in either direction; or it may be directional, that is, the detector is operated by traffic moving over its surface in only one direction.

Magnetic Detector. Installed in or near the roadway, the magnetic detector is actuated by the magnetic disturbance caused by the passage of a vehicle. It is actuated by vehicles traveling at speeds of up to 60 miles per hour, and it responds only to moving vehicles. Therefore, the detector is not rendered inoperative or continuously operative by parked vehicles or other fixed metal objects within its zone of influence.

Radar Detector. This type of detector is installed over the roadway, and is actuated by the passage of a vehicle through its field of emitted radio waves. It is actuated by vehicles traveling from 2 to 70 miles per hour, and its operation is not affected by parked vehicles or other stationary objects in the roadway.[51,52]

Ultrasonic Detector. The ultrasonic detector is installed over, or on the side of, the roadway, and is actuated by the passage of a vehicle through its emitted ultrasonic wave beam.

Infrared Detector. This detector is installed over, or on the side of, the roadway, and is actuated by the passage of a vehicle through its emitted beam of infrared light.

Induction Loop Detector. The induction loop detector is installed in the roadway, and is actuated by the passage of a vehicle which causes a change in the induction of a wire loop.[54,55]

Impact Detector. The impact detector, installed in the roadway, is actuated by the passage of a vehicle which creates a change in air pressure as it passes over the detector.

Photo-Electric Detector. Photo-sensitive cells may be used in conjunction with a light source to detect the presence of vehicles or pedestrians.

Pedestrian Detector. This detector, usually of the push-button type, is installed near the roadway, and is operated by hand pressure.

The placement of detectors with respect to the stop line is dependent on the type and operating characteristics of the controller, vehicle approach speeds, grade and width of the roadway, visibility, and the presence of driveways or channelization. See the MUTCD for recommendations.

Coordination of Traffic-Actuated Signals[56,57]

In addition to the potentially excellent coordination provided by *traffic-adjusted control,* traffic-actuated signals may be coordinated by interconnection, as with pre-timed signals, or by use of synchronous motor control of a series of semi-actuated controllers, or by a series of full-actuated controllers of the volume-density type. *The coordination of full-actuated or volume-density signals means the loss of that flexibility for which the system was installed; such systems will operate like a fixed-time system.*

Traffic-Adjusted Control. The definition of traffic-adjusted control and the relation to digital computer techniques has been discussed. The development of computer control in Toronto, San Jose, Wichita Falls, New York City and other locations is noted in the literature.[31-49] For other applications, both in computer control and in computer simulation, refer to the literature.[58-76]

Interconnection or Synchronous Motor. A master controller is used, interconnected to each secondary controller. The latter can respond to side-street actuation only at the allowable offset point as determined by the time-space diagram. (Refer to Chapter 24.) All green time not required by side-street demand is added to the artery with consequent improvement in the through band. It is possible to insert semi-actuated control at desired locations in a pre-timed coordinated system. This may be a useful device at certain intersections which are poorly spaced for a maintenance of progression, since the traffic-actuated controller can be depended upon to widen the artery through band. A limitation of coordination by interconnection of semi-actuated controllers is that traffic may lose platoon quality with too few calls from minor streets.

Synchronous motors, installed in each semi-actuated controller, can be timed to accomplish coordination in the same manner as by interconnection.

Volume-Density Controllers. A succession of full-actuated controllers on a street will always exhibit a tendency toward facilitating progressive flow thereon, if main-street traffic is significantly greater than side-street movements and exhibits platooning characteristics.

When these controllers are of the volume-density type, the tendency toward progressive platoon movement is substantially intensified. This principle has been successfully applied on streets carrying very heavy traffic volumes with irregular and frequent, short block spacings.

The tendency of full-actuated control to accommodate platoon movement can sometimes be used to advantage in a pre-timed progressive system at points where intersection spacing is such as to cause difficulty in securing a satisfactory through band. Use of volume-density controllers in this manner should be approached with the utmost caution, and only after careful study of flow characteristics at the intersection in question.

A COMPARISON OF PRE-TIMED AND TRAFFIC-ACTUATED CONTROL[77]

With basic pre-timed control, a consistent and regularly repeated sequence of signal indications is given to traffic. By use of attached auxiliary devices or remotely located supervisory equipment, the operation of pre-timed control can be changed within certain limits to meet the requirements of traffic more precisely.

Pre-timed control is best suited to intersections where traffic patterns are relatively stable or where the variations in traffic flow that do occur can be accommodated by a pre-timed schedule without causing unreasonable delay or congestion. Pre-timed control is particularly adaptable to intersections where it is desired to coordinate operation of signals with existing or planned signal installations at nearby intersections on the same street or adjacent streets.

Traffic-actuated control differs basically from pre-timed control in that signal indications are not of fixed length, but are determined by and conformed within certain limits to the changing traffic flow as registered by various forms of vehicle and pedestrian detectors. The length of cycle and the sequence of intervals may vary from cycle to cycle, depending on the type of controller and auxiliary equipment utilized to fit the needs of the intersection. In some cases, certain intervals may be omitted when there is no actuation or demand from waiting vehicles or pedestrians.

Advantages of Pre-timed Control

1. The consistent starting time and duration of intervals of pre-timed control facilitates coordination with adjacent signals and provides more precise coordination than does traffic-actuated control. This factor was substantiated by the results of a study to determine the relative effectiveness of various types of traffic control signals.[8] Average stopped delay per vehicle was shown to be less for pre-timed control than for various types of traffic-actuated control for a series of intersections.
2. Pre-timed controllers are not dependent for proper operation on the movement of approaching vehicles past a detector. Therefore, the operation of the controller is not adversely affected by conditions preventing normal movement past a detector, such as a stopped vehicle or construction work within the area.
3. Pre-timed control may be more acceptable than traffic-actuated control in areas where large and fairly consistent pedestrian volume are present, and where confusion may occur as to the operation of pedestrian push buttons.
4. Generally, the installed cost of pre-timed equipment is substantially less (minimum of one-half) than that of traffic-actuated equipment, and pre-timed equipment is much simpler and more easily maintained.

Advantages of Traffic-Actuated Control

1. Traffic-actuated control may provide maximum efficiency at intersections where fluctuations in traffic cannot be anticipated and programmed for with pre-timed control. This factor was substantiated by a study of delay at signalized intersections,[78] and also in another study of the effect of type of control on intersection delay.[8] In the former study, it was shown that when signalized intersections are spaced so that the signals can be timed for progressive movement, pre-timed control results in the minimum delay.
2. It may provide maximum efficiency at complex intersections.
3. It normally provides maximum efficiency at intersections of a major street and a minor street, by interrupting the major street flow only when required for minor street vehicular or pedestrian traffic, and also by restricting such interruptions to the minimum time required.
4. It may provide maximum efficiency at intersections unfavorably located within progressive, pre-timed systems.
5. It may provide the advantages of continuous stop-and-go operation without unnecessary delay to traffic on the major street, whereas isolated pre-timed signals are sometimes switched to flashing operation during periods of light traffic.
6. It is particularly applicable at locations where traffic signal control is warranted for only brief periods during the day.
7. Traffic-actuated control generally minimizes delay at isolated intersections, particularly where demand is variable.

METHODS OF INTERCONNECTING SIGNALS

Radio Control of Signals

All of the features offered by cable interconnection are provided by radio interconnection. The basic radio coordinated traffic control system consists of the central station equipment, a VHF-FM radio transmitter and antenna, and an intersection unit for each signalized intersection. The central station equipment includes a master controller, a program device, tone generators, and a radio remote unit. Each intersection unit consists of a standard synchronous motor controller, a crystal-controlled VHF-FM radio receiver, an antenna, and tone switches. The switching operation to select the desired function to facilitate traffic is accomplished by the use of dual tone signals transmitted simultaneously for approximately one second. For general background information and reports of applications, consult the literature.[79-86]

Traffic Pacer or Funnel

A means of facilitating traffic flow without stops and increasing roadway capacity is by concentrating vehicles in a traffic lane into platoons timed to arrive at intersections during the green indication. This is accomplished by an illuminated, variable speed sign, placed in advance of each intersection, which advises the motorist how fast to proceed to make the next green indication. The speed signs and traffic signals are interconnected and synchronized throughout the system.[87-91] Consult the literature for examples of how conventional signal equipment (pre-timed) may be modified to operate with actuations.[92-95]

22

Street and Highway Lighting

BENEFITS OF LIGHTING

Accident Reduction

Streets and highways are intended to facilitate the safe movement of vehicular traffic both by day and by night. Nighttime brings increased hazards to users of streets and highways through limited visibility. Night driving is considerably more hazardous than day driving, in both urban and rural areas.

Of the 55,200 fatalities nationally in 1968, 29,300 occurred during the nighttime hours, and 25,900 during the daytime hours. Although motor vehicle deaths at night total only a few thousand more than during the day, death rates at night are significantly greater than day rates; in rural areas, they are over $2\frac{1}{2}$ times higher, as shown in Table 22-1.[1,2]

Artificial lighting should be provided for all locations where proper illumination is a recognized necessity for greater traffic safety and roadway visibility. Before-and-after studies have definitely shown that adequate street lighting has reduced nighttime accident rates.[3-12]

The use of street lighting in urban areas is a commonly accepted practice, particularly where vehicular and pedestrian traffic movements are heavy. Expenditures for street lighting are normally justified, without difficulty, on the basis of road-user benefits, considering reductions in traffic fatalities, injuries, and property damage.[8,11,13]

The continuous lighting of rural highways is not extensively done since results of studies on accident rates are not conclusive, and insufficient facts are available to show economic justification for the expenditure. More extensive studies are required to better understand the safety effects and economics of rural highway lighting.[14-17]

In Great Britain, a Code of Practice outlines the priorities and different areas in which roadways are

TABLE 22-1. DAYTIME VS. NIGHTTIME FATALITY RATES[1,2]

	NUMBER OF DEATHS	*PER CENT OF DEATHS*	*DEATH RATE PER 100 MILLION VEHICLE-MILES*
National			
Day	25,000	47	3.7
Night	28,100	53	9.7
Urban			
Day	7,600	46	2.2
Night	9,100	54	6.2
Rural			
Day	17,400	48	5.2
Night	19,000	52	13.2

(Source: *Accident Facts,* 1968 and 1969 editions, NSC)

provided with lighting installations. Australia has a similar code.[18]

Research by the Central European countries is being undertaken to find a standardization in the best possible combination of lights, spacing, mast heights, etc., for use on their international motorways.[19]

Comfort and Convenience

The driver is subjected to additional strain when driving at night. Adequate roadway lighting will help to provide increased visibility, and thereby decrease driver fatigue and facilitate traffic movement. Highway illumination helps to increase roadway capacity by promoting lateral vehicle placement, vehicle headways, and speeds which are more comparable to daytime conditions.[20-23]

Crime Deterrent

The great majority of crimes are committed at night. Proper street lighting results in a decrease of crime and provides greater protection and convenience for pedestrians. It also aids the police in combatting various types of crime. This benefit applies mostly to urban areas and towns, rather than to isolated rural roads.

Increased Land Values in CBD's

Street lighting aids business and enhances property values. It adds to civic pride and to general safety and pleasant living.[23]

DEFINITIONS

Angle of incident light: Either the angle between the incident ray from a light source and the plane on which it falls, or the angle between the incident ray and the vertical.

Candela: The unit of luminous intensity of a light source.

Candlepower: Luminous intensity expressed in *candelas*. A one-candlepower source will produce a certain illumination on a surface near it. Any other source substituted for this one, which produces the same illumination, is also rated *one candlepower*; if it produces twice as much illumination, it is rated *two candlepower*, etc. It is no indication of the total light output.

Footcandle: The unit of illumination when the *foot* is taken as the unit of length. It is the illumination which falls on a surface *one square foot in area*, on which is uniformly distributed a flux of one lumen. It equals *one lumen per square foot*.

Lumen: The unit quantity of light output. It is the amount of light which falls on an area of *one square foot*, every point of which is one foot distant from a source of *one candela*. A uniform source of light of one candela emits a total of 12.57 lumens, but for practical purposes, a 1 to 10 ratio is used.

Reflectance or reflection factor: The ratio of the light reflected by a surface or object to the light incident upon it.

Brightness: Amount of light emitted by a surface, usually expressed in *footlamberts*.

Footlambert: The uniform brightness of a surface, as observed from a given direction, which reflects *one lumen per square foot*. The average brightness of any reflecting surface, in footlamberts, is the product of the illumination in footcandles by the reflection factor of the surface at the particular angle of observation.

Lamp: The light source employed.

Luminaire: A complete lighting device which directs, diffuses, or modifies the light given out by the illumination source. It consists of a light source and all necessary mechanical, electrical, and decorative parts.

Lighting unit: The assembly of pole or post with bracket and luminaire.

Mounting height: The vertical distance between the roadway surface and the center of the light source in the luminaire.

Spacing: The distance in feet between successive lighting units measured along the centerline of the street.

FUNDAMENTAL FACTORS AFFECTING THE ABILITY TO SEE[24,25]

Discernment by Silhouette

Where the brightness of objects is less than that of the background, discernment is principally by *silhouette*. To enhance discernment by silhouette, brightness of the pavement, and uniformity of brightness along and across the full width of the roadway, is essential.[26] The amount of light reflected, or brightness of the pavement surface, depends upon the *intensity* and *angle* of incident light from the luminaire, pavement *specularity*, and *reflectance* at typical angles of view.

The results of studies of the light-reflecting characteristics of representative pavement surfaces have shown the following:

1. A light tone of pavement surface contributes importantly to the visibility of the pavement itself, and also to discernment of objects by silhouette. The reflectance varies from about 3 to 20 per cent for different types of pavement surfaces. New or clean portland cement concrete surfaces have a reflectance of about 20 per

cent, and asphaltic surfaces range between 3 to 10 per cent, with close-graded asphaltic concrete surfaces at the upper range.[27-29]

2. When pavements are wet, they have a high specular component (mirror reflection), and visibility is greatly enhanced by locating light sources several feet out over the pavement and along the outside of curves.[30]

Discernment by Reverse Silhouette

Where object brightness is greater than that of the immediate background, discernment is by *reverse silhouette.* Typical examples are objects on or adjacent to the roadway, projections above the pavement surface, such as channelizing islands and abutments, and portions of pedestrians and vehicles. Such discernment can be enhanced by increasing the illumination and the reflectance of an object on the sides facing toward the observer.

Discernment by Surface Detail

Where there is a high order of direct illumination on the side of the object toward the observer, discernment is by *surface detail.* The object is seen by variations in brightness or color over its own surface, without general contrast with a background. Under illumination provided for heavy vehicular and pedestrian traffic, discernment by surface detail is predominant. The night legibility of traffic signs depends upon the brightness of surface detail.

Size of Objects

The sizes of objects, and identifying detail which must be recognized, is an important factor in night visibility. Objects of significance to safety may range in size from a small brick to a large truck.[31]

Time Available for Recognition

Another important factor in night visibility is the time available for recognition of an object or traffic situation. The minimum time required for safety would be the PIEV time (for perception and reaction), and the minimum stopping distance time. Therefore, for a driver, the time available for recognition *decreases* with vehicle speed. The time required for recognition *increases* as the level of illumination is decreased, and as the range of brightness in the driver's field of view increases.[26]

Color of Light

Visibility of objects on or near a roadway is substantially the same throughout the range in light color produced by various lamps, when the comparison is on the basis of equal amounts of light similarly distributed.

For best visibility, a sign must be brighter against a night background. Color contrast enhances its brightness effect.[31]

Glare

The effects of glare reduce visibility (veiling glare) and cause ocular discomfort (discomfort glare). The degree of interference with visibility depends upon the illumination produced at the eyes by the glare sources, the incident angles which they make with the line of vision along the roadway, and the fluctuation due to movement, with respect to glare sources.

The interference with seeing caused by glare is diminished by reducing luminaire brightness. This may be accomplished by increasing the effective luminaire area and decreasing the intensity of light at angles higher than that required for pavement brightness.

All glare effects are decreased as the luminaires are removed from the driver's normal line of sight, or the center of his field of vision. Modern practice in luminaire mounting height, and the restriction of light from the luminaire at vertical and lateral angles, where interference with driver visibility is most significant, aids in this respect. The blinding glare in the driver's eyes due to the light source is greatest when the angle between the line of vision and the light source is 20 degrees or less. This effect is reduced by using higher mounting heights, which increase the angle. The relative veiling glare from street lights at different mounting heights (candlepower assumed constant) is shown in Table 22-2.[25] The relative blinding effect with a mounting height of 15 feet is over 10 times greater than that for a mounting height of 40 feet. Increasing the brightness of surroundings, such as the pavement and other areas forming a background in the field of view, reduces glare contrast, and improves visibility and the seeing comfort.[32,33]

Glare effects from approaching upper-beam vehicle headlamps present a serious hazard; however, it is essential to provide sufficient headlamp illumination for adequate sight distance. Various methods have been

TABLE 22-2. RELATIVE VEILING GLARE VS. LUMINAIRE MOUNTING HEIGHT

MOUNTING HEIGHT (MH) (ft)	*RELATIVE VEILING GLARE*
40	1.0
35	1.4
30	1.9
25	2.9
20	5.1
15	10.7

advocated for reducing glare from approaching headlamps. These methods vary from the use of wide medians and glare screens (expanded steel mesh has been used effectively as an anti-glare screen to reduce accident frequency[34]), to the use of polarized headlamps and tinted windshields and special lenses for drivers. However, tinted windshields cause a 30 per cent loss of seeing, which is of concern to night drivers and drivers deficient in red color vision.[35] For some reports of these problems and proposed solutions, consult the literature.[36-48]

To provide for better seeing with vehicle lighting, sealed-beam headlamps are used. Many vehicles are equipped with four headlamps. Improved visibility of vehicle signals is obtained by increasing signal intensity.[49-52] One of the hazards of night driving is encountering vehicles with only one lighted headlamp. Reflex-reflective materials included within a sealed-beam headlamp can help alleviate this problem.[16,53]

QUANTITY OF LIGHT REQUIRED

It would be most desirable to have the same visibility at night as exists in the daytime. Based upon traffic safety, a criterion can be established for the amount of light required. Studies of the ratio of night-to-day fatalities, under various footcandle values, indicate that this ratio is over 7 with 0 footcandles, and that the decrease in the ratio is very rapid until an illumination value of approximately 1.0 footcandle is attained. At this footcandle value, the ratio is slightly over 1. Increasing illumination beyond 1.0 footcandle provides little additional reduction in the ratio of night-to-day fatalities.[54]

Additional criteria that should be considered in determining the amount of light required, in keeping within economically practical limits, are the vehicular and pedestrian minimum requirements. In this regard, the recommendations of the American Standards Association (ASA)[25] provide excellent guides for the level of illumination to be used for different types of streets and roadways located in different areas. An earlier edition of the ASA standard[24] was published in 1953, and discussions of the comparison between the 1953 and 1964 editions are found in the literature.[55-57] Although significant changes have been made in the later edition, based on the findings of research, there is no departure from the basic principles set forth in the earlier edition. Since the pavement surface reflectance has a definite effect on the lighting effectiveness, primarily on streets carrying light traffic where discernment is essentially by silhouette, it also must be considered.[27-29]

Areas are classified as *downtown*, *intermediate*, and *outlying and rural*. Roadways are classified as *major*, *collector*, *local or minor*, and *expressways and freeways*. ASA recommendations for roadway illumination are shown in Table 22-3.

The average values recommended in the table represent average illumination on the roadway pavement when the illuminating source is at its lowest output and when the luminaire is in its dirtiest condition. In the use of these values, the following factors should be considered:

1. At-grade, intersecting roadway areas require higher illumination than that recommended in the table. The illumination in these areas should be at least equal to the sum of the illumination values provided on the roadways which form the intersection.
2. The lowest footcandle value at any point on the pavement should not be less than $\frac{1}{3}$ of the average value. The only exception to this requirement applies to residential roadways, where the lowest footcandle value at any point may be as low as $\frac{1}{6}$ of the average value.
3. It is assumed that good maintenance practices will be followed, as characterized by:
   a. Operation of light sources at their rated current or voltage.
   b. The regular replacement of depreciated lamps.
   c. The periodic cleaning of luminaires.

Lamp output at replacement time can be determined from the published data of lamp manufacturers. Tests

TABLE 22-3. RECOMMENDATIONS FOR AVERAGE HORIZONTAL FOOTCANDLES (Lumens per square foot)

ROADWAYS OTHER THAN EXPRESSWAYS OR FREEWAYS			
Roadway Classification	*Area Classification*		
	Downtown	*Intermediate*	*Outlying and Rural*
Major	2.0	1.2	0.9
Collector	1.2	0.9	0.6
Local or Minor	0.9	0.6	0.2 (residential)

EXPRESSWAYS AND FREEWAYS	
Classification	*Expressways*
Continuous Urban	1.4
Continuous Rural	1.0
Interchange Urban	2.0
Interchange Rural	1.4

(Source: *American Standard Practice for Roadway Lighting,* IES, 1964)

have shown that luminaires on extremely heavily traveled expressways may be reduced in light output by 20 per cent in a six-month period, and approximately 5 per cent for the same period on residential streets or outlying highways. Some downtown streets showed a 10 per cent reduction in light output at the end of six months.

LUMINAIRE LIGHT DISTRIBUTIONS[25,32,58]

To best accomplish the design objective of maximum visibility per dollar, proper distribution of light from the luminaire is essential. The object is to gather all light radiating from the lamp, direct it downward at the desired angle, and then shape a specified pattern upon the roadway. The distribution should be, therefore, downward, so as to utilize a high percentage of the lamp light in illuminating the pavement and adjacent area. The distribution selected should be that which will produce a practical maximum uniformity of pavement brightness, along and over the full width of the roadway, with a practical minimum of glare. The distribution from the luminaire should cover the pavement between curbs, and provide reasonably adequate lighting on adjacent areas extending 10 to 15 feet beyond the pavement edge. This illumination is essential for traffic signs, to detect the presence of movement of pedestrians, and to light the full height of parked vehicles.

Several methods have been developed to indicate the light distribution pattern from a luminaire.

Vertical Light Distributions

Classifications for luminaires are determined by the areas bounded by the following transverse roadway lines (TRL). A luminaire is classified as having one of the following distributions if its maximum candlepower point lies within the specified bounded zone.[59]

Short distribution:	from 1.0 MH TRL to 2.25 MH TRL
Medium distribution:	from 2.25 MH TRL to 3.75 MH TRL
Long distribution:	from 3.75 MH TRL to 6.0 MH TRL

Therefore, the maximum spacing for *short distribution* luminaires would be 4.5 times the mounting height (MH); for *medium*, 7.5 MH; for *long*, 12.0 MH.

To minimize the effects of glare, it is necessary to control the candlepower emitted in the upper portion of the beam above maximum candlepower. This vertical control of the candlepower distribution is divided into three categories:

1. *Cutoff:* When the candlepower beyond the limiting TRL does not numerically exceed 10 per cent of the rated lumens of the light source used. The limiting TRL are:

Short distribution:	3.75 MH
Medium distribution:	6.0 MH
Long distribution:	8.0 MH

2. *Semi-cutoff:* When the candlepower beyond the limiting TRL does not numerically exceed 30 per cent of the rated lumens of the light source used. The limiting TRL are the same as the cutoff category.
3. *Non-cutoff:* When there is no candlepower limitation in the zone above maximum candlepower.

On a semi-rural road with no bright surrounding backgrounds, a cutoff lighting installation provides better revealing power than a semi-cutoff one, because of the effects of reduced glare. The order of preference is (1) cutoff, (2) semi-cutoff, and (3) noncutoff.[60]

Lateral Light Distributions

These are divided into two groups, based on the location of the luminaire as related to the area to be lighted, that is, if the luminaire is mounted over or near the center of the area, or if the luminaire is mounted over or near the side of the area. Each group is subdivided into divisions with regard to the width of the area to be lighted, in terms of the MH ratio. The lateral width boundaries are defined by the one-half maximum candlepower line.[59]

Luminaires At or Near Center of Area. This group of lateral width classifications has similar light distributions on both the *house side* and the *street side* of the reference line, which is the intersection with the roadway of a vertical plane parallel to the curb and passing through the center of the luminaire.

1. *Type I:* This type has a two-way lateral distribution. The two principal light concentrations are in opposite directions along a roadway. The vertical plane of maximum candlepower is parallel with the curbline. The light distribution is similar on both sides of this vertical reference plane, and is generally applicable to a luminaire location near the center of a roadway. It may be used, at extended spacing, on narrow streets for low-intensity lighting, in residential areas, and on very minor traffic routes. Generally, the mounting height (MH) is approximately equal to the roadway width. Therefore, the roadway width should not exceed 2.0 times MH.

2. *Four-way Type I:* This unit has a distribution having four principal concentrations at lateral angles of approximately 90 degrees to one another, each with a width range as described above. This distribution is generally applicable to luminaires located over or near the center of a right-angle intersection.
3. *Type V.* These light distributions have a circular symmetry of candlepower which is essentially the same at all lateral angles. This distribution is intended for luminaire mounting at or near the center of a roadway, in the center islands of parkways, at intersections, and in business districts where it may be desirable to improve its general attractiveness.

Luminaires Near Side of Area. This group of lateral width classifications varies as to the width of the distribution range on the street side of the reference line.

1. *Type II*: This light distribution projects main beams forward into the street and is generally applicable to luminaires located at or near the side of relatively narrow streets of less than about 60 feet wide, or where the roadway width does not exceed about 1.75 MH. It can also be used on wide roadways in opposite arrangements.
2. *Four-way Type II:* In this unit light distributions have four principal light concentrations. The four main beams are bent forward into the street, which makes its installation applicable to luminaire locations near one corner of a right-angle intersection.
3. *Type III:* This unit is similar to Type II, but projects beams farther into the street. This distribution is intended for luminaire mounting at or near the side of medium-width streets (up to about 75 feet), where the width does not exceed about 2.75 MH.
4. *Type IV:* This is similar to Type III. This distribution is intended for luminaire mounting at or near the side of wide streets, where the width exceeds about 2.75 MH.

Photometric Data of Luminaire

The Illuminating Engineering Society has prepared *Guides to Testing Procedures*, which specify methods of measuring the photometric performance of luminaires and provide a form of presenting data. The photometric data sheet contains all the essential information required to compare the performance of luminaires, determine distribution type, and provide necessary information for executing computations. It contains the following:

1. *Candlepower distribution curves.* These are the vertical and lateral distribution curves that were discussed previously, and they are used for establishing luminaire type classifications.
2. *Isofootcandle diagrams.* These depict the manner in which illumination from a single unit, or from a number of units, is distributed on a horizontal plane. All points of equal illumination are connected which results in an *isolux* or *isofootcandle* line. The isolux diagram is made of a series of isolux lines which are similar to contour lines. These diagrams are very useful for determining the uniformity of illumination, proper lamp size, and proper mounting height.
3. *Utilization curves.* These are a measure of the quantities of light striking the horizontal plane in front of and behind the luminaire. They are measures of the actual efficiency of the luminaire. The utilization curve tells how much light falls on the roadway, but reveals nothing of the way in which this light is distributed. It must, therefore, be used in conjunction with an *isolux* diagram, or a distribution curve, in order to picture correctly the true performance of the luminaire. However, the average horizontal footcandles on the street can be determined from the utilization curve by dividing the effective lumens by the area covered by the luminaire.

   The symmetrical unit, such as Types I and V, requires only one curve to represent its distribution. In an asymmetrical unit, such as Types II, III and IV, one curve is required for illumination on the street side and one for illumination on the house side.[59]
4. *Isocandela diagram.* This is another method of presenting the candlepower distribution of the luminaire. The curves represent points of equal candlepower striking on a spherical surface around the luminaire. The points on the surface representing equal illumination are joined by isocandela lines. These lines make up the isocandela diagram and the information obtained therefrom is of the same type as obtained from the isolux diagram.

Examples of these curves and diagrams appear in the literature.[59]

DESIGN OF STREET AND HIGHWAY LIGHTING

Luminaire Mounting Heights

There are two criteria for determining a preferred luminaire mounting height. The first has to do with the previously mentioned relationship of glare to mounting height. A generally recognized means of minimizing the effect of glare on visibility is to raise the luminaire sufficiently above the street level to remove it from the normal line of vision. The second consideration is that of candlepower concentration and distribution.

As mounting heights are increased, proportionately

TABLE 22-4. RECOMMENDED MINIMUM LUMINAIRE MOUNTING HEIGHTS[25]

MAXIMUM LUMINAIRE (CANDLEPOWER)	MINIMUM MOUNTING HEIGHT (ft)		
	Cutoff	Semi-cutoff	Non-cutoff
Under 5,000	20	20	25
Over 10,000	20	25	30
Under 15,000	25	30	35
Over 15,000	30	35	40

(Source: *American Standard Practice for Roadway Lighting,* IES, 1964)

longer spacings between lighting units may be permissible. Wider spacings, of course, require larger lamps to maintain a given level of illumination. Lower mounting heights require closer spacing to preserve a proper distribution of light over the roadway.

Based upon consideration of these factors, the ASA recommends minimum mounting heights as shown in Table 22-4. Recent research has found that lighting systems with luminaires mounted at 40 to 50 feet (high mast) are more economical and effective than those mounted at 30 feet. The former provide safer and more aesthetic lighting.[60,61]

Luminaire Spacing

The spacing of street lighting units is often influenced by the location of existing utility poles, block lengths, and property lot lines. It is good practice to use larger lamps at reasonably long spacings and higher mounting heights than to use smaller lamps at more frequent intervals with lower mountings. This is an economic consideration; also it is usually in the interest of good lighting, provided the spacing-mounting height ratio is within the range of light distribution for which the luminaire is designed.

The use of the street, and the corresponding illuminating values assigned to it, often will influence the spacing between luminaires. On a residential street, it may be permissible to extend the spacing between luminaires so that the main beams barely meet between units. For a major street or business street, where the uniformity of illumination is relatively more important, it may be desirable to close the spacing to provide 50 to 100 per cent overlap of adjacent luminaires.[32,58]

Transverse Location of Luminaire

Luminaire distributions are most effective when suspended over the roadway. The transverse location is, of course, influenced by the classification of the distribution. Types I and V luminaires are intended for locations marking the centerline of the area to be illuminated, while Types II, III and IV luminaires are intended for mounting at or near the curbline. For the latter types of luminaire, it is recommended that the overhang, which is the horizontal distance from the curbline to a vertical axis through the lamp, be a minimum of 4 feet.

Typical Street Lighting Layouts

Typical arrangements of lighting units for roadways are shown in the ASA standards.[25] Additional typical illustrations are shown in the literature.[62]

For additional discussion on roadway lighting design, and the effectiveness of various lighting systems, consult the literature.[9,10,62-79]

A significant design is the one used by the Texas Highway Department. They use Type III 400-W mercury vapor luminaires mounted at 40 feet and spaced 200 feet apart, for roadways with two to three lanes in each direction, and Type III 1000-W luminaires at 50 feet and 300-foot spacing, for three to six lanes.[78]

Situations Requiring Special Considerations

Not all the roadway lighting problems can be fully analyzed and designed by only using the criteria already established. These design data may require modification to provide adequate visibility for certain special situations. These modifications for special situations are treated in Appendix A of ASA.[25] The special situations include: streets with heavy forestation, curves, hills, intersections, railroad grade crossings, alleys, bridges, overpasses, viaducts, plazas, traffic circles, underpasses, tunnels, and highway interchanges.[21,23,62,79-86]

To obtain the best result with lighting, with respect to seeing and general visibility, the following should be remembered:

1. Use the *short cutoff* restriction of the loss of visibility due to DVB (disability veiling brightness or veiling glare). This also improves visual comfort.
2. Adopt *higher mounting heights* to allow for proper spacing-to-mounting-height ratios, such as 3.5 MH, with fewer poles per mile, and to alleviate effects of DVB and discomfort glare.
3. Consider the CIE (international) recommendations for levels of average and minimum visibility to be provided for the handicapped night driver.
4. Consider the ASA recommended practice levels as the minimum visibility for the normal driver.
5. Establish reasonable spacing-to-MH ratio overlapping of the light distributions to provide adequate minimum visibility.

6. Use the new lamps developed to provide higher effectiveness, such as the high-pressure-sodium lamps.[58]

SELECTING THE LAMP TYPE FOR LIGHT SOURCE

The following types of lamps are available for street and highway lighting:

Incandescent or Filament Lamp

This has been, for some time, the most popular light source for roadway lighting. Its compact filament makes it easy to obtain excellent control and high utilization. It has the lowest first cost, and does not require auxiliary equipment or complex circuitry for dependable operation. However, its overall efficiency, economically, is *least* of all the lamps available, with its higher annual lamp replacement and energy costs. These factors will undoubtedly influence its future use for street lighting.

Mercury-Vapor Lamp

This is being used to an increasing extent. Its high efficiency is about twice that of the incandescent lamp. Therefore, the mercury lamp generates about twice as much light as does the incandescent lamp of comparable wattage. It also has a much longer life, which contributes to its advantage in an overall lighting cost analysis. The characteristic bluish-white color of the mercury-vapor lamp also adds to its desirability because of its aesthetic value. Where white light is preferred, the *fluorescent* mercury lamp is recommended. Mercury-vapor lamps require auxiliary equipment and more complex circuitry for dependable operation.

Fluorescent Lamp

This is becoming increasingly more popular for street lighting, particularly in downtown installations. It emits a white light, provides a modern appearance and is approximately $2\frac{1}{2}$ times as efficient as the incandescent lamp. Its white light produces the most natural light to the human eye, and it is the most efficient of the lamps available with regard to energy requirements. However, it does not compare favorably with the incandescent and mercury lamps from the standpoint of lamp size and auxiliary equipment required for protecting the lamp in cold weather. Both fluorescent and mercury-vapor lamps require a transformer, which provides the high starting voltages necessary for lighting the lamps.[81,88]

Sodium-Vapor Lamp

This unit has many disadvantages which outweigh its high efficiency. Yellow light does not penetrate fog much farther than does white light, as was originally claimed, and the sodium light distorts the appearance of complexions to such an extent that it is seldom to be recommended where there are pedestrians.

The readily interpreted yellow light from the sodium lamp has been used to advantage at hazardous locations. However, the 400-W high-pressure sodium-vapor lamps have a "creamy" color and provide high light output with similar color rendering qualities to those of the filament lamps, making them suitable for use in important shopping areas.[58,89]

Economic Lamp Comparison

Lumen output, mean life, and cost data, for various types of lamps, are treated in the literature.[59,82,90,92] It is estimated that over 7 per cent of the occupants of motor vehicles are killed by collision with luminaire supports or poles. Breakaway columns or poles have been developed which could be used to reduce such hazards, but they are not to be used in busy shopping streets.[86,91]

FORMULA FOR COMPUTATION OF ROADWAY ILLUMINATION

As previously mentioned, the values from the utilization curves can be used to obtain the average footcandle (lumens per square foot) value for the entire roadway. The following formula is used to obtain this value:[25,54]

$$\text{Average footcandles (lumens per square foot)} = \frac{(\text{lamp lumens}) \times (\text{coefficient of utilization}) \times (\text{luminaire maintenance factor})}{(\text{spacing between luminaires in feet}) \times (\text{width of roadway in feet})} \qquad (22\text{-}1)$$

Lamp lumens may be any one of the following:

1. Initial lamp lumen rating, if answer is desired in terms of initial illumination.
2. Mean lamp lumen rating, if answer is desired in terms of average illumination during rated life of lamp.
3. Lamp lumen rating at replacement time. When group lamp replacements are made, an appropriate value would be 80 per cent of the rated initial value. Lamp lumen rating at replacement time is available from the lamp manufacturer.

It is a fact that at least 50 per cent of the effectiveness and light output can be lost if a highway lighting system is not properly maintained. Contributing are two principal factors:

1. *Lamp lumen depreciation.* This is a characteristic of light sources. A knowledge of the lumen depreciation curve, and life ratings of the particular lamp being used, is essential. With this information, a relamping plan can be established. The practice of random replacement of lamps as they fail is followed in some instances. However, on highways where the distances to be traveled by service crews may be great, the economics of group replacement of lamps at the end of their useful life is favorable.
2. *Dirt, dust and corrosion.* This is applicable to the luminaire and its components. It would be unrealistic to expect to eliminate the need to clean this equipment.

Correlation of relamping and cleaning schedules is both common and logical. However, even with proper replacement and cleaning schedules, it is good practice to use a *luminaire maintenance factor.* Refer to the section on Quantity of Light Required, to estimate the appropriate maintenance factor.

Spacing between luminaires. Corresponds to longitudinal distance between luminaires, if spaced in a staggered or one-side arrangement, or $\frac{1}{2}$ the longitudinal distance between luminaires, if the luminaires are arranged in opposite spacing.

IV Applications of Control Measures

23 The Intersection

The existing system of streets and highways functions with a most complex set of interrelationships. The operating characteristics of this system are functionally dependent upon the number and kinds of users that require service. This is particularly true at those points where elements of the system come together—the intersection. When traffic volumes increase or change their nature, e.g., an increase in the number of trucks and/or buses, it is usually the intersection that first proves itself incapable of serving the added or altered demand. This is most frequently illustrated by increased congestion and delays, and a worsening accident experience. That this occurs is logical, since the intersection is a concentrated conflict point which generally functions at a lower capacity and level of service than any of the road segments it serves.

When it becomes evident that an intersection has, in fact, become a bottleneck to the satisfactory operation of the adjacent highway system, a detailed study of its physical and operational capabilities and limitations must be made. Based upon that study, appropriate decisions can be made as to the nature and degree of improvement possible or justified.

When dealing with such intersection or other limited area modifications, it is common to refer to them as "point" or "spot" improvements. *Spot* improvements fall into two general classes: minor (limited), and major (extensive). Minor spot improvements usually are those that can be accomplished within the existing right-of-way, and at relatively low cost. Examples of such minor improvements include the installation of traffic control devices, channelization, right-and/or left-turn lanes, parking controls, and roadway illumination. Major spot improvements generally require the purchase of additional right-of-way, and the expenditure of sizable sums of money. Examples of major spot improvements include vehicular and pedestrian grade separations, intersection reconstruction, including widening and straightening of approaches, and purchase of right-of-way to improve sight distance.

INTERSECTION CONTROLS

General Comments

The intersection is a critical area in the effective use of streets and highways. It is the focal point of conflicts and congestion, since it is common to two or more roadways. As the frequency and severity of intersection conflicts increase, regulation and control become more necessary.

Some intersections require controls to designate right-of-way, decrease approach speeds, restrict turning movements, designate lane use, and channelize vehicular and pedestrian traffic.

Proper intersection control is employed to accomplish the following objectives:

1. Increase in intersection capacity
2. Reduction and prevention of accidents
3. Creation and protection of major streets

Increase in Intersection Capacity. Since conflicts of movements occur at intersections, the capacity of the intersection is normally less than that of the approaches. Intersection controls may be applied in a variety of ways to increase intersection capacity. For example, parking restrictions in the vicinity of the intersection provide for more lanes for moving traffic; left-turn restrictions reduce conflicts, with their attendant turbulence in operations; the use of traffic signals to systematically assign the right-of-way to conflicting movements can reduce the effects of these conflicts, thereby increasing capacity.

Reduction and Prevention of Accidents. Crossing, converging, and diverging movements of traffic streams at intersections significantly increase the hazard of accidents. Over one-third of all accidents and traffic fatalities occur at intersections in urban areas. Nearly 25 per cent of all accidents, and about 15 per cent of traffic fatalities, occur at intersections in rural areas.[1] Proper control of speeds, placement of vehicles, and timing of movements can materially aid in reducing many types of intersection accidents, among which are: right-angle, head-on, sideswiping, and pedestrian-vehicular collisions.

Creation and Protection of Major Streets. Intersection controls provide for more continuous movement along major streets at greater speeds and increased safety. The creation of major streets encourages through traffic to use them. This provides greater safety for users of adjacent minor streets by reducing or eliminating the presence of through vehicles on these streets.

The primary function of a through route is to concentrate and facilitate traffic on one route, attracting this type of traffic away from other adjacent traffic ways, which are less suited to high volumes of fast through traffic.[2]

Some *advantages* of through routes are:

1. Delays to through traffic are reduced.
2. Traffic control on adjacent streets is better suited to local traffic, thereby reducing delays to this type of traffic.
3. Accident frequency is reduced on all routes.
4. The capacity of the route is increased, due to more efficient control.

Disadvantages of through routes include:

1. Increased speed of through traffic may result in higher frequency of more severe traffic accidents.
2. Vehicular and pedestrian traffic crossing the through route may be delayed for long periods during peak traffic-flow periods.

No definite volume warrant has been advocated for through route designation. However, a suggested recommendation is that a street is not considered as warranting through-route treatment unless the two-way volume of traffic during the major part of a normal weekday averages at least 300 vehicles per hour, and during the peak hour at least 450 vehicles per hour, over the length of street to be designated.

Types of Intersection Control

The various methods that can be used to control conflicting movements at intersections include the following, presented in order of ascending degree of control and restriction:

Basic Right-of-Way Rule. The basic right-of-way rules for the restriction and control of vehicular and pedestrian traffic at uncontrolled intersections are:

1. The driver of a vehicle approaching an intersection shall yield the right-of-way to a vehicle which has entered the intersection from another approach.
2. When two vehicles enter an intersection from different approaches at approximately the same time, the driver of the vehicle on the left shall yield the right-of-way to the vehicle on the right.
3. The driver of a vehicle intending to turn to the left shall yield the right-of-way to any approaching vehicle from the opposite direction.
4. At unsignalized intersections, the pedestrian has the right-of-way after entering a marked or unmarked crosswalk.

Certainly this type of control is satisfactory only at minor intersections which have adequate sight distance.

YIELD Sign. Every driver approaching a YIELD sign is required to slow down and yield the right-of-way to any vehicle in the intersection, or approaching so closely as to constitute an immediate hazard. Vehicles controlled by a YIELD sign are only required to stop when necessary to avoid interference with other traffic that is given the right-of-way. The YIELD sign was first used in this country in 1951. It serves as an intermediate form of control between the basic right-of-way rule and the STOP sign.

The National Manual on Uniform Traffic Control Devices (MUTCD)[3] adopted, for the first time, warrants for the use of YIELD signs in its 1961 edition. These warrants are essentially unchanged in the current MUTCD, and are discussed in detail therein. The most significant factor concerning the proper application of YIELD signs is the safe approach speed of traffic on the minor (controlled) road. When it is greater than 10 miles per hour, the minor approach will normally be controlled by a YIELD sign; if it is less than 10 miles per hour, a STOP sign is generally used. Some state manuals differ slightly from the National MUTCD. For example, New York State uses 8 miles per hour as the dividing safe approach speed for selecting a YIELD or STOP sign.

Additional warrants have been suggested for the use of YIELD signs.[4] They are based on total approach volumes, restrictive sight distance, and accident hazard.

YIELD signs should not ordinarily be placed to control a major traffic flow. They should not ordinarily be placed on the approaches of more than one of the intersecting roadways. Similarly, they should not be used where STOP signs are on other approaches to the intersection.

Studies on the effectiveness of, and experiences with, YIELD signs are found in the literature.[5-8]

Two-Way Stop. Every driver approaching a STOP sign is required to come to a complete stop before determining whether or not it is safe to enter the intersection. As such, the STOP sign causes substantial inconvenience to motorists. It is important that this form of control only be used where warranted, as indicated in the MUTCD.

STOP sign control is generally used on a minor road, where it intersects a major road. It may also be an effective form of control at an unsignalized intersection in a signalized area. Such factors as high approach speed, restricted sight distance, and serious accident record may indicate the need for STOP sign control.

It has been suggested that a STOP sign is warranted if an average day contains eight or more hours during which the volumes of the main street are such that at least 50 per cent of the minor street volume is delayed.[9] This warrant is complex in its application, since it is necessary to determine the distribution of time gaps on the main street which would permit the safe crossing of a vehicle from the minor street, and hourly volume counts are required on both streets for a substantial part of the day.

Multi-Way Stop. This form of control is considered useful as a safety measure at some locations. It has the disadvantage that all vehicles approaching the intersection are required to come to a complete stop before proceeding through the intersection. Once the vehicle is stopped, the driver's decision to proceed is then governed by the right-of-way rule. Warrants for the application of multi-way stop installations are specified in the MUTCD. Multi-way stops may be used as an interim measure at locations where traffic signals are warranted, but cannot be quickly installed.

Warrants for this kind of control begin to approach, in form, those for traffic signals in that they specify minimum traffic volumes and combined pedestrian and traffic volumes which would justify such control. The more conventional two-way stop does not have such quantitative warrants for use.

A number of experiences with four-way stop control are reported in the literature.[10-17]

Traffic Signals. Next in the scale of increasing restrictiveness is the widely used traffic signal, which alternately assigns the use of the intersection first to one stream of traffic, then to the other. Right-of-way is allocated by time separation of conflicts.

Because the signal systematically assigns the right-of-way to competing movements, it will impose considerable delay on vehicles on all approaches to the intersection. It is most important, therefore, to be sure that a traffic signal is justified before one is recommended. The MUTCD spells out in some detail eight different warrants, at least one of which should be met before a traffic signal is installed. The warrants relate to:

1. Minimum vehicular volume
2. Interruption of continuous traffic
3. Minimum pedestrian volume
4. School crossing
5. Progressive movement
6. Accident experience
7. Systems
8. Combination of warrants

Each of these is discussed briefly below. For greater detail, consult the appropriate sections of the MUTCD.

Minimum Vehicular Volume. This warrant is intended to be applied where the volume of intersecting traffic is the primary reason for consideration of traffic signal control.

Interruption of Continuous Traffic. This warrant applies to operating conditions where the traffic volume on a major road or street is so heavy that cross traffic on the minor street suffers excessive delay or danger in entering onto, or crossing, the major street.

Minimum Pedestrian Volume. This warrant is based on a combination of vehicle and pedestrian volumes which are considered hazardous at a given intersection. Signals installed at isolated intersections under this warrant should be of the pedestrian-actuated type.

School Crossing. A traffic signal may be warranted at an established school crossing if a study shows the gaps in the vehicular traffic, in terms of frequency and size, are not sufficient to safely provide for street crossings by school children. In particular, if gaps of sufficient duration to allow for crossings without interruption of traffic flow occur on the average of less than one per minute, children may become impatient and endanger themselves by attempting to cross the street during an inadequate gap. Under such conditions, some form of traffic control is needed to create adequate gaps in the traffic stream.

Progressive Movement. This warrant is more directly concerned with a signal-controlled intersection as part of a system, rather than as an isolated element. For consistency of approach, however, it seems reasonable to discuss all signal warrants at the same time. Where signals are widely separated, it may be impossible to maintain coordinated control, since vehicle platoons spread out over long distances. Under such circumstances, it may be justifiable to place an intermediate signal at a location where it would otherwise be unwarranted to maintain proper coordination of platoon movements. This topic is discussed in detail in Chapter 24.

Accident Experience. This warrant comes into play when an adequate trial of less restrictive measures, with satisfactory enforcement and observance, has failed to reduce the accident frequency. Under these conditions, the occurrence of a number of accidents involving personal injury or property damage in a twelve-month period may justify the installation of a signal.

Systems. As with street progressions, signal installations under this warrant are designed to encourage concentration and organization of traffic. Where the progressive warrant was concerned with vehicular movement along a street or arterial, the systems warrant is oriented toward benefiting network flow. This subject is discussed in detail in Chapter 24.

Combination of Warrants. In some cases, no single MUTCD warrant may be totally satisfied. If, however, a condition exists where any combination of the first three warrants is satisfied to within 80 per cent of the specified values, a signal may be justified. It is recommended, however, that other remedial measures, which incur less delay to traffic, be tried prior to installation of signals.

Police Officer Control. Control of intersection movements by a police officer is the most expensive method of allocating right-of-way. It may be justified, however, during periods of breakdown at intersections normally controlled by signals, to provide school crossing protection, at locations having brief periods of unusually high volumes of vehicular and/or pedestrian traffic leaving or entering public gathering places (such as race tracks, stadiums, other recreational locations, and shopping centers), and during peak periods of traffic at detour locations.

DEFINITIONS

Major street: At an intersection, the roadway approach or approaches normally carrying the major volume of vehicular traffic.

Minor street: The roadway approach or approaches normally carrying the minor volume of vehicular traffic.

Signal indication: The illumination of a traffic signal lens or a combination of several lenses at the same time.

Time cycle: The number of seconds required for one complete revolution of the timing dial or for one complete sequence of signal indications.

Interval: Any one of the several divisions of the time cycle during which signal indications do not change.

Interval sequence: A pre-determined consecutive order of appearance of signal indications during successive intervals within a time cycle.

Traffic phase (traffic movement): A part of the time cycle allocated to any traffic movement receiving the right-of-way, or to any combination of traffic movements receiving the right-of-way simultaneously during one or more intervals. A traffic movement can signify a vehicular movement alone, a pedestrian movement alone, or a combination of vehicular and pedestrian movements. The sum of all traffic phases is equal to the time cycle.

Clearance interval: The time of display of the signal indication (generally yellow) following the green interval. It is the time allocated for vehicles to clear the intersection after the green interval only.

All-red interval: The time of display of a red indication for all entering traffic. It is sometimes used following a clearance interval to permit vehicles or pedestrians to clear an excessively large intersection before opposing vehicles receive a green indication. It may also be used to create an exclusive pedestrian phase.

SIGNALIZATION OF ISOLATED INTERSECTIONS

Objectives of Signal Timing

The major objectives of a signal timing plan are the minimization of delay and congestion at intersections and within block lengths and series of block lengths, and the increase in safety for all road users. Full utility of traffic control signals is realized only when they are operated so as to satisfy, as nearly as possible, actual traffic requirements.

In general, *short cycle lengths* are desirable, because the delay to standing vehicles is reduced. Cycle lengths ordinarily fall between 30 and 120 seconds, but good practice dictates the selection of a total time cycle in the range of from 35 to 50 seconds for a simple right-angled intersection where the intersection roadways are of average width (30 to 40 feet) and traffic volumes are not extremely heavy. Where intersecting streets are wider, necessitating longer pedestrian crossing time, or volumes extremely heavy, or turning interferences substantial, the cycle will be between 45 and 60 seconds long. Three-street intersection cycles, or three-phase operation, will range from 55 to 80 seconds.

Heavier volumes require longer cycle lengths because the sum of the "Go," or green, intervals must be greater to gain sufficient capacity. The sum of "Go" intervals is a higher percentage of a cycle length with longer cycles since clearance times remain essentially fixed.

A number of excellent reports dealing with optimizing signal operations have been published.[18-21]

Timing a Pre-timed Traffic Signal

Timing of a pre-timed signal requires consideration of several aspects. They are discussed in the following paragraphs.

Traffic Phases. The number of *phases* required for the proper and efficient operation of a signalized intersection varies with the composition and direction of traffic flows, as well as with the number of intersection approaches and the general intersection layout.

The most commonly installed traffic control signal will operate on a two-phase cycle, in which the right-of-way is alternately assigned to each of the two cross movements. Intersections having a large and concentrated volume of left turns, or unusually heavy pedestrian movements, and intersections having more than four approaches for entering traffic, may require a division of the time cycle into more than two phases, to eliminate conflicts between vehicles or between vehicles and pedestrians.

The division of a time cycle into more than two phases should be avoided, where possible, since each additional phase lengthens the overall cycle, thereby increasing delay.

Vehicle Clearance Interval. The purpose of the *amber signal indication* following each green interval is to warn moving traffic facing the signals to come to a stop and, if possible, to do so with safety. It should provide enough time for vehicles to clear the intersection before cross traffic starts to move. Theoretically it should be long enough to permit a vehicle to travel, at normal intersection approach speed, a distance equal to the cross-street width between curbs plus the safe stopping distance. This is necessary because if the approaching vehicle is a few feet less than the safe stopping distance back from the intersection when the signal changes, time should be allowed for the vehicle to travel the stopping distance plus the street width before the cross-street traffic receives the green indication.

If the clearance interval is too long, it may be used as part of the green interval, and thus defeat its own purpose. If it is too short, it may constitute a hazard and increase rear-end collisions. At most urban intersections, a clearance interval of 3 seconds produces good results. Where speeds are high, or streets are exceptionally wide, 4 to 6-second clearance intervals may be warranted.

The *Traffic Engineering Handbook*[2] recommends the use of the following formula that incorporates a safe stopping distance. This formula is derived in the literature.[22]

$$Y = t + \frac{1}{2}\frac{v}{a} + \frac{(w + l)}{v} \tag{23-1}$$

where
- Y = clearance interval, in seconds
- t = perception-reaction time, in seconds, suggested value = 1.0
- v = approach speed of clearing vehicle, in feet per second
- a = deceleration rate of clearing vehicle, in (feet per second^2), suggested value = 15.
- w = intersection width (curb to curb), in feet
- l = length of vehicle, in feet, suggested value = 20.

This formula is based on a coefficient of skidding friction of about 0.5.

While clearance intervals in excess of 6 seconds may be necessary at very wide or complicated intersections, they are likely to cause impatience among drivers awaiting the signal change, which may result in starting through the intersection before the green indication appears. Under such conditions it has been found more satisfactory to supplement the normal amber interval with a short all-red interval of sufficient duration (normally 1 or 2 seconds), immediately after the amber interval, to permit clearance of the intersection before cross traffic is released.

Laws regarding the use of the amber interval vary widely between states. Most states require vehicular traffic to have completely cleared the intersection before the end of the amber phase, while some allow vehicles to enter the intersection legally during the amber phase. Others have laws falling between the two, e.g. "Traffic facing the yellow signal shall stop before entering the nearest crosswalk at the intersection, but if such stop cannot be made in safety, a vehicle may be driven cautiously through the intersection."[23]

In reported British practice nearly all of the amber is considered used as an extension of the green interval. The United States attitude as quoted from Page 406 of the Traffic Engineering Handbook[2] is that, "It is common to assume that no vehicles enter the intersection during the yellow intervals, although under maximum flow conditions this is obviously untrue." Studies (23, 93, 94) have been made of the clearance interval to determine how it is being used by drivers. A recent work (95) has collected this information from which the following regression equation was developed

$$y^1 = 1.32 + 0.050\,V$$

where y^1 = amber effectively used as though it were green, seconds
V = approach speed, MPH

Pedestrian Requirements. As a general principle of traffic signal timing, no vehicle "Go" interval should be shorter than the time required for the waiting group of pedestrians to get started and to cross to a point of safety, unless an exclusive pedestrian interval is also employed. Experiments with signal timing have shown that, insofar as vehicle movements are concerned, excellent efficiency can be attained under certain off-peak conditions with "Go" intervals as short as 15 seconds. Ordinarily, however, they must be somewhat longer to give the pedestrian a safe opportunity for crossing the roadway.

When the pedestrian crossing time runs concurrently with the vehicle "Go" period, which is the usual case, the total "Go" interval should be long enough to allow not less than 5 seconds during which it is indicated that pedestrians may start to cross (pedestrian starting time), and enough additional time to permit pedestrians who have entered the roadway to reach a place of safety, with the additional time provided by the vehicle clearance interval. Thus, if it takes 14 seconds for most pedestrians to cross the roadway or reach a point of safety, and if the vehicle clearance interval is 3 seconds, the total "Go" interval should be at least $5 + 14 - 3 = 16$ seconds. Pedestrian walking speed is usually assumed to average from about *3.5* to *4.0* feet per second in making allowance for crossing the street.

On a street where parking is prohibited in the vicinity of the intersection, it is necessary to provide sufficient time for pedestrians to traverse the entire distance from curb to curb. Where parking is allowed, however, it is only necessary to provide sufficient time for pedestrians to get out of the traveled way. This shortens the minimum required green interval by about 2 seconds.

At many signal installations in urban areas, it is not the traffic volumes but the pedestrian requirements that will determine the minimum intervals at an intersection. This is especially true on wider streets, where vehicular traffic can clear the intersection in less time than a pedestrian can cross. The pedestrian must be taken into consideration in most instances in urban areas.

At locations where pedestrian signals are used (WALK-DON'T WALK signals), a starting or reaction time of 7 seconds is recommended for determining total pedestrian crossing time, and the vehicle clearance time is subtracted from the total crossing time to evaluate the minimum "Go" interval.

On a street with a median at least 4 feet wide, it is only necessary to allow sufficient pedestrian clearance time on a given phase to clear the crossing from the curb to the median. However, if the signal has a pedestrian-actuated phase, it may be necessary to install an additional detector on the median.

Time Spacing of Vehicles Entering a Signalized Intersection. The minimum departure headway represents the time spacing between the instant that successive vehicles, waiting in a single lane, enter the intersection. When a continuous line of vehicles is waiting for a signal to turn green, there is an initial starting delay or time lag for the first few vehicles in the line, followed by approximately equal headways. Average minimum headway values vary according to the number of turning movements, character of traffic, parking conditions, and other factors. The results of a detailed study of departure headways was reported by Greenshields.[24]

TABLE 23-1. AVERAGE DEPARTURE HEADWAYS AT SIGNALIZED INTERSECTIONS

CAR IN LINE (NUMBER)	*GREEN TIME CONSUMED (MEDIAN VALUES)*
1	3.8
2	3.1
3	2.7
4	2.4
5	2.2
6 and over	2.1

(Source: Greenshields et al., *Traffic Performance at Urban Intersections*, Tech. Rep. 1, Yale BHT, 1947)

For vehicles waiting to start on the green indication, it was found that the entrance times shown in Table 23-1 were required per passenger car. Movements through the intersection did not include left-turning vehicles.

It was also found that each left-turning vehicle caused an added delay, not only to the lane from which they turn, but to the opposite lane as well. The study results showed that each left-turning movement, regardless of direction, consumes about 1.3 seconds additional time, and each bus or truck consumed about $1\frac{1}{2}$ as much time as a passenger car. Thus, assuming ten vehicles lined up at the intersection, the third one in line a bus, the eighth in line a truck, and two left-turning movements, the total time required for these ten vehicles to enter the intersection would be

$$(1.3 \times 2) + 3.8 + 3.1 + (2.7 \times 1.5) + 2.4 + 2.2 + (2 \times 2.1) + (2.1 \times 1.5) + (2 \times 2.1) = 29.7 \text{ seconds}$$

This calculation represents only the time for vehicles to enter the intersection. To this amount of green time *must be added the clearance interval*, to permit the last vehicle to clear the intersection before cross traffic receives the "Go" indication.

If it is assumed that all vehicles shown in Table 23-1 proceeded at an equal headway of 2.1 seconds, the time required for the first five vehicles to pass would have been $2.1 \times 5 = 10.5$ seconds, but it was actually 14.2 seconds. Therefore, the additional time of 3.7 seconds may be thought of as representing the starting delay which would be required at the beginning of each green phase. Considering the time requirements in this way is the same as saying that the green time required for n vehicles equals $2.1n + 3.7$. Strictly speaking, this is valid only for $n \geq 5$. For most signals, the green intervals are long enough that at least five vehicles can pass. It is reasonable to use $2.1n + 3.7$ to calculate the necessary green time in cycle length determination.

Other studies of vehicle time spacing[25-30] have reported variations in time spacing of vehicles from those first reported by Greenshields.

Ideally, when developing cycle length requirements, starting time data for the intersection under study should be collected. This is generally not feasible, unless specific conditions warrant the expenditure in time and effort to conduct such a study. In general, however, the values shown in the table are reasonable for most average situations, and allow for the development of acceptable cycle lengths.

Determination of a Satisfactory Cycle Length. Cycle gears for the control of pre-timed signal operation are furnished as standard equipment in multiples of 5 seconds for cycle lengths of up to 90 seconds, and in multiples of 10 seconds from 90 to 120 seconds. Odd cycle lengths require special gears, which are costly. For economic considerations, cycle lengths should be chosen to the nearest 5 or 10 seconds, depending upon cycle length.

Optimum cycle lengths should permit traffic to pass through the intersection with a minimum of delay. In determining the number of vehicles that are to be accommodated by a "Go" interval, it is necessary to not only consider those that accumulate during the red indication for that direction, but also those vehicles which arrive during the green indication for that direction (as well as the clearance intervals); hence, the *full cycle length must be used to determine the number of vehicles to be passed through a "Go" interval.*

In the timing of traffic signals it is extremely desirable to consider the variation in traffic flow during the peak hour. A peak 15-minute period is considered to be the shortest practical time interval to express this variation. The peak-hour factor (PHF) is defined as the ratio of the number of vehicles entering the intersection during the peak hour to four times the number of vehicles entering during the peak 15-minute period. If traffic flow is uniform during the entire peak hour, so that each 15-minute period carries the same amount of traffic, the $\text{PHF} = 4N/4N = 1.0$; at the other extreme, if all the traffic during the peak hour occurs during a single 15-minute period, the $\text{PHF} = N/4N = 0.25$. At most locations the $\text{PHF} = 0.85$. Therefore, traffic flow during the peak 15-minute period $= N/4 \times 0.85 = 0.294N$, say $0.3N$, where N is equal to the peak-hour flow.

Using the time spacings presented in Table 23-1, an approximate average headway per vehicle may be developed. For use in cycle-length determination, Figure 23-1 shows the average headway per vehicle as a function of the number of vehicles served. Figure 23-2 shows the average headway per vehicle as a function of the length of the green interval. From the latter figure, it can be

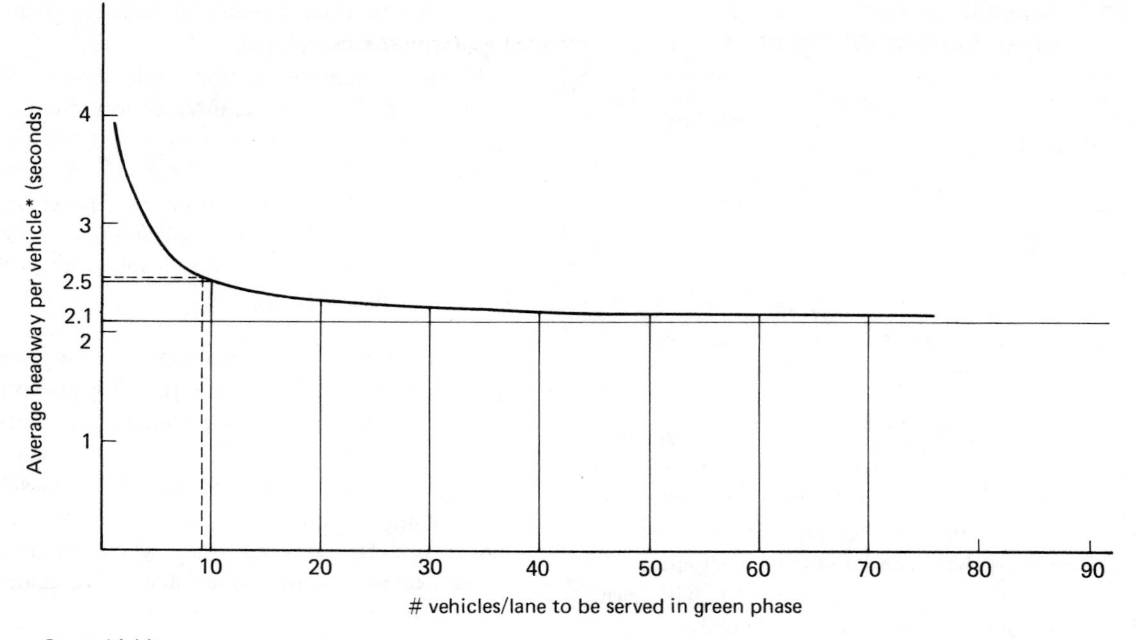

* From Greenshields

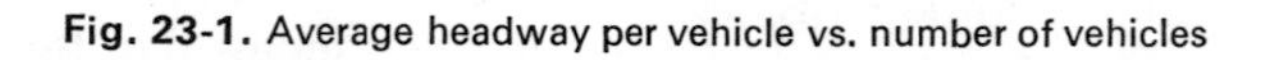

Fig. 23-1. Average headway per vehicle vs. number of vehicles

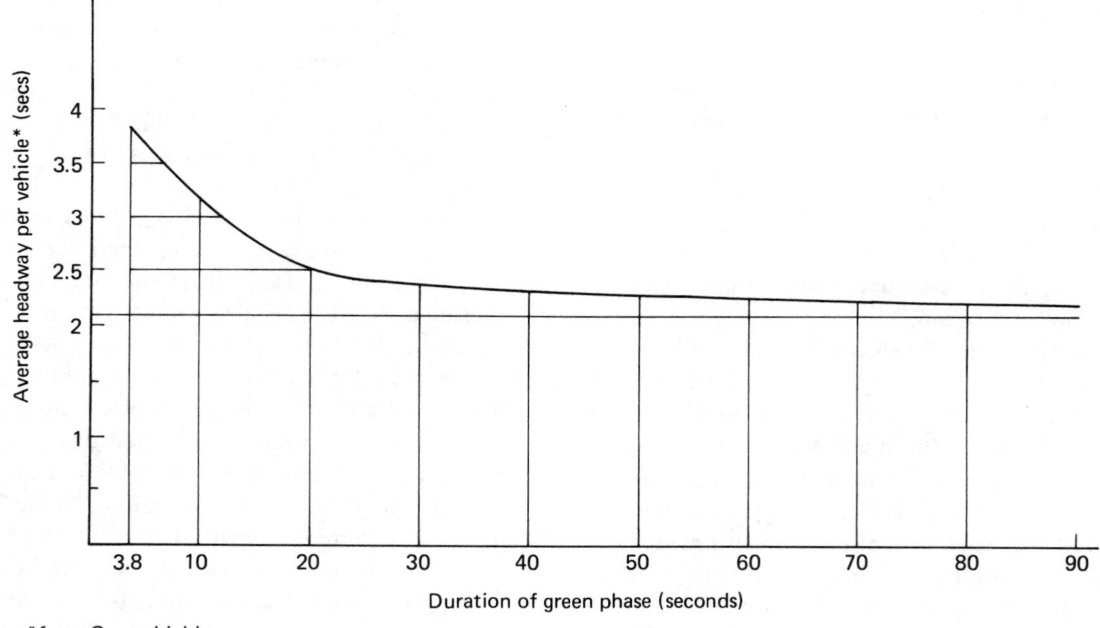

*from Greenshields

Fig. 23-2. Average headway per vehicle vs. length of green phase

seen that for green intervals of 20 to 30 seconds, the average headway per vehicle is 2.4 to 2.5 seconds. If approximately equal splits are assumed, this headway value is reasonable for two-phase signals with cycle lengths on the order of 40 to 60 seconds, and three-phase signals of cycle length 60 to 90 seconds. This is the most common cycle length range in which most signals are timed. If it is known that a much longer cycle length (or green phase) is required, then an appropriately shorter average headway (from Figure 23-2) should be used. For purposes of illustration, an average value of 2.5 seconds per passenger car will be used, unless otherwise noted.

As previously mentioned, left-turning vehicles and commercial vehicles require more green time than do straight-through passenger cars. One method of addressing this problem has been previously treated. That procedure requires the knowledge of the number of vehicles desiring service each cycle. This is a function of the cycle length, and cannot be precisely determined until an approximate cycle length has itself been determined.

The procedure used to account for the detrimental effect of commercial vehicles and turning movements on the time of start-up, or average headway, in the determination of cycle length, is to convert the demand volumes, given in mixed vehicles per hour with percentage left turns and commercial vehicles, to an equivalent volume in straight-through passenger cars. This procedure develops *passenger car equivalents* (*pce*) for all demand volumes. Under this procedure, each passenger car (non-left-turning) is equal to 1 *pce*. Buses and trucks are assumed to be equal to 1.5 *pce*, since their start-up times have been observed to be 1.5 times that of passenger cars.[24] Left-turning vehicles are equivalent to approximately 1.6 *pce*. This is based on the finding that each left-turn movement consumes about 1.3 seconds of additional time. Since the minimum departure headway is 2.1 seconds, it can be shown that

$$1 \text{ left turn} = \frac{2.1 + 1.3}{2.1} = 1.6\, pce$$

It should be noted that this 1.6 expansion factor only holds for those vehicles on the approach which is being used in the determination of minimum acceptable cycle length. Left turns from the opposing direction also require an additional 1.3 seconds. This is the same as considering each opposing left turn as if it were equal to $1.3/2.1 = 0.6\, pce$. If the amount of opposing left-turning traffic is not too great, it is reasonable to ignore it in the computation of an approximate cycle. It should, however, be considered when using the more exact vehicle time requirements in checking the selected cycle. This is the procedure which will be used herein. A very limited amount of data has been collected[19] which indicates that one right-turning vehicle equals approximately 1.4 *pce*. This approximate factor is highly dependent on the curb radius, approach speed, and number of pedestrians present who may conflict with right-turning vehicles. In general, where the number of right-turning vehicles is low, and where pedestrian-vehicle conflicts are minimal, it is common to ignore the extra time which may be required by right-turning vehicles. When pedestrian-right turn conflicts are substantial, approximate values, particularly when based on such limited field data, should not be relied upon. It is preferable to conduct a field study to determine more precisely the actual effect of these conflicts at the location under study.

Considering a simple intersection with four approaches and a two-phase signal cycle, the following procedure may be used to determine an approximate cycle length. It should be noted that the term "approximate" is appropriate. After splitting the cycle into its required phases, the cycle length will be checked against other requirements discussed herein, and adjusted when necessary.

Let N_1 = major street *critical lane volume*, the number of vehicles *in a single lane* entering the intersection on the major street during the peak hour. N_1 is the *largest* of all single-lane volumes of the two major street approaches.

N_2 = minor street *critical lane volume*, the number of vehicles entering the intersection in a *single lane* on the minor street in the same period as N_1. N_2 is the *largest* of all the single-lane volumes of the two minor street approaches.

C = cycle length, in seconds

S_1 = approximate average headway between vehicles entering the intersection among N_1

S_2 = approximate average headway between vehicles entering the intersection among N_2

Y_1 = vehicle clearance interval (amber), in seconds, for the N_1 direction

Y_2 = vehicle clearance interval (amber), in seconds, for the N_2 direction

K = number of signal cycles for a 15-minute period

Then, the *total time required to pass all vehicles* through the intersection *during the peak 15-minute period* is:

$$T = \frac{N_1S_1 + N_2S_2}{4(PHF)} \tag{23-2}$$

And, the total time required for *clearance intervals* is:

$$K(Y_1 + Y_2) \tag{23-3}$$

The sum of these must be *less than or equal to 900 seconds*, the total amount of time available in a 15-minute period:

$$T + K(Y_1 + Y_2) \leq 900 \text{ seconds} \tag{23-4}$$

Thus, as a limiting condition,

$$K = \frac{900 - T}{Y_1 + Y_2} \tag{23-5}$$

But

$$K = \frac{900}{C} \tag{23-6}$$

By substituting Equations (23-2) and (23-6) in Equation (23-5), and solving for C,

$$C_{\min} = \frac{Y_1 + Y_2}{1 - (N_1S_1 + N_2S_2/3{,}600(PHF))} \tag{23-7}$$

The cycle length determined by this relation is intended to maintain *undersaturated operation* during a "typical," or average, peak load, determined by the average peak-hour volume and the peak-hour factor (PHF).

At more complicated intersections, as, for example, one having five or more approaches, or one with a significant left-turning movement, three or more phases of the cycle may be necessary. The above equation can be modified to take this into account. For example, for a three-phase cycle, the equation becomes

$$C_{\min} = \frac{Y_1 + Y_2 + Y_3}{1 - [(N_1S_1 + N_2S_2 + N_3S_3)/3{,}600(PHF)]} \tag{23-8}$$

In general, the expression becomes:

$$C_{\min} = \frac{\sum_{i=1}^{M} Y_i}{1 - \left[\sum_{i=1}^{m} N_iS_i/3{,}600(PHF)\right]},$$

$$M = \text{number of phases} \tag{23-9}$$

The basic expression can also be modified to meet other special requirements. For instance, if an *exclusive pedestrian phase* is to be provided in the cycle, this can be included as follows:

$$C = \frac{Y_1 + Y_2 + P + Y_P}{1 - 0.000333(N_1S_1 + N_2S_2)} \tag{23-10}$$

Where P is the length of the pedestrian "Walk" interval, in seconds, and Y_P is the length of the pedestrian clearance interval.

When the "Walk" interval is shown simultaneously with the green it does not affect the computation for the cycle length, and it does not constitute an additional phase.

Webster[18] has developed an approximate formula for determining the optimum cycle length in terms of minimum delay:

$$C_0 = \frac{1.5L + 5}{1 - \sum_{i=1}^{n} y_i} \tag{23-11}$$

where
- C_o = optimum cycle length, in seconds
- $L = nl + R$ (assumes amber time as green time)
- n = number of phases
- l = average lost time per phase (usually starting delay only), excluding any all-red intervals.
- R = time during each cycle when all signals display red simultaneously
- $y_i = \dfrac{\text{design flow}}{\text{saturation flow}}$, ith phase

The above formulation is a simple approximation of a more complicated relationship that was developed as part of the study mentioned. In this method, all, or nearly all, of the amber time is assumed to be usable for vehicle movement. As such, it generally yields cycle lengths that are somewhat shorter than those of the first method.

Webster's delay analysis, using simulation techniques, was quite comprehensive. The analyses conducted demonstrated a very important fact: Delay is not significantly increased by variation of cycle lengths around the optimum value, as long as they remain in the range $0.75C_o$ to $1.50C_o$. This lack of sensitivity to cycle length over a reasonable range of values (e.g., if $C_o = 60$, the range is 45–90), indicates that it is not necessary to be overly concerned about the need to make simplifying approximations in the determination of *pce* and average headway values.

Division of the Cycle Length. If, during the peak hour traffic flow, the average time spacing (headway) between vehicles on the major and minor approaches is approximately the same, the division of the available green time in the cycle should be made proportional to the *critical lane volumes* on the intersecting streets. For example, consider two intersecting streets A and B, with critical lane volumes N_a and N_b respectively. Assume the departure headway for street A, S_a, is approximately equal to the departure headway for street B, S_b. The total available green time is

$$G_a + G_b = C - (Y_a + Y_b) \tag{23-12}$$

Dividing the green time in proportion to the relative critical lane volumes,

$$G_a = \left(\frac{N_a}{N_a + N_b}\right)[C - (Y_a + Y_b)] \tag{23-13}$$

Simplifying,

$$G_a = \frac{C - (Y_a + Y_b)}{1 + (N_b/N_a)} \tag{23-14}$$

and

$$G_b = C - (Y_a + Y_b) - G_a \tag{23-15}$$

If, however, S_a does not approximately equal S_b, which might occur, for example, if one street were on a significant grade, the available green time should be split in the ratio of the product of the critical lane volume and its average headway. That is to say,

$$\frac{G_a}{G_b} = \frac{N_a S_a}{N_b S_b} \tag{23-16}$$

A similar proportioning of green time may be undertaken for a three-phase (or any multi-phase) cycle as well.

In any case, a time apportionment of this kind ensures that each approach receives essentially "equal" service. The underlying assumption here is that the arrival patterns of traffic on both the major and minor streets are essentially the same. If it is known that one street has a significantly different arrival pattern from the other, even though they have the same average headway time, it may be more efficient to allocate time to the phases by a method other than proportional allocation. Barring such conditions, however, proportional allocation has the obvious benefit of serving everyone "equally."

It should be emphasized at this time that regardless of the results obtained by the above, or other, procedures for division of the cycle length, other considerations, such as the time required for pedestrian crossings, and physical limitations at the intersection, must all be satisfied.

Application of a Cycle Failure Specification. One may define a cycle failure in one of two ways: (1) vehicles

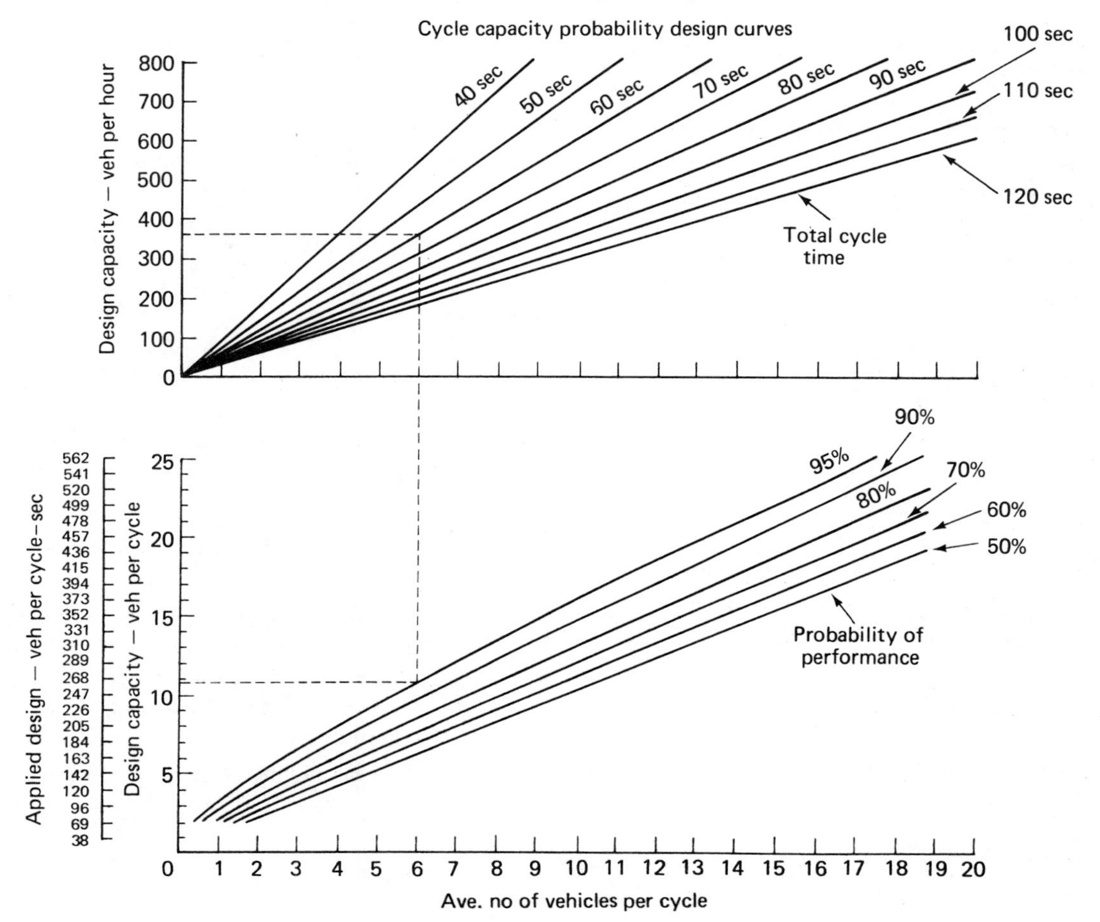

Fig. 23-3. Probability of cycle performance (*Source:* Traffic Engineering Handbook, *ITE, 1965, Figure 11.7*)

arriving in the last cycle time are not cleared in the current green on *at least* one leg, and (2) vehicles so arriving are not cleared on the *critical* leg. A design specification of a 5 per cent or lower probability of cycle failure may be made, and the cycle length C determined by cut-and-try techniques. Figure 23-3 depicts percentage performance versus cycle length, critical volumes. This figure is developed using an assumed Poisson distribution for arriving vehicles.

More often, cycle failure is addressed indirectly by requiring that the peak 15-minute volume for each leg be accommodated in the green time available to that leg in the peak 15 minutes; that is, to require

$$\frac{N_i S_i}{4(PHF)} \leq \frac{900}{C} G_i \tag{23-17}$$

for all approaches.

Sample Problems in Pre-timed Signalization

The following problems illustrate the use of the methodologies discussed in the previous sections.

Problem 1:

This problem deals with a *simple two-phase intersection* (Figure 23-4). The following data is given:

3.5 feet per second pedestrian walking speed
85th percentile approach speed = 25 miles per hour, all approaches
Parking permitted on all approaches
10 per cent trucks
15 per cent left turns
Right turns not considered separately
PHF: 0.85
Volumes as shown

Design a two-phase signalization plan, if possible.

Solution

Computation of critical lane volumes:

By inspection, the critical approach volumes are:

Fig. 23-4.

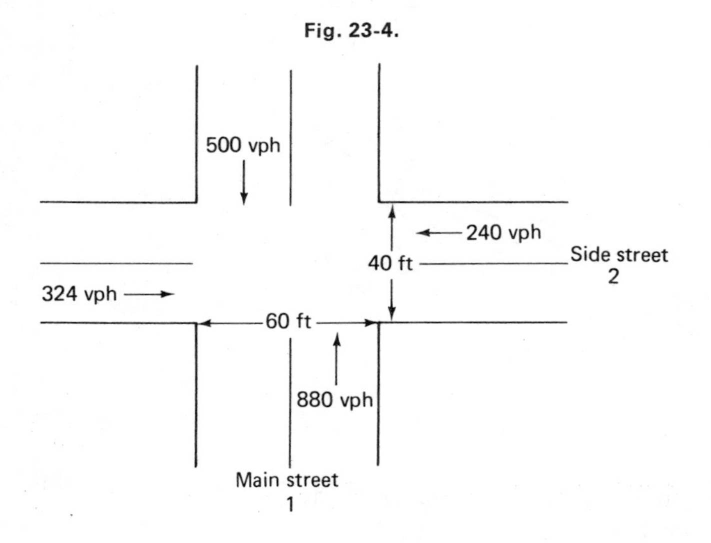

Main street: 880 vehicles per hour
Side street: 324 vehicles per hour

Using a truck factor of 1.5, and a left-turn factor of 1.6 (opposing left turns are ignored at this point), equivalent volumes in *pce* are computed:

Main street: $0.15(1.6)(880) + 0.10(1.5)880 + 0.75(880) = 1{,}000\ pce$

Side street: $0.15(1.6)(324) + 0.10(1.5)324 + 0.75(324) = 370\ pce$

The *main street* is 60 feet wide, with two parking lanes (7 feet each), leaving 46 feet for through traffic, or four 11-foot + lanes, two in each direction. The *side street* is 40 feet wide, with two parking lanes, leaving 26 feet for through traffic, or two 13-foot lanes, one in each direction. Assuming equal distribution among lanes, the critical lane volumes become:

$$N_1 = \frac{1{,}000}{2} = 500\ pce$$

$$N_2 = \frac{370}{1} = 370\ pce$$

Vehicle Clearance Intervals:

Using Equation (23-1),

$$Y = t + \frac{1}{2}\frac{V}{a} + \frac{(W + l)}{V}$$

where $t = 1.0$ seconds
$a = 15$ feet per second^2
$l = 20$ feet

as recommended, and

$V = 25$ miles per hour $= 36.7$ feet per second (both approaches)

W = width of street crossed; for Y_1, $W = 40$ feet and for Y_2, $W = 60$ feet

$$Y_1 = 1 + \frac{36.7}{2(15)} + \frac{40 + 20}{36.7} = 3.9 \text{ seconds}$$

$$Y_2 = 1 + \frac{36.7}{2(15)} + \frac{60 + 20}{36.7} = 4.4 \text{ seconds}$$

Pedestrian crossings:

Pedestrian crossing times determine the *minimum red phase* for the street *being crossed.*

Crossing the main street:

$$R_{1\ \min} = 5 + \frac{60}{3.5} = 22.2 \text{ seconds}$$

where 5 seconds is the assumed pedestrian start-up time, the main street is 60 feet wide, and pedestrians walk at 3.5 feet per second. However, $R_1 = G_2 + Y_2$ under a simple two-phase plan. Therefore, $G_{2\ \min} = 22.2 - 4.4 = 17.8$ seconds.

Crossing the side street:

$$R_{2\ \min} = 5 + \frac{40}{3.5} = 16.4 \text{ seconds} \quad \text{and}$$

$$G_{1\ \min} = 16.4 - 3.9 = 12.5 \text{ seconds}$$

Approximate cycle length:

Equation (23-7),

$$C_{\min} = \frac{Y_1 + Y_2}{1 - (N_1S_1 + N_2S_2)/3{,}600(PHF)}$$

produces a cycle for undersaturated flow during the peak 15-minute period, based on *vehicular considerations only*, of

$$C_{\min} = \frac{3.9 + 4.4}{1 - 500(2.5) + 270(2.5)/3{,}600(0.85)}$$

$$C_{\min} = 29 \text{ seconds}$$

However, recalling the pedestrian requirements, the minimum cycle length can be said to equal $G_{1\ \min} + G_{2\ \min} + Y_1 + Y_2 = 12.5 + 17.8 + 4.4 + 3.9$, or $C_{\min} = 38.6$ seconds. This value obviously controls.

Splitting the cycle:

Should a cycle $C = 40$ seconds be chosen? *Not necessarily.* Consider that the selected cycle should accomplish two purposes. First it must have phases of sufficient length to satisfy *pedestrian requirements*, and second, it should allocate green times to the two phases *in proportion* to the two traffic volumes N_1 and N_2 (assuming $S_1 = S_2$). The following procedure makes use of both facts to arrive at a more desirable cycle length, and more desirable signal splits.

It has been determined that the minimum green time for the side street based on pedestrian requirements = 17.7 seconds. Now the main street and side street volumes are 500 vehicles and 370 vehicles, respectively. If the green time on the two phases is to be allocated proportionally, main street should get

$$G_1 = \text{main street green} = \frac{500}{370}(17.7) = 23.9 \text{ seconds}$$

and the total cycle would be

$$C = G_1 + G_2 + Y_1 + Y_2 = 23.9 + 17.7 + 3.9 + 3.5$$
$$= 50 \text{ seconds}$$

With this cycle length it should be possible to satisfy both the minimum pedestrian time and proportional time allocation requirements.

Checking the timing:

Checking peak 15-minute requirements for each phase:

REQUIRED GREEN TIME (SECONDS)		PROVIDED GREEN TIME (SECONDS)	
$\frac{500}{4(0.85)}(2.5) = 368$	$\leq$	$\frac{900}{50}(23.9) = 430$	OK
$\frac{370}{4(0.85)}(2.5) = 272$	$\leq$	$\frac{900}{50}(17.7) = 319$	OK

Checking per-cycle requirements using vehicle start-up times: There are 72 50-second cycles in an hour. The average number of vehicles per cycle is vehicles per hour/72.

Main street: There are 880 vehicles per hour in two lanes. Of these, 10 per cent are trucks and 15 per cent are left turns. Assume that trucks are evenly divided between the two lanes, and all left turns are made from the left lane. Also assume that total traffic divides equally into two lanes. The critical left lane has 440 vehicles per hour, with 10 per cent trucks and 30 per cent left turns. Additionally, there are $500(0.15) = 75$ left turns in the opposing direction. Per cycle, this yields $440/72 = 6.1$ vehicles per cycle, of which 0.6 are trucks and 1.8 are left turns. There are $\frac{75}{72} \cong 1.0$ opposing left turns each phase. Using the values derived by Greenshields, the total minimum green required is computed below.

Basic time for all vehicles: $2.1(6.1) + 3.7$
$= 16.5$ seconds

Extra time for trucks ($\frac{1}{2}$ vehicle additional): $(2.1)(0.6)$
$= 0.6$ seconds

Extra time for left turns (regardless of direction):
$1.3(1.8 + 1.0) = 3.6$ seconds
$= 20.7$ seconds

Since the required main street green of 20.7 seconds is less than the 23.9 seconds provided, this phase is acceptable.

Side street: The minimum side street green time requirements may be determined in a similar manner. There are 324 vehicles per hour in one lane—10 per cent trucks and 15 per cent left turns. In each cycle, $\frac{324}{72} = 4.5$ vehicles per cycle, of which 0.45 are trucks and 0.68 are left turns. In addition, there are $240(0.15)/72 = 0.50$ opposing left turns each cycle. The required green per cycle is, therefore,

$$[2.1(4.5) + 3.7 + \tfrac{1}{2}(2.1)(0.45)] + [1.3(0.68 + 50)]$$
$$= 15.2 \text{ secs}$$

which is less than the 17.7 seconds provided. Therefore, this phase is also satisfactory. Thus, the final timing is

$$G_1 = 23.9, \quad Y_1 = 3.9, \quad R_1 = 22.2$$
$$G_2 = 17.7, \quad Y_2 = 4.4, \quad R_2 = 27.8$$

Recall, however, that for most pre-timed controllers timing must be to the nearest 1 per cent of the cycle. The actual phasing would have to be shown as in Figure 23-5.

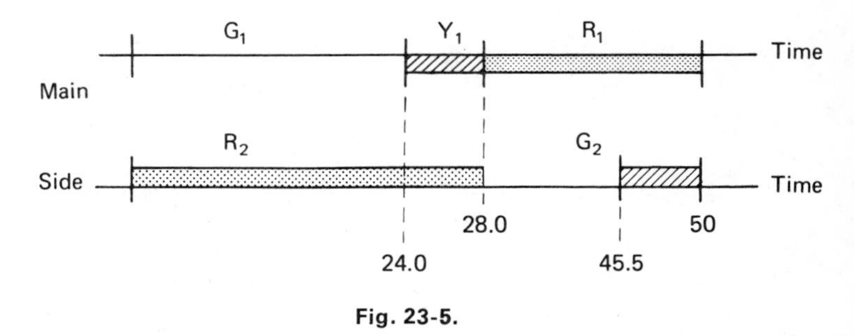

Fig. 23-5.

Problem 2:

This problem deals with a multi-phase T-intersection.

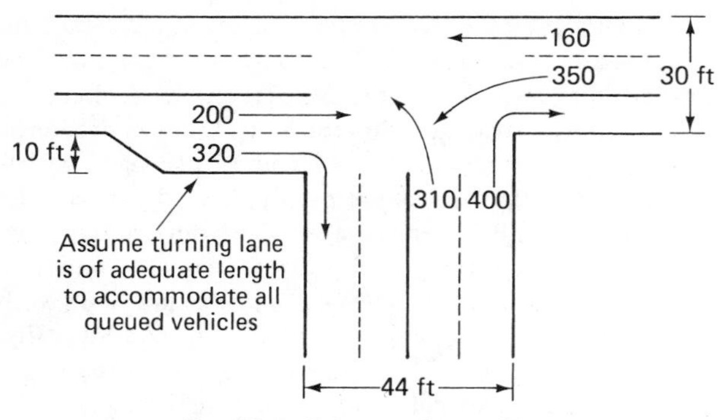

Fig. 23-6. T-intersection problem

Consider the T-intersection, which has significant pedestrian movements, as shown in Figure 23-6. Design an adequate signal timing arrangement. The following data is given:

3.5 feet per second pedestrian walking speed
Use of WALK-DON'T WALK signals (necessitating 7-second initial time assumption)
Approach speeds (85th percentile) = 30 miles per hour, main
10 miles per hour, side
No parking in vicinity of intersection
Negligible number of trucks
PHF = 0.85

Solution:

Phasing:

Phasing this T-intersection is particularly difficult due to the heavy turning movements, which preclude two-phase operation, and due to pedestrian considerations. Note that turning movements from street A, as well as through movements on street B, interfere with pedestrian movement across street B. The phasing shown in Figure 23-7 will eliminate these conflicts.

Fig. 23-7. Phasing for T-intersection

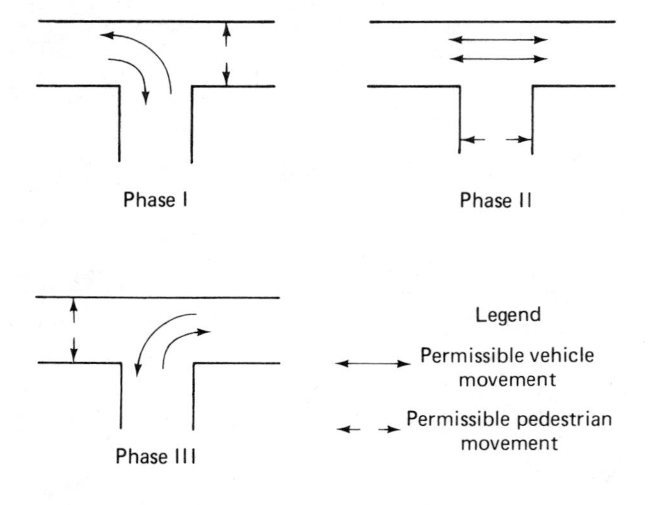

Critical lane volumes:

Because traffic movements are likely to segregate by lane, volumes should not be averaged over all lanes, but considered on a lane-by-lane basis. There are a negligible number of trucks, and left turns are unopposed, no expansion for *pce* is necessary. By inspection,

$$N_1 = 320, \quad N_2 = 200, \quad N_3 = 400$$

Pedestrian requirements:

$$T_1 = G_1 + Y_1 \geq 7 + \frac{30}{3.5} = 15.6 \text{ seconds}$$

$$T_2 = G_2 + Y_2 \geq 7 + \frac{44}{3.5} = 19.6 \text{ seconds}$$

$$T_3 = G_3 + Y_3 \geq 7 + \frac{40}{3.5} = 18.4 \text{ seconds}$$

Note that it would be possible to prohibit crossings on either phase 1 or phase 3, eliminating one of the above requirements, if necessary.

Vehicle clearance interval:

Using values of $t = 1$ second, and $a = 15$ feet per second^2,

$$Y_1 = 1.0 + \frac{1}{2}\frac{14.7}{15} + \frac{20 + 30}{14.7} = 4.9 \text{ seconds}$$

(Note: 10 miles per hour = 14.7 feet per second)

$$Y_2 = 1.0 + \frac{1}{2}\frac{44}{15} + \frac{20 + 44}{14.7} = 3.9 \text{ seconds}$$

(Note: 30 miles per hour = 44 feet per second)

$$Y_3 = 1.0 + \frac{1}{2}\frac{14.7}{15} + \frac{20 + 30}{14.7} = 4.9 \text{ seconds}$$

Note that for phases 2 and 3, the minimum approach speed (10 miles per hour) is used, since this produces more conservative estimates for the clearance interval. Also, because turning vehicles do not have to clear the entire width of the intersection, a reduced value (30 feet) is used.

Approximate cycle length:

$$C_{\min} = \frac{4.9 + 3.9 + 3.9}{1 - \{[320(2.5) + 200(2.5) + 400(2.5)]/3{,}600(0.85)\}}$$
$$= 55 \text{ seconds}$$

Phasing:

$$G_1 + G_2 + G_3 = 55 - 13.7 = 41.3 \text{ seconds}$$

$$G_1 = \frac{N_1}{N_1 + N_2 + N_3}(41.3) = \frac{320}{920}(41.3) = 14.4 \text{ seconds}$$

$$G_2 = \frac{200}{920}(41.3) = 9.0 \text{ seconds}$$

$$G_3 = \frac{400}{920}(41.3) = 18.0 \text{ seconds}$$

Checking the timing:
Checking pedestrian requirements:

	PROVIDED G	REQUIRED G	
G_1	14.4	15.6 − 4.9 = 10.7	OK
G_2	9.0	19.6 − 3.9 = 15.7	NG
G_3	18.0	18.4 − 4.9 = 13.5	OK

As in the previous problem, once one knows that one of the phases is determined by pedestrian requirements, it is possible to directly determine required cycle length. The minimum allowable green time for phase 2, $G_2 = 15.7$ seconds, and $N_1 = 320$, $N_2 = 200$, $N_3 = 400$. Thus,

$$G_1 = G_2\frac{N_1}{N_2} = 15.7\frac{320}{200} = 25.1 \text{ seconds}$$

and

$$G_3 = G_2\frac{N_3}{N_2} = 15.7\frac{400}{200} = 31.4 \text{ seconds}$$

The minimum acceptable cycle length is

$$\begin{aligned} C &= G_1 + G_2 + G_3 + Y_1 + Y_2 + Y_3 \\ &= 25.1 + 15.7 + 31.4 + 4.9 + 3.9 + 4.9 \\ &= 85.9 \text{ seconds} \end{aligned}$$

Using the next highest multiple of 5 seconds gives a cycle of 90 seconds, which will satisfy both pedestrian and vehicular volumes.

To get the values of the three green times for the 90-second cycle, it is necessary to expand the above times as follows: In the 85.9-second cycle, 72.2 seconds are in green time, i.e., the sum of the greens in the three phases. In a 90-second cycle, the available green time would be equal to

$$90 - (Y_1 + Y_2 + Y_3) = 90 - 13.7 = 76.3 \text{ seconds}$$

Each of the three green times previously calculated must be expanded by a factor of 76.2/72.7 = 1.06. The new green time values are

$$\begin{aligned} G_1 &= 25.1(1.06) = 26.6 \text{ seconds} \\ G_2 &= 15.7(1.06) = 16.6 \text{ seconds} \\ G_3 &= 31.9(1.06) = 33.3 \text{ seconds} \end{aligned}$$

Checking the peak 15-minute requirements:

	REQUIRED GREEN TIME (SECONDS)	PROVIDED GREEN TIME (SECONDS)	
G_1:	$\frac{320}{4(0.85)}(2.5) = 235$	$\frac{900}{90}(26.6) = 266$	OK
G_2:	$\frac{200}{4(0.85)}(2.5) = 147$	$\frac{900}{90}(16.6) = 166$	OK
G_3:	$\frac{400}{4(0.85)}(2.5) = 294$	$\frac{900}{90}(33.3) = 333$	OK

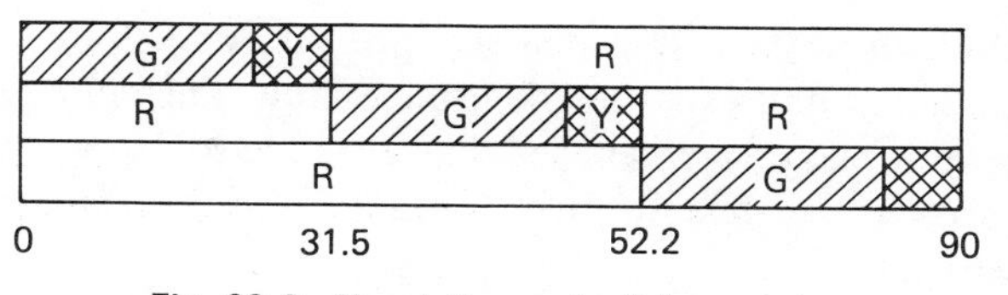

Fig. 23-8. Signal diagram for T-intersection

Individual cycle requirements may be checked, as was done in the previous problem.

Final timing:

$$\begin{aligned} G_1 &= 26.6, & Y_1 &= 4.9, & R_1 &= 58.5 \\ G_2 &= 16.6, & Y_2 &= 3.9, & R_2 &= 69.5 \\ G_3 &= 33.3, & Y_3 &= 4.9, & R_3 &= 51.8 \end{aligned}$$

As indicated previously, signal timing is usually to the nearest 1 per cent of a cycle. The final phasing of the T-intersection to the nearest 1 per cent of a 90-second cycle is shown in Figure 23-8.

DEFINITIONS FOR ACTUATED SIGNALS

Initial portion: The first part of the green interval which is timed out or separately controlled by a traffic-actuated controller before the extendible portion of the interval takes effect.

Extendable portion: The part of the green interval following the initial portion.

Unit extension: The minimum time, during the extendible portion, for which the right-of-way must remain on any traffic phase following an actuation on that phase, but subject to the extension limit.

Extension limit: The maximum time for which actuations on any traffic phase may retain the right-of-way after actuation on another traffic phase.

Minimum period: The shortest period of time that a green interval may be displayed during any traffic phase.

Carryover: The effect which a passing platoon exerts on the controller for the benefit of prompt assignment of the right-of-way for the next following platoon on that traffic phase.

The *initial portion* is timed to permit waiting vehicles, which have accumulated between the detector and the stop line during the red interval, to enter the intersection. The *unit extension* is timed to extend the initial portion to permit a vehicle to travel from the detector to the intersection. The minimum green period is the sum of the initial portion and one unit extension. If there is no vehicle actuation during the first unit extension, the green will be transferred, after the clearance interval, to another approach if there is a demand present. Therefore, the length of a unit extension determines how much of a gap between successive vehicles will be allowed before the controller transfers the green to traffic waiting on another approach.

If a vehicle followed the first, and crossed the detector during the first unit extension, the balance of the unit extension being timed out would be cancelled, and the controller would start timing a new unit extension. As long as vehicles cross the detector within the unit extension, that is, when the time spacing between vehicles is less than the unit extension the green indication will be extended if there has been no detector actuation on the competing approach.

The *extension limit* is the maximum time that an approach is permitted to retain the green interval after there is a vehicle actuation on the other street. The *clearance interval* follows the extension limit, and is timed to permit the last vehicle to clear the intersection.

TIMING A TRAFFIC-ACTUATED SIGNAL

Considerations in Timing a Full-Actuated Signal

Consider the simple two-phase problem (Problem 1) of the previous section. Assume that the peak-hour volumes are as shown, but that the intersection experiences sharp fluctuations in flow during the course of the day. It has been concluded that a full-actuated signal will best respond to the varying demands for service at this intersection.

Detectors are assumed to be located on all four approaches, 120 feet back of the stop line.

The effectiveness of a full-actuated signal largely depends on the timing of the initial interval and the unit extension. The latter is the more important and, thus, shall be discussed first.

Unit Extension. Proper timing of the unit extension depends on how far the detectors are back from the stop line and the average speed of approaching traffic. The unit extension should be no less than the time it takes a vehicle to cross the detector and reach its stop line. In this case that means

$$\frac{120}{36.7} = 3.3 \text{ seconds}$$

(Note: 25 mph = 36.7 feet per second)

This will exactly allow a vehicle, traveling at the average approach speed of 25 mph, to arrive at the stop bar just as the signal turns amber. The vehicle would then clear through the intersection on the amber interval.

If one wished to ensure that a vehicle traveling at 25 mph was completely clear of the intersection, it would be necessary that it not only cover the distance between the detector and the stop line, but also the cross street (30 feet), and its own length (20 feet), to be completely clear of the intersection. This would require a unit extension of

$$\frac{120 + 30 + 20}{36.7} = 4.6 \text{ seconds}$$

In this example, a compromise value of *4 seconds* will be selected. This is enough time for a vehicle traveling at 25 mph to almost reach the far side of the intersection before having the amber appear, a not unreasonable procedure.

Initial Portion. This is the time required to clear those vehicles which may have queued up between the detector and the stop line during the red phase. In actuated systems a minimum green is provided which is the sum of the initial portion plus one unit extension. This minimum green has to be sufficiently long to clear as many vehicles as could possibly be waiting between the stop line and the detector, since actuated control equipment only records that a demand for service exists. It is not known if one, two, or more vehicles are waiting.

In this example, the distance between the stop line and the detector is 120 feet. If each vehicle waiting is assumed to use up 20 feet (its own length plus some small intervehicular gap), then a total of

$$\frac{120}{20} = 6$$

vehicles could be waiting to pass through on the green. Using the time requirements shown earlier, the six vehicles would require a total of 16.3 seconds ($2.1 \times 6 + 3.7$). Thus, the minimum green would be 16 seconds. Now the initial portion equals the minimum green, minus one unit extension. Thus,

$$\text{Initial portion} = 16 - 4 = 12 \text{ seconds}$$

Extension Limit. The way to set the extension limits is to determine the maximum peak-period volumes, and to determine the timing of a pre-timed signal for those conditions. The extension limits can then be set to these values. This is not unreasonable when one considers that in peak periods detector actuations on competing approaches are much more likely to occur at, or very soon after, the initiation of a green, thus causing the signal to operate like a fixed-time device. In this case, the procedure used would be applied to determine the two green requirements. Now, however, what was thought of as the minimum acceptable green under a pre-timed scheme, is here considered to be the maximum allowable green. The underlying philosophy is to keep the possible maximum as short as possible to minimize delay.

In calculating the extension limits by using peak-hour volumes, pedestrian requirements *do not control*. If pedestrians are present in any number, a push-button pedestrian actuator should be provided. The controller will give either the minimum vehicular green time or the minimum required pedestrian time, depending upon the actuation.

Vehicle Clearance Interval. This is computed in the same manner as for a pre-timed signal.

Pedestrian Clearance. Should the need exist to service pedestrians at this location, pedestrian actuators will be provided. Minimum pedestrian green is calculated as for pre-timed control.

Considerations in Timing a Semi-Actuated (Vehicular) Signal

A semi-actuated controller utilizes detectors only on the minor approaches. This type of control is particularly applicable at locations where minor street vehicles may not safely cross a major flow without signals. In many such cases, the minor street flow would not normally warrant a signal. Semi-actuated controllers require the setting of the following times:

1. For the artery or main street:
   a. Minimum green
   b. Clearance interval
   c. Pedestrian clearance (if pedestrian actuator is provided).
2. For the side or minor street:
   a. Initial portion
   b. Unit extension
   c. Maximum green
   d. Clearance interval

The procedure for timing a semi-actuated signal is quite similar to that used in timing a full-actuated signal. Determination of minor street initial portion and unit extension (initial portion + unit extension = minimum green) are identical. While in semi-actuated control there is a maximum green on the minor street, there is no maximum green on the main street, only a minimum green.

The timing of the maximum extension for the side street, and minimum green for the major street, is not as precise a procedure as the items previously discussed. Several considerations must be addressed.

It should be remembered that during periods of peak side street loading the signal may operate as a fixed-time controller, timing out maximum side street and minimum main street green. Under these circumstances, the minimum and maximum green settings must be sufficient to provide adequate capacity for both side and main street loads. Also to be considered is a desirable cycle length and phase split under these conditions.

Platoon movement and the expected size of platoons along the major street should be considered. The minimum main street green should not be so short as to interfere with such movement.

It should be remembered that the main street retains the green at all times, unless the side street is actuated. At the termination of side street actuations, the green automatically returns to the major street.

TURN CONTROLS AT INTERSECTIONS

Turn controls represent one of the most useful traffic regulatory measures. The utilization of proper turn controls can accomplish the following objectives:

1. Elimination or reduction of intersection conflicts involving vehicle versus vehicle and vehicle versus pedestrian
2. Reduction of accident hazard
3. Reduction of delay and increase of intersection capacity

The prohibition of one of the conflicting movements offers great flexibility, both because no important physical changes are required, and because the regulation is easily removed if it proves unsatisfactory or if a better control is found. Since the rerouting of vehicles prohibited from turning at a particular intersection may sometimes create problems at other intersections, it is necessary that each location be thoroughly studied to determine whether or not the benefits of a turn prohibition outweigh the disadvantages. Part-time turning prohibitions are good in theory, and more in keeping with traffic demand, but overall observance by the public is considerably better if the regulations are applied full-time.[31]

Type of Turn Controls

The two basic types of turn controls are left-turn restrictions and right-turn restrictions. There are no commonly accepted warrants for the use of turn controls, but the reports of practices in several cities are useful as guides.

The following warrants were reported for establishing turning restrictions:[31]

1. When the left-turn volume exceeds 20 per cent of the total approach volume
2. When left turns constitute 10 per cent of the total movement on a given street
3. Where left-turn movements interfere with straight-through movements of 15,000 vehicles per day, regardless of number of lanes and at signalized four-way intersections
4. Where a left-turn or right-turn movement interferes with pedestrian crosswalk volumes in excess of 2,000 persons per hour
5. When 600 vehicles conflict with 1,000 or more pedestrians per hour
6. When turning vehicles average 7 per green interval for several successive signal changes
7. Where more than three intersection accidents involving turning vehicles occur within a 12-month period
8. When the number of traffic lanes available at the intersection will accommodate only a single movement in each direction, and there is an appreciable demand for left turns

Left-Turn Restrictions. Left-turn prohibitions are most frequently used at intersections along major streets carrying heavy volumes. However, caution should be exercised in the indiscriminate use of left-turn prohibitions. The effects of restricting a left turn at one intersection must be evaluated with respect to the possibility of creating undue hazard at the other intersections. There should be convenient alternate routes available for left-turning traffic. To determine the availability of such routes, it is necessary to study traffic flow patterns and traffic volumes in the vicinity of the turn restrictions.

Heavy volumes of left-turning traffic should not necessarily be a warrant for prohibiting the turn, but should indicate the need to investigate all possible methods to accommodate the movement. At signalized intersections it is often possible to accommodate a left-turn movement by special phasing of the traffic signal.

Right-Turn Restrictions. Right-turn prohibitions are most commonly used when there is a serious pedestrian-vehicle conflict. The potential accident hazard, and delay and congestion to right-turning traffic, and frequently to straight-through traffic, are factors that may warrant the turn restriction. However, special signal phasing and other less restrictive measures should be studied as possible substitutes for turn prohibition.

Turn Restriction Devices

Turn Prohibition Signs. Signs containing messages of NO RIGHT TURN, NO LEFT TURN, NO TURNS, and NO U-TURN, may be used to inform drivers of the appropriate regulations. When the restricted movement only applies during certain periods of the day, the MUTCD recommends the following alternatives, in order of preference:

1. Internally illuminated signs that are lighted and made legible only during the restricted hours, particularly at signalized intersections
2. Permanently mounted signs, incorporating a supplementary legend showing the hours during which the prohibition is applicable
3. Movable pedestal signs, which are put in place under police supervision only when applicable, and removed at other hours

Traffic Signal Indications. When the turn prohibition is in effect full-time, it is desirable to use green arrows instead of full green lenses.

Pavement Markings. Markings are useful in supplementing signs or signals.

Channelization. When turn prohibition is in effect full-time, channelizing devices may be considered which make it physically impossible to execute turns.

Right Turn on Red

In an effort to ease congestion at intersections where large volumes of pedestrians block right-turning movements, some states and cities allow right turns on a red signal indication *after stopping*. Such a regulation can only be effective without signing when it is applied on a uniform basis within an area, and the vehicle code must require all right-turning traffic to stop before proceeding to turn on a red indication. In some instances, this practice has been eliminated due to the accident hazard, and lack of compliance with the vehicle code.[32,33]

Another regulation legally permits right turns on red without requiring motorists to stop before executing the turn. This is permitted by use of a right-turn green arrow, which is exhibited simultaneously with the red indication. Continuous use of the right lane for turns at a signalized intersection is of great benefit in the reduction of vehicular delay, and in the increase of intersection capacity, but it creates a serious conflict for crossing pedestrians. Conflict between through traffic may also exist if the intersection geometry does not lend itself to proper merging of the two streams. A thorough investigation of pedestrian volumes and right-turning volumes should be made before permitting continuous right turns. For an intelligent approach to the problem, consult the literature.[34]

Lane-Use Control at Intersections

Lane-use controls generally establish lanes for the exclusive use of a certain movement. An exclusive turn lane, left or right, is only warranted when there are heavy turning movements, since it may significantly reduce capacity if it is not fully utilized. Ideally, separate turn lanes should be provided by increasing the intersection's approach width of the pavement, thereby preventing interference with through movements. However, this type of intersection design is seldom possible for most urban intersections. In fact, it is often necessary to permit turns from more than one lane during peak hours of flow. This condition commonly occurs at intersections of one-way streets, and on routes leading to and from freeways.

When turning volumes at peak hours require more than one lane, the lane adjacent to the left- (or right-) turning lane may permit optional movements, such as left (or right) and through movements.

The different methods in use for controlling vehicular movements at the intersection include:

Lane-Use Control Signs. These are signs which indicate, by arrows or word messages, the types of movement that are permitted or required from specific lanes at an intersection. The use of lane-control signs is covered in the MUTCD. The signs may be post-mounted or mounted overhead. Overhead mounting is preferred, because the signs can be placed over the lanes to which they apply.

Pavement Markings. Lane-use control markings should be used on the approach to an intersection to supplement lane-use control signs. The MUTCD gives the recommended design of word markings and arrows applicable to lane-use control. Results of a recent study[35] show that the standards recommended by MUTCD for control of multiple turns at intersections are satisfactory. The conclusion was based on both field and laboratory studies.

For a proposed California standard for marking pavements for two-way left-turn lanes, using striping alone and striping in conjunction with a raised island, consult the literature.[36]

Separate Left-Turn Lane. There are no commonly accepted warrants for the construction of a left-turn refuge. However, the factors that must be considered before a decision can be made should include:[37,38]

1. Accident experience involving left-turning vehicles with opposing traffic, sideswipe collisions, and rear-end collisions
2. Volumes of left-turning traffic and opposing traffic
3. Delay to left-turning and straight-through movements
4. Physical layout of the intersection
5. Cost of construction

Advance Route-Turn Markers. The route turn is often a source of congestion and confusion. A study[39] made of the use of advance route-turn markers, placed ahead of the intersection, has indicated that they have a significantly beneficial effect on locating drivers in the appropriate lane before arriving at the intersection.[40]

Special Phasing of Traffic Signals for Turn Controls

Special signal intervals for turning movements are widely used. Compared with the prohibition of certain turns, this method has the advantage of not requiring rerouting; but it frequently necessitates a lengthening of the signal cycle, which means added delay for other motorists and pedestrians. Special signal phasing, however, is often the superior solution when compared with turn restrictions. The different methods of accomplishing special phasing include the following:

Leading (Advance) and Lagging (Delayed) Green. The application of leading or lagging green intervals in the traffic signal sequence permits left-turning vehicles to clear the intersection without conflict with opposing traffic.

A *leading* green phase consists of permitting all traffic on one approach (say, northbound) to proceed through the intersection unopposed for a period of time (7 to 20 seconds), at the beginning of the northbound green phase. After this initial interval, southbound traffic is given the green indication, and all northbound and southbound traffic proceeds through the intersection, permitting turns in both directions through the opposing traffic, if adequate gaps occur. The green indications for both directions are terminated simultaneously. This is known as giving a leading green phase to northbound traffic.

A *lagging* green phase consists of terminating the

green indication for one approach (say, southbound) for a period of time (7 to 20 seconds) before the green phase ends for northbound traffic, thereby permitting all northbound traffic to proceed through the intersection unopposed. The green indications for both directions are started simultaneously. This is known as giving a lagging green phase to northbound traffic.

The use of either phasing should be carefully evaluated, since the motorist receiving the shorter green tends to continue through the intersection against a red indication, because opposing traffic is moving freely. The following factors should be considered before arriving at a decision to use leading or lagging green.[41]

Advantages of a leading green:

1. It permits higher intersection capacity on restricted-width intersection approaches.
2. It is easier to program in the traffic signal controller.
3. It eliminates conflicts between left-turn and opposing straight-through vehicles, by clearing the left-turn vehicles through the intersection first, thus reducing congestion.
4. Driver reaction time is quicker.
5. It requires only one amber clearance interval, since the green phase is terminated simultaneously for both directions.
6. It is desirable where separate left-turn lanes do not exist.

Disadvantages of leading green:

1. It creates vehicle-pedestrian conflicts during the leading green interval.
2. Left turns may preempt the right-of-way from the opposing straight-through movements as the leading green terminates, which may create an accident potential.
3. Opposing movement makes false starts in an attempt to move with the leading green.
4. It may cause difficulties in timing progressive signal systems because straight-through traffic is released sooner, and may arrive at the next interconnected signalized intersection ahead of the green indication.
5. It does not conform with the right-of-way law and creates entrapment of the left-turn vehicle when the leading green interval expires.

Advantages of lagging green:

1. It conforms to the normal left-turn right-of-way law, and does not create entrapment of left-turn vehicles.
2. It is closer to normal driving behavior.
3. It provides for vehicle-pedestrian separation, since pedestrians cross at the beginning of the green phase.
4. Green phase for both directions starts simultaneously.
5. It allows for greater total approach capacity, because the first left-turning vehicles have moved into the intersection and filter through during the regular portion of the green phase, thereby reducing the time headway for succeeding vehicles.
6. Left turns do not preempt right-of-way from the opposing straight-through movement.
7. It cuts off only platoon stragglers from adjacent signalized interconnected intersections.
8. It is desirable where a separate left-turn lane exists.

Disadvantages of lagging green:

1. It requires two amber clearance intervals, since the green phase is terminated at different times for the two directions.
2. It creates conflicts for opposing left turns at the beginning of the lag interval, since the opposing left-turn drivers think that both movements stop at the same time.
3. It creates an obstruction to through movement during the green interval, where a separate left-turn lane does not exist.

Several sign legends used in conjunction with leads and lags to control both the advanced movement and delayed movement are as follows:[40]

WAIT FOR GREEN LIGHT
ADVANCE GREEN (Lead)—DELAYED GREEN (Lag)
MOVE ON GREEN ONLY
WAIT, DELAYED SIGNAL
OBEY YOUR SIGNAL ONLY

The most popular of these legends is WAIT FOR GREEN LIGHT, where a leading green is employed.

Three-Phase Operation. This type of operation provides an exclusive phase for left-turning traffic. The signal cycle is divided into three phases, with the first permitting straight-through and right-turn movements on opposite approaches of the major street; the second is for the exclusive use of left-turning vehicles on opposite approaches of the major street; the third accommodates all movements on opposite approaches of the minor street. These represent the most common three-phase operation, but there are other variations, depending upon specific needs.

Three-phase operation is normally used to reduce delay to turning vehicles. One study, however, found that the addition of a left-turn phase substantially increased the overall delay at the intersection. Leftturn delay was not reduced significantly compared to delay experienced under two-phase control.

A study[45] of the use and capacity of two-lane left turns, controlled by a separate phase, indicated that each of the two lanes handled very close to equal volumes during peak hours. The capacity of each left-turn lane was found to be somewhat less than that of a single left-turn lane, and it was concluded that approximately a 75 per cent increase in capacity may be expected by adding a second left-turn lane.

Special Pedestrian Phases. These include the methods in which pedestrian signal intervals can be combined with vehicular signal intervals. The four basic combinations will be discussed under Pedestrian Controls. Consult the literature[42-48] for additional comments on special phasing of traffic signals for turn controls.

TRAFFIC CONTROL ISLANDS

A *traffic control island* is a defined area between traffic lanes for the control of vehicle movements or for pedestrian refuge. Vehicular traffic is intended to be excluded from the island area, together with any approach area that is occupied by protective deflecting or warning devices.

Island Classification

Islands may be classed functionally and physically as follows:

1. *Pedestrian Refuge Islands:* A *pedestrian refuge island* is designed for the use and protection of pedestrians. A pedestrian island includes the safety zone, together with the area at the approach end, occupied or outlined by protective deflecting or warning devices.

   The *safety zone* is the area or space officially set apart within a roadway for the exclusive use of pedestrians, and it is protected or marked or indicated by adequate signs so as to be plainly visible at all times. A *loading island* is a pedestrian island especially provided for the protection of transit-vehicle users.
2. *Traffic Divisional Islands:* A traffic divisional island is one, usually elongated and narrow, which follows the course of the roadway to separate traffic moving in the same or opposite directions.
3. *Traffic Channelizing Islands:* These are located in a roadway area to confine specific movements of traffic, usually turning movements, to definite channels.

Functions of Islands[2,3]

Most islands serve many purposes, although they are usually installed for specific needs. Each island type is discussed in the following section.

Pedestrian Refuge Islands. Refuge islands should be used in urban areas on exceptionally wide roadways, or in large, irregularly shaped intersections, where there is a considerable amount of pedestrian traffic and where heavy vehicular volumes make it difficult and dangerous for pedestrians to cross. Such islands may also be desirable on wide streets, where the intersection is controlled by traffic signals, to reduce the necessary clearance periods and expedite the movement of traffic. Pedestrian islands may also be warranted where accident experience shows the occurrence of a number of pedestrian accidents of a type that would probably be eliminated or reduced by the installation of islands.

No refuge island should be installed where it will leave less than two through lanes available for traffic between it and the adjacent curb, or another island, and they should never be located so as to create a hazard for vehicles. It is preferable that refuge islands be raised platforms outlined by barrier curbs so as to provide greater safety for pedestrians.

Divisional Islands. A divisional island, or median, is an important and essential element in the design of major thoroughfares in urban areas to provide adequate safety and capacity. Generally no rural highway designed for four or more lanes in its ultimate stage should be constructed without a median.

The more important functions of a median are as follows:

1. To provide freedom from the interference of opposing traffic which results in greater convenience, comfort and confidence for the motorists, and in a reduction of accidents
2. Where width is sufficient, to provide protection and control for crossing and turning traffic and to provide for the possibility of creating separate turn lanes for the storage and safer maneuvering of turning vehicles, as well as increasing intersection capacity
3. To provide a haven in the case of an emergency
4. To provide a refuge for pedestrians, and decrease the need for the installation of traffic signals
5. To prevent U-turns
6. To define alinement and proper vehicle paths, which make it possible for traffic to move more smoothly and safely at higher operating speeds

Medians should be as wide as feasible, highly visible both day and night, and in definite contrast with the through-traffic lanes. Except for very narrow medians (4 feet or less), and where pedestrian protection is significant, medians should be outlined with mountable curbs where curbing is utilized.

Attention is called to a bibliography of literature

dealing with all aspects of median functions, limitations, and operation.[49] A short abstract is given after each listing. Consult the literature for studies dealing with the results of full-scale tests on median and bridge barriers, and experiences with some of these barriers. The cable-chain link barrier proved to be very effective.[49-55]

Many studies have been undertaken to examine the behavior of traffic on roads with and without median barriers, and to evaluate the effect of the presence of such barriers on accident experience.[56-68]

Channelizing Islands. Channelizing islands are generally used in at-grade intersections to guide traffic into proper channels through the intersection area. Channelization is generally employed for one or more of the following purposes:

1. To separate conflicts
2. To control the angle of conflict
3. To reduce excessive pavement areas
4. To regulate traffic and to indicate the proper use of the intersection
5. To favor predominant turning movements
6. To protect pedestrians
7. To protect and store turning and crossing vehicles
8. To locate traffic-control devices
9. To prohibit specific movements
10. To control speed

The basic warrant for the installation of channelizing islands is that of intersection size, physical characteristics, or complexity of a nature that their use will eliminate or reduce unnecessary or undesirable conflicts and hazards to motorists and pedestrians, as well as disorder and confusion in traffic flow.

Each intersection requires very careful study to determine the appropriate location and shape of islands to accomplish the desired control of traffic movements through the intersection area.

Consult the literature for pertinent data on the design and operational characteristics of various types of channelization. In addition, there is a special report on channelization.[69-71]

General Design Elements

The island design should be carefully planned so that the shape of the island will conform to natural vehicular paths, and so that a raised island will not constitute a hazard in the roadway. A judiciously placed island at an intersection on a wide street may eliminate the need for traffic signal control by channelizing traffic into orderly movements.

Illumination and Reflectorization. Islands should be clearly visible at all times, and be seen sufficiently in advance so that the motorist will not be surprised by its presence.

Refuge islands should not be installed unless they can be adequately reflectorized and illuminated. Illumination of refuge islands, including their approach-end treatment, should be great enough to show the general layout of the island, and the immediate vehicular travel paths, with the greatest concentration of illumination at points of potential danger to pedestrians or vehicles. Reflectorization is required to warn of the presence of the island on nights when illumination may not be in operation.

Divisional and channelizing islands, and the proper lanes of travel along them, should preferably be made clearly visible at night through the use of adequate and properly directed street lighting. If lighting facilities are not available, the islands should be outlined as clearly as possible with high-visibility approach-end devices.

Size and Shape. Islands are generally either narrow and elongated, or triangular in shape, and they are normally situated in areas of the roadway outside the planned vehicle paths. They are shaped and dimensioned to serve as component parts of the roadway or intersection layout.

Actual island size will depend upon specific applications, but they should be large enough to command attention. In rural sections, an island should be at least 75 square feet in area, and a minimum of 50 square feet for very restricted conditions. For urban sections, where speeds are lower, the corresponding areas are 50 and 35 square feet. Whenever possible minimum island sizes should be avoided.

Designation of Area. Generally, islands are delineated by one of the following methods:

1. They may be raised and outlined by curbs, and filled with pavement, turf, or other material.
2. They may be formed by pavement markings, buttons, or raised bars on all paved areas, usually in urban districts where speeds are low and space limited.
3. They may be unsurfaced and flush with the traveled way, and sometimes supplemented by guide posts, stanchions, or other delineators. This treatment is normally used for large islands.

Approach-End Treatment. The approach end of an island should be carefully designed to provide a maximum degree of warning, for approaching traffic, of the presence of the island, and a definite indication of the proper vehicle path or paths to be followed. Motorists may be effectively warned of the island, and guided around it, by a combination of some of the following devices:

1. Pavement markings
2. Object markings
3. Signs
4. Reflecting hazard markers and delineators
5. Flashing yellow beacons
6. Contrasting pavement colors or textures
7. Raised bars, buttons, and blocks
8. Illumination devices

Large mushroom buttons or bars of cast iron or concrete several inches high, with or without reflectors or lights, may be used to outline islands or their approaches, and to assist in channelizing traffic around them. Island-approach warnings may also be effected by jiggle bars, which are raised transverse bars placed on the pavement to make any wheel encroachment within their area obvious to the motorist without loss of control of the vehicle. A technique similar to that of jiggle bars is the use of rumble or rumbler strips, which involves the application of large, angular particles to the surface so that the tires produce an audible hum. Rumble strips are also extremely useful to effectively warn motorists on approaches to dangerous intersections, and to help provide for proper merging operations.[69]

A study indicated that special approach-end treatment to a channelizing island caused traffic to reduce speed at a greater distance before the intersection, and continue at a reduced speed through the channelized section.[70]

For a more detailed treatment of islands and their design, consult HRB Special Report 74 on channelization,[71] the AASHO *Policy on Geometric Design of Rural Highways*, and the AASHO *Policy on Arterial Highways in Urban Areas*.[73]

Island Maintenance

Maintenance methods and procedures are governed to a large extent by the physical makeup of the island. Maintenance operations should include the preservation and maintenance, in original constructed condition of the structural features of the island and its approach-end treatment. These operations may include the repaving of the surface within the island, outlining curbs, straightening of posts, rails or buffer protection, replacing of buttons or reflectors, and periodic painting of curbs and markings. Where islands are sodded, the growth should be mowed for appearance and visibility. Jiggle bars, mushroom buttons, and other protruding pavement inserts may also require resetting or replacement.[74]

INTERSECTION PARKING CONTROLS

The special restriction of parking in or near intersections can have many beneficial results on the operation of the intersection. The elimination or restriction of parking near intersections, particularly during peak periods of flow, reduces the probability of conflicts from cars entering and leaving curb spaces. This additional element of congestion, in an already congested intersection, can often not be tolerated in peak periods.

The complete elimination of parking near intersections will create a right-turning lane, a particularly effective result when heavy movements in that direction are experienced. With no parking near the intersection, there is also extra width for queued vehicle storage, thus helping to prevent long queues from developing.

It should be remembered, however, that the prohibition of parking should be limited to those intersections where the permitting of parking would represent a serious hazard or obstruction to flow.

The subject of curb-parking controls is more fully discussed in Chapter 27.

PEDESTRIAN CONTROLS

Pedestrian-Vehicle Accidents

While the total number of pedestrian traffic fatalities has decreased significantly during the last 30 years, pedestrian deaths still account for about 37 per cent of all traffic fatalities in urban areas.[1]

Furthermore, older people (over 65 years) and younger people (less than 15 years), are more likely to be involved in a traffic fatality than other age groups. The traffic fatalities in the older and younger age groups represented 57 per cent of the total.

Traffic Congestion

At locations where there are heavy volumes of vehicular and pedestrian traffic, as at intersections in central business districts, conflicting movements can create congestion, delay, and danger. This condition can exist even at intersections that are signalized, particularly when heavy turning movements conflict with heavy pedestrian crossings. Proper control of pedestrian and vehicular movements is essential to alleviate and minimize the congestion and delay.

Engineering Aids for Pedestrian Safety and Control

The engineer must determine the need for and the application of, specific types of control or physical protection at locations of major conflict or high accident frequency. Some of these controls and physical measures are discussed in the following paragraphs.

Sidewalks. Sidewalks are a recognized necessity in urban areas, but few are considered necessary in rural areas. However, in many rural areas the need for sidewalks is quite critical because of the high vehicle speeds and general lack of highway lighting. Rural locations where sidewalk construction is extremely desirable include points of community development, such as at schools, meeting halls, churches, local businesses, and industrial plants. Of the total number of pedestrian traffic fatalities in 1967, some 35 per cent occurred on rural highways. This is even more significant when it is realized that although the vehicle-miles of travel are about the same in urban and rural areas, pedestrian density is much greater in urban areas. Therefore, the hazard in rural areas is proportionately much greater.

Table 23-2 shows warrants, based on pedestrian volumes and vehicular volumes and speeds, for the justification of sidewalk construction.

To determine the required sidewalk width, it is assumed that the capacity per 22-inch lane width ranges between 18 and 27 pedestrians per minute, or 1,000 to 1,600 per hour.[2] Another recommendation[73] is that, in general, sidewalk widths should be at least 6 feet wide in built-up districts, and 4 feet wide in residential areas, and preferably these should be 8 to 12 feet and 6 feet, respectively.

A British recommendation regarding the capacity of sidewalks in shopping districts provides for an additional 3 feet of sidewalk width, which is assumed to be not available for pedestrian movement. Therefore, 3 feet must be added to the computed width.[75]

TABLE 23-2. PEDESTRIAN AND VEHICLE VOLUMES FOR WHICH THE CONSTRUCTION OF SIDEWALKS MIGHT BE CONSIDERED

VEHICULAR TRAFFIC	*PEDESTRIANS PER DAY, SUGGESTED FOR CONSTRUCTION OF SIDEWALKS WHEN DESIGN SPEED IS*	
Design Hour Volume	*30 to 50*	*60 to 70*
Sidewalk, one side		
30 to 100	150	100
More than 100	100	50
Sidewalk, both sides*		
50 to 100	500	300
More than 100	300	200

(Source: *Traffic Engineering Handbook,* ITE, 1965, Table 4-10)
* Smaller pedestrian volume may justify two sidewalks, to avoid a considerable amount of pedestrian cross traffic.

Crosswalks. Crosswalks define and concentrate pedestrian walking areas within an intersection, or at other points within the roadway. It is particularly desirable to delineate crosswalks by markings that are visible by day and by night. The width of crosswalks, which are defined by two solid white lines, should be at least as wide as, and preferably wider than, the extended sidewalks. Elimination of left or right vehicular turns at intersections will definitely increase crosswalk capacity.

The use of pavement markings to define crosswalks will help encourage their acceptance by pedestrians and help discourage vehicles from stopping within them. To encourage the proper use of crosswalks in one city, the area between the defining white lines was painted green, and the experiment was quite successful.[76] A study of painted crosswalks revealed that they reduced violation of the pedestrian's right-of-way, and pedestrians tended to use a painted crosswalk in preference to an unpainted one.[77]

The British have found the use of "zebra" crosswalk marking to be effective.[75,78] Zebra crosswalk installation consists of equally spaced white stripes running parallel to the direction of vehicle flow for the width of the crosswalk. The stripes are generally 25 inches wide, as are the areas between them.

There are no accepted warrants for painting crosswalks, but they should be based upon:

1. Pedestrian volume crossing the street
2. Vehicular volume, speed, and turning movements
3. Accident frequency at intersections
4. Width and shape of intersections
5. Use as school crossings

Some articles that present ideas on this subject are found in the literature.[79-82]

Pedestrian Barriers. Barriers prohibit pedestrian crossings of major streets at points where such crossing would result in exceptional hazard, or would delay vehicular traffic movement. Barriers are used to channelize pedestrians into crosswalks, overpasses, underpasses, and at other points where the physical channelization of pedestrians is deemed necessary to prevent hazard or delay. Barriers should be installed only after careful study of vehicular and pedestrian volumes, and of conflicting movements and traffic accident records.

Pedestrian Safety Zones and Islands. These include refuge islands, loading zones, safety zones, and all other areas officially set aside for the exclusive use of pedestrians.

Street and Highway Lighting. Street lighting is successful in reducing night pedestrian accidents, but the extent of this success has not been accurately determined. The location and intensity of highway lighting should be based on vehicular and pedestrian volumes, but all urban streets should be lighted to uniform standards. The subject of street and highway lighting is discussed elsewhere.

In absolute numbers, pedestrian fatalities at night are not significantly different from those that occur during daylight hours. In 1967, some 55 per cent of urban pedestrian deaths, and 53 per cent of rural pedestrian deaths, occurred at night. These percentages are almost identical with those of the overall motor vehicle deaths at night. Even though night-related fatalities total only a few thousand more in absolute numbers, the picture is radically altered if data is converted to a mileage death-rate index. The number of deaths per 100,000,000 vehicle-miles, produces an urban night motor vehicle death rate of nearly 3 times that of the day rate. In rural areas the situation is only slightly improved, with the night rate about 2½ times the day rate.[1]

Pedestrian Tunnels (Underpasses) and Overpasses. Tunnels or overpasses are warranted for high pedestrian and vehicular volumes and extreme hazard. They should be constructed when the problem cannot be solved in some simpler and more economical manner. Their use is normally restricted to special locations, such as factories, schools, sports arenas, railroad and freeway crossings, etc. They should be used in conjunction with pedestrian barriers and constructed with ramps instead of stairs (where space permits) to assure and encourage usage.

Overpasses are preferable to tunnels, since they are less expensive, and require no excavation, forced ventilation, or special drainage facilities. Overpasses are also safer, from the point of view of criminal attack. Tunnels, however, are more esthetically pleasing, require less vertical depth in comparison to the vertical rise of overpasses, and provide greater protection from the elements.

Pedestrian Considerations in the Use of Traffic Signals. One of the accepted warrants justifying the installation of a traffic signal is a minimum pedestrian volume. The details are given in the MUTCD. In addition, the design and operation of special pedestrian signals is treated. Pedestrian signals are traffic signals erected for the exclusive purpose of directing pedestrian traffic at signalized locations. There are several ways in which pedestrian signal intervals can be combined with vehicular signal intervals. The four basic combinations described in the MUTCD are discussed below.

A. Combined pedestrian-vehicular phase

Signal phasing wherein pedestrians may proceed to use certain crosswalks parallel to the through vehicular movement, and wherein vehicles are permitted to turn across the crosswalks, fall into this category. This type of operation is normally used for the following conditions:[83,84]

1. To provide a pedestrian clearance indication to permit pedestrians to clear the crosswalk before the right-of-way is transferred to waiting vehicles
2. To provide an indication for the pedestrian at crosswalks where the view of the vehicle signal is obstructed
3. To signal the pedestrian, by means of the WALK indication, at actuated locations that the pedestrian timing is in effect.

B. Exclusive crosswalk interval

Signal phasing wherein pedestrians may use certain crosswalks, but vehicles are not permitted to cross these walks, is termed an *exclusive crosswalk interval.* Pedestrian signals indicate WALK during such phases. This type of operation is normally applied for the following conditions:[83,84]

1. At locations having heavy vehicular-pedestrian conflict where vehicular turns across the path of the pedestrians are prohibited for part of the signal cycle or for the entire signal cycle
2. At wide intersections having heavy vehicular-pedestrian conflict, but where the time necessary for the exclusive pedestrian phase will be prohibitive
3. At intersections where some, but not all, crosswalks have vehicular-pedestrian conflict

Modifications sometimes used in this type of operation are the control of straight, left, and right vehicular movements by arrow signals. In one modification, pedestrians are held for a short time, giving right-turning movements priority. After this initial interval, both pedestrian and turning movements are allowed. The other modification holds the pedestrians while the turning and straight-through movements take place, and then holds the turns while moving the pedestrians with the straight-through movement. This modification is also known as *Share the Green*. The success of these modifications is reported in the literature.[85]

C. Leading pedestrian interval

Signal phasing wherein an exclusive pedestrian interval is given in advance of the vehicular indication. The indication is a steady WALK, then a flashing WALK when the combined vehicular and pedestrian phase begins.

D. Exclusive pedestrian phase

Pedestrians are given an exclusive phase during which they may cross in any direction. All vehicles are stopped during the phase. The pedestrian indication is a steady WALK. This type of operation is normally applied for the following conditions:[83,84]

1. At locations having particularly heavy vehicular-pedestrian conflict
2. At irregularly shaped intersections
3. At T-intersections, where all entering vehicles must turn and pedestrian volume is relatively heavy

This type of operation can only be successfully applied where streets are fairly narrow (about 50 to 60 feet wide, maximum) as the length of the signal cycle becomes prohibitively long for wider streets. This simply results in undue delay to both motorists and pedestrians. Various reports (mostly negative) on the experiences and characteristics of the exclusive pedestrian phase are given in the literature.[36,81-85]

In different parts of the country there are variations in the wording and color of pedestrian signal indications.[84,86,87] The MUTCD specifies that all pedestrian signals shall be rectangular in shape, and shall contain the lettered messages WALK and DON'T WALK. Pedestrian considerations with respect to the timing of the various phases of signals have been discussed previously.

Legislation and Enforcement

The desirable regulations for cooperative street use between drivers and pedestrians are found in the Uniform Vehicle Code, supplemented by the Model Traffic Ordinance, for application to municipalities.

Since pedestrians and drivers legally share the use of streets, it is appropriate that pedestrians should also be held responsible for any infraction of the regulations governing their use of the roadways. The seriousness of this problem becomes apparent upon examination of accident statistics. Pedestrian traffic accidents still account for over one-third of all traffic fatalities in urban areas on a national basis.

Proper application of selective enforcement and the penalizing of pedestrians for violations are essential if the power of enforcement in inducing pedestrians to accept their responsibilities is to be effective. Pedestrian protection programs must be thoroughly planned before executing them.[88-91]

Education

Prior to the adoption and enforcement of pedestrian regulations it is most important to institute pedestrian education programs. It is essential that these programs be carried out on a continuing basis. Major improvements in pedestrian conditions in the future will be largely dependent on more extensive and continuing traffic eduction programs.

Proper evaluation of accident statistics will clearly show the need for educational programs, and help define their scope. Education programs should include the following points:

1. The public should be acquainted with rules for cooperative street use.
2. Pedestrian responsibilities need to be impressed on the public, and every effort used to make the public accept them.
3. Education must be directed towards areas of major need, such as the elderly, children, nondrivers, and the foreign born, and against those practices which produce the most pedestrian accidents.
4. The public should be acquainted with certain basic information:
   a. They should be aware of the great differences of speed between the walking pedestrian and the moving vehicle.
   b. Pedestrians are relatively invisible to the driver at night.
   c. Pedestrians should realize the distances required for the driver to react and brake a vehicle to a stop.
   d. Pedestrians should realize that driving is a complicated job, and that the driver cannot look out for the pedestrian at all times.

Traffic education programs have no value unless they are disseminated and made to appeal to the public. The various media available for these purposes include:

1. Radio and television
2. Newspapers

3. Schools
4. Outdoor advertising, posters, car cards, leaflets, buttons, street signs, slogans
5. Civic and club groups

School Crossing Protection

There is no traffic accident that arouses greater public comment than one involving a child traveling to and from school. The police and traffic departments constantly receive requests for the use of more signs, traffic markings, signals, and police officers at school crossings.

A detailed procedure is recommended by the Institute of Traffic Engineers for establishing a program for school crossing protection.[92] The 1971 MUTCD[3] also contains a separate chapter treating school area control.

24

Coordination of Signal Systems

To achieve smooth flow on a street or highway system, it is not sufficient to ensure that each point of local conflict is efficiently controlled. Unless the controls at each of these locations are coordinated in some fashion, continuous, smooth flow on the system will be impossible. It is particularly important that signals in close proximity to each other be coordinated, to prevent inefficient stop-and-go flow from developing.

The MUTCD[3] recommends that signals within one-half mile of each other be coordinated on major streets and highways. With greater lengths between signals, vehicles tend to spread out of platoons, making coordination less effective. In these situations, intermediate spacer signals may be justified, for the sole purpose of maintaining platoons. A study of a four-lane, divided highway indicated that coordination was feasible for signal spacings up to 4,600 feet.[2]

Spacer signals, installed at intermediate locations which otherwise do not warrant the use of signals to encourage platoon movement, should be used with care, because of the added vehicular delay caused.

Because signal patterns must repeat themselves each cycle, pre-timed signals are generally used for coordination. Some types of actuated equipment, primarily semi-actuated devices, may also be coordinated, with some loss of the flexibility for which signals are designed. It is also possible, under certain circumstances, to insert a volume-density controller at an isolated location within a coordinated system, where multi-phase or other requirements make pre-timed control inefficient. Under such conditions, platoons from nearby signals will drive the volume-density controller on progression. This, however, only occurs under conditions of heavy platoon flow in one direction only, and such use of the volume-density controller should be made with extreme caution.

There are a number of considerations in determining a timing plan for a given signal system. The prime elements are:

1. *Type of signal system:* One-way arterial, two-way arterial, or network type.
2. *Progressed movements:* On a two-way arterial, one or both movements might be progressed; in a network, preferential paths must be determined and progressed.
3. *Objective:* Progressions must be set with some objective in mind, i.e., maximizing bandwidth ("windows" of green for traveling vehicles), minimizing delay, minimizing a combination of stops and delay, or some other policy.
4. *Special inclusions:* Vehicles queued at intersections may be accounted for; critical intersections, special phasing, and cycle lengths in multiples or submultiples of the system cycle length must be considered.

It should be noted that all signals within the same

signal system must generally have the same cycle length, to make it possible for the pattern of timings to repeat every cycle. Special conditions, however, may make a limited number of multiple or submultiple cycle lengths desirable.

In considering coordinated control, *cycle length* is defined as the time for a complete set of phases to be exhibited at any system intersection, and is truly fixed only for a pre-timed signal with fixed settings. *Offset*, or *signal offset* between any two signals is defined as the time duration between the initiation of the progressed movement (phase) common to them at the two intersections, and is generally measured at the downstream intersection relative to the upstream intersection. For the purpose of minimizing differences from desired values, offsets are best expressed between contiguous intersections, since it is these offsets that have the physical meaning of discomfort or waste to the system user, the driver, and not the offset of the tenth intersection in reference to third, for instance.

Semi-actuated signals may be built into a progressed system only if their function is modified so that they are "blocked out," or "forced off," so that they cannot interfere with the progressed band moving down the arterial. This, however, results in a partial loss of the flexibility for which semi-actuated signals were designed.

Certain conditions can significantly reduce the efficiency of pre-timed signal systems, including:

1. Inadequate roadway capacity
2. Interferences from parking and loading operations
3. Complicated intersections, involving multi-phase control
4. Heavy turning volumes
5. Wide variability in traffic speeds
6. Very short signal spacing (does not affect simultaneous systems)

TYPES OF COORDINATED SIGNAL SYSTEMS

Single-Dial Interconnect

In this system, there is one dial in each intersection controller, allowing only one choice of cycle length and split. One to three offset conditions are allowed. All functions are set at local intersections, but the coordination is kept from a master intersection or from a central master clock.

Three-Dial Interconnect

Three timing mechanisms are present at each intersection, allowing three separate pairs of cycle length and split, and up to three offsets for each pair. The selection of the appropriate pair can be made either with a local time clock or by a remote master or control. Although three cycle lengths are possible at each intersection individually, the system itself is restricted to a common cycle length for all intersections at any given time on the basis of coordination considerations.

A three-dial system has the advantage of allowing separate patterns for an inbound peak, an outbound peak, and a mid-day balance period for a two-way arterial. It could also easily be used for various flow and (queuing) conditions on a one-way arterial.

Traffic Responsive System

A variety of systems exist which set cycle length and timing pattern according to input data from detectors, rather than by time-of-day, but which have the functions either built into hardware or programed into "soft ware." An early and still used type of this equipment, is the PR System, which will be described as an example of the kinds of functions which are commonly performed.

A basic element of the PR system is the *cycle computer*, which is a special purpose analog computer designed specifically for the signal system task. It receives inputs from one to four volume detectors, and specifies a recommended cycle length on the basis of these. The feasible range of volumes is divided into six regions by five thresholds. Depending on the observed volume, a cycle length code (A, B, . . . , F) is specified; this code corresponds to a cycle length, according to some independent dial settings. To avoid recommendations which would lead to oscillation in the cycle length, the threshold for going from C to D, for example, is different than that for going from D to C, so that minor volume fluctuations cannot easily cause continual switching.

The recommendations from competing cycle computers (two directions on a two-way arterial, for instance) are fed into a *system selector*, which selects the longest requested cycle length and the offset pattern on the basis of the *difference* in requested cycle-length codes; the setting can be preferential in either direction, balanced, simultaneous, or "free." Free operation implies no coordination, and holds only when cycle computers request code A (lightest volume).

Given a selected cycle length, a split for each intersection is selected from the three possible splits at each intersection by a predetermined assignment of a particular split to each cycle length. These splits are set locally. For the given offset *pattern*, the particular offset assignment for that pattern is also made locally, and is expressed as a percentage of the cycle length.

To illustrate, an inbound cycle computer request of D and an outbound request of B would result in a cycle length corresponding to code D, and probably (depending on the prior settings), a preferential-inbound offset pattern. At each local intersection, the split coded to cycle length D would be enforced, and the offset percentage set on the dial corresponding to "inbound-preferential" would be enforced.

It should be noted that although this discussion is directed to pre-timed operation, PR systems with semi-actuated features are also implemented.

Computer Controlled

There have been a number of systems implemented in the past decade using a general purpose digital computer to control traffic signals, and there is a definite trend for more systems. This trend is presently being accentuated by the complexity of some of the area and freeway signal systems now being undertaken as unified projects, by the acceptance of computer control as an operational system by funding agencies, and by recent breakthroughs in the cost and size of computers.

In North America, those systems in operation tend to be library-oriented, with many stored solutions, and tend to mimic multiple-dial or PR-type systems in function. This is seen in the systems in Toronto, Wichita Falls, and New York City,[3,4,5] and the experiments in San Jose.[6] These systems have realized significant improvements in traffic operation and substantial benefits in surveillance and in equipment monitoring (signal failures, etc.), but have done this primarily by extensions of existing time-of-day and gross responsiveness signal coodination schemes. Further work is continuing in responsiveness on some of these projects, but the prime current efforts from which highly responsive control formulations can be expected are the UTCS Project in Washington, D.C.,[7] and the project, "Optimizing Flow in an Urban Freeway Corridor," in Dallas, Texas.[8]

In Europe, there has been considerably more emphasis on highly responsive local intersection control within a coordinated framework. This is illustrated by projects in West London[9] and Glasgow.[10] Work is also under way in Munich,[11] and in Madrid and Barcelona.[12]

THE TIME-SPACE DIAGRAM: A BASIC TOOL

On an arterial, or on a route through a network, the basic objective is generally to "keep the vehicles moving and use the green as efficiently as possible." This generally means that the offsets should be timed so that moving platoons of vehicles may sweep along the arterial just as space becomes available for them to do so. This is best seen pictorially by means of a *time-space diagram*. This is simply a two-dimensional representation of (1) the spacing of the various signals along the arterial or path, and (2) the signal indication of each of these signals as a function of time. In drawing this indication, an *amber* is indicated by a shading (//////////), a *red* by a solid (▬▬), and a *green* by an absence of other indications. A simple time-space diagram for four signals is shown in Figure 24-1.

For the indicated signal pattern, note that the offset between signals B and A in the northbound direction is 15 seconds; between C and A it is 0 seconds. This observes a convention that the offset be expressed as less than a cycle length. Also, defining *bandwidth* as the width of the "window" of green in a given direction, note that the northbound bandwidth BW_1 is approximately 22.5 seconds, and the southbound bandwidth BW_2 *happens* to be the same.

Fig. 24-1. A time-space diagram

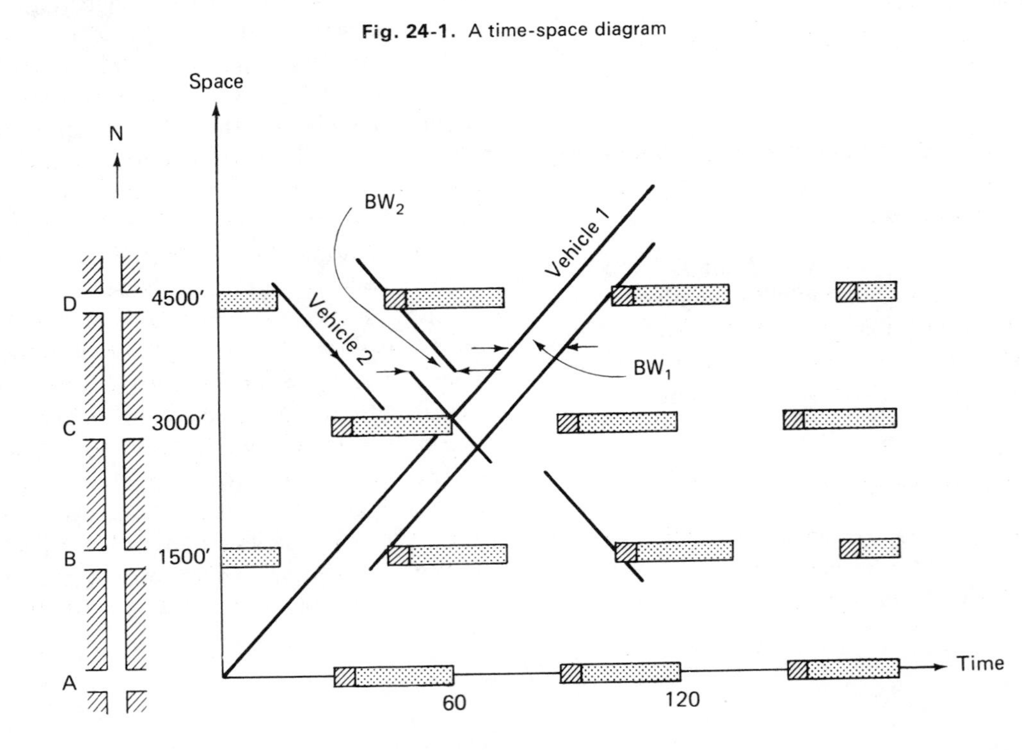

The time-space diagram is also useful for tracing vehicle trajectories and for predicting performance. If vehicle 1 were stopped at signal A at $t = 0$, northbound, and proceeded at 50 feet per second, it would pass through the signal system without stopping. At the same speed, however, southbound vehicle 2 would have to wait until $t = 15$ seconds to start from signal D, and would be stopped at signal C.

PRE-TIMED SIGNAL COORDINATION SCHEMES FOR ARTERIALS

There are several special classifications of signal patterns which were historically among the first implemented, and which still have special application today. They may, however, be regarded as special cases of a more general (or flexible) progression pattern. They are the *simultaneous*, the *alternate*, the *double alternate*, and the *simple* progression.

Simultaneous System

All signals in a simultaneous system have a common indication at all times. Alternatively, all offsets are zero, and all splits are equal. The speed which a vehicle must maintain to just arrive at each signal at green initiation is given by:

$$V = \frac{L}{C} \tag{24-1}$$

where V = speed in feet per second
C = cycle length in seconds
L = signal spacing in feet, usually measured between centers of signalized intersections

Progression is also possible at $\frac{1}{2}$, $\frac{1}{3}$, $\frac{1}{4}$, etc., of the above speed, which results in the vehicle taking 2, 3, and 4 cycle lengths, respectively, to traverse the distance between signals.

Where signal installations are relatively close, or speeds abnormally high, continuous movement is possible. However, in the normal case, continuous movement is not feasible because a very short cycle is necessary. In addition, it is necessary to have signal spacings (or block lengths) fairly uniform, since this distance is assumed constant in the above formula.

The simultaneous system is one of the early types of signal systems, and it has very limited application in modern traffic signal practice. In most applications the simultaneous system has *serious operating disadvantages:*

1. The simultaneous stopping of all traffic along the highway prevents continuous movement of vehicles, and tends to result in high speed between stops, particularly during low-volume periods, with low overall speed.
2. Cycle length and interval proportioning are usually controlled by the requirements of one or two major intersections in the system. This often creates serious inefficiencies at the remaining intersections.
3. When the major street is completely filled with a continuous line of traffic, and this traffic is stopped on a *red* indication, vehicles from the minor street often have difficulty in turning onto, or in crossing, the major street.

The simultaneous system may be *used to advantage* under the following conditions:

1. With closely spaced adjacent signals which are a part of a larger group to be coordinated, simultaneous operation will provide for nearly continuous movement.
2. Under conditions of very heavy traffic a simultaneous system may substantially improve operation of traffic while other systems may break down because of the accumulation of vehicles between signals.

Alternate System

An *alternate system* is a signal system in which alternate signals, or groups of signals, give opposite indications to a given street at the same time. In a *single alternate system* each successive installation shows the opposite signal indication from the previous one at any time. In a *double alternate system* pairs of adjacent installations operate simultaneously, and the signal indications alternate between each succeeding pair; similarly with a *triple alternate system*, wherein a group of three signals operate simultaneously, and the signal indications alternate between groups. To provide for continuous movement in an alternate system on a two-way street, *it is necessary to use a 50-50 split of the cycle.*

The relationship between the speed for continuous movement in both directions, cycle length, and spacing between signal locations in a *single alternate system* is as follows:

$$V\left(\frac{C}{2}\right) = L$$

$$V = \frac{2L}{C} \tag{24-2}$$

where V = speed of progression, in feet per second
C = cycle length, in seconds
L = signal spacing, in feet

Progression is also possible at $\frac{1}{3}$, $\frac{1}{5}$, $\frac{1}{7}$, etc., of the above speed, which results in the vehicle taking 1-$\frac{1}{2}$, 2-$\frac{1}{2}$, 3-$\frac{1}{2}$ cycle lengths, respectively, to traverse the distance between signals.

The relationship between the speed for continuous movement in both directions, cycle length, and spacing between signal locations in a *double alternate* system is as follows:

$$V\left(\frac{C}{2}\right) = 2L$$

$$V = \frac{4L}{C} \qquad (24\text{-}3)$$

Progression is also possible at $\frac{1}{5}$, $\frac{1}{9}$, $\frac{1}{13}$, etc., of this speed, which results in the vehicle taking $1\frac{1}{4}$, $2\frac{1}{4}$, $3\frac{1}{4}$ cycle lengths, respectively, to traverse the distance between signals.

The alternate system is generally an improvement over the simultaneous system, in that continuous movement can be achieved at more reasonable speeds (higher) in the alternate system. However, this can only be achieved if the block lengths, or alternating groups of blocks, are essentially equal, and a 50–50 cycle split is provided. This cycle division may be satisfactory where two major streets intersect, but it gives too much green time to minor streets crossing major arteries.

The alternate system has *limited application* because of the following reasons:

1. It requires substantially equal green phases for both major and minor streets, which is likely to be inefficient at most of the intersections.
2. It is not well adapted to a street having blocks of unequal length.
3. In the double alternate system (or groups of signals), the capacity of the street may be materially reduced during heavy traffic, since the latter part of the vehicle group will be stopped by the second signal in the group when the signal indication changes.
4. Adjustments for changing traffic conditions are difficult to make. In one case reported, a conversion from a triple alternate signal system to a simultaneous system resulted in an *increase* in average speed, a *decrease* in time to traverse the street, and a *reduction* in the number of stops for traffic signals along the street.[13]

Simple Progressive System

In a simple progressive system, all offsets are so arranged that a vehicle entering the system in the progressed direction, just after the green initiation of the first signal, will arrive at all other signals just after the green initiation of that signal. The offset between any two signals (in the progressed direction) is thus simply the *distance* between them divided by the *speed* of the progressed vehicles.

Unlike the previous systems, there is no restriction on the individual splits at the several intersections; these may be determined completely locally. It should be noted, however, that the minimum main street green defines the bandwidth in the progressed direction.

The simple progressive system is normally far more efficient than either the simultaneous or the alternate system, but it does not provide the full flexibility necessary at many locations to meet the variations in traffic flows, such as heavy inbound movement during the morning, and heavy outbound movement during the late afternoon. This type of flexibility is most desirable in a modern traffic signal system.

Flexible Progressive System

In this kind of system the parameters of a simple progression may be redefined several times during a period of operation (for instance, a day), so as to reflect changing needs at individual intersections, along the original progressed direction, or in a new direction (inbound, outbound). This classification reflects a redefinition of the simple progression in force, more than anything else.

General Progressive System

This term is defined herein to denote a system in which the offsets are determined so as to maintain smooth flow of a platoon of vehicles, by accounting for changes in desired vehicle speeds along the arterial, vehicles stored at signals and needing clearance, and other such factors. Unlike the first three systems, the progressed speed of signal-switchings and of vehicle movements can be markedly different, with the former being quite irregular over the arterial so that the latter can be regular.

The general progression, as defined above, has the advantage of being a single formalism which can yield other types of progression as needed. This type also emphasizes the difference between regularity of signal pattern and of vehicle movement, these two concepts being indistinct in other types.

The use of this progression type is illustrated in Figure 24-2, wherein equal per-lane queues are assumed to exist at all intersections. Given that these queues are zero (case A), a simple progression results; given queues of size three (3), with no turning off the arterial (case B), a simultaneous progression results. Further increase naturally yields (case C) a "backward" progression, coming upstream rather than the conventional, or expected sweep downstream. It is interesting to note, however, that the physical progression speed of entering vehicles is the same (40 feet per second) for all cases. Case D illustrates how the signal pattern can become irregular in order to maintain tight and regular pro-

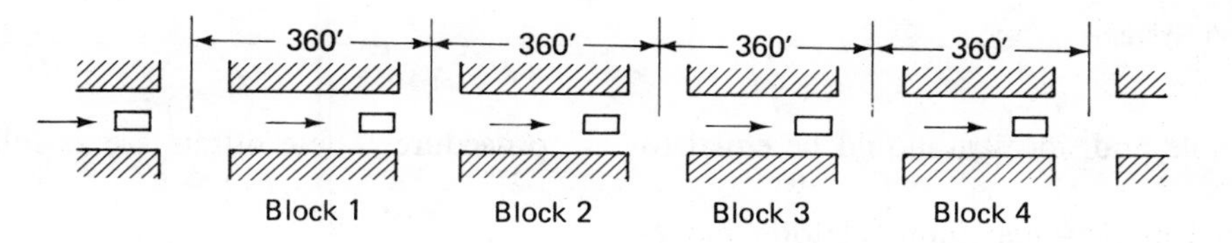

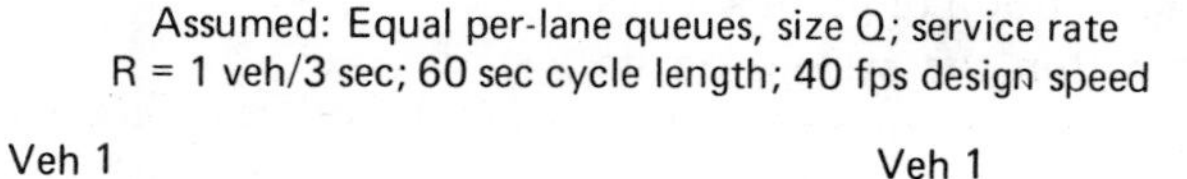

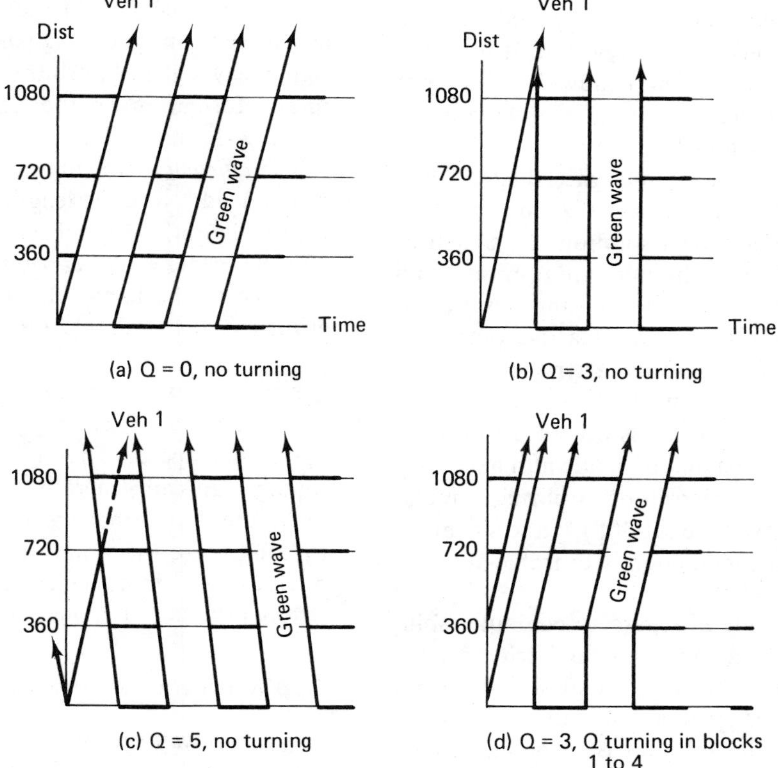

Fig. 24-2. One-way arterial—smooth-flow responses

gression; unequal internal queues, and/or turning, would only accentuate this pattern.

Some other advantages of the general progressive system are:

1. Continuous movement of entire groups of vehicles is possible with a minimum of delay and at an average speed planned for the system.
2. A high degree of efficiency can be obtained by proportioning the intervals to fit the traffic requirements at each intersection.
3. Speeding is discouraged because a vehicle must make frequent stops if it exceeds the speed for which the system is planned.
4. Difference in block lengths can be handled better than with other pre-timed systems.

Progressive systems along major urban streets are generally timed for speeds ranging from 20 to 30 miles per hour; in outlying areas higher system speeds are desirable. In general, design speeds should decrease with: increasing volume, concentration and activity of the area, mixed traffic composition, pedestrian and cross traffic, and decreasing pavement width.

TIMING THE PROGRESSIVE SYSTEM

A common objective in timing a progressive system is to permit a platoon of vehicles to move along the roadway, at a given speed, without stopping. Therefore, the offset at each intersection must be determined so that the first vehicle of the platoon will receive the green indication just as it reaches the intersection. The number of vehicles in the platoon that can pass through the intersection is dependent upon the width of the through band. The *through band* is the time, in seconds, elapsed between the passing of the first and last possible vehicle in a group of vehicles, moving in accordance with the designed speed of a progressive signal system. The through band should be as wide as possible and, ideally, should be equal to the green phase.

The criterion cited above—maximum platoon movement—generally assumes that the platoon of interest does not encounter queued vehicles within the system on its trajectory. Under this condition, the criterion of attaining *maximum bandwidth*, subject to specified balance between the two directions, is quite relevant. It has been shown, however, that maximum bandwidth does not minimize total delay under flow or turning conditions which yield queue formations.[14]

As discussed previously, the time-space diagram is a graphic technique for illustrating signalization schemes. It is the basic tool in graphical solutions of signalization problems. If graphical techniques are to be employed, the maximum bandwidth criterion and a standard technique for achieving it is the only feasible option of achieving this timing for two-way arterials. One-way street solutions are generally trivial.

A significant amount of work has been done on both computer algorithms for maximum bandwidth and other criteria,[15-17] signalization algorithms utilizing simulation subprograms (for example, S1G0P),[18] and signalization algorithms used in conjunction with traffic-adjusted control.[19]

Two points can be made in regard to computer solutions: (1) the underlying assumptions and criteria must be precisely ascertained and accepted, lest the output be simply a collection of meaningless numbers, and (2) a graphical display of the solution is recommended.

Application to a One-Way Street

Considering the case of timing a progressive system for a one-way street, it will become evident as the discussion proceeds that this case is an ideal situation for the timing of a progressive system. Assume that the system is made up of signal locations A through F, as shown in Figure 24-3. The cycle length for the system, based upon the demands at the heaviest traveled intersection, is 50 seconds. The split at each intersection, expressed in per cent, is indicated in Figure 24-3. The desired speed for the system is 18 miles per hour, and the vehicular clearance interval at each intersection is 3 seconds.

For one-way streets, the offset between signal installations is readily determined without any trial and error procedure. These offsets are as follows:

$$T_{\text{off}} = \frac{L}{V}$$

where V is in feet per second (18 miles per hour is equal to 26.4 feet per second). The offsets may be expressed either as between contiguous signals or with reference to a base signal, say signal A, arbitrarily but conventionally. The total information on the signal timing is contained in Table 24-1, and is illustrated in Figure 24-4.

The *efficiency* of a timing plan is defined as the ratio of bandwidth to cycle length, expressed as a percentage. From Figure 24-4 (or Table 24-1), the efficiency is $(\frac{22}{50})$ 100 per cent = 44 per cent, where the amber is not included in the bandwidth determination. This is good efficiency, the desirable range being from 40 to 55 per cent.

For a one-way street, note that any desired speed can be accommodated by simply calculating the offsets based upon that speed. The full efficiency of a progressive timing plan can generally only be realized for one-way operation, since there is no opposing flow of vehicles to complicate the timing of the system. It is therefore obvious why more efficient movement of traffic is a strong argument for one-way operation.

Application to a Two-Way Arterial

For a two-way operation with a simultaneous or alternate system, it is possible to obtain an ideal progression for both directions when the signal installations are evenly spaced and the offset between signal locations is one-half the cycle length, or some multiple of $\frac{1}{2}$ C. From previous discussion of these systems, it is clear that for a given cycle length the speed of progression is directly proportional to the distance between signal installations.

It is extremely important to remember that the timing of two-way signal systems is considerably simplified when the block lengths are essentially equal. Therefore, consideration should be given to this factor in developing any new city plans.

Unfortunately, the normal condition encountered when devising a timing plan is one of unequal signal spacings. The complication can be readily understood if

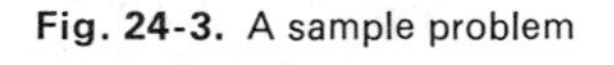
Fig. 24-3. A sample problem

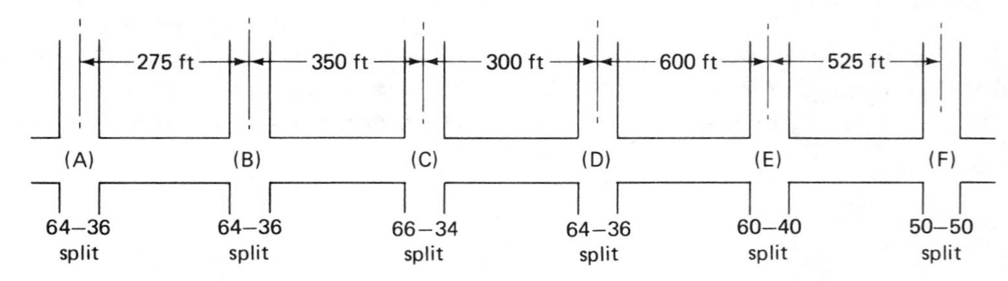

TABLE 24-1. AN ILLUSTRATIVE PROBLEM, C = 50 Seconds

SIGNAL	*SPLIT*	*PHASE 1* *GRN Main*	*PHASE 1* *AMB Main*	*PHASE 2 RED MAIN*	*OFFSET TO UPSTREAM CONTIGUOUS*	*OFFSET TO SIGNAL A*
A	64–36	28	3	19	—	—
B	64–36	28	3	19	10.4	10.4
C	66–34	29	3	18	13.2	23.6
D	64–36	28	3	19	11.4	35.0
E	60–40	26.5	3	20.5	22.7	7.7
F	50–50	22	3	25	19.9	27.6

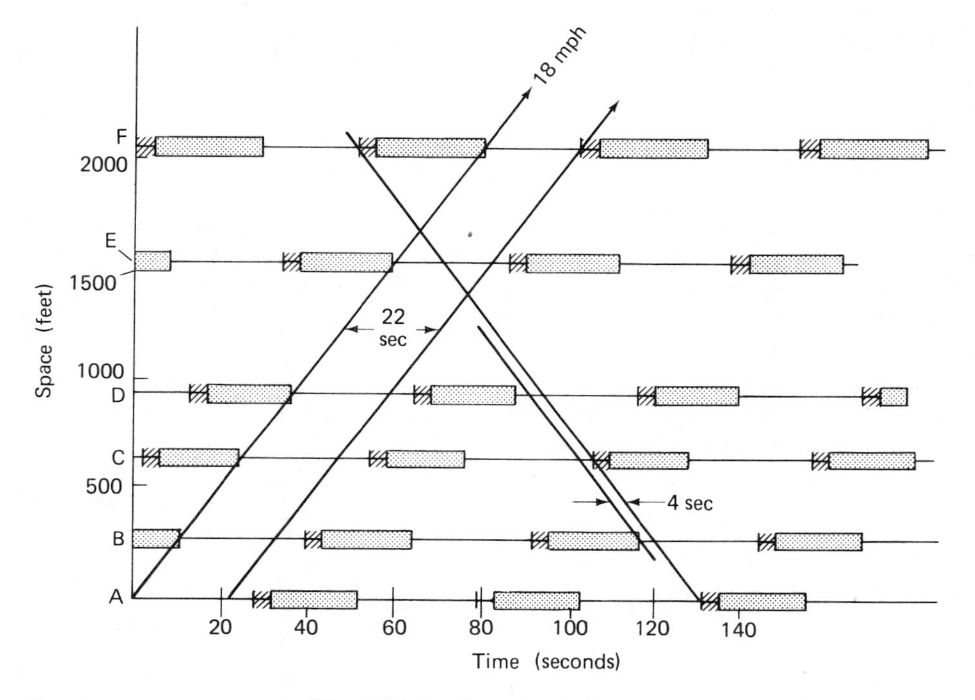

Fig. 24-4. An illustrative problem

Figure 24-4 is assumed to be a two-way street. The system is made up of signal locations A through F, and, as in the one-way progression; the cycle length is assumed to be 50 seconds, with splits as shown in Figure 24-3, a clearance interval of 3 seconds, and a desired speed in both directions of flow of 18 miles per hour.

Three signal plans are most common in considering such a situation: *preferential inbound*, *preferential outbound*, and *balanced.* For a maximum bandwidth criterion, and complete preference (so that for progression design the arterial is effectively one-way), the first plan has already been executed above, and the second may be executed just as easily. Note from Figure 24-4 that only a 4-second bandwidth exists in the nonpreference direction in the first case, but that this can be improved markedly by initiating the main street green *earlier* at signals B, D, and E (if necessary), without affecting the preference bandwidth.

In addition to such visual improvements on the preference plans, one may execute a balanced, or weighted, plan by trial-and-error, by some graphical technique (Bower's technique,[20] for instance), or by one of the computer techniques. Most importantly, it may be prudent to reevaluate the criterion under which the solution is being generated. For instance, should maximum bandwidth be continued with a computer solution, given certain internal queuing or storage patterns?

PRE-TIMED SIGNAL NETWORK COORDINATION

The problem of progressively timing networks of signalized intersections with nonuniform block lengths is extremely complex. The principal reason for this is the closure condition. Given a specified cycle length and phase assignments, and (typically) four intersecting streets, specification of three offsets in the loop thus formed predetermines the fourth offset. The same problem underlies the two-way street (upstream determines downstream), but is not as critical there. Consider the problems of interlocking time-space diagrams which occur when timing a network.

The problem only arises when the design objective is to choose the "best" offsets around such loops. The problem does *not* arise if, for instance, the design objectives were to progressively set all north-south (NS) one-way streets in Manhattan, and *one* east-west (EW) street, say 42nd Street. However, all other EW streets are then determined.

DEFINITIONS

Signal Network: Any two or more intersecting signal systems employing the same cycle length. The basic requirement for coordinating signal systems is that a common cycle must be employed.

Closed Network: A signal network formed by three or more signal systems, each of which has a signal in common with each of two other signal systems. There is a special case of a closed network formed by two signal systems shown in Figure 24-5.

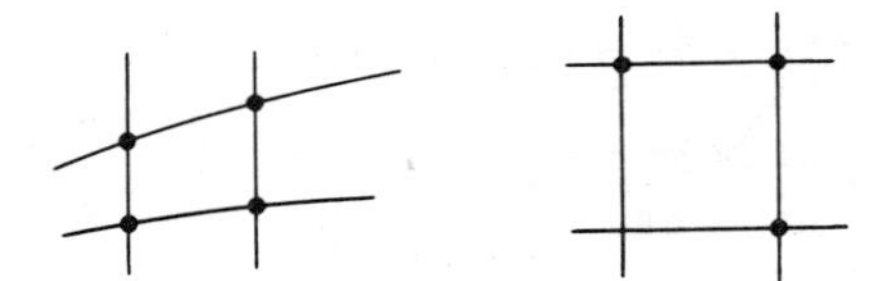

(a) A closed network (b) An open network

Fig. 24-5. Closed network vs. open network

Open Network: A signal network formed by two or more intersecting signal systems that do not create a closed network, shown in Figure 24-5.

Time reference point: In the signal system for any one street, the time-space diagram is normally expressed in terms of the offset between each signal in the system, and one particular signal which serves as the time reference signal. The start of green of the time reference signal on this street is defined as the time reference point.

Network offset: The time relationship between two intersecting signal systems may be expressed in terms of the offset between the time reference points of the two systems. If the offset between the time reference points of any two systems is fixed, then the offset between any signal in one system and any signal in the other system is fixed. Therefore, the offset between the time reference points of two intersecting signal systems expresses the time relationship between the two systems and is defined as the network offset.

The Closure Condition

To illustrate the formal relation of closure, consider the four intersecting one-way streets of Figure 24-6. Define the quantity $T_{D,I}$ as the offset of the green (relative to some reference) on the face in direction D at intersection I. Further define

$$T_{NS,1} = 0$$

and t_i as the offset between adjacent signals as defined in the figure. Thus,

$$T_{NS,2} = 0 + t_1$$

and

$$T_{EW,2} = T_{NS,2} + (\text{grn} + \text{amb})_{NS,1}$$

due to the "cornering" or re-alinement of the observer's viewpoint to continue around the loop. Similarly,

$$\begin{aligned} T_{EW,3} &= T_{EW,2} + t_2 \\ T_{NS,3} &= T_{EW,3} + (\text{grn} + \text{amb})_{EW,3} \\ T_{NS,4} &= T_{NS,4} + t_3 \\ T_{EW,4} &= T_{NS,4} + (\text{grn} + \text{amb})_{NS,4} \\ T_{EW,1} &= T_{EW,4} + t_4 \\ T_{NS,1} &= T_{EW,1} + (\text{grn} + \text{amb})_{EW,1} \end{aligned} \tag{24-4}$$

But this last term can only differ from the original statement $T_{NS,1} = 0$ by an integer multiple of the cycle length. Thus, successively substituting in Equation (24.4),

$$\left\{\begin{aligned} &t_1 + t_2 + t_3 + t_4 + (\text{grn} + \text{amb})_{NS,2} \\ &\quad + (\text{grn} + \text{amb})_{EW,3} + (\text{grn} + \text{amb})_{NS,4} \\ &\quad + (\text{grn} + \text{amb})_{EW,1} \end{aligned}\right\} = MC$$

where M = integer
C = cycle length

There are as many of these equations to be satisfied

Fig. 24-6. Network closure condition

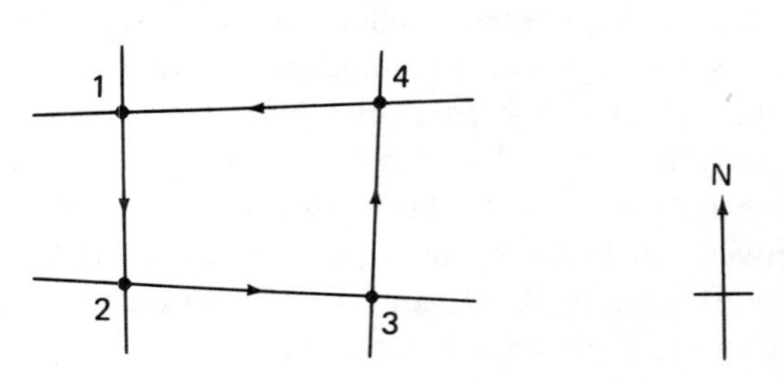

as there are closed loops in a line sketch of the network of interest, assuming leg crossings at nonintersection points can be avoided. This assumption is violated in some, but not all, overpass situations.

It is important to note that the network of interest is *not* the physical street grid, but simply that part of it on which a network progression is to be designed by specification.

Algorithms for Network Timing

The principal reference for manual solution of this type of problem is a 1948 article on an installation in Baltimore.[21] An extensive analysis is described therein for the installation. The technique utilized to time the network was essentially the development of good individual progressions, and a cut-and-try improvement. It contains the only network application reported in such detail in the literature.

Computer algorithms have been developed to treat this problem,[19,22,23] and field applications have been made. A successful and universally well-accepted solution has not yet been achieved, however.

There are at this time no standard warrants for computer control (traffic-adjusted control) of arterials and networks, nor are there standard warrants for determining the conditions under which a collection of streets must be treated as a network in signalization design.

Traffic Signal Optimization Program (SIGOP).[24] SIGOP is a computer program designed to determine optimum cycle lengths, phase splits, and coordination of traffic signal displays in a grid network. The program includes the following functions:

1. It computes and tabulates proper phase splits for each intersection.
2. It computes and tabulates ideal offset differences between each pair of intersections.
3. It determines optimum offset plan which optimizes a weighted sum of stops and delays in the networks.
4. It performs a coarse simulation of traffic operation on the timed network to estimate stops, costs, and delays.
5. It tabulates recommended traffic timing plan and prints selected time-space diagrams for designated arterials.

Volume Priority Method. This algorithm consists of ranking all links in the network in order of descending link volume (two-way). Link offsets are ideally timed, starting with the highest volume link and going on in order of descending volume, until links are reached which are already determined by other previous settings. This method resulted in the ideal timing of about 60 per cent of the network links in the study in which it was used. In this method no consideration is given to the delays on the non "ideally" timed streets. This procedure has the advantage that it minimizes reliance on computers.

Preferential Street Method. This method consists of choosing those streets which should be given preference, and timing these progressively. Streets are ranked in order of decreasing preference.

It must be noted that neither the volume priority method nor the preferential street method is truly a *network* timing plan, but are rather plans for the *decomposition* of networks, so as to effect relatively simple timing plans.

Consult the literature for a review of computer usage in network signalization.[25]

25

Speed Control and Zoning

It is extremely important to control speed in order to provide safety and efficiency for traffic movements. The maximum safe speed at any location will vary as traffic, road, weather, light, and other significant conditions change. Speed control assists the motorist in selecting speeds which are safe for the prevailing conditions.

Speed too great for prevailing conditions is the most frequently reported violation in fatal accident reports. It was a contributing factor in 31.7 per cent of all fatal accidents in 1968; it was a factor in 27.6 per cent of urban fatalities, and 32.9 per cent of rural fatalities. It was reported as a contributing factor in over 15.9 per cent of all traffic accidents; it was a factor in about 10.7 per cent of all urban accidents, and 28.5 per cent of all rural accidents.[1] Severity of accidents at higher speeds is much greater than at lower speeds. Speed too great for prevailing conditions occurs over the entire speed range, and it is the principal factor to be considered in controlling maximum speeds of vehicles for improved traffic safety. An unsafe speed can actually be quite low, in terms of miles per hour, while relatively high speeds can be safe under favorable conditions.[2]

Some principal findings from a detailed study on main rural two-lane and four-lane roads,[3] concerning the relationship between speed and accident involvement, were as follows:

1. The accident involvement, injury, and property damage rates were *highest* at very low speeds, *lowest* at about the average speed of all traffic, and *increased* at the very high speeds, particularly at night. Therefore, the greater the variation in speed of any vehicle from the average speed of all traffic, the greater its chance of being involved in an accident. One of the objectives of speed control is to produce a uniformly moving traffic stream.
2. The severity of accidents *increased* as speed increased, especially at speeds exceeding 60 miles per hour.
3. The fatality rate was *highest* at very high speeds, and lowest at about the average speed.
4. Pairs of passenger car drivers involved in two-car, rearend collisions were much more likely to be traveling at speed differences greatly in excess of those observed for pairs of cars in normal traffic.

Modern highways and vehicles can and should provide fast, safe, and convenient transportation service in all parts of the United States, if proper design and control measures are adopted and if drivers cooperate. Traffic engineers must be well-informed concerning all phases of speed problems. The need today is for uniform laws, speed zoning, and enforcement.

SPEED CONTROL LAWS

Speed control is controversial because of differences of opinion among engineers, enforcement officers, motorists, and people living along the street or highway concerning the solution of the speed problem, and the relative influence of each group.

While there is considerable variance among states and municipalities as to the extent and nature of their speed regulations, there are three principal types of speed control laws:

Basic Speed Rule

The basic speed rule is that no person shall drive a vehicle on a highway at a speed greater than is reasonable and prudent under existing conditions, giving heed to the actual and potential hazards.

Prima Facie Speed Limit

This type of speed control law provides that any vehicle speed in excess of a designated numerical limit is presumed to constitute a violation of the basic speed rule. The burden of proof falls upon the driver to demonstrate that he was not traveling at an unsafe speed. Evidence that a person was driving faster than the designated speed limit is sufficient to authorize the court to find him guilty of violating the basic speed rule, but does not compel such verdict. If he chooses to do so, the defendant may introduce evidence to prove that the speed at which he was driving was not improper under the existing conditions. Therefore, *prima facie* limits allow for the fact that no particular rate of speed is necessarily safe or unsafe at all times.

The main disadvantages of this type of speed limit are that its meaning is often not understood, and enforcement is comparatively difficult.

Absolute Speed Limit

A speed limit above which it is always illegal to drive, regardless of conditions. This type of speed limit allows no question as to whether or not the driver was proceeding at a safe speed when he exceeded the speed limit.

Advantages of absolute speed limits are their clarity of meaning, and the comparative ease with which verdicts can be reached. Absolute speed limits tend to be set somewhat higher than *prima facie* limits, as a result of their nature and requirements.

LEGISLATION FOR SPEED CONTROL

The nature and provisions of speed laws vary widely among states. Every state has a law equivalent to the basic speed rule. The Uniform Vehicle Code, and Model Traffic Ordinance, provide the models for state speed laws, and city ordinances, respectively. Consult the literature[4] for the speed laws that were in existence in the various states as of December 1, 1963.

Absolute Speed Limits

The Code speed law is the basic speed rule, supplemented by absolute speed limits (Section 11-801):

1. 30 miles per hour in any urban district
2. 60 miles per hour in other locations during the daytime
3. 55 miles per hour in other locations during the nighttime

Daytime means from a half hour before sunrise to a half hour after sunset. *Nighttime* means any other hour.

State Speed Zones

The Code permits the establishment of state speed zones (Section 11-802).

Whenever the state determines, upon the basis of an engineering and traffic investigation, that any maximum speed is greater or less than is reasonable or safe under the conditions found to exist at any intersection or other location on the highway system, the state may determine and declare appropriate safe maximum speeds which shall be effective after proper signs have been posted. Such a maximum speed limit may be declared to be effective at all times or at such times as are indicated upon the posted signs; and different limits may be established for different times of the day, different type of vehicles, varying weather conditions, and other factors bearing on safe speeds.

Maximum Limits

The Code permits local authorities to alter maximum limits (Section 11-803).

Whenever local authorities determine, on the basis of an engineering and traffic investigation, that any maximum speed is greater or less than is reasonable and safe under the conditions found to exist upon a highway, or part of a highway, the local authority, with the approval of the state, may determine and declare an appropriate safe maximum limit thereon, which does one of the following:

1. Decreases the limit at intersections
2. Increases the limit within an urban district, but not to more than 60 miles per hour during daytime or 55 miles per hour during nighttime
3. Decreases the limit outside an urban district, but not to less than 35 miles per hour

Any altered limit shall be effective at all times, or during hours of darkness, or at other times as may be determined after proper signs have been posted.

Minimum Speed Limits

The Code provides for the establishment of minimum speed limits (Section 11-804).

The minimum speed limit is based upon that speed which impedes the normal and reasonable movement of traffic, except when reduced speed is necessary for safe operation, or in compliance with law. Determination of the minimum speed limit is based on engineering and traffic investigation. Regulation of minimum speeds has resulted from the fact that speeds much lower than those of normal traffic contribute to the accident hazard because they impede the free movement of traffic and often result in passing maneuvers being made by other motorists when it is not safe to do so.

Truck Speed Limits

Since the brake performance for different types of vehicles varies, the stopping distance for these vehicles will also differ. Trucks generally require longer distances to come to a stop in an emergency situation than do passenger cars. In view of the differences in braking characteristics between passenger cars and trucks, several states have adopted speed laws wherein speed limits for trucks are lower than the limits for passenger cars.

The then BPR tested various types of vehicles on level grades in 1955, to determine their stopping ability from a speed of 20 miles per hour.[5] The requirements of the Code (Section 12-302) on the performance ability of brakes have, for the most part, been based on the results of the BPR tests.

The Code requires that every motor vehicle, at all times and under all conditions of loading, upon application of the service brake, shall be capable of the following:

1. Developing a braking force that is not less than the percentage of its gross weight tabulated for its classification
2. Decelerating to a stop from not more than 20 miles per hour, at not less than the rate tabulated for its classification.
3. Stopping from a speed of 20 miles per hour, in not more than the distance tabulated for its classification, such distance to be measured from the beginning of brake application

These standards are shown in Table 25-1.

Predominant Speed Limits

Predominant speed limits, among the states, are 60 to 65 miles per hour for daytime passenger-car travel on rural roads, 25 to 30 miles per hour in residential districts, and 20 to 25 miles per hour in business districts. More than 30 states have established lower daytime speed limits on rural roads for trucks, while about 20 states have reduced speed limits for nighttime travel. About half of the states now have laws providing for the establishment of minimum speed limits, and this type of speed control is being utilized on certain roads and streets in about one-fourth of the states.

TABLE 25-1. BRAKING CHARACTERISTIC STANDARDS

CLASSIFICATION OF VEHICLES	*BRAKING FORCE (% of Gross Weight)*	*DECELERATION (ft/sec/sec)*	*STOPPING DISTANCE FROM 20 mph(ft)*
Passenger Vehicles	52.8	17	25
Motorcycles	43.5	14	30
Single Units below 10,000 lbs	43.5	14	30
Single Units, two-axle, over 10,000 lbs	43.5	14	40
Combinations with Gross Weight of Trailer not over 30,000 lbs and Buses without Weight Rating	43.5	14	40
All other Vehicles	43.5	14	50

(Source: *Uniform Vehicle Code,* National Committee on Uniform Traffic Laws and Ordinances, 1968)

EDUCATION FOR EFFECTIVE SPEED CONTROL

General public and driver education through all publicity channels and by all interested organizations play a very important part in the dissemination of the meaning of speed laws, the importance of observing posted limits for the general welfare of all, how to adjust speeds to conditions, and engineering and enforcement steps necessary to obtain speed control. As drivers understand and become familiar with traffic regulations, the more respect they will have for them.

ENGINEERING ASPECTS OF SPEED CONTROL

Highway Design

Design considerations, including such elements as sight distance, curvature, grades, cross-section dimensions, and other features, represent one limitation on safe speeds. Whenever it is feasible to correct hazards through improvement of the roadway or conditions surrounding it, that method should be preferred to speed restriction.

Speed Zoning

Establishment of safe and reasonable speed limits for certain special zones or sections of street or highway, where the general state-wide statutory speed limits do not fit the road and traffic conditions, is often necessary. Various locations at which speed zoning is commonly warranted include:

1. Transitional stretches of roadway, from rural to urban conditions.
2. Unusual roadway conditions, or other features which make it advisable to establish speed limits different from those applicable under general laws. Such conditions would include winding sections of roadway, sharp curvature, steep downgrades, restricted sight distance or view obstructions, restricted lateral clearance, poor surface conditions, and other dangerous locations.
3. Intersection approaches, particularly where view obstructions exist.
4. Highways and streets which have design standards considerably higher or lower than other highways in the state or community involved. Such examples are freeways, expressways, urban extensions of primary state highways, major streets and arterials.
5. Temporary zoning at construction sites and school zones.

The following factors should be considered in establishing speed zones and the maximum speed limit to be applied therein.[6]

Prevailing Vehicle Speed. Prevailing vehicle speed is a significantly important factor in speed zoning. If speed zoning is to be effective, speed limits must be generally consistent with speeds that drivers feel are safe and proper. This involves making spot speed studies to determine average, median, 15th percentile speed, 85th percentile speed, and pace. The criterion most generally used and preferred in determining the specific maximum speed limit from speed studies is the 85th percentile speed. Another criterion that is used in selecting a proper speed limit is the 10 miles per hour pace, as the numerical limit should not be set at a value below the lower limit of the pace.

Physical Features. Physical features of the section of roadway should be considered in determining whether or not a speed zone is desirable and what the numerical limits should be. These features may indicate the necessity to select a speed limit different from that based on prevailing speeds. Test runs should be made with a *ball-bank indicator* to determine the maximum comfortable speed on curves. The maximum speed at which a curve may be traveled comfortably is indicated by a reading of 10 degrees on a ball-bank indicator. For isolated sharp curves, *advisory speed signs* (which are not speed limits) will generally be used.

The spacing between intersections should be measured, since roadways with many intersections require lower speed limits because of the increased number of potential conflicts. The relationship of these to the appropriate speed limit is given in Table 25-2.

The number of roadside businesses per mile should be determined, because business establishments with highway access increase the roadside friction encountered by traffic using a highway. The relationship of these to the appropriate speed limit is shown in Table 25-2.

Roadway surface conditions and characteristics influence the maximum safe speed, and they should be reviewed when a judgment is made on whether or not a proposed speed limit is appropriate. These factors include slipperiness of pavement, roughness of pavement, presence of transverse dips and bumps, presence and condition of shoulders, and presence and width of median.

Accident Experience. Accident experience should be reviewed with respect to frequency, severity, type, and cause. Lowering of the speed limit will not necessarily reduce vehicle speeds, or result in fewer accidents. Colli-

TABLE 25-2. CHECK SHEET FOR SPEED ZONES

Part 1

HIGHWAY CONDITIONS (THREE OR MORE MUST BE SATISFIED)				*PRELIMINARY ESTIMATE OF MAXIMUM SPEED (mph)*
Design Speed (mph)	*Minimum Length of Zone Equals or Exceeds (miles)*	*Average Distance between Intersections Equals or Exceeds (feet)*	*Number of Roadside Businesses does not Exceed per mile*	
20	0.2	no min.	no max.	20
30	0.2	no min.	no max.	30
40	0.3	125	8	40
50	0.5	250	6	50
60	0.5	500	4	60
70	—	1000	1	70

Part 2

SPEED CHARACTERISTICS (TWO OR MORE MUST BE SATISFIED)			*MAXIMUM PROPOSED SPEED LIMIT (mph)*
85th Percentile Speed Between (mph)	*Limits of 10-mph Pace Between (mph)*	*Average Test Run Speed Equals or Exceeds (mph)*	
under 22.5	under 25	17.5	20
22.5–27.5	11–29	22.5	25
27.5–32.5	16–34	27.5	30
32.5–37.5	21–39	32.5	35
37.5–42.5	26–44	37.5	40
42.5–47.5	31–49	42.5	45
47.5–52.5	36–54	47.5	50
52.5–57.5	41–59	52.5	55
57.5–62.5	46–64	57.5	60
62.5–67.5	51–66	62.5	65
67.5 or over	over 55	67.5	70

(Source: *Traffic Engineering Handbook,* ITE, 1965, Figure 14-2)

sion frequencies and accident rates have often been reduced by raising speed limits to realistic levels. Particular attention should be given to those accidents in which unreasonable speeds appear to have been a causative or severity factor.

Traffic Characteristics and Control. Traffic characteristics and control should be considered when determining the appropriateness of a proposed speed limit. These factors include traffic volumes during peak and off-peak periods, parking and loading and other vehicle operations adjacent to travel lanes, proportion of commercial vehicles in traffic stream, turning movements and controls, traffic signals and other traffic control devices, and vehicle-pedestrian conflicts. Speed differentials are not objectionable on low-volume highways, but they become more hazardous as volumes increase.

Changes of Posted Speed Limits. It is impractical to change posted speed limits at intervals of less than 1,000 feet. This is reflected in the minimum length of a speed zone permitted, shown in Table 25-2. The minimum length of speed zone permitted increases as speed increases.

Setting the Limit

The "Check Sheet for Speed Zones," Table 25-2, will help the engineer to make an evaluation of the measurable conditions along the highway being considered for speed zoning. If three or more of the conditions stated are met, a speed zone which has a maximum as given may be appropriate. The value of the maximum limit thus determined is only a preliminary estimate, and must be considered in conjunction with the actual speed characteristics measured in the field.

In the selection of the numerical *value of the speed limit*, it is necessary to measure the actual vehicular speed characteristics. Spot speed studies are made to determine the 85th percentile speed and the pace. In addition, as a check on the speed values determined from the spot speed study, the average test-run speed should be found.

The average test-run speed is most simply determined by using a passenger car with a calibrated speedometer, from which the speed is recorded at 0.1-mile intervals over the length of the highway being studied. Runs are made under light traffic conditions, to insure that the conditions of the road and its environment, rather than traffic conditions, govern the driver's speed. Two runs in each direction are generally sufficient.

Maximum speed limits based on the 85th percentile speed and the 10 miles per hour pace are usually reasonable by all standards. The upper limit of the pace usually approximates the 85th percentile speed. It is recommended that the maximum speed limit selected should not be more than 3 miles per hour below the upper limit of the pace or the 85th percentile speed, whichever is the lower.

When the above values have been determined, a more accurate estimate of the speed limit to be set can be found by entering the second part of the "Check Sheet." The speed limit which satisfies two or more of the conditions shown is the limit which should be set for the speed zone.

Area Speed Limits

It is frequently advisable to establish speed limits on an area basis, rather than an individual highway basis. Many jurisdictions have laws setting forth specific speed limits on all roads and streets in areas or districts of a particular class. Localities where speed zoning of this nature is often desirable include:

1. Business and commercial districts
2. Residential districts
3. Industrial districts
4. Large school and other public institution areas
5. Park and recreational areas
6. Public assembly areas
7. Areas of much more intensive development than surrounding or adjacent areas

Drivers will expect and accept greater speed restrictions in the above areas. The factors to be considered in speed zoning on an area basis are identical to those used in speed zoning along individual highways. However, such speed limits should not preclude the establishment of higher or lower speed limits on individual streets within an area where warranted.

When speed limit selections are based on the type of locality or place involved, it is advisable to consider whether or not the speed restrictions need to be in effect at all times, or only during certain times or seasons.

Posting of Limits

Signs indicating the statutory speed limit (*prima facie* or absolute) should be erected at the entrance of the state and at boundaries of metropolitan areas. Whenever the speed limit is altered, a limit sign should be located at the beginning of each section where the speed is altered and at appropriate intermediate locations. At the end of each such section a limit sign should be posted, giving the next speed limit. A sign should be posted prior to a speed zone in a rural area at a distance of not less than 300 feet, nor more than 1,000 feet, in advance of the speed zone to which it applies.[7-9]

For additional comments about the theory of and need for speed zoning and speed control, and speed characteristics, consult the literature.[10-27] One study of speed zoning practices[26,27] reported the following results:

1. There is a strong relationship between the rate of occurrence of accidents and the speed distribution on rural state highways. The accident rate is significantly higher where the speed distribution is nonnormal, and the accident rate is reduced when the distribution is changed to a normal one.
2. The best parameter to use in determining nonnormality is the skewness of the distribution.
3. Changing the speed distribution from nonnormal to normal results in an accident rate reduction which is about twice that found under any other set of before-and-after conditions.
4. Warrants for speed zoning should be established which include the speed distribution as a factor.
5. The "before" speed distribution alone is not adequate as a warrant for speed zoning.

Speed Control Signals

When high vehicular speeds present problems at hazardous locations, speed control signals may prove to be useful.

Nonintersection. Nonintersection signals control the speed of vehicles at such locations as curves, bridges, and school zones. The signal indication is normally *red*, and the signal is actuated by a vehicle detector in advance of the signal. The interval of delay is set so that the signal will turn *green* to permit the passage of a vehicle traveling at or below the maximum allowable speed.

Intersection. Used at intersections where speeds on the artery approach present a particular hazard. There is full traffic-actuated control, and all signals normally show *red* in the absence of approaching traffic. A single vehicle, approaching the intersection and crossing the detector, will receive the *green* indication at the end of a vehicle-approach period. If the vehicle is traveling at or below the speed designated as safe for the location, the *green* indication will come on in time to allow passage of the vehicle without its being required to stop.

An additional method of utilizing traffic signals to effectively regulate vehicular speed is in a progressive signal system. The driver must drive at the speed for which the system is timed if he wishes to pass through a series of intersections without stopping.

Critical Speed Computations

Horizontal Curves. The maximum safe speed for horizontal curves must be based on the curvature, the rate of superelevation, and the safe coefficient of side friction.[28] The defining relationship is:

$$V^2 = 15R(e + f) \qquad (25\text{-}1)$$

where V = speed, in miles per hour
R = radius of the curve, in feet
e = rate of superelevation, in feet per foot
f = safe coefficient of side friction

Intersection Approach Speeds. Safe intersection approach speeds must be determined when a sight restriction exists. There must be unobstructed sight along both roads at an intersection, and across their included corner, for distances sufficient to allow the drivers of vehicles approaching simultaneously to see each other in time to prevent collision at the intersection, providing that there is no intersection control such as a YIELD sign, STOP sign, or traffic signal.

The space-time-velocity relations define the minimum sight triangle or visibility triangle (see Figure 25-1) required to be free of obstructions, or, if a sight triangle below the desirable minimum must be used, they fix the necessary modifications in approach speeds. Where an obstruction to a driver's sight distance exists, and the obstruction fixes the vertices of the visibility triangle at points that are less than the safe stopping distances from the intersection, vehicles may be brought to a stop (after sighting other vehicles on the intersecting road) only if they are traveling at a speed appropriate to the available sight distance. If the speed of a vehicle on one of the roads is assumed (for example, the design speed), the critical corresponding speed on the other road can be evaluated in terms of this assumed speed and the dimensions to the known obstruction.

The critical speed for vehicles approaching the intersection on the minor street can be calculated and then used as the basis for establishing a special speed limit at that point. The speed of vehicles on the major street can be taken as either the legal speed limit, the design speed for the highway, or a percentile speed, such as the 85th percentile speed.

Other uses for calculations of critical speeds at street intersections include:

1. Determining whether there is a view obstruction for the prevailing speeds at an intersection
2. Determining to what extent a view obstruction must be removed to make further speed restrictions unnecessary
3. Determining to what extent curb parking should be eliminated to remove the obstruction to view caused by such vehicles
4. Deciding whether there is sufficient view obstruction to require a YIELD sign or STOP sign
5. Establishing criteria for avoiding view obstruction hazards in the design of new streets or highways, or improvement of existing ones

There are several methods for computing the critical approach speeds for vehicles at intersections where the view is obstructed.[4,5,28,29] Each method is based upon slightly different assumptions regarding driver and vehicle performance. In each of the methods described below, the location of points A and B of the visibility triangle, Figure 25-1, are determined in the same manner. The critical speed for vehicle B is computed after determining the distance from B to collision point. Referring to Figure 25-1, in the typical case V_a is known (or assumed), and a and b are the known distances to the sight obstruction from the respective paths of vehicles A and B. Triangle 123 is similar to triangle 145, therefore:

$$\frac{D_b}{D_a} = \frac{a}{D_a - b} \quad \text{or} \quad D_b = \frac{aD_a}{D_a - b} \qquad (25\text{-}2)$$

National Safety Council Method. This method is applicable to streets intersecting at any angle, if the measurements a and b are taken parallel to the paths of vehicles B and A, respectively.

The assumption is made that vehicle B is traveling at a sufficiently low uniform speed so that when both vehicles reach their critical sight points at the same time, vehicle B can continue through the intersection at speed V_b and pass behind vehicle A, which also continues at a uniform speed. The distance D_a is taken equal to the driver safe stopping distance for the speed of vehicle A, plus a 16-foot clearance distance.

The approximate braking distance of a vehicle on level roadway is determined by use of the standard formula:[28]

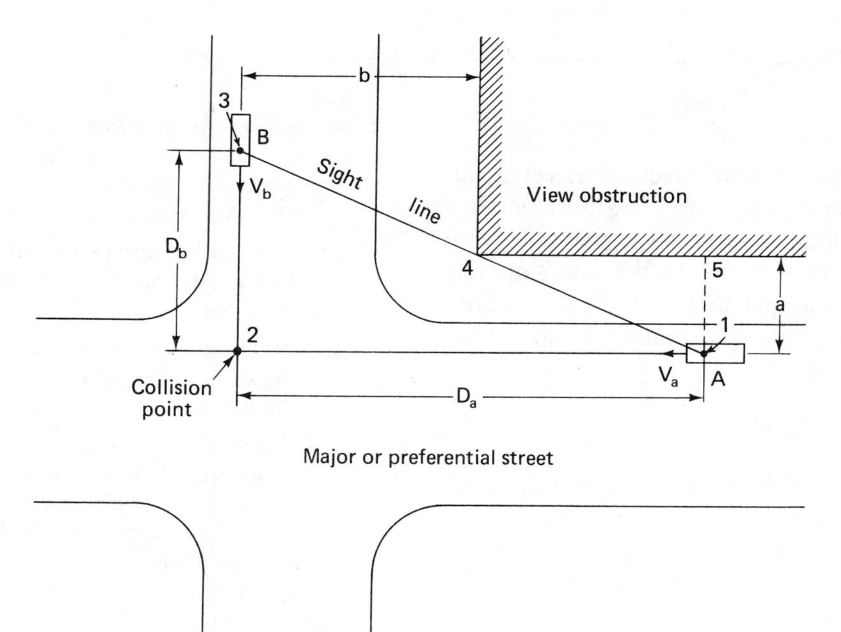

Fig. 25-1. Visibility triangle

$$d = \frac{V^2}{30f} \qquad (25\text{-}3)$$

where d = braking distance, in feet
V = initial speed, in miles per hour
f = coefficient of skidding friction

In the NSC method, f is assumed to be 0.53; then

$$d = \frac{V^2}{30 \times 0.53} = 0.063V^2 \qquad (25\text{-}4)$$

The safe stopping distance is then determined by using the formula

$$S = 0.063V^2 + 1.47tV \qquad (25\text{-}5)$$

where S = safe stopping distance, in feet
t = perception-reaction time, in seconds

V (mph):	0	10	20	30	40	50	60
t (sec):	0.76	0.80	0.90	1.00	1.22	1.55	2.00

Therefore,

$$D_a = 0.063V_a^2 + 1.47tV_a + 16$$

and since

$$D_b = \frac{aD_a}{D_a - b}$$

from Equation (25-2), the time for vehicle A to clear collision point

$$t_a = \frac{D_a + 12}{1.47V_a}$$

where the 12 represents half a car length. This same time should be assumed for vehicle B to travel a distance of $D_b - 16$, since vehicle B should not get closer to the collision point than the clearance distance of 16 feet; therefore

$$\frac{D_a + 12}{1.47V_a} = \frac{D_b - 16}{1.47V_b}$$

and the critical approach speed for vehicle B

$$V_b = \frac{(D_b - 16)V_a}{D_a + 12}$$

AAA method

The formulas given below for this method are applicable to right-angle intersections only. In this method, the distance D_a is based on the safe stopping distance, where f is taken equal to 0.50, plus a clearance distance of 15 feet. The perception-reaction time is assumed to be 1.0 second throughout all ranges of speed. Therefore,

$$D_a = 0.067V_a^2 + 1.47V_a + 15$$

V_a is assumed to be the legal speed limit, or the 85th percentile speed. The critical speed for vehicle B is based on the distance from B to the collision point, and this distance is assumed to be equal to the safe stopping distance for vehicle B, plus a clearance distance of 15 feet.

$$D_b = \frac{aD_a}{D_a - b} = 0.067V_b^2 + 1.47V_b + 15$$

and

$$V_b = -10.9 + 2.73\sqrt{2D_b - 14}$$

AASHO method

This method is applicable to streets intersecting at any angle, if the measurements a and b are taken parallel to the paths of vehicles B and A, respectively. In this method, the distance D_a is based on the safe stopping distance, where f is taken equal to 0.40, and a perception-reaction time of 2.0 seconds is assumed throughout all ranges of speed. Therefore,

$$D_a = 0.083V_a^2 + 2.93V_a$$

V_a is assumed to be the design speed, with a range of 30 to 70 miles per hour. The critical speed for vehicle B is based on the distance from B to the collision point, and this distance is assumed to be equal to the safe stopping distance for vehicle B.

$$D_b = \frac{aD_a}{D_a - b} = 0.083V_b^2 + 2.93V_b$$

and

$$V_b = -17.6 + \sqrt{12.1D_b + 310}$$

Illustrative example for the three methods

Assume $V_a = 30$ miles per hour, $a = 50$ feet, $b = 60$ feet.

By NSC method:

$$D_a = 0.063(30)^2 + 1.47 \times 1.0 \times 30 + 16 = 116.8 \text{ feet}$$

$$D_b = \frac{50 \times 116.8}{116.8 - 60} = 102.8 \text{ feet}$$

$$V_b = \frac{(102.8 - 16) \times 30}{116.8 + 12} = 20.2 \text{ miles per hour}$$

By AAA method:

$$D_a = 0.067(30)^2 + 1.47 \times 30 + 15 = 119.4 \text{ feet}$$

$$D_b = \frac{50 \times 119.4}{119.4 - 60} = 100.5 \text{ feet}$$

$$V_b = -10.9 + 2.73\sqrt{2 \times 100.5 - 14} = 26.5 \text{ miles per hour}$$

By AASHO method:

$$D_a = 0.083(30)^2 + 2.93 \times 30 = 162.6 \text{ feet}$$

$$D_b = \frac{50 \times 162.6}{162.6 - 60} = 79.2 \text{ feet}$$

$$V_b = -17.6 + \sqrt{12.1 \times 79.2 + 310} = 18.0 \text{ miles per hour}$$

Consult the literature for a method of obtaining the safe approach speed for a vehicle at a railroad crossing where the line of sight is obstructed.[30]

Summary of Major Points[18]

1. Extensive transition zoning, based upon engineering and traffic investigations, is extremely desirable.
2. Where critical speeds on curves are below the speed limit for the open highway, adequate posting of signs is essential.
3. Avoid zoning if limits cannot be established on the basis of engineering and traffic studies, or cannot be enforced.
4. If more than about 15 per cent of the vehicles exceed the numerical limit by more than 5 miles per hour after the establishment of a speed zone, the zone should be restudied to determine whether the limit should be raised or whether there are other contributing factors such as inadequate posting, or lack of enforcement or education.
5. The majority of drivers drive properly most of the time.
6. Most drivers select safe and proper speeds based on roadway and traffic conditions, and drivers have little respect for, and rarely obey, unreasonable speed limits.
7. Reasonable speed limits will have the greatest effect in slowing the faster driver, speeding up the slow drivers, and increasing the number traveling at or near the same speed.
8. A speed limit should seem too fast to at least 85 per cent of the drivers, or it is not a maximum limit.
9. Unreasonable speed limits cannot be enforced with reasonable enforcement. Speed limits should be for the most part self-enforcing.
10. Forcing drivers to drive more slowly does not mean that they are driving more safely.
11. There is no such thing as a safe speed limit. Traffic accidents occur at all speeds, as no single limit is effective for all driving conditions. The slow drivers are often a greater hazard on the roadway than the fast drivers.

ENFORCEMENT OF SPEED LIMITS

Legislation is only as effective as its enforcement. Proper enforcement can only be achieved with adequate trained personnel. It is essential that sufficient attention be given to speed control, with appropriations for personnel, training, and equipment. Training of police in the meaning and effective application of speed laws is extremely important.

Methods of Enforcement

Pacing Speed Violators. This is the most common procedure in use, and it involves having an officer follow the violator at a sufficient distance to ascertain his speed. In some cases, ordinances require pacing before arrests can be made. This method may be wasteful of personnel, and hazardous when traffic is heavy or when the violator is traveling at a high speed. The distance required for pacing is, in most cases, left to the discretion of the officer.

Distance Versus Time. This involves the determination of speed by measuring the time it takes for a vehicle to traverse a measured distance, such as is done in spot-speed studies. Various types of meters are used to determine the time.

Radar Meter. This device is effective, and decreases the number of personnel required for enforcement. The most common technique used in apprehending violators is the use of two vehicles. An officer in one vehicle, equipped with radar and radio-telephone, identifies the violator and transmits this information to a second officer, stationed some distance ahead in another vehicle. The officer in the second vehicle observes approaching traffic, and stops the violator. The use of radar could also prove to be unfair if it is not enforced with a proper tolerance. This method is unpopular, and is sometimes used in "speed traps."

Tolerance

Many factors enter into the accurate determination of speeds at which vehicles are traveling. When one vehicle is pacing another, it is difficult to accurately ascertain speed because of acceleration, deceleration, differences in speedometers, and other factors (such as traffic conditions). When meters are used to measure speed, it is necessary to take into consideration the limitations and accuracy of the meter, and techniques of speed measurement.[31] Because of these factors it is good practice for enforcement personnel to allow tolerances, whereby motorists are permitted to exceed speed limits by reasonable values, before apprehending violators. The unreasonableness of some speed limits, and the desire to maintain public support of the program, are additional reasons for tolerances. Public support is sometimes difficult to obtain unless it is properly appraised of the objectives of speed control. In most cases, a 5-miles per hour tolerance is considered reasonable.

Enforcing a law pertaining to slow drivers proved to be extremely difficult in one case that was reported.[32] Such laws, in general, are difficult to enforce because of the judgment required in determining that a motorist is driving at such a slow speed as to impede or block the normal and reasonable movement of traffic.

Studies indicate that the effect of patrol units on speeds of vehicles in the vicinity of the patrol unit is not influential over great distances.[33-39]

EXPERIENCES WITH SPEED ZONING

Many studies have been done with regard to speed zoning and are discussed briefly.

Illinois

In Illinois, several before-and-after studies were conducted to determine the effectiveness of a state-wide change in the speed law.[40] A study of 30 locations where speed zones had been increased from 30 miles per hour to 40 miles per hour indicated, for a typical zone, that the 85th percentile speed was 38.9 miles per hour before the zone was changed, and 38.4 miles per hour after the zone was changed. The selection of a reasonable speed for the zone was beneficial, since the observance was significantly better and accidents were reduced.

At 11 other locations where new speed zones were established, the results of the study showed a lowering of all speed characteristics in the after study. The 85th percentile speed decreased 3.6 miles per hour. The new speed limits were observed by 88 per cent of the drivers, which again illustrates the reasonableness of the limit.

At four other locations where speed zones were lowered, the results showed that the behavior of the traffic followed a pattern very similar to the earlier pattern, and the new posted speed had little apparent influence.

Speed studies were conducted at nine stations along one U.S. route, five of which were set up in rural areas, and four in small built-up towns.[20] The speed limit varied along the route, with a 65-miles per hour absolute limit in rural sections, and 45, 35, 35, and 40-miles per hour, respectively, in the towns. The results of the study indicated that the posted speed limits in the towns were not effective as a traffic control device in regulating the speed of traffic.

Nashville

Three streets were selected to determine the effectiveness of speed limit signs. The conclusion drawn from the study was that numerical speed limits are not realistic, since drivers will select their speeds based on road conditions, and it is not possible to determine a numerical speed that would satisfy all conditions.[41]

Pennsylvania

The effectiveness of speed limit signs were studied at two locations on roads approaching Allentown.[42] Both roads

TABLE 25-3. RESULTS OF BEFORE AND AFTER SPEED STUDIES

LOCATION	*POSTED SPEED LIMIT*		*AVERAGE SPEED*			*85th PERCENTILE SPEED*		
	Before	*After*	*Before*	*2–3 yrs. After*	*4–5 yrs. After*	*Before*	*2–3 yrs. After*	*4–5 yrs. After*
1A	none	35	42.6	33.1	34.1	48.7	41.3	36.0
1B	none	40	44.0	44.5	38.8	49.4	51.8	44.2
1C	none	40	45.4	43.3	45.1	51.6	50.7	52.1
1D	none	50	49.3	48.8	47.3	57.0	58.1	54.5
2	none	35	36.4	39.7	36.2	43.2	45.4	42.0
3A	none	45	46.6	40.7	40.1	53.7	48.1	47.1
3B	none	45	44.6	43.5	40.8	51.5	50.2	48.0
3C	none	45	43.4	43.8	39.7	49.5	49.6	46.4
5	none	50	43.3	42.7		51.8	49.3	

(Source: Mohr, "Results of Speed Zoning on Rural Highways," *Proc. HRB*, 1954)

had similar characteristics with respect to adjacent land use, with both sides of the roads built up with service stations, restaurants, motels, etc. The erection of 35-miles per hour speed limit signs, in a previously designated 50-miles per hour zone, had a significant effect on the speed of traffic, even though a sizable percentage of traffic did not observe the 35-miles per hour limit. At one location, the per cent of total traffic traveling at or less than 35 miles per hour increased from 22 to 46 per cent after erection of signs; and at the other location the corresponding percentages were 13 and 39 per cent, respectively. The 85th percentile speed was reduced from 45 to 40 miles per hour at one location, and reduced from 49 to 42 miles per hour at the other. The per cent of vehicles in the pace was increased from 58 to 71 per cent at one location and increased from 59 to 64 per cent in the other location.

Increase in the number of vehicles traveling within the pace is highly desirable because the accident potential is decreased, and capacity is increased.

Wisconsin

Several studies were made at rural zones of speeds before and after erection of speed limit signs.[43] The conclusions drawn from the investigation were:

1. A substantial reduction may be expected in the frequency and severity of motor vehicle accidents when the speed zone limits are determined on the basis of engineering and traffic investigations, and when adequate speed limit signs are provided.
2. When speed limits are reasonably lowered through properly applied speed zoning, there is generally a substantial reduction in the average and 85th percentile speeds. See Table 25-3.

These reductions took place without the benefit of special enforcement activities.

Five Suburban Locations

A study of the effectiveness of speed limit signs on speed at five suburban locations revealed that the drivers paid little attention to the posted speed limits.[44] Most of the drivers selected speeds that they considered to be safe and proper for the prevalent roadway and traffic conditions, regardless of regulations. At one location the results shown in Table 25-4 were obtained.

St. Paul

Studies were conducted on 11 arterial streets to determine the effects of raising the usual speed limit of 30 miles per hour to 35 miles per hour, and in some cases to 40 miles per hour.[45] The new limits were set substantially in accordance with the 85th percentile speed. The speed limits for the before and after conditions were posted. Results indicated that the 85th percentile criteria for speed zoning is satisfactory for use on urban arterial

TABLE 25-4.

SPEED CHARACTERISTIC	*NONE*	*40 mph*	*35 mph*	*30 mph*
Mean Speed	31.99	33.8	33.91	34.31
85th percentile speed	37.38	38.44	38.42	38.49
10-mph pace	26–35	28–37	28–37	30–39
Per cent in pace	73.0	72.8	74.6	78.9
Per cent exceeding limit by 5 mph or more		1.6	10.1	37.8

(Source: Elmberg and Michael, "Effect of Speed Limit Signs on Speeds on Suburban Arterial Streets," *HRBB 303*, 1961)

streets. After raising the limits, it was found that the mean, median, modal, and 85th percentile speeds remained very close to those occurring before the change.

Connecticut

A progressive signal system proved to be very effective as a speed control on a U.S. route that passed through several built-up areas.[46] The road was posted for 40 and 45-miles per hour limits, but the majority of drivers did not observe these limits. After the installation of signals, which were progressively timed for 45 miles per hour, there was a minimum of vehicles traveling above that speed. Although the frequency of accidents during the first month after the installation of signals was more than twice the monthly average prior to installation, after four months the monthly frequency was less than prior to installation.

New Mexico

The state highway department realistically approached the problem of lack of observance of speed limits.[47] Lack of observance was primarily due to the setting of speed limits too low, and irregular, haphazard posting of limits. The results of several studies dispelled the belief that if drivers exceed a posted limit significantly they will increase their speed to an even greater extent if the limit is raised. It was found that as a result of raising speed limits, the 85th percentile speed was increased by not more than 2 miles per hour. In fact, the 85th percentile speed generally drops when an unreasonably low speed limit is raised.

Nebraska

Hundreds of speed studies have indicated that speed signs do not change average speed characteristics, that enforcement does not change average speed characteristics, that motorists drive according to their own appraisal of roadway and traffic conditions, and that traffic accidents can be reduced by adopting reasonable speed limits.[48]

California

In 1965, California established minimum speed limits, by lane, on multi-lane highways. The desired results of (1) decreased travel time for the fast driver, (2) less frustration to a driver being delayed, and (3) increased safety, were *not* realized. The minimum speed limit by lane did not keep drivers on the right, and passing vehicles on the left. It caused, instead, (1) a reduction of mean speeds for vehicles traveling in the left lanes, (2) increased passing on the right, and (3) increased, rather than reduced, travel time. Therefore it is concluded that minimum speed by lane would only add clutter to the highways, with definite operational and safety disadvantages.[49]

Miscellaneous Studies

A study of the use of STOP signs to control speeds revealed that the STOP sign only affected speeds for a short distance. Within 100 feet of the STOP sign, speeds were within 3 miles per hour of their maximum level.[50]

A study to compare the effect of signal lens size on approach speeds at a signalized rural intersection revealed that the 12 inch lens significantly reduced approach speeds in comparison to an 8 inch lens.[51] This was obviously due to the increase in target value provided by the 12-inch lens.

In a comparative study of driver performance in response to messages received by sign displays alone, and radio communication plus sign displays revealed that drivers of vehicles equipped with radio receivers reacted more safely in control of their vehicle speed when warned of some impending hazard.[52]

A study to determine the effect on vehicle spacing of communicating with the rear driver by means of signal displays on the lead vehicle, revealed that variability in headways was substantially reduced when the driver of the rear car had advance information on the speed changes.[53]

26

One-Way Streets, Unbalanced Flow and Reserved Transit Lanes

Congestion in urban areas is often aggravated by the number of conflicts on the street system. Impediments to the movement of traffic include conflicting turning movements, vehicle-pedestrian conflicts, conflicts between curb parking and through-movement, and conflicts between types of vehicles (transit, trucks, passenger cars). When it becomes clear that these conflicts cannot be controlled effectively with existing control mechanisms, serious consideration must be given to the prevention and elimination of such conflicts through flow regulation.

In earlier chapters, such spot-flow regulations as turning prohibitions and pedestrian considerations were discussed. Chapter 27 treats the subject of curb-parking controls. This discussion is directed toward major flow control over significant lengths of an arterial, street, or network. These include the provision of one-way streets, unbalanced flow, and/or reserved transit lanes. They are provided with the intent of increasing the efficiency of flow through minimization of conflicts, and more efficient use of roadway space.

The types of flow control which may be considered under these categories include:

1. *Normal one-way street:* Traffic is permitted to move in one direction only at all times.
2. *Reversible one-way street:* A one-way street on which the direction of flow is reversed at different times of the day to meet demand.
3. *Partial reversible one-way street:* A reversible one-way street on which flow is one-way during peak periods (in the direction of heavy demand), and two-ways at other times.
4. *Unbalanced flow street:* A two-way street on which the number of lanes allotted to each direction is changed in the morning peak, mid-day, and the evening peak, to suit demand.
5. *One-way preference street:* A two-way street permanently marked to favor flow in one direction.
6. *Reserved transit lanes:* A lane on a street or arterial reserved for the sole use of transit vehicles.

It is obvious from Table 26-1 that one-way streets under 50 feet wide, with parking on both sides, are not being operated in an efficient manner so as to obtain their maximum potential capacities. Even though the most favorable situation has been assumed for two-way operation, there is a definite indication of the possibility that traffic control on some one-way streets is not as efficient as it should be.

A set of 1962 data appears to substantiate this conclusion, as shown by the comparison in Table 26-2.[6] This table indicates the same trend with respect to one-way streets with parking. However, it also shows that

TABLE 26-1. COMPARISON OF CAPACITY* ON A PAIR OF STREETS (1) (VPHG)

STREET WIDTH CURB TO CURB (ft)	CURB PARKING BOTH SIDES			STREET WIDTH CURB TO CURB (ft)	CURB PARKING PROHIBITED		
	Two-Way	One-Way	% Change		Two-Way	One-Way	% Change
40	3,320	2,400	−26.5	30	3,400	4,420	+33.0
45	3,550	2,920	−17.7	35	4,080	5,460	+33.8
50	3,860	3,640	− 5.7	40	4,650	6,860	+47.5
55	4,210	4,600	+ 9.3	45	5,670	8,810	+55.3
60	4,650	5,900	+26.9				
65	5,130	7,450					

(Source: "Current Intersection Capacities," HRB Correlation Service Circular 376, December, 1958)
* Practical capacity, as per the 1950 *Highway Capacity Manual.*

the full capacity potential is not being obtained on wider one-way streets without parking. In this respect, the two tables give conflicting and seemingly unrealistic results.

The *Model Traffic Ordinance* provides for the designation of one-way streets and alleys (Section 7-2), and for reversible one-way streets and unbalanced flow (Section 7-3). The traffic engineer is given the authority, and is required, to erect and maintain signs indicating the direction of lawful traffic movement at every intersection where movement of traffic in the opposite direction is prohibited (Section 7-1). Reserved transit lanes are not specifically authorized, but sections allowing weight and vehicle class restrictions may be interpreted to allow reserved transit lanes. Many municipalities, though, have specifically amended their traffic ordinances to permit such lanes.

ONE-WAY STREETS

Advantages[1-4]

Increase in Capacity. One-way streets usually attract traffic from nearby parallel streets because of greater freedom of movement. Due to the reduction of conflicts at intersections and more efficient timing of traffic signals, one-way streets can accommodate a larger volume of traffic than the same streets operated as two-way streets.

TABLE 26-2. COMPARISON OF CAPACITY ON A PAIR OF STREETS (2) (VPHG)

STREET WIDTHS CURB TO CURB (ft)	CURB PARKING BOTH SIDES % CHANGE FROM TWO-WAY TO ONE-WAY	CURB PARKING PROHIBITED % CHANGE FROM TWO-WAY TO ONE-WAY
35	−20.0	+23.0
40	−12.3	+16.4
45	− 5.1	+10.6
50	− 1.7	+ 5.1
55	+11.7	+ 1.9
60	+23.1	− 2.9

(Source: Normann, "Variation in Flow at Intersections as Related to Size of City, Type of Facility, and Capacity Utilization," in HRBB 352, 1962)

The advantage of one-way streets over two-way streets will depend upon the directional distribution on the two-way streets, the relative number of turning movements involved, width of the streets, and parking conditions. To illustrate these influences, assume two parallel, signalized, two-way streets, with average conditions of 10 per cent commercial vehicles, 10 per cent left turns, and 10 per cent right turns and no bus stop; compare practical capacities considering that the same streets are operated as one way. The computations are based on curves in a 1958 study.[5] A 50–50 distribution of directional movement for traffic on the two-way streets is assumed, which results in the most favorable condition for the two-way operation. With the exception of streets under 50 feet wide, with parking on both sides, the capacity of a pair of one-way streets is higher than the capacity of a pair of two-way streets.

The extremely beneficial effects of proper traffic control can be seen from a study[7] which was conducted on an individual street. The street is operated one-way inbound during the morning peak period, and one-way outbound during the evening peak period; for the remaining time it operates as a two-way street with parking on one side. The results of the investigation are shown in Table 26-3, and they indicate that the capacity of the major street (40 feet wide) is significantly increased under one-way operation; however, the elimination of parking on the major street had a much greater effect on its capacity than changing from two-way to one-way operation. In addition, the capacity of the cross streets, whether operated as one-way or two-way, is higher when the major street operates as one-way than when it operates as two-way. This is primarily due to the reduction in turning conflicts.

While these earlier studies produced somewhat conflicting results, curves shown in the 1965 *Highway*

TABLE 26-3. CHANGING FROM TWO-WAY TO ONE-WAY OPERATION

CHANGE IN CONDITION	% INCREASE IN MAJOR STREET	POSSIBLE CAPACITY* MINOR STREET
On Major Street with no parking		
With two-way traffic on minor street	25	12
With one-way traffic on minor street	11	20
On Minor Street		
A. With two-way traffic on major street		
Parking both sides of major street	0	15
Parking one side of major street	8	15
No parking on major street	12	15
Elimination of Parking from Major Street (two-way in all cases)		
A. From one side		
At one-way minor streets	48	0
At two-way streets	37	0
B. From both sides		
At one-way minor streets	97	0
At two-way minor streets	75	0

(Source: French, "Capacities of One-Way and Two-Way Streets with Signals and Stop Signs," *HRBB 112*, 1956)
* 1950 HCM definition.

TABLE 26-4. COMPARISON OF CAPACITY ON A PAIR OF STREETS (3) (VPHG)

STREET WIDTHS CURB TO CURB	CURB PARKING BOTH SIDES Two-Way	One-Way	% Change	CURB PARKING PROHIBITED Two-Way	One-Way	% Change
35	3,680	4,220	+14.7	5,600	6,680	+19.3
40	4,380	5,240	+19.6	6,460	7,740	+19.8
45	5,080	6,380	+25.6	7,320	8,800	+20.2
50	5,860	7,480	+27.6	8,220	9,900	+20.4
55	6,640	8,650	+29.8	9,100	11,000	+20.9
60	7,360	9,783	+32.9	9,960	12,120	+21.7

TABLE 26-5. NUMBER OF CONFLICTING MOVEMENTS

STREET A	STREET B	BASIC MOVEMENTS	NUMBER OF CONFLICTS
two-lane, two-way	two-lane, two-way	12	24
two-lane, one-way	two-lane, two-way	7	11
two-lane, one-way	two-lane, one-way	4	6

Capacity Manual reflect more consistent results with respect to conversion from two-way to one-way operation. Table 26-4 represents comparative capacities for streets located within the CBD of a city with a population of 1,000,000, assuming a load factor of 0.70, and a peak-hour factor of 0.85, and average conditions with respect to turning movements and commercial vehicles.

Reduction in Potential Conflicting Movements. The number of potential vehicular conflicts at intersections is considerably reduced when two-way operation is converted to one-way. This is primarily due to the removal of opposed turning movements (Table 26-5).

The number of conflicts between lanes is reduced. One-way streets eliminate the head-on conflict between opposing traffic streams.

The number of pedestrian-vehicular conflicts is reduced. Pedestrians need look one way, or at the most two ways, instead of four ways. Due to the greatly reduced number of turning movements at intersections when one or both of the streets are one-way, there is much less interference with pedestrians when they are in the crosswalk.

Increased Safety. One-way streets have a number of characteristics which contribute to highway safety. There are fewer potential conflicts at intersections between vehicles and between vehicles and pedestrians.

With no opposing streams of traffic, the possibility of head-on collisions and sideswipe collisions is practically eliminated. The hazard of headlight glare is also eliminated.

The long-range effect of conversion from two-way to one-way is a reduction in the accident frequency. In a comprehensive report on highway safety[8] it was concluded that one of the most effective measures to reduce accident hazards is a well-planned, integrated system of one-way streets.

Facilitated Timing of Signals. It is possible to time signals for perfect progression for traffic in each direction on a pair of one-way streets. Progressive signal timing is simple on one-way streets, even though it is sometimes extremely difficult to obtain progressive timing for each direction on the same streets when they are two-way.

The smoother flow of traffic which results from an efficient progressive system is conducive to traffic safety. Sudden stops are reduced, and progressive timing of signals provides an effective method for controlling speeds.

Improvement in Parking Conditions. When curb parking is of considerable importance to abutting or nearby properties, one-way operation may postpone, eliminate, or decrease the need for elimination of parking on heavily traveled streets. Conversion from two-way to one-way makes it possible to retain one side for parking, which would not be possible under two-way operation. Many narrow streets can be made one-way with parking on one side, whereas they would be much less effective if operated as two-way streets without parking.

Increase in Average Traffic Speed. The elimination of opposing streams of traffic, the more efficient timing of traffic signals, and reductions in delay and congestion due to a decrease in conflicts, all contribute to increase the average speed of vehicular traffic.

Improved Traffic Operation. One-way operation permits the passing of slow-moving vehicles on streets which might not otherwise permit passing maneuvers. One-way operation allows economic benefits through improved speed of traffic, and reductions in costly delays and accidents. It permits better utilization of streets having an odd number of lanes for traffic movement. One-way operation makes it simpler to provide reserved transit lanes.

Effect on Business. In general, businessmen oppose the conversion to one-way operation. However, their opposition is unfounded, for the most part. Such opposition usually disappears after one-way operation has been in effect for a trial period. The fear of the unknown is usually the reason for the businessmen's initial opposition. It is an understandable reaction; a certain combination of factors has enabled a businessman to show a profit at the end of a number of successive years, and any proposal to disturb the status quo is apt to distress him considerably.[9] The public and businessmen must be well informed as to the objectives for the conversion to one-way operation, and the benefits to be derived from the conversion. Public support is normally enthusiastic. Since businessmen are extremely sensitive to public reactions and relations, they will more readily submit to a trial period.

One-way streets invariably result in increased volumes of business, and consequently increased property values, in the downtown area. These benefits are primarily due to improved accessibility to, and reduction of congestion, within the business district. It is often possible to retain curb parking where it would have to be eliminated under continued two-way operation.

Conversion to one-way operation may adversely affect some types of business that cater to traffic from a particular direction, such as parking lots or garages, food stores, filling stations, restaurants, and drive-in establishments. When these businesses suddenly are located on a one-way street leading *out of* the business district rather than *into* the business district, their volume of sales may be temporarily reduced.

A study[10] made in Denver, Colorado, concluded that one-way conversion did not have any overall adverse effects on business located along the converted streets.

The broad beneficial effect of conversion to one-way operation is that the movement of all vehicular traffic is improved. There is additional space for all types of vehicles, including passenger cars, commercial vehicles, taxicabs, and buses. Mass transportation vehicles are of particular importance, because they carry a large number of shoppers into the business district. Since one-way operation results in reduced congestion and travel time, bus schedules are significantly improved.

Disadvantages[1-4]

Increased Travel Distance. Motorists must travel farther to reach certain destinations, and must travel around the block to get to their destination. This effect accumulates in proportion to the number of one-way streets in the system and the block lengths. This situation is confusing to strangers and annoying to some local drivers, particularly during hours of the day when traffic volumes are low and one-way operation is not required. One-way operation may be a cure for a 4-hour problem, but a disadvantage during the remaining 20 hours.

Possible Reduction in Transit Capacity. Transit vehicles normally operate best in a single lane because of the frequency of stops to load and discharge. A system of streets with eight north-south streets would provide for eight lanes for transit in each direction as two-way streets, and only four right-hand curb lanes if they are made one-way streets. If the eight lanes are operating at or near capacity under two-way operation, then the four lanes available under one-way operation would not be sufficient to handle the transit load. This problem can be partially alleviated by alternating bus stops, with half the bus lines stopping in one block and the other half stopping in the next block.

Walking distance to transit lines and between transfer points may be increased, because a line, normally operating in both directions on one two-way street, has to be divided to operate on two streets with one-way operation. If the block lengths between a pair of one-way streets is excessively long, as in Manhattan, the increase in walking distance for passengers becomes significant. Changes in transit routing are particularly distressing to residents within an area as they suddenly find themselves confused with respect to direction and location of appropriate transit lines. However, this confusion is only temporary, and it can be minimized to a great extent by fully acquainting the public with the changes before they are put into operation.

Adverse Effects on Some Businesses. Certain types of business establishments, as mentioned above, may be adversely affected where it becomes more difficult to reach the business location. Businesses that depend heavily on inbound traffic may suffer loss in sales if the street is made one-way outbound, and vice versa. An excellent example of the adverse business effect is an establishment located at a bus stop, depending heavily on drop-in trade. If the bus stop is removed, business may suffer significantly.

Occasionally, bus lines may suffer substantial financial loss because of an extreme drop in passengers. This could result when access to bus lines has been made extremely difficult for residents because of the rerouting that is necessary with conversion to one-way operation.

Reduction in Turning Capacity. The number of opportunities to execute right and left turns may be reduced by $\frac{1}{4}$ when a grid network of two-way streets is converted to a grid network of one-way streets. When there are heavy turning movements, delay to traffic or delay to transit operations may result. One method used to alleviate this problem is to permit right turns from two adjacent lanes, with the traffic in the curb lane *required* to turn right and traffic in the next lane *permitted* to turn right. Two adjacent lanes may also be utilized for left turns.

It is inherent in a system of one-way streets that there are nearly always more turns than there would be if the same streets were two-way, because of increased circulation of traffic.

Additional Control Devices Necessitated. Additional signs are required, since one-way signs must be erected at every intersection. Other special signs may be needed, such as turn prohibition signs, DO NOT ENTER signs, and lane control signs, when left or right turns are permitted from two lanes.

Difficult Pedestrian Crossings on Wide Streets. On wide one-way streets (four or more lanes), pedestrian refuge islands cannot be introduced at the center of the pavement, as is possible on two-way streets. There may be additional hazard on one-way streets because of the inability of pedestrians to break a good safety habit, that of looking first to the left and then to the right as one crosses the street. Looking to the left as one leaves the curb is hazardous on a one-way street, with traffic approaching from the right.

Warrants for One-Way Streets[1,11]

Before instituting a one-way street system, all advantages and disadvantages should be carefully weighed. It should be remembered that such measures, which cause some degree of inconvenience, should be used only when less restrictive measures are infeasible or unable to solve the problem at hand. Specific warrants for initiation of one-way streets are published in the *Traffic Engineering Handbook*, and elsewhere.

Certain conditions warranting the use of one-way streets have already been treated, i.e., reduction in conflicting movements for which lesser methods of control are impractical. Other situations may require one-way streets, such as expressway frontage roads and ramp connections, rotary roadways, and narrow streets which make any significant two-way flow hazardous.

Street pairs should be studied carefully before initiating their operation as a one-way couple. The pairs should serve approximately the same origins and destinations, have about equal capacity, and have convenient terminals for transition from one-way to two-way operation. Grids should have similar characteristics if effective one-way operation is to be instituted.

Individual one-way streets of a minor nature can also be used to simplify complex intersections and their control. For example, a five-leg intersection will require at least three phases if signalized and all legs are two-way.

If the minor leg is made one-way leaving the intersection, normal two-phase signalization may be used.

Before instituting a one-way street plan, the following information should be studied in detail:

1. Origins and destinations of traffic
2. Flow (Volumes) on each link of the system: peak hour, classified, with turning movements
3. Travel time and delay along typical routes during peak and off-peak periods
4. Detailed street and intersection capacities: one-way versus two-way
5. Complete physical layout of system
6. Economic evaluation of added travel distance (as a result of one-way operation) versus travel time decreases
7. Location of major traffic generators
8. Possible effects on businesses, loading zones and other access functions, accidents, transit routes, emergency vehicle access, and pedestrian movements

Installing the System

To obtain the full benefits of the one-way plan it is preferable to install the entire system at the same time. However, if approval cannot be obtained on this basis, the alternative of a step-by-step installation should be properly planned to demonstrate as early as possible the advantages to be gained from conversion to one-way operation.

A trial period may be necessary when there are differences of opinion as to the need for, and benefits of, a proposed one-way plan. A minimum trial period of three months is required, and preferably longer, to prove or to disprove the worth of the one-way plan. During the trial period traffic studies should be conducted along with other related studies to determine the effects of the one-way plan.

Signs and markings for one-way streets should be installed in accordance with the recommendations of the MUTCD.

Proper advance publicity must be provided to acquaint the general public, motorists, and pedestrians with the plan and the reasons for its inception. Press, radio, TV, and talks before interested civic groups should be utilized as much as possible. An educational program in conjunction with the policy and civic groups should be planned. Prior to the adoption of a comprehensive system of one-way streets, it is extremely desirable to institute a continuous educational program which should last several months.

Adequate studies should be conducted after installation of the system to determine the effects of the one-way plan. These studies should include volume, speed and delay, signal timing, accidents, public acceptance, and the effects on business and mass transit.

UNBALANCED, REVERSIBLE, AND PREFERENTIAL FLOW

Reversible and Partially Reversible One-Way Streets

Both of these types of operation are only warranted under the condition that a heavy, unbalanced, directional distribution of traffic exists at different periods, and there is only one suitable street available to carry the peak flows. Usually, the partial reversible type of operation is used for situations where traffic is one-way inbound during the morning peak period, one-way outbound during the evening peak period, and two-way during other periods when traffic flow is more evenly balanced. This type of control is generally used on arterial streets leading to and from a business district where inbound flow is heavy in the morning and outbound flow is heavy in the evening.

Reversible operation should definitely be restricted to those cases where peak period directional flow is in excess of street capacity under two-way operation, and there is no suitable parallel street to establish normal one-way operation. However, it is necessary that a nearby parallel street exists to accommodate the smaller traffic flow in the opposite direction.

The temptation may arise to use reversible control rather than normal one-way operation, assuming it is feasible, because of the extra travel distance that may be required for both motorists and pedestrians under one-way operation. Another impetus is the criticism that one-way operation may be a 24-hour cure for a 4-hour problem. However, the disadvantages of reversible flow should discourage its use in place of normal one-way operation when it is feasible:

1. Control devices are difficult to set up to establish proper regulation of traffic
2. It is difficult for motorists, especially strangers, to understand and comply with regulations. This results in many inadvertent violations of the regulation by motorists attempting to travel in the wrong direction.
3. Accidents may increase unless the regulation is carefully enforced.

The MUTCD presents no standard devices for controlling reversible flow. Most commonly used are signs with arrows and time of day information and regulation.

Lane-use control signals may also be used. Supplementary turn regulation signs are desirable. Most importantly, an extensive public education program should be conducted when initiating such operation.

Off-Center Lane Movement (Unbalanced Flow)

The purpose of this technique is to provide for more efficient use of existing street width when there are periods of heavy unbalanced traffic movements. Off-center lane operation is applied to overcome the inefficiency of two-way street operation, when peak periods of flow are primarily unidirectional, with traffic movement in the opposite direction significantly less than the capacity of half the street width. Under such conditions, roadway space is allocated in proportion to the directional distribution of traffic movements. For example, four of six lanes would be provided in the morning peak period for inbound traffic, and four of six lanes would be provided in the opposite direction during the evening peak period for outbound traffic, and three lanes in each direction would be provided for other periods.

Advantages. The advantages of off-center lane operations include the following:

1. It provides additional capacity in appropriate direction when required.
2. It eliminates possible extra travel distance for motorists and inconvenience for pedestrians under normal one-way operation.
3. It provides some of the advantages of one-way operation without the necessity of having a suitable parallel street.
4. Rerouting of mass-transit vehicles may not be required.

Disadvantages. Some disadvantages are:

1. It may result in inadequate capacity for minor traffic flow.
2. It may necessitate the prohibition of left turns from and bus stops along the minor direction of flow.
3. Costs for providing adequate control devices may be high.
4. It may result in increased accident frequency and severity.

Warrants. Specific warrants for unbalanced flow have been developed in the literature.[12] Such operation is justified under a variety of circumstances, often on bridges and tunnels with heavily directional demand patterns. Arterials which demonstrate periodic congestion, and for which normal one-way streets are either not feasible or not justifiable, are also candidates for unbalanced flow.

Extensive study should be conducted to ensure the need for unbalanced flow. Often a lesser form of regulation will suffice. Measures such as peak-hour parking prohibitions, turn restrictions, elimination of loading during peak periods, improved signalizations, and better lane markings should all be carefully considered.

On narrow streets (three to four lanes), unbalanced flow is generally not practical. However, under certain conditions it may provide a useful solution to a problem. In general, speeds and density on the street should be low, and all stopping in the single lane of minor flow should be prohibited.

Methods of Control.[12,13] There is considerable variation in the use of traffic control devices to indicate proper lane usage. The primary reason for the wide differences in application of control devices is economic. The three basic systems of controls include:

1. Physical barriers
2. Lane-direction signals
3. Signs

These three systems are not entirely independent of each other. Almost all systems use some components of the others. However, each of the systems differs in the amount of flexibility and control afforded, and in the economics of their establishment and operation.

Physical barrier. These include permanent installations, temporary barriers, and lane markings. One type of permanent installation utilizes mountable curbs to establish reversible lane usage. The upper deck of the George Washington Bridge, with its eight lanes, was an example of this type of installation. The middle two lanes were segregated from the other lanes by mountable curbs on either side, so that the roadway presented a 3-2-3 lane arrangement. During periods of balanced traffic flow the roadway was used in a 3-3 or 4-4 lane split, according to the traffic demand. During peak periods, with unbalanced traffic flow, the lanes were split 5-3. Police at the extremities of the roadway effected changes by placing channelizing stanchions and traffic cones. This operation was replaced by a single, fixed steel divider down the center of the bridge, in the interests of safety.

Another type of permanent installation utilizes movable fins which raise and lower. One portion of an eight-lane expressway in Chicago provides for reversible lane movement with continuous fins which are raised (8 inches) or lowered by hydraulic action. By raising or lowering the lines of fins, the roadway can be arranged in three different combinations of traffic lanes providing two, four, or six lanes in either direction. Another type of permanent installation uses medians and/or guardrails to separate express, reversible roadways on freeways from the two conventional roadways. Access to the reversible roadway is provided for relatively long distances. The use of permanent installations normally involves a large initial expenditure and lower daily labor maintenance requirements.

Temporary physical barriers such as traffic cones, stanchions, barricades, and sign pedestals are more commonly used to effect reversible-lane usage. The use of temporary barriers provides much less positive separation, reduces the initial expense, and increases the daily labor maintenance costs.

Lane lines are used to designate reversible-lane usage by variations in width, striping, color, or by a combination of these methods.

Lane direction signals. A very popular technique to establish reversible-lane operation is the use of overhead signal lights to indicate the direction usage of traffic lanes. This system offers no physical barrier for separation; however, overhead lane signals have been very successfully applied. The signal lights can be either automatically or manually operated, which makes their use extremely simple. There is a high initial expense involved in the installation of the system, but there is practically no daily labor maintenance requirements.

Details about the design, location, and operation of lane direction signals are found in the MUTCD.

Signs. Informative signs for the control of lane usage are either post-mounted at the curb, or mounted overhead. The marginal display of signs provides the minimum amount of control. They should only be used where the roadway predominantly serves local traffic, so as to insure rapid familiarity with the system. It would be advisable to adopt a color code for both the informative signs and the reversible-lane markings, to distinguish them from conventional use. It is essential to provide strict enforcement during the initial application period.

Overhead signs are extensively used to indicate the direction of usage of traffic lanes. In a general way, the remarks made about lane direction signals also apply to overhead signs. Word messages on signs vary in different cities. For example, in Detroit, the following overhead sign messages provided effective control. The signs were mounted at intervals of about 1,000 feet over each of the two center reversible lanes of a six-lane roadway, and alternate signs facing inbound traffic read as follows:

USE BOTH CENTER LANES 7:00—9:00 A.M. Monday Thru Friday	KEEP OFF BOTH CENTER LANES 4:30—6:30 P.M. Monday Thru Friday

Alternate signs facing outbound traffic read as follows:

USE BOTH CENTER LANES 4:30—6:30 P.M. Monday Thru Friday	KEEP OFF BOTH CENTER LANES 7:00—9:00 A.M. Monday Thru Friday

Preferential Streets

The use of permanent off-center operation, wherein a two-way street is permanently marked for unbalanced flow, could conceivably incorporate most of the advantages of one-way and two-way streets, while eliminating some of the disadvantages of both types of operation.

Permanent off-center operation would eliminate all of the difficulties associated with control of reversible unbalanced flow operation as well as the expense involved in such control. There would be no additional expense involved as the lanes would be permanently marked. It also results in very efficient use of street width which provides for an odd number of moving lanes. For example, a fire-lane street could be marked for three lanes in the preferred direction and two lanes in the opposite direction, instead of two lanes in each direction under normal two-way operation.

Permanent off-center operation also provides an excellent intermediate step between conversion from two-way streets to one-way streets. Consult the literature for a suggested method to effect a transition from two-way streets, to one-way preference streets, to normal one-way streets.[14]

EXPERIENCES WITH ONE-WAY STREETS AND UNBALANCED FLOW

New York City established its first one-way street in 1907, in the Park Row section of Manhattan. By 1939, 85 per cent of Manhattan's streets were one-way.[15] Today, in New York City, one-way streets total well in excess of 1,000 miles.

The Traffic Control Plan for Manhattan's north-south streets was initially introduced in 1951, and it had as its objectives the following:

1. Reduction of accidents
2. Reduction of congestion
3. Increase in street capacity.

To accomplish this, the plan involved the following:

1. One-way operation on the north-south avenues
2. Retiming of traffic signals to provide progressive signal timing on the avenues and major crosstown streets
3. Re-allocation of signal periods to reduce waiting time, and distribute available time more equitably between avenues and crosstown streets

The first pair of north-south avenues to be converted to one-way was First and Second Avenues. This conversion took place in June, 1951, and it resulted in a significant improvement in the movement of north-south traffic.[16]

On May 21, 1959, the Department of Traffic officially submitted a plan to the Board of Estimates for the completion of one-way operation of the north-south streets in Manhattan. This plan extended *one-way operation* to include Fifth Avenue, Madison Avenue, Lexington Avenue and Third Avenue, which would leave Park Avenue as the only remaining north-south street with two-way operation. At this meeting reports and remarks were also submitted by the opponents and proponents of one-way operation. The Department of Air Pollution was a proponent of the plan. Its support was based on the fact that less air pollution results under one-way operation as motor vehicle idling time is reduced 65 to 70 per cent, and stops are reduced 65 to 72 per cent. A motor vehicle produces its smallest concentration of pollutants when it is in motion at a uniform speed. It produces twice the amount of pollution when it is decelerating to a stop (and approximately four times as much pollution when it is stopped and idling), as it does when it is in motion.[17] (See Figure 26-1.)

The Board of Estimate approved this important plan in 1960. In January, 1966, Fifth and Madison Avenues were converted to one-way operation, thereby completing the plan submitted to the Board of Estimates. This action alleviates congestion on Fifth and Madison Avenues. The magnitude of this congestion is reflected by studies of traffic movements on the avenues. It was found that during peak hours of traffic, average travel time on Fifth Avenue, for a 30-block trip from 33rd to 63rd Streets, was 25 minutes; that a vehicle is forced to come to a complete stop on the average of 23 times; that the average delay caused by congestion amounted to 86 per cent of the total trip time; that stopped-time delay consumed 71 per cent of the total; and that the average speed was 3.5 miles per hour.[18]

TABLE 26-6. NEW YORK CITY—IMPROVEMENTS ACHIEVED BY ONE-WAY OPERATION

ONE-WAY Improvements	*MINIMUM IMPROVEMENT ACHIEVED THROUGH CONVERSION (%)*
Pedestrian Accidents Reduced	20
Congestion Relief as measured by	
1. Total Trip Time Reduced	22
2. Stopped Time Reduced	60
3. Number of Stops Reduced	65
Crosstown Delay reduced	40
Crosstown Capacity Increased	20

(Source: *Annual Report,* City of New York, Dept. of Traffic, 1965)

One-way streets in New York City have proved their value. Studies clearly indicate that one-way operation has produced significant results in improving the safe and efficient movement of traffic. Before-and-after studies conducted by the Department of Traffic and other interested groups (notably AAA) amply prove this fact.

Studies conducted before and after several one-way conversions reveal the *minimum* improvements shown in Table 26-6. Similar results were obtained from studies on the other one-way avenues.

One-way operation has contributed to enhanced property values on Manhattan's avenues, as reflected by the comparisons shown in Table 26-7. The amounts shown are the totals for all real property parcels on both sides of the avenues from 42nd to 59th Streets. The figures reflect the assessed valuations of real property alone and do not include the assessed valuations of the building improvements constructed on the property.

One of the important features of the one-way plan is

Fig. 26-1. New York City one-way street plan

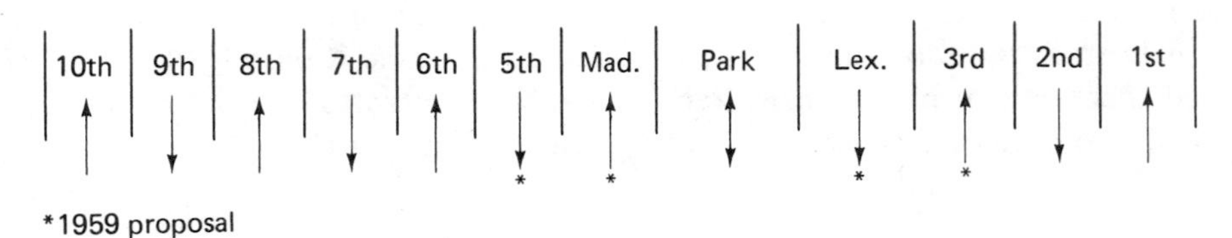

TABLE 26-7. NEW YORK CITY—INCREASED LAND VALUE UNDER ONE-WAY OPERATION

AVENUE	*VALUATION IN FISCAL YEAR PRIOR TO CONVERSION*	*VALUATION IN FISCAL YEAR 1965*	*% INCREASE IN VALUATION*
Third	$26,380,000	$ 57,409,000	*118
Lexington	38,635,000	45,000,000	16
Americas	86,627,000	135,517,000	56

(Source: *Annual Report*, City of New York., Dept. of Traffic, 1965)
* Tearing down of 3rd Avenue El may have contributed to this figure.

the improvement of *crosstown flow*. As already indicated, gains had been realized with only part of the plan in effect.

Crosstown traffic delay is reduced, and traffic capacity increased, through rescheduling the signal timing. One-way operation makes the avenues so much more efficient that it is possible to give a greater proportion of the green signal time to the cross streets. For example, under two-way operation, a 90-second cycle was required, with a 60–30 split, or 33 per cent to the side street. Under one-way operation, a 60-second cycle was feasible, with a 35–25 split, or 41 per cent to the side street.

With a 90-second cycle, cross-street traffic gets the green signal $3,600/90 = 40$ times an hour and therefore, for a total of 40×30 seconds per cycle $= 1,200$ seconds of green per hour. While under a 60-second cycle, it gets the green signal 60 times an hour, or a total of $60 \times 25 = 1,500$ seconds of green per hour. Thus, while cross-street traffic gets less green time per cycle, because the signal changes more frequently, and a greater proportion of the cycle is assigned to it, it gets 25 per cent more green time per hour.

Increased crosstown capacity can also be demonstrated by sound traffic engineering principles which verify the results obtained in the before-and-after studies. This analysis is presented in the following paragraphs.

The time required for cars in line to clear an intersection controlled by a signal decreases for each car in line, down to a certain minimum value. The values reported in "Traffic Performance at Urban Street Intersections" (*Technical Report No. 1*, Yale Bureau of Highway Traffic) are for ideal conditions, but they will serve to illustrate the basic principle (Table 26-8).

While the latter part of a signal phase is more efficient, it is clear that the same number of cars will clear the intersection in the first 25 seconds of a 60-second cycle, as will in the first 25 seconds of a 90-second cycle. This, of course, applies only if basic traffic conditions are the same, which must be assumed to make a valid comparison.

Thus:

1. Under a 60-second cycle, during the 25 seconds allotted, the signal should pass approximately 10 vehicles.
2. Under a 90-second cycle, during the 30 seconds allotted, the signal should pass approximately 12.5 vehicles.

But on an hourly basis:

1. The 60-second cycle will pass 10 vehicles each cycle of the 60 cycles per hour, or 600 vehicles per hour.
2. The 90-second cycle will pass 12.5 vehicles each cycle of the 40 cycles per hour, or 500 vehicles per hour.

Therefore, the 60-second cycle under one-way operation is 20 per cent more efficient for cross streets than the 90-second cycle, which is part of two-way operation.

In Washington, D.C., three parallel, two-way streets were converted (December, 1961) to a system of a one-way streets on each side of the center two-way street. The center street was designed to facilitate mass bus transportation movements by improved traffic signal timing and creation of exclusive bus lanes during peak periods.[16] Although there was only a slight change in traffic volume, traffic movement on the two one-way streets was considerably improved, as reflected by an 11.5 per cent decrease in travel time, less driver tension, and 33 per cent fewer stops.

In Los Angeles, California, unbalanced flow operation has been extensively used since 1928. One important installation is a broad, six-lane (10-foot lane widths) arterial traversing the city. This roadway is marked with a continuous, double, centerline, and with intermittent companion lane lines designating a split of three traffic lanes in each direction for off-peak periods. During the morning and evening peak periods, a 4-2 split is effected by the use of rubber traffic cones and movable pedestal signs. The four, 10-foot lanes average over 900 vehicles per hour during peak periods, and these volumes include a large proportion of bus traffic.[19,20]

The Seattle Freeway (Interstate Highway No. 5) in

TABLE 26-8. INTERSECTION START-UP TIMES

CAR-IN-LINE	*GREEN TIME CONSUMED (sec)*	*ACCUMULATIVE TIME (sec)*
1	3.8	3.8
2	3.1	6.9
3	2.7	9.6
4	2.4	12.0
5	2.2	14.2
$N \geq 6$	2.1	$14.2 + 2.1(N + 5)$

(Source: Greenshulds, et al., "Traffic Performance at Urban Street Intersections," *Technical Report No. 1*, Yale Bureau of Highway Traffic, 1947)

Seattle, Washington, has been designed to provide for four reversible lanes, 7 miles long. The 12-lane facility was completely engineered to permit reversible flow on four lanes, inbound in the morning and outbound in the evening. There is no interchange of traffic between the reversible lanes and other through lanes, except at the terminals. However, interchanges permit access from city streets at intermediate points. Over most of the length, the reversible lanes are beneath the other through lanes. Control of directional flow on the reversible lanes is accomplished by actuating automatic gates which open and close the access to the reversible lanes. The control system is keyed to a closed-circuit television system. A control panel, showing the position of each gate, supplements the television screens.[21] Consult the literature for more accounts and studies concerning one-way streets and unbalanced flow.[22-33]

RESERVED TRANSIT LANES

A *reserved transit lane* is that portion of the roadway devoted entirely to the use of mass transit vehicles that are in motion, in the act of loading or unloading passengers, or stopped in response to street traffic controls. Other vehicles may enter, leave, and cross this lane only when permitted, but may not substantially interfere with the transit vehicle movement in the lane. Transit vehicles may use other street lanes on a street where a transit lane is established, but under no circumstances may they make service stops outside the transit lane.[34]

The purpose of the transit lane is to segregate transit traffic from other vehicles and to prevent interference of one by the other. The reserved transit lane is an attempt to eliminate the friction and conflicts of purpose of the various classes of traffic that make up the traffic stream. Other methods that are extensively used to minimize or eliminate conflicts between elements of the traffic stream include one-way operation, unbalanced-flow operation, parkways, and exclusive turn lanes.

Reserved lanes for buses on freeways and expressways in urban areas have also gained attention. A single, reserved bus lane can theoretically carry 60,000 persons per hour, as compared to about 3,000 persons per hour in private cars. This value is somewhat unrealistic, requiring 1,200 buses per hour. For short distances this is feasible, but for long-haul runs, the labor requirement of 1,200 drivers per hour is unrealistic. Demand of this magnitude (60,000 persons per hour) is more suited to rapid transit. Notwithstanding, the potential to carry smaller volumes conveniently on reserved transit lanes on freeways deserves, and has received, serious study in several areas.[35]

The major *advantages* of reserved transit lanes include: the elimination of friction between transit and other vehicles; decreased transit travel times; the possibility of street capacity increase with corresponding decrease in delay to other vehicles (if large numbers of transit vehicles disrupt several lanes normally); reduction in accident frequency.

Disadvantages include the blocking of land access by curb lanes, conflicts with turning movements, the necessity of special loading islands for center lanes, and the additional control devices needed for operation.

Warrants for curb and center transit lane, full-time and part-time, have been established.[34,36]

Operating Criteria for a Curb Transit Lane

1. Turning movements of other vehicles should not interfere with the operation of the transit lane. Right turns from the subject street by other than transit vehicles may be provided for, to the extent necessary for traffic circulation over the intersecting streeets.
2. At times when the transit lane is in effect, no other vehicles may use the lane except for turns and taxicab pickup and discharge.
3. Bus stops should be located on the far side of the intersection, or mid-block, to permit the use of the transit lane for right turns.
4. No transit vehicle should be permitted to leave the transit lane except to make a turn or, in case of emergency, to pass a stalled vehicle.
5. Two curb transit lanes should not be operated simultaneously on a two-way street. For streets wide enough to sustain two curb transit lanes, center transit lanes are recommended.
6. Provision should be made for adequate enforcement of regulations to insure free-flowing traffic.

Operating Criteria for a Center Transit Lane

1. On one-way streets other vehicles may use the transit lane within an intersection, or after passing through the intersection, for the purpose of weaving from one side of the transit lane to the other, provided such a maneuver does not interfere with the operation of the transit lane.
2. If two-way center transit lanes are established, left turns should be prohibited to the degree required for proper operation.
3. Nearside stops for transit vehicles should be used exclu-

sively. Safety islands must be provided at the stopping place for passengers boarding and alighting. Access to the safety islands should only be permitted at the near-side pedestrian crosswalk.

4. No transit vehicle should be permitted to leave the transit lane except to make a turn or, in case of emergency, to pass a stalled vehicle.
5. If the lane is operated part-time, transit vehicles should continue to operate in the center lane (shared) in the off-hours), and make stops at the designated loading islands.

Since reserved transit lanes deprive other vehicles of the use of part of the roadway, and block (curb lanes) access to abutting lanes, as well as interfere with turning vehicles, they should be implemented only when transit traffic is such that overall movement and convenience is benefited. Consult published warrants[34,36] for specifications of conditions considered to meet this requirement.

Transit lanes have been implemented in Chicago, Atlanta, Dallas, Baltimore, New York, and Washington, D.C. Consult the literature for descriptions.[36]

27

Curb-Parking Controls

CURB-PARKING RESTRICTIONS

The severity of the parking problem in any urban area is first recognized in the CBD, when the available curb space is no longer able to accommodate the parking demand. Congestion is an inevitable consequence of this situation. Physical characteristics of the CBD, e.g., size limitation and intensity of land use, combine together to aggravate the situation.

When the severity of congestion so caused becomes great, some form of curb-parking restriction is needed. The following points should be considered.

Traffic Versus Parking

Whenever demand for public service becomes greater than the supply, priorities must be established. Streets intended primarily for travel are needed for free movement of traffic and must not be used for parking. However, no reasonable use of streets for parking should be prohibited or restricted until traffic demands become pressing, and parking should be permitted and regulated wherever and for as long periods as possible. Traffic and parking are in the public welfare, and neither should be unduly benefited by group or private interests.

Street Capacity

The capacities of streets in the downtown area are significantly affected by curb parking. At signalized intersections, where parking is prohibited on both sides of two-way streets, the capacity is approximately double that with parking permitted on both sides. Similarly, at signalized intersections where parking is prohibited on both sides of one-way streets, the capacity is about two and one-half times as great as that with parking permitted on both sides.[1-3]

Accidents

On-street parking is a major factor in traffic accidents. Vehicles entering or leaving curb spaces, vehicles improperly or illegally parked, and pedestrians entering the street from between parked vehicles, are serious and common causes of accidents. In 1960, over 10 per cent of pedestrian fatalities, and 25 per cent of pedestrian injuries in New York City were caused by pedestrians coming from behind parked vehicles.[4]

In a study conducted in 1965, it was found that 18.3 per cent of all accidents involved street parking. Of these, 90 per cent were directly related, while 10 per cent were indirectly related, such as pedestrians stepping from between parked vehicles.[5]

Business in the Downtown Area

Decentralization and loss of business to merchants is partly due to congestion and an inadequate supply of

TABLE 27-1. MINIMUM CURB BUS LOADING AREAS

APPROXIMATE BUS SEATING CAPACITY	APPROXIMATE BUS LENGTH	ONE-BUS STOP			TWO-BUS STOP		
		Near Side	Far Side	Mid-block	Near Side	Far Side	Mid-block
30 and less	25	90	65	125	120	90	150
35	30	95	70	130	130	100	160
40–45	35	100	75	135	140	110	170
51	40	105	80	140	150	120	180

(Source: *Standards for Street Facilities and Services,* PMTA, NCUT, PAS, 1958)

TABLE 27-2. ITE RECOMMENDED LENGTHS FOR A ONE-BUS STOP*

NEAR SIDE	FAR SIDE**	MID-BLOCK
105	80	140

(Source: *Proper Location of Bus Stops,* ITE Recommended Practice, December, 1967)

* In each case, an additional 45 feet of length should be provided for each additional bus expected to stop simultaneously.

** A far-side bus stop after a right turn, for a single bus, should be 140 feet in length.

parking spaces. Several studies have shown that downtown areas have experienced declining proportions of total metropolitan area sales.[6-8] Downtown establishments are losing some business to shopping centers, and are not gaining business as fast as the suburbs, but they are gaining in absolute volume. They need better parking facilities, but it is impossible for them to compete with suburban shopping centers in that respect. In many cases, the provision of off-street parking has significantly helped downtown areas.[9-12]

Emergency Vehicles

Police and fire departments are importantly concerned with curb-parking conditions, because curb parking can interfere with their duties. Curb parking constitutes a serious fire hazard, by obstructing apparatus and fire hydrants. Adequate parking restrictions in the vicinity of fire houses and fire hydrants, sufficient street width for maneuvering public-service vehicles, and space for laying fire hose, are public safety requirements that must be considered when street-parking regulations are recommended.

Loading Zones

Provisions must be made for curb-loading zones for buses, commercial vehicles, and taxis.

Bus Loading Zones. To permit bus drivers to pull in and out with reasonable ease and to park parallel to the curb, out of the flow of traffic, the minimum curb bus loading zones given in Table 27-1 are recommended.[13]

ITE recommendations[14] for the minimum curb bus loading zones for a 40-foot bus are given in Table 27-2.

Bus stops are generally located at intersections. This location provides for the optimum convenience for passengers, since the stop can be reached from four directions and, where bus lines cross, transfer distance is at a minimum. This location is also the most efficient from the point of view of street capacity.

Truck Loading Zones. An essential purpose of streets in the CBD is to provide access to commercial activities in the area. This includes movement of merchandise into and out of the CBD, and loading and unloading of merchandise. In many cases, loading and unloading take place at the curb. A truck handling merchandise must be parked near its destination, and trucks play a major part in double parking and obstructing the flow of traffic. These practices lead to excessive parking violations by truck drivers. The need for truck loading zones should be determined by study of parking conditions, availability of off-street or alley loading space, frequency of delivery or pickup, size of load, type of vehicle, and number of adjacent establishments being serviced. The zone length should not be less than 25 feet, and 30 feet, or multiples thereof, are preferable.

Warrants for truck curb-loading zones should include the following provisions:

1. No alley or off-street space available for loading use
2. No curb loading space within 100 feet of proposed zone without crossing a street or alley, except in areas of concentrated activity
3. A minimum of 10–15 stops per day for pickup and/or delivery per business or combination of businesses proposed to use the zone
4. Weight, quantity, and time requirements of loads handled

The minimum length of zone that should be provided for single-unit trucks is 30 to 40 feet. For practices of cities in providing truck loading zones, consult the literature.[15-17]

Taxi Zones. Standing curb zones for taxis should be provided in downtown areas for the convenience of the public. The number and length of such zones will depend upon occupancy and demand for service. Practices in different cities are reported in the literature.[18,19]

The suggested minimum length for a taxicab zone with a high turnover is 30 to 40 feet.

TYPES OF CURB PARKING

There are two basic ways in which a vehicle can park at the curb: *angle* parking and *parallel* parking.

Angle parking accommodates more vehicles per unit of curb space than parallel parking. This advantage increases as the angle increases, until at 90 degrees nearly two and one-half times as many spaces are possible. As the angle increases, so do the requirements of roadway width used for parking, and additional width needed for maneuvering into and out of spaces. Usually, 60 degrees is the maximum practicable, and 45 degree stalls generally give best results. Angle parking should only be considered on streets over 70 feet wide.

Angle parking simplifies and speeds up the act of parking. However, the act of leaving an angle stall is more dangerous than unparking from a parallel stall. Many studies, of the before-and-after type, have shown that angle parking at the curb is a common factor in accidents, and a change to parallel parking brings a definite reduction.[20]

Stalls should be clearly marked on the pavement. The most commonly used dimensions for stalls are $8\frac{1}{2}$ by 22 feet.

ITE recommendations regarding minimum street widths where parallel parking is permitted are shown in Table 27-3.

CONDITIONS WARRANTING PARKING PROHIBITION

Many locations, such as fire hydrants, crosswalks, driveways, structures, intersections, excavation sites, and others, receive automatic parking prohibitions because of safety considerations. Traffic conditions will often warrant the posting of parking prohibitions for other purposes (bus stops, loading zones, through traffic.).

The Uniform Vehicle Code and Model Traffic Ordinance define three terms for use in parking prohibitions:

1. *Parking:* Leaving an *unattended* vehicle at curbside with motor off.
2. *Standing:* An *attended* vehicle stopped at curbside for an extended period with motor running.
3. *Stopping:* A vehicle stopping momentarily at curbside to pick up or discharge a passenger or parcel.

NO STOPPING regulations bar all stopping, standing and parking. NO STANDING regulations bar all standing and parking. NO PARKING regulations bar all parking. Consult the literature for recommended applications of such controls.[22,23]

ESTABLISHING TIME CONTROL

Parking should be prohibited only for the length of time for which prohibition is warranted. Time restrictions may range from none at all in the smallest cities, where existing space meets the demand, to the other extreme in the largest cities, where it may become necessary to prohibit all curb parking in the CBD during peak hours of traffic flow.

Time Restrictions

These are developed from an analysis of these parking characteristics as related to existing and forecasted traffic volume patterns. For most cities, the following time limits are satisfactory where parking can be permitted:

1. *One hour* is the normal time for the central part of the downtown area. The majority of the motorists do not care to stay in downtown districts for more than one hour.[24]
2. *Two-hour* limits in the fringe areas of downtown dis-

TABLE 27-3. MINIMUM STREET WIDTHS FOR PARALLEL PARKING[21]

	COMMERCIAL PARKING LESS THAN 15%		*COMMERCIAL PARKING OVER 15%*	
	Two Side Parking	*One Side Parking*	*Two Side Parking*	*One Side Parking*
Two-way	36	28	38	29
One-way	26	18	28	19

(Source: Wermyer, "Progress Report of Committee 6-E," Proc. ITE, 1955)

tricts will normally accommodate parkers who desire to park longer and are willing to walk further.

3. *15- to 30-minute limits* in the vicinity of post offices, banks, and public utility offices will normally be of sufficient duration for parkers to accomplish their desires.

ENFORCEMENT

Regulatory signs and markings should indicate parking regulations and restrictions. Specifications for standard signs and markings is found in the MUTCD. To fully serve their purpose, parking regulations should be strictly enforced. Since great competition exists for the available curb spaces, time limits require enforcement to provide for the accommodation of the greatest number of parkers and orderly traffic flow. When parking is completely prohibited for any period, it is essential that the restriction be enforced to provide for additional roadway capacity.

Large personnel and equipment requirements are necessary to enforce time and space regulations. Many methods have been developed for enforcing such regulations, but in most cities the police are usually in the difficult position of attempting to enforce regulations which are contrary to public desires. Enforcement of curb parking regulations is not only one of the largest jobs of city police, but one of the most unpopular. In New York City in 1967, over one million summonses were issued for parking violations. The different methods used to enforce parking regulations include foot patrol, motor patrol, mounted police, and towing and impounding of parked vehicles.

For further discussion of these factors, consult the literature.[11,15,16,25,26-31]

PARKING METERS

Parking meters originated in 1935, and they play a major role in the control of curb parking. A most complete study on the use of parking meters is found in the literature.[32] The study was purely a factual survey of existing practices relating to the number, revenue, and use of parking meters in municipalities. Some of the important findings and major trends in the development of parking meters obtained from this investigation include:

Extent of Parking Meter Use

It was estimated that by the end of January, 1952, there were 1,113,000 parking meters in the United States in over 2,800 localities. New York City first installed meters at the end of 1951, and by the end of 1962 there were 47,000 curb meters in operation including 13,000 in Manhattan, 15,000 in Brooklyn, 8,100 in the Bronx, 10,000 in Queens, and 900 in Staten Island.[33] By 1966 the total was up to nearly 62,000 meters. Many of the larger cities are providing metered off-street parking facilities. By the end of 1966 there were some 12,100 off-street metered spaces in New York City in 44 facilities serving about seven million parkers annually.

Public Attitude Toward Meters

Community attitudes toward meters before their installation involve uncertainty, or fear of the unknown but these disappear once the advantages of meter use are demonstrated. Over 95 per cent of the cities covered in the study indicated public approval of meters after installation.

Advantages of Meters

The parking meter, when accompanied by adequate length of stalls, time restrictions appropriate to the demand, and proper provision for loading zones, produces the following benefits:

1. Provides an accurate time check on parking, simplifying detection of overtime parking and discouraging all-day parkers.
2. Reduces overtime parking, increases turnover, and makes parking available for more motorists.
3. Aids merchants in metered areas by increasing turnover.
4. Reduces personnel required for parking enforcement. Some communities use meters primarily for this purpose.[34]
5. Reduces double parking.
6. Aids traffic flow by reducing congestion.
7. Aids in the financing of traffic control and off-street parking facilities.

Disadvantages of Meters

To provide the above advantages and to be fully effective, meters must be adequately used, thoroughly supervised, and enforced. Some possible disadvantages include the following:

1. If used where not warranted, they arouse resentment.
2. Unless properly enforced, motorists learn that they can park overtime without receiving a summons.
3. Unless frequently checked, some motorists will park

overtime for long periods by feeding coins into the meter.

4. After meters have been installed, the desire to continue the revenue may discourage the elimination of curb parking when traffic demands indicate a need for it.
5. On streets where parking is prohibited during peak hours of traffic flow, the presence of meters increases the difficulty of enforcement.

Legality of Meters

Ordinances relating to parking meters in some communities may be deficient, rendering portions of them invalid if challenged in court. A common circumstance is to find meter installations placed by the authority of some city official, or agency, but without the enacted and recorded ordinance required to give them legal status. An excellent model for a parking meter ordinance is found in the Model Traffic Ordinance.

Litigation is sometimes necessary to settle many questions relating to parking. For example, a court upheld the right of the city to finance off-street parking with meter revenues, rather than placing them in the general fund.[35]

Enforcement of Meters

In more than 75 per cent of the places reporting, the police department properly had this function. The average number of meters served per policeman ranged from 88 in the smaller cities (population under 2,500) to more than 300, averaging 277 per policeman for all cities.

Enforcement in New York City is carried out by "meter maids" and "meter men" who are employed by the Department of Traffic.

Cost and Service Life of Parking Meters

The average cost per meter in 1951 was $61.00, and based on available data the estimated service life ranged from 6 to 15 years.

Meter Revenues and their Disposition

In general, the larger the city, the greater was its average annual revenue per meter in 1951. The estimated average annual revenue per meter was $70.48, and the range was from $42.28 per meter, for places with populations under 2,500, to a maximum of $89.67 per meter. Total gross revenue was about $76 million.

A more recent survey[36] showed that the average annual revenue per meter was $65.88 for off-street meters, and $74.07 for curb meters. Meter revenues for 1962 in New York City totaled $6,500,000, which averaged to over $138 per meter. It is interesting to note that the total parking revenues ($8,250,000) received by the Department of Traffic constituted almost 76 per cent of the department's total operating expenses.[23] On-street meter revenues for 1965 in New York City totaled $8,590,400. Off-street parking revenues amounted to $2,116,000.[23]

The need for parking accommodations is so great that any diversion of revenues to nonparking purposes is looked upon as undesirable in the public interest as well as conflicting with the legal justification for parking meters. Some cities have wisely earmarked meter revenue for providing off-street parking facilities, and for traffic improvement.

In 1951, approximately 35 per cent of gross meter revenues was spent for curb and off-street parking accommodations. Administration and amortization of meters totaled 23.1 per cent but only 8.6 per cent was used for off-street parking facilities. In New York City, the law prescribes that all meter revenues be paid into a special traffic improvement fund.

Effectiveness of Meters[37,38]

The basic objective in using meters is to reduce overtime parking and increase turnover at the curb. However, the effectiveness of parking meters, as compared to unmetered time restrictions, should be judged by considering several factors:

1. Proportion of overtime parkers
2. Proportion of available time used by them
3. Average parking duration of these violators
4. Parking turnover

An improvement found in a metered area in any one of these factors may be credited to the meters, but may not constitute any net improvement in conditions unless accompanied by improvement in several other or all of the other conditions. An improvement in all of the above factors is reasonably conclusive evidence of meter effectiveness.

In 15- and 30-minute zones, meters are not markedly effective unless strongly enforced. Enforcement in all zones is desirable, but meters have a definitely beneficial effect in one- and two-hour zones, whether enforced or not, and these zones include 96 per cent of all metered zones.

Warrants for Meters[39, 40]

An information report by the ITE contains a statement of description and purposes of parking meters and installation details, as well as recommended warrants for their use. Curb-parking meters are warranted as an aid in the control of parking in a given block side if all of the following conditions exist:

1. Parking time limits exist and are enforced.
2. High demand is shown by usage of at least 70 per cent of the available space-hours during the time of limited parking.
3. The block side is within walking distance of the generator of high short-time parking demands, such as stores and office buildings.
4. Observation shows a need for greater turnover. This is indicated, for example, by:
   a. High average parking duration
   b. High level of enforcement required to prevent parking violations
   c. Cruising by drivers desiring parking space

In addition, the report contains recommendations on design, marking, and operation of parking meters, and comments on control, enforcement, and use of revenues.

28

Special Applications of Traffic Control to Limited-Access Facilities

The freeway poses many special problems to the traffic engineer. One of these is the regulating and guiding of vehicles at relatively high speeds. Because of these speeds, the proper use and placement of signs is extremely important, since drivers will have traveled several hundred feet in the time it takes to simply read the sign.

Another problem of special interest is the interface between freeway ramps and local streets. Where the diamond interchange configuration is used, special consideration in signalization becomes necessary.

In urban areas, where freeways often operate at Levels of Service E and F, consideration is being given to ramp metering and freeway surveillance. This relatively recent control technique attempts to prevent continuous interruption of through traffic by entering ramp vehicles, which would cause operation at Level of Service E to break down to Level of Service F.

These items are discussed in the sections that follow.

EXPRESSWAY SIGNING

For the purpose of this book, and according to the MUTCD, *expressways* include all types of high-speed highways which are characterized by three distinctive features: divided roadways, controlled-access, and grade separations at intersections.

Expressways are normally built to high design standards developed through research and experience, including numerous observations and analyses of traffic performance and driver behavior. These high design standards are reflected in the use of adequate lane widths, adequate shoulder width and clearance to vertical obstructions from the edge of the pavement, proper lengths of acceleration and deceleration lanes, gentle side slopes and back slopes, vertical and horizontal alinement consistent with the design speed, and other geometric design features to help satisfy the needs and desires of motorists. An expressway is a facility that, potentially, can offer a maximum degree of safety, since it is designed within practical limits to satisfy the existing and future patterns of human behavior and vehicle development.

Studies comparing accident rates on highways with controlled access, with those without access control, illustrate the importance of location and design to traffic safety. The BPR for several years analyzed data from 30 states on this subject. Accident and fatality rates on fully controlled-access highways have consistently been only $\frac{1}{3}$ to $\frac{1}{2}$ as great as those on highways with no control of access. This is not due wholly to the control-of-access feature, but also to grade separation of intersections, provision of separate roadways for opposing directions of traffic, and other design features customarily employed in conjunction with access control.[1]

To a large degree, traffic control is incorporated in the design of expressways, and the need for traffic control devices is reduced to a minimum. Since traffic safety has, to a great extent, been incorporated in the design, the need for regulatory and warning signs is considerably reduced; however, it is imperative to provide adequate guide signs if motorists are to leave the expressway with convenience and without interference with other traffic. The high speeds and high volumes normally associated with expressway facilities, particularly in urban areas, dictate the need for increased emphasis on guidance.

Variable Signs[2]

Signs which at special times display emergency warnings, as of fog or icy surfaces ahead, or impose special regulations as to speed, turn prohibitions, or lane movements, are becoming increasingly more important on high-speed expressways. Such signs may be changed manually, by remote control, or for some purposes by automatic controls using sensing elements for the conditions that require special sign messages. Signs on the New Jersey Turnpike are being remotely operated by radio to provide speed and reliability in alerting motorists of vital road conditions.

The sign may have a single message, made legible only when required, or several messages, of which one or more can be made legible at any time. The various methods of accomplishing this are discussed in Chapter 20.

Design and Location of Signs

An ultimate goal for expressway signing should be the standards promulgated for the interstate system. Therefore, the comments about and requirements for the interstate system made in Chapter 20 (Traffic Signs) should be adopted, in general, for expressway signing.

Operational characteristics of certain expressways, such as very heavy traffic volumes, reduced speeds, and the location and spacing of interchanges, may not warrant or make possible the complete adoption of interstate standards. These conditions very often exist on urban expressways, but signing for expressways that permit high speeds should conform to a major degree with the interstate standards.

Regulatory and Warning Signs[2,3]

On expressways, these signs must ordinarily be larger than those on other highways, to provide quicker recognition. Otherwise, standard shapes, colors, and legends should be used where they are applicable. A few special messages are required to deal with particular features of expressway design, primarily dealing with proper placing and merging of traffic. Such regulations as may be required must be clearly posted to prevent improper or unexpected behavior in fast or high-volume traffic. With an increasing mileage of expressways in use, motorists can be expected to become familiar with expressway controls, and regulatory and warning signs should only be used where there is a definite need for them.

Interstate standards require regulatory and warning signs to be in general conformance with the established principles prescribed in the MUTCD. In all cases, the background should be reflectorized or illuminated. There is fairly general agreement that internally illuminated signs are more desirable under certain conditions, including inclement weather, high-intensity background illumination, points of multiple decisions, and lack of reflectance from vehicle headlamps when using low beams. For some of the thoughts dealing with specifications for internally illuminated signs for expressways see the literature.[4]

Because of greater demands for legible signing on the interstate system, all regulatory and warning signs should equal or exceed in their dimensions the minimum specifications of the MUTCD. *Warning* signs should have a *black* legend and border on a *yellow* background, and *regulatory* signs should have a *black* legend and border on a *white* background. For both types of signs, the size of the lettering or symbol must determine the size of the sign, and all legends should be in capital letters which, with few exceptions, should be a minimum of 10 inches in height.

Some of the pertinent regulatory and warning signs required to deal with special design features of the interstate system include:

1. *MERGING TRAFFIC:* This warning sign is to be erected on the appropriate side of the through roadway in advance of where another roadway enters.
2. *KEEP OFF MEDIAN:* This regulatory sign should be erected within the median wherever there is a tendency for drivers to enter or cross, and at random intervals as required.
3. *EMERGENCY PARKING ONLY* or *EMERGENCY STOPPING ONLY:* This regulatory sign should be used at random intervals as required.
4. *SLOWER TRAFFIC KEEP RIGHT:* This regulatory sign should be used to the extent necessary to maintain orderly use of lanes.
5. *SPEED LIMIT:* These regulatory signs should be of standard design, and the legend in accord with the respective state laws.

6. *YIELD:* This regulatory sign is to be used on entrance at locations where a traffic engineering study indicates that the sign will be conducive to safe and orderly merging entrance movement. Where full-length acceleration lanes exist, the YIELD sign will not normally be required.
7. *EXIT* (speed) *MPH*: Where it is necessary to post a lower speed on an exit ramp, an advisory warning exit speed sign should be used. It should have a *black* legend and border on a *yellow* background, and should be mounted on the right-hand side of the ramp roadway just beyond the gore.

Guide Signs[2,3]

The major problems concerned with the development of expressway signing relate to the design and application of guide signs. Expressway guide signs must be large enough and legible enough to be read from vehicles moving at high speeds, or from vehicle in high traffic volumes. At interchanges, drivers must often depend completely on signs in choosing the proper roadways, and since in some cases interchanges are relatively far apart, an error in turning could possibly add miles to a driver's trip. It may also be necessary to inform the traveler where he can reach certain service facilities that are not directly visible or accessible from the expressway.

While expressway guide signs are, in principle and in function, the same as other guide signs, the characteristic design of expressways requires special types of directional signs. The distinctive feature of expressway guide signs is that they incorporate all directional information required—route numbers, destinations, and directions—into single, large signs, which are so placed in sequence as to give the traveler advance information concerning diverging roadways at an interchange, and confirming information at the actual point of divergence.

It is extremely important that expressway signing and geometric design be integrated to provide for maximum safety and efficiency on the expressway. For a coverage of principles of directional signing, relationship between geometric design and signing, and correlation of geometric design and signing, see the literature.[5]

Basic Principles. A study was conducted to determine the desires of motorists to help to establish principles in developing expressway directional signing.[6] The following six basic principles were recommended as a guide to be used in the design, installation, and maintenance of directional signing:

Interpretation. All possible interpretations and misinterpretations must be considered in phrasing sign messages (words and symbols). Messages must be complete and clearly stated.

Continuity. Each sign must be designed in context with those which precede it, so that continuity is achieved through relatively long sections of highway. The driver should not be expected to evaluate more than one new alternative at any advance sign. At the decision point he should never be given new information about either the through route or the turnoff.

Advance notice. Signing must prepare the driver ahead of time for each decision he has to make. A single advance sign is usually not sufficient to overcome the element of surprise on the motorist's part when he reaches the point of decision. Therefore, it is advisable to use *two* advance signs when feasible.

Relatability. Sign messages should be in the same terms as information available to the driver from other sources, such as touring maps and addresses given in tourist information and advertising.

Prominence. The size and position, as well as the number of times a sign or message is repeated, should be related to the competition from other demands on the driver's attention. These demands can come from other visual aids, other signs or parts of the message, as well as the task of driving. When the roadway is wide, traffic volume high, and there are numerous competing commercial signs or buildings (as is typical of a downtown urban expressway), the directional signs must be very large, well illuminated, and properly placed. This usually requires an overhead installation.

Unusual Maneuvers. Signing must be specially designed at points where the driver has to make a movement which is unexpected or unnatural.

Size of Lettering. On expressways the principal legend on a guide sign, mounted at the roadside, on a through lane, should be in letters at least 10 inches high if capitals, or at least 8 inches if lower case. Much larger lettering is necessary for destination signs at, and in advance of, major interchanges.

Standards for the interstate system require that all names of places and highways on guide signs should be composed of lower-case letters with initial capitals. Letter sizes are prescribed in detail for each of the signs specified for use on the interstate system.

Amount of Legend. Expressway guide signs, in particular, should be limited to three lines of principal legend on roadside signs, and two on overhead signs. Principal legend includes only destinations—place names, route numbers, and highway and street names. Symbols, exit instructions, and cardinal directions may make up other lines of legend. The message dimensions should be determined first, and these in turn will determine the overall sign dimensions. For the interstate system, principal legend on both roadside and overhead mounted major signs is limited to two destination names.

Use of Arrows. Proper aiming of arrows is important to provide motorists with accurate alinement information. On all signs used to direct traffic into an exit roadway from the main traffic lanes the arrows should point upward at an angle to fit the alinement as well as possible. On roadside signs for through traffic, the arrows should ordinarily point vertically upward. On overhead installations, interstate standards require directional arrows to be pointed downward, and should designate the lane or lanes for the specific routes or destinations shown on the sign.

Signing for Interchanges. An interchange on an expressway usually connects with a numbered or named highway or street. In addition, the intersected highway will in most cases, lead to some destination or destinations of interest in one or both directions. Therefore, in general, both route numbers and destinations must be shown on interchange signs. In urban areas, a street name may be all that can be shown without reference to any particular destination. The designation of interchanges by name may be advantageous under some circumstances, but generally an interchange can be identified by the name or number of the intersected highway or the primary destination thereon. A special interchange designation, such as numbered exits, only adds to the legend that must appear on the interchange signs. Where a large number of exits are used, numbering should be from south to north and from west to east. On the New Jersey Turnpike, serial numbers were chosen as the primary designation of exits, and they were emphasized with 30-inch numerals on all major signs.

On expressway guide signs the emphasis should always be placed on the information that will be most helpful to the strangers unfamiliar with the route.

The major signs at interchanges and on their approaches are of three general types, discussed in the following paragraphs.

Advance guide signs

The advance guide sign notifies the driver well in advance of the intersecting route (or routes), the principal destinations served by the interchange being approached, and the distance to that interchange. Use of one or more such advance guide signs may be necessary. Sometimes a *NEXT EXIT* sign may be necessary. Where the distance to the next interchange beyond the one for which the advance guide sign is posted exceeds 5 miles, it may be desirable to use a supplementary panel mounted below the advance guide sign.

Exit direction signs

The exit direction sign repeats the route and destination information displayed on the advance guide sign, and provides drivers with more specific information concerning the action that should be taken to reach the interchange exit. For any given exit, no more than one exit direction sign, located on the immediate approach to the interchange, is used.

Where there is an adequate overhead sign at the gore giving the required directional information, or where the distance between interchanges is short, or where there are at least two advance exit guide signs, the exit direction sign may be omitted.

Gore signs

The gore sign is erected at the point of departure where drivers leave the through-traffic lanes. Depending upon specific conditions, the gore sign carries the word EXIT, with an appropriate directional arrow, or repeats route and destination information previously displayed on the advance guide signs and exit direction sign.

Details of design, placement, and application of all the above signs are given in the MUTCD and the Interstate Manual. In addition, the Interstate Manual contains sketches showing application of the principal guide signs and some regulatory signs. Typical interchanges shown include the following:

1. Interchange of two interstate routes
2. Cloverleaf interchange
3. Diamond interchange
4. Urban interchange with a city street
5. Two closely spaced diamond interchanges

A survey conducted by the BPR soliciting advice on current ramp terminal signing from the 50 states revealed that the standard signs in the MUTCD appeared to be adequate, when properly applied and located.[8] However, 15 per cent of the states said that wrong-way movements on exit ramps was a problem.

A proposed revision of the direction of arrows at interchanges[9] was successfully used on expressways leading to the New York World's Fair of 1964.[10] A 2-o'clock arrow was used to indicate an advance movement to the right, and where positive action was required in front of or at the sign, a 5 or 7 o'clock arrow was used.

Other Expressway Guide Signs.

Rest area signs. There should be set aside special areas, with properly designed access roadways, to serve the comfort and convenience of travelers. For this purpose, a guide sign should be placed one or two miles in advance of the rest area, and should carry the appropriate legend. At, or in advance of, the beginning of the deceleration lane, there should be placed a sign similar to the exit direction sign, and within the gore there should be a final sign with an arrow pointing in the appropriate direction.

Service signs. As a general rule, access to vehicle service stations, restaurants, overnight lodgings, and other commercial establishments is only available at interchanges of expressways, or where a special area or service roadways have been provided. It is often necessary to indicate by signs where these facilities can be reached. However, such signing is normally required only in rural areas.

Other directional signs. Other signs may be used where they can provide additional useful directions. Nearby destinations not shown on expressway interchange advance signs may be shown on a special sign carrying up to three additional names, and the legend Next Exit, Next Right, or Second Right, erected about $1\frac{1}{2}$ miles in advance of the interchange.

Sign Installations and Highway Safety

In 1968, 7.7 per cent of all fatal motor vehicle accidents were single-car accidents involving collisions with roadside objects.[7] A large portion of these objects are believed to be posts or structures supporting traffic control devices. It is the traffic engineer's responsibility to the motoring public, when installing signs, especially on high-speed roadways, to consider not only the operational aspects, but also the safety of the situation. Signs should be located to eliminate or minimize the exposure of traffic to structures supporting signs, and such structures should be designed to minimize the impact on colliding vehicles. See Chapter 17 of this book for further studies on this subject.

Signalization of Diamond Interchanges[11]

Phasing. The signalization of diamond interchanges presents an interesting challenge to the traffic engineer. Where heavy turning movements prevail, vehicle storage between ramps becomes a major problem. Consider the two conventional phasing schemes shown in Figure 28-1. Consider signalization (a). If turning volumes from the ramps are high, a significant storage problem arises in phase 3. Signalization (b) adds the problem of storing vehicles turning onto the ramps in phase 1. In both cases, the storage is between ramps, an area of severely restricted length. Should a back-up across either ramp occur, the entire signalization will collapse.

A four-phase signalization, with two overlaps, has been developed which eliminates the storage of vehicles between the ramps. This is shown in Figure 28-2.

Capacity Calculations. Consider a single lane on a single approach. Let n = number of vehicles that can clear the signal during one green phase from this lane.

$$n = \left\{\frac{G - D}{H} + 2\right\} \text{ vehicles per cycle} \qquad (28\text{-}1)$$

where
G = length of green
D = starting delay of first vehicle
H = headway of remaining vehicles

Assuming

$$D = 5.8 \text{ seconds}$$
$$H = 2.0 \text{ seconds}$$

Fig. 28-1. 4-phase signalization

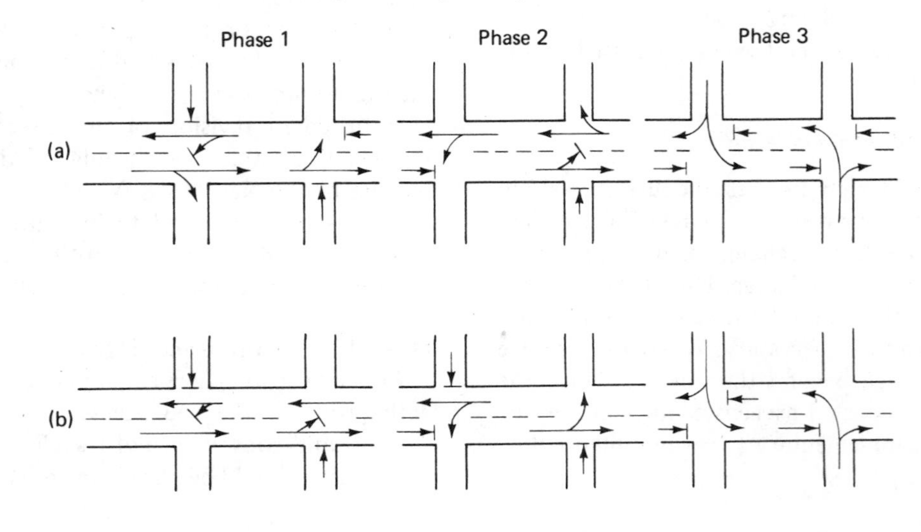

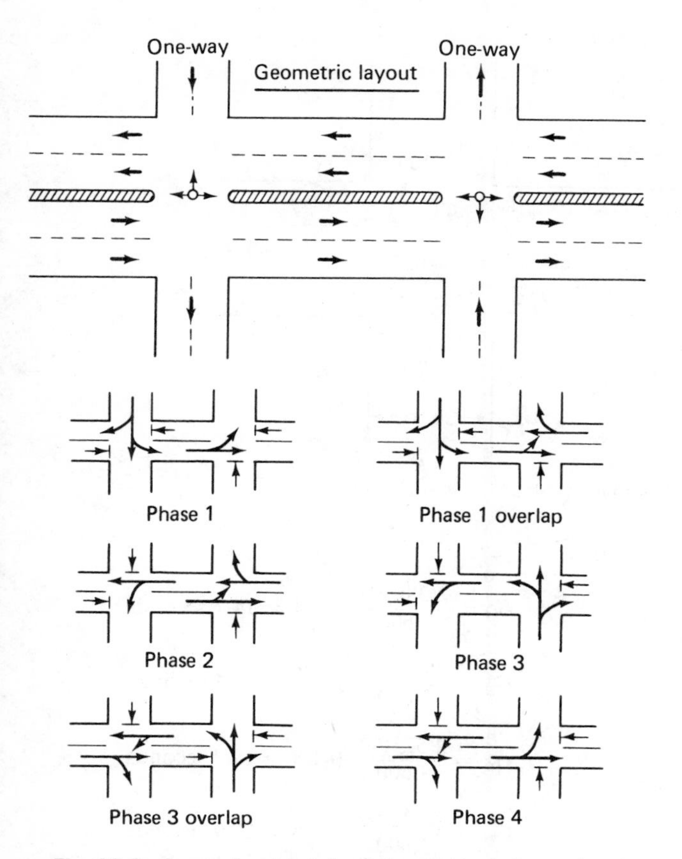

Fig. 28-2. Conventional two-level diamond interchange signal phasing (*Source: "Urban Interchange Design as Related to Traffic Operation," Pinnell and Burh, TE, March, 1966, Figure 4*)

then

$$n = \frac{G - 5.8}{2.0} + 2.0 = \frac{G - 1.8}{2.0}$$

Multiplying by the number of cycles per hour produces N, the number of vehicles per hour through one lane:

$$N = \frac{G - 1.8}{2.0}\left(\frac{3{,}600}{C}\right) \text{ vehicles per hour}$$

Under four-phase signalization, there are four critical volumes (see Figure 28-3). Thus,

$$N_C = \sum_1^k N_i = \sum_1^k \left(\frac{G_i - 1.8}{2.0}\right)\left(\frac{3{,}600}{C}\right)$$

$$= \frac{\left(\sum_1^k G_i\right) - 1.8k}{2.0}\left(\frac{3{,}600}{C}\right)$$

Cycle length C is equal to total green time plus total amber time *minus* total overlap time ϕ.

$$C = \sum G_i + A - \phi \qquad (28\text{-}2)$$

$$\sum G_i = C - A + \phi$$

Substituting

$$N_C = \frac{C - A + \phi - 1.8k}{2.0}\left(\frac{3{,}600}{C}\right) \qquad (28\text{-}3)$$

for k = four phases, $A = 12$ seconds (3 per phase),

$$N_C = 1{,}800 - (19.2 - \phi)\left(\frac{3{,}600}{2C}\right)$$

A desirable condition exists if $\phi > 19.2$. If $\phi > 19.2$, N_C gets larger as the cycle length gets smaller.

Timing the Four-Phase Diamond Interchange.

Example (see Figure 28-4).

Critical volumes:

$$N_A = 388$$

$$N_B = \frac{786 + 150}{2} = 468$$

$$N_C = 525$$

$$N_D = \frac{1{,}050 + 513}{2} = 782$$

$$\text{Total} = 2{,}163$$

Capacity:

$$N = \frac{C - A + \phi - 1.8k}{2.0}\left(\frac{3{,}600}{C}\right)$$

As long as $\phi \geq 20$ seconds, N increases as C decreases. Therefore, a short cycle length of 60 seconds will be used. However, $N_C \geq 2{,}163$. Knowing C, A, K, and N, we may solve for the necessary ϕ to accommodate N.

$$2{,}200 = \frac{60 - 12 + \phi - 7.2}{2}\left(\frac{3{,}600}{60}\right)$$

$$\phi = \frac{2{,}200}{30} - 60 + 19.2$$

$$\phi = 32.5$$

Therefore, if the diamond is spaced to permit ϕ phases 16.25 seconds or more in each direction, the phasing is

Fig. 28-3. 4-phase movements

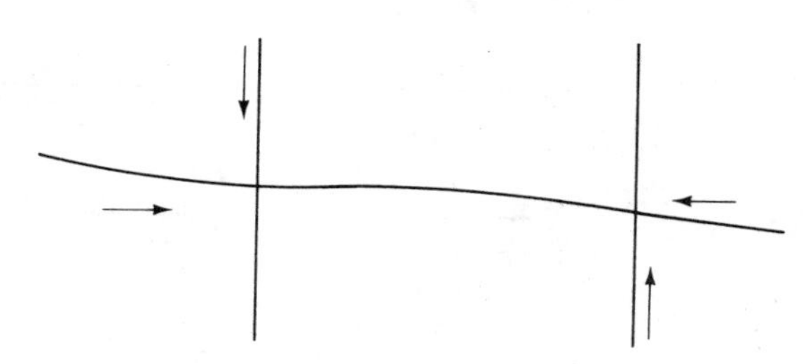

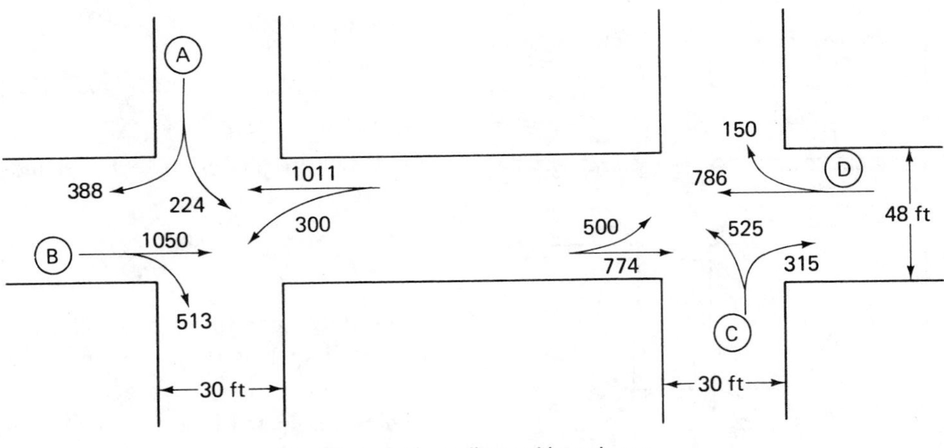

Fig. 28-4. 4-phase diamond interchange

feasible. If the diamond is spaced too closely for ϕ's of 16 seconds, the diamond may not be properly timed.

Timing:

Assume 16.25 seconds, if possible. 450 to 500 feet spacing is necessary. Note that

$$N = \frac{G - 1.8}{2.0}\left(\frac{3{,}600}{C}\right)$$

which implies

$$G = \frac{2N \times C}{3{,}600} + 1.8$$

Now,

$$G_A = \frac{2(388)60}{3{,}600} + 1.8 = 14.7 \text{ seconds}$$

$$G_B = \frac{2(468)60}{3{,}600} + 1.8 = 17.4 \text{ seconds}$$

$$G_C = \frac{2(525)60}{3{,}600} + 1.8 = 19.3 \text{ seconds}$$

$$G_D = \frac{2(782)60}{3{,}600} + 1.8 = 27.9 \text{ seconds}$$

Fig. 28-5. Phasing of diamond interchange

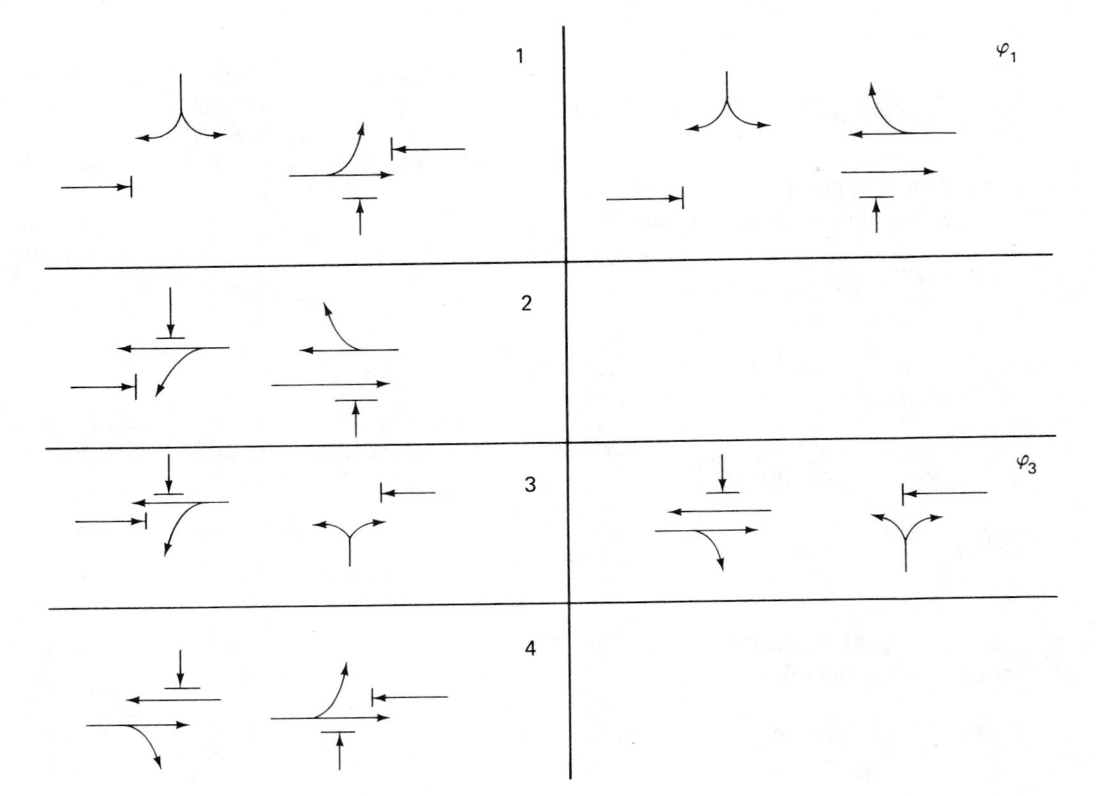

Phasing

See Figure 28-5. Phasing follows the four-phase systems illustrated in Figure 28-2.

Requirements:

PHASE	*REQUIREMENT*
$1 + \phi_1$	14.7 seconds
$\phi_1 + 2$	17.4 seconds
$3 + \phi_3$	19.3 seconds
$\phi_3 + 4$	27.9 seconds

Phase timing:

See Figure 28.6. The timing plan meets all the above requirements, and is accomplished through trial and error, guided by computed requirements.

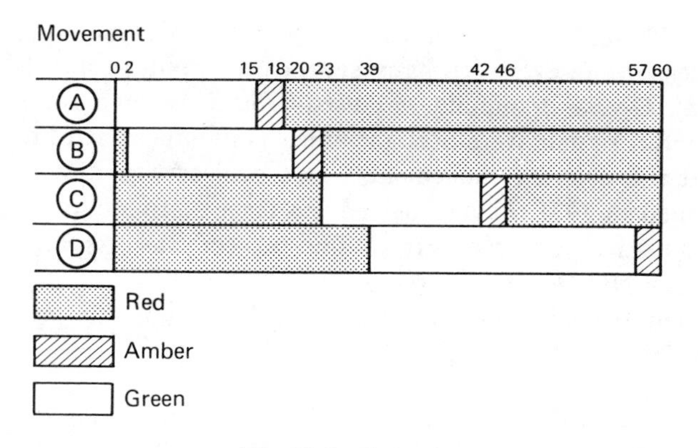

Fig. 28-6. Phase timing

RAMP AND FREEWAY CONTROLS

Many limited-access facilities in urban areas experience a breakdown in traffic flow during peak hours of travel, because the demand exceeds the capacity of the freeway. There are two basic alternatives: (1) increase the capacity by physical improvements or additional facility construction, or (2) regulate the demands the given facility can experience. The latter alternative has been used more frequently in the past several years than ever before, and has advanced from the research stage to operational acceptance. There are two aspects: individual ramp control, and freeway (facility) surveillance and control.

The fact that there are operational systems for facility surveillance and control does not at all imply that development in this area is complete; there are a number of unsolved problems, particularly in regard to proper practice for freeway management policies over the onset and duration of the peak period. For instance, how much service should be allocated to various ramps, and how should this vary with time? And there are questions in regard to allocation among facilities—freeways, service roads, and arterial streets—during the onset of a breakdown. Systems now in operation generally have only relatively empiric allocations, specific in intent and capability. The best of the existing systems, the Gulf Freeway in Houston,[12] has the capability of accepting future management policies as they are developed, by the simple expedient of reprogramming. This is one marked advantage of digital computer control over systems in which the policy is physically built into the hardware.

Freeway Surveillance and Control

The various projects in existence have been noted in Chapter 12 in regard to the prevention of breakdowns at capacity flow levels. These are principally the projects in Chicago,[13] Detroit,[13] and Houston.[12] There is also a major project in Dallas, Texas,[14] in which an expressway is controlled as part of a traffic corridor, including the expressway frontage (service) road and arterial streets and grids.

In addition to these major projects, there has been considerable work on surveillance and control in the tunnels linking New York and New Jersey by the Port of New York Authority.[15,16] Although these are not freeways, results have significantly advanced the state of knowledge about surveillance and control policies, both on policy and implementation levels. There have also been more limited but significant projects in Chula Vista, California,[17] and Los Angeles.[18] A pilot study for control of the Van Wyck Expressway in New York City was also undertaken.[19]

Detroit Project. Detroit established the first freeway surveillance and control project which initially incorporated 3.2 miles of outbound freeway in 1960. Control was subsequently extended to 8.5 miles at the beginning of 1967.

On the John C. Lodge Freeway in Detroit, traffic was first monitored by a 14-camera closed-circuit television and by an electronic sensing system consisting of overhead-mounted ultrasonic detectors which provide inputs to determine speed and lane occupancy. The detector information was then transmitted to a CDC 8090 digital computer which output various traffic characteristics on punched tape, printout sheets, and display panels. The data provided by television and detector systems were visually interpreted by observers who institute control measures by use of variable-speed signs, lane control signals, and entrance ramp control signals. Eight entrance ramps were controlled over the 8.5-mile section of freeway.

In a latter phase of this research project, television monitoring and on-freeway advisory signing were discontinued. The eight controlled ramps, all of which were northbound, were controlled by a simple metering release (one every so many seconds, computer-controlled). Diversion signing was used extensively, and some arterial signals were brought under computer control to aid the diverted movements. This project was discontinued in 1971, having passed the term of funding. An exhaustive survey report of this project was prepared under NCHRP sponsorship.

The benefits which have been derived from this project include the following:

1. Travel time through the 8.5-mile section during the peak period has been decreased from 29 to 17 minutes.
2. Volume on the freeway has decreased.
3. Volume on the arterial streets increased, but changes made in traffic signal timing resulted in reduced travel times.
4. Congestion was reduced.

Chicago Project. Chicago initiated the application of surveillance and control techniques in 1961 on a 5-mile section of the Eisenhower Expressway. Traffic was monitored by overhead-mounted ultrasonic detectors which transmitted data to a GE 4040 digital computer. The information was analyzed to determine vehicle speeds, volumes, and lane occupancy, at a number of points along the section. The computer automatically implemented a ramp metering system, which consisted of traffic signals and detectors located on entrance ramps. Ramp vehicle metering rates were automatically established by the detector-computer system, which decreased the volume of traffic entering the freeway in the vicinity of impending congestion. In an attempt to discourage traffic from using the congested expressway, informational signs symbolically displaying existing delays at entrance ramps and on the expressway were installed at appropriate locations on parallel arterials. The computer also executed the control functions on a 4-mile section of the Dan Ryan Expressway.

The results accomplished with these projects include the following:

1. Volume on the Eisenhower Expressway remained unchanged.
2. Travel time through the 5-mile section during the peak period was decreased from 15 to 12 minutes.
3. Accidents decreased by about 16 per cent.
4. Travel time through the 4-mile section of the Dan Ryan Expressway during the peak period decreased from 20 to 10 minutes.

The success achieved with these projects provoked a planned expansion of these control techniques wherever need indicated on Chicago's 75-mile expressway system. An initial phase of this program provided coverage in both directions of the Eisenhower Expressway for a length of 10 miles. A total of 22 entrance ramps would be under control in this section.

Port of New York Authority Project. An extensive surveillance and control system was installed in the Lincoln Tunnel between New York and New Jersey in 1964. Traffic was monitored by vehicle detectors with traffic-condition display and control devices and closed-circuit television. A special computer received traffic data and implemented controls to prevent congestion from occurring. Two-way radio communication was also maintained between control centers and police in the tunnel, and a rail-guided rubber-tired vehicle was installed to speedily dispatch police to any point in the tunnel.

The improvements achieved with this project include the following:

1. Peak period volume increased 5 per cent.
2. Travel time through the tunnel during the peak period decreased 12 per cent.
3. Air pollution concentration decreased 29 per cent.

Houston Project. The Gulf Freeway in Houston has one of the most sophisticated surveillance and control projects. Control was instituted on eight entrance ramps along a 6.5-mile section of inbound freeway. The project was formulated with the basic objective of developing criteria for the design and operation of automatic surveillance and control systems, which would permit the attainment of acceptable freeway levels of service during peak periods of demand. In addition to the more standard ramp metering controls, analog prototype ramp control equipment was installed, capable of operating the ramp control signal in a traffic-responsive manner. The ramp signal was actuated by the detection and projection of acceptable gaps in the shoulder lane. An IBM 1800 digital computer system was used to integrate the local entrance-ramp controllers, and selected through-lane traffic characteristics to provide overall freeway system control. Inputs were provided by loop detectors. A 14-camera closed-circuit television system was employed in control studies for visual surveillance.

The astounding success achieved with this project was evidenced by significant improvements in safety and reductions in congestion. The results accomplished include the following:

1. Peak period volume increased 10 per cent.

2. Speeds on the test section during the peak period increased 30 per cent.
3. Travel time over a 5-mile section of the freeway during the peak period decreased from 16 to 11 minutes.
4. The number of accidents occurring during the peak period decreased from about 145 to 75 per year, over the 6.5-mile section.
5. In the two years after the first prototype controller was installed on an entrance ramp, no accidents occurred on the ramp.

Chula Vista Project. The California Division of Highways, in July 1968, instituted ramp metering at four locations on the northbound Route 5 freeway in the vicinity of Chula Vista, a community just south of San Diego.

At each location a solid-state, fixed-time controller was installed. Each controller provided for cycle length adjustments from 6 to 20 seconds, green phase adjustments from $\frac{1}{2}$ to 2 seconds, and amber phase adjustments from $\frac{1}{2}$ to 2 seconds. The controllers had the ability to change cycle lengths four times during the A.M. peak period (the only one under control).

Available results, all qualitative, are as follows:

1. Freeway speeds *up.*
2. Net travel times *down.*
3. Vehicle-miles traveled on the freeway *up.*
4. Traffic density and speeds on parallel city streets *unchanged.*
5. Public reaction *favorable.*
6. Accident frequency *down.*

According to the California Highway Patrol, and local police officers, *reported* plus *unreported* A.M. peak accidents were five to ten per month fewer than before metering. Eliminating these minor accidents and the delays they cause increases travel time savings.

Los Angeles Project. The California Division of Highways implemented two freeway control projects in the Los Angeles area. The Hollywood Freeway project was very small (only two ramps). The other project, on the Harbor Freeway, was far more extensive and more successful. Only the Harbor Freeway ramp-metering system was reported on in the literature. This latter system involved the metering and/or closing of a number of southbound ramps along a five-mile section of the freeway during the P.M. peak period. Metering was through use of fixed-time signals. Use of a three-dial controller provided for up to three different rates. When releasing single vehicles, the green and amber phase was fixed and very short (about 2 seconds). The variable rate was obtained by varying the red phase. When platoon-metering was used (necessary when entering rates were greater than could be handled one at a time), the green and amber phase was still fixed, but at a time duration sufficient, on the average, to service the desired number per cycle. Ramp closure was effected through use of hand-placed barricades, because signs alone were found to be ineffective.

Initial results of a typical day's controlled operation yielded reduced freeway travel time, on the order of 1,000 vehicle-hours. Some individual drivers saved up to 9 minutes on a 5-mile trip. Delays to drivers diverted or in ramp queues, and a small amount of other surface street traffic, was about 130 vehicle-hours.

New York City Pilot Study. The Department of Traffic of the City of New York, in conjunction with the Polytechnic Institute of Brooklyn, undertook a pilot study of the southbound Van Wyck Expressway (in the direction of Kennedy International Airport) in 1969. One entrance ramp was closed, and two other ramps were metered with a simple fixed-time metering scheme. The service road signals were retimed and set to more traffic. Advisory signs were installed.

The pilot study resulted in a significant reduction in travel time through the controlled section, increased the average speeds, and provided considerably improved entrance patterns, since "bursts" of vehicles from a nearby service road signal were metered in, providing needed spacing. The installation was continued in operation after the test period, and a decision was made to install a complete surveillance and control system. Although the evaluation period was short, a marked improvement in the accident record was observed.

Dallas Corridor Project. The latest freeway surveillance and control project implemented was one in Dallas, Texas, in June of 1971. This project was part of a larger corridor control project in which freeway, arterial, and grid traffic would be controlled for the purpose of enhancing traffic flow in an urban corridor. This overall project was undertaken by the Texas Transportation Institute of Texas A & M University, with the sponsorship and support of the Federal Highway Administration, the Texas Highway Department, and the City of Dallas.

The controlled facility was the North Central Expressway. Control was implemented on an IBM 1800 digital computer. The control mode used was gap acceptance control, with a fixed-rate override to assure certain minimum volumes. This was the same approach used in Houston. Television surveillance was also used.

Individual Ramp Control

Individual ramps have been studied intensively in recent years with regard to entrance control of vehicles by ramp signalization and a governing control algorithm. These algorithms have taken four prime forms: (1) simple metering, (2) demand-capacity responsive metering, (3) gap acceptance injection, and (4) paced or "follow-the-rabbit" injection. Forms (1) through (3) have been in use in both operational and research installations; form (4) was tested in a research program.

Simple Metering. An application of simple metering is illustrated in the next section; this form consists of the proper setting of a pre-timed signal (with extremely short cycle) at the ramp entrance, so as to limit the hourly flow into the freeway at the controlled ramp. It has been used effectively in New York City,[19] Chula Vista,[17] and Los Angeles.[18] It has also been used to effect reductions in rear-end collisions in Atlanta.[20]

Demand-Capacity. Demand-capacity responsive metering is the response of the metering rate to known or measured flow levels. A treatment of such a volume control (pre-programmed) is given in the literature.[21] The second problem in the next section illustrates a restricted form of this control.

Gap Acceptance. Gap acceptance injection addresses the problem of optimally timing a vehicle's release so as to maximize the probability that it will meet (match) an observed acceptable gap in the freeway shoulder lane. This technique was used on the Gulf Freeway in Houston,[12,22] and was implemented in the project "Optimizing Flow in an Urban Freeway Corridor," in Dallas.[14] Refer to Figure 28-7.

Pacer System. The pacer, or "follow-the-rabbit" control, was studied by the Raytheon Company for the United

Fig. 28-7. Illustration of gap acceptance mode of ramp control (*Source: "Gap Acceptance and Traffic Interaction in the Freeway Merging Process," Phase II Final Report, TTI*)

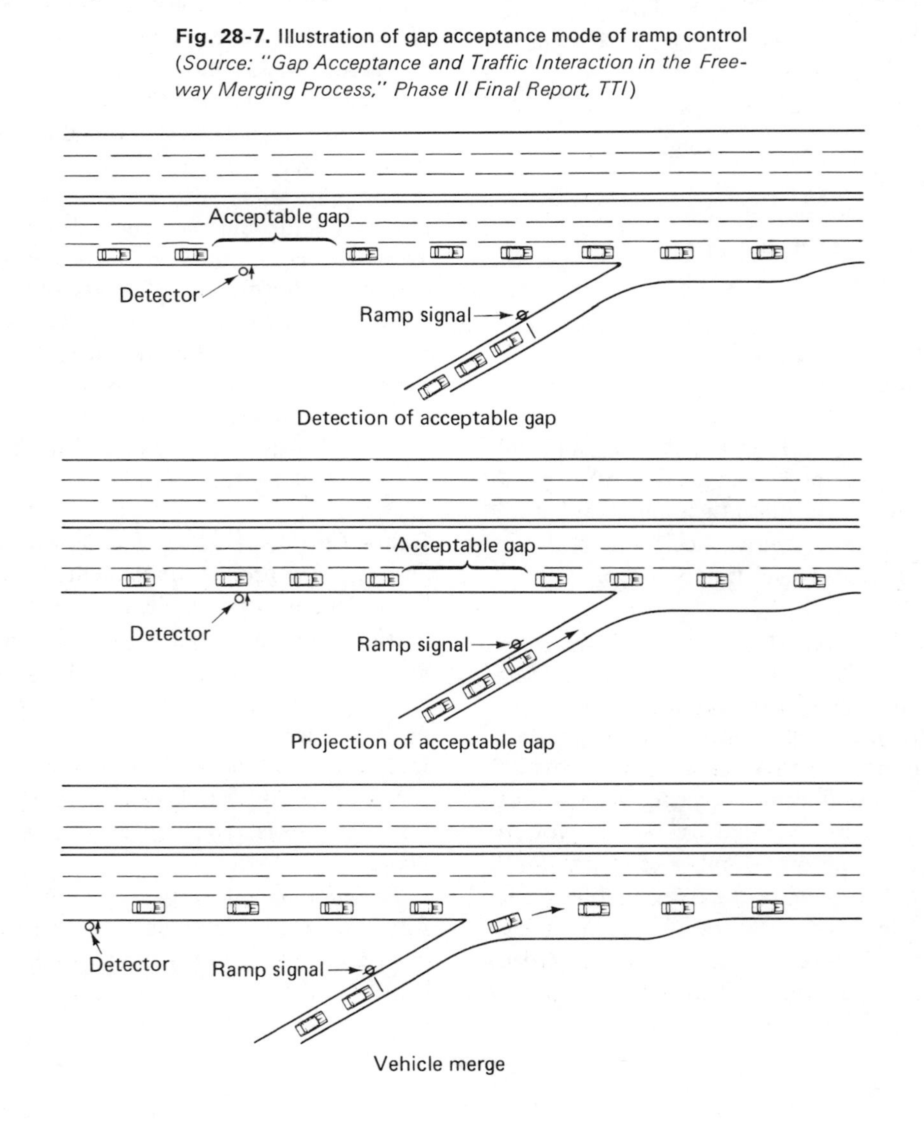

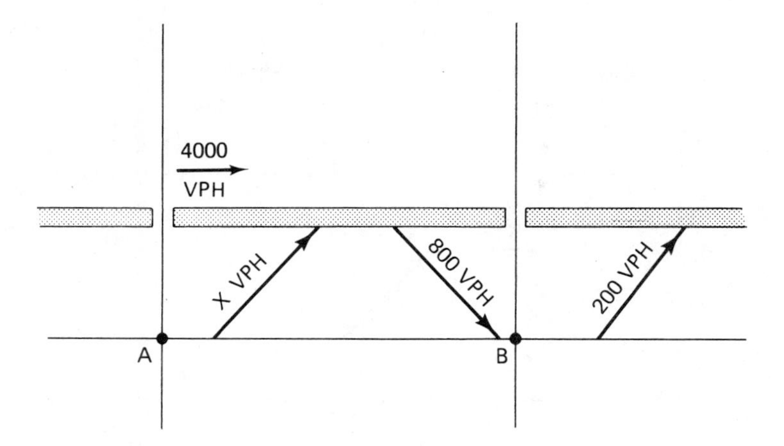

Fig. 28-8. Flows for sample ramp metering problem

States Bureau of Public Roads. It involves an entering vehicle following, or keeping pace with, a positive indication (on) in a string of signals, or in a band of light (both were tested), whose movement is determined by the changing movement of an acceptable gap on the freeway.

Sample Problems

The purpose of these sample problems is to illustrate some of the considerations of simple ramp metering, including demand-capacity calculations.

Problem 1:

Given a ramp with a demand of 900 vehicles per hour for entrance, and the flows as indicated in Figure 28-8. Given also that the demand must not exceed 4,500 vehicles per hour in order to avoid breakdown. Determine the type of ramp control and the flow X to achieve nonbreakdown operation. Identify probable problem areas and incidental benefits.

Solution:

From the given information, it would seem that a simple metering ramp control would suffice. This controller should have a cycle length of $3{,}600/500 = 7.2$ seconds. A 2-second green followed by a 5.2-second red is recommended, so that a single vehicle release per cycle is assured. (Actually, 2.016 and 5.184 seconds, because of the 100-unit division on the timing dial). This will allow 500 vehicles per hour on at the controlled ramp, with 400 vehicles per hour diverted downstream along the service road (frontage road), to enter at the next on-ramp, increasing its flow to $200 + 400 = 600$ vehicles per hour. It is recommended that the signal at intersection B be retimed to accommodate the increased flow and that it be offset relative to A to best handle the increased load.

Probable Problem Areas:

1. It must be established that the service road can handle the increased volume without breakdown itself.
2. It must be anticipated that vehicles may queue at the metered ramp, backing up into and blocking intersection A. This is particularly the case if the service road breaks down.
3. It must be recognized that the area downstream of the controlled ramp is a weaving section, and that the figure of 4,500 vehicles per hour may be high for its capacity.
4. It must be established that the upstream 4,000 vehicles per hour was a *demand*, and not a flow at breakdown conditions, lest removal of the (weaving section) breakdown by metering allows this to rise, causing the same breakdown for a different reason.
5. It may be that some of the 900 vehicles per hour (actually, the 400 vehicles per hour diverted) shall choose to divert *upstream*, thus increasing the 4,000 vehicles per hour and decreasing the permissible 500 vehicles per hour metered.

Incidental Benefits

1. The ramp control shall smooth out the injection of "batch" or "pulsed" arrivals from signal A into the freeway, thus relieving a periodic load by spreading it out.
2. Accidents will probably decrease, particularly rear-end stop-and-go types on the freeway in breakdown, and on the ramp in "batch" arrivals.

Problem 2:

This problem deals with traffic conditions at a controlled ramp. Traffic volume on the ramp upstream and peak flow demand for the ramp are given in Figure 28-9. Determine (1) an appropriate fixed (single-dial) ramp metering rate for the indicated capacity, (2) the duration of the build-up on the ramp, (3) the maximum delay for an individual vehicle, and (4) a measure of the total delay.

Solution:

Ramp metering is to be fixed at a rate which allows 500 vehicles per hour. The demand at the ramp increases linearly to a peak of 750 vehicles per hour. Ramp metering starts at 5: 10 P.M., and is maintained *until the backlog on the ramp* is cleared. This does *not* occur at 5: 50 P.M.,

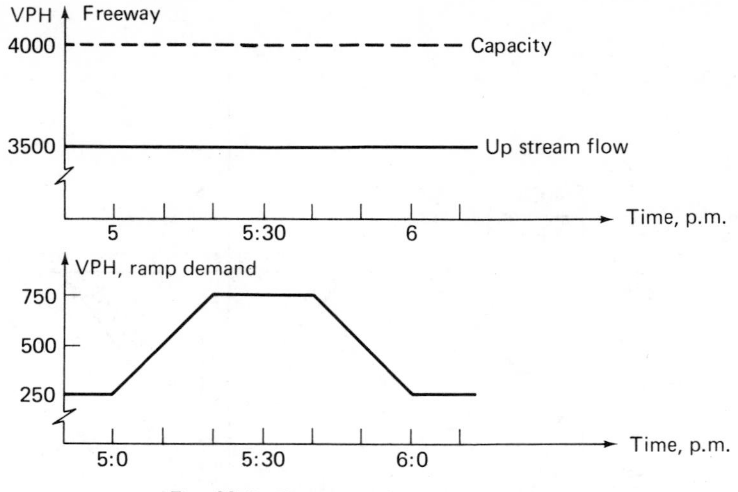

Fig. 28-9. Freeway and ramp demand flow

since some vehicles delayed prior to 5:50 P.M. are still present, even though the new arrivals have fallen below 500 vehicles per hour.

To achieve a flow rate of 500 vehicles per hour, the metering can be set at 1 vehicle per 7.2 seconds. This is accomplished with a single-dial, pre-timed signal with either a nonstandard 7.2-second cycle, 2-second green, or with standard timing gears, by setting the multiple-timing key position in (1) a 30-second timing gear with keys for four repetitions of the pattern, with a cycle for a rate of $3{,}600(\frac{4}{30}) = 480$ vehicles per hour; (2) a 35-second timing gear with keys for five repetitions of the pattern, within a cycle for a rate of $3{,}600(\frac{5}{35}) = 514$ vehicles per hour. This complication was not considered in the first example.

Note that the average arrival rate between 5:00 P.M. and 5:10 P.M. is $(250 + 500)/2 = 375$ vehicles per hour. In this 10-minute period ($\frac{1}{6}$ hour), $375(\frac{1}{6}) = 62.5$ vehicles arrive. Similarly, 104 vehicles arrive between 5:10 and 5:20 P.M., and 125 vehicles between 5:20 and 5:30 P.M.

A number of the quantities desired can be determined easily, in this particular case, from the graphs of Figure 28-9. For instance, by manipulating total vehicles, as in Figure 28-10(a), one may see that the duration of the build-up is from 5:10 P.M. to 6:25 P.M. This is so because, between 5:10 and 5:50 P.M., an extra, i.e., above 500 vehicles per hour, (125 vehicles per hour) ($\frac{1}{6}$ hour) + (250 vehicles per hour) ($\frac{1}{3}$ hour) + (125 vehicles per hour) ($\frac{1}{6}$ hour) = 125 vehicles arrive. Between 5:50 and 6:00 P.M. there is the ability to service (125 vehicles per hour)

Fig. 28-10. Operations at metered ramp

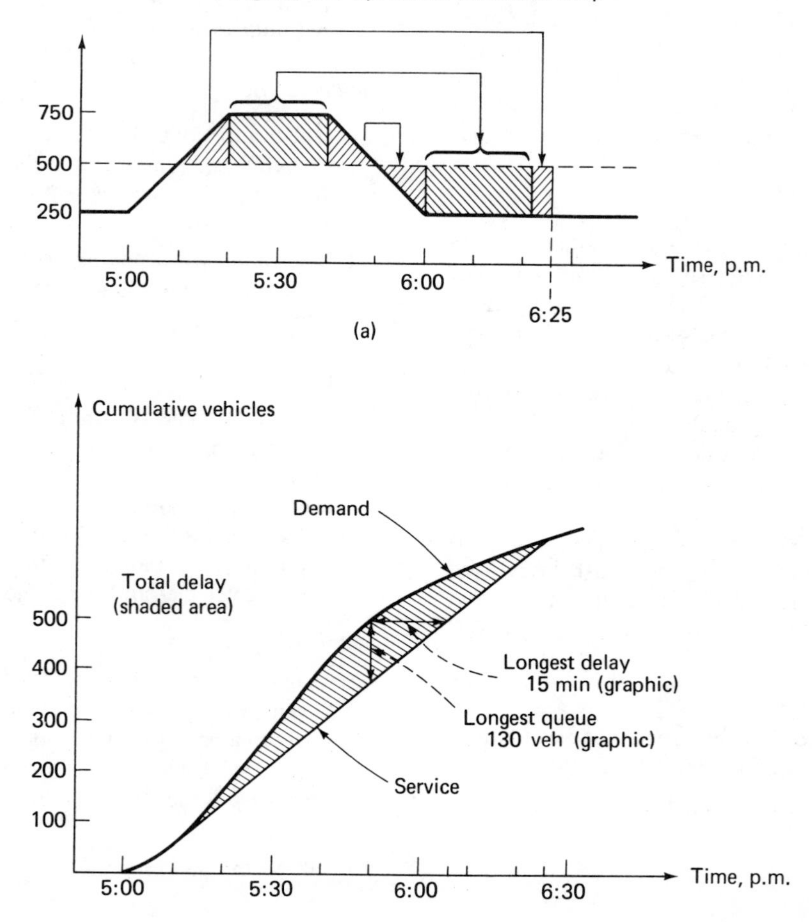

($\frac{1}{6}$ hour) $\cong$ 21 vehicles net above the number that arrive. Beyond 6:00 P.M., a net of 250 vehicles per hour beyond arrivals (500 versus 250 vehicles per hour) may be serviced. With 125 − 21 = 104 vehicles to service, this takes ($\frac{104}{250}$) = 0.416 hours = 25 minutes. Thus, the end of the ramp build-up is at 6:25 P.M.

Rather than pursue this avenue, attention is now directed to a more general methodology. This method, shown graphically, is easy to apply to (1) total delay, (2) simultaneous variations in ramp flow, and upstream flow and/or downstream capacity, and (3) more complicated flow patterns than straight-line pieces.

To determine the delays and queue length, the cumulative demand and service versus time are drawn (which are easily obtained by integration of demand and service rates). Integration can be easily performed graphically, by adding the total vehicle arrivals by successive 10-minute periods. For instance, from 5:00 to 5:10 P.M., 62.5 vehicles arrived. By 5:20 P.M., the total is 62.5 + 104 = 166.5 vehicles. By 5:30 P.M., it is 166.5 + 125 = 291.5 vehicles, and so forth.

The total delay at the ramp is shown by the shaded areas in Figure 28-10(b). The case shown assumes no diversion (everyone waits at the ramp), but it can also be similarly evaluated to account for diversions.

The same approach outlined in this example may be used to design a responsive, demand-capacity control setting. This setting may be implemented by using a variable rate, or by selecting the closest of several pre-set rates (multiple-dial selection). Such policies are used to advantage when either the freeway arrival flow, or capacity, varies with time, rather than remaining fixed (or approximately so), as illustrated in Figure 28-9. This variation may be known from historical data and the metering rate variation may be pre-planned, or detectors may be used to monitor freeway flow with the metering rate being periodically recomputed based on these measurements. See the literature[21] for an example and discussion.

Choice of a System

Given that a facility or a section of a facility is a candidate for a management and control project, several things must be ascertained about the facility as a whole:

1. The capacity of the system as defined.
2. The aggregate demands and the particular usage patterns: Collector funneling to CBD, distributor, mixed short trips in urban area, etc.
3. Diversion routes for limited traffic must be established, including service road usage, arterial paths, etc.
4. Existence of present *de facto* diversion due to poor facility performance, and probable return to improved facility and ancillary routes must be anticipated.
5. The "system" or area to be controlled must be redefined as necessary to accommodate the above points.
6. The common bottleneck locations, the regularity of, and duration of, composite demands, and the frequency of disruptive incidents.

Among other things, the above needs will generally imply that origin and destination patterns must be established, either by land use knowledge or by survey. This is particularly true of shorter facilities located principally within or on the fringe of urban areas, as for instance the Van Wyck Expressway in New York City. It is a significantly less acute problem for a rather lengthy, suburban collector facility terminating or distributing into a CBD, where patterns are more readily identified.

The above knowledge will allow formulation of an overall management policy (current sophistication or better), estimation of critical control and surveillance points, and allocation of control limitations. It will also allow some determination of the sophistication necessary for individual ramp controls (simple metering, simple metering with remote selection of multiple dials, demand-responsive, etc.). This general or overall analysis must be supplemented with the particulars of individual ramps:

1. Storage capabilities on the ramp proper and on the service road prior to upstream signals
2. Adequacy of geometrics (e.g., sight distances, grade, length), and the need to aid injection process
3. Type of arrival pattern: regular, burst, etc.
4. Relation ships to other ramps, and the occurrence of weaving and other behavior

This information, together with the above, will aid in determining the type of local control and surveillance required. Indeed, this analysis and care is as applicable in determining that a simple closure-fixed-metering solution to a ramp pair is in fact that simple, as it is in determining a complete freeway project. The quantity of data required will of course vary by project, as will the detail. It may be evident, for instance, that a rather local problem with an acute accident history exists and must be treated,[20] and the analysis will be so tempered.

29

Applications for Highway Safety

There are many approaches[13] taken in the war on traffic accidents: safer cars; better "packaging" of the occupants; bigger and brighter signs and signals; pavement marking improvements; better radii on better highways; automated highways; safety slogans. The list is endless, and in many respects directionless.

There is a simultaneous attempt to alleviate the results of accidents, reduce the occurrence of accidents, improve the quality of roads, tighten the qualifications for drivers' licenses, and other considerations. Some of these desires are at cross purposes. The better highways become (from a speed-maintaining point of view), the bigger and costlier the signs become, the worse the effects of accidents become, and the more difficult the creation of a "safe" car and the packaging of its occupants become. The newest of superhighways are so super that signing must be so huge and so spaced, tens of miles in advance of exits, that the slightest misunderstanding or indecision by the driver will result in either a terrible accident or the missing of a destination by many miles.

True *accidents* cannot be eliminated as long as human beings are not kept completely immobile. But *traffic* accidents, as has been noted many times, are not really accidents. There have been many proposals that they not be *called* accidents, that this misnomer lulls the public and the expert into the dead-end acceptance of "a certain number" of these "accidents" as being inevitable, even under the best of conditions. Traffic accidents, it has been said, should perhaps better be called "incidents," all ascribable to some *causative* factor. The word accident refers to an unexpected, a chance occurrence. We all know that certain types of traffic "accidents" recur at certain locations under similar conditions: rear-end accidents at signals, sideswipes at merge points. The condition of chance and unexpectedness does not exist. *If a driver is not careful, a known type of accident will occur*. The conclusion is clear: Traffic accidents are *not* inevitable, therefore *they can be eliminated.*

THE HUMAN EQUATION

There is no equation, no mathematical expression, that describes the activity of the simplest living organism, much less the human being. Students of animal behavior, for instance, have developed what is known as the "Harvard Law": *When stimulation is repeatedly applied under conditions in which environmental factors are precisely controlled the animal will react exactly as it pleases.*

The attempts to develop theories of traffic flow are, therefore, paradoxical. Molecules of gases, liquids, or solids have no individual capacity for decision. This is the reason that elimination of the traffic accident requires removal of decision-making in control of individual vehicles from the individual driver. Erection of the first STOP sign, the first signal, was a step in this direction. Single-direction roads or tunnels limit freedom of choice and action in the interests of safety. But there remains the individual's freedom to obey that signal or not, as he chooses, or to enter the one-way road at the wrong end.

Ultimate control would mean the end of the individually driven car. It means the change from automated highways to fully automated, complete systems, with programmed destinations and travel patterns to be punched into a control board at the beginning of any trip, short or long, rural or centrally-located urban. Complete automation, complete control, a completely accident-free situation, also means complete separation of modes of transportation—there is *no* rationalization for pedestrian accidents (or railroad-crossing accidents)—and complete separation of directions.

A LOGICAL PROGRESSION

Control Through Intervehicular Communication:

The relaying of desires and intentions from one driver to another atrophied with the brake light and the turn signal. The brake light goes on when the brake is *hit*, which, under conditions of close following or high speed or both is usually too late. Drivers need to be told more explicitly that the driver ahead is going to slow or stop; possibly they need to be told his *destination*, within reason, to be prepared for his movements.

Control through Knowledge of Traffic Conditions:

Under today's crowded conditions, which can only be expected to worsen, the driver needs to know *what to expect*, in terms of congestion within the next ten miles, weather or road conditions ahead, alternate routing possibilities ahead, etc., to a much greater degree than at present.

Control through Roadway design:

Channelization and one-way streets and ramps must be taken further. Decision-making at bifurcations must be removed from high-speed locations.

Control through Mechano-Electrical Developments:

Means must be developed to make it impossible for one vehicle to strike the rear of another. Or to strike the guardrail alongside it. Or to absorb most of the shock of such encounters.

Control through Limitation:

The optimum volume of traffic can be found for any stretch of roadway. Access to that roadway must be cut off when it reaches such a point. The overloaded "express"-way is of little value; the on-ramps can be barricaded and the next motorist can be told the best available alternate. Certain areas of the cities were never meant to accommodate automobiles at all. Vehicles can be prevented from forcing themselves into these areas. Access can be limited on city streets.

There is no excuse for wrong-way driving on the exit ramps of major arterials. Physical barriers activated by direction sensors can be developed.

Optimum capacity is also developed at certain known or obtainable combinations of speed and spacing. Hardware or electronic means of controlling these variables must be developed.

In facilities with limited width, physical means of making it impossible to leave the correct path must be found.

Control through Maintenance of Flow:

The knowledge and ability is available to create means of maintaining flow by the removal of disabled vehicles as quickly as possible from traffic streams. Mechanical means, bridging the roadway and traveling quickly on rollers have been envisioned.

Control through the Upgrading of Present Control Measures:

Wherever pavement markings can be replaced by positive barriers such as dividers and islands, they should be; where they cannot, the entire geometrics of the road should be restudied so that more positive control can be included. Flexibility can perhaps be retained through the use of changeable, removable, or otherwise alterable devices. But "flexibility" is the direct reason for much of the accident problem.

CURRENT PROCEDURES FOR THE STUDY OF HIGH ACCIDENT-FREQUENCY LOCATIONS

A study of high accident-frequency locations is essential to identify and help develop corrective measures for these locations. The study should also involve a comprehensive analysis of every potential high accident-frequency location in an area.

Applications

The results of the analysis of causes, effects, and remedies for accidents, and the priority of corrective measures required, may be applied in the following ways:

1. To determine a logical plan of accident reduction measures, based on treating locations in the order of severity rating
2. To determine definite ways that accident frequency can be reduced through simple engineering measures
3. To justify certain recommendations which necessitate large expenditures or marked changes in geometric design of streets and highways
4. To aid in planning a street and highway improvement program
5. To reveal and prove the need for additional enforcement or police supervision
6. To reveal certain driver or pedestrian actions causing accidents which might be prevented through public education
7. To point out a need for better maintenance of streets, highways, and control devices
8. To assist in developing a program of signal, sign, or pavement marking installation
9. To assist in developing a curb-parking program
10. To determine priority of need for roadway lighting
11. To aid in developing a speed-zoning program
12. To determine the need for sidewalk construction

Identification of Hazardous Locations

Hazardous locations can be broadly divided into those which are in urban areas and those which are in rural areas. In *urban areas,* accident spot maps are used to select the locations with the highest accident frequency and list them with total number of fatal accidents, personal injury accidents, and property damage accidents. In general, an urban location at which there are four or more injuries (including fatal and personal injuries) within a year warrants investigation. Based on property damage accidents alone, the number of accidents warranting an investigation of a location would normally be higher, such as five accidents at a residential intersection, or ten at an arterial street intersection. Locations with a high ratio of *night* to *day* accidents warrant investigation of the adequacy of lighting at that location.

In *rural areas,* high accident-frequency locations are generally identified from accident rate computations and accident profiles. Rate computations are based on accidents per mile, vehicle-miles, population, and registrations. Accident profiles may show widths of highway, curvature, grade, intersections, volumes, speeds, and the locations of accidents. It is also very desirable to distinguish between day and night accidents and between single-vehicle and multiple-vehicle accidents.

Several methods have been developed to identify hazardous locations systematically and objectively, based on accident experiences.[2]

1. *Rating method:* A point rating is assigned, based on such factors as traffic volumes, number of lanes, and accidents.
2. *Number method:* This involves a simple comparison of the number of accidents at intersections or on segments of roadway.
3. *Accident rate method:* The numbers of accidents per vehicle-mile, or per vehicle, are compared for each intersection or segment of a roadway.
4. *Number-rate method:* Number and accident rate methods are applied in combination.
5. *Quality control method:* Comparison is made by accident rate. However, the location is eliminated if the accident rate for that location is not significantly greater than the mean rate of a route or an area.

Study of High Accident-Frequency Locations

A general outline of a typical study of a high accident-frequency location is as follows:

1. The first step is to obtain copies of complete reports for accidents occurring at the locations under study. These reports should cover all accidents occurring within the past one to four years.
2. Prepare collision diagram for each high accident-frequency location.
3. Prepare condition diagram for each high accident-frequency location.
4. Prepare accident summary for each high accident-frequency location to help establish any predominant accident characteristics.
5. Obtain other traffic data as may be required. These data might include vehicular volume counts, pedestrian volume counts, approach speeds, driver observance of traffic signals and signs, and pedestrian observance of traffic signals and signs.
6. Compute the economic costs of accidents if feasible to do so. (See Chapter 17.)
7. Utilize the above data to select the most appropriate remedies for the high accident-frequency locations, considering available funds and the relative importance of the locations.

When several alternate remedies are available for one

location, or priorities have to be established among various locations, benefit-cost analysis should be made. Application of benefit-cost analysis to highway safety programs is discussed further in the following section.

METHODS OF SOLUTION

Selecting Corrective Measures

The relatively recent, greatly intensified efforts of the Federal Government in initiating and participating in various safety programs and studies are indicative of national concern over the significance of the highway safety problem.[3-6] Many comprehensive studies are in progress, and when the results are published, they will provide valuable data and recommendations for selecting appropriate corrective measures for various aspects of the traffic accident problem.

There are several national organizations actively engaged in conducting investigations or sponsoring research in one or more of the three elements (road user, vehicle, highway) involved in traffic accidents,[8] and the services and reports issued by these organizations significantly contribute to the strengthening of traffic safety efforts at all levels of government. In addition, the President's Committee for Traffic Safety has published a series of basic booklets on street and highway safety to serve as an aid to local communities and states in establishing an action program for highway safety. Their recommendations are consolidated in a separate summary booklet.[8]

A listing[21] of simple engineering remedies for patterns of predominant accident types is reproduced under the heading Specific Accident Remedies and appears at the end of these notes.

In *Getting Results Through Traffic Engineering*,[9] there are case studies of various accident locations presenting pertinent data and before-and-after studies with the solutions used. Numerous accident studies and remedies are reported in various issues of *Traffic Engineering* and other publications.[10-12] However, it should be understood by all practicing traffic engineers that there is no general theory of accident occurrence. This makes all studies of the past useful only to the understanding of the particular location at which the study was made.

Benefit-Cost Study

The objective of cost-effectiveness analysis in traffic safety programs is to assist in the rational allocation of remedial efforts. Cost-effectiveness analysis may be used to select one of several alternate remedies for one location, or to establish priority among several different locations. The traditional benefit-cost analysis has been simplified, normally, to include only the costs and benefits related to highway construction and maintenance, vehicle operating costs, and costs of travel time. Costs of accidents are seldom included in such analysis because of their unreliability, the complexity of determining accident costs, and the lack of uniform procedures to determine such costs.

There are four essential elements required in such an analysis, and the soundness of cost-effectiveness analysis will be largely dependent upon the accurate determination of these elements:

1. Identification of hazardous locations
2. Forecasting results of remedial efforts
3. Costs of accidents and costs of remedies
4. Selection of analysis method

Identification of Hazardous Locations. This was discussed in a previous section.

Forecasting Results of Remedial Efforts. Once high accident-frequency locations are identified and types of remedial action are determined, it is necessary to estimate quantitatively the accident reduction benefits that may be realized at each individual location. These benefits may be of both a direct and an indirect nature. Indirect benefits might be a decrease or increase in travel time or operating costs, resulting from improvement directed toward a reduction of accidents. Relatively extensive work has been done in evaluating such benefits.[14,15]

Extensive work has also been done on the accident characteristics of all types of freeways and expressways, related to types of facility, traffic volumes, degree of access control, and other operating characteristics of roadways. Therefore, it is possible, by utilizing these data, to determine the reductions in accident occurrence and severity resulting from major projects such as the addition of a lane, the installation of a median barrier, etc.

Present methods used to predict accident reductions resulting from minor improvements are very limited in scope and use. The methods were based on analysis of various before-and-after studies of minor improvements made in the past.[2,16]

Costs of Accidents and Costs of Remedies. Costs of accidents were discussed in a previous section. Since cost estimates of improvements, both physical and control devices, are well established and usually obtained from other sources, they are not discussed here.

Safety improvements always affect some other aspect of traffic characteristics and, therefore, the entire cost of a safety improvement cannot be considered as directed solely toward the reduction of accidents. As an example, if a section of guardrail is installed to reduce the severity of collisions involving vehicles running off the road, the entire cost of the guardrail could be considered directed at the reduction of the severity of such accidents. But, when a signal is installed to reduce angle collisions, it may reduce the overall delay at the intersection, and thus some of the cost should be assigned to nonsafety items. Since cost-effectiveness analysis is usually made for both safety and operational improvements together, the need for such differentiation of the cost will not usually arise.

Selection of Analysis Method. The purpose of a cost-effectiveness analysis is to optimize in an economic sense, i.e., to get the maximum return from a fixed investment or to achieve a fixed level of performance at minimum cost. Thus, in many cost-effectiveness studies, it is simply a comparison of the benefits of a specific investment and the cost of the investment that determine the final rankings of the options.

The application of cost-effectiveness analysis to highway economics is well explained and synthesized in *Road User Benefit Analysis for Highway Improvements*, by AASHO, and *Economic Analysis for Highways*, by Winfrey. Three methods are most frequently used: *cost-benefit ratio*, *rate of return*, and *annual cost*.

Evaluation of Corrective Measures

Upon taking any type of corrective measures to reduce the frequency or severity of accidents, evaluation of the effectiveness of such measures is an essential part of accident analysis. Careful evaluation through before-and-after studies should be conducted to measure effectiveness of such corrective measures and to detect at the earliest time possible if such measures create any adverse effects on accident occurrence. If any adverse effect were found, the corrective measures should be immediately removed or modified.

Various techniques, including a simple comparison of frequency and severity rates, and statistical methods, have been developed.[17-21] Differences in average severity rates and in accident occurrence per unit time or mileage (accident rates) may be potentially evaluated using a variety of statistical techniques.

The *central limit theorem*, which states that the distribution of average values (rates) approaches the normal distribution for large sample sizes, regardless of the underlying distribution, enables many such evaluations to be simply performed using the *normal approximation*. This test is simply manipulated and easily applied utilizing standard statistical tables of the normal distribution.

In accident analysis, where sample sizes (total numbers of accidents considered) may be small, other statistical methods must be applied. Where the underlying distribution can be shown or assumed to be normal, differences between small samples may be evaluated using the *t-distribution*. This distribution is easily manipulated, but assumes not only normalcy, but equal variances. This latter assumption must be tested using an *F-distribution*, and if it proves false, the *t*-distribution may not be used.

Where normalcy or equivalency of variances may not be assured, a number of nonparametric distributions, which make no assumptions as to underlying distribution or variance equivalency, are of potential utility. Differences may be evaluated directly using *run* or *rank* tests, whose basis is nonparametric.

Alternatively, *goodness of fit tests* may be applied to evaluate differences. Generally useful in fitting data to a theoretic distribution, these tests may be used to evaluate two data sets, where one is considered a "theoretic" distribution to be fit by the other. The chi-square (χ^2) test and the Kolmogorov-Smirnov (K-S) test are examples of such tests. These tests are of utility in evaluating differences between *average* accident rates and severities. It is also useful to evaluate the possible difference in the range, or spread, of accident rates and severities. This may be accomplished through the use of the *F*-distribution or order statistics.

Specific Accident Remedies

The patterns of predominant accident types shown by the collision diagrams will usually give a clue to the remedies needed. The following remedies should be used only as a guide. Additional guides of great value are copies of *Getting Results Through Traffic Engineering* prepared by the Association of Casualty and Surety Companies.[9] These are case studies of various accident locations presenting before-and-after studies with the solution used.

However, as previously stressed, it should be understood that there is no general theory of accident occurrence, and studies of the past can only help in the understanding of the particular location at which the study was made. Various types of accidents and specific remedies which proved to be effective are outlined below.

Right Angle and Rear-End Collisions at Intersections

1. Removal of view obstructions, such as foliage, bushes, billboards, or parking at curb
2. Installation of warning signs, if speeds are high and the element of surprise is present
3. Installation of stop signs, if view is obstructed to such an extent that safe approach speed is 8 miles per hour or less, if one street is an approach street, or no other remedy reduces accident frequency
4. Installation of traffic signals if minimum warrants are met
5. Continuing operation of traffic signals during certain light traffic hours when signals are normally off
6. Provision of proper clearance interval in signal cycle
7. Relocation, repair, or other means of providing better visibility of signs or signals
8. Better street lighting
9. Provision of pedestrian cross-walk marking and/or pedestrian barriers
10. Rerouting of through traffic onto specially designated and protected through streets
11. Creation of one-way streets
12. Provision of traffic signal system time for progressive movement
13. Speed zoning to safe approach speed

Head-On, Left-Turn Collisions at Intersections.

1. Provision of turning guide lines
2. Prohibition of left turns (provided such movement is of little importance)
3. Provision of channelizing islands
4. Provision of protected turning interval, via traffic signal control
5. Installation of STOP signs (provided no other remedy works)
6. Elimination of view obstructions
7. Creation of one-way street
8. Routing of turning traffic via an alternate route (with proper signs) to eliminate left turn

Pedestrian-Vehicular Collisions at Intersections.

1. Installation of pedestrian cross-walk lines
2. Erection of pedestrian barriers
3. Installation of traffic signals
4. Provision of pedestrian refuge islands
5. Prohibition of curb parking
6. Provision of adequate street lighting
7. Creation of one-way street
8. Rerouting of through traffic to specially designated and protected through streets
9. Addition of pedestrian indications and pedestrian actuation features to existing traffic signals

Sideswiping Collisions.

1. Installation of painted pavement lane lines
2. Installation of channelizing islands, if at intersections
3. Installation of advance warning signs to warn drivers of proper lane for certain destinations
4. Speed zoning
5. Provision of acceleration or deceleration lanes at intersections
6. Widening of pavement
7. Creation of one-way street
8. Elimination of marginal obstructions such as caused by parked vehicles or other bottlenecks

Head-On Collisions.

1. Same remedies as for side-swiping collisions
2. Installation of "no-passing" zone at curves or other points with restricted view
3. Installation of center dividing strip

Vehicles Running off Roadway.

1. Installation of pavement centerline
2. Installation of warning reflectors, guardrail, or white posts at curve
3. Installation of advance warning signs
4. Installation of roadside delineators
5. Speed zoning
6. Street lighting
7. Skid-proofing slippery black top pavement, improving shoulder maintenance, and prompt ice treatment and snow removal

Collision with Fixed Objects.

1. Application of paint and reflectors to fixed object
2. Use of pavement guide lines to guide traffic around obstruction
3. Street lighting
4. Reduction of the number of fixed objects
   a. Place signs that must be in the median back-to-back wherever possible.

   b. Remove unnecessary sign posts (consolidate signs).
   c. Combine signs and light poles where possible.
   d. Utilize existing structures for posting signs.
   e. Use sign bridges where possible rather than gore signs.
5. Reduction of exposure to fixed objects
   a. Place signs and light poles on the *right* side of pavements rather than in the median or gore areas if feasible. (Reducing exposure to total traffic.)
   b. Use sign bridges where possible rather than gore signs.
6. Minimizing of the hazards of fixed objects
   a. Provide guardrail in front of those objects.
   b. Use prows and other methods wherever guardrail is not suitable.
   c. Use breakaway sign supports and light poles.

Collisions with Parked Cars.

1. Parking prohibitions.
2. Change from angle to parallel parking.
3. Rerouting of through traffic to less congested, specially protected through streets.
4. Creation of one-way streets.

Appendix

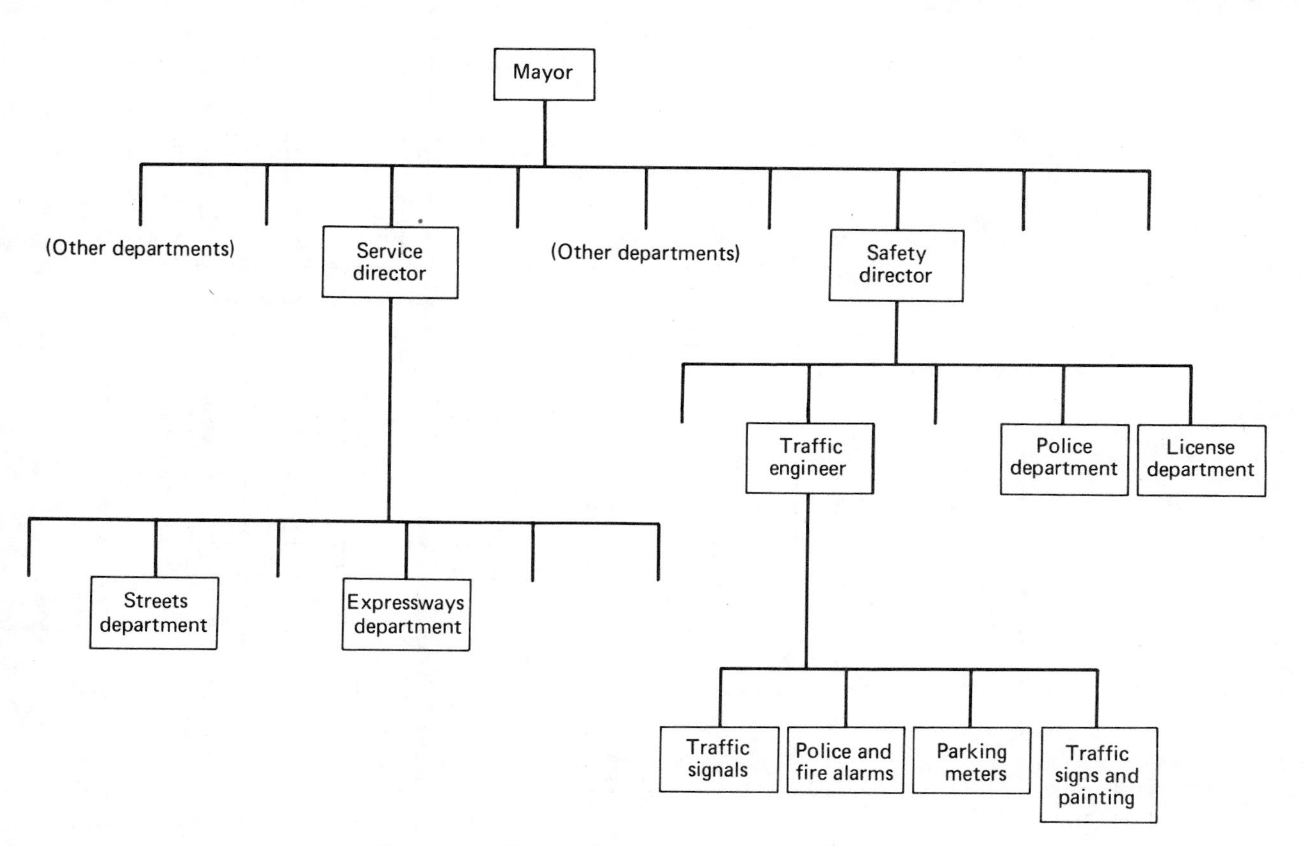

Fig. A-1. Traffic and related areas under separate branches

Fig. A-2. Traffic engineering dept. of approximately 400,000 people

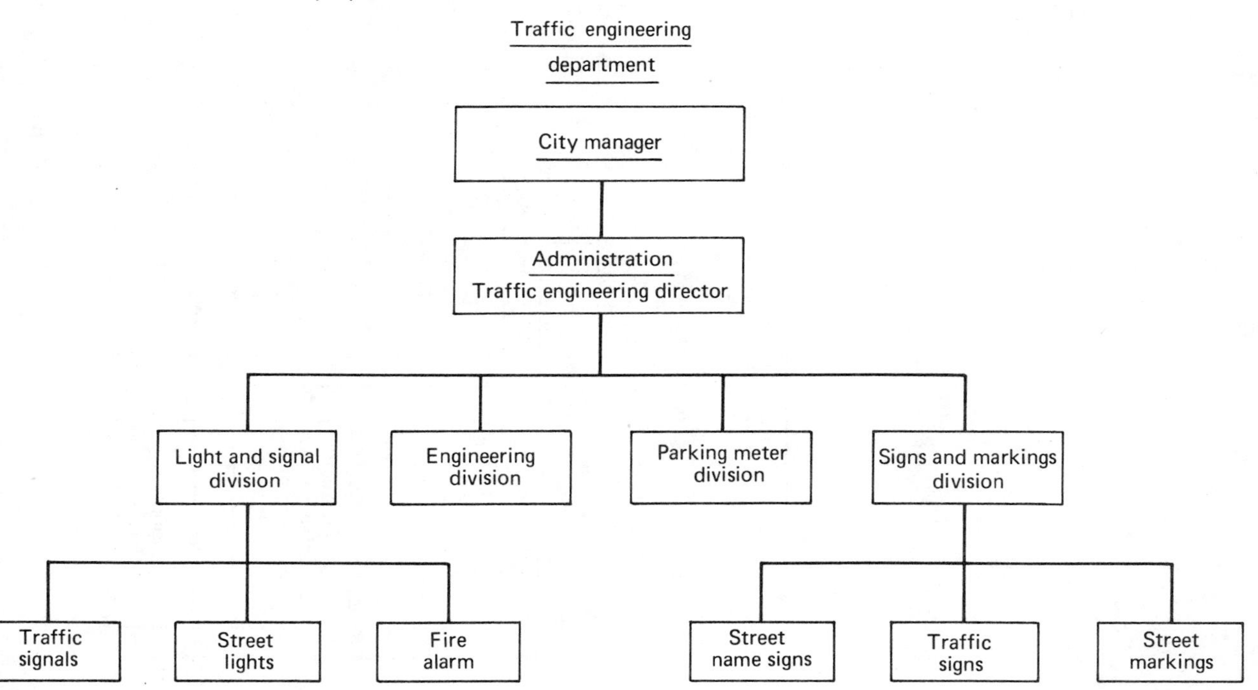

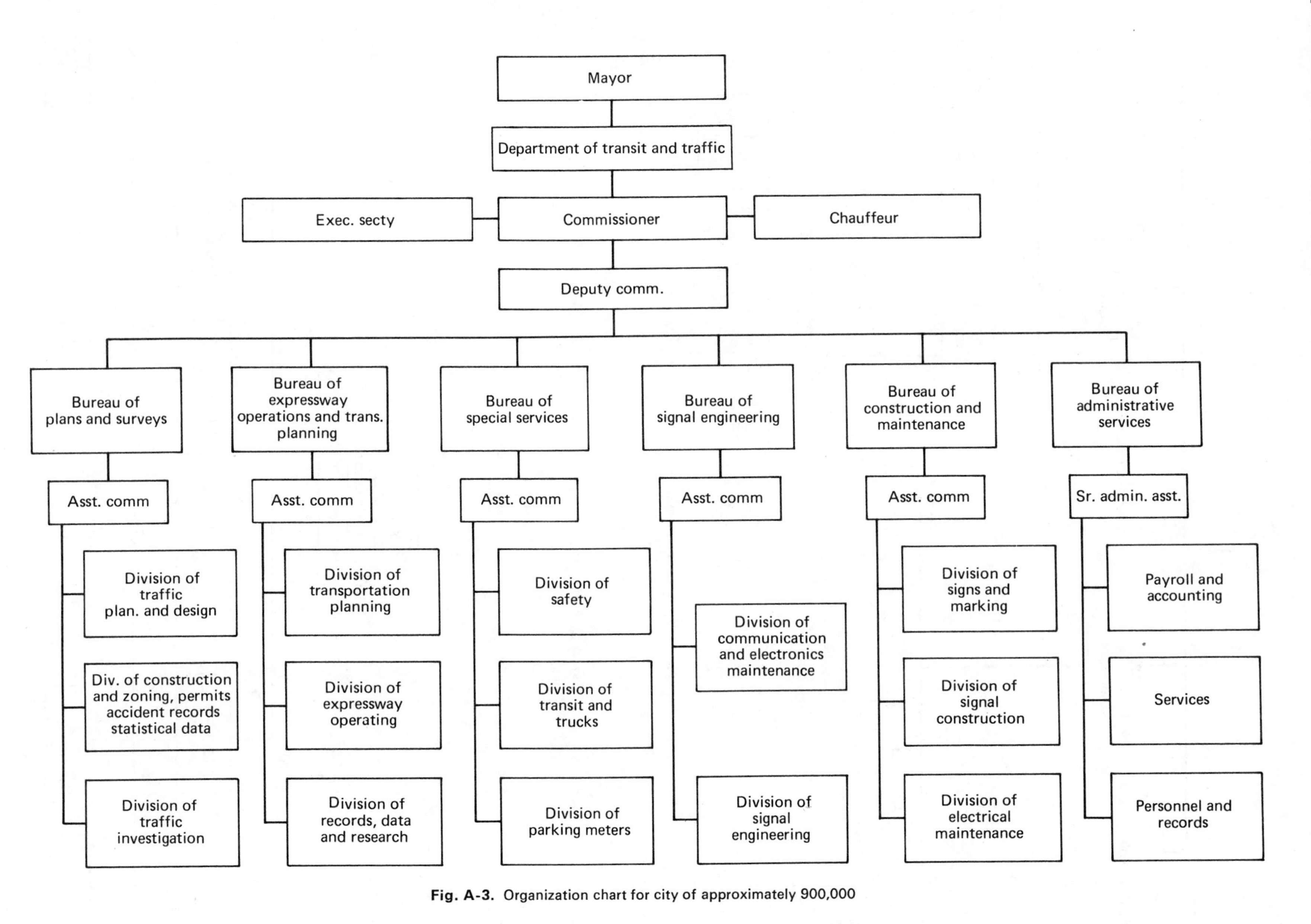

Fig. A-3. Organization chart for city of approximately 900,000

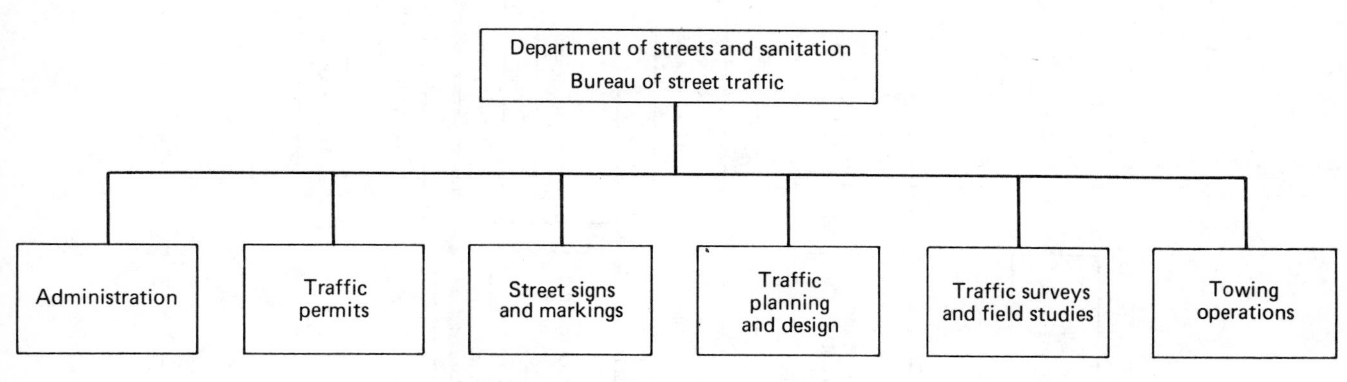

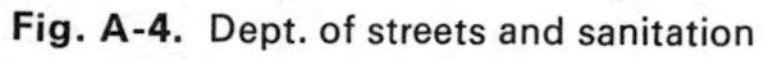
Fig. A-4. Dept. of streets and sanitation

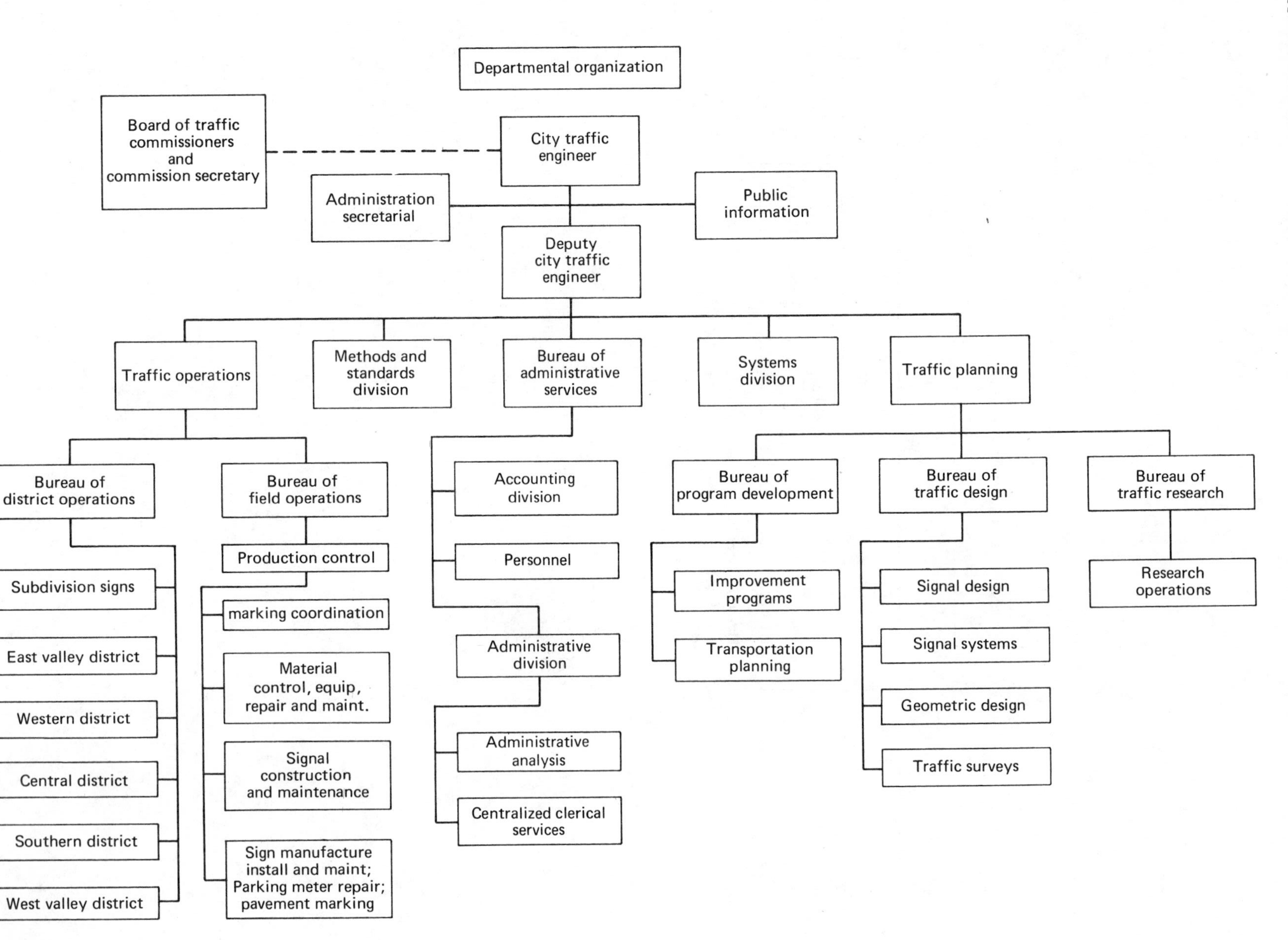

Fig. A-5. Organization of Los Angeles, Calif. Dept. of Traffic

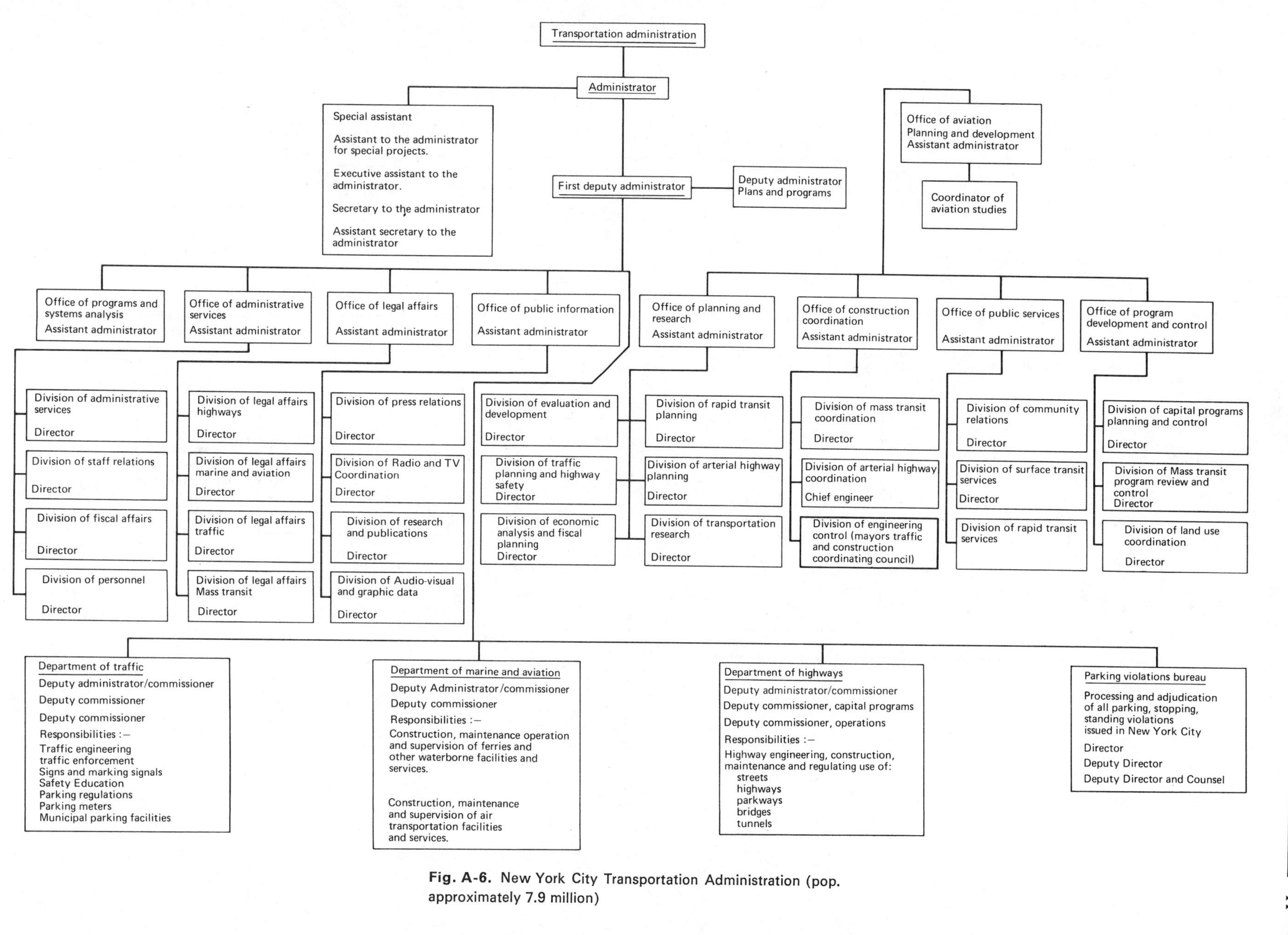

Fig. A-6. New York City Transportation Administration (pop. approximately 7.9 million)

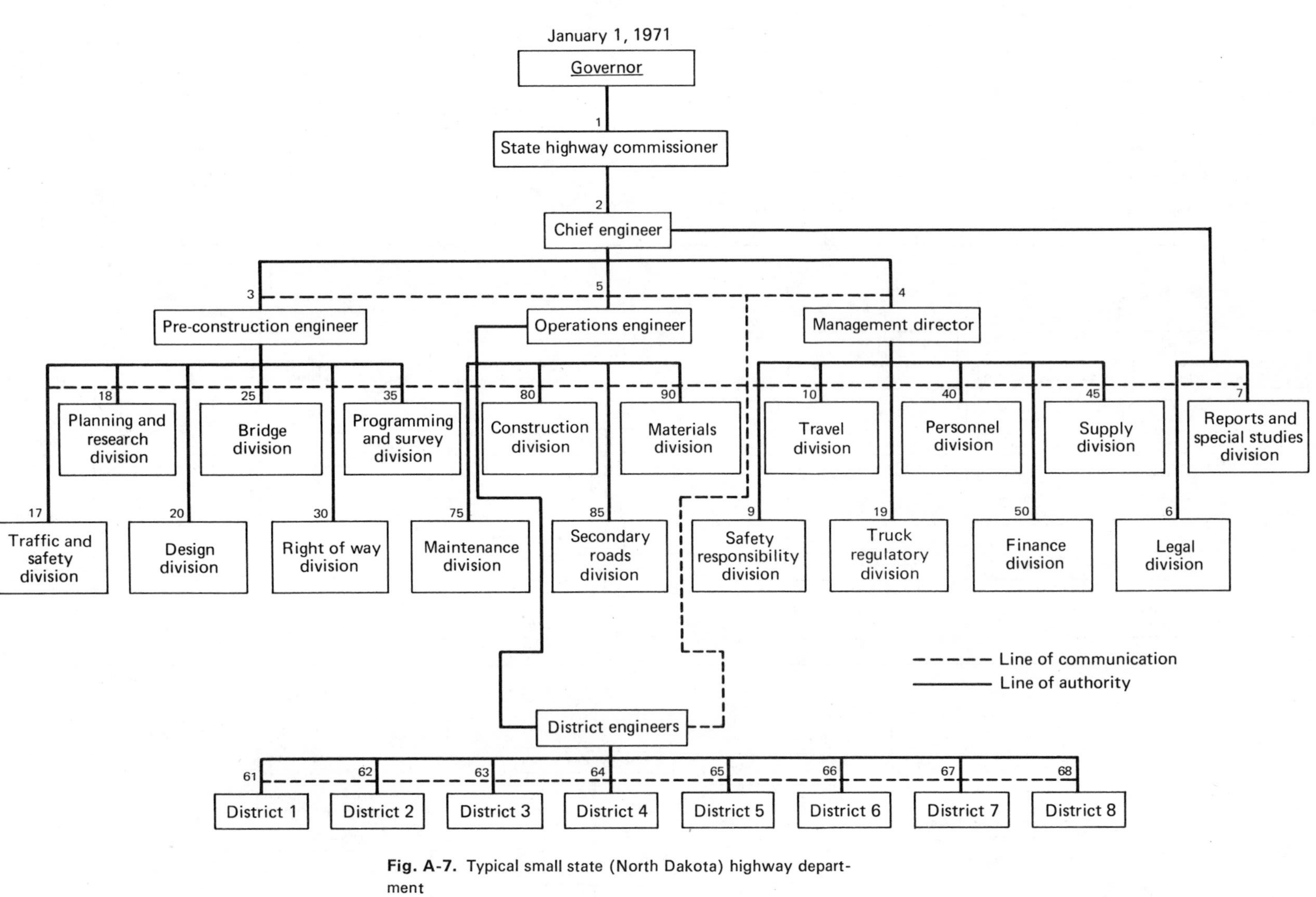

Fig. A-7. Typical small state (North Dakota) highway department

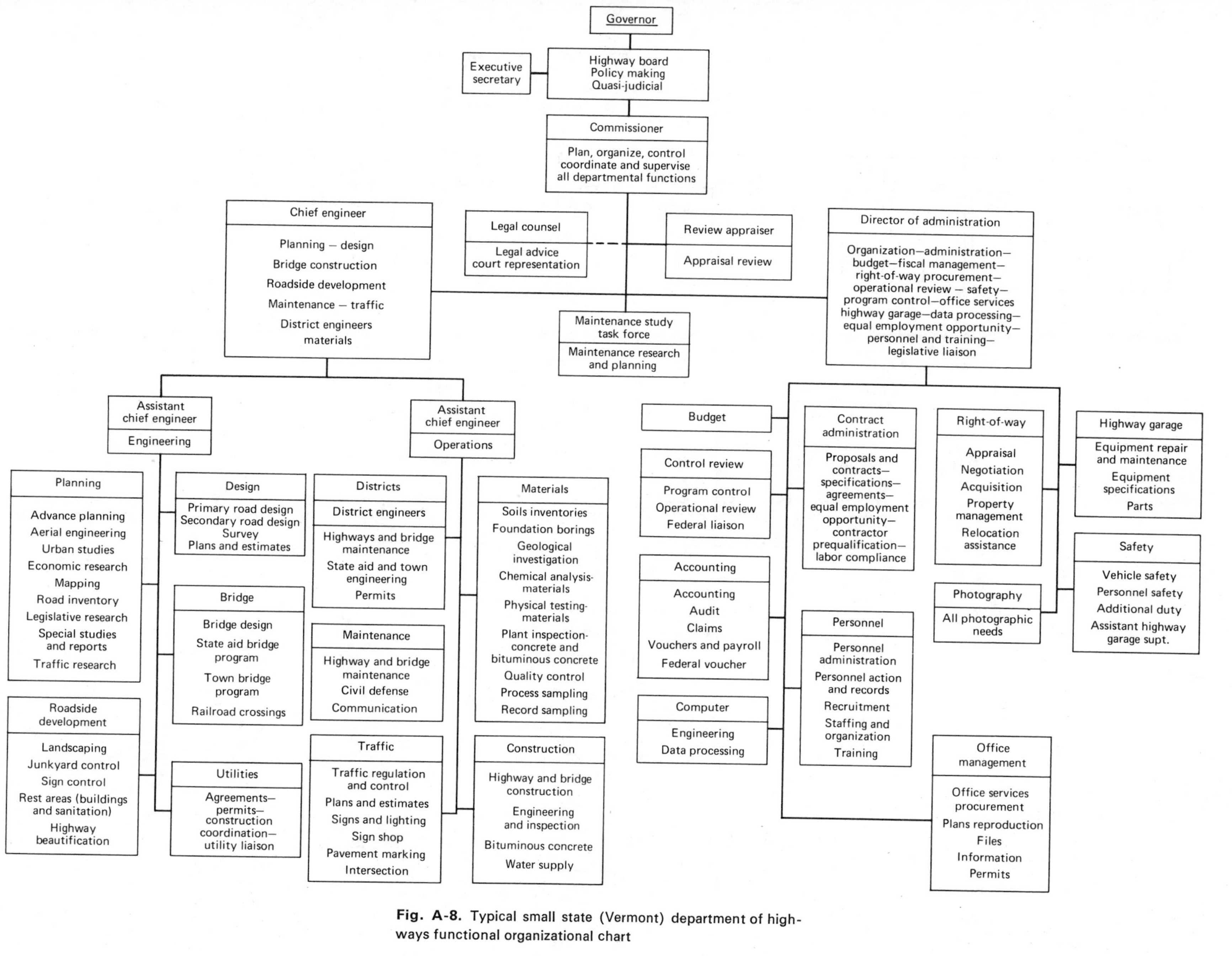

Fig. A-8. Typical small state (Vermont) department of highways functional organizational chart

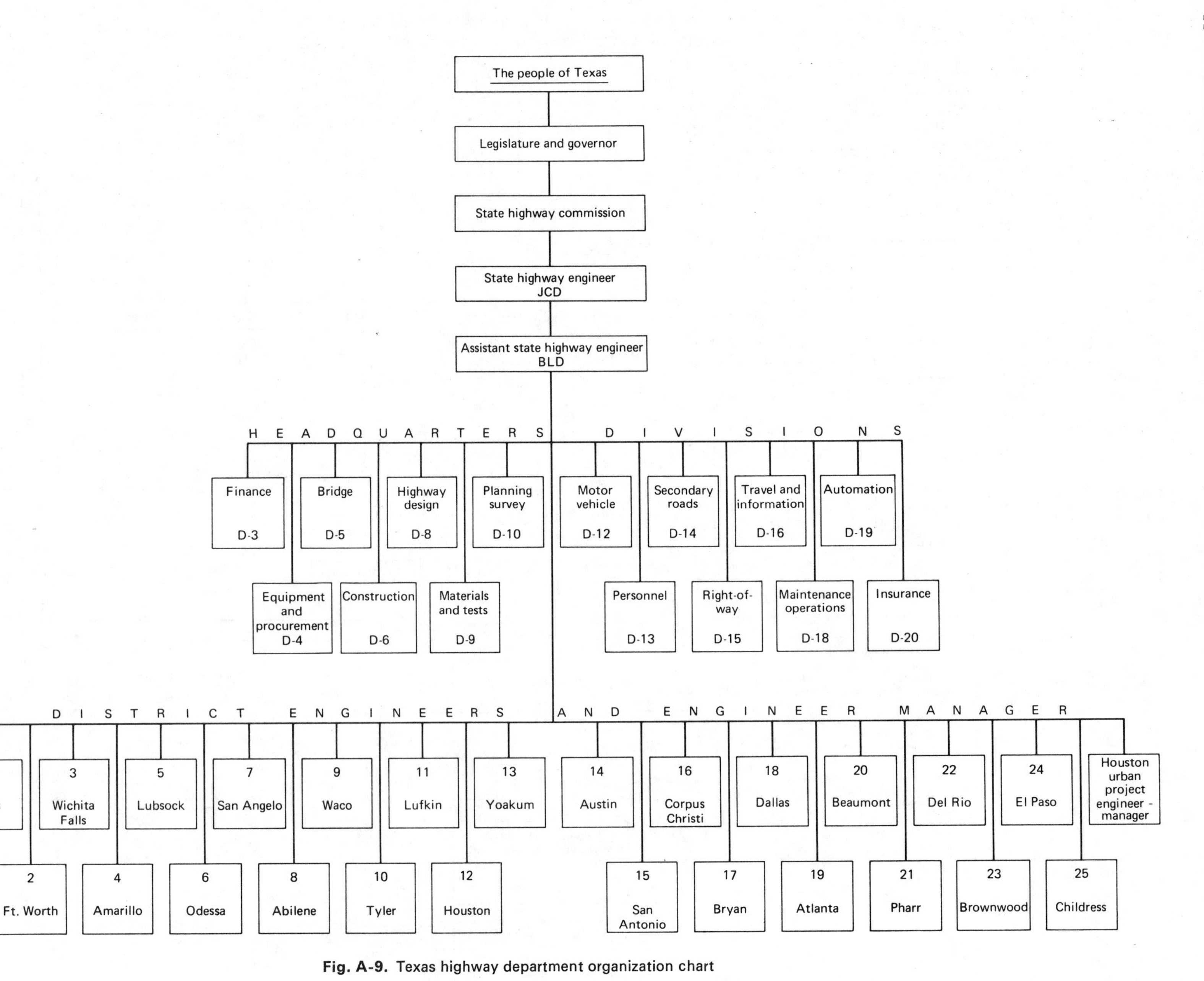

Fig. A-9. Texas highway department organization chart

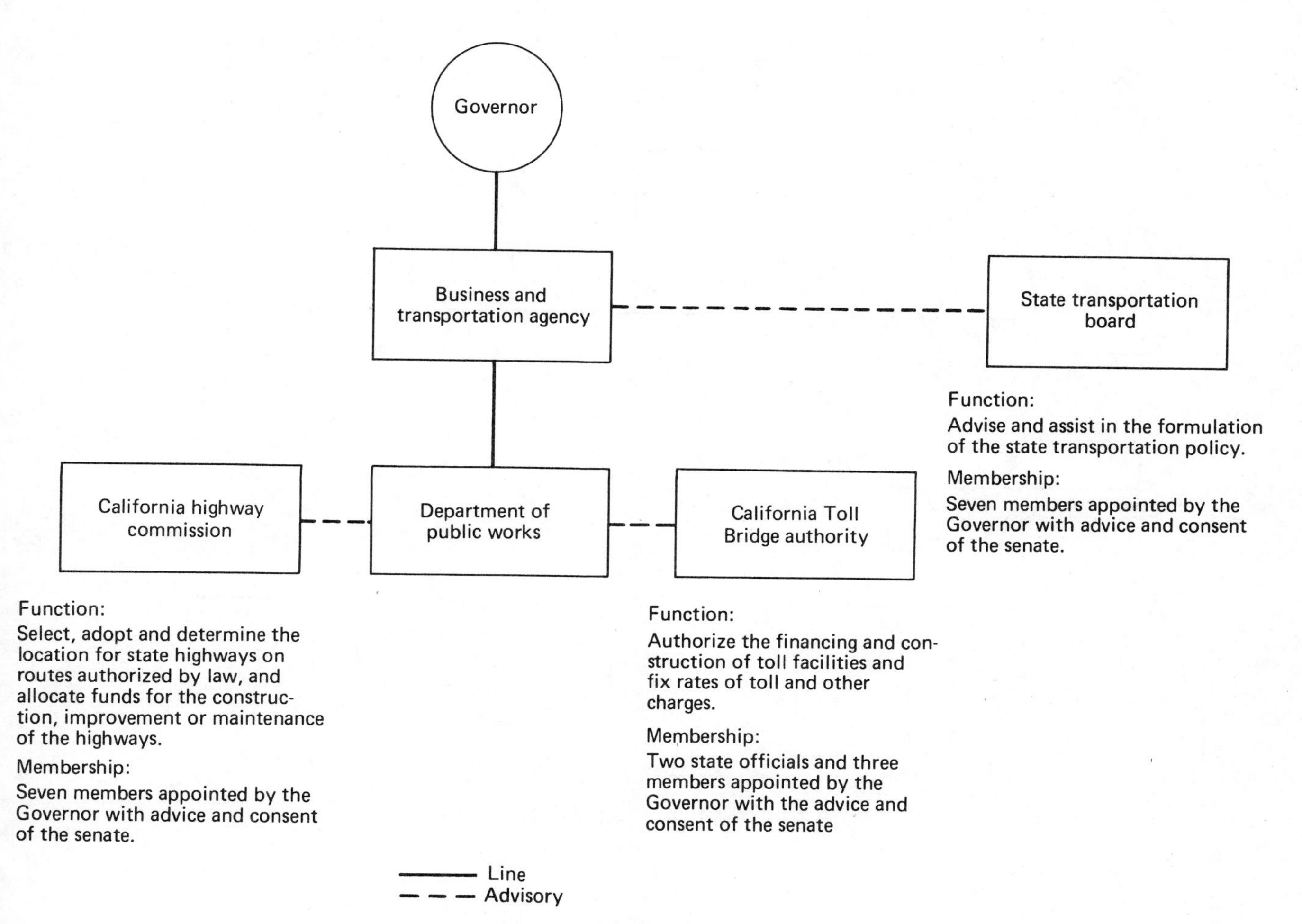

Fig. A-10. State of California Transportation Agencies

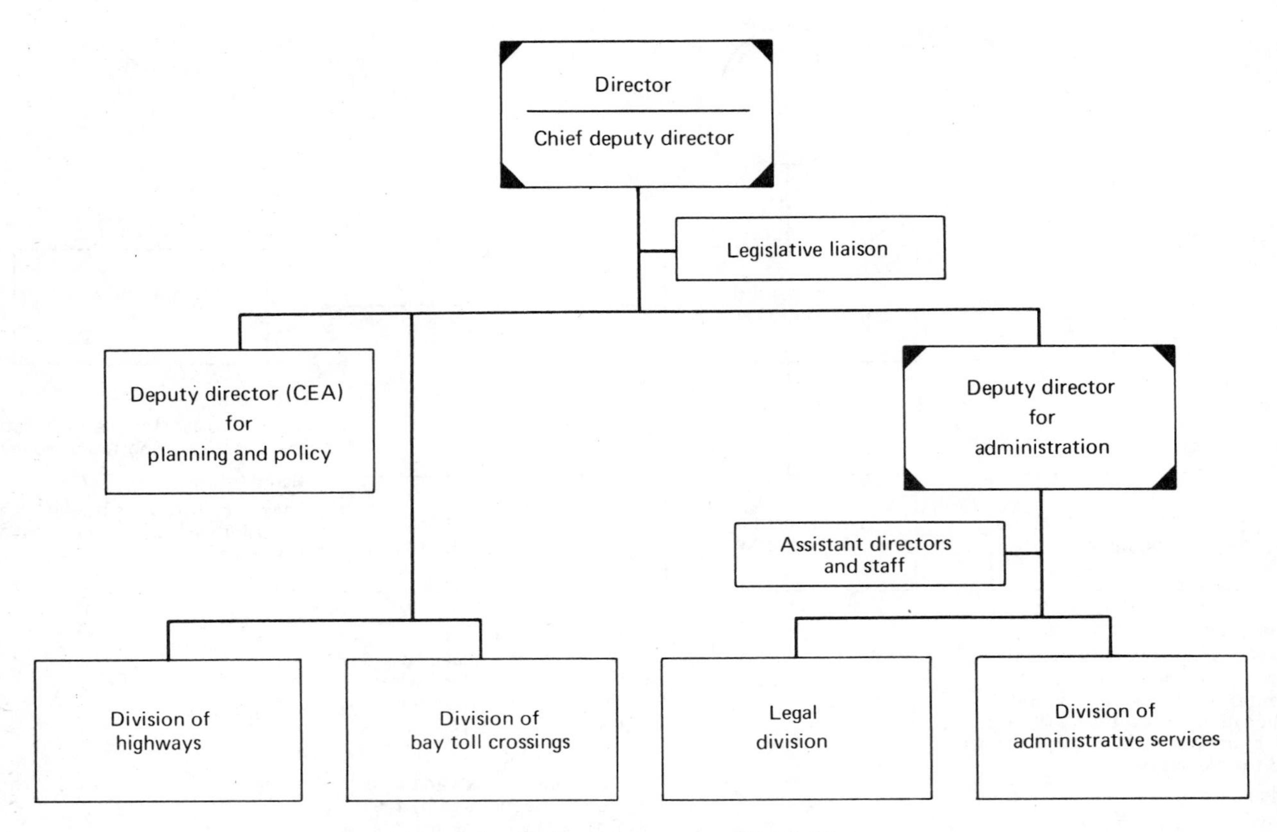

Fig. A-11. California Department of Public Works

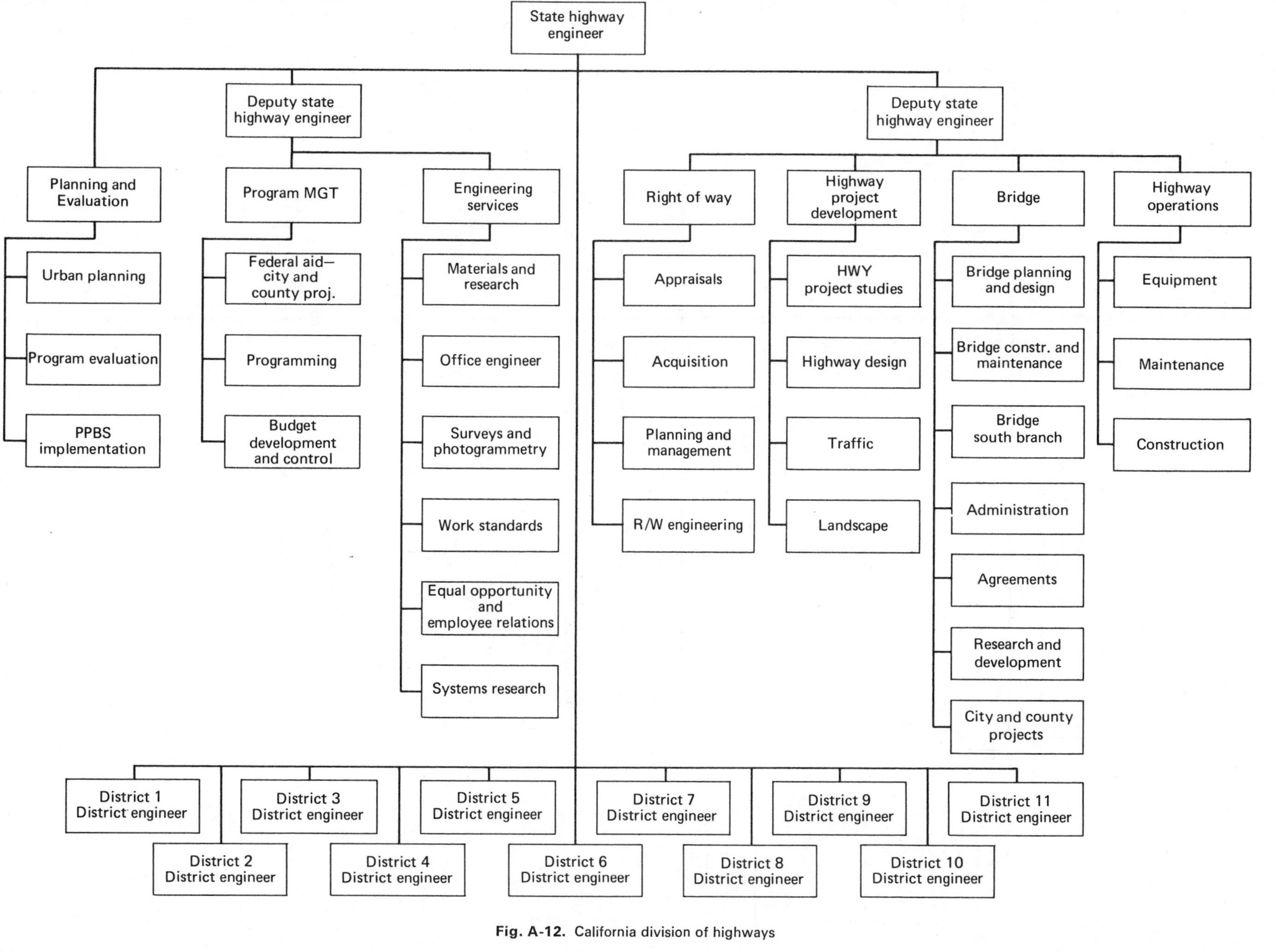

Fig. A-12. California division of highways

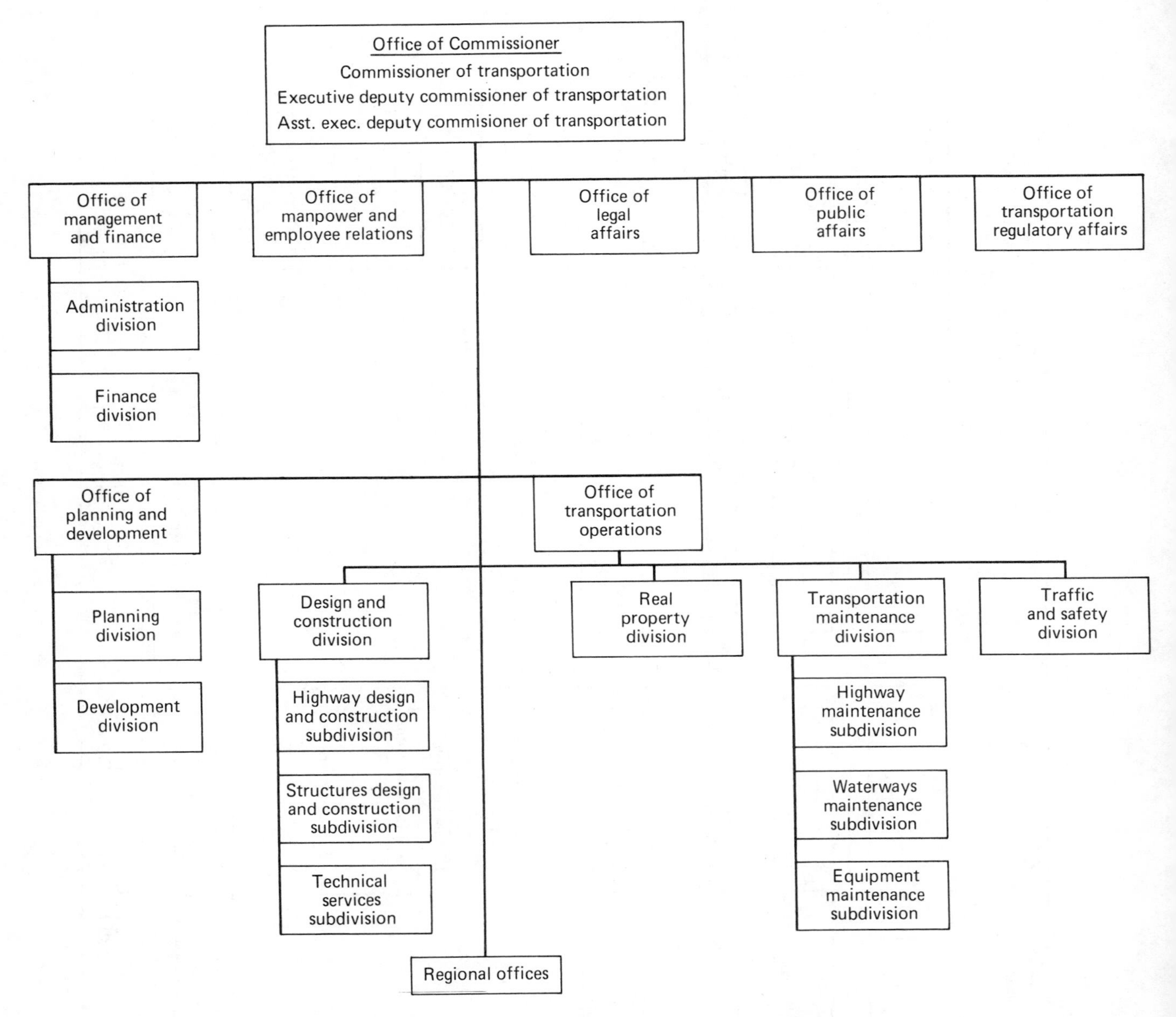

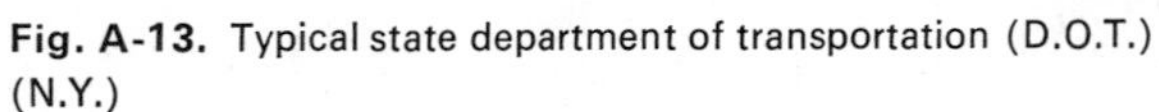

Fig. A-13. Typical state department of transportation (D.O.T.) (N.Y.)

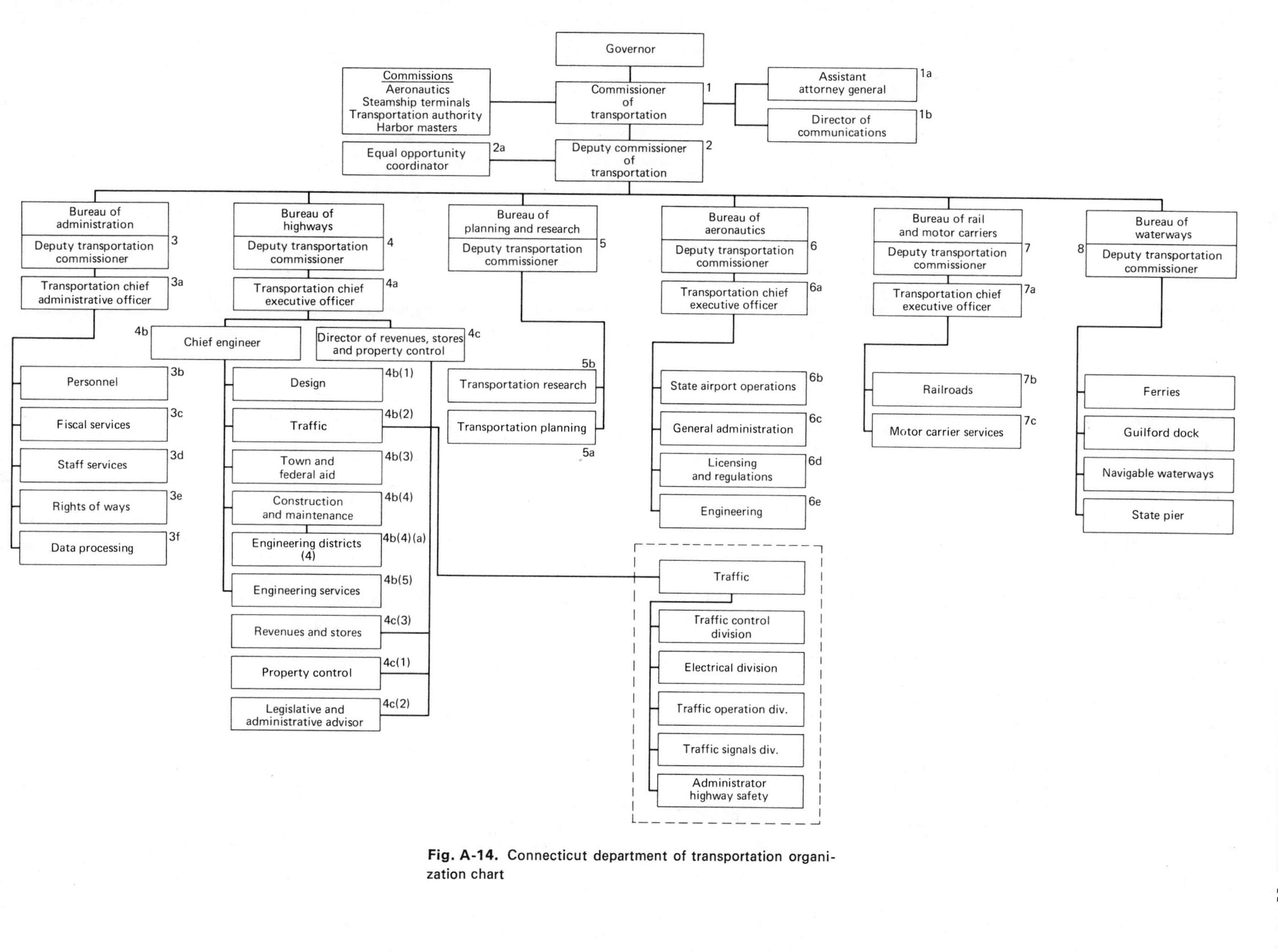

Fig. A-14. Connecticut department of transportation organization chart

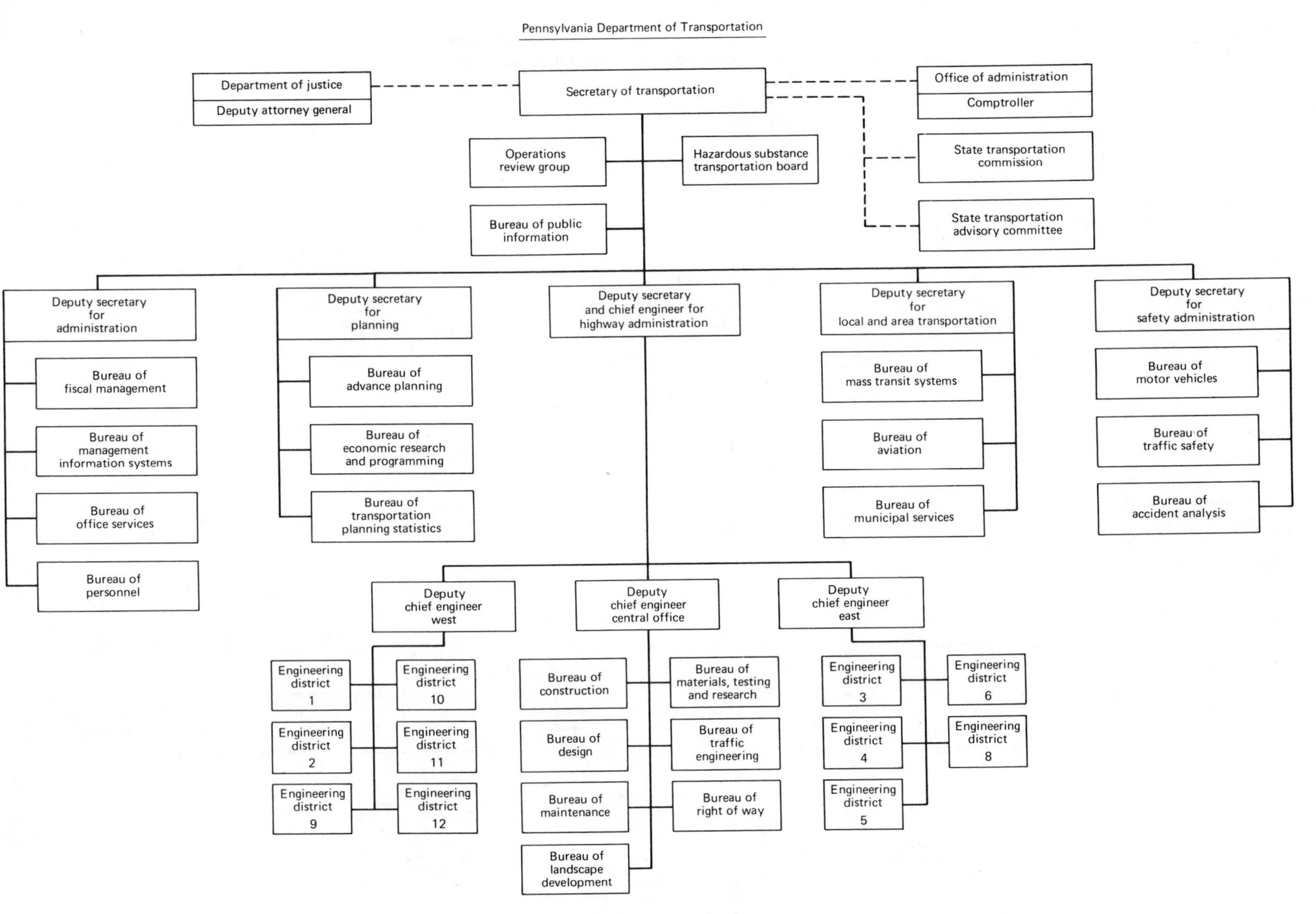

Fig. A-15. Typical state D.O.T. (Penn.)

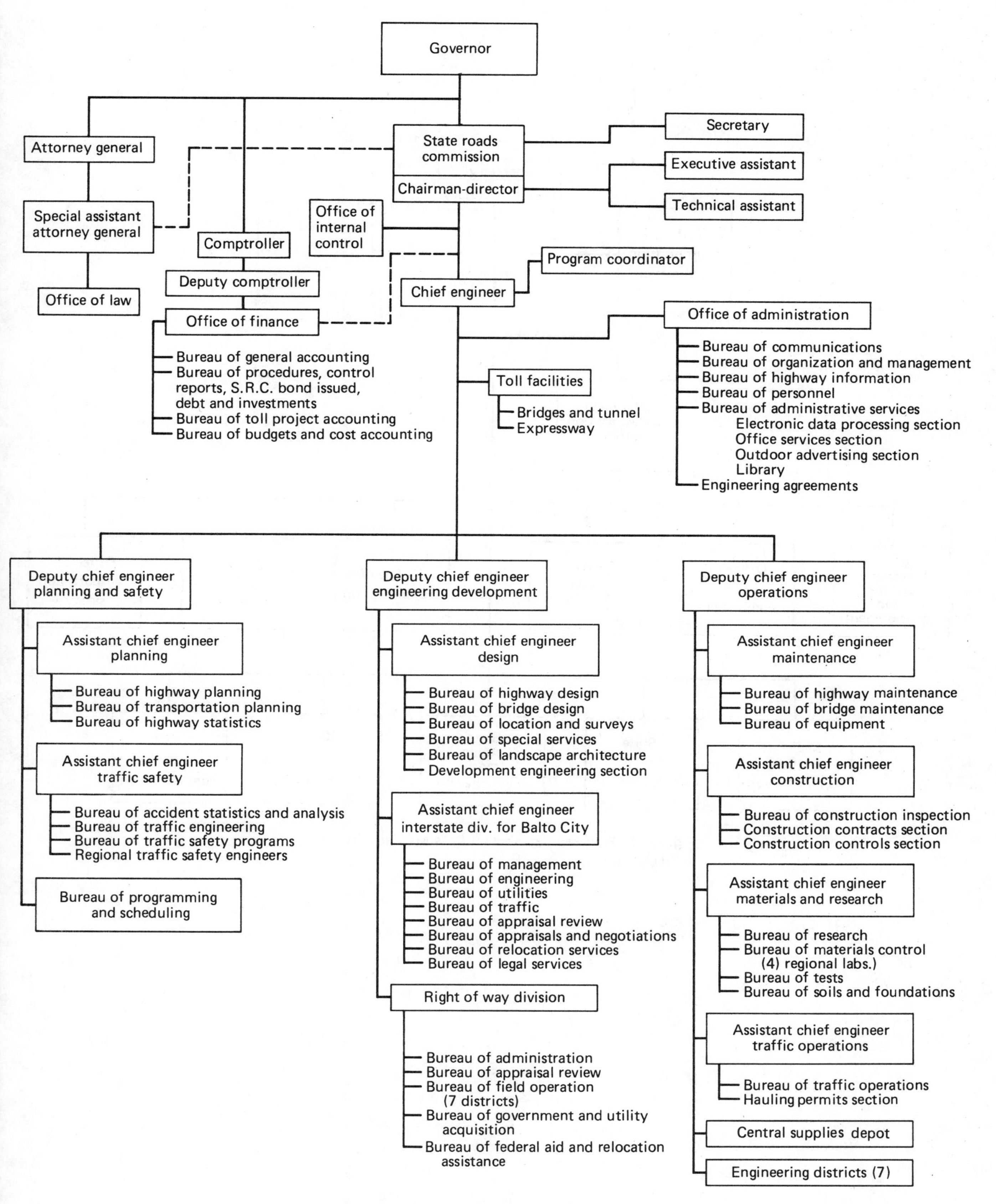

Fig. A-16. State Roads Commission of Maryland

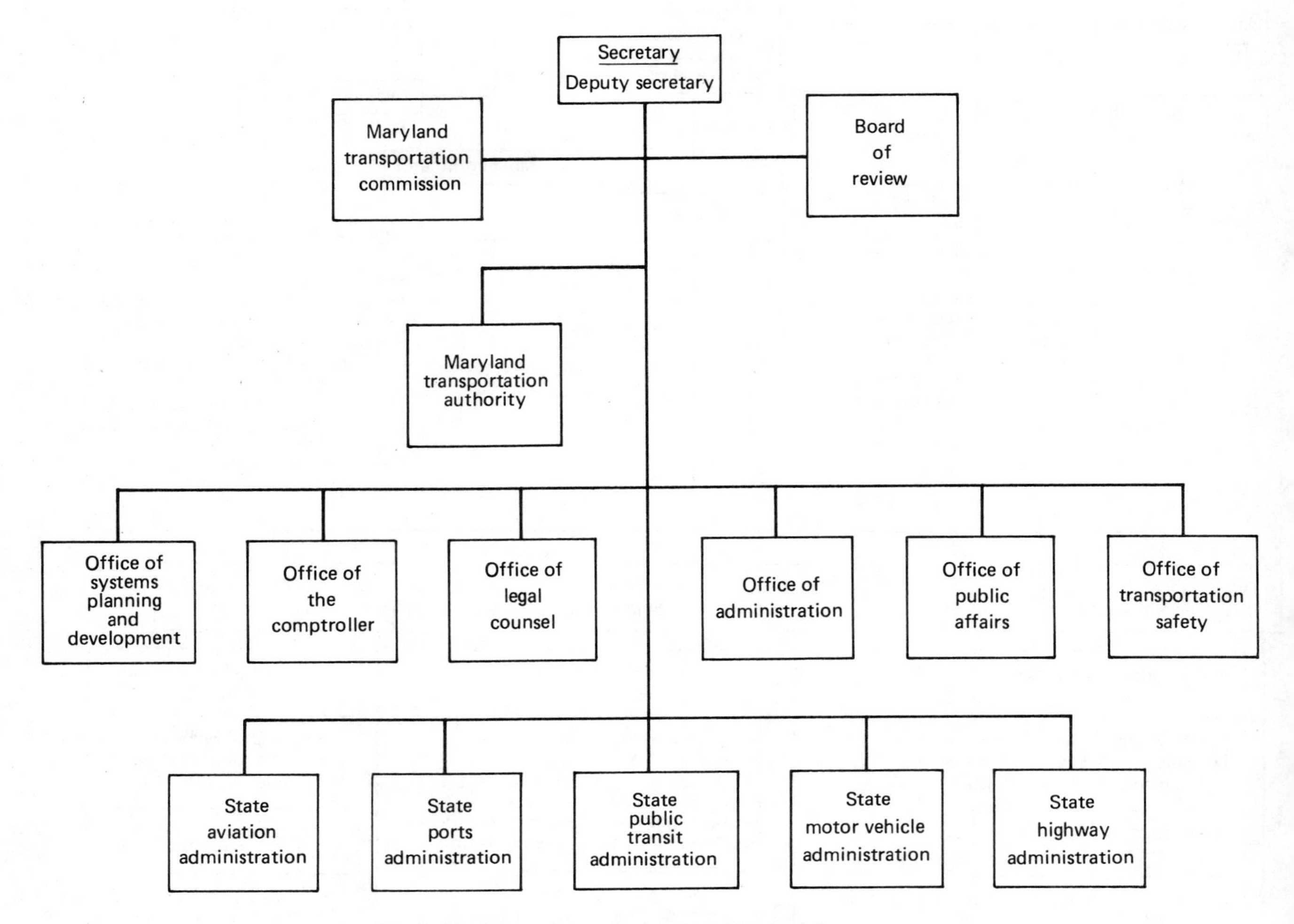

Fig. A-17. Proposed organization for a Maryland Department of Transportation

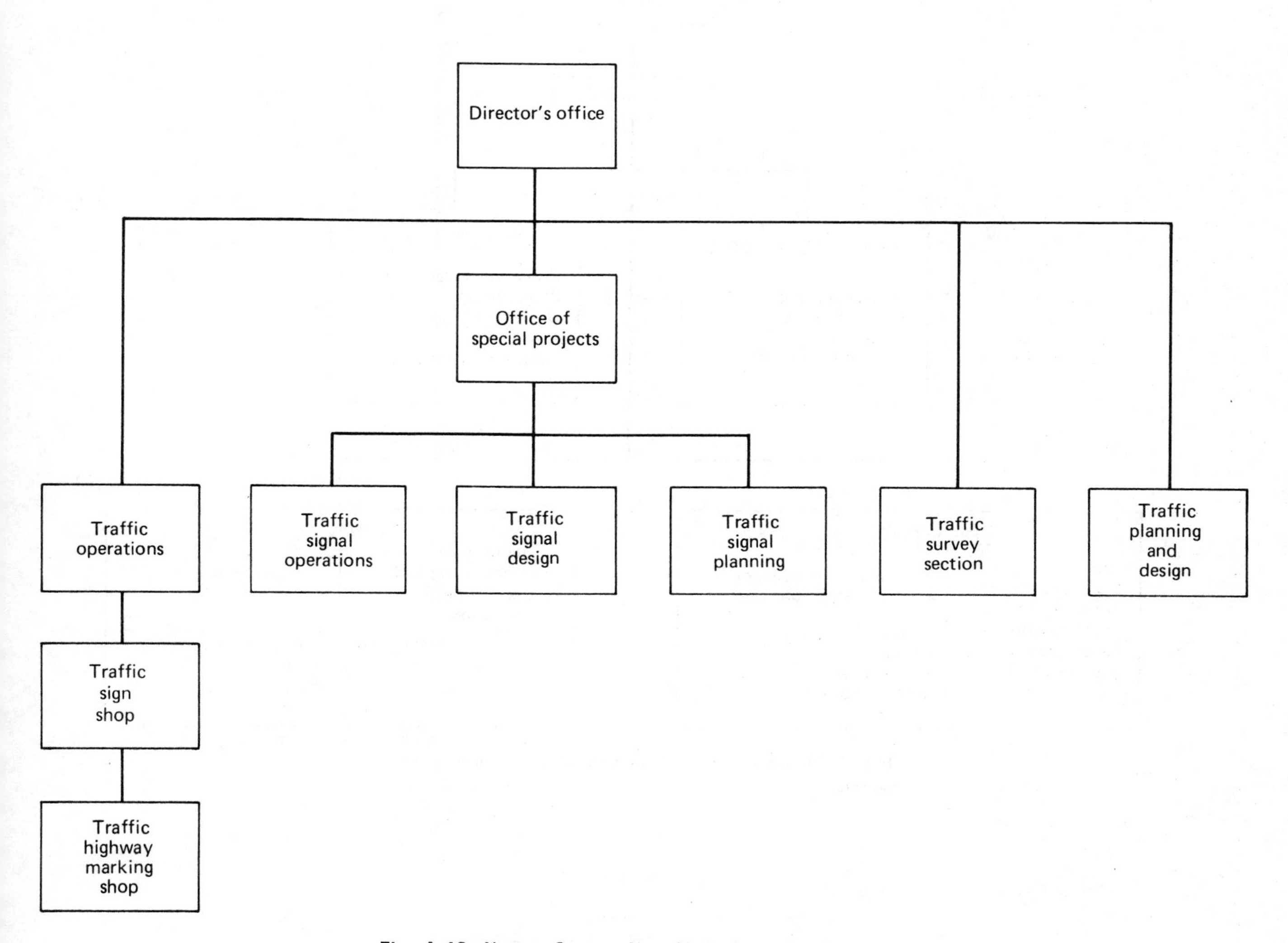

Fig. A-18. Nassau County, New York, Dept. Public Works Traffic Engineering Division

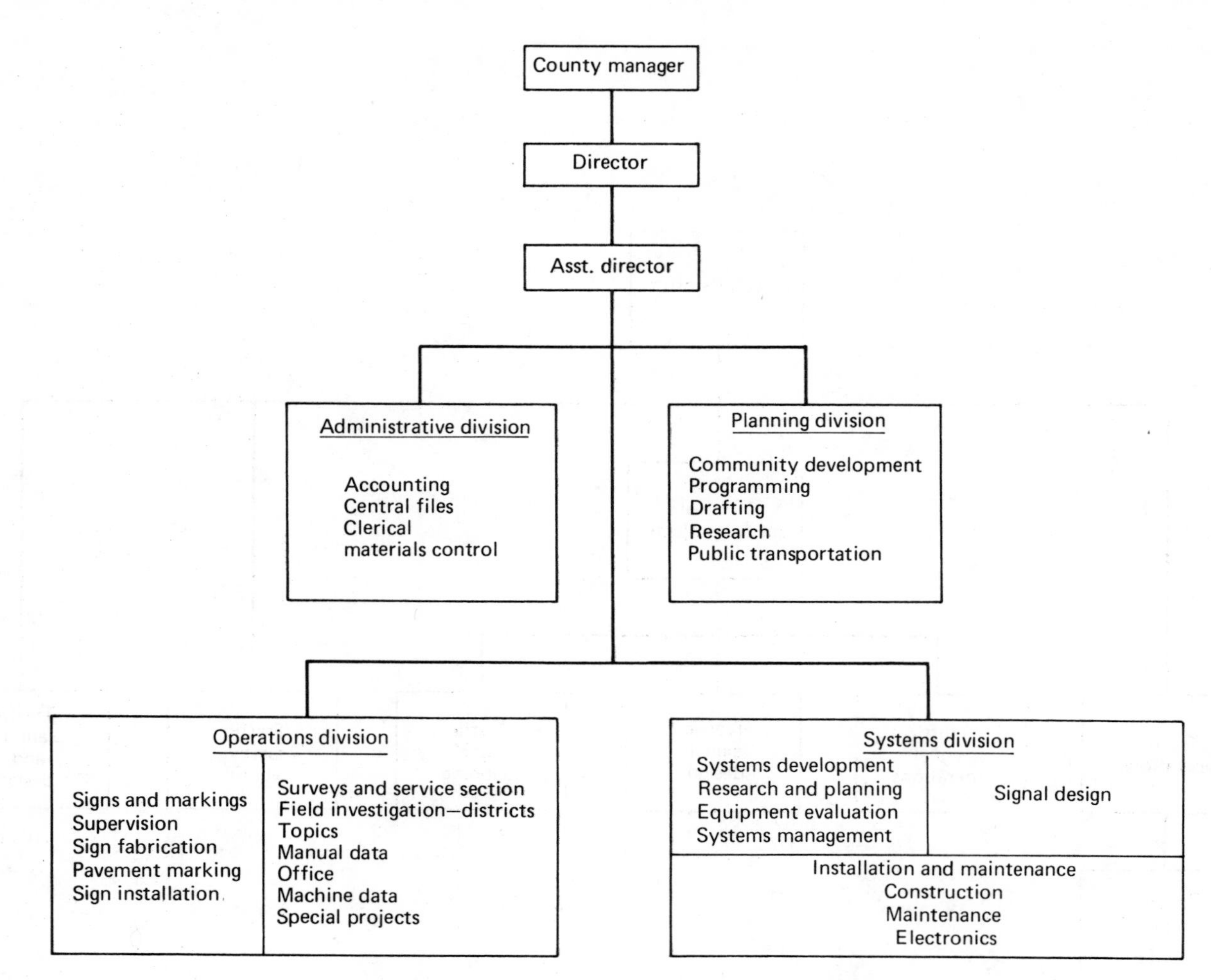

Fig. A-19. Dade County, Florida, Dept. of Traffic and Transportation

Fig. A-20. U.S.D.O.T.

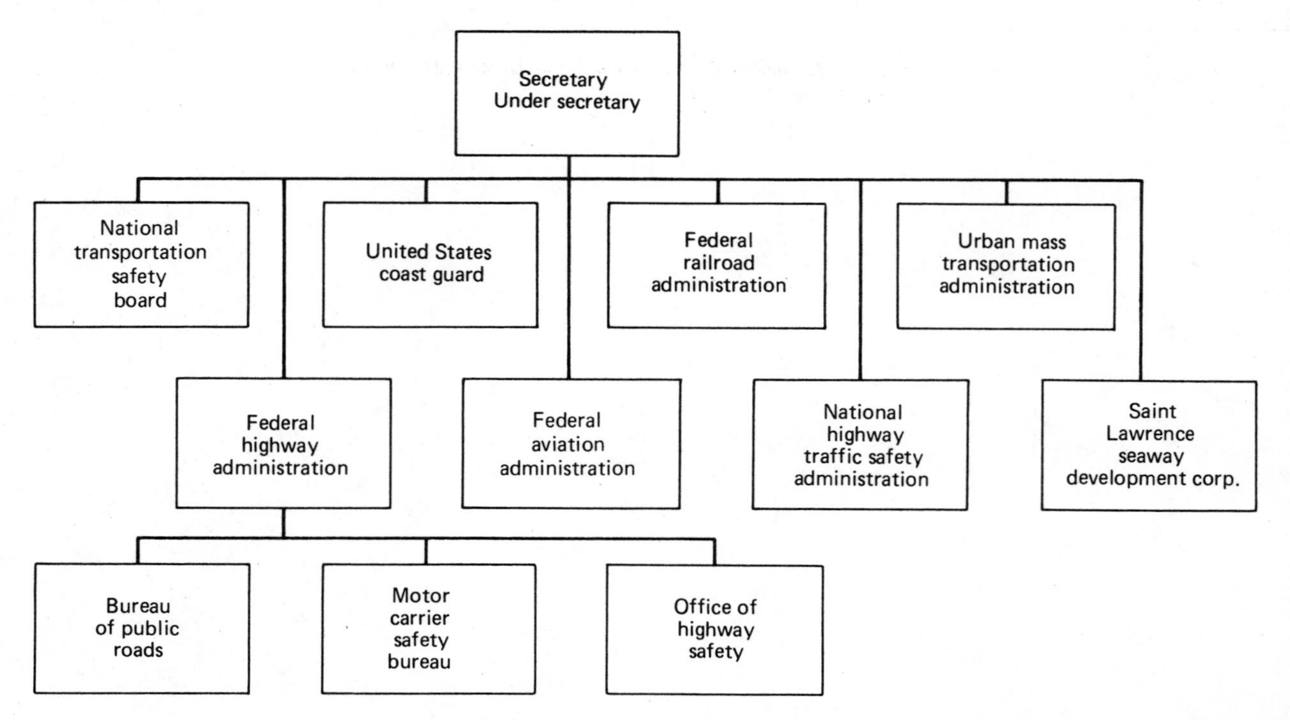

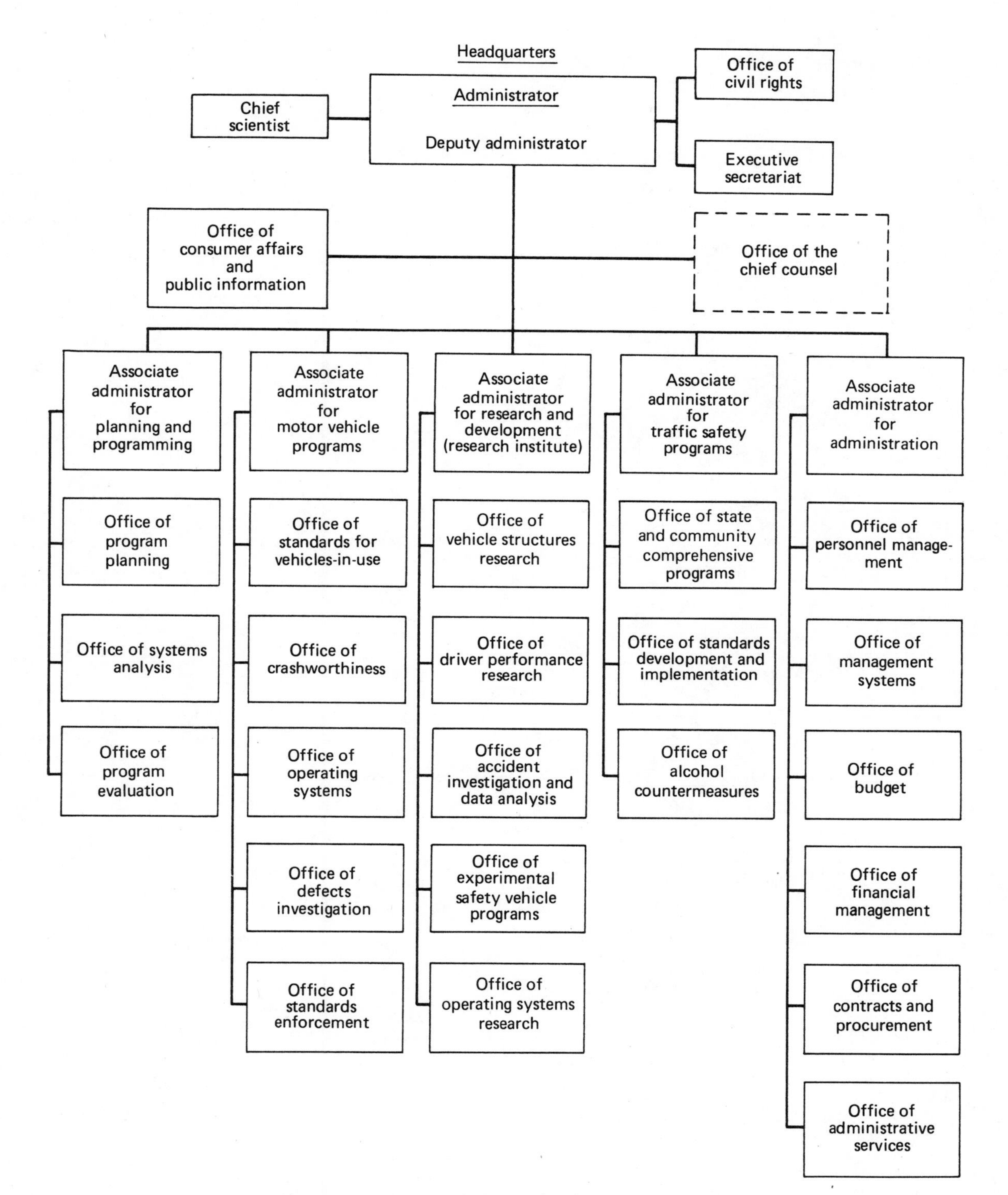

Fig. A-21. National Highway Traffic Safety Administration

Reference Bibliography

KEY TO REFERENCE ABBREVIATIONS AND ADDRESSES OF SOURCES

AASHO = American Association of State Highway Officials
341 National Press Building
Washington, D.C. 20004

AH = *American Highways*, AASHO

AMA = Automobile Manufacturers Association
320 New Center Building
Detroit, Michigan 48202

ARB = *American Road Builder*
ARBA Building
525 School Street, S.W.
Washington, D.C. 20024

ARRBB = Australian Road Research Board Bulletin
60 Denmark Street
Kew, Victoria, Australia

ASCE = American Society of Civil Engineers
345 East 47th Street
New York, N.Y. 10017

ASF = Automotive Safety Foundation
now *Highway Users Federation for Safety & Mobility*
1776 Massachusetts Ave., N.W.
Washington, D.C. 20036

ASTDJ = *Aero-Space Transport Division Journal*, ASCE

BHT = Bureau of Highway Traffic
Pennsylvania State University
State College, Pa.
(formerly at Yale University)

BPR = Bureau of Public Roads, USDOT

BST = Bureau of Safety and Traffic, NJDT

CATS = Chicago Area Transportation Study
230 North Michigan Avenue
Chicago, Illinois 60601

CDJ = Construction Division Journal, ASCE

CE = *Civil Engineering* Magazine, ASCE

CR = Council Report, ITE

Dover = Dover Publications
180 Varick Street
New York, N.Y. 10014

DVRPC = Delaware Valley Regional Planning Commission
Philadelphia, Penna.

Eno = Eno Foundation for Transportation, Inc.
Saugatuck, Connecticut 06880

ER = Engineering Report
FHWA = Federal Highway Administration, USDOT
HD = House Document (Congress of the United States), USGPO
HDJ = Highway Division Journal, ASCE
HF = *Human Factors*, John Hopkins Press
Baltimore, Maryland 21218
HMSO = Her Majesty's Stationery Office
York House
Kingsway, London, W.C. 2, England
HPTR = Highway Planning Technical Report
HRB = Highway Research Board, National Research Council
National Academy of Sciences
2101 Constitution Avenue
Washington, D.C. 20418
HRBB = HRB Bulletin, HRB
HRBBib = HRB Bibliography, HRB
HRBC = HRB Circular, HRB
HRBR = HRB Record, HRB
HRC = Highway Research Circular, HRB
HRN = Highway Research News, HRB
HRR = Highway Research Record of the HRB
ICE = Institution of Civil Engineers
Great George Street
Westminster, London S.W. 1, England
IDH = Illinois Division of Highway
Springfield, Illinois 62706
IE = *Illuminating Engineering*, IES
IEEE = Institute of Electrical & Electronics Engineering
345 East 47th Street
New York, N.Y. 10017
IEES = Iowa Engineering Experiment Station
Iowa State University
Ames, Iowa 50010
IES = Illuminating Engineering Society
345 East 47th Street
New York, N.Y. 10017
ITE = Institute of Traffic Engineers
2029 K Street, N.W.
Washington, D.C. 20006
ITTE = Institute of Transportation & Traffic Engineering
University of California
Berkeley, California
MIT = Massachusetts Institute of Technology
Cambridge, Mass. 02139
NCHRPR = National Cooperative Highway Research Program Report, HRB
NCUTLO = National Committee on Uniform Traffic Laws and Ordinances
NCUT, PAS = National Committee on Urban Transportation
Public Administration Service
1313 East 60th Street
Chicago, Illinois
NHUC = National Highway Users Conference, Inc.
Now Highway Federation for Safety and Mobility
1776 Massachusetts Avenue, N.W.
Washington, D.C. 20036
NJCUTCD = National Joint Committee on Uniform Traffic Control Devices
NJDT = New Jersey Department of Transportation
1035 Parkway Avenue
Trenton, N.J. 08625
NSC = National Safety Council
425 N. Michigan Avenue
Chicago, Illinois
NYSDMV = New York State Department of Motor Vehicles
Office of Public Information
462 Central Avenue
Albany, N.Y. 12206
NYSDOT = New York State Department of Transportation
1220 Washington Avenue
Albany, N.Y. 12226
OR = *Operations Research*
428 East Preston Street
Baltimore, Md. 21202
OSHC = Orgon State Highway Commission
State Highway Building
Salem, Oregon 97310
OSS = Office of Statistical Standards
Bureau of the Budget
Executive Office of the President
Washington, D.C.
OSU = Ohio State University
Columbus, Ohio
Pelican = Pelican Books
3300 Clipper Mill Road
Baltimore, Md.
PIB = Polytechnic Institute of Brooklyn
333 Jay Street
Brooklyn, N.Y. 11201

PJTS = Penn-Jersey Transportation Study
51 Street & Parkside Avenue
Philadelphia, Pa.

PM = Procedure Manual, NCUT

PONYA = The Port of New York Authority
111 Eighth Avenue
New York, N.Y. 10011

PR = Public Roads
Superintendent of Documents
U.S. Government Printing Office
Washington, D.C. 20402

Proc = Proceedings

Proc P = Proceedings Paper

RN = Research News

RP = Road Paper

RRL = Road Research Laboratory
Department of the Environment
Crowthorne, Berkshire, England

SD = Senate Document, USGPO

SR = Special Report

SRI = Standard Research Institute
Menlo Park, Calif.

SRR = Student Research Report

TE = *Traffic Engineering* Magazine, ITE

TEC = *Traffic Engineering & Control*
Morley House
26–30 Holborn Vinduct
London OCIA 2BP

Tech Bull = Technical Bulletin

TEJ = Transportation Engineering Journal, ASCE

TI = The Traffic Institute
Northwestern University
Evanston, Illinois 1963

TM = Technical Memorandum

TQ = *Traffic Quarterly*, Eno

TTI = Texas Transportation Institute
Texas A & M University
College Station, Texas

TTR = Texas Transportation Research

UIP = University of Illinois Press
Urbana, Illinois

USDOT = United States Department of Transportation
Bureau of Public Roads
800 Independence Ave. S.W.
Washington, D.C. 20590

USGPO = Superintendent of Documents
United States Government Printing Office
Washington, D.C. 20402

WS = Wilbur Smith and Associates
155 Whitney Avenue
P.O. Box 993
New Haven, Connecticut 06504

Introduction

1. *Standard Metropolitan Statistical Areas*, OSS.
2. *1970 Yearbook*, ITE.
3. MORRIS, "The Story of the Center Line," in *TE*, November, 1967.
4. STARK, "Houston's Coordinated Traffic Signals," in *TE*, April, 1968.
5. *Traffic Engineering Handbook*, 2nd ed., ITE, 1950.
6. "Special Legislative Report—The 1968 Federal Aid Highway Act," in *ARB*, August, 1968.
7. *Public Law 91-605*, HR 19504, 91st Cong., USGPO, December 31, 1970.
8. *ITE Membership Survey—1959*, ITE.
9. "First Report on ITE Study Involving Salary, Education Median for Members Released," in *TE*, June, 1965.
10. McCLURE, "The Scope of Traffic Engineering in City Administration," in *TE*, September, 1958.
11. DOYLE, "Possible Traffic Help for Small Towns," in *TE*, January, 1960.
12. "Traffic Engineering Functions in Cities of 80,000 to 20,000 Population," in *TE*, July, 1963.
13. "What Are Organization Methods?", in *TE*, October, 1963.
14. ROSEVEARE, "Unique Role of a County Traffic Engineer," in *TE*, May, 1964.
15. WEBB, "The Organization and Administration of a State Traffic Department," in *TE*, February, 1965.
16. BAERWALD, "Why Aren't Traffic Engineering Decisions Made by Professionals?", in *TE*, April, 1967.
17. HOMBURGER, "Traffic Engineering Organization and Staffing in California Cities and Counties—1963," in *ITE SR*, May, 1963.
18. BRANDES, "County Traffic Engineering," in *TQ*, April, 1957.
19. LAYTHAM, "Traffic Engineering Features of Missouri State Highways in Metropolitan St. Louis," in *TE*, September, 1967.
20. HENRY, "Modern Building Benefits Traffic Division Operations," in *TE*, September, 1967.
21. CONNER, "Your New Traffic Engineer—Handle With Care," in *TE*, March, 1968.
22. SELTZER, "Los Angles County Traffic Engineering Advisory Services to Cities," in *TE*, August, 1969.
23. BAERWALD, "The Decision-Making Role of the Traffic Engineer," in *TE*, July, 1970.
24. POSTAS, "The Role of Public Relations in Traffic Engineering," in *TE*, July, 1970.
25. WATSON, "Traffic Engineering in Houston," in *TE*, September, 1970.
26. KEEFER, "The Role of the Traffic Engineer in the Urban Transportation Study," in *TE*, February, 1962.
27. BENSON, "The Traffic Engineer and Urban Transportation," in *TE*, September, 1962.

28. Braff, "New Life for Cities: Traffic Engineers' Role," in *TE*, August, 1964.
29. Keefer, "City Traffic Engineer and Urban Transportation Study," in *TE*, April, 1965.
30. Mickle, "The Expanding Role of Traffic Operations in Cities and States," in *TE*, February, 1958.
31. Mickle, "Traffic Engineering in Tomorrow's Transportation," in *TE*, January, 1959.
32. Cherniack, "Transportation—A New Dimension of Traffic Engineering," in *TE*, March, 1964.
34. Sielski, "Coordinating City Planning and Traffic Engineering," in *TE*, May, 1959.
35. Webber, "The Engineer's Responsibility for the Form of Cities," in *TE*, October, 1959.
36. "The Professional Responsibility of City Planners and Traffic Engineers in Urban Transportation," in *TE*, February, 1961.
37. Willey, "Highway Administration and the Traffic Engineer," in *TE*, December, 1962.
38. Gruen, "Renewing Cities for the Automobile Age," in *TE*, May, 1957.
39. Ehlers, "Highway Planning and Urban Renewal," in *TE*, July, 1958.
40. Englen, "The Opportunity is Here—For Coordination of Renewal, Transportation Planning," in *TE*, April, 1962.
41. "Traffic Planning in Urban Renewal Projects," in *TE*, January, 1964.
42. McGrath, "The Traffic Engineer's Role in Rebuilding Cities," in *TE*, June, 1964.
43. Edwards, "Transportation Planning and Urban Renewal," in *Proc. ITE*, 1964.
44. Barkeley, "Factors in Urban Renewal Plans," in *TQ*, January, 1959.
45. Steiner, "Traffic Improvement, Urban Renewal," in *TQ*, January, 1959.
46. Claire, "Urban Renewal and Transportation," in *TQ*, July, 1959.
47. Mueller, "Traffic Engineering and Cooperative Research," in *TE*, October, 1964.
48. Blackburn, "The Civil Engineer and Urban Transportation," in *Proc. P2957* ASCE, September, 1965.
49. Fratar, "Civil Engineering in Urban Transportation Planning," in *Proc. RA500* ASCE, December, 1965.
50. Lovelace, "The Civil Engineer and Urban Planning," in *Proc. P4809* ASCE, May, 1966.
51. Talbot, "Role of the City Engineer in Urban Planning," in *Proc. P4821* ASCE, May, 1966.
52. Branch and Boelter, "Civil Engineering in City and Regional Planning," in *Proc. P4827* ASCE, May, 1966.
53. Cummings, "Traffic Requirements for Construction Projects," in *TE*, November, 1967.
54. Tipton and Gordon, "Newport News Traffic Operation Plan," in *TE*, December, 1967.
55. Edwards, "Urban Design—The Traffic Engineer's Role," in *TE*, July, 1968.
56. Dasch, "A Working Community Relations Program," in *HRN 35*, Spring, 1969.
57. Smith, "Transportation Role for Civils," in *TEJ* ASCE Vol. 95, February, 1964.
58. Liebrand, "City Planning and Traffic Engineering," in *TQ*, October, 1968.
59. Werner, "The Urban Design Concept Team," in *ARB*, November, 1967.
60. "Team Concepts for Urban Highways and Urban Designs," in *HRR 220*, 1968.
61. "The Traffic Engineer's Role in Environmental Design," in *CR TE*, June, 1969.
62. Pikarsky, "The Chicago Crosstown Expressway," in *SR 104 HRB*, 1969.
63. "Chicago's Crosstown Expressway," in *ARB*, January, 1970.
64. Holmes, "To Grow with the Needs," in *TE*, December, 1970.
65. "Joint Development and Multiple Use of Right-of-Way," in *SR 104 HRB*, November, 1968.
66. "Multiple Use of Lands Within Highway Rights-of-Way," *NCHRPR 53*, 1969.
67. Slavis and Pignataro, "Utilization of Air Rights over Highway Right-of-Way," in *TQ*, January, 1969.
68. "National Survey of Transportation Attitudes and Behavior," *NCHRPR 249*, 1968.
69. Rash and Hille, "Public Attitudes toward Transport Modes: A Summary of Two Pilot Studies," in *HRR 233*, 1968.
70. Matson, Smith, and Hurd, *Traffic Engineering*, McGraw-Hill, New York, 1955.
71. Seburn and Marsh, *Urban Transportation Administration*, Yale BHT.

Chapter I: The Road User

1. Gordon and Michaels, "Static and Dynamic Visual Fields in Vehicular Guidance," in *HRR 84*, 1965.
2. Burg and Coppin, "Visual Acuity and Driving Record," in *HRR 122*, 1966.
3. Burg, "Vision and Driving: A Summary of Research Findings," in *HRR 216*, 1968.
4. Wolf, "Effects of Age on Peripheral Vision," in *HRBB 336*, 1962.
5. Wolf, "Studies on the Shrinkage of the Visual Field With Age," in *HRR 164*, 1967.
6. Salvatore, "Estimation of Vehicular Velocity Under Time Limitation and Restricted Conditions of Observation," *HRR 195*, 1967.
7. Forbes, "Driver Characteristics and Highway Operations," in *Proc. ITE*, 1953.
8. Wolf, "Glare Sensitivity in Relation to Age," in *HRBB 298*, 1961.
9. Peckham and Hart, "The Association Between Retinal Sensitivity and the Glare Problem," in *HRBB 255*, 1960.
10. Kinney and Connors, "Recovery of Foveal Dark Adaptation," in *HRR 70*, 1965.
11. Mortimer, "The Effect of Glare in Simulated Night Driving," *HRB*, 1967.
12. "Night Visibility—Selected References," in *HRBBib* 45, 1967.
13. "Headlight Glare," in *HRBBib* 46, 1968.
14. Coppia and Peck, "The Totally Deaf Driver in California," in *HRR 79*, 1965.
15. Konz and Daccarett, "Controls for Automotive Brakes," in *HRR 195*, 1967.
16. *A Policy on Geometric Design of Rural Highways*, AASHO, 1965.
17. Michaels, "Human Factors in Highway Safety," in *TQ*, October, 1961.
18. Michaels, "Tension Responses of Drivers Generated on Urban Streets," in *HRBB 271*, 1960.
19. Michaels, "Effect of Expresswsy Design on Driver Tension Responses," in *HRBB 330*, 1962.

20. Michaels, "Attitudes of Drivers Toward Alternative Highways and Their Relation to Route Choice, in *HRR 122*, 1966.
21. "Human Variables in Traffic Accidents—A Digest of Research and Selected Bibliography," in *HRBBib 31*, 1962.
22. *The Motor-Vehicle Driver: His Nature and Improvement*, Eno, 1949.
23. Coppin *et al*, "The Effectiveness of Short Individual Driver Improvement Sessions," in *HRR 195*, 1967.
24. Waller and Goo, "Accident and Violation Experience and Driver Test Score," in *HRR 225*, 1968.
25. Brody, "Personal Characteristics of Chronic Violators and Accident Repeaters," in *HRBB 152*, 1957.
26. Rainey *et al*, "An Investigation of the Role of Psychological Factors in Motor Vehicle Accidents," in *HRBB 212*, 1959.
27. Heath, "Relationships Between Driving Records, Selected Personality Characteristics, and Biographical Data of Traffic Offenders and Nonoffenders," in *HRBB 212*, 1959.
28. Uhlaner and Drucker, "Selection Tests—Dubious Aid to Driver Licensing," in *HRR 84*, 1965.
29. Ferdun *et al*, "The Teen-Aged Driver," in *HRR 163*, 1967.
30. Peck *et al*, "Driver Record by Age, Sex, and Marital Status," in *HRR 163*, 1967.
31. Marsh *et al*, "A Reevaluation of Group Driver Improvement Meeting," in *HRR 163*, 1967.
32. Silver, "Performance Criteria—Direct or Indirect," in *HRR 55*, 1964.
33. Mast *et al*, "Effects of Fatigue on Performance in a Driving Device," in *HRR 122*, 1966.
34. Dobbins *et al*, "Human Factors Research Reports—AASHO Road Test: I. Field Study of Vigilance Under Highway Driving Conditions," in *HRBB 330*, 1962.
35. Dobbins *et al*, "Human Factors Research Reports—AASHO Road Test: II Prediction of Vigilance," in *HRBB 330*, 1962.
36. Platt, "A New Method of Measuring the Effects of Continued Driving Performance," in *HRR 25*, 1963.
37. Greenshields, "Investigating Traffic, Highway Events in Relation to Driver-Action," in *TQ*, October, 1961.
38. Greenshields, "Driving Behavior and Related Problems," in *HRR 25*, 1963.
39. "Special Purpose Traffic Survey Devices," Report of ITE Project Committee 7F (62), *TE*, February, 1966.
40. Heimstra *et al*, "Effects of Fatigue on Basic Processes Involved in Human Operator Performance: 1. Simple Vigilance and Target Detection," in *HRR 55*, 1964.
41. Schlesinger and Safren, "Perceptual Analysis of the Driving Task," in *HRR 84*, 1965.
42. Greenshields, "Changes in Driver Performance with Time in Driving," in *HRR 122*, 1966.
43. Safford and Rockwell, "Performance Decrement in Twenty-Four Hour Driving," in *HRR 163*, 1967.
44. Zahavi, "Traffic Signals for the Blind in Israel," in *TE*, November, 1963.
45. Hoel, "Pedestrian Travel Rates in Central Business Districts," in *TE*, January, 1968.
46. Lindsay and Rockwell, "Freeway Illumination and Driving Performance," in *TE*, March, 1969.
47. Farber, Silver, and Landis, "Knowledge of Closing Rate Versus Knowledge of Oncoming Car Speed As Determiners of Driver Passing Behavior," in *HRR 247*, 1965.
48. Gorden and Mast, "Drivers' Decision in Overtaking and Passing," in *HRR 247*, 1968.
49. Farber and Silver, "Behavior of Drivers Performing a Flying Pass," in *HRR 247*, 1968.
50. Silver and Farber, "Driver Judgment in Overtaking Situations," in *HRR 247*, 1968.
51. Cassel and Janoff, "A Simulation Model of a Two-Lane Rural Road," in *HRR 257*, 1968.
52. Hutchenson *et al*, "An Evaluation of the Effectiveness of Televised, Locally Oriented Driver Re-education," in *HRR 292*, 1969.
53. Lindsay and Rockwell, "Freeway Illumination and Driving Performance," in *TE*, March, 1969.
54. Hoffman, "Note on Detection of Vehicle Velocity Changes," in *HF*, April, 1966.
55. Gorden, "Experimental Isolation of the Driver's Visual Input," in *HF*, April, 1966.
56. Johannson and Rumar, "Drivers' Brake Reaction Times," in *HF*, Feburay, 1971.
57. Borg, "Vision and Driving: A Report on Research," in *HF*, Feburary, 1971.
58. Davies and Watts, "Further Investigations of Movement Time Between Brake and Accelerator Pedals in Automobiles," in *HF*, June, 1970.
59. Gordon and Mast, "Drivers' Judgments in Overtaking and Passing," in *HF*, June, 1970.
60. Schuster, "Attitudes Toward Driving Safety and their Modification" in *TE*, June, 1970.
61. Konz and McDougal, "The Effect of Background Music on the Control Activity of an Automobile Driver," in *HF*, June, 1968.
62. Salvatore, "The Estimation of Vehicular Velocity as a Function of Visual Stimulation," in *HF*, February, 1968.

Chapter 2: The Vehicle

1. *Automobile Facts and Figures*, AMA, 1969.
2. Webb, "The Organization and Administration of a State Traffic Department," in *ITE*, Feburary, 1965.
3. *Motor Truck Facts*, AMA, 1969.
4. "Relation Between Vehicle Characteristics and High Design," in *HRBB 195*, 1958.
5. Haber and Witheford, "Correlation of Vehicle Design and Highway Design," in *Proc P2529*, ASCE, June, 1960.
6. "Motor Vehicle Size and Weights Limits," in *PR*, August, 1965.
7. *A Policy on Arterial Highways in Urban Areas*, AASHO, 1957.
8. "Directional Channelization and Determination of Pavement Widths," in *HRBB 72*, 1953.
9. Clark, "Truck Turning Path Study," in *TE*, September, 1958.
10. Jindra, "Track Widths of Vehicles on Curves," in *TE*, September, 1962.
11. Foxworth, "Determination of Oversized Vehicle Tracking Patterns by Adjustable Scale Model," in *Proc. HRB*, 1960.
12. Stevens *et al*, "Offtracking Calculation Charts for Trailer Combinations," in *HRR 159*, 1967.
13. Normann, "Braking Distances of Vehicles from High Speeds," in *Proc. HRB*, 1953.
14. Stonex, "Review of Vehicle Dimensions and Performance Characteristics," in *Proc. HRB*, 1960.
15. Stonex, "Correlation of Future Vehicle and Highway Design," in *Proc. P3281 ASCE*, September, 1962.
16. Baker, *Traffic Accident Investigator's Manual for Police*, TI, 1963.
17. "Vehicle Characteristics," in *HRB 334*, 1962.

18. Carmichael, "Motor-Vehicle Performance and Highway Safety," in *Proc. HRB*, 1953.
19. Wilson, "Deceleration Distances for High Speed Vehicles," in *Proc. HRB*, 1940.
20. Petring, "Stopping Ability of Motor Vehicles Selected from the General Traffic," in *PR*, June, 1959.
21. Tignor, "Braking Performance of Motor Vehicles, in *PR*, October, 1966.
22. "Pavement Slipperiness Factors and Their Measurement," in *HRBB 186*, 1958.
23. "Road Roughness and Skidding Measurements: 1960," in *HRBB 264*, 1960.
24. "Skid Testing: 1961," in *HRBB 302*, 1961.
25. "Pavement Roughness: Measuring Technique and Changes," in *HRBB 264*, 1960.
26. "Skidding Measurement Techniques: 1962 Developments," in *HRBB 348*, 1962.
27. "Roadway Surface Properties," in *HRR 28*, 1963.
28. Anderson *et al*, "Road Resistance of Large Transport Trucks," in *HRR 49*, 1964.
29. Moore, "Prediction of Skid-Resistance Gradient and Drainage Characteristics for Pavements," in *HRR 131*, 1966.
30. McCullough and Hankins, "Skid Resistance Guidelines for Surface Improvements on Texas Highways," in *HRR 131*, 1966.
31. "Studded Tires," in *HRR 136*, 1966.
32. "Studded Tires," in *HRR 171*, 1967.
33. *Reprint No. 34*, Virginia Council of Highway Investigation and Research, April, 1962.
34. Rizenbergs and Ward, "Skid Testing With An Automobile," in *HRR 189*, 1967.
35. "Tentative Skid-Resistance Requirements for Main Rural Highways," in *NCHRPR 37*, 1967.
36. Csathy *et al*, "State of the Art of Skid Resistance Research," in *HRBSR 95*, 1968.
37. "Surface Properties of Pavements and Vehicle Interaction," in *HRR 214*, 1968.
38. Hegman and Meyer, "The Effectiveness of Antiskid Materials in Cars," in *Proc. HRB*, 1960.
39. McFarland and Domey, "Human Factors in the Design of Passenger Cars," in *Proc. HRB*, 1960.
40. King and Sutro, "Dynamic Visual Fields," in *HRBB 152*, 1957.
41. Wolf *et al*, "Influence of Tinted Windshield Glass on Five Visual Functions," in *HRBB 255*, 1960.
42. McFarland *et al*, "Dark Adaptation as a Function of Age and Tinted Windshield Glass," in *HRBB 255*, 1960.
43. Richards, "Tinted Contact Lenses—A Handicap for Night Driving," in *HRR 25*, 1963.
44. Kilgour, "Some Results of Cooperative Vehicle Lighting Research," in *HRBB 255*, 1960.
45. Kilgour, "Cooperative Research in Vehicle Lighting," in *TQ*, January, 1962.
46. Roper and Meese, "More Light on the Headlighting Problem," in *HRR 70*, 1965.
47. Hanson and Palmquist, "Effectiveness of Reflectorized Headlamps," in *HRR 164*, 1967.
48. Cheeseman and Voss, "Motor Vehicle Headlight Beam Usage on a Section of Interstate Highway 90," in *HRR 164*, 1967.
49. Allen, "Vision, Vehicles, and Highway Safety," in *HRN 25*, 1966.
50. "Evaluation of Studded Tires" in *NCHRPR 61*, 1969.
51. *A Policy on Geometric Design of Rural Highways*, AASHO, 1965.
52. "Headlight Glare," in *HRBBib 46*, 1968.

Chapter 3: The Road and Geometric Design

1. *A Policy on Geometric Design of Rural Highways*, AASHO, 1965.
2. *A Policy on Arterial Highways in Urban Areas*, AASHO, 1957.
3. *Traffic Engineering Handbook*, 3rd ed., ITE, 1965.
4. *An Introduction to Highway Transportation Engineering*, ITE, 1968.
5. Ritter and Paquette, *Highway Engineering*, 3rd ed., Ronald Press, New York, 1968.
6. Drew, *Traffic Flow Theory and Control*, McGraw-Hill, New York, 1968.
7. *Better Transportation for Your City*, National Committee on Urban Transportation, Public Administration Service, 1958.
8. *A Policy on Design Standards-Interstate System*, AASHO, 1965.
9. *Geometric Deisn Standards for Highways other than Freeways*, AASHO, 1961.
10. *Report by the Special Freeway Study and Analysis Committee*, AASHO, 1960.
11. *Highway Design and Operational Practices Related to Highway Safety*, AASHO, 1967.

Chapter 4: Introduction to Transportation Planning

1. *CATS*, Vols. 1, 2, 3.
2. *PJTS*, Vols. 1, 2, 3, Philadelphia, Pa., 1964.
3. *Plan Report No. 5*, DVRPC, Philadelphia, Pa, 1969.
4. Martin *et al*, *Principles and Techniques of Predicting Future Traffic Demand for Urban Area Transportation*, MIT Report No. 3, MIT Press, 1961.
5. R. L. Creighton, *Urban Transportation Planning*, University of Illinois Press, Urbana, Ill., 1970.
6. *Traffic in Towns* (The Buchanan Report), HM Stationery Office, 1963. Also available in paperback from Penguin Books, Inc.
7. From the United States Department of Transportation's Bureau of Public Roads, Washington D. C., Manuals on:
   (a) *Trip Generation*
   (b) *Trip Distribution*
   (c) *Modal Split*
   (d) *Traffic Assignment*
8. *Urban Transportation Planning, General Information and Introduction to system 360*, U.S. DOT FHWA, BPR, June, 1970.
9. *Future Highways and Urban Growth*, Wilbur Smith and Associates, New Haven, Conn., 1961, for the AMA.
10. *Automobile Facts and Figures*, AMA, 1969.
11. Silver and Stowers, "Population, Economic, and Land Use Studies in Urban Transportation Planning," *BPR*, July, 1964.
12. Fratar, "Forecasting Distribution of Interzonal Vehicular Trips by Successive Approximations" in *HRB*, Vol. *33*, 1954.

Chapter 5: Origin and Destination Studies

1. *Procedure Manual 2B—Conducting a Home Interview Origin—Destination Survey*, NCUT, PAS, 1958.
2. MacLachlan, "The Coordinate Method of O & D Analysis," in *Proc. HRB*, Vol. 29, 1949.

3. BARKLEY, "Origin-Destination Surveys and Traffic Volume Studies," in *HRBBib 11*, 1951.
4. "Traffic Surveys by Post Cards," in *HRBB 41*, 1951.
5. WINFREY, "Postcard Method of Obtaining Origin and Destination of Traffic and Comparison with Roadside-Interview Method," in *HRBB 76*, 1953.
6. MURRAY, "A Comparative Study of Origin and Destination Data Obtained by Home Interview and Controlled Post Card Methods," in *Proc. HRB*, 1957.
7. CLEVELAND (ed.), *Manual of Traffic Engineering Studies*, ITE, 1964.
8. KRYGER and OTTESEN, "Can the License-Plate Method Be Used for Traffic Surveys?", in *TQ*, July 1956.
9. BEBEE, "Novel Traffic Survey Method Utilizes Vehicle Lights," in *HRBB 224*, 1959.
10. GOODMAN *et al*, "The 'Headlights On' Traffic Survey Technique," in *Proc. ITE*, 1964.
11. MAYER and WALLACE, "A New Method of Obtaining Origin and Destination Data," in *HRBB 347*, 1962.
12. EDENS, "Origin and Destination Surveys by Telephone," in *TE*, April, 1963.
13. STOCKTON, "Dallas Traffic-Survey Methods and Cost Analysis," in *HRBB 76*, 1953.
14. MCGRATH and GUINN, "Simulated Home Interview by Television," in *HRR 41*, 1963.
15. WYNN, "Studies of Trip Generation in the Nation's Capital" in *HRBB 230*, 1959.
16. CURRAN and STEGMAIER, "Travel Patterns in 50 Cities," in *HRBB 203*, 1958.
17. SCHMIDT and CAMPBELL, *Highway Traffic Estimation*, Eno, 1956.
18. WYNN, "Intracity Traffic Movements," in *HRBB 119*, 1956.
19. CARRIL, "Traffic Forecast Based on Anticipated Land Use and Current Travel Habits," in *Proc. HRB*, Vol. 31, 1952.
20. SMITH, "Gravity Model Theory Applied to a Small City Using a Small Sample of Origin-Destination Data"; BEN *et al*, "An Evaluation of Simplified Procedures for Determining Travel Patterns in a Small Urban Area " in *HRR 88*, 1965.
21. BOSTICK and TODD, "Travel Habits in Cities of 100,000 or More," in *PR*, February, 1966.
22. SMITH, "Analyzing and Projecting Travel Data," in *Proc.* ASCE, Vol. 86, No. HW2, June, 1960.
23. SHULDINER, "Trip Generation and the Home," in *HRBB 347*, 1962.
24. HALL, "Travel Characteristics of Two San Diego Subdivision Developments"; SHARPE, *et al*, "Factors Affecting Trip Generation of Residential Land-Use Areas," in *HRBB 203*, 1958.
25. JANES, "Method for Estimating Potential Increases in Traffic Volumes Based on O-D Survey Data from a Mid-Western City," in *HRR 88*, 1965.
26. WALKER, "Social Status of Head of Household and Trip Generation from Home," in *HRR 114*, 1966.
27. KOLE, "Variation of Work-Home Trip as a Function of Travel Time," in *TE*, December, 1964.
28. HOWE, "A Critical Analysis of an Origin-Destination Survey," in *HRR 41*, 1963.
29. HOWE, "A Theoretical Prediction of Work Trips in the Minneapolis-St. Paul Area," in *HRBB 347*, 1962.
30. HOWE, "A Theoretical Prediction of Work-Trip Patterns," in *HRBB 253*, 1963.
31. LAPIN, "Report on Analysis of Urban Work Trips" in *HRBB 224*, 1959.
32. LAPIN, "The Analysis of Work-Trip Data: Relationships and Variances," in *TQ*, April, 1957.
33. DEARINGER, "Auto Trips to the Central Business District," in *Proc. ASCE*, Vol. 88, No. HW2, September, 1962.
34. THABIT and LOBE, "Shopper Origins and Destinations," in *TQ*, January, 1957.
35. "Shopping Habits and Travel Patterns," in *HRBSR 11-B*, 1955.
36. "Travel to Commercial Centers," in *HRBB 79*, 1953.
37. WRIGHT, "Traffic, Traffic Generators in Central Business District," in *TE*, March, 1965.
38. SILVER and HANSEN, "Characteristics of Travel to a Regional Shopping Center," *BPR*, December, 1960.
39. HARPER and EDWARDS, "Generation of Person Trips by Areas Within the Central Business District," in *HRBB 253*, 1960.
40. TRUEMAN *et al*, "The Effect of Building Space Usage on Traffic Generation and Parking Demand," in *Proc. HRB*, Vol. 28, 1948.
41. WRIGHT, "Relationships of Traffic and Floor Space Use in Central Business District," in *HRR 114*, 1966.
42. HORN, "Impact of Industrial Development on Traffic Generation in Rural Areas of North Carolina," in *HRBB 347*, 1962.
43. BLACK, "Comparison of Three Parameters of Nonresidential Trip Generations," in *HRR 114*, 1966.
44. SULLIVAN, "Variations in Personal Travel Habits by Day of Week"; Crevo, "Characteristics of Summer Weekend Recreational Travel," in *HRR 41*, 1963.
45. COLEMAN, "Evaluation of Some Elements of Auto-Driver Trip Productions," in *HRR 41*, 1963.
46. HOEL, "Truck Travel in the Los Angeles Metropolitan Area," in *TQ*, October, 1964.
47. VOORHEES *et al*, "Traffic Patterns and Land-Use Alternatives," in *HRBB 347*, 1962.
48. WIANT, "A Simplified Method for Forecasting Urban Traffic," in *HRBB 297*, 1961.
49. WYNN and LINDER, "Tests of Interactance Formulas Derived from O-D Data," in *HRBB 253*, 1960.
50. VOORHEES and MORRIS, "Estimating and Forecasting Travel for Baltimore by Use of a Mathematical Model," in *HRBB 224*, 1959.
51. BEN *et al*, "Simplified Procedures for Determining Travel Patterns Evaluated for a Small Urban Area," in *PR*, February, 1965.
52. *Better Transportation for Your City*, NCUT, 1958.
53. HANSEN, "Traffic Approaching Cities," in *PR*, April, 1961.
54. CARROL and JONES, "Interpretation of Desire Line Charts Made on a Cartographatron," in *HRBB 253*, 1960.
55. *Procedure Manual 2A—Origin-Destination and Land Use*, NCUT, PAS 1958.
56. O'FLAHERTY, "Simplified Methods for Travel Studies in Smaller Cities," in *Proc. HRB*, Vol. 27, 1947.
57. *Procedure Manual 3D—Conducting a Comprehensive Parking*, NCUT, PAS, 1958.

Chapter 6: Highway Economy Studies

1. GRANT and IRESON, *Principles of Engineering Economy*, 4th ed., Ronald Press, New York, 1950.
2. THUESEN, *Engineering Economy*, Prentice-Hall, Englewood Cliffs, N.J., 1950.
3. DEGARMO, *Engineering Economy*, 4th ed., Macmillan, New York, 1967.

4. *Road User Benefit Analyses for Highway Improvements*, AASHO, 1960.
5. CLAFFEY, "Characteristics of Passenger Car Travel on Toll Roads and Comparable Free Roads," in *HRBB 306*, 1961; also in *PR*, June, 1961.
6. THOMAS, "Value of Time for Commuting Motorists," in *HRR 245*, 1968.
7. HANCY, "Use of Two Concepts of the Value of Time," in *HRBRec 12*, 1963.
8. FLEISCHER, "Effect of Highway Improvement on Travel Time of Commercial Vehicles, A Twenty-Five-Year Case Study," in *HRBRec 12*, 1963.
9. St. CLAIR and LIEDER, "Evaluation of Unit Cost of Time and Strain-and-Discomfort of Non-Uniform Driving," in *HRBSR 56*, 1960.
10. LAWTON, "Evaluating Highway Improvements on Mileage and Time Cost Basis," in *TQ*, January, 1950.
11. WEST, "Economic Value of Time Savings in Traffic," in *Proc. ITE*, 1946.
12. VASWANI, "Value of Automobile Transit Time in Highway Planning," in *Proc. HRB*, 1958.
13. St. CLAIR *et al*, "The Measurement of Vehicular Benefits," in *HRR 138*, 1966.
14. CURRY, "Use of Marginal Cost of Time in Highway Economy Studies," in *HRR 77*, 1965.
15. HANING and WOOTAN, "Value of Commerical Motor Vehicle Time Saved," in *HRR 82*, 1965.
16. "Values of Time Savings of Commercial Vehicles," in *NCHRPR 33*, 1967.
17. LISCO, "Value of Commuters' Travel Time—A Study in Urban Transportation," in *HRR 245*, 1968.
18. MICHAELS, "Attitudes of Drivers Toward Alternative Highways and Their Relation to Route Choice," in *HRR 122*, 1966; also in *PR*, December, 1965.
19. WACHS, "Relationships Between Drivers' Attitudes Toward Alternative Routes and Driver and Route Characteristics," in *HRR 197*, 1967.
20. WINFREY, *Highway Engineering Handbook*, Section 3, ed. Woods, McGraw-Hill, New York, 1960.
21. LIEDER, "Passenger Car Fuel-Consumption Rates," in *PR*, December, 1962.
22. FIREY and PETERSON, "An Analysis of Speed Changes for Large Transportation Trucks," in *HRBB 334*, 1962.
23. SAWHILL and FIREYS, "Predicting Fuel Consumption and Travel Time by Motor Transport Vehicles," in *HRBB 334*, 1962.
24. STEVENS, "Line-Haul Trucking Costs in Relation to Vehicle Gross Weights," in *HRBB 301*, 1961.
25. KENT, "Fuel and Time Consumption for Highway User Benefit Studies," in *HRBB 276*, 1960; also in *PR*, April, 1960.
26. CLAFFEY, "Time and Fuel Consumption for Highway User Benefit Studies" in *HRBB 276*, 1960; also in *PR*, April, 1960.
27. SAWHILL and FIREY, "Motor Transport Fuel Consumption Rates and Travel Time," in *HRBB 276*, 1960.
28. GLAZE and VAN MIEGHEM, "Washington Motor Vehicle Operating Cost Survey" in *Proc. HRB*, 1957.
29. SCHWENDER, "Vehicle Operating Characteristics on the West Virginia Turnpike and Alternative Routes" in *Proc. HRB*, 1957.
30. SAAL, "Operating Characteristics of a Passenger Car on Selected Routes," in *PR*, August, 1955; also in *HRBB 107*, 1957.
31. WINFREY, "Gasoline Consumption, Weight, and Mileage of Commercial Vehicles," in *HRB 92*, 1954.
32. SAAL, "Time and Gasoline Consumption in Motor Truck Operation as Affected by the Weight and Power of Vehicles and the Rise and Fall in Highways," in *HRBRR 9A*, 1950.
33. MOYER, "Motor Vehicle Power Requirements on Highway Grades" in *Proc. HRB*, 1934.
34. "The Effect of Highway Design on Vehicle Speed and Fuel Consumption," in *Tech. Bull. 5 OSHC*, 1937.
35. MARCELLIS, "An Economic Analysis of Traffic Movement at Various Speeds," in *HRR 35*, 1963.
36. SAWHILL and CRANDAIL, "Some Measurable Qualities of Traffic Service Influenced by Freeways," in *HRR 49*, 1964.
37. WINFREY, "Research on Motor Vehicle Performance Related to Analyses for Transportation Economy," in *HRR 77*, 1965.
38. BEVIS, "Estimating a Road-User Cost Function from Diversion Curve Data," in *HRR 100*, 1965.
39. "Running Cost of Motor Vehicles as Affected by Highway Design," in *NCHRPR 13*, 1965.
40. BEVIS, "Automobile Operating Costs," in *CATS RN*, September, 1957.
41. JOSEPH, "Automobile Operating Costs," in *CATS RN*, November, 1959.
42. STEVENS, "Line-Haul Trucking Costs Upgraded, 1966," in *HRR 127*, 1966.
43. JAN DE WEILLE, "Quantification of Road User Savings," in *World Bank Staff Occasional Papers No. 2*, 1966, distributed by The Johns Hopkins Press (available at UN Bookstore).
44. CURRY and HANEY, *A Manual for Conducting Highway Economy Studies*, SRI, August, 1966.
45. *Maximum Desirable Dimensions and Weights of Vehicles Operated on the Federal-Aid Systems*, HD No. 354, 88th Cong. USGPO, 1964.
46. SAAL, "Relation Between Gross Weights of Motor Trucks and Their Horsepower," in *PR*, October, 1957.
47. WRIGHT and TIGNOR, "Relation of Gross Weights and Horsepowers of Commercial Vehicles," in *PR*, Octomber, 1966.
48. EVANS and TREADWAY, "Economic Analysis of Truck Climbing Lanes on Two-Lane Highways," in *HRR 245*, 1968.
49. GIBBONS and PROCTOR, "Economic Costs of Traffic Congestion," in *HRBB 86*, 1954.
50. LAPIN, "Costs of Excessive Congestion in Philadelphia," in *TE*, September, 1957.
51. HAIKALIS and JOSEPH, "Economic Evaluation of Traffic Networks," in *HRBB 306*, 1961.
52. RUITER and SHULDINER, "Operating Costs at Intersections Obtained from the Simulation of Traffic Flow," in *HRR 89*, 1965.
53. *Accident Facts—1967 Edition*, NSC, 1967.
54. DUNMAN, "The Economic Costs of Motor Vehicle Accidents," in *TE*, November, 1956.
55. DUNMAN, "The Economic Costs of Motor Vehicle Accidents," in *HRBB 208*, 1959; also in *PR*, June, 1958.
56. McCARTHY, "Economic Cost of Traffic Accidents in Relation to the Vehicle," in *HRBB 263*, 1960; also in *PR*, June, 1960.
57. JOHNSTON, "Economic Cost of Traffic Accidents in Relation to Highway Planning," in *HRBB 263*, 1960; also in *PR*, June, 1960.
58. DRAKE and KRAFT, "Motor Vehicle Accident Costs in the Washington Metropolitan Area," in *HRR 188*, 1967.

59. Flanakin, "The Washington Area Motor Vehicle Accident Cost Study," in *TE*, June, 1967.

60. Billingsley and Jorgenson, "Direct Cost and Frequencies of 1958 Illinois Motor-Vehicle Accidents" in *HRBR 12*, 1963; also in *PR*, August, 1963.

61. Jorgenson, "A Brief Description of the Illinois Accident Cost Study" in *CATS RN*, December, 1961.

62. Smith and Tamburri, "Direct Costs of California State Highway Accidents," in *HRR 225*, 1968.

63. *Traffic Control and Roadway Elements*, ASF, 1963.

64. "Accident Rates as Related to Design Elements of Rural Highways," in *NCHRPR 47*, 1968.

65. *The Federal Role in Highway Safety*, HD No. 93, 86th Cong. USGPO 1959.

66. Byington, "Interstate System Accident Research," in *PR*, December, 1963.

67. "Turnpike Accident and Fatality Statistics 1960–1961," in *TE*, March, 1962.

68. Frye, "The Effect of an Expressway on an Area-Accidents" in *CATS RN*, December, 1961.

69. Frye, "Effect of an Expressway on Distribution of Traffic and Accidents," in *HRR 21*, 1963.

70. Hoch, "Chicago's Accident Experience on Arterials and Expressways," in *TQ*, July, 1960.

71. Solomon, *Accidents on Main Rural Highways Related to Speed, Driver, and Vehicle*, BPR, USGPO, July, 1964.

72. Twombly, "Economic Cost of Traffic Accidents in Relation to the Highway," in *HRBB 263*, 1960; also in *PR*, June, 1960.

73. Jorgenson. "Accident Costs and Rates on Chicago Area Streets and Highways," in *CATS RN*, March, 1962.

74. *Final Report of the Highway Cost Allocation Study*, HD No. 72, 86th Cong. USGPO, 1961.

75. Grant, "Interest and the Rate of Return on Investments," in *HRBSR 56*, 1960.

76. Glancy, "Utilization of Economic Analysis by State Highway Departments," in *HRR 77*, 1965.

77. Grant, "Concepts and Applications of Engineering Economy," in *HRBSR 56*, 1960.

78. Winfrey, "Concepts and Applications of Engineering Economy in the Highway Field," in *HRBSR 56*, 1960.

79. "Know Your Highway Costs," in *HRBSR 13*, 1953.

80. *Highway Statistics-1966*, BPR, 1968.

81. Balkins and Srour, "Limited-Access Highway Construction Costs in the Tri-State Region," in *HRR 252*, 1968.

82. Winfrey and Farrel, "Life Characteristics of Surfaces Constructed on Primary Rural Highways," in *PR*, March, 1941; also in *Proc. HRB*, 1940.

83. Farrell and Paterick, "Life Characteristics of Highway Surfaces," in *PR*, June, 1956; also in *Proc. HRB, 1956*.

84. Gronberg and Blosser, "Lives of Highway Surfaces—Half Century Trends," in *PR*, June, 1956; also *Proc. HRB*, 1948.

85. Winfrey and Howell, "Highway Pavements—Their Service Lives," in *HRR 252*, 1968.

86. *Highway Maintenance*, NHUC Research Department, October, 1964.

87. Lang, "Turnpike Maintenance," in *ARB*, November, 1960.

88. "Maintenance Practices," in *HRBR 11*, 1963.

89. "Repairs of Concrete Pavements," in *HRBB 322*, 1962.

90. *Highway Research News No. 4*, HRB, 1963.

91. "Iowa State Highway Maintenance Study," in *HRBSR 65*, 1961.

92. "Supplement I, Iowa State Highway Maintenance Study," in *HRBSR 65*, 1961.

93. "Maintenance Costs and Performance—1961," in *HRBC 478*, 1962.

94. Larsen, "Iowa County Highway Maintenance Practices" in *Proc. HRB*, 1961.

95. Betz, "Highway Maintenance Costs—A Consideration for Developing Areas," in *HRR 94*, 1964.

96. Lang, "Expressway Maintenance—Costs and Problems in Urban Areas" in *CE*, September, 1964.

97. "Highway Maintenance Studies," in *HRBB 155*, 1957.

98. "Repair of Structures and Pavement by Thin Concrete Patching," in *HRR 93*, 1965.

99. "Pavement Condition Evaluation," in *HRR 40*, 1963.

100. "Maintenance Practices—1963, Administration, Methods, and Materials," in *HRR 61*, 1964.

101. "Maintenance Practices—1963, Administration, Methods and Materials," in *HRR 61*, 1964.

102. Cardone, "Maintenance Cost of Rest Areas in Michigan," in *HRR 93*, 1965.

103. "Evaluation of Pavements by Deflection Studies for Maintance Purposes," in *HRR 129*, 1966.

104. "Pavement and Bridge Maintenance," in *HRR 173*, 1967.

105. "Maintenance and Equipment," in *HRR 173*, 1967.

106. "Highway Maintenance, A Survey of State Laws," in *HRBSR 84*, 1965.

107. "Non-Chemical Methods of Snow and Ice Control on Highway Structures," in *NCHRPR 4*, 1964.

108. Tinney and Carr, "Sign Maintenance in California," in *HRN 3*, May, 1963.

109. "Annual Progress Report," Joint Committee on Maintenance Personnel, in *HRN 9*, November, 1963.

110. "Reports of Committee on Maintenance Costs," in *HRN 15*, November, 1964.

111. Edwards, "Development of a Maintanance Cost Formula and Its Application," in *HRN 16*, December, 1964.

112. Mann, "Predicting Highway Maintenance Costs," in *HRN 19*, June, 1965.

113. Gills, "Salvaging of Old Pavements to Meet Interstate Standards by Resurfacing with Asphaltic Concrete"; Hewitt, "Cost Analysis of Methods for Interstate Guide Sign Maintenance," in *HRN 21*, 1965.

114. Arman, "Better Use of Engineering Manpower in Maintenance" in *HRN 28*, Summer, 1967.

115. Sloane and Stackhouse, "Melting Snow on a Mountain Pass Highway by Using Coal Dust," in *HRN 29*, Autumn, 1967.

116. Gladstone and Cooper, "State Highway Patrols—Their Functions and Financing," in *HRR 138*, 1966; also in *PR*, December, 1966.

117. Cron, "Snowdrift Control Through Highway Design," in *PR*, December, 1967.

118. Records, "Horizons in Highway Maintenance," in *PR*, December, 1967.

119. Parisi, "Cheap Paint is Costly," in *TE*, December, 1967.

120. "Progress Report of HRB-AASHO Joint Committee on Maintenance Personnel," in *HRC 11*, December, 1965.

121. *An Information Guide for Physical Maintenance of Pavements*, AASHO, 1963.

122. *An Information Guide for Methods and Procedures in Contract Maintenance*, AASHO, 1963.
123. NUSBAUM, "Operating and Maintaining Urban Expressways," in *ARB*, June, 1964.
124. GLENDENING, "Maintaining City Streets," in *ARB*, July, 1967.
125. BAKER, "Maintenance Ahead—Needed: Materials, Equipment and Ideas," in *ARB*, August, 1967.
126. CRAWFORD, "Highway Maintenance in Louisiana," in *ARB*, December, 1967.
127. *HRC Nos. 17*, *33*, *42*, *57*, *58*, 1966, and *63*, *64*, 1967.
128. "Interstate Highway Maintenance Requirements and Unit Maintenance Expenditure Index," in *NCHRPR 42*, 1967.
129. BYRD, "Interstate Maintenance," in *ARB*, January, 1969.
130. "Maintenance Management—1967," in *HRR 241*, 1968.
131. "Maintenance Practices—1967," in *HRR 254*, 1968.
132. "Maintenance Management," in *HRBSR 100*, 1968.
133. "Progress Report of HRB-AASHO Joint Committee on Maintenance Personnel," in *HRC 90*, 1968.
134. RECORDS, "Highway Maintenance Research—An Overview," in *PR*, October, 1968.
135. DOUGLAS, "Optimum Life of Equipment for Maximum Profit," in *CDJ*, ASCE, Vol. 94, No. CO1, January, 1968 (*Proc P5745*).
136. "An Analysis of Techniques Used in Highway Economy Studies by Various States," in *HRN 31*, Spring, 1968.
137. OGLESBY and GRANT, "Economic Analysis—The Fundamental Approach to Decisions in Highway Planning and Design," in *Proc. HRB*, 1958.
138. GRANT and OGLESBY, "Economy Studies for Highways" in *HRBB 306*, 1961.
139. GRANT and OGLESBY, "A Critique of Some Recent Economic Studies Comparing Alternate Highway Locations," in *Proc. HRB*, 1960.
140. MANHEIM, "Data Accuracy in Route Location," in *TQ*, January, 1961.
141. MACDORMAN, "Case Study in Sensitivity of Highway Economy Factors," in *HRR 100*, 1965.
142. STEINER, "Engineering Economy and Economic Penetration Roads," in *Proc P 5609*, HDJ, ASCE, Vol. 93, No. HW2, November, 1967.
143. BELLOMO and PROVOST, "Two Procedures to Improve the Economic Evaluation of Alternative Highway Systems," in *HRR 224*, 1968.
144. EVANS, "Economic Studies for Highways," in *TQ*, October, 1968.
145. LEE and GRANT, "Inflation and Highway Economy Studies," in *HRR 100*, 1965.
146. MAYER and LAMUS, "A Study of Annual Costs of Flexible and Rigid Pavements for State Highways in California" in *HRR 77*, 1965.
147. BALDOCK, "The Annual Cost of Highways," in *HRR 12*, 1963.
148. WORK, "An Economic Replacement Model for Highway Surface Determination," in *HRR 12*, 1963.
149. *Economics of Asphalt and Concrete for Highway Construction*, SRI, 1961.
150. *Highway Capacity Manual*, HRBSR 87, 1965.
151. SMITH, "Benefit-Cost Ratios: A Word of Caution," in *HRR 12*, 1963.
152. HOCH, "Benefit-Cost Methods for Evaluating Expressway Construction," in *TQ*, April, 1961.
153. HANSEN, "A Procedure for Determining the Annual Cost of a Section of Rural Highway," in *PR*, April, 1951.
154. WOHL and MARTIN, "Evaluation of Mutually Exclusive Design Projects," in *HRBSR 92*, 1967.
155. BROWN and HARRAL, "Estimating Highway Benefits in Underdeveloped Countries," in *HRR 115*, 1966.
156. SOBERMAN, "Economic Analysis of Highway Design in Developing Countries," in *HRR 115*, 1966.
157. NEWCOMB, "New Approach to Benefit Cost Analysis," in *HRR 138*, 1966.
158. HILL, "A Method for the Evaluation of Transportation Plans," in *HRR 180*, 1967.
159. FRANKLIN, "Benefit-Cost Analysis in Transportation Plans," in *HRR 180*, 1967.
160. SHORTREED and BERRY, "Spacing of Grade Separations on Rural Freeways," in *HRR 224*, 1968.
161. *Cost-Effectiveness in Traffic Safety*, Arthur D. Little, Inc., Frederick A. Praeger Publishers, New York, 1968.
162. CRIBBINS, "Investment Return Analysis: A New Approach for Scheduling Improvements at Hazardous Highway Locations," in *HRBB 320*, 1962.
163. OGLESBY and SARGENT, "An Economy Study Aimed at Justifying a Secondary Road Improvement," in *HRBB 320*, 1962.
164. HEAD and BLENSLY, "Total Annual Cost Analysis," in *HRBB 320*, 1962.
165. MORGALI and OGLESBY, "Procedures for Determining the Most Economical Design for Bridges and Roadways Crossing Flood Plains," in *HRBB 320*, 1962.
166. WINFREY, "Cost Comparison of Four-Lane vs. Stage Construction on Interstate Highways," in *HRBB 306*, 1961.
167. GILBERT, "Economic Balance of Transportation Modes," in *TE*, October, 1963; also in *Proc. ITE*, 1963.
168. HOCH, "Chicago's Accident Experience on Arterials and Expressways," in *TQ*, July, 1960.
169. CARSTENS and CSANYI, *Economic Analyses for Highway Improvements in Under-Developed Countries*, ER 44, EES, November, 1964.
170. STAFFORD, "Economics of Parking Structures at Airports," in *Proc. P 4627* JASTD ASCE, January, 1966.
171. CAMPBELL, "Social and Economic Factors in Highway Locations," in *Proc. P 4926*, HDJ, ASCE, Vol. 92., No. HW2, October, 1966.
172. BETZ, "Interpretation of Costs" in *Proc P 4937*, HDJ ASCE, Vol. 92, No. HW2, October, 1966.
173. NIEBUR, "Preliminary Engineering Economy Analysis of Puget Sound Regional Transportation Systems," in *HRR 180*, 1967.
174. WANG, *et al*, "Toward the Solution for the Optimal Allocation of Investment in Urban Transportation Networks," in *HRR 238*, 1968.
175. HARVEY, "A Method of Network Evaluation Using the Output of the Traffic Assignment Process," in *HRR 238*, 1968.
176. SHAW and MICHAEL, "Evaluation of Delays and Accidents at Intersections to Warrant Construction of a Median Lane, in *HRR 257*, 1968.
177. LAGO, "Cost Functions and Optimum Technology for Intercity Highway Transportation Systems in Developing Countries," in *TQ*, October, 1968.
178. LANG *et al*, "An Evaluation of Techniques for Highway User Cost Computation," in *HRBB 320*, 1962.
179. ROBERTS and SUHRBIER, "Link Analysis for Route Location," in *HRR 77*, 1965.

180. SUHRBIER and ROBERTS, "Engineering of Location: The Selection and Evaluation of Trial Grade Lines by an Electronic Digital Computer," in *HRR 100*, 1965.

181. JOHNSON, "Use of Traffic Volume Data in Evaluation of Highway User Costs for Economic Analysis," in *HRR 100*, 1965.

182. SPENCER, "An Approach to Planning and Programming Local Road Improvements Based on a Network-Wide Assessment of Economic Consequences," in *HRR 224*, 1968.

183. WINFREY, *Economic Analysis for Highways*, International Textbook Company, Scranton, Pa., 1969.

184. "Running Cost of Motor Vehicles as Affected by Road Design and Traffic," in *NCHRPR 111*, 1971.

Chapter 7: Travel Time and Delay Studies

1. ROTHROCK, "Urban Congestion Index Principles," in *HRBB 86*, 1954.
2. SCHROEDER, "A Suggested Congestion Rating for Urban Highways," in *SRR* No. 5, 1954.
3. GREENSHIELDS, "Quality of Traffic Transmission," in *Proc. HRB*, 1954.
4. GREENSHIELDS, "The Quality of Traffic Flow," *in Quality and Theory of Traffic Flow*, Symposium, Yale University, 1961.
5. HALL and GEORGE, "Travel Time—An Effective Measurement of Congestion and Level of Service," in *Proc. HRB*, 1959.
6. DEEN, "The Use of Travel Time as a Factor in Rating Urban Streets," in *TE*, January, 1960.
7. SCHWART, "Quality of Traffic Service," in *TQ*, January, 1966.
8. WARDROP, "Some Theoretical Aspects of Road Traffic Research," in *Road Paper No. 36*, *ICE*, 1952.
9. DRAKE *et al*, "A Statistical Analysis of Speed Density Hypothesis," *in HRR 154*, 1967.
10. BERRY, "Evaluation of Techniques for Determining Overall Travel Time," in *Proc. HRB*, 1949.
11. BERRY and GREEN, "Techniques for Measuring Overall Speeds in Urban Areas," in *Proc. HRB*, 1949.
12. *Determining Travel Time*, PM No. 3B, NCUT, 1958.
13. WALKER, "Speed and Travel Time Measurement in Urban Areas," in *HRBB 156*, 1957.
14. CRIBBINS *et al*, "Development and Use of Maximum-Car Technique For Measuring Travel Time," in *HRBB 303*, 1961.
15. HORN *et al*, "Effects of Commercial Roadside Development on Traffic Flow in North Carolina," in *HRBB 303*, 1961.
16. MUELLER, "Recent Speed and Delay Insturments," in *TE*, December, 1954.
17. YOUNGS, "The Traffic Chronograph," in *TE*, May, 1953.
18. RODGERS, "A New Traffic Delay Measuring Device," in *TE*, Februray, 1957.
19. MAY and KANEKO, "Comparative Study of Two Vehicle Operating Characteristics Instruments," in *Proc. HRB*, 1958.
20. FOOTE, "Special Purpose Traffic Survey Devices," in *Proc. ITE*, 1963.
21. BERRY and VAN TIL, "A Comparison of Three Methods for Measuring Delay at Intersections," in *Proc. California Street and Highway Conference*, 1954; also in *TE*, December, 1954.
22. BERRY, "Field Measurements of Delay at Signalized Intersections," in *Proc. HRB*, 1956.
23. ATHOL, "Interdependence of Certain Operational Characteristics Within a Moving Traffic Stream," in *HRR 72*, 1965.
24. "Urban Travel Time Measurement by Taxicab Speed Studies," in *TE*, June, 1961.
25. WARDROP and CHARLESWORTH, "A Method of Estimating Speed and Flow of Traffic From a Moving Vehicle," in *Paper No. 5925*, *ICE*, 1954.
26. BELMONT, "A Note on the Use of the Wardrop-Charlesworth Method," in *TM No. B-29*, *ITTE* 1954.
27. MORTIMER, "Moving Vehicle Method of Estimating Traffic Volumes and Speeds," in *HRBB 156*, 1957; also in *TE*, September, 1956.
28. BLENSLY, "Moving Vehicle Method of Estimating Traffic Volumes," in *TE*, December, 1956.
29. HALL, "Intersection Delay-Signal vs. Four-Way Stop," in *Proc. ITE*, 1952.
30. RAFF, *A Volume Warrant for Stop Signs*, Eno, 1950.
31. BELLIS, "Traffic Report Before and After Improvement at Intersection of Routes 1 and 25," in *Proc. HRB*, 1950.
32. SOLOMON, "Accuracy of the Volume-Density Method of Measuring Travel Time," in *TE*, March, 1957.
33. GIBBONS and PROCTOR, "Economic Costs of Traffic Congestion," in *HRBB 86*, 1954.
34. *Standards for Street Facilities and Services*, PM 7A, NCUT.
35. ROTHROCK and KEEFER, "Measurement of Urban Traffic Congestion," in *HRBB 156*, 1957.
36. COLEMAN, "A Study of Urban Travel Times in Pennsylvania Cities," in *HRBB 303*, 1961.
37. HIXON, "Analysis of Urban Travel Times and Traffic Volume Characteristics," in *HRBB 303*, 1961.
38. HALEY *et al*, "Travel Time—A Measure of Service and a Criterion for Improvement Priorities," in *HRR 35*, 1963.
39. ATHOL, "Headway Groupings," in *HRR 72*, 1965.
40. PLATT, "A Proposed Index for the Level of Traffic Service," in *TE*, November, 1963.
41. THARP and HARR, "A Quantitative Evaluation of the Geometric Aspects of Highways," in *HRR 83*, 1965.
42. GEORGE, "Measurement and Evaluation of Traffic Congestion," in *Quality and Theory of Traffic Flow*, Symposium, Yale University, 1961.
43. DREW *et al*, "Freeway Level of Service as Described by an Energy-Acceleration Noise Model," in *HRR 167*, 1967.
44. KEESE *et al*, "Highway Capacity—The Level of Service Concept," in *Proc. ITE*, 1965.
45. "Quality of Traffic Service," in *HRBBib 36*, 1963.
46. MATSON, SMITH, and HURD, *Traffic Engineering*, McGraw-Hill, New York, 1955.
47. BONE, "Travel-Time and Gasoline-Consumption Studies in Boston," in *Proc. HRB*, 1952.
48. SIELSKI, "Effect of Northwest Expressway on Alternate Arterial Streets," in *HRR 21*, 1963.
49. VOLK, "Effect of Type of Control on Intersection Delay," in *Proc. HRB*, 1956.
50. FRENCH, "Capacities of One-Way and Two-Way Streets with Signals and with Stop Signs," in *PR*, February, 1956.
51. KENEIPP, "Efficiency of Four-Way Stop Control at Urban Intersections," in *TE*, January, 1951.
52. GURNETT, "Intersection Delay and Left Turn Phasing," in *TE*, June, 1969.
53. WEBSTER, "Traffic Delay on an Urban Arterial Street as a Result of Curb Parking Maneuvers," in *HRR 27*, 1969.
54. KENNEDY *et al*, *Fundamentals of Traffic Engineering*, 7th ed., ITTE, UCLA, 1967.
55. TREADWAY and OPPENLANDER, "Statistical Modeling of Travel Speeds and Delays on a High Volume Highway," in *HRR 199*, 1967.

56. Christensen, "Use of a Computer and Vehicle Loop Detectors to Measure Queries and Delay at Signalized Intersections," in *HRR 211*, 1967.
57. Greenshields, "The Measurement of Highway Traffic Performance," in *TE*, April, 1969.
58. Votan and Levinson, *Elementary Sampling for Traffic Engineers*, Eno, 1962.

Chapter 8: Spot Speed Studies

1. "Spot Speed Survey Devices," in *TE*, May, 1962.
2. Greenshields and Weida, *Statistics with Applications to Highway Traffic Analysis*, Eno, 1952.
3. Schwar and Puy-Huarte, *Statistical Methods in Traffic Engineering*, OSU, August, 1964.
4. Votaw and Levinson, *Elementary Sampling for Traffic Engineers*, Eno, 1962.
5. Dixon and Massey, *Introduction to Statistical Analysis*, McGraw-Hill, New York, 1951.
6. Moroney, *Facts from Figures*, Pelican, Baltimone, Md., 1956.
7. Crow, Davis, and Maxfield, *Statistics Manual*, Dover, New York, 1960.
8. Kennedy, Kell, and Homburger, *Fundamentals of Traffic Engineering*, 6th ed., UCLA, 1966.
9. Oppenlander *et al*, "Sample Size Requirements for Vehicular Speed Studies," in *HRBB 281*, 1961.
10. Oppenlander, "Sample Size Determination for Spot-Speed Studies at Rural, Intermediate, and Urban Locations," in *HRR 35*, 1963.
11. "Traffic Speed Trends," in *BPR*, March, 1958.
12. "Highway Statistics—1963," in *BPR*, March, 1965.
13. Forbes, "Speed, Headway, and Volume Relationships on a Freeway," in *Proc. ITE*, 1951.
14. May and Wagner, "Headway Characteristics and Interrelationships of Fundamental Characteristics of Traffic Flow," in *Proc. HRB*, 1960.
15. Rowan and Keese, "A Study of Factors Influencing Traffic Speeds," in *HRBB 341*, 1962.
16. "Report of Speeds at High Speed Locations on Illinois Rural State Highways," in *IDH*, February, 1956.
17. Whitby, "Small-Car Speeds and Spacings on Urban Expressways," in *HRBB 351*, 1962.
18. Lefeve, "Speed Characteristics on Vertical Curves," in *Proc. HRB*, 1953.
19. Taragin, "Driver Performance on Horizontal Curves," in *Proc. HRB*, 1954.
20. Powers and Michael, "Effects on Speed and Accidents of Improved Delineation at Three Hazardous Locations," in *HRBB 303*, 1961.
21. Taragin and Rudy, "Traffic Operations as Related to Highway Illumination and Delineation," in *PR*, August, 1960.
22. Taragin, "The Effect on Driver Behavior of Center Lines on Two-Lane Roads," in *Proc. HRB*, 1947.
23. Jouzy and Michael, "Use and Design of Acceleration and Deceleration Lanes in Indiana," in *HRR 9*, 1963.
24. Conklin, "A Comparison of Vehicle Operating Characteristics Between Parallel Lane and Direct Taper Types of Freeway Off-Ramps," in *TE*, December, 1959; also in *Proc. ITE*, 1959.
25. Berry *et al*, "A Study of Left-Hand Exit Ramps on Freeways," in *HRR 21*, 1963.
26. Covault and Roberts, "Influence of On-Ramp Spacing on Traffic Flow Characteristics on Atlanta Freeway and Arterial Street System," in *HRR 21*, 1963.
27. Covault and Kirk, "Influence of On-Ramp Spacing on Traffic Flow Characteristics on Atlanta Freeway and Arterial Street System," in *HRR 59*, 1964.
28. Gervais, "Optimization of Freeway Traffic by Ramp Control," in *HRR 59*, 1964.
29. Oppenlander, "Multivariate Analysis of Vehicular Speeds," in *HRR 35*, 1963.
30. Oppenlander and Dawson, "Criteria for Balanced Geometric Design of Two-Lane Rural Highways"; Tharp and Harr, "A Quantitative Evaluation of the Geometric Aspects of Highways," in *HRR 83*, 1965.
31. Wortman, "A Multivarate Analysis of Vehicular Speeds on Four-Lane Rural Highways," in *HRR 72*, 1965.
32. Sacks, "Effect of Guardrail in a Narrow Median Upon Pennsylvania Drivers," in *HRR 83*, 1965.
33. Taragin, "Driver Behavior as Affected by Objects on Highway Shoulders," in *Proc. HRB*, 1955; also in *PR*, June, 1955.
34. Taragin, "Driver Behavior Related to Types and Widths of Shoulders on Two-Lane Highways," in *PR*, August, 1957.
35. Taragin and Eckhardt, "Effect of Shoulder on Speed and Lateral Placement of Motor Vehicles," in *Proc. HRB*, 1953.
36. Walker, "Speed and Travel Time Measurement in Urban Areas," in *HRBB 156*, 1957.
37. Wilson, "California's Reduced Visibility Study Helps Cut Down Traffic Accidents When Fog Hits Areas," in *TE*, August, 1965.
38. Lefeve, "Speed Habits Observed on a Rural Highway," in *Proc. HRB*, 1954.
39. Webster, "Driver Opinions and Characteristics Related to Rural Speed," in *TE*, July, 1966.
40. Snider, "Capability of Automobile Drivers to Sense Vehicle Velocity," in *HRR 159*, 1967.
41. Salvatore, "Vehicles Speed Estimation from Visual Stimuli," in *PR*, February, 1967.
42. Shumate and Crowther, "Variability of Fixed-Point Speed Measurements," in *HRBB 281*, 1961.
43. "Motor Vehicle Speeds," in *HRBBib 27*, 1960.
44. "Variables Influencing Spot-Speed Characteristics—Review of Literature," in *HRBSR 89*, 1966.
45. Taragin and Hopkins, "A Traffic Analyzer: Its Development and Application," in *PR*, December, 1960; also in *Proc. ITE*, 1960.
46. Covault, "Time-Lapse Movie Photography Used to Study Traffic Flow Characteristics," in *TE*, March, 1960.
47. Wohl, "Vehicle Speeds and Volumes Using Sonne Stereo Continuous Strip Photography," in *TE*, January, 1959.
48. Sickle, "Continuous Strip Photography—An Approach to Traffic Studies," in *TE*, July, 1959.
49. "Special Purpose Traffic Survey Devices," in *TE*, February, 1966.
50. Stern, "Traffic Flow Data Acquisition Using Magnetic-Loop Vehicle Detectors," in *HRR 154*, 1967.
51. Christensen, "Use of a Computer and Vehicle Loop Detectors to Measure Queues and Delays at Signalized Intersections," in *HRR 211*, 1967.
52. Sagi, "Multi-Parameter Vehicle Flow Detection" in *TE*, January, 1968.
53. Treiterer and Taylor, "Traffic Flow Investigations by Photogrammetric Techniques," in *HRR 142*, 1966.
54. Herd, "An Aerial Photographic Technique for Presenting Displacement Data," in *HRR 232*, 1968.
55. Rib, "Remote Sensing Applications to Highway Engineering," in *PR*, June, 1968.

56. *Highway Statistics, Summary to 1965*, USGPO, March, 1967.
57. *1967 Annual Speed Survey*, PR7-70200-01, NYSDOT, December, 1967.
58. Vey and Ferreri, "The Effect of Lane Width on Traffic Operation," in *TE*, August, 1968.
59. Treadway and Oppenlander, "Statistical Modeling of Travel Speeds and Delays on a High-Volume Highway," in *HRR 199*, 1967.
60. Salvatore, "Estimation of Vehicular Speed Under Time Limitation and Restricted Conditions of Observations" in *HRR 195*, 1967.
61. Farber and Silver, "Knowledge of On-Coming Car Speed as Determiner of Driver's Passing Behavior," in *HRR 195*, 1967.
62. *Manual of Traffic Engineering Studies*, 3rd ed., ITE, 1964.
63. May, "Traffic Characteristics and Phenomena on High Density Controlled Access Facilities," in *TE*, March, 1961.
64. Ryan and Breuning, "Some Fundamental Relationships of Traffic Flow on a Freeway," in *HRB 324*, 1962.
65. Webb and Moskowitz, "California Freeway Capacity Study—1956," in *Proc. HRB*, 1957.
66. Keese *et al*, "A Study of Freeway Traffic Operation," in *HRBB, 235*, 1960.
67. Malo *et al*, "Traffic Behavior on an Urban Expressway," in *HRBB 235*, 1960.
68. Underwood, "Speed, Volume, and Density Relationships," in *Symposium on Quality, Theory of Traffic Flow*, BHT, Yale, 1961.
69. *Highway Capacity Manual*, HRBSR 87, 1965.
70. Wagner and May, "Volume and Speed Characteristics at Seven Study Locations," in *HRBB 281*, 1961.

Chapter 9: Volume Studies

1. *Highway Capacity Manual*, HRBSR 87, 1965.
2. Greenshields, "The Density Factor in Traffic Flow," in *TE*, March, 1960.
3. *Manual of Traffic Engineering Studies*, 3rd ed., The Accident Prevention Department, Association of Casualty and Surety Companies, 1953 (out of print).
4. *Manual of Traffic Engineering Studies*, 3rd ed., ITE, 1964.
5. "Volume Survey Devices," Report of Technical Committee 7-G of ITE, in *TE*, March, 1961.
6. Huppert, "Familiarity With Loop Detectors Will Aid Application, Installation," in *TE*, August, 1965.
7. O'Connell, "Radio Controlled Traffic Counting," in *TE*, July, 1957.
8. Andrews, "Use of Non-Recording Traffic Counters," in *TE*, June, 1965.
9. Auer, "A System for the Collection and Processing of Traffic Flow Data by Machine Methods," in *HRBB 324*, 1962.
10. Dickins, "Sky Count—New System Developed for Traffic Data Acquisition," in *TE*, December, 1964.
11. Jordan, "Development of the Sky Count Technique for Highway Traffic Analysis," in *HRR 19*, 1963.
12. Jordan, "The Sky Count Program for Highway Systems Management," in *Proceedings Conference on Traffic Surveillance, Simulation, and Control*, BPR, Wash,. D.C., September, 1964.
13. Jordan, "Sky Count of Traffic Congestion and Demand, in *TEC*, September, 1965.
14. *Pedestrian Traffic Study—World Trade Center*, Report of Operations Standards Division, PONYA, April, 1965.
15. "Traffic Counting, Classification, and Weighing," in *BPR*, 1957.
16. *Measuring Traffic Volumes*, PM 3A, NCUT.
17. *Determining Street Use*, PM 1A, NCUT, 1958.
18. *Measuring Transit Service*, PM 4A, NCUT, 1958.
19. *Highway Capacity Manual*, BPR, 1950.
20. May, "Traffic Characteristics and Phenomena on High Density Controlled Access Facilities," in *TE*, March, 1961.
21. Ryan and Breuning, "Some Fundamental Relationships of Traffic Flow on a Freeway," in *HRBB 324*, 1962.
22. Webb and Moskowitz, "California Freeway Capacity Study—1956," in *Proc. HRB*, 1957.
23. Keese *et al*, "A Study of Freeway Traffic Operation," in *HRBB 235*, 1960.
24. Malo *et al*, "Traffic Behavior on an Urban Expressway," in *HRBB 235*, 1960.
25. Underwood, "Speed, Volume, and Density Relationships," in *Symposium on Quality and Theory of Traffic Flow*, BHT Yale, 1961.
26. Wagner and May, "Volume and Speed Characteristics at Seven Study Locations," in *HRBB 281*, 1961.
27. Pearson *et al*, "Operational Study—Schuylkill Expressway," in *HRBB 291*, 1961.
28. Moskowitz and Newman, "Notes on Freeway Capacity," in *HRR 27*, 1963.
29. Hess, "Capacities and Characteristics of Ramp-Freeway Connections," in *HRR 27*, 1963.
30. May *et al*, "Development and Evaluation of Congress Street Expressway in Pilot Detection System," in *HRR 21*, 1963.
31. Normann, "Variations in Flow at Intersections as Related to Size of City, Type of Facility, and Capacity Utilization," in *HRBB 352*, 1962.
32. Drew and Keese, "Freeway Level of Service as Influenced by Volume and Capacity Characteristics," in *HRR 99*, 1965.
33. Howie and Young, "The Traffic Counting Program in Cincinnati," in *Proc. HRB*, 1957.
34. Darrell *et al*, "Minnesota Experience in Counting Traffic on Low-Volume Roads," in *Proc. HRB*, 1958.
35. Dimmick, "Traffic and Travel Trends, 1955," in *PR*, December, 1950.
36. Dickerson, "Estimated Travel by Motor Vehicles in 1960," in *PR*, April, 1962.
37. French and Dickerson, "Travel by Motor Vehicles in 1963, 1964, and 1965," in *PR*, February, 1967.
38. Muranyi, "Estimating Traffic Volumes by Systematic Sampling," in *HRBB 281*, 1961.
39. Walker, "Trends in the 30th–Hour Factor," in *HRBB 167*, 1957.
40. Bellis and Hones, "30th Peak Hour Factor Trend," in *HRR 27*, 1963.
41. Heiol, "A Method for Estimating Design Hourly Traffic Volume," in *HRR 72*, 1965.
42. Pelz and White, "*Short Count Results*," in *TE*, August, 1948.
43. Adams, "Five Minute Cluster Sampling for Determining Urban Traffic Volumes," in *Proc. HRB*, 1955.
44. Gilbert, "Evaluation of the Six-Minute Sample Count Procedure," in *TE*, November, 1962.
45. Wilson, "Maximum Info At Minimum Cost Provided by Unique Traffic Count," in *TE*, July, 1965.
46. *Traffic Engineering Handbook*, 2nd ed., ITE, 1950; 3rd ed., 1965.

47. Petroff and Blensly, "Urban Traffic Volume Patterns in Tennessee," in *Proc. HRB*, 1958.
48. Petroff, "Experience in Application of Statistical Method to Traffic Counting," in *PR*, December, 1956.
49. Petroff and Kancler, "Urban Traffic Volume Patterns in Tennessee," in *Proc. HRB*, 1958.
50. *Guide for Traffic Volume Counting Manual*, 2nd ed., BPR, February, 1965.
51. Bodle, "Evaluation of Rural Coverage Count Duration for Estimating Annual Average Daily Traffic," in *HPTR No. 5*, BPR, October, 1966.
52. Morris, "Standard Deviation and Coefficient of Variation of Automatic Traffic Recorder Counts," in *Proc. HRB*, 1950.
53. Breuning and Wagner, "Analysis of 24-Hour Counts by Digital Computer," in *TE*, April, 1961.
54. Head and Blensly, "Application of Mechanical Analysis to Traffic Recorder Data," in *HRBB 303*, 1961.
55. Haigh, "Statistical Digital Computer Methods for Traffic Count Analysis," in *HRN 13*, June, 1964.
56. Phillips and Woolman, "Systems Analysis Approach to Processing of Vehicular Traffic Records from Continuous-Count Stations," in *HRN 13*, June, 1964.
57. Clark and Cribbins, "Traffic Volume Measurement Using Drivometer Events," in *HRR 230*, 1968.
58. *1966 Traffic Volumes on State Routes*, Subdivision of Transportation Planning and Programming, N.Y.S. Department of Transportation, September, 1967.
59. Melton, *30th Peak Hour Factor Trend—1964*, BST, NJDT, April, 1966.
60. Reilly and Radics, *30th Peak Hour Factor Trend*, BST, NJDT, August, 1966; also see *HRR 199*, 1967.
61. *Comparison of Monthly and Annual Average Daily Traffic at Continuous Traffic Counting Stations, 1964–1967*, Planning Division, N.Y.S. DOT, April, 1968.
62. Parrish *et al*, "Georgia's Program for Automated Acquisition and Analysis of Traffic-Count Data" in *HRR 199*, 1967.
63. Falcocchio, *Highway Traffic Patterns in the Delaware Valley*, DVRPC Draft Report, October, 1968.
64. Maltayan and Pallat, "A Multistation, Centralized Digital Traffic-Counting System," in *HRR 10*, 1963.
65. Crawford *et al*, "Automated Traffic Counters," in *HRR 10*, 1963.
66. "Detector Locations," An ITE Informational Report, in *TE*, February, 1969.
67. *A Policy on Geometric Design of Rural Highways*, AASHO, 1965.
68. *Automobile Facts and Figures*, AMA, 1969.

Chapter 10: Traffic Theory: Flow and Control

1. Gazis and Edie, "Traffic Flow Theory," in *Proc. IEEE*, April, 1968.
2. Drew, *Traffic Flow Theory and Control*, McGraw-Hill, New York, 1968.
3. Ashton, *Theory of Road Traffic Flow*, Methuen, London 1967.
4. *Analytic Models of Unidirectional Multi-Lane Traffic Flow: A Survey of the Literature*, SDC Technical Memo, January, 1969.
5. Herman, *Theory of Traffic Flow*, Elsevier, Amsterdam 1961.
6. Herman *et al*, "Traffic Dynamics: Studies in Car Following," in *OR*, Vol. 7, 1959.
7. McShane and Yagoda, "Effects of Vertical Geometry on Vehicular Stream Dynamics," in *High-Speed Ground Transportation Journal*, May, 1967.
8. "Improved Criteria for Traffic Signals at Individual Intersections," in *NCHRPR 32*, 1967.
9. Greenshields, "A Study in Highway Capacity," in *Proc. HRB*, 1934.
10. Morrisons, "The Traffic Analogy to Compressible Fluid Flow," in *Advanced Research Engineering Bulletin*, 1964.
11. Harr and Leonhards, "A Theory of Traffic Flow for Evaluation of the Geometric Aspects of Highways," in *HRBB 308*, 1961.
12. Lighthill and Whitham, "On Kinematic Waves: II. A Theory of Traffic Flow on Long Crowded Roads," in *HRB 79*, 1964.
13. Prigogine, "A Boltzmann-like Approach to the Statistical Theory of Traffic Flow," in *Theory of Traffic Flow*, ed. R. Herman, Elsevier, Amsterdam, 1971.
14. Rashba, *Freeway Management and Control*, Dissertation, Polytechnic Institute of Brooklyn, Brooklyn, N. Y. 1970.
15. Beckmann *et al*, *Studies in the Economics of Transportation*, Yale, New Haven, Conn., 1956.
16. Newell, "Queues for a Fixed Cycle Traffic Light," in *Annals of Mathematical Statistics*, 1960.
17. Castoldi, "Queue Alternance and Traffic Flow at a Crossroad," in *Bollettino del Centro per la Ricerca Operativa*, 1957.
18. Haight, "Overflow at a Traffic Light," in *Biomertrika*, 1959.
19. Gardwood, "The Application of the Theory of Probability to the Operation of Vehicular Controlled Traffic Signals," in *Journal of the Royal Statistical Social Supplement*, 1940.
20. "An Introduction to Traffic Flow Theory," in *HRB SR 79*, 1964.
21. *Traffic Engineering Handbook*, ITE, 1965.
22. *Highway Capacity Manual*, HRB SR 87, 1965.
23. *Weaving Area Operations Study*, Final Report, PIB, 1970.
24. Yagoda and Pignataro, "The Analysis and Design of Freeway Entrance Ramp Control Systems," in *HRR 303*, 1970.
25. Worrall and Bullen, *Lane-Changing or Multilane Highways*, Final Report, Northwestern University, 1969.
26. Jones and Potts, "The Measurement of Acceleration Noise—A Traffic Parameter" in *OR*, Vol. 10, No. 6, 1958.
27. *Gap Acceptance and Traffic Interaction in the Freeway Merging Process*, Final Report, TTI, BPR.
28. Weiner, *Some Theoretical Considerations in the Design and Operation of Gap Acceptance Ramp Control Systems for Freeways*, Dissertation, Polytechnic Institute of Brooklyn, Brooklyn; N. Y. 1968.
29. Hodgins, "Effects of Volume Controls on Freeway Traffic Flow—A Theoretical Analysis," in *PR*, February, 1968.
30. Rosen *et al*, *An Electronic Route Guidance System for Highway Vehicles*, FHWA, October, 1969.
31. Moore, paper presented at the International Symposium on the Theory of Switching, 1957.
32. Almond, "Traffic Assignment with Flow Dependent Journey Times," in *Vehicular Traffic Science Proceedings*, 3rd International Symposium on the Theory of Traffic Flow, 1967.
33. "Optimum Control of a System of Oversaturated Intersections," in *OR*, Vol. 12, 1964.
34. *System Analysis Methodology in Urban Traffic Control Systems*, Final Report, June, 1969.
35. *Advanced Control Technology in Urban Traffic Control System: Volume 1, System Description*, UTCS, Sperry, October, 1969.

36. HELLY and BAKER, "Acceleration Noise in a Congestion Signalized Environment," in *Vehicular Traffic Science Proceedings*, 3rd International Symposium on the Theory of Traffic Flow, 1967.

Chapter 11: Highway Capacity: Introduction and Background

1. *Highway Capacity Manual*, HRBSR 87,1965.
2. *Highway Capacity Manual*, BPR,1950.
3. NEWMAN, *Introduction to Capacity*, California Division of Highways, unpublished.
4. GERLOUGH and SCHUHL, *Poisson and Traffic*, Eno. 1955.
5. GREENSHIELDS and WEIDA, *Statistics with Applications to Highway Traffic Analyses*, Eno, 1952.
6. RAFF, *A Volume Warrant for Stop Signs*, Eno, 1950.
7. PEARSON *et al*, "Operational Study—Schuylkill Expressway," in *HRBB 291*, 1961.
8. GREENSHIELDS, *Traffic Performance at Urban Street Intersections*, Yale BHT, TR No. 1, 1947.
9. MOSKOWITZ, "Waiting for a Gap in a Traffic Stream," in *Proc. HRB*, 1954.
10. GERLOUGH, "Traffic Inputs for Simulation on a Digital Computer," *Proc. HRB*, 1959.
11. GLICKSTEIN *et al*, "Application of Computer Simulation Techniques to Interchange Design Problems," in *HRBB 291*, 1961.
12. PERCHONOK and LEVY, "Application of Digital Simulation Techniques to Freeway On-Ramp Traffic Operations," in *Proc. HRB*, 1960.
13. HAIGHT *et al*, "New Statistical Method for Describing Highway Distribution of Cars," in *Proc. HRB*, 1961.
14. KELL, "Analyzing Vehicular Delay at Intersections Through Simulation," in *HRBB 356*, 1962.
15. OLIVER and THIBAULT, "A High-Flow Traffic-Counting Distribution" in *HRBB 356*, 1962.
16. GREENSHIELDS, "The History of an Idea on the Spacing of Vehicles on the Highway," in *TE*, July, 1961.
17. HAIGHT, *Mathematical Theories of Road Traffic*, ITTE, March, 1960.
18. MAY, "Gap Availability Studies," in *HRR 72*, 1965.
19. ATHOL, "Headway Groupings," in *HRR 72*, 1965.
20. LEWIS, "A Proposed Headway Distribution for Traffic Simulation Studies," in *TE*, February, 1963.
21. DREW *et al*, "Gap Acceptance in the Freeway Merging Process," in *HRR 208*, 1967.
22. GRECCO and SWORD, "Prediction of Parameters for Schuhl's Headway Distribution," in *TE*, February, 1968.
23. BUHR, "Gap Stability and Its Application to Freeway Merging Control Systems" in *TE*, March, 1968.
24. FORBES, "Speed, Headway, and Volume Relationship on a Freeway," in *Proc. ITE*, 1951.
25. MAY and WAGNER, "Headway Characteristics and Interrelationships of Fundamental Characteristics of Traffic Flow," in *Proc. HRB*, 1960.
26. CAMPBELL *et al*, "A Method for Predicting Speeds Through Signalized Street Sections," in *HRBB 230*, 1959.
27. ROTHROCK and KEEFER, "Measurement of Urban Traffic Congestion," in *HRBB 156*, 1957.
28. UNDERWOOD, "Speed, Volume, and Density Relationships," in *Symposium on Quality and Theory of Traffic Flow*, Yale BHT, 1961.
29. PALMER, "The Development of Traffic Congestion," in *Symposium on Quality and Theory of Traffic Flow*, Yale BHT, 1961.
30. EDIE *et al*, "Analysis of Single-Lane Traffic Flow," in *TE*, January, 1963.
31. HAYNES, "Some Considerations of Vehicular Density on Urban Freeways," in *HRR 99*, 1965.
32. GARCIA and JONES, "High Density Vehicular Flow-Density Relationships" in *ASCE*, Vol. 91, No. HW2, December, 1965.
33. BENROFF and MAGHADELAS, "Stopped Vehicle Spacing on Freeways," in *TE*, Febuary, 1970.
34. KEESE *et al*, "Highway Capacity—The Level of Service Concept" in *Proc. ITE*, 1965.
35. DREW and KEESE, "Freeway Level of Service as Influenced by Volume and Capacity Characteristics," in *HRR 99*, 1965.
36. DREW *et al*, "Freeway Level of Service as Described by an Engergy-Acceleration Noise Model," in *HRR 162*, 1967.
37. DRAKE *et al*, "A Statistical Analysis of Speed Density Hypothesis," in *HRR 154*, 1967.
38. HODGKINS, "Effect of Buses on Freeway Capacity," in *HRR 59*, 1964.

Chapter 12: Highway Capacity: Freeways and Expressways

1. *Highway Capacity Manual*, HRBSR 87, 1965.
2. NORMANN, "Operation of Weaving Areas," in *HRBB 167*, 1957.
3. HESS, "Capacities and Characteristics of Ramp-Freeway Connections," in *HRR 27*, 1963.
4. HESS, "Ramp-Freeway Terminal Operation as Related to Freeway Lane Distribution and Adjacent Ramp Influence," in *HRR 99*, 1965.
5. MOSKOWITZ and HEWMAN, "Notes on Freeway Capacity," in *HRR 27*, 1963.
6. PEARSON *et al*, "Operational Study—Schuylkill-Expressway," in *HRBB 291*, 1961.
7. WEBB and MOSKOWITZ, "California Freeway Capacity Study—1956," in *Proc. HRB*, 1957.
8. PINNELL and KEESE, "Traffic Behavior and Freeway Ramp Design," in *HDJ*, *Proc. ASCE*, Vol. 86, No. NW3, September, 1960.
9. LESSIEU, "Operational Characteristics of High Volume On-Ramps," in *TE*, December, 1957.
10. MOSKOWITZ, "Research on Operating Characteristics of Freeways," in *Proc. ITE*, 1956.
11. GLICKSTEIN *et al*, "Application of Computer Simulation Techniques to Interchange Design Problems," in *HRBB 291*, 1961.
12. NEWMAN, "Traffic Operation at Two Interchanges in California," in *HRR 27*, 1963.
13. BERRY *et al*, "A Study of Left-Hand Exit Ramps on Freeways," in *HRR 21*, 1963.
14. WORRALL *et al.*, "Study of Operational Characteristics of Left-Hand Entrance and Exit Ramps on Urban Freeways," in *HRR 99*, 1965.
15. LUNDY, "The Effect of Ramp Type and Geometry on Accidents," in *HRR 163*, 1967.
16. BURH *et al*, "A Nationwide Study of Freeway Merging Operations," in *HRR 202*, 1967.
17. JOHNSON and NEWMAN, "East Los Angeles Interchange Operation Study," in *HRR 244*, 1968.
18. DREW *et al*, "Determination of Merging Capacity and Its Applications to Freeway Design and Controll," in *HRR 244*, 1968.

19. OOKERT and WALKER, "Criteria to be Used in Developing Warrants for Interchanges on Rural Expressways," in *HRR 244*, 1968.
20. DREW, "Taking the Scourge out of the Merge," in *TE*, August, 1967.
21. PFEFER, "Two-Lane Entrance Ramps" in *TE*, November, 1968.
22. WOODIE *et al*, "A Computer Program for Weaving Capacity," in *TE*, 1969.
23. *A Policy on Geometric Design of Rural Highways*, AASHO, 1965.
24. PINNELL and TULT, "Evaluation of Frontage Roads as an Urban Freeway Design Element," in *HRR 9*, 1963.
25. *A Policy on Arterial Highways in Urban Areas*, AASHO, 1957.
26. CONKLIN, "A Comparison of Vehicle Operating Characteristics Between Parallel Lane and Direct Taper Types of Freeway Off-Ramps," in *Proc. ITE*, 1959; also in *TE*, December, 1959.
27. *Geometric Design Standards for the National System of Interstate and Defense Highways*, AASHO, 1956.
28. LEISCH, "Adaptability of Interchanges on Freeways in Urban Areas," in *Proc. ASCE*, Vol. 84, No. HW 1, January, 1958.
29. LEISCH, "Spacing of Interchanges on Freeways in Urban Areas," in *Proc. ASCE*, Vol. 85, No. HW 4, December, 1959.
30. PETERSON, "Freeway Spacing in an Urban Freeway System," in *Proc. ASCE*, Vol. 85, No. HW 4, December, 1959.
31. PINNELL and KEESE, "Traffic Behavior and Freeway Ramp Design," in *Proc. ASCE*, Vol. 86, No. HW 3, September, 1960.
32. LOUTZENHEISER, "New Concepts for Urban Freeway Interchanges," in *Proc. ASCE*, Vol. 88, No. HW 1, May, 1962.
33. JONES, *The Geometric Design of Modern Highways*, Wiley, New York, 1961.
34. JOUZY and MICHAEL, "Use and Design of Acceleration and Deceleration Lanes in Indiana," in *HRR 9*, 1963.
35. COVAULT and ROBERTS, "Influence of On-Ramp Spacing on Traffic Flow on Atlanta Freeway and Arterial Street System," in *HRR 21*, 1963.
36. COVAULT and KIRK, "Influence of Off-Ramp Spacing on Traffic Flow Characteristics on Atlanta Freeway and Arterial Street System," in *HRR 59*, 1964.
37. FUKUTOME and MOSKOWITZ, "Traffic Behavior and Off-Ramp Design," in *HRR 21*, 1963.
38. GERN and JOYNER, "Crossroute Access Design in Interchange Areas," in *HRR 59*, 1964.
39. MCDERMOTT and MCLEAN, "Improving Traffic Flow at Transfer Roadways on Collector-Distributor Type Expressways," in *HRR 59*, 1964.
40. ALEXANDER and MANHEIM, "The Design of Highway Interchanges, An Example of a General Method for Analyzing Engineering Design Problems," in *HRR 83*, 1965.
41. TIPTON and PINNELL, "An Investigation of Factors Affecting the Design Location of Freeway Ramps," in *HRR 152*, 1967.
42. WATTLEWORTH *et al*, "Operational Effects of Some Entrance Ramp Geometrics on Freeway Merging," in *HRR 208*, 1967.
43. BERRY and MCCABE, "Development and Use of Models in the Design of Highways," in *HRR 172*, 1967.
44. SATTERLY and BERRY, "Spacing of Interchanges and Grade Separations on Urban Freeways," in *HRR 172*, 1967.
45. AHLBORN *et al*, "A Computer Program for Ramp Capacity," in *TE*, December, 1968.
46. NEWMAN, *Introduction to Capacity*, California Division of Highways, unpublished.
47. KERMODE and MYYRA, "Freeway Lane Closures," in *TE*, February, 1970.
48. DREW *et al*, "Multilevel Approach to the Design of a Freeway Control System," in *HRR 279*, 1969.
49. HAACK *et al*, "Traffic Control on a Two-Lane, High-Speed, High-Volume Freeway Entrance," in *HRR 279*, 1969.
50. BREWER *et al*, "Ramp Capacity and Service Volume as related to Freeway Control," in *HRR 279*, 1969.
51. BUHR, *et al*, "Traffic Characteristics for Implementation and Calibration of Freeway Merging Control Systems," in *HRR 279*, 1969.
52. BUHR *et al*, "A Moving Vehicle Merging Control System," in *HRR 279*, 1969.
53. BUHR *et al*, "Design of Freeway Entrance Ramp Merging Control Systems," in *HRR 279*, 1969.
54. BHUR *et al*, "Some Design Considerations of Digital Computer Systems," in *HRR 279*, 1969.
55. RUSSELL, "Ramp Control of Freeways in California," in *HRR 279*, 1969.
56. WATTLEWORTH *et al*, "Development and Evaluation of a Ramp Metering System on the Lodge Freeway," in *HRR 244*, 1968.
57. NEWMAN *et al*, "Freeway Ramp Control—What It Can And Cannot Do," in *TE*, June, 1969.
58. THOMAS, "Simple Ramp Metering Device Reduces Rear-End Collisions," in *TE*, June, 1969.
59. COURAGE, "Evaluation and Improvement of Operations in a Freeway Corridor," in *TE*, March, 1970.
60. *Weaving Area Operations Study, Final Report*, Dept. of Transportation Planning and Engineering, Polytechnic Institute of Brooklyn, NCHRP Project 3-15, unpublished, 1971.
61. GWYNN, "Truck Equivalency," in *TQ*, April, 1968.
62. COLLINS and MAY, "A Computer Program for Freeway and Highway Capacity," in *TE*, April, 1968.

Chapter 13: Highway Capacity: Urban Streets and Arterials

1. *Highway Capacity Manual*, HRBSR 87, 1965.
2. MILLER, "Capacity of Signalized Intersections in Australia," in *ARRBB 3*, March, 1968.
3. *Intersection Capacity Study*, Study 68-2, Missouri State Highway Department, Division of Planning, unpublished, August, 1969.
4. MAY and GYAMFI, "Extension and Preliminary Validation of a Simulation of Load Factor at Signalized Intersections," in *TE*, October, 1969.
5. SAGI and CAMPBELL, "Vehicle Delay at Signalized Intersections —Theory and Practice," in *TE*, February, 1969.
6. MAY and PRATT, "A Simulation Study of Load Factor at Signalized Intersections," in *TE*, February, 1968.
7. LEISCH, "Capacity Analysis Techniques for Design of Signalized Intersections," Installment No. 1, in *PR*, August, 1967.
8. LEISCH, "Capacity Analysis Techniques for Design of Signalized Intersections," Installment No. 2, in *PR*, October, 1967.
9. LEISCH, "Design Capacity Charts for Signalized Street and Highway Intersections," in *PR*, February, 1951.
10. MAY *et al*, "A Computer Program for Intersection Capacity," in *TE*, January, 1968.
11. MAY and HOM, "Intersection Capacity of Exclusive Turning Lanes," in *TE*, March, 1968.

12. NORMANN, "Variations in Flow at Intersections as Related to Size of City, Type of Facility, and Capacity Utilization," in *HRBB 352*, 1962.
13. BELLIS, "Capacity of Traffic Signals and Traffic Signal Timing," in *HRBB 271*, 1960.
14. LEISCH, "Design Geometries for Diamond Interchanges," in *Proc. ITE*, 1960.
15. LEROY and CLINTON, "Signalization of Diamond Interchanges," in *Proc. ITE*, 1960.
16. CAPALLE and PINNELL, "Capacity Study of Signalized Diamond Interchanges," in *HRBB 291*, 1961.
17. CAPELLE and PINNELL, "Operational Study of Signalized Diamond Interchanges," in *HRBB 324*, 1962.
18. PINNEL and TULT, "Evaluation of Frontage Roads as an Urban Freeway Design Element," in *HRR 9*, 1963.
19. PINNEL, "The Value of Signal Phase Overlap in Signalized Intersection Capacity," in *TE*, December, 1962.
20. DREW, "Design and Signalization of High-Type Facilities," in *TE*, July, 1963.
21. DREW and PINNELL, "A Study of Peaking Characteristics of Signalized Urban Intersections as Related to Capacity and Design," in *HRBB 352*, 1962.
22. RYAN, "Intersection Capacity," in *HRBB 352*, 1962.
23. REILLY and SIEFERT, "Capacity of Signalized Intersections," in *HRR 321*, 1970.
24. *Australian Road Capacity Guide*, ARRBB, No. 4, June, 1968.
25. *Final Report on Intersection Traffic Flow*, C.E.I.R. Consultants, unpublished, December, 1960.
26. CARTER, "Increasing the Traffic-Carrying Capability of Urban Arterial Streets," in *BPR*, May, 1962.
27. BERMAN, "Appendices to the Original Wisconsin's Ave. Study," in *BPR*, May, 1962.
28. BERMAN and CARTER, "Increasing the Traffic-Carrying Capability of Urban Arterial Streets," in *HRBB 271*, 1960.
29. HODGKINS, "Effect of Traffic Improvements on Operation of an Urban Arterial Street," in *HRBB 308*, 1962.
30. FRENCH, "Capacities of One-Way and Two-Way Streets with Signals and with Stop Signs," in *PR*, February, 1956.
31. STANHAGEN and MULLINS, "Application of Police Power and Planning Controls to Arterial Streets," in *HRBB 271*, 1960.
32. TIDWELL, JR., and HUMPHREYS, "Relation of Signalized Intersection Level of Service to Failure Rate and Average Individual Delay," in *HRR 321*, 1970.

Chapter 14: Highway Capacity: Rural Highways Without Access Control

1. *Overtaking and Passing on Two-Lane Rural Highways*, literature review, USGPO, June, 1967.
2. *Highway Capacity Manual*, HRB SR 87, 1965.
3. "Statements of Needed Research in Capacity and Freeway Operations," in *HRC 81*, July, 1968.
4. MAY, "Initial Experiences With the New Highway Capacity Manual," in *TE*, December, 1967.

Chapter 15: Pedestrian Studies

1. *Traffic Engineering Handbook*, ITE, 1965.
2. "A Program for School Crossing Protection-A Recommended Practice of the Institute of Traffic Engineers," in *TE*, October, 1962.
3. FRUIN, *Designing for Pedestrians—A Level of Service Concept*, Dissertation, Polytechnic Institute of Brooklyn, Brooklyn, N. Y., January, 1970.
4. *Planned Pedestrian Program*, American Automobile Association, 1968.
5. HOEL, "Pedestrian Travel Rates in Central Business Districts," in *TE*, January, 1968.
6. NAVIN and WHEELER, "Pedestrian Flow Characteristics," in *TE*, June 1969.
7. *Highway Capacity Manual*, HRBSR 87, 1965.
8. OLDER, "Movement of Pedestrians on Footways in Shopping Streets," in *TEC* 10(4), 1968.
9. *A Policy on Geometric Design of Rural Highways*, AASHO, 1965.
10. "Geometric Design of Loading Platforms and Bus Runways for Local and Suburban Terminals," Reference Notes, in *TE*, January, 1958.
11. *Traffic in Towns*, the Buchanan Report, Penguin Books, Inc., Baltimore, Md., 1963.

Chapter 16: Parking Studies

1. BURRAGE and MOGREN, *Parking*, Eno, 1957.
2. *Conducting a Comprehensive Parking Study*, PM 3D, NCUT, PAS.
3. *Conducting a Limited Parking Study*, 3C, NCUT.
4. *Parking in the City Center*, Wilbur Smith and Associates, AMA, 1965.
5. "Urban Travel Patterns for Airports, Shopping Centers, and Industrial Plants," NCHRPR 24, 1966.
6. "Parking Requirements for Shopping Centers," in *Tech Bull* No. *53*, Urban Land Institute, 1965.
7. "Parking Facilities for Industrial Plants," in *Informational Report of the ITE*, September, 1969.
8. "Survey of Ground Access Problems of Airports," in *TEJ*, *Proc.* ASCE, February, 1969.

Chapter 17: Accident Study

1. *The State of the Art of Traffic Safety*, Arthur D. Little, Inc., June, 1966.
2. *National Highway Safety Standards*, USDOT, 1967.
3. *Highway Safety Program Management*, ASF, 1968.
4. *International Road Safety Research Directory*, 2nd ed., Organization for Economic Co-operation and Development, Paris, 1966.
5. *Hearings on Traffic Safety* HR 13228, 1966.
6. *Hearings on Roadside Hazards*, 1967.
7. *Hearings on Traffic Safety*, 53005, 1966.
8. *First Annual Report on the Administration of the Highway Safety Act of 1966 for pd Sept. 66–Dec. 67*, House Document 311, USGPO, 1968.
9. *Research into Road Safety*, Organization for European Economic Cooperation; Report on 1st International Meeting on Research into Road Safety, 1960.
10. *Highway Safety Manpower & Training* (*Training and Manpower Needs*), Traffic Education and Training Committee—Traffic Conference, 1968.
11. *Traffic Safety—Strategies for Research and Action*, Conference Report, The Travelers Research Centre, Inc., 1967.

12. *Guidelines for Accident Reduction Through Programming of Highway Safety Improvements*, BPR.
13. *Highway Design and Operational Practices Related to Highway Safety*, AASHO, 1967.
14. Blatnik and Prisk, *Roadside Hazards*, Eno, 1968.
15. *Accident Facts—1969 Edition*, published annually by NSC.
16. *Accident Facts—1969* NYS DMV, Albany, N.Y., 1969.
17. Solomon, "Accidents on Main Rural Highways Related to Speed, Driver and Vehicle," in *BPR, USSPC*, July, 1964.
18. Cantilli, "Daylight 'Lights-On' Program," in *TE*, December, 1965.
19. Cantilli, "Daylight 'Running Lights' Reduce Accidents," in *TE*, February, 1969.
20. *Manual of Traffic Engineering Studies*, Association of Casualty and Surety Companies 1953; New edition, 1960, Institute of Traffic Engineers.
21. *Uses of Traffic Accident Records*, Eno, 1947.
22. Halsey, *State Traffic Safety*, Eno, 1953.
23. *The Federal Role in Highway Safety*, HD 93, 86th Cong., USGPO, 1959.
24. *Manual on Classification of Motor Vehicle Traffic Accident Statistics*, Committee on Uniform Traffic Accident Statistics, NSC, 1962.
25. Cantilli, "Computer Applications to the Urban Traffic Accident Problems," *5th Annual Urban Symposium on the Application of Computers to the Problems of Urban Society*, 1970.
26. *Automobile Facts and Figures—1968*, AMA, 1968.
27. Cantilli, "Statistical Evaluation of Traffic Accident Severity," in *HRBB 208*, 1958.
28. *Traffic Control and Roadway Elements*, ASF, 1963.
29. *Accident Rates as Related to Design Elements of Rural Highway*, NCHRPR 47, 1968.
30. Byington, "Interstate System Accident Research," in *PR*, December, 1963.
31. Prisk, "Life Saving Benefits of the Interstate System" in *PR*, December, 1961.
32. "How Many Lives Will Controlled-Access Highways Save?", NSC, in *TE*, December, 1959.
33. Moskowitz, "Accidents on Freeways in California," in *Theme IV*, World Traffic Engineering Conference, 1961.
34. Raff, "Interstate Highway-Accident Study," in *HRBB 74*, 1953.
35. Belmont, "Effect of Average Speed and Volume on Motor Vehicle Accidents on Two-Lane Tangents," in *Proc. HRB*, 1953.
36. Cope, "Traffic Accident Experience—Before and After Pavement Widening," in *TE*, December, 1955.
37. Billion and Stohner, "A Detailed Study of Accidents as Related to Highway Shoulders in New York State," in *Proc. HRB*, 1956.
38. Stohner, "Relation of Highway Accidents to Shoulder Width on Two-Lane Rural Highways in New York State," in *Proc. HRB*, 1956.
39. Musick, "Effect of Pavement Edge Marking on Two-Lane Rural State Highways in Ohio," in *HRBB 266*, 1960.
40. Basile, "Effect of Pavement Edge Marking on Traffic Accidents in Kansas," in *HRB 308*, 1962.
41. Thomas, "Pavement Edge Lines on Twenty-Four Foot Surfaces in Louisiana," in *HRBB 178*, 1958.
42. Willey, "Arizona's Dashed Shoulder Stripe," in *TQ*, April, 1955.
43. Williston, "Effect of Pavement Edge Markings on Operator Behavior," in *HRB 266*, 1960.
44. *Traffic Accident Experience Before and After Pavement Edge Painting*, Springfield, Illinois, Division of Highways, 1959.
45. "Highways with a Narrow Median," in *HRBB 35*, 1951.
46. Hurd, "Accident Experience with Traversable Medians of Different Widths," *HRBB 137*, 1956.
47. Billion, "Effect of Median Dividers on Driver Behavior," in *HRBB 266*, 1960.
48. Moskowitz and Schaefer, "California Median Study: 1958," in *HRBB 266*, 1960.
49. Crosby, "Cross-Median Accident Experience on the N.J. Turnpike," in *HRBB 266*, 1960.
50. Billion *et al*, "Effect of Parkway Medians on Driver Behavior —Westchester County Parkways," in *HRBB 308*, 1962.
51. Billion and Parsons, "Median Accident Study—Long Island, New York," *HRBB 308*, 1962.
52. Telford and Israel, "Median Study (California)", in *Proc. HRB*, 1953.
53. Webb, "Median Study," and Benton and Field, "Impact Tests," in *TE*, December, 1959.
54. Stonex, "Roadside Design for Safety," in *Proc. HRB*, 1960.
55. Kohn, "Cross Median Accident Experience on New Jersey's Garden State Parkway," in *TE*, December, 1961.
56. Bowman, "Ohio Turnpike Accident Study," in *TE*, June, 1958.
57. Pinnell and Keese, "Traffic Behavior and Freeway Ramp Design," in *JHD, Proc. ASCE*, Vol. 86, No. HW3, September, 1960.
58. Brenning and Bowe, "Interchange Accident Exposure," in *HRBB 240*, 1960.
59. Fisher, "Accident and Operating Experience at Interchanges," in *HRBB 291*, 1961.
60. Pinnell," Driver Requirements in Freeway Entrance Ramp Design," in *TE*, December, 1960.
61. Sadal, "Accident Analysis of Freeway Interchanges," in *TE*, March, 1961.
62. Lind and Horq, "Traffic Accident Study on Milwaukee Expressway," in *JHD, Proc. ASCE*, Vol. 91, No. HW1, January, 1965.
63. Kipp, "Final Report on the Minnesota Roadside Study," in *HRBB 55*, 1952.
64. Staffeld, "Accidents Related to Access Points and Advertising Signs in Study," in *TQ*, January, 1963.
65. Wallen, "Landscaped Structures for Traffic Control," in *TE*, January, 1961.
66. Syrek, "Accident Rates at Intersections," in *TE*, May, 1955.
67. Solomon, "Traffic Signals and Accidents in Michigan," in *PR*, October, 1959.
68. Warmier, "The Effectiveness of Flashing Beacons," in *TE*, January, 1951.
69. Kell, "Applications of 'Yield-Right-of-Way' Signs," in *TE*, July, 1958.
70. Ray, "Six Months Use of 'Yield Right-of-Way' Signs," in *TE*, October, 1957.
71. Kell, "Yield Signs: Warrants and Applications," in *TE*, April, 1960.
72. Webb, "The Relation Between Accidents and Traffic Volumes at Signalized Intersections," in *Proc. ITE*, 1955.
73. Vey, "Effect of Signalization on Motor Vehicle Accident Experience," in *Proc. ITE*, 1933.

74. McMonagle, "Relation of Traffic Signals to Intersection Accidents," in *HRBB 74*, 1953.
75. Ckydem, "Michigan Study Indicates Signals Increase Accidents," in *TE*, November, 1964.
76. McEachron, "A Study of Railroad Grade Crossing Protection in Houston," in *Proc. ITE*, 1960.
77. Johnson, "Maximum Safe Vehicle Speeds at Railroad Grade Crossings," in *TE*, June, 1958.
78. Cantilli and Lee, "Upgrading Highways for Safety—Systematically," in *TE*, February, 1968.
79. Wyatt and Lozano, "Effect of Street Lighting on Night Traffic Accident Rate," in *HRBB 146*, 1956.
80. Ives, "Does Highway Illumination Affect Accident Occurence?", in *TQ*, April, 1962.
81. "Night Visibility—1961," in *HRBB 298*, 1961.
82. "Night Visibility—1962," in *HRBB 336*, 1962.
83. Blythe, "Highway Lighting and Accidents in Indiana," in *HRBB 146*, 1957.
84. *Personal Characteristics of Traffic-Accident Repeaters*, Eno, 1948.
85. "Driver Personality and Behavior Characteristics," in *HRBB 285*, 1961.
86. "Driver Characteristics", in *HRB 330*, 1962.
87. *A Selected Bibliography of Highway Traffic Safety with Annotations (1956–1960)*, Whitelaw, Michigan State University, 1961.
88. "Human Variables in Traffic Accidents," in *HRBBib 31*, 1962.
89. "Driver Characteristics and Accidents," in *HRBB 73*, 1953.
90. Brenner and Hulbert, "Psychology of Trip Geography," in *HRBB 91*, 1954.
91. "Traffic Accidents and Violations," in *HRBB 120*, 1956.
92. "Driver Characteristics," in *HRBB 152*, 1957.
93. "Investigating and Forecasting Traffic Accidents," in *HRBB 161*, 1957.
94. "Characteristics of Vehicle Operators," in *HRBB 212*, 1959.
95. Levonian *et al*, "Prediction of Recorded Accidents and Violations Using Non-Driving Predictors," in *HRR 4*, 1963.
96. "Driver Characteristics, Night Visibility, and Driving Simulation," in *HRR 25*, 1963.
97. "Driving Simulation: 1963," in *HRR 55*, 1964.
98. "Road-User Characteristics: 1963," in *HRR 84*, 1965.
99. Fox, "The Problem of Countermeasures in Drinking and Driving," in *TQ*, July, 1965.
100. Haddon and Klein, "Assessing the Efficiency of Accident Countermeasures," in *TQ*, July, 1965.
101. Blommer, "Perceptual Defense and Vigilance and Driving Safety," in *TQ* October, 1961.
102. McGlade, "Traffic Accident Research: Review and Prognosis," in *TQ*, October, 1962.
103. Michaels, "Human Factors in Highway Safety," in *TQ*, October, 1961.
104. Sherman, "Seeing Habits and Vision, A Neglected Area in Traffic Safety," in *TQ*, October, 1961.
105. Greenshields, "Attitudes, Emotions, Accidents," in *TQ*, April, 1959.
106. Malfetti, "Scare Techniques and Traffic Safety," in *TQ*, April, 1961.
107. Danman, "Economic Costs of Motor Vehicle Accidents," in *HRBB 208*, 1959.
108. Johston, "Economic Costs and Traffic Accidents in Relation to Highway Planning," in *HRBB 263*, 1960.
109. Billingsley and Jorgenson, "Direct Costs and Frequencies of 1958 Illinois Motor Vehicle Accidents," in *HRR 12*, 1963.
110. Smith and Tamburri, "Direct costs of California State Highway Accidents" in *HRR 225*, 1968.
111. Wilbur Smith and Associates, *Motor Vehicle Accident Costs—Washington Metropolitan Area*, 1966.
112. Flanakin, "The Washington Area Motor Vehicle Accident Cost Study," in *TE*, June 1967.
113. Drake and Kraft, "Motor Vehicle Accident Cost in the Washington Metropolitan Area," in *HRB 188*, 1967.
114. *Running Cost of Motor Vehicles As Affected by Road Design and Traffic, NCHRPR 111*, 1971.
115. *Traffic Engineering Handbook*, ITE, 1965.

Chapter 18: Traffic Laws and Ordinances

1. *Automobile Facts and Figures—1969 Edition*, AMA, 1969.
2. *Accident Facts—1969 Edition*, NSC, 1969.
3. *Uniform Vehicle Code*, NCUTLO, 1968.
4. *Model Traffic Ordinance*, NCUTLO, 1968.
5. "General Description of Significant Changes in the Uniform Vehicle Code," in *Proc. ITE*, 1959.
6. Mitton, "Improvements in Traffic Laws and Ordinances Necessary to Meet the Needs of Traffic Operations on the Interstate System," in *Proc. ITE*, 1959.
7. Berry and Jordan, "The Uniform Vehicle Code and Traffic Engineering," in *Proc. ITE*, 1963.
8. "TDER Committee Compares Manual, Code," in *TE*, January, 1965.
9. *Traffic Laws Annual 1964*, NCUTLO, 1964.
10. "A Colloquy on Motor Vehicle and Traffic Law," in *HRBSR 86*, 1965.
11. "Report of Technical Committee on Traffic and Safety," in *TE*, October, 1966.
12. Hennessee and Kerrick, "Licensing the Driver," in *TQ*, July, 1961.
13. Carmichael, "Driver Control Through Licensing," in *TQ*, January, 1962.
14. "BPR Will Maintain Federal Driver Register," in *TE*, October, 1960.
15. Williams, "A New Dimension in Highway Safety," in *TQ*, January, 1963.
16. *The National Driver Register*, BPR, USGPO, September, 1965.
17. Richards, "Motorist Vision and the Driver's License," in *TQ*, January, 1966.
18. "New Device Tests Driver Skills in Minutes," in *TE*, January, 1964.
19. Platt, "A Proposed Index for the Level of Traffic Service," in *TE*, November, 1963.
20. *1964 Annual Report*, NUS DMV, March, 1965.
21. Reilly and Schmitz, "Building Safety Into Motor Vehicles: Part I", in *The Highway User*, NHUC, November, 1965.
22. Reilly and Schmitz, "Building Safety Into Motor Vehicles: Part II," in *The Highway User*, NHUC, December, 1965.
23. Reilly and Schmitz, "Building Safety Into Motor Vehicles: Part III," in *The Highway User*, NHUC, January, 1966.
24. *Automotive Air Pollution*, Report of the Secretary of Health, Education, and Welfare to U.S. Congress, SD No. 7, 89th Cong, USGPO, January 15, 1965.
25. *Automotive Air Polution*, Second Report, SD No. 42, 89th Cong., USGPO, July 15, 1965.

26. *Automobile Industry Program to Control Emissions from Motor Vehicles*, AMA, June, 1964.
27. *Senate Bill 306 to Amend the Clean Air Act*, AMA, April, 1965.
28. BUSH, "Urban Atmosphere Pollution," in *CE*, May, 1965.
29. DOCKERTY and BAYLEY, "Carbon Monoxide Pollution of Urban Air," in *TEC*, April, 1970.

Chapter 19: Introduction to Traffic Control Devices

1. *Uniform Vehicle Code*, NCUTLO, 1968.
2. *Model Traffic Ordinance*, NCUTLO, 1968.
3. BERRY and JORDAN, "The Uniform Vehicle Code and Traffic Engineering," in *Proc. ITE*, 1963.
4. "TDER Committee Compares Manual, Code," in *TE*, January, 1965.
5. *Manual on Uniform Traffic Control Devices*, BPR, 1961.
6. *Manual of Uniform Traffic Control Devices*, State of New York, State Traffic Commission, July, 1958.
7. ROBERTS, "Traffic Control—The Elternal Triangle," in *TE*, December, 1962.
8. "The Relation of the National Highway Safety Program to Traffic Engineers," in *SR TE*, May, 1968.
9. MARSH, "Why Adhere to the Manual on Uniform Traffic Control Devices?", in *TE*, January, 1963.
10. DARRELL, "New Signing Standards," in *Proc. ITE*, 1960.
11. WEBB, "New Marking Standards," in *Proc. ITE*, 1960.
12. HOWIE, "New Signal Standards," in *Proc. ITE*, 1960.
13. "Manual on Uniform Traffic Control Devices—Errata," in *TE*, February, 1962.
14. *Manual for Signing and Pavement Marking of the National System of Interstate and Defense Highways*, AASHO, 1961; revised ed., 1962.
15. *Revised Manual on Uniform Traffic Control Devices*, ARBA, February, 1961.
16. "Questions and Answers on the Manual of Uniform Traffic Control Devices," in *Proc. ITE*, 1962.
17. "Application of Traffic Control Devices," in *Proc. ITE*, 1962.
18. *Proceedings of the National Conference on Uniform Traffic Control Devices*, NJCUTED, November, 1962.
19. "How Far Should We Go in Striving for Uniformity?", A Panel Discussion, in *Proc. ITE*, 1958.
20. SMITH, "Progress and Current Activity on Uniformity of Traffic Control Devices," in *Proc. ITE*, 1950.
21. GALLOWAY, "How Far Should We Go For Uniformity?", in *TE*, June, 1959.
22. MICHAEL, "Uniformity of Traffic Control: Has It Been Achieved?", in *TQ*, January, 1961.
23. ISBELL and GUYUN, "Promoting Uniformity in Traffic Control Devices," in *TE*, May, 1964.
24. REILLY and WOODS, "The Driver and Traffic Control Devices," in *TE*, June, 1967.
25. HILTS, "World-Wide Standards for the Devices of Traffic Control," in *TQ*, January, 1953.
26. USBONE, "International Standardization of Road Traffic Signs," in *TE*, July, 1968.
27. MASSON, "United Nations Conference on Road Traffic (1968)," in *TE*, October, 1968.
28. GROTTEROD and LIAVAAG, "The New Norwegian Signing Code," in *TE*, July, 1967.
29. FRAVEL, "A Brief Look at Traffic and Traffic Control Around the World," in *Proc. ITE*, 1963.
30. ELIOT, "Symbology on the Highways of the World," in *TE*, December, 1960.
31. *Uniform Traffic Control Devices for Canada*, Joint Committee on Uniform Traffic Control Devices for Canada, January, 1960.
32. SHOAF, "Traffic Signs," in *HRBSR 93*, 1968.
33. "New Traffic Signals, Their Effect on Street Utilization," in *HRBSR 93*, 1967.
34. "National Joint Committee Meeting," in *TESR*, February, 1968.
35. "Public Support for Uniform Signs, Signals, and Marking," in *TE*, January, 1963.
36. "Public Support for Uniform Signs, Signals, and Markings," in *TE*, February, 1963.
37. "Report of the Technical Vice-President," 1969 ITE Annual Business Meeting, in *TE*, November, 1969.
38. CONNER, "The National Joint Committee on Uniform Traffic Control Devices," in *TE*, May, 1969.
39. ZUNIGA, "International Effort Toward Uniformity on Signs, Signals, and Markings" in *TE*, May, 1969.
40. WILSON, "United Nations Conference on Road Traffic," in *TE*, May, 1969.
41. *Manual on Uniform Traffic Control Devices*, FHWA, 1971.
42. *Manual for Signing and Pavement Marking of the National System of Interstate and Defense Highways*, AASHO, 1970.

Chapter 20: Traffic Signs and Markings

1. *Manual on Uniform Traffic Control Devices*, BPR, 1961.
2. *Manual of Uniform Traffic Control Devices*, State of New York, State Traffic Commission, July, 1958.
3. *Manual on Uniform Traffic Control Devices*, FHWA, 1971.
4. "Report on Preparation of Performance-Type Specification for Reflectorized Pavement Marking Paint," H.R.B. Committee on Coating, Signing and Marking Materials, in *HRN 15*, November, 1964; also Moore, "Committee Develops Guide for Preparing Paint Specs," in *TE*, April, 1965.
5. SPRUNGMAN, "Progress Report of Committee 4C," in *Proc. ITE*, 1955. (Parking meters.)
6. BOTTS, "Traffic Paint Development in California"; PEED, "Physical Properties of Traffic Paint"; BAUMANN, "Traffic Paint Tests"; VANNOY; SMITH, "Road Tests of Traffic Paints"; MINOR and CODY, "Application of Plain and Beaded Traffic Paints"; SHELBURNE *et al* and BURCH, "Field Studies of Traffic Paint" and "Performance Test Pavement Marking Materials" in *HRBB 57*, 1952.
7. WATERS, "Methods and Application Procedures for Pavement Marking," and Ashman, "Present Preferences for Traffic Paint," in *HRBB 36*, 1951.
8. "Model Performance Specification for Purchase of Pavement Marking Paint," in *TE*, May, 1960.
9. PARISI, "Cheap Paint is Costly", in *TE*, December, 1967.
10. "Development of Improved Pavement Marking Materials," *in NCHR PR 45*, 1967.
11. BOWMAN, "Pavement Marking Economics on Ohio Turnpike," in *TE*, July, 1960.
12. HAYES, "Use of Airless Spray Equipment for City Pavement Markings," in *TE*, May, 1962.

13. COLBY, "Pre-Lining for Better Street Marking," in *TE*, March, 1957.
14. KEESE and BENSON, "Thermoplastic Striping Compounds," in *HRBB 57*, 1952.
15. HARRISON, "Thermo-Plastic Pavement Marking—A Solution to Staggering Maintenance Costs?", in *TE*, December, 1961.
16. PEED, "Physical Properties of Traffic Paints"; Pocock and Rhodes, "Principles of Glass-Bead Reflectorization"; and Shelburne, "Field Studies of Traffic Paints," in *HRBB 57*, 1952.
17. MATHEWS, "Reflectorized Pavement Markings in the Western States," in *Proc. ITE*, 1948.
18. STAPLES, "Reflectorized Traffic Line Markings," in *AH*, April, 1953.
19. HARRINGTON and JOHNSON, "An Improved Instrument for Measurement of Pavement Marking Reflective Performance," in *HRBB 336*, 1962.
20. LYON and ROBINSON, "A Study of Glass Beads for Reflectorizing Traffic Paint," in *Proc. HRB*, 1949.
21. TREMPER and MINOR, "Experience With Reflectorized Traffic Paint," in *Proc. HRB*, 1948.
22. WARNER, "The Visibility of Highway Markings in the Atlantic Seaboard States," in *TQ*, April, 1958.
23. *Policy on Maintenance of Safety and Traffic Control Devices and Related Traffic Services*, AASHO, August, 1954.
24. SUMMERFIELD, GIESA, TORNBERG, SCHRAM, and TRONCON, "Carriage Markings in Europe," in *TEC*, April, 1968.
25. DUFF, JAMES, GIESA, SCHRAM, and YU, "Focus on Road Marking and Materials," in *TEC*, March, 1970.
26. TARAGIN, "The Effect on Driver Behavior of Center Lines on Two-Lane Roads," in *Proc. HRB*, 1947.
27. RICE, "Effect of Lane Line on Lane Usage," in *Proc. ITE*, 1960.
28. GORDON, "Experimental Isolation of Drivers' Visual Input," in *PR*, February, 1966.
29. FAULKNER, "Experimental Carriageway Markings at Uxbridge Circus Roundabout," in *TEC*, May, 1968.
30. PRISK, "Effect of Barrier-Line Location at No-Passing Zones," in *Proc. HRB*, 1952; also in *PR*, June, 1952.
31. BARTELS, "No-Passing Zone Procedures," in *TE*, April, 1958.
32. CORDER, "The No-Passing Zone," in *TE*, October, 1959.
33. "Increased Use of Edgelining Reported," in *TE*, May, 1962.
34. THOMAS and TAYLOR, "Effect of Edge Striping on Traffic Operations," in *HRBB* 244, 1960.
35. WILLISTON, "Effect of Pavement Edge Markings on Operator Behavior," in *HRBB 266*, 1960.
36. JORAL, "Lateral Vehicle Placement as Affected by Shoulder Design on Rural Idaho Highways," in *Proc. HRB*, 1962.
37. MUSICK, "Effect of Pavement Edge Marking on Two-Lane Rural State Highways in Ohio," in *HRBB 266*, 1960.
38. BASILE, "Effect of Pavement Edge Markings on Traffic Accidents in Kansas," in *HRBB 308*, 1962.
39. RICE, "Effectiveness of Lane Markings on Urban Turning Movements," in *TE*, July, 1950.
40. KOLTONOW, "Two-Way Left Turn Lanes Are Proving Successful," in *TE*, October, 1964.
41. FAILMEZGER, "The Effectiveness of Painted Channelization," in *TE*, February, 1966.
42. TERRY and KASSAN, "Effects of Paint Channelization on Accidents," in *TE*, December, 1968.
43. PINNELL, "Driver Requirements in Freeway Entrance Ramp Design," *Proc. ITE*, 1960; also in *TE*, December, 1960.
44. TARAGIN and RUDY, "Traffic Operations as Related to Highway Illumination and Delineation," in *HRBB 255*, 1960; also in *PR*, August, 1960.
45. DARRELL and DUNNETTE, "Driver Performance Related to Interchange Marking and Night-time Visibility Conditions," in *HRBB 255*, 1960.
46. FITZPATRICK, "Unified Reflective Sign, Pavement, and Delineation Treatments for Night Traffic Guidance," in *HRBB 255*, 1960.
47. HUBER, "Traffic Operations and Driver Performance as Related to Various Conditions of Night-time Visibility," in *HRBB 336*, 1962.
48. HUBER, "Night Visibility and Drivers," in *TQ*, January, 1961.
49. ROTH and DERUSE, "Interchange Ramp Color Delineation and Marking Study," in *HRR 105*, 1966.
50. "Color Zoning an Intersection," in *ARB*, December, 1963.
51. WALASCHEK, "Color Pavement," in *ARB*, November, 1965.
52. MCDERMITT and MCLEAN, "Improving Traffic Flow at Transfer Roadways on Collector-Distributor Type Expressways," in *HRR 59*, 1964.
53. DIER, "Pavement Marking Templates," in *TE* July, 1963.
54. NASCIMENTO, "Priority Road Rules in Europe Help Drivers in Each Nation," in *TE*, September, 1965.
55. TWISS, "Reflective Painting Pays Dividends," in *TE*, June, 1962.
56. FINCH, "Roadway Delineation with Curb Marker Lights," in *HRB 336*, 1962.
57. "Warning Device to be Tested," in *TE*, January, 1961.
58. ROWAN, "Approach-End Treatment of Channelization—Signing and Delineation," in *HRR 31*, 1963.
59. DART, "A Study of Roadside Delineator Effectiveness on an Interstate Highway," in *HRR 105*, 1966.
60. HUTCHINSON and LACIS, "An Experiment with Evergreen Trees in Expressway Medians to Improve Roadway Delineation," in *HRR 105*, 1966.
61. "Report of National Joint Committee," in *TE*, May, 1967.
62. DECKER, "Highway Sign Studies—Virginia 1960," in *Proc. HRB*, 1961.
63. ELIOT, "Symbology on the Highways of the World," in *TE*, December, 1960.
64. "Priority Road Rules in Europe Help Drivers in Each Nation," in *Proc. ITE*, 1960; also in *TE*, December, 1960.
65. MAYOR, "Mexico's Sign Program Aimed Toward International Markings," in *TE*, November, 1965.
66. HULBERT and BURG, "The Effects of Underlining on the Readability of Highway Destination Signs," in *Proc. HRB*, 1957.
67. STEPHENS and MICHAELS, "Time-Sharing Between Compensatory Tracking and Search-and-Recognition Tasks," in *HRR 55*, 1964; also in *PR*, December, 1964.
68. DESRASIERS, "Moving Picture Technique for Highway Signing Studies—An Investigation of its Applicability," in *PR*, April, 1965.
69. FORBES and HOLMES, "Legibility Distances of Highway Destination Signs in Relation to Letter Height, Letter Width, and Reflectorization," in *Proc. HRB*, 1959.
70. NEAL, "Legibility of Highway Signs," in *TE*, September, 1947.
71. LAUER, "Certain Structural Components of Letters for Improving the Efficiency of the Stop Sign," in *Proc. HRB*, 1964.
72. FORBES *et al*, "A Comparison of Lower Case and Capital Letters for Highway Signs," in *Proc. HRB*, 1950.
73. SOLOMON, "The Effect of Letter Width and Spacing on Night Legibility of Highway Signs," in *TE*, December, 1956; also in *PR*, April, 1956 and *Proc. HRB*, 1956.

74. CORNOG *et al*, *Legibility of Alphanumeric Characters and Other Symbols, A Permuted Title Index and Bibliography*, NBS Miscellaneous Pub. 262-1, December, 1964, available from USGPO.
75. FORBES, "Factors in Highway Visibility, in *TE*, September, 1969.
76. PAIN, "Brightness and Brightness Ratio as Factors in the Attention Value of Highway Signs," in *HRR 275*, 1969.
77. DOUGHTY, "Application of Internally Illuminated Signs to City Streets," in *Proc. ITE*, 1960.
78. YOUNG, "Overhead Land Direction Signs in Cincinnati," in *Proc. ITE*, 1956.
79. BOUMAN, "The Use of Yield Signs and Illuminated One-Way Signs in San Diego," in *TE*, December, 1958.
80. NEU, "Internally Illuminated Traffic Signs," in *TQ*, April, 1956.
81. STRICKLAND and HART, "Standards for External and Internal Illumination of Traffic Signs," in *Proc. ITE*, 1959.
82. "Traffic Sign Illumination with External Fluorescent Fixtures," Report of Technical Committee 7-C of ITE, in *TE*, March, 1960.
83. BAUER, "Some Solutions of Visibility and Legibility Problems in Changeable Speed Command Signs," in *HRBB 330*, 1962.
84. WHITE, "Use of Neon Highway Signs," in *TE*, March, 1956.
85. CLYDE and COOL, "Electrical Animation," in *TE*, May, 1963.
86. PECKHAM and HART, "A Hypereffective Visual Signal for Night Driving Warning Device," in *HRR 25*, 1965.
87. "Sight and Sound Alert Motorists," in *TE*, September, 1964.
88. GOODMAN, "Interchange Design to Eliminate Wrong-Way Entry," in *TE*, October, 1969.
89. STRAUB and ALLEN, "Sign Brightness in Relation to Position, Distance, and Reflectorization," in *HRBB 146*, 1957.
90. ALLEN, "Night Legibility Distances of Highway Signs," in *HRBB 191*, 1958.
91. ELSTAD *et al*, "Requisite Luminance Characteristics for Reflective Signs," in *HRBB 336*, 1962.
92. BLACKWELL *et al*, "Illumination Requirements for Roadway Visual Tasks," in *HRBB 255*, 1960.
93. POWERS, "Effectiveness of Sign Background Reflectorization," in *PR*, June, 1965; also in *HRR 70*, 1965.
94. BEERS and HULBERT, "Nighttime Effectiveness of Two Types of Reflectorized Stop Signs," in *TE*, November, 1964.
95. FORBES *et al*, *A study of Traffic Sign Requirements—II An Annotated Bibliography*, Michigan State University, East Lansing, Mich. August, 1964.
96. HONG, "Collapsible Signs," in *TE*, 1967.
97. CLEVELAND and KEESE, "Intersections at Night," in *TQ*, July, 1961.
98. BAUCH *et al*, "Sign Supports—Characteristics, Materials, Design, Criteria, and Costs," in *Proc. ITE*, 1959.
99. CHANDLER, "Sign Erection Techniques and Maintenance," in *Proc. ITE*, 1953.
100. RIEGELNEIER, "Rehabilitating Signs," in *Proc. ITE*, 1942.
101. STRAUB, "Causes and Costs of Highway-Sign Replacement," in *Proc. HRB*, 1955.
102. TINNEY and CARR, "Sign Maintenance in California," in *HRN 3*, May 1963.
103. HEWITT, "Cost Analysis of Methods for Interstate Guide Sign Maintenance," *HRN 21*, November, 1965.
104. VOLK, "Wisconsin's Sign Shop Serves Tate Trunk System," in *TE*, February, 1965.
105. MCQUAID, "The Signs in Maine are Plainly a Highway Gain," in *HU*, October, 1965.
106. ROCKEY, "Liability of State or Municipality for Failure to Erect or Maintain Stop Signs," in *TE*, November, 1963.
107. *Manual for Signing and Pavement Marking The Natural System of Interstate and Defense Highways*, NASHO, 1970.
108. FORSTER, "Holosigns," in *TE*, April, 1968.
109. BERG and HULBERT, "Predicting the Effectiveness of Highway Signs," in *HRBB 324*, 1962.
110. POWERS, "Advance Route-Turn Markers on City Streets," in *PR*, April, 1962; also in *Proc. HRB*, 1962.
111. ZIMMERMAN, "Reversible Sign Technique," in *TE*, August, 1954.
112. CYSEWSKI, "A New Signing Approach," in *TE*, November, 1955.
113. CROSBY, "A Case for Overhead Signing," in *TE*, August, 1959.
114. LANER and MCMONAGLE, "Do Road Signs Affect Accidents?" in *TQ*, July, 1955.
115. MCCAMMERT, "New Kansas Curve Signs Reduce Deaths," in *TE*, February, 1959.
116. ORRETT, "Effective Traffic Signing Gains Approval in Monterey," in *TE*, July, 1965.
117. HANSON, "Uniform Street Sign Program for St. Louis," in *TE*, July, 1959.
118. SCHWAR, "Dublin, Ohio—A Model System of Traffic Signs and Markings," in *TE*, September, 1963.
119. "How Many Signs Are Enough?", in *TE*, October, 1957.
120. DARRELL, "Use and Misuse of Traffic Signs," in *TQ*, October, 1961.
121. *Report on Traffic Control Devices Workshop*, November, 1966.
122. DILLON, "A Country Updates its Signs," in *ARB*, May, 1964.
123. DARELL, "Traffic Signing and Marking," in *ARB*, August, 1964.
124. SHOAF, "Traffic Signs," in *HRSR 93*, 1968.

Chapter 21: Traffic Signals

1. *Manual on Uniform Traffic Control Devices*, BPR, 1961.
2. *Manual on Uniform Traffic Control Devices*, FHWA, 1971.
3. "A Standard for Traffic Signal Lamps," revised ITE Standard, in *TE*, March, 1968.
4. HILL, "Speed Control by Traffic Signals," in *TQ*, January, 1954.
5. HENRY, "Route Signal Systems," in *HRB 93*, 1967.
6. YOUNG, "New Traffic Signals—Their Effect on Street Utilization," *HRBR 93*, 1967.
7. MALO, "Signal Modernization," in *HRBS 93*, 1967.
8. VOLK, "Effect of Type of Control on Intersection Delay," in *Proc. HRB*, 1956.
9. HALL, "Intersection Delay-Signal vs. Four-Way Stop," in *Proc. ITE*, 1952.
10. FRENCH, "Capacities of 1-Way and 2-Way Streets with Signals and with Stop Signs," in *HRBB 112*, 1956.
11. DEROSE, "Reversible Center-Lane Traffic System—Directional and Left-Turn Usage," in *HRR 151*, 1966.
12. "A Tentative Standard for Traffic Signal Lamps," in *TE*, April, 1966.
13. "Tentative Revised ITE Standards for Adjustable Face Vehicle Traffic Control Signal Heads," in *TE*, April, 1969.
14. MANN, "Signal Maintenance," in *TE*, April, 1959.
15. WARR, "Pin-Pointing Traffic Signal Troubles," in *TE*, May, 1956.

16. SMITH and OPPENLANDER, "Economic Model for the Maintenance of Traffic Signals," in *HRR 254*, 1968.
17. BUTTERFIELD, "Philadelphia's New Program for Traffic Signal Modernization," in *TQ*, April, 1956.
18. WYNN "Establishing a Priority Basis for New Traffic Signals in Cities," in *TQ*, October, 1951.
19. LANTEN, "New York Signal Formula," in *TQ*, July, 1955.
20. HORN, "The Oakland Signal Formula," in *TQ*, January, 1957.
21. PARKER and BARTTE, "Signalization Priority in Los Angeles County," in *TE*, June, 1954.
22. BELLES, "Selecting the Intersection Type by Traffic Volume: A Chart," in *TQ*, January, 1951.
23. GALLOWAY, "Expanded Signal Volume Warrants," in *TE*, September, 1958.
24. BLEYT, "A Method for Estimating the Satisfaction of Traffic Signal Warrants Based on ADT Volumes," in *TE*, October, 1963.
25. ALROTH, "Assembly, Analysis, and Application of Data on Warrants for Traffic Control Signals," in *TE*, January, 1968.
26. *Highway Capacity Manual*, HRBSR 87, 1965.
27. "Pre-Timed, Fixed Cycle, Traffic Signal Controllers," ITE Report, in *TE*, September, 1958.
28. GERLOUGH, "Operation of the 'Volume-Density' Vehicle-Actuated Traffic Signal Controller," in *TE*, July, 1953.
29. BEAUBIEN, "The GE Volume-Density Controller," in *TE*, December, 1954.
30. WEBB, "Prunedale Signals," in *TE*, May, 1962.
31. HAUGH, "The Field of Modern Interconnected Traffic Control Signals," in *Proc. ITE*, 1953; also in *TE*, April, 1954.
32. HOOSE, "Direct Time Cycle Control by Traffic Density in Charlotte, North Carolina," in *TE*, April, 1956.
33. BARKER, "A Traffic Actuated Cycle and Offset Selector System," in *Proc. ITE*, 1949.
34. READING, "Traffic Provides its Own Progression," in *Proc. ITE*, 1950.
35. BARNES, "Denver Installs a Modern Signal System," in *Proc. ITE*, 1951.
36. FLYE, "Traffic Signal Systems for Small Cities," in *TE*, June, 1962.
37. HIMELHOCK, "Case History," in *TE*, March, 1958.
38. HEWTON, "Metropolitan Toronto Traffic Surveillance and Control System," in *CE*, February, 1969.
39. WILSHIRE, "The Benefits of Computer Traffic Control," in *TE*, April, 1969.
40. KARAGHEUZOFF, *A Perspective on Electronics in Traffic Control*, Presented at IEEE: NEREM, Boston, Mass., November 6, 1969.
41. *San Jose Traffic Control Project*, Final Report, 1967.
42. GORDON, "Urban Traffic Control," in *Sperry Rand Engineering Review*, 22.1, 1969.
43. MCCASLAND and CARVELL, *Optimizing Flow in an Urban Freeway Corridor*, Annual Report: Fiscal Year 1968–69, TTI, BPR Contract No. FH-11-6931.
44. HILLIER and HOLROYD, "The Glasgow Experiment in Area Traffic Control," in *TE*, October, 1969.
45. BOLKE *et al*, "Munich's Traffic Control Center," in *TEC*, August, 1967.
46. FERRATE, "Requirements of an Advanced Hierarchy System, Using Sub-Masters, For Computer Area Traffic Control," presented at the *1967 IFAC Control Conference*, Haifa, Israel.
47. VALDEZ and DE LA RICO, "Area Traffic Control by Computer in Madrid," in *TEC*, July, 1970.
48. HUDDART and CHANDLER, "Area Traffic Control for Central London," in *TEC*, September, 1970.
49. COBBE, "Traffic Control for West London," in *Electronics and Power*, April, 1967.
50. "Traffic-Actuated, Traffic Signal Controllers and Detectors," in *TE*, October, 1958.
51. LEMARCH, "Seattle's Radar Signal Installations," in *TE*, November, 1955.
52. JORDAN, "Norfolk Installs Unique Signal Control System," in *TE*, November, 1964.
53. *Policy on Maintenance of Safety and Traffic Control Devices and Related Traffic Services*, AASHO, August, 1954.
54. CLAIBORNE, "Induction Vehicle Detectors for Traffic-Actuated Signals," in *TE*, December, 1962.
55. HUPPER, "Familiarity With Loop Detectors Will Aid Application, Installation," in *TE*, August, 1965.
56. *Traffic Engineering Handbook*, 3rd ed., ITE, 1965.
57. RAUS, "Traffic-Actuated Signal Control Without the Initial and Vehicle Interval," in *TE*, February, 1964.
58. PINNELL, "Utilization of Digital Computer for Real Time Traffic Control, Part II," in *TE*, July, 1968.
59. WARDELL and MURRAY, "Computerized Traffic Signal System Maximizes Roadway Network Capacity and Improves Pedestrian Safety At J. F. Kennedy International Airport," in *TE*, September, 1968.
60. HANSON, "Modernizing the Traffic Signal System of the Nation's Capital," in *TE*, May, 1967.
61. "Central Area Traffic Control in Liverpool," in *TEC*, August, 1968.
62. BOLKE, "Munich's Traffic Control Center," in *TEC*, August, 1967.
63. PAVEL, "Report from Munich on Centralizer Computer Control of Traffic Signals," in *TEC*, September, 1967.
64. TILLOTSON, "Traffic Signal Control Using Television," *TEC*, September, 1968.
65. WEINBERG, GOLDSTEIN, MCDADE, and WAHLEN, "Digital-Computer-Controlled Traffic Signal System for a Small City," in *NCHRPR 29*, 1966.
66. GERLOUGH and WAGNER, "Improved Criteria for Traffic Signals at Individual Intersections," in *NCHRPR* 32, 1967.
67. FOODY and TAYLOR, "An Analysis of Flashing Systems," *HRR 221*, 1968.
68. BLEYL, "A Practical Computer Program for Designing Traffic-System Timing Plans," in *HRR 211*, 1967.
69. CASS, "Signal Networks," in *HRB SP 93*, 1967.
70. FRIEDLANDER, "Computer Controlled Vehicular Traffic," in *IEEE Spectrum*, Vol. 6, No. 2, February, 1969.
71. KOHNERT and MUSIAL, "Traffic Signal Control by Computer," in *Control Engineering*, Vol. 16, No. 6, June, 1969.
72. OSBORNE, "Buffalo Weds Computer and Traffic Signal System," in *TE*, July, 1964.
73. RAYNOR, "Charleston's Computerized Traffic Control System," in *TEC*, May, 1969.
74. HILLIER, "Area Traffic Control by Computer-Equipment on the Glasgow Experiment," in *TEC*, February, 1968.
75. HILLIER, "Glasgow's Experiment in Area Traffic Control," in *TEC*, December, 1965, and January, 1966.
76. GORDEN, "Urban Traffic Control," in *Sperry Rand Engineering Review*, Vol. 22.1, 1969.
77. FRAVEL, "A Brief Look at Traffic and Traffic Control Devices Around the World," in *Proc. ITE*, 1963.
78. BERRY, "Field Measurement of Delay at Signalized Intersections," *Proc. HRB*, 1956.

79. HILLER, "The Tampa Signal System," in *TE*, July, 1962.
80. MICHEL, "Radio Co-ordinated Traffic Control Signals," in *Proc. ITE*, 1955.
81. RADZIKOWSKI and DILLON, "Report on Expansion in the uses of Radio and Availability of Radio Frequencies for Highway Department Purposes as a Result of Recent FCC Actions," in *TE*, September, 1958.
82. "Radio Frequency Allocation Sought from FCC for Traffic Control," in *TE*, February, 1959.
83. KELLER, "The FCC and Radio Control of Traffic Devices," in *TE*, May, 1959.
84. KAVANAUGH, "Radio Control for Reversible Flow," in *TE*, February, 1959.
85. "Chicago Controls Traffic Signals with Radio," in *TE*, February, 1956.
86. TWISS, "Centralized Control of a Metropolitan Signal System," *TE*, July, 1962.
87. MORRISON *et al*, "Traffic Pacer," in *HRBB 338*, 1962.
88. BRENNUG, "Intersection Traffic Control Through Coordination of Approach Speed," in *HRBB 324*, 1962.
89. RANABAUER, "Remote Control of Traffic on the Autobahn," in *TE*, July, 1967.
90. BRISSON and VILLE, "Traffic Control for the 10th Winter Olympic Games at Grenoble," in *TE*, October, 1968.
91. DARE, "The Traffic Actuated Speed Signal Funnel," in *TE*, November, 1969.
92. HEWTON, "Modification of Fixed Time Traffic Signal Controller to Give Semi-Actuated Operation," in *TE*, March, 1957.
93. ROGERS, "More Flexibility with Fixed Time Equipment," in *TE*, December, 1955.
94. CLARKE and SAGI, "Circuit Design for Railway Actuated Signals at Three Intersections," in *TE*, March, 1962.
95. FREED, "Highway Traffic Detour Uses Live Railroad Loop Track," in *TE*, September, 1964.

Chapter 22: Street and Highway Lighting

1. *Accident Facts—1968 Edition*, NSC, 1968.
2. *Accident Facts—1969 Edition*, NSC, 1968.
3. "Effect of Adequate Street Lighting in Reducing Night Traffic Accidents," in *TE*, May, 1958.
4. CULVER, "Traffic Safey Lighting," in *TE*, July, 1952.
5. SCHRENK, "The Importance of Adequate Fixed Lighting for Maximum Safety on Expressways," in *TE*, December, 1950.
6. SEBURN, "Relighting a City," in *Proc. ITE*, 1948.
7. BLYTHE, "Highway Lighting and Accidents in Indiana," in *HRB B 146*, 1957.
8. WYATT and LOZANO, "Effect of Street Lighting on Night Traffic Accident Rate," in *HRB B 146*, 1957.
9. CHRISTIE, "An Experimental Low-Cost Lighting System for Important Highways in Rural Areas," in *HRBB 255*, 1960.
10. REX and FRANKLIN, "Visual Comfort Evaluations of Roadway Lighting," in *HRBB 255*, 1960.
11. CORK, "Impact of Street Lighting on Road Accidents," in *TEC*, October, 1967.
12. SIELSKI, "Proper Lighting Saves Lives," in *ARB*, July, 1968.
13. BERRY, "Influence of Lighting on Cost of Road Accidents," in *TEC*, January, 1968.
14. ARGRAVES, "Lighting the Connecticut Turnpike," in *TQ*, July, 1959.
15. IVES, "Does Highway Illumination Affect Accident Occurrence?", in *TQ*, April, 1962.
16. BERRY, "Progress in Street Lighting," *TEC*, May, 1967.
17. WILLIAMS, "Roadway Lighting," in *TE*, March, 1970.
18. FISHER, "Visibility, Accidents and the S.A.A. (Standards Association of Australia) Street Lighting Code," in *ARRBB*, December, 1967.
19. BOEREBOON, "Lighting of Public Thoroughfares in Europe," in *ITE*, 1965.
20. BOX, "Effect of Highway Lighting on Night Capacity," in *TE*, January, 1958.
21. BAKER, "Lighting for Single-Level Road Junctions Including Roundabouts," in *TEC*, June, 1968.
22. COLUIN, "Lighting Road Junctions and Roundabouts," in *TEC*, September, 1967.
23. WILCOCK, "Influence on Current and Future Street Lighting Practice," in *TEC*, March, 1968.
24. *American Standard Practice for Street and Highway Lighting*, IES, 1953.
25. *American Standard Practice for Roadway Lighting*, IES, 1964.
26. HAZLETT and ALLEN, "The Ability to See a Pedestrian at Night: Effects of Clothing, Reflectorization, and Driver Intoxication," in *HRR 216*, 1968.
27. KING and FINCH, "A Laboratory Method for Obtaining Pavement Reflectance Data," in *HRR 216*, 1968.
28. FINCH and KING, "A Simplified Method for Obtaining Pavement Reflectance Data," in *HRR 179*, 1967.
29. POHLENZ and STOLZENBERG, "An Efficient Installation for Measuring Road-Surface Luminance," in *Lichttechnik*, Helios-Verlag, Berlin, Germany, March, 1968 (in German).
30. BAILLIF, "Lighting of French Motorways," in *TEC*, April, 1968.
31. FORBES *et al*, "Color and Brightness Factor in Simulation and Full-Scale Traffic Sign Visibility," in *HRR 216*, 1968.
32. REX, "Roadway Lighting for the Motorist," in *IE*, February, 1967.
33. SAUR, "Influence of Luminance and Geometry on Glare Impression," in *Optical Society of America Journal*, June, 1968.
34. COLEMAN and SACKS, "An Investigation of the Use of Expanded Metal Mark as an Anti-Glare Screen," in *HRR 179*, 1967.
35. ALLEN, "Automobile Windshields, A New Car Study—1966 Models," in *Optometry Weekly*, 57 (28, pt. 2), 1966.
36. FRIES and ROSS, "Headlight Glare vs. Median Width," in *HRBB 298*, 1961.
37. HOFER, "Glare Screen for Divided Highways," in *HRBB 336*, 1962.
38. LAND, "The Use of Polarized Headlights for Safe Night Driving," in *TQ*, October, 1948.
39. CHUBB, "Polarized Light for Auto Headlights," in *TE*, April, 1950.
40. LAND and CHUBB, "Polarized Light for Auto Headlights," in *TE*, July, 1950.
41. "Against the Adoption of Polarized Headlights at This Time," Engineering Liaison Committee, in *TE*, October, 1950 and November, 1950.
42. WOLF *et al*, "Influence of Tinted Windshield Glass on Five Visual Functions," in *HRBB 255*, 1960.
43. MCFARLAND *et al*, "Dark Adaptation as a Function of Age and Tinted Windshield Glass," in *HRBB 255*, 1960.
44. BRYAN, "Lenses for Night Driving," in *HRBB 336*, 1962.
45. POWERS and SOLOMON, "Headlight Glare and Median Width: Three Exploratory Studies," in *HRR 70*, 1965; also in *PR*, April, 1965.
46. MORTIMER, "The Effect of Glare in Simulated Night Driving," in *HRR 70*, 1965.

47. SCHWAB, "Night Visibility for Opposing Drivers with High and Low Headlight Beams," in *HRR 70*, 1965.
48. LAGELSTOM (ed.), *Lighting Problems in Highway Traffic*, Macmillan, New York, 1968.
49. ROBER, "Four Headlamps for Better Seeing," in *TE*, January, 1957.
50. KILGOUR, "Some results of Cooperative Vehicle Lighting Research," in *HRBB 255*, 1960.
51. KILGOUR, "Cooperative Research in Vehicle Lighting," in *TQ*, January, 1962.
52. ROPER and MEESE, "More Light on the Headlighting Problem," in *HRR 70*, 1965.
53. HANSON and PALMQUIST, "Effectiveness of Reflecterized Headlamps," in *HRR 164*, 1967.
54. *Westinghouse Street Lighting Engineering Guide*, B-5460, Westinghouse Electric Corp.
55. BOX and EDMAN, "Roadway Lighting," in *TE*, April, 1964.
56. SEBMAN, "Roadway Lighting," *CE*, October, 1964.
57. SEBURN, "New IES Roadway Lighting Standards," in *Proc. ITE*, 1963.
58. REX, "Light Distribution for the Motorist," in *TE*, January, 1967.
59. *Outdoor Lighting Photometric Data and Sample Calculations*, GET-3100C, General Electric.
60. WALKER, "Cut-off and Semi-cut-off street Lighting Installations," in *TEC*, February, 1967.
61. DUNCAN, "Progress Report on High Mast Lighting," in *TEC*, January, 1967.
62. *1963 Designer's and Buyer's Guide—Roadway Lighting*, General Electric.
63. REX, "Modern Lighting for Residential Streets," in *Proc. ITE*, 1945.
64. REX, "Improving Seeing Efficiency with Roadway Lighting," in *TE*, August, 1956.
65. REX, "Principles and Figures of Merit for Roadway Lighting as an Aid to Night Motor Vehicle Transportation," in *HRBB 146*, 1957.
66. REX, "New Developments in the Field of Roadway Lighting," in *Proc. ITE*, 1959; also in *TE*, March, 1960.
67. REX, "Advancement in Roadway Lighting," in *HRBB 255*, 1960.
68. REX, "Comparison of Effectiveness Rating—Roadway Lighting," in *HRBB 298*, 1961.
69. BLACKWELL *et al*, "Illumination Requirements for Roadway Visual Tasks," *HRBB 255*, 1960.
70. DE BEER, "Road Surface Luminance and Glare Limitation in Highway Lighting," in *HRBB 298*, 1961.
71. SEBURN, "Lighting our Streets and Highways," in *TQ*, January, 1956.
72. BOLLONG, "Modern Street Illumination," in *TE*, March, 1950.
73. SCHWANHAUSSER, "Highway Lighting—A Feature of Modern Roadway Design," in *TE*, December, 1958.
74. "Non-Lighting Considerations in the Location of Light Supports," in *TE*, November, 1962.
75. BLACKWELL *et al*, "Illumination Variables in Visual Tasks of Drivers," in *PR*, December, 1965.
76. IVES, "Highway Illumination Warrants-Design-Maintenance-Costs," in *AH*, AASHO, July, 1961.
77. "Economic Study of Roadway Lighting," in *NCHRPR 20*, 1966.
78. ROWAN and WALTON, "Optimization of Roadway Lighting Systems," in *HRR 216*, 1968.
79. CLEVELAND, "Driver Tension and Rural Intersection Illumination," in *TE*, October, 1961.
80. SCANLON and REID, "Street Lighting Correlated with Tree Management," in *TQ*, October, 1947.
81. GONSETH, "Effectiveness of Holland Tunnel Transitional Lighting during the Winter Months," in *HRBB 255*, 1960.
82. "Lighting Plays Key Role in New George Washington Bridge Roadway," in *TE*, February, 1963.
83. *An Informational Guide for Lighting Controlled Access Highways*, AASHO, 1960.
84. FALK, "New Technique for Illuminating Underpasses," in *TE*, August, 1965.
85. GWYNN, "Low Level Bridge Lighting," in *TE*, December, 1965.
86. ROWAN *et al*, "Floodlighting Highway Interchange Areas," in *TTR*, July, 1968.
87. CARTER, "A New Highway Luminaire for Intersection Lighting," in *TE*, May, 1952.
88. DIMITRI, "Fluorescent Lighting for a New York City Parking Area," in *TE*, December, 1956.
89. BERRY, "Road Lighting Review," in *TEC*, May, 1968.
90. *Traffic Engineering Handbook*, 3rd ed. *ITE*, 1965.
91. SIELSKI, "Relationship of Roadway Lighting and Traffic Accidents," in *HRBS 93*, 1967.
92. WITHEFORD, "The Economic Analysis of Freeway Lighting," in *TQ*, April, 1967.

Chapter 23: The Intersection

1. *Accident Facts—1970 Edition*, National Safety Council, 1970.
2. *Traffic Engineering Handbook*, 3rd ed., 1965.
3. *Manual on Uniform Traffic Control Devices*, FHWA, 1971.
4. KELL, "Yield Right-of-Way Signs: Warrants and Applications," in *Proc. ITE*, 1959.
5. KELL, "Applications of Yield Right-of-Way Signs," in *TE*, July, 1958.
6. RAY, "Six Months Use of Yield Right-of-Way Signs," in *TE*, October, 1957.
7. BERRY and KELL, "Use of Yield Signs," in *TE*, January, 1956.
8. RICE, "The Yield-Right-of-Way Sign," in *TQ*, January, 1952.
9. RAFF, *A Volume Warrant for Urban Stop Signs*, Eno, 1950.
10. MARKS, "Warrants for Four-Way Stop Signs," in *Proc. ITE*, 1959.
11. HANSON, "Are There Too Many Four-Way Stops?", in *TE*, November, 1957.
12. HALL, "Intersection Delay-Signal vs. Four-Way Stop," in *Proc. ITE*, 1952.
13. MCEACHERN, "A Four-Way Stop-Sign System at Urban Intersections," in *TQ*, April, 1949.
14. HERBERT, "A Study of Four-Way Stop Intersection Capacities," in *HRR 27*, 1963.
15. WILKIE, "58,732 Motorists Checked at Stop Signs," in *TE*, April, 1954.
16. KENEIPP, "Efficiency of Four-Way Stop Control at Urban Intersections," in *TE*, January, 1951.
17. HARRISON, "Four-Way Stops," in *TE*, February, 1949.
18. WEBSTER, *Traffic Signal Settings*, Road Research Technical Paper No. 39, RRL, 1961.
19. BONE *et al*, *The Selection of a Cycle Length for Fixed Time Traffic Signals*, Joint Highway Research Project, MIT., Cambridge, Mass., April, 1964.

20. Gerlough and Wagner, "Improved Criteria for Traffic Signals at Individual Intersections, Interim Report," in *NCHRPR 3*, 1964.
21. Kell, "Results of Computer Simulation Studies as Related to Traffic Signal Operation," in *Proc. ITE*, 1963.
22. Gazis, Herman, and Maradad, "The Problem of the Amber Signal Light in Traffic Flow," in *TE*, July, 1960.
23. May, "Clearance Interval at Traffic Signals," in *HRR 221*, 1968.
24. Greenshields, Shapiro, and Ericksen, *Traffic Performance at Urban Intersections*, Technical Report No. 1, Yale BHT, 1947.
25. Bartle *et al*, "Starting Delay and Time Spacing of Vehicle Entering Signalized Intersection," in *HRBB 112*, 1956.
26. Webb and Moskowitz, "Intersection Capacity," in *TE*, January, 1956.
27. Leroy, "Signal Timing to Meet Peak Loads of Traffic Demands," in *TQ*, January, 1949.
28. Capelle and Pinnell, "Capacity Study of Signalized Diamond Interchanges," in *HRBB 291*, 1961.
29. Helm, "Saturation Flow of Traffic at Light-Controlled Intersections," in *TE*, February, 1962.
10. Wildermuth, "Average Vehicle Headways at Signalized Intersections," in *TE*, November, 1962.
31. *Turn Controls in Urban Traffic*, Eno, 1951.
32. Ray, "Experience With Right-Turn-on-Red" in *Proc. ITE*, 1956; also in *TE*, December, 1957.
33. Rankin, "Report on Results from Right Turn on Red Light Questionnaire Survey," in *TE*, October, 1955.
34. Kuehl, "Warrants for Right-Turn Arrows," in *TE*, June, 1952.
35. Berry *et al*, "Evaluating Effectiveness of Lane-Use Control Devices at Intersections," in *Proc. HRB*, 1962.
36. Koltnow, "Two-Way Left Turn Lanes Are Proving Successful," in *TE*, October, 1964.
37. Failmezger, "Relative Warrant for Left-Turn Refuge Construction," in *TE*, April, 1963.
38. Ray, "Installation of a Two-Way Median Left Turn Lane," in *TE*, March, 1961.
39. Powers, "Advance Route-Turn Markers on City Streets," in *PR*, April, 1962; also in *Proc. HRB*, 1962.
40. Clyde and Smith, "Route Turn Beacon," in *TE*, January, 1963.
41. Hawkins, "A Comparison of Leading and Lagging Greens in Traffic Signal Sequence," in *Proc. ITE*, 1963.
42. Ray, "Two-Lane Left Turns Studied at Signalized Intersections," in *TE*, April, 1965.
43. Bruening, "Separating Walkers and Right-Turners," in *TE*, October, 1957.
44. Schwanhausser, "A Turn to the Left," in *TE*, April, 1957.
45. Lema, "Presence Detector for Left-Turning Vehicles," in *Proc. ITE*, 1960.
46. Sansom, "Use of Flashing Green Signal to Indicate Advance Green Phase," in *Proc. ITE*, 1960.
47. Webb and Moskowitz, "Intersection Capacity," in *TE*, January, 1956.
48. Exnicios, "The Problem of the Signalized Intersection—The Left Turn," in *Proc. ITE*, 1963.
49. "Medians of Divided Highways," in *HRBBib 34*, 1963.
50. Beaton and Field, "Impact Tests," in *TE*, December, 1959.
51. Beaton and Field. "Dynamic Full-Scale Tests of Median Barriers," in *HRBB 266*, 1960.
52. Beaton *et al*, "Median Barriers: One Year's Experience and Further Controlled Full-Scale Tests," in *Proc. HRB*, 1962.
53. Nordlin *et al*, "Dynamic Full-Scale Impact Tests of Bridge Barrier Rails, *HRR 83*, 1965.
54. Lundstrom, "A Bridge Parapet Designed for Safety," in *HRR 83*, 1965.
55. McAlpin *et al*, "Development of an Analytical Procedure for Prediction of Highway Barrier Performance," in *HRR 3*, 1965.
56. Moskowitz and Schaefer, "California Median Study: 1958," in *HRBB 266*, 1960.
57. Crosby, "Cross-Median Accident Experience on the New Jersey Turnpike," in *HRBB 266*, 1960.
58. Sacks, "Effect of Guardrail in a Narrow Median Upon Pennsylvania Drivers," in *HRR 83*, 1965.
59. Johnson, "Effectiveness of Median Barriers," in *HRR 105*, 1966.
60. Billion *et al*, "Effect of Parkway Medians on Driver Behavior —Westchester County Parkways," in *HRBB 308*, 1962.
61. Billion and Parsons, "Median Accident Study—Long Island, New York," in *HRBB 308*, 1962.
62. Hutchinson *et al*, "History of Median Development in Illinois," in *HRR 105*, 1966.
63. Fries and Ross, "Headlight Glare vs. Median Width," in *HRBB 298*, 1961.
64. Hofer, "Glare Screen for Divided Highways," in *HRBB 336*, 1962.
65. Powers and Solomon, "Headlight Glare and Median Width: Three Exploratory Studies," in *HRR 70*, 1965; also in *PR*, April, 1965.
66. Hanna, "Median Dividers on Urban Streets," in *TQ*, October, 1963.
67. Himelhoch, "Urban Roads Given High Capacity Through Medians and Signals," in *TQ*, October, 1957.
68. Lipp, "Lincoln Road Mall, Miami Beach," in *TQ*, July, 1961.
69. Kermit and Hein, "Effect of Rumble Strips on Traffic Control and Driver Behavior," in *Proc. HRB*, 1962.
70. Rowan, "Approach-End Treatment of Channelization— Signing and Delineation," in *HRR 31*, 1963.
71. "Channelization—The Design of Highway Intersections at Grade," *HRBSR 74*, 1962.
72. Fruin, *Design for Pedestrians*, MAUDEP Press, New York, 1971.
73. *A Policy on Arterial Highways in Urban Areas*, AASHO, 1957.
74. *Policy on Maintenance of Safety and Traffic Control Devices and Related Traffic Services*, AASHO, August, 1954.
75. *Research on Road Traffic*, RRL Department of Scientific and Industrial Research, HMSO, London, 1965.
76. Storey, "New Street Marking System Controls Pedestrian Traffic," in *TE*, October, 1954.
77. *Planned Pedestrian Program*, AAA Foundation for Traffic Safety, Washington, D. C. 1958.
78. *Research on Road Safety*, RRL, 1963.
79. Rotman, "The Toronto Pedestrian Crossover Program," in *TE*, February, 1961.
80. Roer, "Pedestrian Crossovers," in *TE*, August, 1961.
81. Massey, "Mathematical Determination of Warrants for Pedestrian Crossings," in *TE*, September, 1962.
82. Clearwater, "Los Angeles Marks Crosswalks with Plastic," in *TE*, August, 1964.
83. Gove, "Pedestrian Signal Warrants," in *Proc. ITE*, 1958.
84. Rudden, "Warrants For and Experience With Pedestrian Intervals at Signalized Intersections," in *Proc. ITE*, 1959.
85. Dier, "Pedestrian 'Scramble' Control," in *TE*, August, 1954.
86. Lawton, "The Lawton Pedestrian Signal," in *TE*, September, 1955.

87. HOFFMAN, "Pedestrian Signal," in *TE*, September, 1955.
88. MURPHY, "Planned Pedestrian Programs in New York City," in *TE*, November, 1961.
89. MARCH, "What About the Pedestrian," in *Proc. TE*, 1957.
90. LEE, "Portland Protects its Pedestrians," in *TQ*, July, 1952.
91. MILLER, "Pedestrian Protection and Control," in *TQ*, July, 1950.
92. "A Program for School Crossing Protection," in *TE*, October, 1962.
93. OLSON and ROTHERY, "Driver Response to Amber Phase of Traffic Signals," HRB Bull. 330, 1962.
94. JENKINS, "A Study of Selection of Yellow Clearance Intervals for Traffic Signals," Report TSD-TR-104-69, Michigan Dept. of State Highways, 1969.
95. CARSTENS, "Some Traffic Parameters at Signalized Intersections," in *TE*, August, 1971.

Chapter 24: Coordination of Signal Systems

1. *Manual on Uniform Traffic Control Devices*, FHWA, 1971.
2. RUDY, "Limitations of Signal Spacings in a Coordinated System on a High Speed, Dual Highway," in *TE*, October, 1957.
3. HEWTON, "Metropolitan Toronto Traffic Surveillance and Control System," in *CE*, February, 1969.
4. WILSHIRE, "The Benefits of Computer Traffic Control," in *TE*, April, 1969.
5. KARAGHEUZOFF, *A Perspective on Electronics in Traffic Control*, presented at IEEE: Nerem, Boston, Mass., November 6, 1969.
6. *San Jose Traffic Control Project*, Final Report, 1967.
7. GORDON, "Urban Traffic Control," in *Sperry Rand Engineering Review*, Vol 22.1, 1969.
8. MCCASLAND and CARVELL, "Optimizing Flow in an Urban Freeway Corridor," *Annual Report: Fiscal Year 1968–69*, TTI, BPR Contract No. FH-11-6931.
9. COBBE, "Traffic Control for West London," in *Electronics and Power*, April, 1967.
10. HILLIER and HOLROYD, "The Glasgow Experiment in Area Traffic Control," in *TE*, October, 1969.
11. BOLKE and WEREIGH, "Munich's Traffic Control Center," in *TEC*, August, 1967.
12. FERRATE, "Requirements of an Advanced Hierarchy System, Using Sub-Masters, For Computer Area Traffic Control," presented at the *1967 IFAC Control Conference*, Haifa, Israel.
13. HANNA *et al*, "A Simultaneous vs. Triple Alternate Signal System," in *TE*, March, 1957.
14. BAVAREZ and NEWELL, "Traffic Signal Synchronization on a One-Way Street," in *Transportation Science*, Vol. 1,No. 2, 1967.
15. LITTLE, "The Synchronization of Traffic Signals by Mixed-Integer Linear Programming," in *OR*, 1966.
16. YARDENI, "Algorithms for Traffic-Signal Control," in *IBM Systems Journal*, Vol 4: Nos. 148–161, 1965.
17. MORGAN and LITTLE, "Synchronizing Traffic Signals For Maximal Bandwidth," in *OR*, 1964.
18. IRWIN, "Development of a Method for the Optimum Timing of Traffic Signals in an Urban Street Network," in *Proc. BPR Program Review Meeting*, Gaithersburg, Md., December, 1966.
19. *San Jose Traffic Control Project*, Final Report, 1967.
20. BOWERS, "Progressive Timing for Traffic Signals," in *Proc. ITE*, 1947.
21. MURPHY, "Development of a Signal Timing Plan for a Network of Signal System," in *Proc. ITE*, 1948.
22. CHANG, "Synchronization of Traffic Signals in Grid Networks," in *IBM Journal of Research and Development*, July, 1967.
23. IRWIN, "The Toronto Computer-Controlled Traffic Signal System," *Traffic Control, Theory, and Instrumentation*, Plenum Press, New York, 1965.
24. *SIGOP Traffic Signal Optimization Program Users Manual*, Peat, Marwick, Livingston and Co., New York, December, 1968.
25. FRIEDLANDER, "Computer-Controlled Vehicular Traffic," *IEEE Spectrum*, February, 1969.

Chapter 25: Speed Control and Zoning

1. *Accident Facts—1969 Edition*, NSC, 1969.
2. PRISK, "The Speed Factor in Highway Accidents," in *TE*, August, 1959.
3. SOLOMON, *Accidents on Main Rural Highways Related to Speed, Driver, and Vehicle*, BPR, July, 1964.
4. *Traffic Engineering Handbook*, 3rd ed., 1967.
5. PETRING, "Stopping Ability of Motor Vehicles Selected from the General Traffic," in *PR*, June, 1957.
6. "An Informational Report on Speed Zoning," in *TE*, July, 1961.
7. CARSTEN, "Inform the Driver," in *TE*, January, 1958.
8. SMITH, "Control of Speeds with Signs and Markings," in *Proc. ITE*, 1939.
9. *Manual on Uniform Traffic Control Devices*, FHWA, 1971.
10. BAERWALD, "Theory of Speed Zones in Developed Areas," in *TE*, December, 1957.
11. OPPENLANDER, "A Theory on Vehicular Speed Regulation," in *HRBB 341*, 1962.
12. SIELSKI, "What Should the Maximum Speed Limit Be?", in *TE*, September, 1956.
13. SEBURN, "Reduced Speed for Safe Night Driving," in *Proc. ITE*, 1939.
14. DARRELL, "Speed Zoning on Rural Highways," in *Proc. ITE*, 1941.
15. KUNZ, "Can Speed Zoning be Applied at Isolated Intersections," in *Proc. ITE*, 1941.
16. MCMONAGLE, "The Need and Application of Speed Zoning," in *Proc. ITE*, 1946.
17. DARRELL, "What About Speed Limits?," in *Proc. ITE*, 1950.
18. JOHNSON, "Speed Control and Regulation," in *Proc. ITE*, 1955.
19. LEFEVE, "Speed Habits Observed on Rural Highway," in *Proc. HRB*, 1954.
20. OGAWA *et al*, "Driver Behavior Study—Influence of Speed Limits on Spot Speed Characteristics in a Series of Contiguous Rural and Urban Areas," in *HRBB 341*, 1962.
21. ROWAN and KEESE, "A Study of Factors Influencing Traffic Speeds," in *HRBB 341*, 1962.
22. WRIGHT, "Highway Speed Zoning and Control as Practiced in State of Utah," in *TQ*, April, 1949.
23. LEWIS, "The Case for Speed Zoning," in *TQ*, July, 1950.
24. WILSON, "Open Discussion—Are Rural Speed Zones Effective?", in *TE*, October, 1951.
25. BAERWALD, "300-Horsepower Roads—Horse and Buggy Speed Laws," in *TE*, June, 1964.
26. TAYLOR, "A New Concept of Speed Zoning," in *TE*, September, 1964.
27. TAYLOR, "Speed Zoning: A Theory and its Proof," in *TE*, January, 1964; also in *Proc. ITE*, 1964.

28. *A Policy on Geometric Design of Rural Highways*, AASHO, 1965.
29. *Traffic Engineering Handbook*, ITE, 1950.
30. JOHNSON, "Maximum Safe Vehicle Speeds at Railroad Grade Crossings," in *TE*, June, 1958.
31. STACK *et al*, "A Survey of the Uses of Radar in Speed Control Activities," in *TQ*, October, 1954.
32. SCHEIDT, "Enforcing Law on Slow Driving," in *TQ*, July, 1954.
33. BAKER, "Effect of Enforcement on Vehicle Speeds," in *HRBB 91*, 1954.
34. WILSON and CHENN, "Effect of Color and Location of Observation Car When Conducting Speed Checks," in *TE*, January, 1960.
35. KENNEDY *et al*, *Fundamentals of Traffic Engineering*, UCLA, Berkeley, Calif., 1960.
36. MICHAELS, "The Effects of Enforcement on Traffic Behavior," in *PR*, December, 1960.
37. "Wyoming Study Shows Highway Maintenance, Law Enforcement Yield Big Safety Gains," in *HRN* 27, Spring, 1967.
38. DESROSIERS, "Speed Estimation on Residental Streets," in *PR*, August, 1962.
39. SALVATORE, "Estimation of Vehicular Velocity Under Time Limitation and Restricted Conditions of Observation," in *HRR 195*, 1967.
40. KESSLER, "The Effect of Speed Zone Modifications Occasioned by the Illinois Speed Law," in *TE*, July, 1959.
41. DEEN, "Effectiveness of Speed Limit Signs," in *TE*, April, 1959.
42. COLEMAN, "The Effect of Speed Limit Signs," in *TE*, January, 1957.
43. MOHR, "Results of Speed Zoning on Rural Highways," in *Proc. HRB*, 1954.
44. ELMBERG and MICHAEL, "Effect of Speed Limit Signs on Speeds on Suburban Arterial Streets," in *HRBB 303*, 1961.
45. AVERY, "Effect of Raising Speed Limits on Urban Arterial Streets," in *HRBB 244*, 1960.
46. HILL, "Speed Control by Traffic Signals," in *TQ*, January, 1954.
47. CONRADT, "Speed Limits in New Mexico," in *TQ*, October, 1953.
48. JOHNSON, "How About Vehicle Speeds?", in *TQ*, July, 1951.
49. WINGERD, "Minimum Speed Limit on Freeways," in *HRR 244*, 1968.
50. "Stop Sign Speed Control," in *TE*, September, 1964.
51. CHEGIN, "Effect of Different Sign Displays on Speed," in *TE*, November, 1964.
52. COVAULT and BOWES, "A Study of the Feasibility of Using Roadside Radio Communications for Traffic Control and Driver Information," in *HRR 49*, 1964.
53. MICHAELS and SOLOMON, "The Effect of Speed Change Information on Spacing Between Vehicles," in *HRBB 330*, 1962; also in *PR*, February, 1962.

Chapter 26: One-Way Streets, Unbalanced Flow and Reserved Transit Lanes

1. *Traffic Engineering Handbook*, 3rd ed., ITE, 1965.
2. *A Policy on Arterial Highways in Urban Areas*, AASHO, 1957.
3. BEVINS, "Facilitating Traffic Flow With One-Way Streets," in *TE*, October, 1949.
4. "One-Way Business Streets," Transportation and Communication Dept., Chamber of Commerce of the U.S., July, 1955.
5. "Current Intersection Capacities," *HRB Correlation Service Circular 376*, December, 1958.
6. NORMANN, "Variations in Flow at Intersections as Related to Size of City, Type of Facility, and Capacity Utilization," in *HRBB 352*, 1962.
7. FRENCH, "Capacities of One-Way and Two-Way Streets with Signals and with Stop Signs," in *HRBB 112*, 1956; also in *PR*, February, 1956.
8. *The Federal Role in Highway Safety*, HD No. 93, 86th. Cong. USGPO, 1959.
9. DENNIS, "The Businessman's Viewpoint on One-Way Streets," in *TE*, April, 1953.
10. THOMAS, "The Effect of One-Way Traffic Operation on Arterial Business," in *TE*, July, 1967.
11. MCCRACKEN, "Warrants for One-Way Streets and Their Values," in *Proc. ITE*, 1941.
12. "Off-Center Lane Movements," Report of Technical Committee 3F of ITE, in *TE*, November, 1958.
13. "Off-Center Lane Movements," Report of Technical Committee 3F of ITE, in *TE*, October, 1958.
14. WEIGARTEN, "The One-Way Preference Street," in *TE*, March, 1958.
15. ROCHESTER, "Relieving Congested Districts by Systems of One-Way Streets," in *Proc. ITE*, 1939.
16. WILEY, "One-Way Avenues, Re-timed Signals Improve Manhattan Traffic," in *TQ*, January, 1952.
17. CHANG, "Syncronization of Traffic Signals in Grid Networks" in *IBM Journal of Research and Development*, July, 1967.
18. *Annual Report—1962, 1963, 1964, 1965, 1966, 1967*, City of New York, Department of Traffic.
19. DORSEY, "Use of the Off-Center Lane Movement in Los Angeles," in *TQ*, July, 1948.
20. DORSEY, "Off-Center Aids the City Solution," in *Proc. ITE*, 1951.
21. HAMMOND, "Seattle's Foolproof Reversible Highways," in *Construction Craftsman*, October, 1965.
22. DIER, "One-Way Street Experience of Cities," in *TE*, January, 1950.
23. SMITH and HART, "A Case Study of One-Way Streets," in *TQ*, October, 1949.
24. FOWLER, "One-Way Grid System of Portland, Oregon," in *TE*, April, 1953.
25. FAUSTMAN, "Improving the Traffic Access to Sacramento's Business District," in *TQ*, July, 1950.
26. EASTMAN, "Ann Arbor Discontinues One-Way Operation of Two Streets," in *TE*, December, 1950.
27. TODD, "Reversing Flow in Center Lane of 3-Lane Roads," in *Proc. ITE*, 1949 (shorter treatment: Todd, "Effects of Reversible Lane Movement Signalization of Three Lane Highways," *PHRB*, 1950.)
28. EXNICIOS, "Report on Unbalanced Traffic Flow Operations," in *Proc. ITE*, 1960.
29. STILL, "4½ Miles of Lane Control Signals," in *TE*, September, 1961.
30. HOOSE, "Planning Effective Reversible Lane Control," in *TQ*, July, 1963.
31. BRUENING, "Lane Control in Freeway Terminal Area," in *Proc. ITE*, 1964.
32. BRUENING, "6-Lane Street—8-Lane Performance," in *TE*, July, 1964.
33. WAKE, "Reversible Lanes Reduce Congestion at Little Cost," in *TE*, May, 1964.

34. "Reserved Transit Lanes," Report of Committee 3-D of ITE, in *TE*, July, 1959.
35. MORAN and REAGAN, "Reserved Lanes for Buses and Car Pools," in *TE*, July, 1969.
36. RUSSO, *Reserved Mass Transit Lanes*, M.S. Thesis, PIB, Brooklyn, N. Y., June, 1962.

Chapter 27: Curb Parking Controls

1. "Current Intersection Capacities," in *HRB Correlation Service Circular 376*, December, 1958.
2. BARTLE, "Effect of Parked Vehicle on Traffic Capacity of Signalized Intersection," in *HRBB 112*, 1956.
3. *Highway Capacity Manual*, HRBSR 87, 1965.
4. *Accident Facts, 1967*, N. Y. State DMV.
5. *Zoning Applied to Parking*, Eno, 1947.
6. STEGMAIER, "Parking as a Factor in Business," in *Proc. ITE*, 1955.
7. "Parking as a Factor in Business," in *HRBR 11*, 1953.
8. EVANS, "Parking and Its Importance to the Downtown Business District," in *Proc. ITE*, 1953.
9. BEVINS and SLIDERS, "Atlanta Solves Its Downtown Parking Problem," in *TE*, December, 1957.
10. HALL, "The Parking Problem—Detroit Finds an Answer," in *TE*, August, 1957.
11. GITTENS, "Relief for Pittsburgh's Traffic Congestion Through Curb Parking Regulations," in *TE*, July, 1955.
12. *Parking in the City Center*, Wilbur Smith and Associates, *AMA*, May, 1965.
13. *Standards for Street Facilities and Services*, PM7A, NCUT, PAS, 1958.
14. *Proper Location of Bus Stops*, An ITE Recommended Practice, December, 1967.
15. "Zoning for Truck-Loading Facilities," in *HRBB 59*, 1952.
16. GREEN, "Loading Experience as a Basis for Zoning Requirements," in *TQ*, October, 1953.
17. HUNNICUTT, "Nashville's Loading Zone Policy," in *TQ*, April, 1959.
18. SOLER, "A Study on Taxicab Stands," in *TE*, November, 1949.
19. ANDREWS, "Taxi Stands," in *TE*, November, 1957.
20. FISHER, "Angle vs. Parallel Parking," in *TE*, October, 1950.
21. OVERMYER, "Progress Report of Committee 6-E", in *Proc. ITE*, 1955.
22. "A Tentative Recommended Practice for Special Parking Prohibitions in Connection with Snow Storm Emergencies," ITETR, in *TE*, January, 1966.
23. "Local Street Parking Criteria," ITE Project Committee 3F(64), in *TE*, March, 1967.
24. BAKER and THOMAS, "Saturday Parkers in Business Areas of Richmond, Virginia," in *TE*, October, 1961.
25. *Parking*, Eno, 1957.
26. GOODWIN, "Truck Loading Facilities as Related to Parking," in *TE*, February, 1950.
27. MATSON, "Basic Factors in the Formulation of Parking Policy," in *TE*, November, 1951.
28. CHERNIAK, "The Zoning Ordinance—An Attack on the City Parking Problem," in *Proc. ITE*, 1946.
29. MCCRACKEN, "Commercial Vehicle Loading to Minimize Traffic Hazard and Congestion," in *Proc. ITE*, 1946.
30. *Parking—Legal, Financial, Administrative*, Eno, 1956.
31. *Access and Parking for Institutions*, Eno, 1960.
32. LEVIN, "Parking Meters," in *HRBB 81*, 1954.
33. *Annual Report—1962, 1963, 1964, 1965, 1966, 1967*, City of N. Y. Department of Traffic.
34. MILTON, "Parking Meters as an Aid to Enforcement," in *Proc. ITE*, 1949.
35. "Court Upholds City's Parking Program," in *TE*, August, 1954.
36. KERSTETTER, "Mechanized Hitching-Post, 1960," in *TQ*, October, 1960.
37. BURRAGE, "The Effectiveness of Parking Meters," in *TQ*, April, 1955.
38. BORROWMAN, "Winnipeg's Parking Meters—Before and After," in *TE*, December, 1950.
39. "Parking Meters," in *TE*, August, 1957.
40. SPRUNGMAN, "Progress Report of Committee 4C," in *Proc. ITE*, 1955. (Parking meters).

Chapter 28: Limited—Access Facilities

1. MUSICK, "The Federal Role in Highway Safety," HD No. 93, in *Proc. ITE*, 1960.
2. *Manual on Uniform Traffic Control Devices*, State of New York, State Traffic Commission, July, 1958.
3. *Manual for Signing and Pavement Marking of the National System of Interstate and Defense Highways*, AASHO, 1961; revised ed., 1962.
4. MUSICK, "Internally Illuminated Signs for Interstate Highways," in *Proc. ITE*, 1960.
5. WEBB, "Correlation of Geometric Design and Directional Signing," in *Proc. P 1627 ASCE*, May, 1958.
6. SCHOPPERT *et al*, "Some Principles of Freeway Directional Signing Based on Motorists Experiences," in *HRBB* 244, 1960.
7. CANTILLI and LEE, "Upgrading of Highways for Safety Systematically," in *TE*, March, 1970.
8. "Exit Ramps, Wrong-Way Traffic Problem for U. S.," in *TE*, January, 1965.
9. ERNSTEIN, "Guide Sign Revisions May Eliminate Confusion," in *TE*, August, 1964.
10. RICKER, "Traffic Signs for an Express Highway," in *Proc. ITE*, 1952.
11. PINNELL and BOHR, "Urban Interchange Design as Related to Traffic Operation," *TE*, March, 1966.
12. *Gap Acceptance and Traffic Interaction in the Freeway Merging Process, Phase II*, Final Report, TTI.
13. BUHR, "Freeway Ramp Control," in *CE*, June, 1968.
14. MCCASLAND, and CARVELL, "Optimizing Flow in an Urban Freeway Corridor," *Annual Report: Fiscal Year 1968–69*, TTI, BPR Contract No. FH-11-6921.
15. FOOTE, CROWLEY, and GONSETH, "Instrumentation for Improved Traffic Flow," in *Proc. ITE*, 1960.
16. FOOTE and GAZIS, "Surveillance and Control of Tunnel Traffic by an On-Line Digital Computer," in *Transportation Science*, Vol 3, No. 3, August, 1969.
17. *Ramp Metering in Chula Vista*, California Division of Highways, unpublished report, January, 1969.
18. NEWMAN, DUNNET, and MEIS, "An Evaluation of Ramp Control on the Harbor Freeway in Los Angeles," in *HRR 303*, 1970.
19. *Study Summary*, unpublished, PIB, Brooklyn, N. Y., 1969.
20. THOMAS, "Simple Ramp Metering Device Reduces Rear-End Collisions," in *TE*, June, 1969.

21. HODGINS, "Effects of Column Controls on Freeway Traffic Flow—A Theoretical Analysis," *PR*, February, 1968.
22. YAGODA and PIGNATARO, "The Analysis and Design of Freeway Entrance Ramp Control Systems," in *HRR 303*, 1970.

Chapter 29: Highway Safety

1. *Manual of Traffic Engineering Studies, 1953*, Association of Casualty and Surety Companies, ITE, 1964.
2. JORGENSEN and ASSOCIATES and WESTAT RESEARCH ANALYSTS, *Evaluation of Criteria for Safety Improvements*, 1966.
3. WHITTON, "Entire Bureau of Roads Program Aimed Towards Improved Safety," in *TE*, June, 1965.
4. SAAL, "Public Roads Research Concentrates on Three Critical Problems," in *Proc. ITE*, 1964.
5. WHITTON, "The Role of Highway and Traffic Engineering and Highway Safety," in *TQ*, January, 1965.
6. WILLIAMS, "A New Dimension in Highway Safety," in *TQ*, January, 1963.
7. *Traffic Safety Services—Directory of National Organizations*, Office of Highway Safety, BPR, July, 1963.
8. *Highway Safety Action Program*, President's Committee for Traffic Safety, 1962.
9. *Getting Results Through Traffic Engineering*, Booklets I, II, III, and Case Studies 73 through 140, Accident Prevention Department of the Association of Casualty and Surety Companies (no longer available).
10. The following are some articles from *Traffic Engineering* dealing with traffic accidents:
    (a) CONNER and MCMILLEN, "Traffic Engineering Reduces Accidents in Ohio"; Stover, "Locating Slippery Highway Surfaces Through Accident Report Analysis," November, 1962.
    (b) TWISS, "Reflective Painting Pays Dividends," June, 1962.
    (c) BURCH and ZOOK, "Expressway Accident Rates in North Carolina," April, 1962.
    (d) MCMULLINS, "Improving Freeway Traffic Accident Reporting," December, 1960.
    (e) FARSH, "Aging and Driving," November, 1960.
    (f) PRISK, "The Speed Factor in Highway Accidents," August, 1959.
    (g) GRAVES, "Safety and Economic Aspects of Expressway Construction in Michigan"; "Toll Road Accident and Fatality Statistics," May, 1959.
    (h) BATTEY, "The Measurement of Exposure to Motor-Vehicle Accidents," March, 1959.
    (i) WEBB, "Accident Study," October, 1958.
    (j) PRISK, "The Relationship of Engineering Science to Highway Safety," June, 1958.
    (k) PRISK, "How Access Control Affects Accident Experience," March, 1956.
    (l) WEBB and ISRAEL, "Fatalities on California Freeways," March, 1957.
    (m) DUNMAN, "The Economic Costs of Motor Vehicle Accidents," November, 1956.
11. The following are some articles in *Traffic Engineering* dealing with this subject:
    (a) WILSON, "California's Reduced Visibility Study Helps Cut Down Traffic Accidents When Fog Hits Area," March, 1965.
    (b) WILLIAMS, "Evaluating Safety," March, 1965.
    (c) MATTHEWS, "Spot Icing Major Culprit in Winter Death, Injury Toll," February, 1965.
    (d) KNISELY, "Wood Break-Away Posts Provide Safety for Motorists," August, 1964.
    (e) MARCONI, "Punch Cards Code City Traffic Accidents," July, 1964.
    (f) MAY, "A Determination of an Accident Prone Location," February, 1964.
    (g) ROTMAN, "Coding Accident Locations," February, 1964.
    (h) BALDWIN, "Traffic Accident Records," July, 1963.
    (i) MICHALSKI, "Traffic Safety in Chicago," February, 1960.
12. The following are some articles dealing with this subject:
    (a) MITCHELL, "Accident Analyses for Program Planning," in *Proc. HRB*, 1949.
    (b) KELEHER *et al*, "New Developments in Accident Reporting and Analysis in Chicago," in *Proc. ITE*, 1964.
    (c) "Traffic Accident Studies—1958," in *HRBB 208*, 1959.
    (d) "Highway Accident Studies," in *HRBB 240*, 1960.
    (e) RUDY, "Operational Route Analysis," in *HRBB 341*, 1962.
    (f) JORGENSEN, "Programming Highway Accident Reduction," in *HRR 12*, 1963.
    (g) SAWHILL and NEUZIL, "Streets with Two-Way Median Left-Turn Lanes," in *HRR 31*, 1963.
    (h) DEROSE, "An Analysis of Random Freeway Traffic Accidents and Vehicle Disabilities," in *HRR 59*, 1964.
    (i) "Highway Safety 1963 and 1964," in *HRR 79*, 1965.
    (j) CAMPBELL, "Highway Traffic Safety—Is It Possible?", in *TQ*, July, 1965.
    (k) ALLAN, "Accident Prevention in Arizona," in *TQ*, October, 1960.
    (l) HOBACK, "Accidents on Oklahoma Turnpikes," in *TQ*, October, 1959.
    (m) ALLEN, "Correcting Traffic Difficulties," in *TQ*, October, 1953.
    (n) ELLISON, "Traffic Safety," in *TQ*, January, 1963.
    (o) ELLISON, "Traffic Safety in the Nation's Capital Moves Steadily Ahead," in *TQ*, July, 1960.
    (p) EHRMAN, "Causes of Highway Accidents, United States Experience," in *TQ*, January, 1958.
    (q) DAMON, "North Dakota Highway Safety Study," in *TQ*, April, 1955.
    (r) RICKER, "Fog on the Jersey Turnpike," in *TQ*, July, 1953.
13. CANTILLI, "A Philosophy for Accident Prevention," in *TE*, May, 1965.
14. *Road User Benefit Analysis for Highway Improvements*, AASHO, 1960.
15. *Running Cost of Motor Vehicles as Affected by Road Design and Traffic*, NCHRPR 111, 1971.
16. TAMBURRI and HAMMER, *Evaluation of Minor Improvements*, California Div. of Highways, Parts 1 to 5, 1967.

Index

A

B

C

D

E

F

G

H

I

J

L

M

N

O

P

Q

R

S

T

U

V

W

Y